Dynamics of Structures

Levon Gregory Petrosian

Dynamics of Structures

Oscillations/Vibrations

Springer

Levon Gregory Petrosian
District Department of Transportation
Project Delivery Administration
Washington, DC, USA

ISBN 978-3-031-63538-0 ISBN 978-3-031-63539-7 (eBook)
https://doi.org/10.1007/978-3-031-63539-7

This Springer imprint is published by the registered company Springer Nature Switzerland AG
The registered company address is: Gewerbestrasse 11, 6330 Cham, Switzerland

To my Angel my Beloved Wife
NATALIE

Preface

The book is a textbook for students of technical and engineering specialties in the study of structural mechanics. The book can also be used in other non-engineering colleges and universities as a textbook and could become a practical guide/tool for designers, undergraduate, graduate, and postgraduate students and for PhD research programs, researchers, structural designers, and engineers who study and use the dynamics of structures in their practical work.

This book presents both methodological and practical purposes of dynamic analysis of engineering structures. Therefore, accounting of all methods is accompanied by solutions to specific problems, which not only are illustrative materials but may also represent independent interest in the study of various technical issues.

The textbook offers efficient solutions for readers looking to understand basic concepts and methods of dynamic analysis of engineering structures and structural elements; explains foundational concepts and principles needed to understand structural vibrations under various dynamic loads; describes objectives of the study of dynamic response of engineering structures and the importance of such study in their design.

The book is designed for a wide range of readers who have mastered the basics of structural mechanics and strength of materials in the volume of courses of higher technical schools: students and graduate students at technical universities, as well as engineers engaged in design and research work. Additionally, the textbook can serve as a guide for students of technical colleges in the study of relevant sections for courses in dynamics of structures. Each chapter covers a particular topic. Chapters dealing with related topics are grouped together.

The book contains a large number of most typical problems with detailed solutions, which are preceded by a fairly complete and detailed presentation of theory. The text explains structural responses from dynamic loads and modelling vibration systems and calculations of dynamic responses of structural systems. The book presents the classical vibration theory in a clear and systematic way, detailing original work on rod systems, beams, arches, trusses, frames, plates, membranes, foundations, combined bridge systems such as rigid beam and flexible arch, and much more. The book presents analysis and methodologies adopted by the civil, structural, and mechanical engineering society/community for solving structural dynamics problems.

The author is deeply grateful to those users of the present edition of the book who kindly provide feedback and suggestions for an improved second edition.

I want to thank Springer Nature and Executive Editor Michael Luby for their effective cooperation, and their great care in preparing this edition.

Washington, DC, USA Levon Gregory Petrosian

Introduction

This book is a textbook for students of technical and engineering specialties in the study of structural mechanics: dynamics of structures. This textbook could also become a practical guide for many undergraduate, graduate, and postgraduate students, researchers, structural designers, and civil engineers who study and use the dynamics of structures in their practical work, among other users.

The book is compiled in such a way that the level of complexity of the presented course increases as the material is presented. The problems are arranged in the order of gradual complexity and can be used to study the subject, or as final individual assignments or coursework or for in-depth study of theory in research seminars, conferences, and so on. The problems range from extremely simplified and most transparent formulations of individual test problems to very complex.

The first fifteenth chapters of the book are based on the materials of a course of lectures and practical classes conducted by the author at the Armenian National University of Architecture and Construction in Yerevan, where he served as the Chair of the Department of Structural Mechanics. The book was drafted, accounting for modern achievements in the field of structural mechanics, in particular: the dynamics of structures, as well as related areas of applied mechanics. Hence, the first part of the book (the first 15 chapters) is ideal as a textbook for advanced undergraduates, graduates, or postgraduate students taking their first course in structural dynamics. This textbook can be used for two semesters. In addition, the book will serve as a primary reference for practicing engineers and research workers, as well as a self-study guide for students, researchers, and professional engineers.

The second part of the book (16th chapter onward) is intended mainly for professionals and specialists in the field of dynamics of structures and related areas. In preparing the book, the author tried to make it accessible not only for scientists, researchers, graduate, and postgraduate students in the field of structural mechanics, dynamics of structures and related areas, but also for engineers working in design centers and organizations.The author offers methods and applications (not always a simple task considering boundary-value problems), in a sufficiently simple and clear way in order to make it easily accessible for readers with mathematical and technical background at university level. However, for comprehension of the second part of book initial knowledge in such areas, of mathematics as the theory of the generalized functions, transforms, theory of linear integral equations is necessary. The specified mathematical apparatus in recent decades has widely penetrated the technical sciences, especially structural mechanics, and knowledgeable and sophisticated reader should be familiar with it.

Key Features

- Offers efficient solutions for readers looking to understand basic concepts and methods of dynamic analysis of engineering structures and structural elements.
- Explains the foundational concepts and principles needed to understand structural vibrations under various dynamic loads.
- Describes the objectives of the study of dynamic response of engineering structures and the importance of such a study in their design.
- Presents analysis and methodologies adopted by the civil, structural, and mechanical engineering society/community to solve structural dynamics problems.
- The book presents the classical vibration theory in a clear and systematic way, including recent global research and scientific developments in the dynamics of structures and applied mechanics.
- Presents methods of dynamic design of structural elements. The method of separation variables, variational and energy calculation methods, integral equations and integral transforms methods, the finite element method (FEM), the finite difference method (FDM), the boundary element method (BEM), the delta-transform method (DTM), the spectral method of boundary elements (SMBE), problems, and solutions are discussed in this book. The application of the presented methods of analysis and calculation to the problems of the dynamics of structures is proposed. The presented book includes extensive practical examples and problems.
- Presents a method of calculating the Hilbert transformation and allows for a comparatively easy analysis of the aprioristic models of dynamic systems based on certain complex rigidity or pliability of a system.
- Contains a large number of problems with solutions, which allows the reader to assimilate the presented material most effectively.
- The Appendix:
  - Certificate of Essential Building Data (CEBD). CRBD is a systematized, concise record of the most important structural features of a building, including basic geometry, structural system type, foundation type, basic dynamic characteristics, etc.
  - Amplitudes of dynamic reactions
  - Values of the functions those are included in the formulas of the displacement method
  - Correction functions and their numerical values

* * *

Many of the impacts on engineering structures have a distinctly expressed dynamic nature. Dynamics of structures is one of the special sections of structural mechanics; it is engaged in the development of principles and methods of calculation of structures to the action of dynamic loads. The special feature of dynamic loads is the fact that, in most cases, they cause vibrations of structures.

A significant difference between dynamic and static methods of calculation is the introduction of a new variable—time, which participates in the equations either explicitly or in the form of derivatives of unknown functions on time. Thus, dynamics means time varying, and as the nature suggests, the applied loads vary with time and consequently induce time varying responses (displacements, reaction forces, stresses, velocities, accelerations, etc.). One of the most notable differences in the dynamic analysis is the explicit consideration of the inertial forces developed in a structure when it is excited by time varying loads. They cause vibrations of structures in connection with which there are forces of inertia causing additional internal forces and stresses. Inertial forces arise, causing additional internal forces and stresses.

The study of oscillations/vibrations is a very complex and extremely interesting task. Free oscillations are possible due to the fact that mechanical systems have two important properties. Firstly, they have mass, and, therefore, during oscillations, a system can accumulate kinetic energy. Secondly, the system can accumulate potential energy when it deviates from the state of equilibrium. When the structure oscillates, the variable stresses arise and with sufficiently intense oscillations can lead to a destruction of the structure.

Dynamic behavior of structures is an important subject in structural mechanics and adjacent fields. In civil engineering, understanding of structural dynamics is important in the design of structures, the behavior, and modifications of dynamic models of structures, including reinforcements of structures to withstand severe dynamic loads from earthquakes, hurricanes, blasts, as well as production dynamic loads from technological processes, and strong winds. Another important reason for the thorough understanding and analysis of structural dynamics is to identify the occurrence and location of damage within an existing structure. In general, structural response to any dynamic load is expressed basically in terms of displacements of structures.

The purpose of this textbook is to identify the basic regularities in the behavior of models of buildings, structures, and structural elements while solving problems of dynamics. The calculation methods presented in the book will give reliable estimates of the actual behavior of the system. It must be understood, however, that any computational model is an abstraction that reflects basic properties of an object's behavior. No one calculation model, no matter how accurate it is from a mathematical point of view, is able to give a solution whose accuracy exceeds the accuracy of the original data.

The task of the dynamics of structures is the calculation of structures on the action of dynamic loads. Currently, due to various reasons, successful design of buildings and structures is impossible without considering dynamic effects. There are many reasons for the foregoing. The most obvious reason is the growth of dynamic loads caused by machines, mechanisms with unbalanced moving parts, cranes and other equipment, the widespread use of vibrations, shock, impacts, industrial explosions as elements of the technological process, the construction of flexible hanging, suspension structures for which it is necessary to take into account the dynamic effect of the wind, etc.

The purpose of dynamic calculation is to check the strength of structures and to determine dynamic displacements, velocities, and accelerations, whose values under the influence of vibrations on people and certain types of precision equipment, such as measuring devices, instruments, high-precision machines, equipment of microelectronics production, etc., should not exceed permissible limits. Therefore, the increased attention to dynamics is also associated with the development of precise technological processes that require reduction of vibration levels, use of precise measuring instruments, and special laboratory equipment in scientific research. One of the most important technical tasks is to ensure a level of vibration, which is acceptable from a sanitary and hygienic point of view.

Dynamic loading is a load that causes the movement of the body and is accompanied by the appearance of inertia forces. Under a dynamic load, in addition to all usual external and internal forces, the inertia forces also act on each particle of a deforming body. In the case of dynamic loading, the loading depends on the time factor, and the loading rate for a deforming body does not coincide with the growth rate of internal forces. As a result, the equilibrium of the deforming body is disturbed and its movement occurs, accompanied by the inertial forces for weighty bodies, and dependent on the time factor.

The forces of inertia are associated with the motion of the points of the body with their accelerations. Therefore, the forces of inertia of a deforming body depend on its deformation, even when the deformed scheme of the system is not taken into account, so the problem of determining the forces is statically indeterminate.

In addition, dynamic loading affects mechanical properties of a body: the phenomena of material fatigue and elastic hysteresis increase and the physical law changes. Hysteresis, which also occurs under static loading, in the case of dynamic loading acquires a special importance. In this case, the hysteresis depends on the speed of the body and affects the damping of vibrations due to energy dissipation.

Modern developments in various industries and transportation make mechanical oscillations, especially significant in the operating conditions of engineering structures. New types of machines and mechanisms are being introduced; their speeds and masses are increasing as well as traffic intensity, while the relative masses of structures are decreasing due to the use of high-strength building materials. The creation of ultra-high-speed ground transport is one of the most urgent tasks of modern times. From the foregoing, it follows that the methods for calculating engineering structures must be updated and improved over time constantly. In this regard, the problem of dynamics of engineering structures that are affected by moving loads, vibration, impact/shock/impulse loads, and combined dynamic loads is an urgent task of structural mechanics.

The book is compiled in such a way that the level of complexity of the presented course increases as the material is presented. The problems are arranged in the order of gradual complexity and can be used to study the subject, or as final individual assignments or coursework or for in-depth study of theory in research seminars, conferences, and so on. The problems range from extremely simplified and most transparent formulations of individual test problems to very complex.

The first fifteen chapters of the book are based on the materials of a course of lectures and practical classes conducted by the author at the Armenian National University of Architecture and Construction in Yerevan, where he was the Chair of the Department of Structural Mechanics. The book was drafted accounting for modern achievements in the field of structural mechanics, in particular: the dynamics of structures, as well as related areas of applied mechanics.

This book illustrates modern methods of static and dynamic analysis of structures and the methods for solving boundary value problems of structural mechanics and soil mechanics. The application of spectral method of boundary elements (SMBE) to the vibration of the plates on an elastic foundation is examined. It features extensive use of the generalized functions for describing the impacts on structures, and substantiations of the methods of the apparatus of the generalized functions. It describes methods of dynamic calculation of structures using frequency-independent damping models.

The book presents analysis and methodologies adopted by the civil, structural, and mechanical engineering society/community to solve structural dynamics problems. The book also analyzes the impact of seismic influences on regular frame structures. Further, the book offers a method of physical realization of the dynamic systems based on the examples of an elastic-viscous foundation with internal friction. The internal friction in the foundation is described by the model of Kelvin-Voigt and Sorokin.

The book begins with the basic response of single-degree-of-freedom system to various excitations. Chapter 1 gives an overview of structural dynamics, including classification of oscillatory systems and vibration process, general provisions cover the linear and non-linear systems, stationary and non-stationary systems; conservative and non-conservative systems; generalized forces; free vibration; forced vibrations; dissipative system; self-oscillating/vibrating systems; autonomous systems. Types of dynamic load include vibration loads, impact and impulse loads, moving loads, as well as arbitrary loads with excitation as a force or displacement. The methods of discretization of dynamic problems, including dynamic degree-of-freedom, dynamic design models (DDM), generalized displacements including how to formulate equations of motion, and vibration analysis of a single-degree-of-freedom system are discussed.

Chapter 2 deals with the fundamentals of the theory of calculation of elastic systems and variation principles of structural mechanics. Discussion focuses on: potential field of forces; potential energy of masses and definition of force; virtual work of external and internal forces; the principle of virtual displacements; virtual work of external forces or potential deformation energy; total potential energy; extreme properties of potential energy; the principle of virtual displacements; the methods of drawing up equations of motion;

Lagrange variation principle and Lagrange formula; the principle of virtual changes in the stress state and formulation of the principle; Castigliano variation principle; the principle of least work; Hamilton Principle; and the principle of reciprocity of work. The variation principles of structure mechanics discussed in this chapter present a conceptually new approach to studying the course dynamics of structures. Knowledge and the study of the basic theorems of elastic systems and variation principles of structural mechanics are the necessary bases for the successful study of general and special courses in the dynamics of structures. From the methodological point of view, methods and materials presented in the second chapter allow for a smooth and organic transition to the study of energy calculation methods, which are very effectively used in structural and applied mechanics, in the theory of elasticity, as well as in the dynamics of structures.

Chapter 3 presents energy methods of structural dynamics. Differential equations of motion can be obtained using energy principles using the calculus of variations, i.e., the same differential equations that are obtained using the static method. The main goal of the energy principles is to find very effective approximate solutions of dynamic problems. Energy methods are associated with the integration operations. In this case, the approximate solutions can have greater accuracy as compared to the use of the differentiation operations. Thus, the conditions for applying integration are better as compared to differentiation. Energy methods can also be considered as methods of approximating functions, since their application solves the problem of the best approximation of given functions in advance to the actual ones. Energy methods belong to the category of applied mathematical methods and can be used in many fields and problems of mechanics. Derivation of equations of motion using the D'Alembert principle, static method, energy method, Rayleigh, Ritz, and Bubnov-Galerkin methods, and applicable problems and solutions are discussed.

Chapter 4 presents the free vibrations of a system with single-degree-of-freedom (SDOF). Analysis of free vibration, components of the basic dynamic system such as degree-of-freedom, mass, elastic properties of the system, equation of the motion for (SDOF), damping that includes critically damped system, over-critically damped system, under-critically damped system, no damped system, as well as the total energy of the vibrating mechanical system, and problems and solutions are discussed.

Chapter 5 presents forced vibrations of a system with single-degree-of-freedom (SDOF). Analysis of forced vibrations, differential equation of forced vibrations, dynamic coefficient, beat phenomenon, and problem and solutions are discussed. Chapter also deals with forced harmonic vibrations of a damped system with single-degree-of-freedom (SDOF). Analysis of forced harmonic vibrations, differential equation of motion, resonance phenomenon and definition, problems, and solutions are presented.

Chapter 6 presents the motion of an elastic system with single-degree-of-freedom (SDOF) under the action of force varying by arbitrary law. Reactions of a system with single-degree-of-freedom (SDOF) under impulsive and suddenly applied loads, the action of a sudden applied force of constant value, the action of force growth according to linear law, the action of a suddenly applied force of constant value and suddenly terminating its action, the short-term impulse (momentum) of force, the action of periodic instantaneous impulses, and problems and solutions are presented.

Chapter 7 is devoted to impact loading. An impact occurs when two or more bodies collide. The impact could be caused by a weight falling on the design object. An important characteristic of an impact load is the generation of relatively large forces at the point of contact for a relatively short period of action time. Impact load is defined as the force provided by a sudden load on the design object. Analysis of impact, dynamic deflection, and numbers of problems and solutions are discussed.

Chapters 8 is devoted to free vibrations of a system with multi-degrees-of-freedom. The natural transverse vibrations of beams and frames are considered as systems with an n-th number of concentrated (lumped) masses. When composing equations and choosing computational dynamic model, the classical methods of calculating rod system are used: the method of forces and the method of displacements. The chapter includes extensive practical examples and problems with solutions.

Chapter 9 presents forced vibrations of a system with multi-degree-of-freedom. An elastic system with lumped masses acting on the vibrational forces is considered. Bending moments as well as shear and longitudinal forces will vary over time when a harmonic vibrational load on an elastic system with many degrees of freedom. The values of vibration amplitudes depend on the frequencies of the perturbing forces. The chapter includes extensive practical examples and problems with solutions.

Chapter 10 devoted to vibration of a system with finite number of-degrees-of-freedom. The equation of motion in a matrix formulation is discussed. The determination of displacements of rod systems in a matrix form is presented. General provisions, Simpson formula, pliability matrix, quasi-diagonal compliance (pliability) matrix of whole system, canonical equations in matrix form, vector matrix, etc., are discussed. Chapter presents a matrix form of determining the frequencies of free vibrations of a system with concentrated (lumped) masses. Derivation of resolving equations, matrix form of determining of structure's frequencies on vibration load, calculation of frames on vibration load by the force method, calculation of frames on vibration load by the displacement method, derivation of resolving equations, and number of problems and solutions are discussed.

Chapter 11 is devoted to systems with an infinitely large number-of-degrees-of-freedom. Free vibrations of beams with uniformly distributed mass, problems and solutions, fundamental functions of the equation of vibration of beams, longitudinal vibrations of rods with distributed mass, frequencies of free vibrations of trusses, and applicable problems and solutions are discussed in presented chapter.

Chapter 12 presents vibrations of beams under a moving load and discusses two cases. Case 1: The differential equation of motion is based on the static method. The equations of dynamic equilibrium differ from the equations of static equilibrium by additionally taking into account the forces of inertia in the form of product of masses or their moments of inertia and accelerations, that is, the second derivatives of linear or angular displacements over time with respect to time. Case 2: The differential equation of motion or Lagrange equations of motions is based on the energy method. The energy method is based on the law of conservation of energy, according to which for a conservative system that vibrates without energy loss. The problems and solutions are discussed for two cases in the presented chapter.

In Chap. 13, the static and dynamic prerequisites for the calculation of frames and rods with distributed mass as systems with an infinite number of degrees of freedom are defined. In the calculation of such frames by the displacement method, the frame is converted into a system made up of single-span beams with different boundary conditions at the nodes by introducing additional links. The brief information from the theory of dynamic calculation of frames by the displacement method is provided, the calculation sequence is presented, and the table of correction coefficients for support reactions is given. An example of calculating frames using the displacement method for vibration loads is given. At the end of the book, the appendix contains tables with numerical values of correction coefficients for the dynamic calculation of frames.

Chapter 14 presents the methods of dynamic design of structural elements. This chapter shows the main methods used in the dynamic analysis of rods, plates, and shells. Historically, one of the first methods of solving equations of mathematical physics is the method of separation of variables. The application of this method is described in the examples of solving problems of free vibration of rectangular and circular membranes. The range of problems which can be solved accurately, in a closed form, is limited, and, therefore, there are often variation methods used in practice. The variation methods are discussed.

The use of integral equations, in particular the reduction of problems with vibration of one-dimensional structures to the integral equations of Volterra of the second kind and Fredholm of the second kind, is demonstrated. The methods used for numerical solutions of problems of dynamics are described. These are the finite element method (FEM) and the finite difference method (FDM). In either the FEM or FDM, the whole domain of the partial differential equations (PDEs) requires discretization. The FEM is presented in an example of a plane problem of the theory of elasticity. A diagram of solving a problem of non-stationary vibration/oscillations of a cantilever plate is shown. The vibrations of a compressed rod are solved using the finite difference method (FDM). This chapter describes both a general approach to the construction of various types of boundary equations and a spectral method of boundary elements (SMBE), which allows to obtain effective solutions of various boundary value problems of structural mechanics. The application of SMBE in this book is further illustrated by solving several problems on the structural analysis of the design of construction on the static and dynamic impacts. The generalized functions in structural mechanics, generalized method of integral transformation (GMIT), delta-transform method (DTM), and compensating load method (CLM) are also presented in this chapter.

Chapter 15 is devoted to plates' vibrations. Technical theory of bending; small vibrations of elastic plates; application of the refined theory of bending in plate vibrations; anisotropic plates; flexible plates; methods of solving equations of vibrations of plates; free vibrations of rectangular plates; forced vibrations of plates; vibration of plates on elastic foundation; forced vibrations of a rectangular plate on an elastic foundation, problems, and solutions are provided. As an example of the application of the spectral method of boundary elements (SMBE) to dynamic problems, the oscillations of a plate of an arbitrary form, which lies on the elastic Winklerian base, under the action of a load $q(x, y, t)$ are presented. In the present chapter, the SMBE is used to solve the problem of the harmonic oscillations of the rectangular isotropic plate lying on an elastic foundation of the Winklerian type.

Chapter 16 describes free vibrations of combined systems, such as "rigid beam-flexible arch," considering the dynamic thrust when the axis of the arch is outlined by a square parabola and circular arch. Applicable problems and solutions are discussed. The chapter shows the methodology and results of dynamic testing models of rigid beam-flexible arch. The chapter describes an experimentally theoretical method for calculating the combined systems, taking into account the joint dynamic work of the main span and adjacent bridge trestles using factor "increasing" frequency. Presented study and analysis are devoted to the oscillations of the combined system, taking into account its extent at a given harmonic motion base. The chapter shows a method of seismic analysis of combined systems with regard to their length, using real seismograms of powerful earthquakes. There are general equations of motion for a system with a finite number of degrees-of-freedom in a spatial deformation structure.

Chapter 17 determines reactions of frames, accounting for the initial phase of passage of a seismic wave. Chapter discusses the vibrations of multi-story, multi-span frame structures of considerable long span. A dynamic design model is accepted as the system with finite number of degrees of freedom. The equations of seismic oscillations of an elastic frame system considering the extent can be obtained on the basis of the equations of the displacement method with the static horizontal shifting of supports and taking into account the inertial forces in accordance with Dalembert's principle. In the equations of motions, it is assumed that the energy absorption of the first floor at asynchronous oscillations columns is the same as with the synchronous oscillations. Equations of motion, Laplace-Carson transformation, and equation of motion of a single-span frame, taking into account its length with arbitrary movement of the base, are discussed.

Chapter 18 describes features for solving dynamic methods of boundary equations. It is known that most building materials have damping properties, characterized by the frequency's independent loss. However, there is a certain dependency of damping parameters (coefficients loss, decrement) on the frequency of the deformation for soils in the experiments. Frequency-dependent losses are typical for a number of modern materials, such as certain plastics. It is, therefore, a very important task to describe the dynamic behavior of soil and structural materials possessing both frequency-independent and frequency-dependent internal friction. This chapter presents a new way of calculation of Hilbert transform through Fourier transformation of corresponding functions, which simplifies the problem, due to extensive Fourier transforms tables. The presented method of calculating the Hilbert transform allows to comparatively and easily analyze the aprioristic models of the dynamic systems based on the creation of complex rigidity or complex pliability of system. This method makes it possible to analytically determine one of the components of the complex rigidity in accordance with the experimentally defined other components. A few examples of the calculation of the Hilbert transform of delta-functions, Heaviside function, and a power function are examined. The above-described method of calculating the Hilbert transform is applied to the analysis of dynamic models bases of internal friction, described by the hypothesis of Voigt, Sorokin, and Schlippe-Boc. This chapter is devoted to construction of Green's function for an unlimited plate, lying on Winklerian-type elastic-viscous base with an arbitrary dependence of parameters of the complex coefficient of bed on the frequency. In this chapter, the problem of construction of Green's function for single-dimensional dynamic systems with damping, described by different model frequency-independent and frequency-dependent internal friction, is examined and solved.

Chapter 19 presents nonlinear oscillations of a system with a single-degree-of-freedom (SDOF), with multi-degrees-of-freedom (MDOF), with distributed mass and oscillations of the beam with variable cross-sections. The values of the periods of free vibration are obtained by various methods, such as direct integration of the equations of motion and the use of asymptotic methods. A study of free vibrations of a beam with an arbitrary law of variation of the cross-section is also discussed in the nineteenth chapter.

Chapter 20 deals with vibrational damage, criteria and limits, short-term and long-term vibrations, vibration control measures, and reduction of dynamic impact on the structure. Ways to reduce vibrations and resonant phenomena, are discussed.

The presented book has been written to serve as a textbook not only for college and university students but also for postgraduate studies including PhD programs and for research works. In addition, this book will serve as a primary reference material for practicing engineers.

* * *

Modern development of science, technology, structural mechanics, civil, aerospace, naval, automotive engineering, and construction industry require the immediate addition of a chapter on dynamics of structures in the section Topic V: Structural in the book *Civil Engineering Reference Manual for the PE Exam* (Author Michael R. Lindeburg). This book serves as the main reference source and basis for passing PE exam, which is required for all professional engineers.

Here, I would like to share my thoughts on the requirements and administration of the PE exam. I believe it would be most expedient and more effective to allow American universities to conduct the exam as a requirement for awarding a highly qualified professional engineer degree PE in the final graduating year of study for students with civil engineering major. Universities must be responsible for their graduates, for conducting the PE exam, and for the PE licensing and certification procedures. Moreover,

I strongly believe that PE licensing candidates must have a minimum of a master's degree in civil/construction engineering to be admitted for the exam, as undergraduate coursework does not sufficiently prepare engineers for the rigors of the PE exam and civil engineering practice.

The above-mentioned will provide an opportunity for graduates who wish to take the exam and obtain a PE license at the university itself before starting their professional careers.

Fundamental university education is the basis for the successful development of modern civilized society. Nowadays, fundamental education is one of the most important parts of life.

Here we are talking only about graduates of American universities. For specialists, engineers, graduates of colleges and universities of other countries, who work in the USA and are PE licensing candidates, the certification and licensing procedure should be continued and carried out according to the current practice of conducting PE exam and licensing.

Universities' ultimate commitment to graduates is their professional success after graduation.

* * *

I would like to express my gratitude to my teachers and colleagues for the knowledge and experience that I gained working on various problems of structural mechanics, theory of elasticity, dynamics of structures, and stability.

I wish to take this opportunity to express my appreciation to my former teachers and colleagues for the knowledge and experience that I gained working with them during the different time on various problems of structural mechanics, theory of elasticity, methods of static and dynamic analysis of structures, methods for solving boundary-value problems of structural mechanics and soil mechanics based on the application of boundary equations, dynamics of structures, stability of structures, design structures on elastic foundation, foundation design, seismic resistance of buildings and structures, structures protection, etc. In particular, I gratefully acknowledge indebtedness to professors A.I. Tseitlin, A.R. Rzhanitsin, B.G. Korenev, N.N. Leontiev, V.I. Travush, D.N. Sobolev, A.A. Babloyan, V.A. Ilichev, E.E. Khachiyan, V.A. Ambartsumian, V.A. Smirnov, M.A. Dashevski, C. Gavarini, A.S. Veletsos, and A.W. Taylor.

And as always, I am most grateful to my wife Natalie, and my daughters, Galina and Irina, for their love and support.

Finally, I want to thank Springer's Natural Publishing Company and the staff for their effective cooperation and their great care in preparing this edition of the book. And a separate thank you to the production coordinator Mr. Brian Halm, to the production editor Dinesh Praveen, and to the production supervisor Mario Gabriele.

Contents

About the Author

Levon Gregory Petrosian is a professor of Structural Engineering and Structural Mechanics. He received his BSCE and MS degrees from the Armenian National University of Architecture and Construction. He received a PhD from the Moscow State University of Civil Engineering (MGSU) (1980) and the Doctor of Technical Science degree from the Moscow Research Center of Construction, Research Institute of Building Structures (TSNIISK) (1990). Dr. Petrosian has taught undergraduate, graduate, and postgraduate structure mechanics, structural dynamics, and stability classes for 20 years. Dr. Petrosian has made significant contribution to the fields of structural analysis and applied mechanics, structural dynamics, analysis of various engineering structures, including structures on elastic foundations and foundation models, incorporating contemporary methods of analysis and calculation: boundary element method (BEM), generalized integral transformation method (GITM), delta-transformation method (DTM), spectral method of boundary elements (SMBE), as well as application of generalized functions in solving numerous structural mechanics problems. Dr. Petrosian is the author of more than 70 scientific works and publications, including monographs, books, and textbooks. Dr. Petrosian has held professorship at the University of Rome, Institute of Building Structures, and research fellowship of International Center of Structural Models Studies (ICSMS), Bergamo, Italy (1982–1983). Dr. Petrosian has held professorship as a guest researcher at the National Institute of Standards and Technology, USA (1995–1996). Dr. Petrosian participates and provides the leadership in caring out number of engineering project (more than 250 science-research reports). He has been associated with transportation and structural engineering industries for over 40 years as a research engineer, scientist, teacher, and consultant. Levon Gregory Petrosian is a renowned expert on the analysis of engineering structures.

Prior to his move to the United States, Professor Petrosian was the chairman of Department of Structural Mechanics at the Armenian National University of Architecture and Construction as well as the executive director and head of the Armenian Earthquake Engineering Research Institute. Currently, Dr. Petrosian serves as the chief of the Plan Review Division at the District of Columbia Department of Transportation.

List of Symbols

Roman Symbols

a	Constant, some value
a	Parameter
a	Initial stiffness of system
a	Distance; deflection
a_0	Initial amplitude of free vibrations
a_i	Constants, coefficients
a_{ir}	Generalized rigidity of the i th floor with the nonlinear deformation with the parameter r
a_0, a_k	Fourier's coefficients
a_{mn}, b_{mn}	Coefficients of series
a, b, c	Dimensions
a_k	Coefficient
A	Area; cross-sectional area
A	Work of internal forces
A	Potential energy of internal forces
δA	Total possible work of internal forces
A	Amplitude of vibrations
A_{mn}	Constants
A_δ	Displacement matrix
A_i	Pliability matrix, $A_i = EI/EI_i$
$\boldsymbol{A}$	Matrix of coefficients
$\boldsymbol{A^T}$	Transposed matrix $\boldsymbol{A}$
$[\boldsymbol{A}]^{-1}$	Inverse matrix
A	Square matrix of unit displacements
AE	Stiffness at axial deformation
AG	Stiffness at shear deformation
A, *and* B	Constants
b	Distance
b	Constant, some value
b_0, b_k	Fourier's coefficients
b_i	Sub-matrices
b_i	Pliability matrix for particular section

$\boldsymbol{B}$	Quasidiagonal compliance matrix
B_i	Arbitrary constant
$\boldsymbol{B}_i$	Pliability matrix
c	Damping coefficient or coefficient of proportionality
c_i	Displacement
c_c	Critical damping coefficient
C_1, C_2, C_3, C_4	Arbitrary coefficients
$\boldsymbol{C}$	Stiffness matrix for internal forces
C_{ir}	Normalized values of displacements
$\boldsymbol{C}$	Euler's constant
c_n	Fourier's coefficients, coefficient depending on load
c_{mn}	Coefficients
c	Wave propagation velocity
c	Constant
C_1, C_2	Constants of integration
d	Distance or diameter
d	Displacement, deflection
D	Domain
D	Constant of integration
D	Plate/shell flexural rigidity
D	Square matrix
D^{-1}	Inverse matrix with respect to D
e	Eccentricity, coefficient of restitution, natural logarithmic base
E	Modulus of elasticity
$\boldsymbol{E}$	Unit matrix
E_a	Modulus of elasticity of the arch
E_b	Modulus of elasticity of the beam
EI	Bending stiffness
E_{total}	Total energy
f	Height of the arch
f	Technical frequency of natural vibrations
$\|f\|$	Finite element
$f_i(x)$	Approximating functions
$f(x)$	Integrable basic function
F	Force
F_d	Damping force, resistance force
F_x, F_y, F_z	Vector components
F^a, F^b, F^c	Values of the integrals at the beginning, at the end, and in the middle of a segment
F_a	Arch cross-sectional area
F_n	Coefficient of decomposition
g	Acceleration of gravity
G	Shear modulus of elasticity
$G(x, \xi)$	Green's function
h	Height, thickness
$H(t)$	Heaviside function

H	Static thrust
I_i	Inertial force
I	Moment of inertia, impulse
$I^*_{mx}, I^*_{my}, I^*_{mz}$	Moment of inertia of masses relative to the coordinate axes
I_b	Beam cross-section moment of inertia
I	Functional
I	Force of inertia
I_r	Polar moment of inertia
$\vec{I}$	Vector of amplitude of inertial forces
Im	Designation for imaginary part
I_n, Y_n	Bessel functions of nth order of the first and second kind, respectively
j	Integers
n	Numbers of degree-of-freedom
n	Integer
n	Numbers of floor
n	Technical frequency of natural vibrations/oscillations
n	Coefficient of damping
n_q	Normal to Γ' at point q
N	Normal force, longitudinal force
$\vec{N}$	Vector of internal forces
$N(\xi_1)$	Rayleigh function
K	Kinetic energy of masses, kinetic energy of the system
$\boldsymbol{K}$	Single inverse operator
$K(\mathrm{t})$	Impulse transfer function of the system with frequency-independent internal Friction
k	Coefficient of stiffness
k_0	Recovery factor for impact
k	Numbers of floor columns
$k(\omega)$	Complex coefficient of bed
$k'(\omega)$	Elastic component of the complex rigidity
$k''(\omega)$	Dissipative component of the complex rigidity
l	Length, span length
L	Linear operator
L_j	Length of jth span
L_m	Influence matrix
L^0_m	Matrix of influence of bending moments
L^T_{ml}	Matrix transposed with respect to L_{ml}
M	Moment of force, bending moment
M_s	Matrix column
$\overline{M}$	Rectangular matrix
$\vec{M}_p$	Matrix of bending moments
M_{jk}	Minor of element δ_{kj}
$\overline{M}'_{ik}$	Transposed matrix
$\vec{M}_{ip}$	Vector of bending moments of the load diagram

$\vec{M}_{\text{dyn}}$	Vector of bending moment
M_{xy}	Torque moments
m	Mass, number of members
$\overline{m}$	Linear value of the distributed mass
$\vec{m}$	Vector of masses
$m(x)$	Distributed mass with intensity
P	Force, external force
p_i	Circular frequency
$p_{1,\,2}$	Roots of the characteristic equation
p_{mn}	Frequency of natural vibrations of the plate
$q(t)$	External impact /excitation
q_i	Generalized coordinates
q_r	Generalized coordinates
$q_i(t)$	Generalized coordinates depending on time
$q(x, y)$	Intensity of harmonic load
Q	Force, load
Q	Shear force
Q_i	Generalized forces
r	Reaction force
r_{ik}	Unit reactions
r	Polar coordinate
R	Resultant force, reaction, radius
R_s	Restoring force
$R(t)$	Relaxation kernel
R_λ	Resolvent of the self-adjoint operator
S	Static moment,
S_i	Generalized elastic forces
S_i	Seismic force
S	Impulse of impact force
S_0	Initial single instantaneous impulse
S	Tension of the string
s	Number of floor columns
t	Time
t	Thickness
T	Period of vibrations/oscillations
T_p^0	Short period of time (period of action)
T_p	Moment of termination of the force
T	Tension of membrane
$\boldsymbol{T_n}$	Matrix differential operator of stresses
u	Displacement, displacement component
u_0	Initial deflection
U	Potential energy of deformations, total potential energy, potential energy of elastic forces
U'	Functional (additional energy)
$U(\xi, \tau)$	Transform of displacement
v_0	Initial velocity
v	Velocity, linear velocity

v_i	Velocity of the concentrated masses m_i
V	Potential energy of external forces, virtual work of external forces, potential energy of the masses
V'	Additional work
w	Deflection
W	Actual work of internal forces
W	Work of internal forces
$\overline{W}_\delta(\xi, \eta, t)$	Impulse transient function
w_δ	Impulse transient response function
x, y, z	Orthogonal coordinates
x	Linear displacements in direction of X axis
x	Variable
x, y	Cartesian coordinates
y_0	Initial displacement
y	Deflection, displacement
$\ddot{y}$	Mass acceleration
$\dot{y}$	Velocity
$y(x)$	Function
$y^{(n)}(x)$	Function of high orders
$y^{[k]}(x)$	Quasi-derivatives
$y_i(t)$	Generalized displacements, generalized coordinates
y_{st}	Static deflection, static displacement
$y_\delta(t)$	Impulse response function
z	Coordinate
z	Linear displacements in direction of Z axis

Greek Symbols

α	Damping coefficient
(α, β)	Interval of continuity
α, β	Constants
α	Angle of the front of seismic excitation to the longitudinal axis of the bridge
α, β, γ	Angles of rotation relative to the coordinate axes
β	Coefficient
β	Spectral curve
γ	Weight density
γ, δ	Damping parameters
γ	Loss coefficient
γ	Shift angle
δ	Decrement of vibrations/oscillations
δ	Displacement, deflection
δ_{st}	Static deflection, static displacement
δ_{ik}	Unit displacement
$\lvert\delta\rvert^e$	Matrix of the displacements of nodal points
$\delta(x)$	Delta function

$\delta_{\lambda\mu}$	Kronecker's symbol
δ_{q_i}	Infinitesimal displacement
Δ	Increment of a quantity, deflection
Δ	Laplace operator in rectangular coordinates
Δ_r	Laplace operator in polar coordinates
Δ	Displacement, total displacement
Δ	Matrix of unit displacements
Δ_y	Absolute shift
Δ_i	Generalized displacements
$\vec{\Delta}_p$	Column matrix
ΔH	Dynamic part of the thrust
$\Delta B_j w(a)$, $\Delta B_j w(b)$	Jump functions in the sections $x = a$, and $x = b$
$\chi,\ \varepsilon,\ \gamma$	Deformations from corresponding internal forces M, N, and Q
ε	Arbitrary small value
ε	Relative deformation
ε	Coefficient of non-linearity of deformation
ε_{xt}	Relative elongation
ε	Damping coefficient per Voight
ζ	Damping ratio (fraction of critical damping)
θ	Circular frequency of excitation force
θ	Polar coordinate
α	Damping coefficient per Voight hypothesis
η	Dynamic deflection
η	Modes of natural vibrations
η_{ir}	Coefficient of modes of oscillations taken by building codes
λ	Constant
λ_i	Characteristic number
λ	Intermediate value of displacement
$\vec{\lambda}$	Deformation vector
λ, μ	Lame parameters
μ	Dynamic coefficient
μ	Coefficient characterizing the section of the structure
ν	Poisson's ratio
ν	Variable
ξ	Variable
ξ	Variable
ρ	Density of the material
ρ	Mass of the system per unit length
τ	Current coordinate of time
τ	Period
τ	Shear stress
$\tau(\lambda)$	Spectral function of distribution
τ_i	Normalizing numbers of discrete decomposition
φ_0	Initial phase of free vibrations/oscillations
φ_{ij}	Angle of rotation of the node with the coordinates i and j

$\varphi(x, y)$	Eigenfunction
$\ni$	Total potential energy of the system
σ	Normal stress
$\sigma_x, \sigma_y, \tau_{xy}\ \tau_{xz}$	Plate's stress components
σ'	Weighting function
$\psi(x, y)$	Eigenfunction
ψ	Absorption coefficient
Φ	Basic functions or functions of modes
$\Phi(\omega)$	Transfer function
$\varphi(x)$	Some arbitrary functions
$\varphi_i(x)$	Fundamental functions
Ω	Extended domain
Γ	Border
$\Gamma'(1)$	Derivative of the gamma function
ω_0	Natural angular frequency
ω	Frequency of excitation force
ω	Frequency of free oscillations/vibrations
ω_n	Natural frequency
ω_d	Damped natural frequency
∇^2	Laplace operator in Cartesian coordinates
∇^*	Matrix differential operator

Special Constants

e	Natural base of logarithms $= 2.71828182845\ldots$
i	Imaginary unit $= \sqrt{-1}$
π	$3.14159265358\ldots$

General Concept and Principles of Structural Dynamics 1

Abstract

This chapter is an overview of structural dynamics and covers the following: classification of oscillatory systems and vibration process, general provisions covering linear and nonlinear systems, stationary and nonstationary systems; conservative and nonconservative systems; generalized forces; free oscillations/vibration; forced oscillations/vibrations; dissipative system; self-oscillating/vibrating systems; and autonomous systems. Types of dynamic loads include vibration loads, impact and impulse loads, moving loads, and arbitrary loads with excitation as a force or displacement. Furthermore, the chapter discusses methods of discretization of dynamic problems, including dynamic degree of freedom, dynamic design models, generalized displacements, including how to formulate equations of motion, and vibration analysis of a single degree-of-freedom system.

1.1 Overview of Structural Dynamics

A great contribution to the development of the theory and analysis of Structural Dynamics in the last several decades was made due to the fundamental works of the following scientists and research studies whose input should be acknowledged: S.P. Timoshenko [1], C.Y.Harris, C.E. Crede [2], R.W. Clough, and Penzien J. [3], A.K. Copra [4], A.M. Kabe, B.H. Sako [5], H.R. Busby, G.H. Staab [6], J.T. Katsikadelis [7] A.P. Smirnov [8], V.V. Bolotin [9], Y.S. Uflyand [10], I.I. Vorovich and V.A. Bobeshko [11], R.R.Jr. Craig, and A.J. Kurdila [12], B.G. Korenev [13], A.I.Tseitlin [14], J.C. Anderson, F. Naeim [15], Zhihui Zhou, and Ying Wen, et al.[16], MarioPaz, and Young Hoon Kim [17], F Ladislav [18], I.b. Czeslaw [19], F.Y. Cheng [20], J.F. Doyle [21], W. Goldsmith [22], N. Cristescu [23], N.A.Klichevski [24], J Norman [25], I.M.Gelfand, and S.V. Fomin [26], V. Slivker [27], W. Wunderlich, W.D. Pilkey [28], and many others who were not highlighted here.

It is well known that a mechanical system under disturbance is able to make free oscillations. Oscillation is the variation, usually with time, of the magnitude of a quantity with respect to a specified reference when the magnitude is alternately greater and smaller than the Ref. [2]. Free oscillations in themselves are usually of little interest because, as a result of damping, they quickly disappear. Free oscillation/vibration of the system is the oscillation that occurs in the absence of forced oscillation/ vibration. However, in the oscillatory process of mechanical systems free oscillations have an important role. Free oscillations characterize the behavior of a mechanical system under different

L. G. Petrosian, *Dynamics of Structures*, https://doi.org/10.1007/978-3-031-63539-7_1

conditions of oscillations; in other words, they characterize the dynamic individuality of a system, construction, structure, etc. Free oscillations are possibly due to the fact that mechanical systems have two important properties. First, mechanical systems have mass and, therefore, during the motion/oscillations/vibrations such mechanical systems can accumulate kinetic energy. Second is the ability of a mechanical system to accumulate potential energy when it deviates from the state of equilibrium.

Mechanical systems have a number of modes of free oscillation. These modes are called natural modes of free oscillation. Each natural mode of free oscillations corresponds to a certain frequency and rate of damping of oscillations. The fundamental mode of oscillation of a system is the mode having the lowest natural frequency. Natural frequency is the frequency of free oscillation/vibration of a system. Vibration is an oscillation wherein the quantity is a parameter that defines the motion of a mechanical or structural system. Vibrations of a mechanical or structural system are oscillations of the system about an equilibrium position.

Thus, the natural frequencies, natural modes, and damping rates of a mechanical system or a structure are its main dynamic characteristics, and they are not related or dependent on external excitations. The dynamic response of the structure, or the dynamic effect, or in other words, the reaction of a mechanical system to dynamic excitations strongly depends on the natural frequency of the structure, its basic mode(s) of oscillations, as well as the damping properties of the structure itself and its environment. Since the main dynamic characteristics have a particularly important role in oscillations/vibrations, we will consider each of them separately as the material is presented.

It is also known that increasing the mass of a mechanical system or structure leads to a decrease in the frequency of the system's natural oscillations (system reduces the frequency of the system's natural oscillations), while increasing the rigidity of a mechanical system leads to an increase in all its natural frequencies. Note here that different frequencies vary to varying degrees. However, if the rigidity of a mechanical system is proportional to its mass, the increase in the mass does not change its natural frequency of system oscillations.

Thus, when calculating structures for dynamic action, it is necessary to know their individual features that are the basic dynamic characteristics of the structure: the natural frequencies of the structure, the corresponding modes of oscillations, as well as damping properties of the structure and the environment.

1.2 Classification of Oscillatory Systems and Oscillation/Vibration Processes

1.2.1 General Provisions

The analytical dependence between the vector of the system state and the vector of influence of the system from the environment for the oscillatory mechanical system can be written in the following form:

$$\boldsymbol{L}\boldsymbol{w}(t)=\boldsymbol{q}(t) \tag{1.1}$$

where $\boldsymbol{L}$ is the operator of the system, which includes all the equations and additional conditions necessary for an unambiguous description of the behavior of a system $\boldsymbol{w}(t)$ under external impact (excitation) $\boldsymbol{q}(t)$. For mechanical systems, Eq. (1.1) is reduced to the system of some differential equations with boundary and initial conditions:

Linear and Nonlinear Systems The system is called linear if its operator is linear, i.e., it satisfies the condition:

$$\boldsymbol{L}(\alpha_1 \boldsymbol{w}_1 + \alpha_2 \boldsymbol{w}_2) = \alpha_1 \boldsymbol{L} \boldsymbol{w}_1 + \alpha_2 \boldsymbol{L} \boldsymbol{w}_2, \tag{1.2}$$

for any admissible laws of change of state condition $\boldsymbol{w}_1(t)$ and $\boldsymbol{w}_2(t)$ and any numerical multipliers/coefficients α_1 and α_2. In other words, a system is linear if for every element in the system, the response is proportional to the excitation. The dynamic properties of each element in the system can be represented by a set of linear differential equations with constant coefficients. Formula (1.2) expresses the principle of superposition for linear systems. If under the external action (excitation) $\boldsymbol{q}_1(t)$ the behavior of the system is described by the vector $\boldsymbol{u}_1(t)$ and with under the external action (excitation) $\boldsymbol{q}_2(t)$ the behavior of the system is described by the vector $\boldsymbol{u}_2(t)$, then the behavior of the system under the external action $\boldsymbol{q}_1(t) + \boldsymbol{q}_2(t)$ will be described by the vector $\boldsymbol{w}_1(t) + \boldsymbol{w}_2(t)$. Superposition principle is one of the most important properties of linear system. If the condition (1.2) is not met, then the system is called nonlinear.

Stationary and Nonstationary Systems If the properties of a system are not changed at a given interval of time, then the system is called stationary at that time interval. If the properties of a system change over time, then the system is called nonstationary. The processes occurring in stationary systems are described by differential equations with constant coefficients. The processes occurring in nonstationary systems are described by differential equations with variable coefficients.

In stationary problems, the time parameter can be excluded. An example of such stationary problem can be the uniform movement of a load along an infinite beam.

Conservative and Nonconservative Systems The system is called conservative if its total mechanical energy remains constant during oscillations/vibrations. Otherwise, the system is called *nonconservative*.

Generalized Forces Generalized forces are linear combinations of simple forces acting on the system. There could be a total of n linearly independent generalized forces in a system with n degrees of freedom.

Free Oscillations/Vibrations Free oscillations of a mechanical system are oscillations that occur in the absence of excitation. Excitation is an external force (or other input) applied to a system that causes the system to respond in some way.

Forced Oscillations/Vibrations The oscillations of a system are forced if the response is imposed by the excitation. If the excitation is periodic and continuing, the oscillation is steady-state.

Dissipative System A system is called dissipative if its total mechanical energy decreases with any movement of the corresponding autonomous system.

Self-Oscillating/Vibrating System A system is called self-oscillating/vibrating system if it is stationary and autonomous. In addition, self-excitation of oscillations is possible under certain circumstances. Self-oscillating/vibrating systems are characterized by the presence of a non-oscillatory energy source in them. In this case, the incoming energy is regulated by the movement of the system itself.

Autonomous Systems In the operator, Eq. (1.1) for autonomous systems should be taken $q \equiv 0$. Oscillatory processes in autonomous systems can occur only at the expense of internal energy sources, or energy reported (communicated) in the form of initial perturbation. The rest of the systems are called *nonautonomous*.

Many impacts on the engineering structures have a distinctly expressed dynamic character. Dynamics of structures is one of the special sections of structural mechanics; it is engaged in the development of principles and methods of calculation of structures to the action of dynamic loads.

Dynamics is a branch of mechanics concerned with the motion of bodies under the action of forces. As these forces vary over time, they induce varying responses such as displacements, reactionary forces, stresses, velocities, accelerations, etc. One of the most notable differences between static and dynamic analysis is the explicit consideration of inertial forces from varying loads that change within a dynamic structure over time. These time-varying loads cause vibrations of structures in connection with which inertial forces appear, causing additional internal forces and stresses.

A dynamic load is defined as any load that changes in value, position, direction, or place of application over time and creates tangible inertial forces on a structure. These inertial forces produced by masses of both the dynamic load and the structure under the dynamic load play a significant role in dynamic structure analysis.

Under the action of a dynamic load, the development of system deformations and the occurrence of displacements in the system represents a process that varies over time.

Under a dynamic load, the inertial forces impact each particle of the structure in addition to external and internal forces within the structure.

In dynamics of structures, it is common to use the word "mass" in a general sense to designate a rigid body. The mass of a rigid body in a mechanical system can be considered as a property of the body responding with acceleration under applied force. The oscillating processes occurring in these systems are called mechanical oscillations. A mass of a rigid body has an acceleration, according to Newton's law, which is proportional to the resultant of all forces acting on the mass.

For a deforming body, the rate of dynamic loading does not correspond to the rate of growth of internal forces; therefore, the equilibrium is disrupted, and the body comes into motion.

At the same time, the speeds and especially the accelerations of the moving system can reach dangerous values both for individual parts and for the entire structure.

Thus, a significant difference between dynamic and static methods of calculation is the introduction of a new variable—time, which participates in the equations either explicitly or in the form of derivatives of unknown functions over time.

The special feature of dynamic loads is the fact that in most cases they cause vibrations of structures.

The term vibration is used where vibrations have a relatively small amplitude and not too low frequency of oscillations. In technology, especially in mechanical engineering, the term vibration is widely used. For our purposes, the term vibration is synonymous with the terms mechanical oscillations or oscillations of a mechanical system.

A dynamic load that varies in magnitude with time and repeats itself at regular intervals is called a periodic load. With a periodic repetition of small dynamic impacts under certain conditions, a system can accumulate energy which is expressed in a gradual increase in the range of oscillations, and, with it, the intensity of inertial forces up to very large sizes. This is explained by the ability to make elastic oscillations of various frequencies. A structure is responsive to such periodic actions of a load of a certain specific frequency. This explains why a small periodic load of prolonged action is able to shake a structure to dangerous limits.

This phenomenon is called resonance and further explains the danger of destruction by low-impact forces on a structure that is otherwise strong enough to withstand usual static loads.

The purpose of dynamic calculation is to check the strength of structures as well as to determine the dynamic displacements, velocities, and accelerations whose values under the influence of vibrations on people and certain types of precision equipment, such as measuring instruments, high-precision machines, and equipment for microelectronics production, should not exceed the permissible limits.

If a system is under the action of a statically applied load, i.e., if the load is applied to the body sufficiently slowly, monotonically, without noticeable accelerations, the internal forces can only be determined when the system satisfies the conditions of static equilibrium.

A static problem has only one solution and one response which is either a displacement or deflection, whereas a dynamic problem has mainly three responses such as displacement, velocity, and acceleration and an infinite number of solutions that are time dependent in nature. Therefore, it is necessary to choose a sequence of solutions corresponding to all moments of time which are of interest in determining the reactions of the system. Thus, dynamic analysis is more complex and time-consuming than static analysis.

Usually, where the static problem is solved using ordinary algebraic or transcendental equations, the corresponding dynamic problem requires the solution of differential equations with the derivatives over time.

The displacements of systems under dynamic load are determined not only by external forces but also by inertia forces of the system itself.

The presence of forces of inertia, which seem to resist the acceleration of a system, is a very important and distinguishing feature of the problems of dynamics of structures. In addition, the dynamic effects affect the mechanical properties of a system, including the phenomenon of fatigue of the material increase, elastic hysteresis, and the physical law of deformation.

Hence, when distinguishing the dynamics of structures from the static of structures it is necessary to take into account the presence of resistance forces. The nature of the resistance forces is complex and diverse; therefore, the inclusion of the resistance forces is more complicated than the account of the forces of inertia.

It is known that hysteresis takes place at static loading and in the case of dynamic loading it becomes of particular importance.

Hysteresis (Hysteresis loop) depends on the speed of motion of a system and affects the damping of vibrations as a result of energy dissipation. The question about the influence of dynamic effects on the mechanical properties of the material is very important. At the same time, accounting for these phenomena significantly complicates the construction of equations of motion for deforming system.

Thus, the dynamic calculation is considerably more complex and labor-intensive than static. Dynamic calculations involve not only the determination of maximum dynamic forces in the elements of structures but also the study of modes of vibrations of elastic systems. As mentioned before, the term vibration is most often used where oscillations have relatively small amplitude and without too low a frequency.

An engineer must study measures directed toward the decrease of dynamic effects and extinction of harmful vibrations. Undoubtedly, the most unpleasant consequence of vibration may be the fatigue of material. This type of structural failure usually is so catastrophic as it is unexpected. Often, fatigue failure plays very insidious role, since usually it is not preceded by any signals of impending danger.

One of the conditions for the existence of free oscillations is the accumulation of energy by a mechanical system when it is taken out of equilibrium. There are two types of dumping: (1) natural damping, associated with the natural frictional forces of the system itself, the structure, the construction, and the environment, and (2) artificially introduced damping. The latter is used to reduce the impact of dynamic loads on structures. Prior to installing the artificial damping into a structure, the structure and the damping system itself should be subjected to a thorough theoretical analysis and calculation.

When solving problems of the dynamics of structures, engineers need to know the values of the natural frequencies of the system's vibrations and the corresponding modes of vibrations. The average density of the natural frequency distribution increases with the number of the corresponding vibration modes.

In structures with low friction, we expect more intense vibrations as compared to high-friction structures. Energy dissipation occurs in any oscillating/vibrating system. For example, building structures should have significant damping, as it is extremely important in terms of the behavior of buildings under vibration loads and seismic impacts.

Experiments have shown that the increase in friction does not lead to noticeable changes in the frequencies and modes of free oscillations of a system. Therefore, in practical calculations in order to determine the natural frequencies and modes of free oscillations of a given system, i.e., a system with damping, it is possible to accept the frequencies and modes of free oscillations of the same system, but without taking into account the damping. This approach avoids complexity in determining frequencies and modes of oscillations. Thus, the frequencies and modes of free oscillations of the real system, where damping (friction) is taken into account, differ little from the frequencies and modes of free oscillations of the same imaginary but nonetheless the same system without damping.

Natural vibration/oscillation modes of a system create a certain sequence of vibrations/oscillations modes and each such created mode differs from all others. That is, each mode of the system's natural vibrations/oscillations is orthogonal to all other modes of oscillations.

The frequencies and modes of natural oscillations of a system are determined by the values of masses and stiffnesses of structures, as well as their distribution within the structure.

Each natural mode corresponds to a certain natural frequency. In any structure, experimentally, it is possible to excite free vibrations with frequency and mode close to the found frequency and mode of free vibrations of an imaginary system. Slight differences in these characteristics are due to the presence of friction/damping.

Thus, the mass of the system, its rigidity (stiffness), and damping properties of both the system and the environment are the main factors that determine the system's free oscillations. All of these characteristics can vary under the influence of numerous causes. One of these causes is the effect of damping. For example, with high damping, the frequencies and modes of free oscillations/vibrations noticeably change under the action of damping, and they can no longer be considered the same as in a system without damping.

Thus, our assumption that the vibrations/oscillations of each of the natural modes are independent of the vibrations/oscillations of all other modes is valid ***only*** when the damping is very small. For example, with significant damping, if the initial deformation of the structure is set to correspond to one of its natural modes, the subsequent free oscillations occur also under the influence of other modes of oscillations. That is the damping forces arising from free vibrations lead to the mutual influence of their natural modes of vibration. The latter circumstance greatly complicates the study of vibrations of engineering structures.

Solving dynamic problems is simplified under the conditions of instantaneous propagation of deformation throughout the body or structure, and in the case of forced vibrations, it is assumed that natural vibrations are damped. These simplifications are not always applicable. In earthquakes, destruction occurs in the initial period of time (first 1–2 s). That is, when the free oscillations do not have time to damp out. Therefore, the natural oscillations caused by the action of the disturbing force must be taken into account in the transient process.

The instantaneous propagation of deformations throughout the body or throughout the system creates a more dangerous situation for it than the case when the deformation does not have enough time at the moment of loading or impact to spread throughout the body or throughout the system. For example, when a bullet is fired, flying through a glass, it forms only a hole in it at the point of impact. The stone shatters the glass when it is struck.

In the first case, i.e., the bullet example, the time of impact of the bullet on the glass is so short that the deformation does not have time at the moment of impact to spread over the glass. Therefore, the glass receives only a small local destruction. In the second case, the impact time of the stone on the glass is longer, as a result, the deformation has time to spread to a large area of the glass and this causes its general destruction.

1.3 Types of Dynamic Load

1.3.1 General Provisions

When dynamic loads act on a structure, the inertial forces of masses of these loads and masses structures arise and play a significant role. The most important characteristic of dynamic load is its intensity. Dynamic loads are divided into periodic, impulsive, random, and random and nondeterministic loads. Dynamic loads are divided into the following main types: *vibration load; impact load; moving loads; and arbitrary loads*. In the process of operation, constructions undergo various kinds of dynamic influences. These include:

1. *Vibration Load*. Periodic loads exhibit a time variation that is repeated at regular intervals, called periods. Periodic loads act on certain points of construction and are created by stationary machines and mechanisms with moving parts, such as electric motors, turbine generators, and machine tools. Rotating machines that are not fully balanced around the center of rotation will give rise to excitation forces that vary with time. The loads of this type are almost independent from the properties of structures they act on, but they are the basic sources of vibrations of these constructions.
2. *Impact Load*. These are short-term-impulsive loads, created by falling loads and falling parts of power plants: hammers, pile drivers, etc. These loads are characterized by a short duration of action and depend on the elastic and inertial properties of a structure, receiving the impact.
3. *Moving Loads*. Moving load is a load of either constant or changing intensity over time, the position of which in the span of structures also varies with time. These moving loads can be accelerating or deaccelerating, for example, the loads of rolling stock of railroads, electric locomotives, streetcars, trucks, buses, cars, mobile cranes, etc.
4. *Arbitrary Loads (Seismic Load)*. Seismic loads occur in the form of random movements of the ground, causing forced displacements of structural foundations. Seismic loads vary in time according to complex laws and, as a result, cause complex vibrations of structures. When earthquake occurs, a construction undergoes a dynamic motion. Since earthquake motions vary with time and inertia forces vary with time and direction, seismic loads are not constant in terms of time and space.
5. *Arbitrary Loads (Wind Load)*. Wind loads are a dynamic and random phenomenon in both time and space. Wind loads are quite obviously nondeterministic. Wind loads are produced by the flow of wind around structures. Wind load magnitudes vary in proportion to the distance from the base of a structure, peak wind speed, type of terrain, and side and shape of the structure. The dynamic response involving deformations is usually vibratory/oscillatory in its nature. Structures under wind loads vibrate about a configuration of a stable equilibrium. For example, suppose a building structure is in a state of a static equilibrium under the gravity loads acting on it. When subjected to wind, the structure will vibrate/oscillate about this position to the state of equilibrium.

Thus, dynamic loads also include wind, load from blast wave, influence of sea wave on constructions, and others. Dynamic loads can also be combined, for example: an impulse periodic or shock-vibration from the pile driver of a periodic action. In this book, we will mainly consider the effects of a vibrating load in the form of periodically acting forces varying in time according to the harmonic law and sine or cosine waves.

Dynamic loads are time-varying loads. Dynamic loads can be grouped into one of the following categories:

- Type of excitation:
  - Free vibration/oscillation
  - Forced vibration/oscillation
    - Periodic loads
      - Harmonic loads
        (Harmonic loads vary according to a time function with or without a certain phase shift. When an applied load varies as a sine or a cosine function, it is called harmonic loading. Example of harmonic oscillations is that of a mathematical pendulum)
      - Nonharmonic loads
        (Example of nonharmonic oscillations: sea wave motion)
    - Transient loads (end structural response to dynamic loads: deterministic excitation and nondeterministic, stochastic excitation, and random excitation)
      - Impact loads
        (Example of impact loads: falling parts, equipment dropped on a structure, sudden collapse of load-bearing structure's elements, crash, collision, etc.)
      - Impulse/impact/shock or blast loads
        (Example of impulse or shock load: force impulse due to explosion.)
      - Arbitrary loads
        (Example of arbitrary load: seismic load and wind load)

The deterministic representation implies comprehensive knowledge of the behavior of a load with time. The stochastic excitation only implies the knowledge of the statistical distributions of a load's behavior. Seismic and wind loads in general can be conveniently represented by the stochastic category of excitation. A dynamic load that varies in a highly irregular fashion with time is referred to as a random load.

1.4 Methods of Discretization of Dynamic Problems

The calculational methods at the present stage of development of structural mechanics can give reliable estimates of the actual behavior of a system. However, it should always be borne in mind that any computational model is an abstraction that reflects the basic properties of a given structure's behavior. Not one single calculational model, no matter how accurate it is from a mathematical point of view, can give a solution whose accuracy exceeds the accuracy of the original data.

1.4.1 Dynamic Degree of Freedom

The complexity of dynamic calculations of an elastic system depends primarily on the degrees of freedom of the system. The number of independent parameters required to define the displacements of all the masses from their reference positions is called the number of degrees of freedom. That is, the number of independent geometric parameters determines the position of all masses at any virtual elastic displacements of the system.

Let me remind that the term "Virtual" means that an infinitely small displacement is made intentionally in any kinematic acceptable direction.

The number of degrees of freedom of an elastic system is composed of the degrees of freedom of various masses located in it. The position of any concentrated mass of an elastic system on the plane is determined by its three displacements: two progressive in the vertical and horizontal directions and by the turning of the mass itself. Hence, this mass has three degrees of freedom.

The number of degrees of freedom of an elastic system is convenient to define as the number of links that are necessary to add to the system, so that its masses are not able to move. The calculations of the degrees of freedom in the kinematic analysis of static problems are produced without accounting for its own deformations of disks and rods, which are considered as absolutely rigid.

In the problems of dynamics of structures when determining the degrees of freedom of an elastic system, the system's elastic or elastic-plastic deformations are considered. Any distributed mass on the elastic deformed system can be represented as an infinitely large number of infinitely small concentrated masses. Thus, such a system has an infinite number of degrees of freedom.

1.4.2 Dynamic Design Models (DDM)

A number of degrees of freedom depends on the type of dynamic design (calculation) model (DDM) (DCM) by which the actual structure is schematized. The calculations of structures for any dynamic loads always begin with the choice of the calculated dynamic model. Strictly speaking, any (DDM) (DCM) approximates the real properties of structures. At the same time, the design of a dynamic model should take into account the basic, most essential properties of structures. The choice of a model depends on the nature of external influences, estimated level of tension, oscillations of the structure, and their planning, design, and constructive solutions.

According to the number of degrees of freedom the dynamic design models of structures are subdivided into schemes with finite and infinite degrees of freedom. In general, the dynamic design model can be categorized as either discrete or continuous. Structures with discrete distribution of masses are considered as systems with a finite number of degrees of freedom. Structures with a continuous distribution of masses represent systems with an infinite number of degrees of freedom. However, in some cases, it is possible to adopt a system with a finite number of masses concentrated in some characteristic points of the system, by calculating the dynamic impact of these masses resulting in a calculated dynamic model of the original structure. The places of the concentration of masses are usually chosen in such a way as to combine their position with the locations of the largest vertical loads.

For example, when choosing a calculated dynamic model to determine the horizontal oscillations of a building, its mass is assumed to be concentrated at the level of overlap of each story. In the absence of such loads, concentrated masses (lumped masses) are evenly spread along or at the height of the structure. In frame structures, the positions of the concentrated masses are usually combined with the position of frame nodes. In dynamic calculations, any engineering structure loaded with a large, concentrated load compared to which its own weight of structure is small and can be neglected, can be considered as a system with single degree of freedom (SDOF).

Let us first consider a system with concentrated masses. In such systems, the inertia forces of the rods can be neglected and take into account only forces of inertia arising from the concentrated masses.

For example, the weightless beam with one concentrate mass shown in Fig. 1.1 has a single degree of freedom, since the position of this mass is determined by one parameter y.

The weightless frame with one concentrate mass shown in Fig. 1.2 has also a single degree of-freedom since the position of this mass is determined by one parameter x.

The weightless beam with three concentrated masses shown in Fig. 1.3 has three degrees of freedom despite the infinite number of possible forms of elastic line. The position of the three masses is defined by three deflections. In this case, the number of degrees of freedom of a beam equals to the number of concentrated masses.

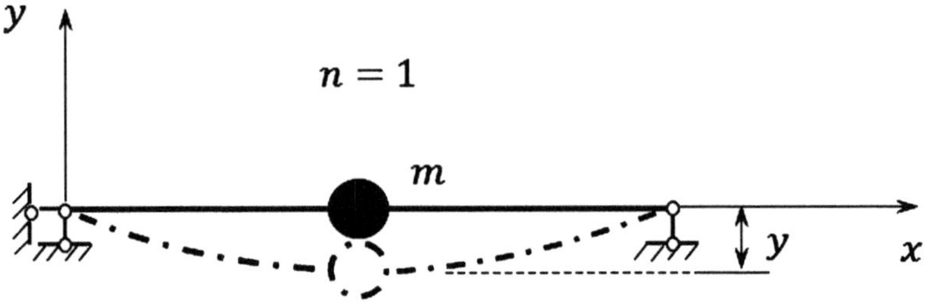

Fig. 1.1 Weightless beam with one discrete mass

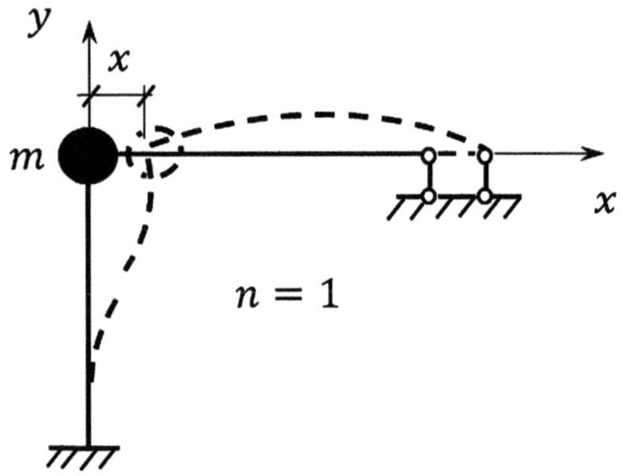

Fig. 1.2 Weightless frame with one discrete mass

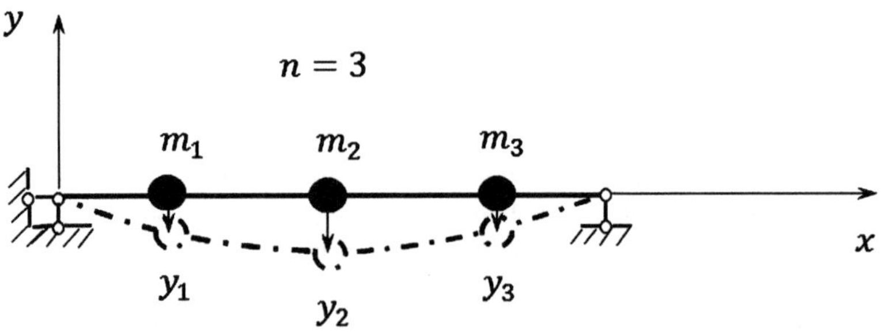

Fig. 1.3 Weightless beam with three discrete masses

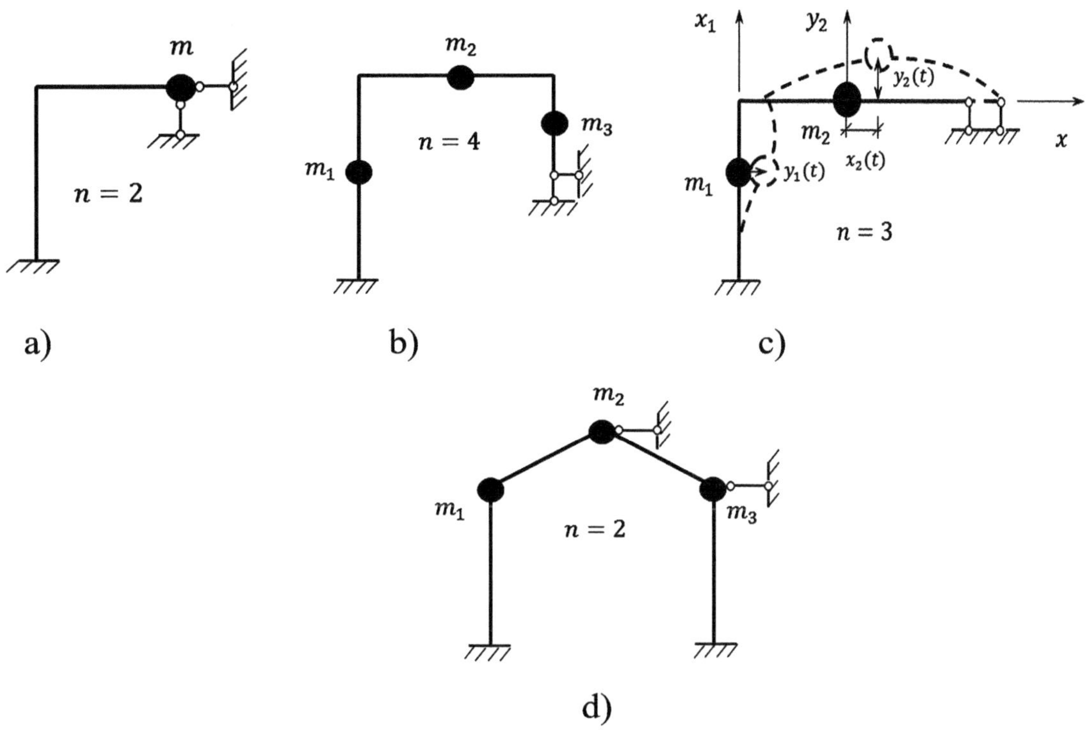

Fig. 1.4 Weightless frames with discrete masses

If in the frames shown in the figure, the deformations from the longitudinal forces are neglected, in the frame shown in Fig. 1.4a–c, the number of degrees of freedom is greater than the number of masses, and in the frame shown in Fig. 1.4d, the number of degrees of freedom is less than the number of masses.

If we take into account the longitudinal deformation of the rods, the frame shown in Fig. 1.4b, should be considered as a system with six degrees of freedom, since the masses m_1 and m_3 in this case can have not only horizontal but also vertical movements. However, in most cases, mass oscillations associated with longitudinal deformations of the rods can be neglected.

When studying oscillations of trusses deformation of elements arise only from longitudinal deformations of rods, therefore, they should not be neglected.

Let us now consider a hinge-less arch, shown in Fig. 1.5a. The dead weight (eigen mass) of the arch is represented by the three masses, concentrated respectively in the quarters m_1, m_3 and in the middle of the arch's span— m_2. Six links should be added to completely eliminate the displacements of the concentrated masses of the arch (Fig. 1.5b) thus, the system has six degrees of freedom.

Let us consider the scheme of the combined system shown in Fig. 1.6. The combined system consists of a rigid beam and a flexible arch. The weight of the beam, arch, and vertical elements of arches is distributed continuously along the span length. The dynamic calculation scheme of the bridge is taken as a system with a finite number of degrees of freedom. The application of a continuous calculation scheme is preferable, but in the case of forced oscillations, this leads to very labor-consuming computational operations.

Sufficient accuracy of calculation can be obtained if the masses of all elements are concentrated in the places of connection of rigid beams to the posts (racks, vertical elements). The value of each mass

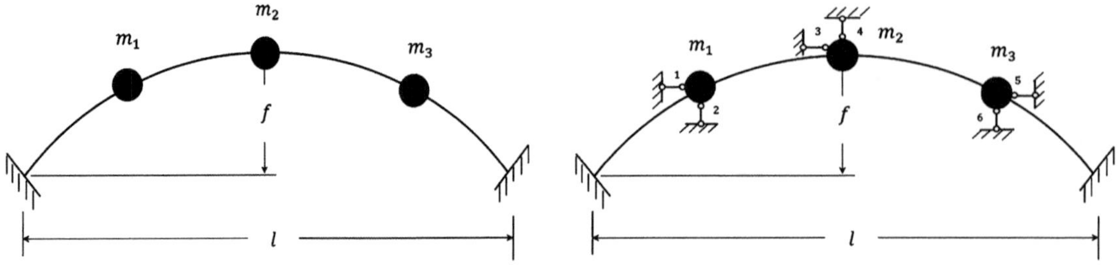

Fig. 1.5 Weightless arch with three discrete masses and the corresponding six links preventing the movement of arch system

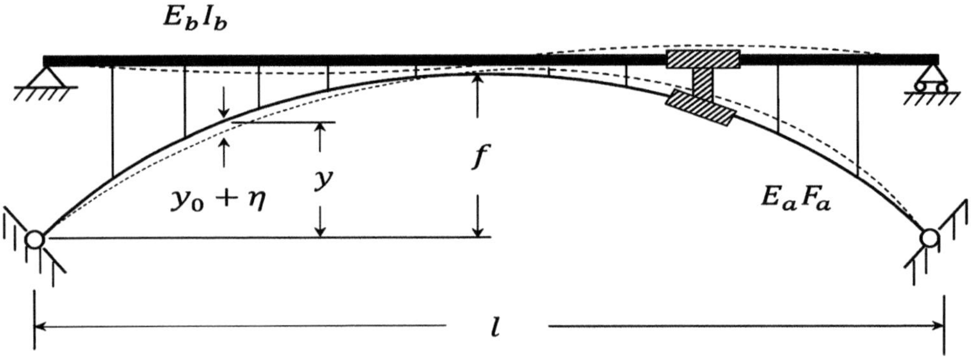

Fig. 1.6 Preliminary design scheme of the bridge structures

is composed of the mass of one panel of the beam, post (rack), and arch parts. For example, the mass in the third node (from the right side) will be equal to the sum of the masses of adjacent elements shown by the hatched surface in Fig. 1.6.

The dynamic design scheme (model) of a long-span combined bridge system of the type: rigid beam-flexible arch is taken as a system with a finite number of degrees of freedom as shown in Fig. 1.7. In this case, each concentrated mass possesses three degrees of freedom, and the position of i—th mass in the space is characterized by the displacements of the X_i, Y_i in the plane of bridge, and displacement Z_i in the direction perpendicular to the plane of bridge.

Figure 1.8a, b shows discrete and continuum dynamic models of the structure, respectively.

The inertial properties of the structure are given by the values of discrete masses m_k and the ordinates of the points of their fixation y_k ($k = 1, 2, \ldots, n$). The deformation properties can be described by unit displacements δ_k at the points of lumped masses. The stress–strain state of a discrete model at each (any) moment of time is determined by a set of relative displacements $X_k(t)$ of lumped masses from the equilibrium state. The number of degrees of freedom is equal to the number of lumped masses.

Let us now consider a continuum dynamic model (Fig. 1.8b). In this case, the system has an infinitely large number of degrees of freedom. The stress–strain state of the system is described by the function $x(y, t)$. This function determines the relative displacement of the points of the system from the equilibrium state. The inertial properties of the system are determined by the function $m(y)$. This function expresses the law of variation of the intensity of mass. The deformation properties of the system can be specified using a linear differential operator L over the function x on the variable y.

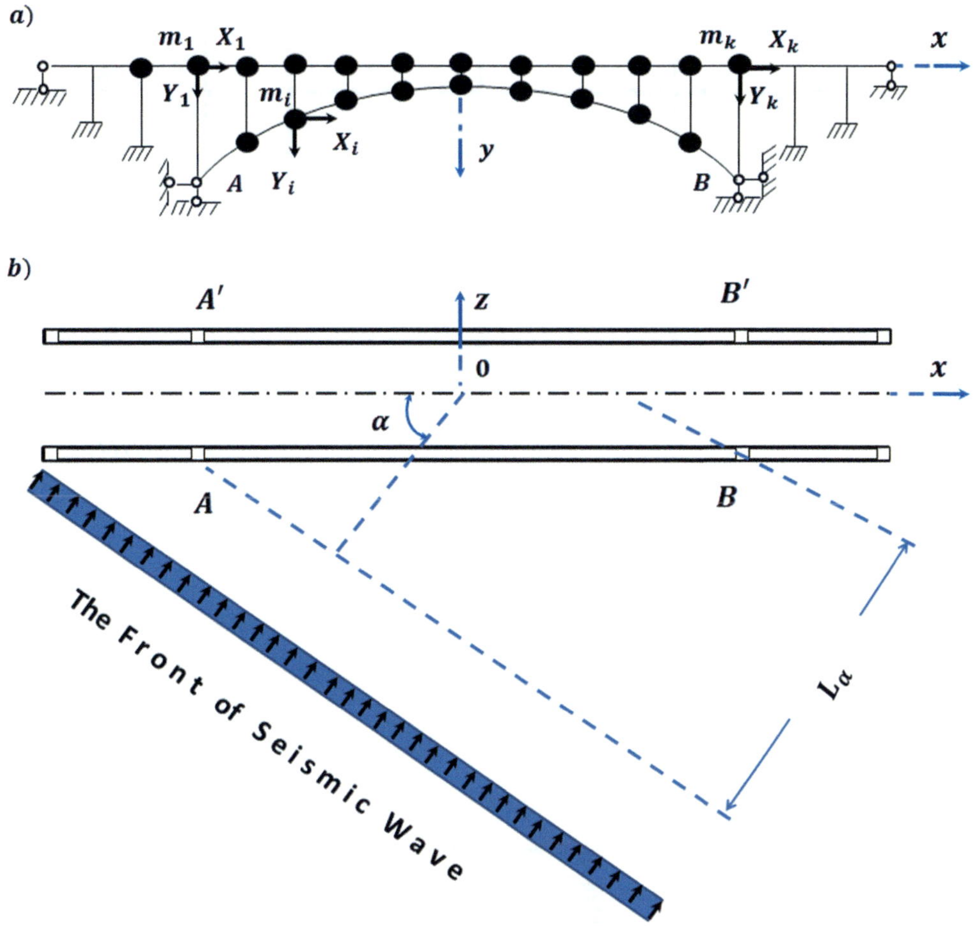

Fig. 1.7 The dynamic design scheme and the scheme of seismic excitation

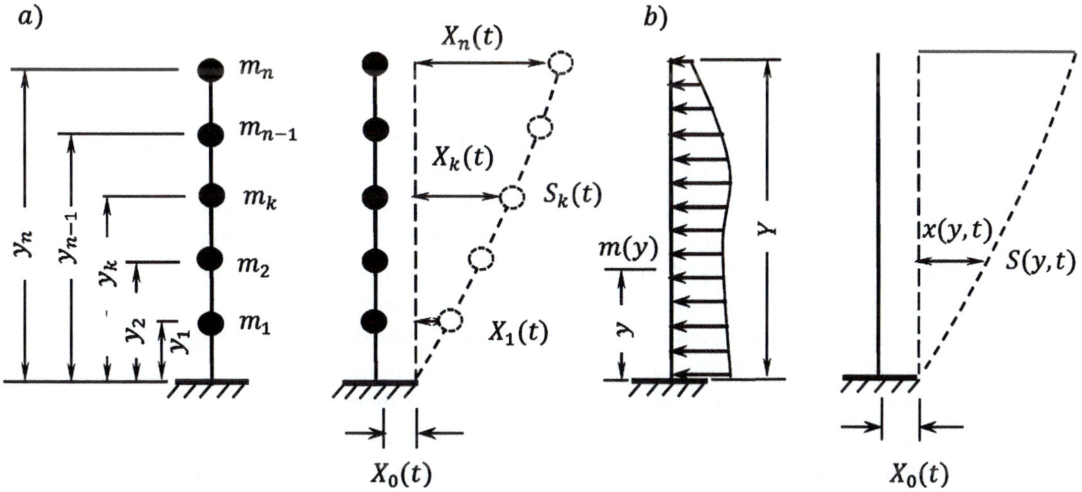

Fig. 1.8 (**a**) Discrete dynamic design scheme. (**b**) Continuum dynamic design scheme (dynamic model with distributed parameters)

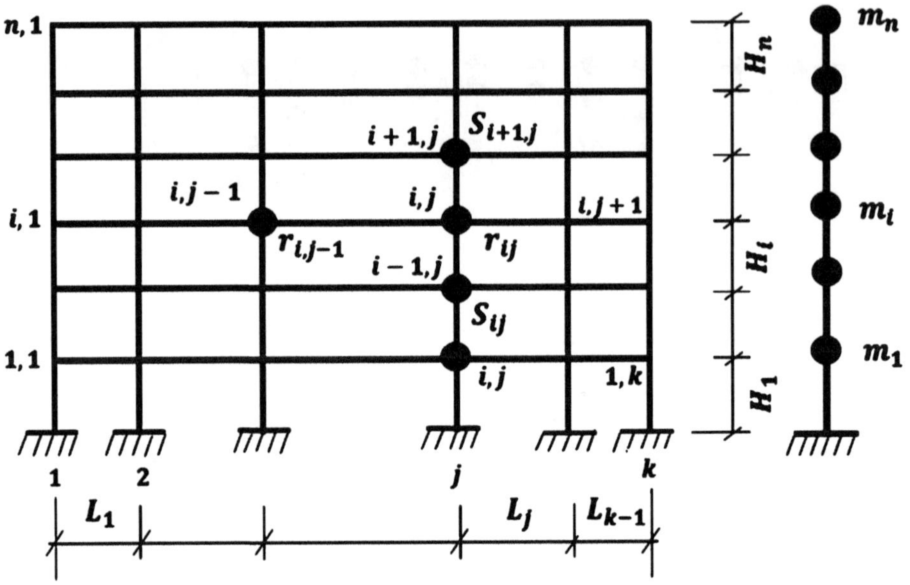

Fig. 1.9 Dynamic design scheme of the multi-story, multi-span frame structure *(Lumped mass model of n-story shear building)*

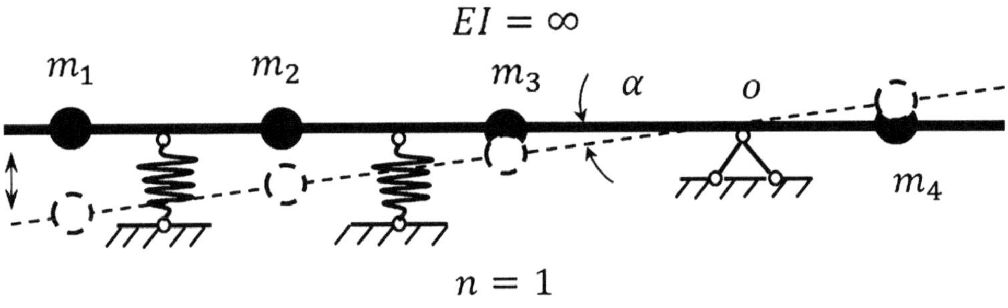

Fig. 1.10 Absolutely rigid beam with four discrete masses

Figure 1.9 shows the dynamic design scheme of the multi-story, multi-span frame structure. The dynamic design model is accepted as a system with a finite number of degrees of freedom, which is called multi-degree of freedom (MDOF)

If an oscillating system is absolutely rigid, for example, an absolutely rigid beam (Fig. 1.10) with the number $k = m_1$, m_2, m_3, and m_4 of the concentrated masses lying on spring's pliable supports rotating around a simple hinge support that is a system with a single degree of freedom, regardless of the number of concentrated masses and the number of elastic spring supports.

Hence, the position of all masses of the system is determined by one parameter, namely, the angle α of rotation of an absolutely rigid beam relative to the center of hinge support.

If in a structural system it is impossible to neglect rod inertia, then the system has an infinitely large number of degrees of freedom. For calculations of such systems, it is necessary to compile and solve systems of partial differential equations, since the inertial forces in this case depend on both the x coordinate and the time t.

If we take into account the own distributed mass of rods of elastic system, then the degrees of freedom of frames shown in the drawings Figs. 1.2 and 1.4 should be considered as a system with an infinitely large number of degrees of freedom. Since all structures have a distributed mass, they are always systems with a degree of freedom equal to infinity. However, in many cases, it is possible to reduce the calculation of such systems to calculation of a system with a final number of degrees of freedom.

To carry out a dynamic calculation, it is necessary to determine the frequencies and modes of natural vibrations of the structure, as well as the disturbing load. Following this, dynamic forces and displacements can be determined. Based on the results obtained, the question of whether there is a need to implement measures to reduce vibrations is considered.

Dynamic calculation is carried out in the following sequence:

1. Dynamic loads are determined and classified by frequency level: low-frequency, medium-frequency, and high-frequency;
2. Amplitudes of dynamic displacements are determined;
3. Fulfillment of physiological and technological requirements to limit the level of oscillations/vibrations is checked;
4. Strength calculations are carried out;
5. Amplitudes of internal forces in structures (bending moments and shear forces) are determined and strength and endurance calculation is carried out.

For dynamic calculation of structures, the method of decomposition by natural oscillations/vibration modes is most widely used. The accuracy of the calculation according to this method in a large part depends on the number of considered oscillations/vibration modes. The smaller the number of frequencies included in the frequency specter of natural oscillations, the smaller the number of terms in the decomposition according to the forms of natural oscillations. In the case of harmonic loads, it is sufficient to take into account n first modes of oscillations/vibrations in the calculations. Here n is the number of the smallest natural frequency of oscillations, which exceeds the frequency of forced oscillations. Thus, if the frequency of the main tone of oscillations is greater than the frequency of the disturbing load, only the first mode of natural oscillations can be taken into account. In practical calculations of building structures, the main case is the calculation for resonance. In this case, it is believed that the main contribution to the solution is made by the resonant mode of oscillations and therefore the influence of other modes can be neglected without a large loss of accuracy in the calculation.

The results of harmonic load calculations at or near resonance are highly dependent on changes in the dynamic characteristics of the system. For example, even a small change in the natural frequency of oscillations can increase or decrease the amplitude of oscillations many times, especially at low values of the inelastic resistance coefficient.

1.4.3 Generalized Displacements

Another method of discretization is to use generalized displacements and the corresponding basic functions. This method assumes that the form of bending the structure can be expressed as the sum of simple modes of displacement. Then these modes become the coordinates of the displacements of the structure. Assume, for example, that it is necessary to describe oscillations of a beam, which has the distributed mass with intensity $m(x)$ and bending (flexural) rigidity—$EI(x)$.

Deflections of the beam depend on two arguments: from x and from time t, i.e.,

$$y = y(x, t). \tag{1.3}$$

The unknown function (1.3) can be represented as a special set of basic (beam) functions $\Phi_i(x)$, ($i = 1, 2, \ldots n$), that satisfy the boundary conditions of the problem.

At the hinge support of the beam, the deflection can be represented as a trigonometric series of sine, and, in a general case, in the form of any other functions satisfying the boundary conditions of the problem. Instead of (1.3) let us write down a truncated series:

$$y = y_1(\mathrm{t})\ \Phi_1(x) + y_2(\mathrm{t})\ \Phi_2(x) + \ldots + y_{n-1}(\mathrm{t})\ \Phi_{n-1}(x) + y_n(\mathrm{t})\ \Phi_n(x) \tag{1.4}$$

Here, the values of y_1(t), y_2(t), ..., y_{n-1}(t), y_n(t) are the generalized displacements or generalized coordinates. The bending of the beam is determined by the generalized displacements according to formula (1.4). Under the term generalized force, we understand any force action applied to the system: concentrated force, bending moment, group of forces, etc. Under the term generalized displacement or generalized coordinates we understand a type of displacement on which the generalized force performs work. By using expression (1.4) instead of (1.3), we switched from a continuous dynamic model to a discrete model with n degrees of freedom. The advantage of this approach lies in the possibility to approximate well the form of bending of a beam or any other bending system by limiting the terms of series of independent harmonic. So, for example, a system with three degrees of freedom will be described by only three terms of the trigonometric series, and so on.

In generalized coordinates, expression (1.4) may be written in the form:

$$y = \sum_{i=1}^{n} q_i(t)\Phi_i(x) \tag{1.5}$$

where q_i—generalized coordinates depending only on time; Φ_i—basic functions or functions of forms (modes) depending only on spatial coordinates.

Based on the analysis we can see that the number of degrees of freedom depends not only on the geometry of the system but also on the accuracy with which we desire to obtain the final result of the calculation. Thus, in each individual case, it is necessary to first formulate the basic assumptions and then determine the number of degrees of freedom. The number of degrees-of-freedom is one of the main characteristics of a system in dynamic calculations. The greater the number of degrees of freedom, the more difficult the dynamic calculation is. Let us also note that the number of degrees of freedom does not depend on whether the system is statically determinable or statically indeterminable.

References

1. Timoshenko, S. P. (1967). *Oscillations in engineering* (444p). Nauka.
2. Harris, C. Y., & Crede, C. E. (2002). *Shock and vibration* (5th ed., 1456p). McGraw W-Hill Book Company.
3. Clough, R. W., & Penzien, J. (1993). *Dynamics of structures* (2nd ed., 738p). McGraw-Hill College.
4. Copra, A. K. (2016). *Dynamics of structures* (992p). Pearson.
5. Kabe, A. M., & Sako, B. H. (2020). *Structural dynamics fundamentals and advanced applications* (Vol. I, 1st ed., 970p). Academic Press.
6. Busby, H. R., & Staab, G. H. (2017). *Structural dynamics: Concepts and applications* (568p). CRC Press.
7. Katsikadelis, J. T. (2020). *Dynamic analysis of structures* (1st ed., 922p). Academic Press.
8. Smirnov, A. P. (1974). *Construction stability and oscillations* (571p). Transzheldorizdat.

9. Bolotin, V. V. (1976). *The dynamic stability of elastic systems* (600p). Gostekhteoretizdat. (State Publishing House of Technical and Theoretical Literature).
10. Uflyand, Y. S. (1967). *Integral transformations in the problems of the theory of elasticity* (420p). Nauka.
11. Vorovich, I. I., & Babeshko, V. A. (1979). *The dynamic mixed problems of the theory of elasticity for non-classical domains* (320p). Nauka.
12. Craig, R. R., Jr., & Kurdila, A. J. (2011). *Fundamentals of structural dynamics* (2nd ed., 752p). John Wiley Sons, Inc.
13. Korenev, B. G. (1960). *Some tasks of the theory of elasticity and thermal conductivity solving in bessel functions* (458p). Fizmatgiz.
14. Tseitlin, A. I. (1984). *The applied methods of solving the boundary-value problems of structural mechanics* (334p). Stroyizdat.
15. Anderson, J. C., & Naeim, F. (2012). *Basic structural dynamics* (280p). John Wiley & Sons, Inc.
16. Zhou, Z., Wen, Y., et al. (2021). *Fundamentals of structural dynamics* (286p). Elsevier.
17. Paz, M., & Kim, Y. H. (2018). *Structural dynamics: Theory an computation* (6th ed., 634p). Springer.
18. Fryba, L. (1999). *Vibration of solids and structures under moving load. Institute of Theoretical and Applied Mechanics Academy of Sciences of Czech Repablic* (495p). Thomas Telford.
19. Bajer, C. I., & Dyniewicz, B. (2012). *Numerical analysis of vibrations of structures under moving inertial loads* (300p). Springer.
20. Cheng, F. Y. (2001). *Matrix analysis of structural dynamics applications and earthquake engineering* (1016p). CRC Press.
21. Doyle, J. F. (1991). Static and dynamic analysis of structures with an emphasis on mechanics and computer matrix methods. In *Solid mechanics & its application* (442p). Springer.
22. Goldsmith, W. (2001). Impact. In *The theory and physical behavior of colliding solids* (390p). Dover Publications, Inc.
23. Cristescu, N. (1967). *Dynamic plasticity* (*Series in applied mathematics and mechanics*) (Vol. 4, 620p). North-Holland Publish Company.
24. Kilchevcki, N. A. (1976). Dynamic contact compression of solid bodies. *Impact*. Naukova Dumka. Kyiv, 319p.
25. Jones, N. (2012). *Structural impact* (2nd ed., 600p). University of Liverpool. Cambridge University Press.
26. Gelfand I. M., & Fomin, S. V. (2000). *Calculus of variations* (240p) (R. A. Silverman, Trans.). Dover Publications Inc.
27. Slivker, V. (2007). Basic variational principles of structural mechanics. In *Foundations of engineering mechanics theory and applications* (700p). Springer-Verlag.
28. Wunderlich, W., & Pilkey, W. D. (2019). *Mechanics of structures. Variational and computational methods* (2nd ed., 912p). CRC Press & Francis Group.

2 The Fundamentals of the Theory of Calculation of Elastic Systems and Variation Principles of Structural Mechanics

Abstract

This chapter deals with the basic fundamentals of the theory of calculation of elastic systems and variation principles of structural mechanics. This chapter discusses the following: potential field of forces; potential energy of masses and definition of force; virtual work of external and internal forces; the principle of virtual displacements; virtual work of external forces or potential deformation energy; total potential energy; extreme properties of potential energy; the principle of virtual displacements; the methods of drawing up equations of motion; Lagrange variation principle, and Lagrange formula; the principle of virtual changes in the stress state and formulation of the principle; Castigliano variation principle; the principle of least work; Hamilton principle; and the principle of reciprocity of work.

The variation principles of structural mechanics discussed in this chapter present a conceptually new approach to studying the course of dynamics of structures. Thorough knowledge of the basic theorems of elastic systems and variation principles of structural mechanics are the bases for a successful study of general and special courses in the dynamics of structures. From a methodological point of view, the methods and materials presented in the second chapter allow a smooth and organic transition to the study of energy calculation methods, that are very effectively used in structural and applied mechanics, in the theory of elasticity, as well as in the dynamics of structures.

2.1 The Concept of Functional and a Necessary Condition of Extreme

Variational methods of mechanics are unified and unifying principles for calculating problems of structural mechanics. Variational principles are the bases for calculating any elastic systems and allow studying problems of statics, stability and dynamics of structures from a common unified position. The basic theorems of elastic systems form the bases for using all modern methods for calculating problems of structural mechanics.

Among the numerous publications of the recent decades devoted to the application of variational principles to solving problems of structural dynamics, let me highlight the works of the following authors: S.P. Timoshenko, J.M. Gere [1], S.P. Timoshenko [2], N.N. Leontiev, D.N. Sobolev [3], A.R. Rzhanitsin [4], R.W. Clough, J. Penzien [5], M.I. Gelfand, S.V. Fomin [6], V. Slivker [7], J.C. Anderson, F. Naeim [8], A.K. Copra [9], W. Wunderlich, W.D. Pilkey [10], A.M. Kabe, B.H. Sako [11], J.T. Katsikadelis [12], and others.

L. G. Petrosian, *Dynamics of Structures*, https://doi.org/10.1007/978-3-031-63539-7_2

Let us recall that the main problem of the calculus of variations is the problem of finding the extremum of a given functional. The law according to which each function from a certain class of functions corresponds to the value of a certain numerical parameter is called a *functional*. The simplest example of a functional is a definite integral.

Let us consider some definite integral:

$$I = \int_a^b F(x, y, y')dx, \tag{2.1}$$

with boundary conditions

$$y(a) = \alpha, \quad y(b) = \beta. \tag{2.2}$$

At the same time, we will not give the physical meaning of the integrand function $F(x, y, y')$ as well as the variables x and y. Here, we will only note that the value of the integral will vary depending on the type of function $y(x)$. Expressions of type (2.1) in the calculus of variations are called functional. The search for conditions under which a functional takes stationary values is the main task of the calculus of variations. For example, if the rate of change of a function at some point in any direction is zero, i.e., when the first partial derivatives with respect to all independent variables are zero, then the function at that point has a stationary value.

Let there be some function $y(x)$, which gives the integral (2.1) a stationary value. In order to show that this value is indeed stationary, let us consider some function $\overline{y}(x)$, which is different from the function $y(x)$, but at the same time is infinitely close to it (Fig. 2.1). This condition can be represented as:

$$\overline{y}(x) = y(x) + \varepsilon\varphi(x), \tag{2.3}$$

where ε is an arbitrarily small value, $\varphi(x)$ is some arbitrary function that satisfies the conditions of continuity and differentiability. It follows from (2.3) that the function $\overline{y}(x)$ can be changed to an arbitrarily small value if ε is changed. The difference $\overline{y}(x) - y(x) = \varepsilon\varphi(x)$ is called the variation of the function $y(x)$.

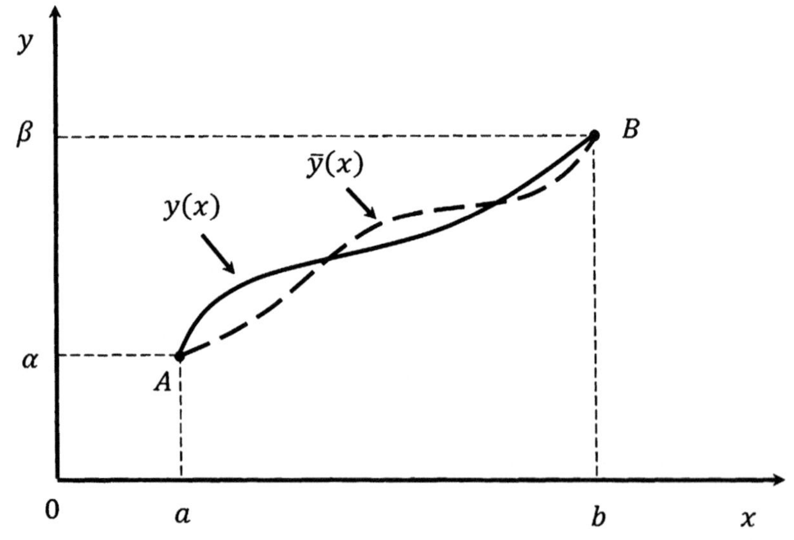

Fig. 2.1 Two functions that are close to each other

In the calculus of variations, the variation of a function $y(x)$ has a special notation δy introduced by Lagrange. Here, let us note the fundamental difference between the notations δy and dy. Both operations mean an infinitesimal change in the function $y(x)$; dy means a change due to an infinitesimal increment of the function dx of the independent variable x, while δy is such an infinitesimal change of function $y(x)$, that creates a new function $y + \delta y$. If the function $y(x)$ is given the meaning of displacement, then it would be appropriate and correct to call δy a possible displacement. It follows that the variation of the independent variable $\delta x = 0$. Moreover, at the boundaries at $x = a$ and $x = b$, we have:

$$\begin{aligned} \delta y(x)|_{x=a} &= \bar{y}(x)|_{x=a} - y(x)|_{x=a} = 0, \\ \delta y(x)|_{x=b} &= \bar{y}(x)|_{x=b} - y(x)|_{x=b} = 0, \end{aligned} \tag{2.4}$$

Hence, the values of $y(x)$ and $\bar{y}(x)$ at the boundaries coincide. Now let us return to functional (2.1) and find the variation of the integrand function $F(x, y, y')$, which is due to the variation of $y(x)$. At the same time, we assume that F is a given function of the variables x, y, and y'. Then we can write the following expression:

$$\delta F(x,y,y') = F(x, y + \varepsilon\varphi, y' + \varepsilon\varphi') - F(x,y,y') \tag{2.5}$$

If the difference on the right side of expression (2.5) is decomposed into a Taylor series of powers ε, and the linear terms are retained, then given that $\varepsilon \to 0$, we obtain:

$$\delta F(x,y,y') = \varepsilon\left(\frac{\partial F}{\partial y}\varphi + \frac{\partial F}{\partial y'}\varphi'\right) \tag{2.6}$$

It is known from calculus of variations that the operations of variation and differentiation, variation, and integration can be interchanged, i.e., we can use the commutability property. Therefore, for functional (2.1) we can write the following expression:

$$\delta I = \delta \int_a^b F(x,y,y')dx = \int_a^b \delta F(x,y,y')dx = \int_a^b \varepsilon\left(\frac{\partial F}{\partial y}\varphi + \frac{\partial F}{\partial y'}\varphi'\right)dx$$

if we divide the obtained expression by ε, then we obtain:

$$\frac{\delta I}{\varepsilon} = \int_a^b \left(\frac{\partial F}{\partial y}\varphi + \frac{\partial F}{\partial y'}\varphi'\right)dx. \tag{2.7}$$

The integral of the second summand can be calculated by integration by parts:

$$\int_a^b \frac{\partial F}{\partial y'}\varphi' dx = \frac{\partial F}{\partial y'}\varphi\bigg|_a^b - \int_a^b \frac{d}{dx}\left(\frac{\partial F}{\partial y'}\right)\varphi dx$$

The first summand in the right-hand side disappears because the function $\varphi(x)$ is zero at the ends of the interval $[a, b]$ and expression (2.7) is reduced to the form:

$$\frac{\delta I}{\varepsilon} = \int_a^b \left[\frac{\partial F}{\partial y} - \frac{d}{dx}\left(\frac{\partial F}{\partial y'}\right)\right]\varphi dx. \tag{2.8}$$

Since, the function $y(x)$ must provide a stationary value to the functional (2.1), the variation δI, which was obtained as a result of an infinitesimal change of $y(x)$, must be equal to zero. In the calculus of variations, it is proved that this is possible for an arbitrary function $\varphi(x)$ only in the case:

$$\frac{\partial F}{\partial y} - \frac{d}{dx} \cdot \left(\frac{\partial F}{\partial y'} \right) = 0, \qquad x \in [a, b]. \tag{2.9}$$

The presented differential equation expresses the necessary and sufficient condition for the stationary of the integral (2.1) under given boundary conditions. This equation was obtained by Lagrange. In the calculus of variations, the Lagrange equation plays a fundamental role. Differential Equation (2.9) can be written in expanded form:

$$\frac{\partial F}{\partial y} - \frac{\partial^2 F}{\partial x \partial y'} - \frac{\partial^2 F}{\partial y \partial y'} y' - \frac{\partial^2 F}{\partial y' \partial y'} y'' = 0. \tag{2.10}$$

For functional of high orders, i.e., for functional that depend on the derivatives of functions $y^{(n)}(x)$,

$$I = \int_a^b F\left(x, y, y', y'', \ldots, y^{(n)}\right) dx, \tag{2.11}$$

Lagrange equation will have the following form:

$$\frac{\partial F}{\partial y} - \frac{d}{dx} \cdot \frac{\partial F}{\partial y'} + \frac{d^2}{dx^2} \cdot \frac{\partial F}{\partial y''} + \cdots + (-1)^{(n)} \frac{d^n}{dx^n} \cdot \frac{\partial F}{\partial y^{(n)}} = 0, \tag{2.12}$$

$$x \in [a, b].$$

2.2 Potential Field of Forces

The concepts of force and field of forces. The field of forces exists by itself, as an objective reality, such as gravitational, electric, or magnetic fields. As soon as an object appears in this field, this object will be under the action of this field. As a result, this object, depending on the nature of the field, that is, the force acting on this object, can begin to move in a certain direction.

Let us consider the potential field of forces in the static calculation of elastic systems. The potential field of forces can be characterized by a scalar function F of the coordinates of some, for example, three-dimensional space x, y, z

$$F = F(x, y, z) \tag{2.13}$$

The function F receives an increment when the coordinates change:

$$dF = \frac{\partial F}{\partial x} dx + \frac{\partial F}{\partial y} dy + \frac{\partial F}{\partial z} dz. \tag{2.14}$$

The values of dx, dy, and dz are considered arbitrary, and partial derivatives $\frac{\partial F}{\partial x}$, $\frac{\partial F}{\partial y}$, and $\frac{\partial F}{\partial z}$ form a three-dimensional vector. This vector is called the gradient of the potential field. The gradient of the potential field indicates the direction in which the function F increases more intensively. If we define a gradient at each point of the field, we obtain a vector field. At each point of this field, a vector with the following components is given:

$$F_x = \frac{\partial F}{\partial x}; \quad F_x = \frac{\partial F}{\partial y}; \quad F_z = \frac{\partial F}{\partial z}. \tag{2.15}$$

From Eq. (2.15), it follows that:

$$\frac{\partial F_x}{\partial y} = \frac{\partial^2 F}{\partial x \partial y} = \frac{\partial F_y}{\partial x}; \quad \frac{\partial F_y}{\partial z} = \frac{\partial F_z}{\partial y}; \quad \frac{\partial F_z}{\partial x} = \frac{\partial F_x}{\partial z}. \tag{2.16}$$

These conditions are met only if there is a scalar function F that satisfies the conditions (2.15). And vice versa, every vector field with vector components F_x, F_y, and F_z can be shown to satisfy conditions (2.16) has a scalar function F. This function is called the field potential and can be defined up to an arbitrary constant. An example of a potential vector field is the gravitational field of forces of gravity. Near the earth's surface, this field is expressed analytically:

$$F_x \equiv F_y \equiv 0; \quad F_x = g = \text{constant},$$

where g is the acceleration of gravity on earth.

It is obvious that the conditions (2.16) are satisfied in this field, and the potential function has the following form:

$$F = gz + C, \tag{2.17}$$

where C is a constant, which depends on the reference level of the coordinate z.

The potential field can also be in the space of many variables

$$F = F(u_1, u_2, \ldots, u_n), \tag{2.18}$$

and its derivatives

$$Q_1 = \frac{\partial F}{\partial u_1}, Q_2 = \frac{\partial F}{\partial u_2}, \ldots, Q_n = \frac{\partial F}{\partial u_n}, \tag{2.19}$$

can be considered as components of an n dimensional vector. This vector forms a potential vector field, which is characterized by the following equality:

$$\frac{\partial Q_i}{\partial u_k} = \frac{\partial Q_k}{\partial u_i} = \frac{\partial^2 F}{\partial u_k \partial u_i} \tag{2.20}$$

2.3 Potential Energy of Masses and Definition of Force

If there are objects in the potential field that have mass properties, then such objects should respond with accelerations to the applied forces. The magnitude (value)

$$V = \sum m_i F_i \tag{2.21}$$

is the potential energy of the masses, which is a function of their coordinates. In formula (2.21), F_i is the value of the function F at the location of the mass m_i.

If we change one of the coordinates of any mass, for example, the coordinate x_i of mass m_i, we can find the partial derivative of the potential energy of the mass by the coordinate x_i. This derivative, taken with an inverse sign, can be defined as a force that acts on the mass m_i in the direction of the x_i coordinate.

Thus, the force P_{x_i}, which acts on the mass in the direction x_i, is equal to the decrease in the potential energy of the masses when the mass m_i and is moved by an infinitesimal unit segment $dx_i = 1$:

$$P_{x_i} = -\frac{\partial V}{\partial x_i} = -\frac{m_i \partial F_i}{\partial x_i} \tag{2.22}$$

If the masses are interconnected by rigid or pliable bonds, then the masses form a certain system. Then the masses of this system cannot move freely in all directions. In this case, the position of the masses is determined by the coordinates $u_1, u_2, \ldots, u_n$, these coordinates represent independent displacements and the number of degrees of freedom of the system.

In this case the position of the masses is determined by the coordinates u, v, and m, these coordinates represent independent displacements and the number of degrees of freedom of the system n. Potential mass energy (2.21) is a function of only n parameters, namely, displacements u_i.

The derivatives of the potential energy of the masses for each parameter u_i, taken with the opposite sign, represent the external forces P_i. These forces act in the direction of displacements u_i and are called conjugate:

$$P_i = -\frac{\partial V}{\partial u_i}(u_1, u_2, \ldots, u_n). \tag{2.23}$$

The following equalities must be met.

$$\frac{\partial P_i}{\partial u_k} = \frac{\partial P_k}{\partial u_i} = \frac{\partial^2 V}{\partial u_i \partial u_k}. \tag{2.24}$$

The totality of all external forces that act on the system will be called the vector of external forces $\vec{P}$. The components of the vector of external forces in coordinates $u_1, u_2, \ldots, u_n$ are equal to: $P_1, P_2, \ldots, P_n$:

$$\vec{P} = (P_1, P_2, \ldots, P_n). \tag{2.25}$$

The potential energy of the masses that enter the system will be called the potential energy of the external forces.

2.4 Virtual Work of External Forces

Let us assume that the system has received some small displacements $\delta u_1, \delta u_2, \cdots, \delta u_n$. Then the potential energy of external forces will receive the following increment:

$$\delta V = \frac{\partial V}{\partial u_1}\delta u_1 + \frac{\partial V}{\partial u_2}\delta u_2 + \cdots + \frac{\partial V}{\partial u_n}\delta u_n = -P_1\delta u_1 - P_2\delta u_2 - \cdots - P_n\delta u_n \tag{2.26}$$

The presented expression for the increment is the work of external forces, produced by these forces on the displacements δu_i. In this case, the work of external forces is considered with an opposite sign. Since the displacements under consideration can have any small values, the magnitude of

$$-\delta V = \sum_{i=1}^{n} P_i \delta u_i \tag{2.27}$$

is called the virtual work of external forces on infinitesimal displacements.

In linearly deformable systems, infinitesimal displacements can be replaced by finite ones, and then the virtual work of external forces will be expressed by the following formula:

$$-V = \sum_{i=1}^{n} P_i u_i \tag{2.28}$$

The generalized displacement is the type of displacement on which the generalized force produces work. Let us also recall that we distinguish between the virtual and the actual work of the forces. At the same time, virtual work is understood as the work that these forces perform on displacements that are caused by some (any) extraneous influences (or forces). If the force has a constant value and direction of action, then such work is determined by the product of the force on the path traveled. Work is called real/actual if it is performed by these forces on displacements that are created by the forces themselves in the process of deformation of a structure. It is assumed that the forces and the displacements caused by them are continuously and slowly increasing (without causing acceleration in the systems) from zero to some finite values.

Such work for the case when the dependence between the generalized force and the generalized displacement is nonlinear is determined by the following expression (Fig. 2.2a):

$$V = \int_0^{\Delta_k} P(\Delta) d\Delta = \eta P_k \Delta_k \tag{2.29}$$

where η is the diagram completeness factor: $P = P(\Delta)$.

The difference between $P_k\Delta_k$ and V is called additional work:

$$V' = P_k\Delta_k - V = \int_0^{P_k} \Delta(P) dP \tag{2.30}$$

From formula (2.30) and Fig. 2.2a, it follows that the "additional work" is defined as the area of the diagram $\Delta = \Delta(P)$, which complements the hatched area to a rectangle. This rectangle represents the virtual work of a constant force P on the displacement Δ. This circumstance explains the use of the term "additional work." For a linearly elastic body or linearly deformable system (Fig. 2.2b),

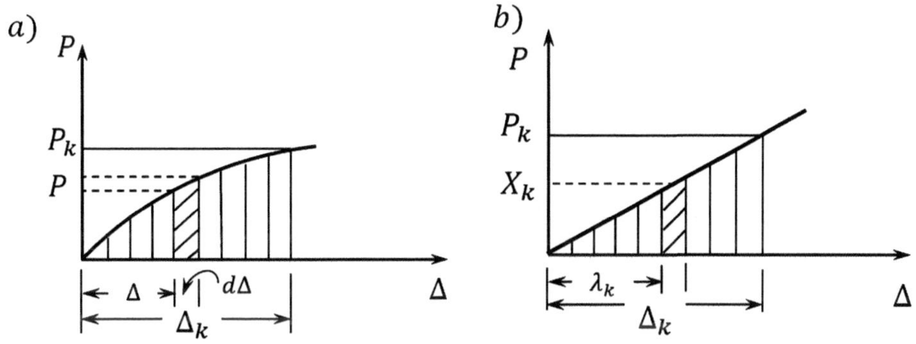

Fig. 2.2 Dependence between the generalized force and the generalized displacement: (**a**) nonlinear; (**b**) linear

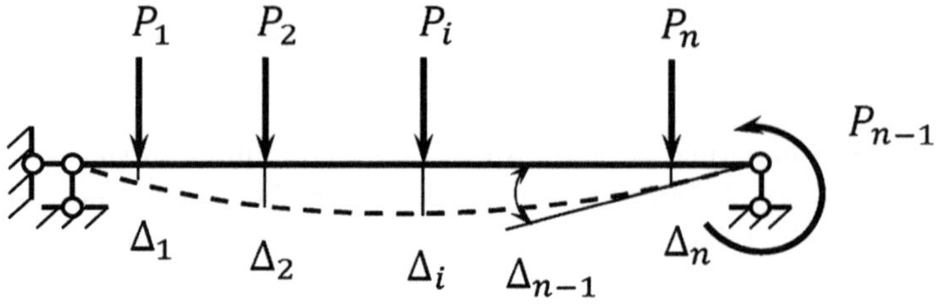

Fig. 2.3 Displacements of a certain type from generalized forces

displacements and deformations are expressed as linear homogeneous functions of external forces P_1, $P_2, \ldots, P_n$ (Fig. 2.3)

$$\Delta = P_1\Delta_1 + P_2\Delta_2 + \cdots + P_i\Delta_i + \ldots + P_{n-1}\Delta_{n-1} + P_n\Delta_n \tag{2.31}$$

where

Δ is displacements of a certain type (vertical, horizontal, angular);
$\Delta_1, \Delta_2, \ldots, \Delta_i, \ldots, \Delta_n$ are displacements of the same type in the same place only from forces, respectively: $P_1 = 1, P_2 = 1, \ldots, P_i = 1, \ldots, P_{n-1} = 1, P_n = 1$

Expression (2.31) is called a generalized Hooke's law. Systems that obey the law (2.31) are called linearly deformable systems. For such systems, the principle of independent action of forces is applicable. According to this principle, the effect of the joint action of several forces is equal to the sum of the effects of each force separately.

If $P_1 = P_2 = \cdots = P_i = \cdots = P_{n-1} = P_n = 0$, then $\Delta = 0$. Thus, there are no residual deformations in linearly deformable systems. If $P_2 = \cdots = P_i = \cdots = P_{n-1} = P_n = 0$, then $\Delta = P_1\Delta_1$. That is, the displacements are proportional to the force.

Suppose that the force acts statically from zero to its finite value P_k. The displacement caused by the action of this force increases accordingly from zero to its final value Δ_k. The variable value of the force is denoted by X_k, and the corresponding intermediate value of displacement through λ_k. Let us give λ_k an increment $d\lambda_k$. On this increment, the force X_k will do the elementary work

$$dV = X_k d\lambda_k \tag{2.32}$$

The displacement λ can be expressed as follows: $\lambda_k = X\delta_{kk}$, where δ_{kk} is the same displacement as λ_k, when $X_k = 1$. It follows that: $d\lambda_k = \delta_{kk}dX_k$. This value is substituted in (2.32), and we obtain the expression for elementary work: $dV = X_k \cdot \delta_{kk} \cdot dx$. The formula for determining the work of a statically applied external force is obtained by integrating this expression within the limits of the total change of force from zero to its final value.

$$V = \int_0^{P_k} \delta_{kk} X_k dx = \frac{\delta_{kk} X_k^2}{2}\bigg|_0^{P_k} = \frac{\delta_{kk} P_k^2}{2} \tag{2.33}$$

Since $\Delta_k = \delta_{kk} \cdot P_k$, the resulting formula can be represented as:

$$V = \frac{P_k \Delta_k}{2} \tag{2.34}$$

Thus for a linearly elastic body (Fig. 2.2b), the following relationship is valid:

$$V = V' = \frac{P_k \Delta_k}{2} \tag{2.35}$$

That is, the actual and additional work are equal to each other and each makes up the product of the final value of the force by the final value of the displacement Clapeyron's Theorem.

Thus, if there are several generalized forces acting on the structure (Fig. 2.3), in this case the work of external forces will be expressed by the following formula:

$$V = \frac{P_1 \Delta_1}{2} + \frac{P_2 \Delta_2}{2} + \cdots + \frac{P_i \Delta_i}{2} + \cdots + \frac{P_{n-1} \Delta_{n-1}}{2} + \frac{P_n \Delta_n}{2} \tag{2.36}$$

or

$$V = \frac{1}{2} \sum_{i=1}^{n} P_i \Delta_i \tag{2.37}$$

2.5 Virtual Work of Internal Forces

Deformed links and elements of an elastic system have some potential energy. Here, the role of masses is played by molecules that are in the field of intermolecular forces. The work of internal forces is called the potential energy W. This energy accumulates as a result of deformations of all elements and links of the system. For each element, the increment of the potential energy at infinitesimal deformation j th of element λ_j.

$$\delta W_j = N_j \delta \lambda_j \tag{2.38}$$

Hence it follows that the internal force N_j can be defined as the increment of the energy of the j-th element with a unit infinitesimal deformation of this element $\lambda_j = 1$.

The total possible work of internal forces of the system at small deformations

$$\delta W = \sum_{j=1}^{m} N_j \delta \lambda_j \tag{2.39}$$

In case of linearly deformable systems, the work of internal forces can be represented similarly to (2.28) by the following formula:

$$W = \sum_{j=1}^{m} N_j \lambda_j. \tag{2.40}$$

Here m is the number of elements and links that are subjected to deformation. It follows from (2.40) that

$$\frac{\partial W}{\partial \lambda_j} = N_j \tag{2.41}$$

2.6 The Principle of Virtual Displacements

The total potential energy of an elastic system U is the sum of the potential energy of external forces V and the potential energy of internal forces W:

$$U = V + W \tag{2.42}$$

The minimum of the total potential energy of the system is reached when the system is in equilibrium, i.e., we have the following conditions:

$$U = \min; \qquad \frac{\partial U}{\partial u_1} = \frac{\partial U}{\partial u_2} = \cdots = \frac{\partial U}{\partial u_n} = 0, \tag{2.43}$$

In this case, the infinitely small increment δU of the total potential energy of the system is equal to zero:

$$\delta U = \frac{\partial U}{\partial u_1}\delta u_1 = \frac{\partial U}{\partial u_2}\delta u_2 + \cdots + \frac{\partial U}{\partial u_n}\delta u_n = 0, \tag{2.44}$$

hence it follows:

$$\delta V + \delta W = 0 \quad \text{or} \quad -\delta V = \delta W \tag{2.45}$$

Equality (2.45) according to (2.27) and (2.39) can be represented as:

$$\sum_{i=1}^{n} P_i \delta u_i = \sum_{j=1}^{m} N_j \delta \lambda \tag{2.46}$$

Thus, the work of external forces acting on the equilibrium system at any possible/virtual infinitesimal displacements is equal to the virtual work of internal forces at these same displacements. This position is called the Principle of Virtual Displacements.

For linearly deformable systems, we have the equality:

$$\sum_{i=1}^{n} P_i \delta u_i = \sum_{j=1}^{m} N_j \lambda_j \tag{2.47}$$

2.7 Virtual Work of External Forces or Potential Deformation Energy

The work of external forces on the displacements caused by them can be expressed through internal forces, namely: through bending moments, longitudinal and transverse forces that arise in all deformable elements and bonds of a structure. The internal forces arising in a deformable system perform the work. This work will always be negative because the directions of internal forces and deformations do not coincide. The work of internal forces is the potential energy V accumulated by all elements and bonds during their deformation. Since in an elastic system, there is no energy loss due to overcoming friction, heat generation, etc., the actual work of external forces V is equal and opposite in sign to the work of internal forces W.

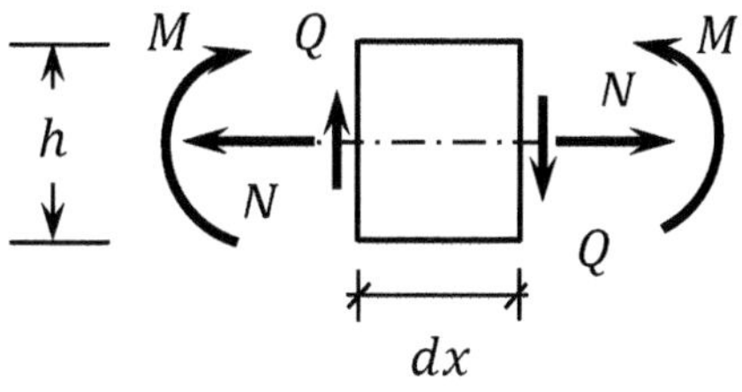

Fig. 2.4 Infinitesimal element of length dx

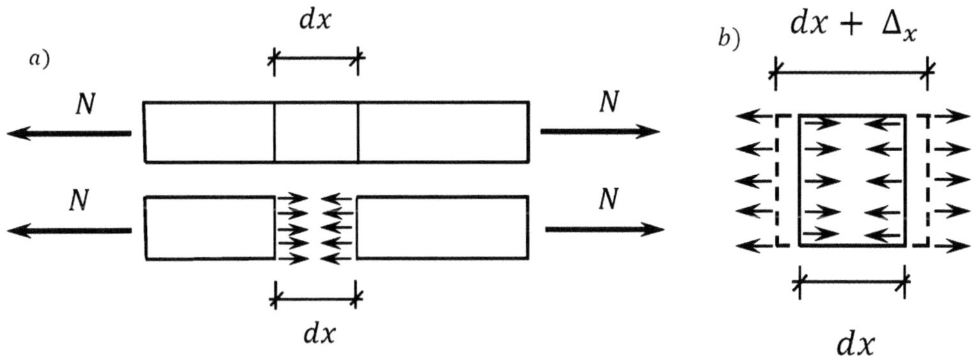

Fig. 2.5 (**a**) Element dx is under the action of only longitudinal forces N; (**b**) Internal forces in the element dx prevent tensile deformations

$$V + W = 0 \text{ or } V = -W \tag{2.48}$$

The work of the internal forces W, taken with the inverse sign, characterizes the deformation energy and is called the potential deformation energy of the elastic system:

$$U = V = -W \tag{2.49}$$

Let us isolate an infinitesimal element of length dx from a rectilinear rod of the system with two sections perpendicular to its axis (Fig. 2.4) to determine the potential energy of deformations of the elastic system. Let us apply to this element the forces M, N, and Q acting on it from the discarded parts of the system.

The deformations caused by these forces will be considered. The actual work of the internal forces of the element, which are equal in magnitude and inverse in sign to the forces M, N, and Q, will be:

$$dW = dW_M + dW_N + dW_Q = -\frac{1}{2}\left(M\varphi + N\Delta_x + Q\Delta_y\right). \tag{2.50}$$

As noted above, the minus sign indicates that the direction of the internal forces of the element dx is opposite to the direction of its deformation. The internal forces try to inhibit the deformation of the elastic system. Therefore, the work of the internal forces of the elastic system is always negative. Let us consider the effect of each of forces M, N, and Q separately:

The element dx is under the action of only longitudinal forces N, Fig. 2.5.

The magnitude of the axial deformation will be:

$$\Delta_x = \varepsilon dx = \frac{Ndx}{AE} \tag{2.51}$$

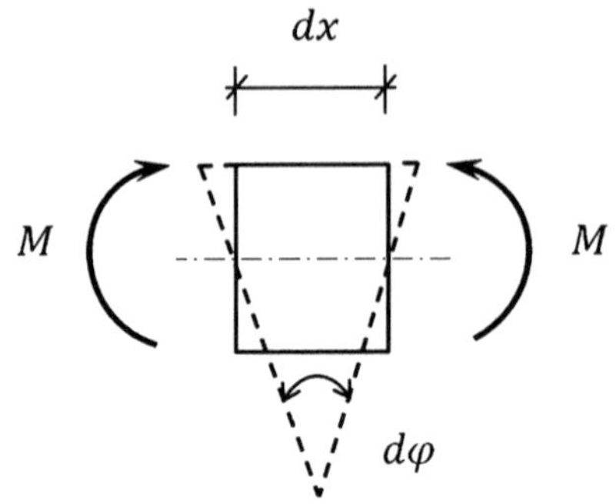

Fig. 2.6 Element dx is under the action of only bending moments M

where A and E are the cross-sectional area and modulus of elasticity, respectively, of the considered rod.

The internal forces N on this displacement will perform the work that within the element dx will be:

$$dW_N = \frac{N\Delta_x}{2} = \frac{N^2 dx}{2AE} \tag{2.52}$$

where AE is the stiffness at axial deformation.

The element dx is under the action of only bending moments M, Fig. 2.6.

The mutual angle of rotation of the end sections of the element dx will be equal:

$$d\varphi = \frac{Mdx}{EI} \tag{2.53}$$

The moment M on this angular displacement will perform work which within the element dx will be:

$$dW_M = \frac{Md\varphi}{2} = \frac{M^2 dx}{2EI} \tag{2.54}$$

where EI is the bending stiffness.

The change in the bending moments along the length of the rob is accompanied by the appearance of shear forces and the shear (tangential) shear stresses caused by them.

$$Q = dM/dx.$$

The element dx is under the action only of shear forces Q, Fig. 2.7.

In this case, shear deformations in the element dx occur. The resultant of the shear forces τdA is the shear force Q. Let us assume that shear stresses τ are uniformly distributed over the entire cross-sectional area A, that is: $=Q/A$. Then the displacement Δ_y (Fig. 2.7a), which is caused by the action of the shear force Q, is a shift of the edge sections of the element dx relative to each other. The shear displacement is determined from the expression:

$$\Delta_y = \gamma dx = \frac{Qdx}{AG}, \tag{2.55}$$

where γ is a shift angle, Δ_y is an absolute shift, and where G is the modulus of elasticity of the second kind.

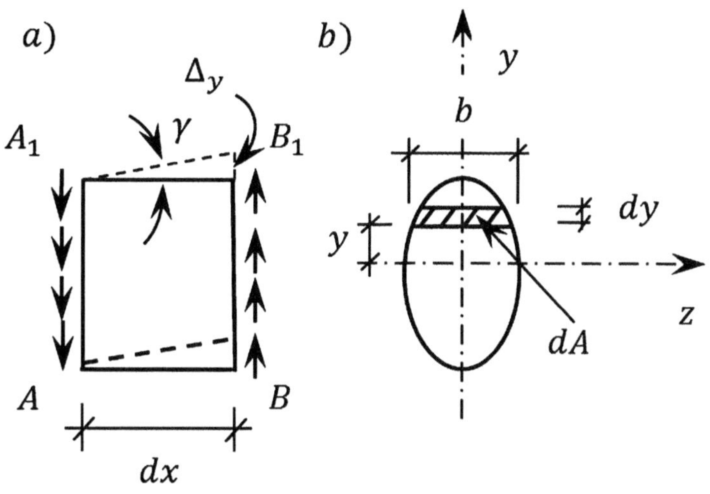

Fig. 2.7 The element dx is under the action only of shear forces Q. (**a**) Shear deformations in the element dx; (**b**) elementary strip with a thickness dy

The work of shear force Q on this displacement will be:

$$dW_Q = \frac{1}{2}Q\Delta_y = \frac{1}{2}Q\frac{Q}{AG}dx. \tag{2.56}$$

In fact, the shear stresses τ over the cross-sectional area are unevenly distributed. This fact is taken into account by introducing a correction factor μ. Therefore, expression (2.56) can be represented as follows:

$$dW_Q = \frac{1}{2}Q\frac{Qdx}{AG}\mu \tag{2.57}$$

The value of the shear force varies with the height of the section. Therefore, first, we find the work of these forces that are applied to the elementary area dA (in Fig. 2.7b) it is shown as an elementary strip with a thickness dy). As is known, the force τdA will perform work on displacements equal to $\Delta_y = \gamma dx$. This work according to Clapeyron's theorem is equal to $\tau dA\gamma dx/2$. According to Hooke's law in shear $\gamma = \tau/G$, the shear forces/stresses in bending are determined by the formula:

$$\tau = \frac{QS_z}{Ib}, \tag{2.58}$$

where S_z is the static moment relative to the $z - z$ axis of the part of the section located above or below the x axis. The mutual linear shift of the two corresponding strips, which are taken at both ends (AA_1 and BB_1) of the dx element, will be:

$$\gamma dx = \frac{\tau dx}{G}. \tag{2.59}$$

In this regard, the work of an infinitesimal tangential force τdA on the displacement γdx will be determined by the following product:

$$\frac{1}{2}\tau dA\gamma dx \tag{2.60}$$

Let us integrate the expression for the work of elementary forces within the entire cross-section A. We obtain the work of all shear forces that act in the cross-section.

$$dW_Q = \int_A \frac{1}{2}\tau\gamma dx dA = \int_A \frac{\tau^2 dx}{2G} dA = \int_A \frac{Q^2 S_z^2}{I^2 b^2}\frac{dx}{2G} dA = \frac{Q^2}{2}\frac{dx}{AG}\frac{A}{I^2}\int_A \frac{S_z^2}{b^2} dA \tag{2.61}$$

Denote

$$\mu = \frac{A}{I^2}\int_A \frac{S_z^2}{b^2} dA$$

the dimensionless coefficient, which depends only on the shape of the cross-section. Then:

$$dW_Q = \mu \frac{Q^2 dx}{2AG} \tag{2.62}$$

If the longitudinal force, bending moment, and transverse force act simultaneously on a selected element, the work of each of them on the displacements caused by the other forces is equal to zero. Therefore, the expression for the total work is:

$$dW = -\left(\frac{N^2 dx}{2AE} + \frac{M^2 dx}{2EI} + \mu \frac{Q^2 dx}{2AG}\right) \tag{2.63}$$

Integrating the obtained expression within the length of each section, each rod/bar and summing over all sections, rods/bars of the system, we determine the actual work of internal forces for the entire system:

$$W = -\frac{1}{2}\left[\sum \int_0^l \frac{M^2 dx}{2EI} + \sum \int_0^l \frac{N^2 dx}{2AE} + \sum \int_0^l \mu \frac{Q^2 dx}{2AG}\right] \tag{2.64}$$

Summing up the elementary work of the internal forces along the length of the rod, and then over all the rods of the system, we finally obtain the work of the internal forces of the system.

Analyzing expression (2.64), one can conclude that the potential energy of deformation U is always positive, since it is expressed in terms of the squares of internal forces. For this reason, the principle of independence of the action of forces is not applicable to the calculation of potential energy, that is, the potential energy caused by the action of a group of forces is not equal to the sum of the potential energies caused by each of the force separately.

In an elastic system, the total work of internal and external forces is determined only by the final state of the system and does not depend on how the system came to this condition. Since the values of the internal forces M, N,and Q of the system do not depend on the intermediate states.

The potential energy of deformations can also be expressed through the work of external forces (Fig. 2.3).

$$U = V = \frac{P_1\Delta_1}{2} + \frac{P_2\Delta_2}{2} + \cdots + \frac{P_i\Delta_i}{2} + \cdots + \frac{P_{n-1}\Delta_{n-1}}{2} + \frac{P_n\Delta_n}{2} \tag{2.65}$$

each of the total displacements Δ_i included in the formula can be expressed in terms of unit displacements in the same direction δ_{ik} caused separately by forces ($P_1 = 1$, $P_2 = 1$, . . ., $P_{n-1} = 1$, $P_n = 1$) as follows:

$$\Delta_i = \delta_{i1}P_1 + \delta_{i2}P_2 + \cdots + \delta_{ii}P_i + \ldots + P_{n-1}\delta_{i-1,n-1} + P_n\delta_{in} \tag{2.66}$$

after substituting formula (2.66) into (2.65) we can obtain:

$$U=\frac{1}{2}\sum_{i=1}^{n}\delta_{ii}P_i^2+\sum_{i=1(k=1)}^{n}\delta_{ik}P_iP_k \quad (i\neq k, \delta_{ik}=\delta_{ki}) \tag{2.67}$$

Thus, the potential energy of deformations (2.67) or the work of external forces is represented in the form of a homogeneous algebraic polynomial of the second degree with respect to external forces. Such polynomials are called quadratic. Since the potential energy is always positive, the quadratic format of external forces (2.67) cannot be negative for any values of the variables: $P_1, P_2, \ldots, P_{n-1}, P_n$, as a result of which it is definitely positive. The potential energy can also be represented as a function of the given displacements $\Delta_1, \Delta_2, \ldots, \Delta_{n-1}, \Delta_n$, which are included in formula (2.65).

In order to represent forces P_i as functions of displacements in formula (2.65), we can write n formulas of the form (2.65) and solve a system of n algebraic equations with respect to the quantities P_i. Thus we obtain:

$$P_i=r_{i1}\Delta_1+r_{i2}\Delta_2+\cdots+r_{i,n-1}\Delta_{n-1}+r_{in}\Delta_n \tag{2.68}$$

where r_{ik} is the reaction in the bond, i (the force applied at the point i, arising from a unit displacement of the bond k ($\Delta_k=1$).

The potential energy of deformations can now be represented as a quadratic form from the displacements, after substituting formula (2.68) into (2.65), we obtain:

$$U=\frac{1}{2}\sum_{i=1}^{n}r_{ii}\Delta_i^2+\sum_{i=1(k=1)}^{n}r_{ik}\Delta_i\Delta_k \quad (i\neq k, r_{ik}=r_{ki}) \tag{2.69}$$

If as independent variables in formula (2.65) we take part of the force and part of the displacement, then we can obtain the expression of potential energy in in the form of a mixed quadratic form.

2.8 Total Potential Energy and Extreme Properties of Potential Energy

When solving many problems of structural mechanics, which relate to the statics, dynamics, and stability of structures, the functional is introduced. The functional is introduced to study the general laws of behavior of an elastic system under load. This functional contains simultaneously the work of both internal and external forces and is called the total potential energy of an elastic system and is defined as

$$\ni = U - W \tag{2.70}$$

The total energy of an elastic system can be interpreted as the work done by internal and external forces during the transition of the system from the deformed state to the initial non-deformed state. In this case, the internal forces will perform actual positive work, where $U=-W$, and the external forces will perform possible negative work V, which is calculated as the total product of the force on the path it has traveled: $V=2W=P\Delta$.

Thus, the potential energy is a function of the system displacements, and, therefore, does not depend on the path of development of displacements from zero to the given equilibrium state. At the same time, in an elastic system, each vector of motions corresponds to a certain vector of external forces, according to the formula (2.66). Therefore, the potential energy of the system also does not depend on the way the system is loaded by external forces. In order to calculate the potential energy of an elastic system when it is loaded by forces $P_1, P_2, \ldots, P_{n-1}, P_n$, it is sufficient to determine this energy for any one loading path.

Fig. 2.8 The beam deforms (deflects) under the force P

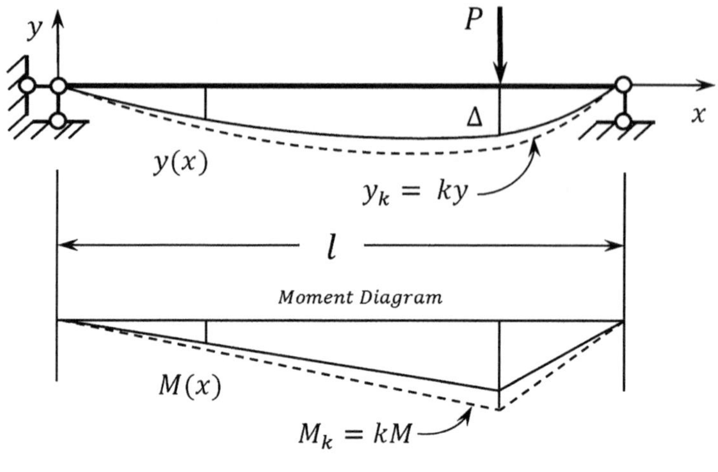

Thus, the total energy of an elastic system ∋ is equal in magnitude and inverse in sign to the potential energy of deformation U, i.e., $\ni = -U$. The expression of total energy contains both external and internal forces and establishes a certain relationship between them. This circumstance makes it possible to identify many common patterns of structural mechanics.

The potential energy of an elastic system has remarkable properties of an extreme nature, which are effectively used for the approximate solution of complex problems of structural mechanics. Let us show this with the example of a simple beam under the action of force P (Fig. 2.8).

Let us assume that we have found the beam deflection function $y = y(x)$ and the bending moment diagram $M(x)$. Then, the total potential energy of bending of the beam can be represented in the form of neglecting the work of transverse shear forces Q:

$$\ni = U(M) - W = \frac{1}{2}\int_0^l \frac{M^2}{EI}dx - P\Delta = \frac{1}{2}P\Delta - P\Delta \tag{2.71}$$

The total potential energy of bending of the beam can be represented in the following form:

$$\ni = U(y) - W = \frac{1}{2}\int_0^l EIy''^2 dx - P\Delta = \frac{1}{2}P\Delta - P\Delta \tag{2.72}$$

Here, the bending moment is expressed through the second derivative of the deflection function of the beam.

Let us now imagine that we want to vary the bent axis of the beam by an arbitrary law. For example, let's increase all the ordinates of the bent axis of the beam by k times: $y_k = ky$, $\Delta_k = k\Delta$ (Fig. 2.8).

$$\ni_k(y) = = \frac{k^2}{2}\int_0^l EIy''^2 dx - kP\Delta = k^2\frac{P\Delta}{2} - kP\Delta = P\Delta\left(\frac{k^2}{2} - k\right) \tag{2.73}$$

Thus, the potential energy is represented by a quadratic function of k. From the graph of the quadratic function, which is shown in Fig. 2.9, the potential energy has a minimum value at $k = 1$, that is, in the actual equilibrium state of the beam.

This leads to an important conclusion that out of all possible deformed states of a beam, only that state exists that gives the potential energy its minimum value.

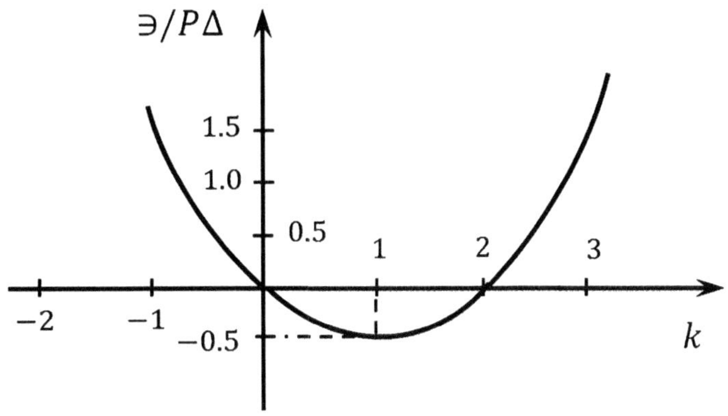

Fig. 2.9 Graph of the potential energy of the curved beam axis

Let us try to vary the stress state of the beam, i.e., the bending moment diagram in expression (2.72), instead of the bent axis of the beam. For example, increase the ordinates of the bending moment diagram k times (Fig. 2.8), then the expression for the potential energy will again look like (2.73). And in this case, that is at $k = 1$, the potential energy will have the minimum value in the equilibrium state of the system. Hence, it follows, that of all stress states (of all possible diagrams of bending moments passing through given points at the ends of the beam), there is that stress state that gives the potential energy a minimum value actually taking place.

The extreme properties of potential energy, shown by the example of a simple beam, are manifested in the most complex cases of the equilibrium state of elastic systems, as well as in those cases where the system makes elastic motions. These properties are a consequence of the fundamental principles of structural mechanics. A presentation of these principles will be shown in the following paragraphs.

2.9 The Principle of Virtual Displacements

The principle of possible displacements is formulated as follows: if a system is in a state of equilibrium, then the sum of works of all external and internal forces on every (any) possible infinitesimal change of displacements is equal to zero. In addition, we remind that such changes of displacements, which do not violate external and internal connections anywhere, i.e., they satisfy the conditions of compatibility of deformations, are considered as possible. In other words, the kinematically admissible variations of displacements are the possible changes of displacements. It is assumed that in the process of virtual displacements all forces that act on the system retain their magnitudes and directions of action, i.e., remain unchanged. Therefore, when calculating the work of these forces, the full value of the product of each force and the corresponding displacement is taken.

As an example, consider any system, for example, a beam (Fig. 2.10) which is in equilibrium under the action of generalized forces P_i.

The internal forces of the beam M, N, Q, and their corresponding deformations $\varkappa$, ε,and γ are shown in Figure (Fig. 2.10). The infinitesimal possible variation of displacements has been represented by a dotted line in the figures (Figs. 2.10 and 2.11). In accordance with the Principle of Possible Displacements, we obtain:

$$\sum_{i=1}^{n} P_i \delta\Delta_i - \delta U = 0, \tag{2.74}$$

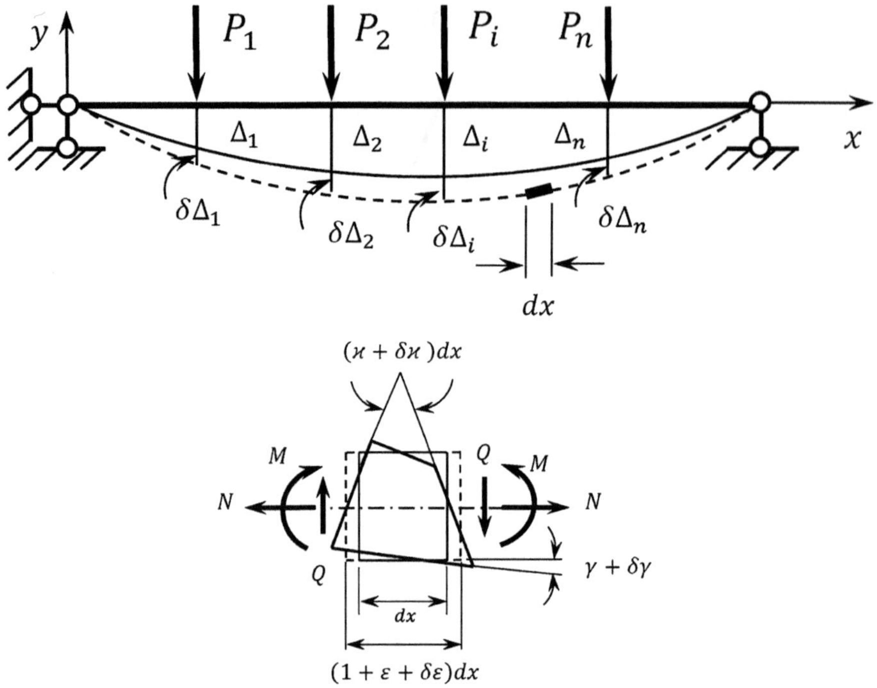

Fig. 2.10 A beam is in equilibrium under the action of generalized forces P_i

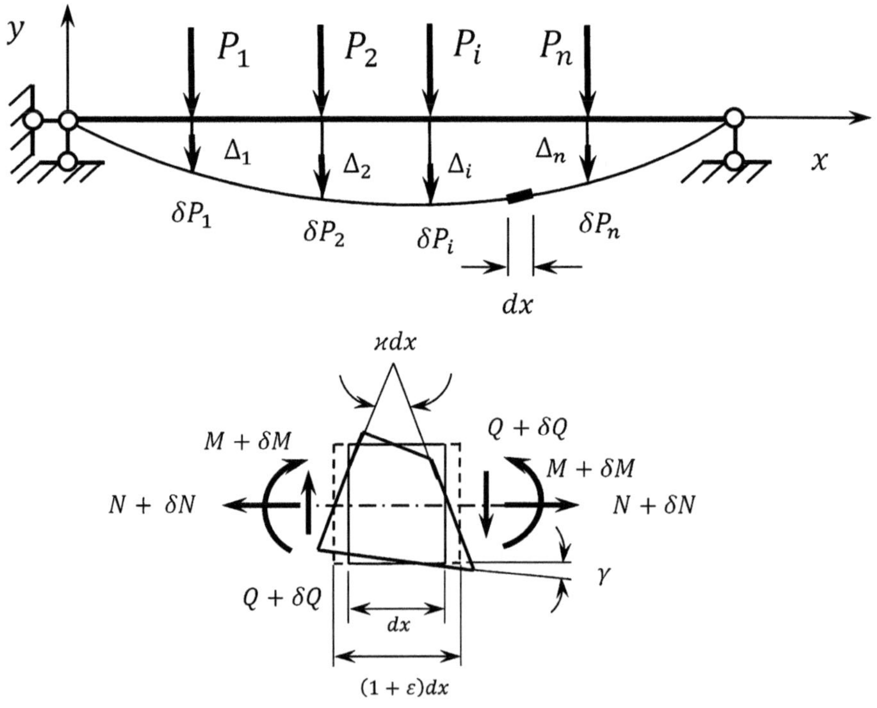

Fig. 2.11 A beam in a deformed position has an additional infinitesimal variation in displacements

Here δU is a variation of the potential energy of system deformations. This deformation is caused by a possible change in the displacements or deformations of the system.

$$\delta U = \sum \left[\int M\delta\varkappa dx + \int N\delta\varepsilon dx + \int Q\delta\gamma dx \right] \tag{2.75}$$

The values of deformation of the element in accordance with Hooke's Law are determined by the following formulas:

$$\varphi = \varkappa dx = \frac{Mdx}{EI}; \quad \Delta_x = \varepsilon dx = \frac{Ndx}{AE}; \quad \Delta_y = \gamma dx = \mu \frac{Qdx}{AG} \tag{2.76}$$

Formula (2.74) is a mathematical notation of the Principle of Possible Displacements for a one-dimensional rod system. The Principle of Possible Displacements can be formulated differently if the initial conditions and the consequence are reversed, that is: if the sum of works of all external and internal forces of the system on every possible infinitesimal displacement is equal to zero, then the system (each part of it) is in a state of equilibrium.

The Principle of Virtual Displacements is the most general principle of statics and is the basis of one of the methods for calculating statically indeterminate systems of the displacement method. The resolving equations of the method of displacements express the system's equilibrium conditions. The Principle of Virtual Displacements can be applied to any systems with elastic, viscous, and plastic properties. A feature and peculiarity of applying the Principle of Virtual Displacements to elastic systems is that all work of external and internal forces is reduced to the formation of deformations that is to the accumulation of potential energy of deformations.

Historically, the Principle of Virtual Displacements was initially used in solving static problems. The extension of the Principle of Virtual Displacements to dynamics became possible by the work of the outstanding French mathematician D'Alembert. He proposed to introduce inertial forces into consideration. The inertial forces were defined in accordance with Newton's basic law of motion as the product of mass and acceleration. This meant that inertial forces, in addition to active forces, should be applied to a system that is in motion, and the equilibrium condition of the system should be used. As a result, the problems of dynamics were reduced to the problems of statics, i.e., the equations of dynamics became possible to deduce from the equations of equilibrium of statics. Thus, the d'Alembert principle states that any system of forces is in equilibrium if we add to the applied forces, forces of inertia. It can be said that a body or system moves in such a way that the additional inertial forces which are caused by motion (vibration, oscillation) bring the balance of work to zero. Hence it follows that the total virtual work of all applied forces and inertial forces is equal to zero on displacements that are compatible with the given kinematic conditions.

D'Alembert's principle gives a complete solution to the problems of dynamics. The equations of motion of an arbitrary mechanical system follow from the D'Alembert principle.

2.10 The Methods of Drawing Up Equations of Motion

Let us remind that the virtual are such changes of displacements that do not disrupt external or internal connections of the system, i.e., satisfy the conditions of the compatibility of deformations. In other words, the kinematic virtual variation of displacements is kinematically allowable variations of displacements. According to the Principle of Virtual Displacements, the system of forces is in equilibrium if at any virtual displacement δr_i the work of all forces is zero:

$$\delta W_i = \delta W_1^{\text{elastic}} + \delta W_i^{\text{external}} = 0 \qquad (i = 1, \ldots, n), \tag{2.77}$$

where the first term denotes work of internal forces of elasticity, and the second external forces including forces of inertia on the displacement δr_i.

The state of deformable systems is determined by n generalized coordinates $q_1, q_2, \ldots, q_n$. The kinetic energy of masses of the system K and potential energy of elastic forces U can be presented in the form of some dependencies on the coordinates q_i and velocities $\dot{q}_i$:

$$K = K(q_1, \ldots, q_n, \dot{q}_1, \ldots, \dot{q}_n); \qquad U = U(q_1, \ldots, q_n). \tag{2.78}$$

We assume that the work of external forces Q on the variations of displacements $\delta q_1, \ldots, \delta q_n$ linearly depends on these variations, that is

$$\delta W_Q = Q_1 \delta q_1 + \ldots + Q_n \delta q_n \tag{2.79}$$

From the theoretical mechanics, it is known that the generalized forces of inertia I_i and the generalized elastic forces S_i corresponding variations of displacement δq_i are expressed through derivatives of the function K and U as follows:

$$I_i = -\frac{d}{dt}\left(\frac{\partial K}{\partial \dot{q}_i}\right) + \frac{\partial K}{\partial q_i}; \qquad S_i = -\frac{\partial U}{\partial q_i}. \tag{2.80}$$

Now, let us write the conditions of equilibrium of the system using the Principle of Virtual Displacements (2.77):

$$\delta W_i = (I_i + S_i + Q_i)\delta q_i = 0 \qquad (i = 1, \ldots, n). \tag{2.81}$$

Substituting Eq. (2.80) in Eq. (2.81) and reducing on arbitrary variation δq_i, we obtain n Lagrange equations of motion:

$$\frac{d}{dt}\left(\frac{\partial K}{\partial \dot{q}_i}\right) - \frac{\partial K}{\partial q_i} + \frac{\partial U}{\partial q_i} = Q_i \qquad (i = 1, \ldots, n). \tag{2.82}$$

The value Q_i is called the generalized force corresponding to the coordinate q_i. For conservative systems, Eq. (2.82) will take the form:

$$\frac{d}{dt}\left(\frac{\partial K}{\partial \dot{q}_i}\right) - \frac{\partial K}{\partial q_i} = \frac{\partial U}{\partial q_i} \qquad (i = 1, \ldots, n). \tag{2.83}$$

Equation (2.83) determines in particular the free oscillations of the system in the absence of resistances.

The Principle of Virtual Displacements states that a given mechanical system will be in equilibrium only if and when the total virtual work of all applied forces is equal to zero. Describing the equilibrium conditions of mechanical system, it should be noted that the Principle of Virtual Displacement refers primarily to the static of structures. The extension of the Principle of Virtual Displacements to the dynamics of structures became possible due to the works of French mathematician D'Alembert. D'Alembert proposed to introduce forces of inertia, determined in accordance with the basic Newton's law of motion as the product of mass and acceleration. It means that in a system of motion, in addition to the active forces, it is necessary to apply forces of inertia, and then use the equilibrium conditions of the system. As a result, the dynamics was reduced to the statics, i.e., it became possible to derive the dynamic equations from the consideration of static equations. Thus, D'Alembert's principle states that any system of forces is in equilibrium, if the forces of inertia are added to the applied forces. The difference of applied forces and inertial forces for a dynamic system does no virtual work.

$$\delta W = \sum_i (Q_i - m_i a_i) \cdot \delta r_i = 0 \tag{2.84}$$

where

δW is the virtual work done by the total and inertia forces together through an arbitrary virtual displacement.
Q_i is the applied force.
δr_i is the virtual displacement of the system, consistent with the constrains.
m_i is the mass of the particles in the system.
a_i is the accelerations of the particles in the system.
$m_i a_i$ are the inertial forces that result from the total forces.
i is an integer used to indicate a variable corresponding to a particular particle.

The additional inertial forces resulting from the movement of a body or system lead the balance of work of all forces to zero. Hence it follows that the total virtual work of all applied forces and inertial forces is equal to zero on displacements that are compatible with the given kinematic conditions. Based on D'Alembert's principle, the equations of motion of an arbitrary mechanical system are derived. Particularly, the energy conservation law follows from D'Alembert's principle, which states that during oscillations of a mechanical system, the sum of the kinetic and potential energy remains constant.

2.11 Lagrange Variation Principle and Lagrange Formula

The Principle of Virtual Displacements can be written down in the following form based on the concept of total potential energy.

$$\delta \ni = \delta(U - V) = 0 \tag{2.85}$$

Expression (2.85) is a stationary condition for the functional R, i.e., of all virtual equilibrium states of a system that is under the action of external forces, the real is the one where the total energy of the system takes a stationary value. The variational principle in this formulation is called Lagrange. From the condition of stationarity of the functional, in a general case, as was shown in this chapter, Sect. 2.1, the differential equilibrium equations are obtained (the Euler–Lagrange equations), as well as the equilibrium conditions at the boundary of the domain under consideration.

If the potential energy of deformations U, which is included in formula (2.74), is expressed through the generalized displacements, i.e., it is a functional of Δ_i, then its first variation can be defined as:

$$\delta U = \sum_{i=1}^{n} \frac{\partial U}{\partial \Delta_i} \delta \Delta_i \tag{2.86}$$

By substituting Eq. (2.86) into Eq. (2.74), we obtain:

$$\sum_{i=1}^{n} P_i \delta \Delta_i - \sum_{i=1}^{n} \frac{\partial U}{\partial \Delta_i} \delta \Delta_i = 0 \quad \text{or} \quad \sum_{i=1}^{n} \left(P_i - \frac{\partial U}{\partial \Delta_i} \right) \delta \Delta_i = 0,$$

from where, due to the arbitrariness of the value $\delta\Delta_i$, we will have:

$$-\frac{\partial U}{\partial \Delta_i}=0 \quad \text{or} \quad P_i=\frac{\partial U}{\partial \Delta_i}. \tag{2.87}$$

This result can also be obtained from considering expression (2.69). Indeed, the partial derivative of one of the generalized displacements Δ_i from the potential energy of deformation U is:

$$\frac{\partial U}{\partial \Delta_i}=r_{ii}\Delta_i+\sum_{k=1}^{n} r_{ik}\Delta_k=r_{i1}\Delta_1+r_{i2}\Delta_2+\cdots+r_{ii}\Delta_i+\cdots+r_{in}\Delta_n \qquad (i\neq k).$$

Taking into account expression (2.68), we obtain:

$$\frac{\partial U}{\partial \Delta_i}=P_i \tag{2.88}$$

Thus, the partial derivative of the potential energy of deformations with respect to the generalized displacement is equal to the generalized force corresponding to this displacement. Thus, the Lagrange formula is: valid for any linear and nonlinear deformable system.

2.12 The Principle of Virtual Changes in the Stress State

2.12.1 Formulation of the Principle

The Principle of Virtual Changes in the stress state is formulated as follows: if the compatibility of the deformations of the system is observed, then the sum of virtual works that are performed by virtual infinitesimal increments of all external and internal forces on the actual displacements of the system, which are caused by the forces themselves, is equal to zero.

Let us look at the beam shown in Fig. 2.12 to illustrate the above. Suppose that the external and internal forces of the beam received virtual/passible variations δP_i, δM, δN, and δQ in which the system did not experience any changes in deformations.

In accordance with the Principle of Virtual Changes in the stress state, we obtain:

$$\sum_{i=1}^{n}\Delta_i\delta P_i-\delta U'=0 \tag{2.89}$$

where $\delta U'$ is the variation in the strain energy, which is due to the variation in internal forces:

$$\delta U'=\sum\int(M\delta\varkappa+N\delta\varepsilon+Q\delta\gamma)dx. \tag{2.90}$$

The functional U' is called the additional energy and can be represented in the following form:

$$U'=\sum_{i=1}^{n}\Delta_iP_i-U, \tag{2.91}$$

The presented formula (2.91) is an analog of formula (2.50) (Fig. 2.2a). For a linearly deformable system: $U'=U$.

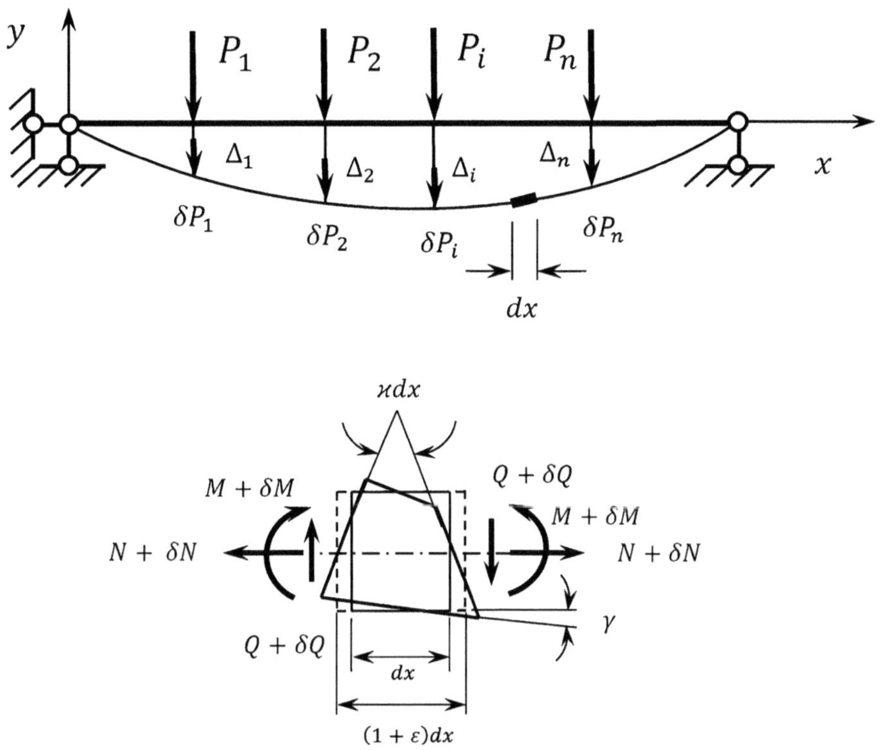

Fig. 2.12 A beam in a deformed position has an additional infinitesimal variation in displacements

The Principle of Virtual Changes in the stress state of a system can also be formulated in the following form: if the sum of virtual work performed by virtual infinitesimal changes of all external and internal forces on the actual displacements of a system is equal to zero, then the deformation of the system is consistent with all existing external and internal links, i.e., the compatibility of system's deformations is observed (i.e., observance of conjoint deformations of the system). Thus, the Principle of Virtual Changes of stresses is the most general principle of kinematics. From this principle, as a consequence, the equations of continuity of deformations arrive: the Saint-Venant equations. As a result, the Principle of Virtual Changes in the stress state of a system underlies the second basic method of calculating statically indeterminate systems, namely, the method of forces. The equations of the method of forces express the conditions for the continuity of the system.

2.12.2 Castigliano's Variational Principle

Among the methods of solving problems of structural mechanics and the theory of elasticity, variational methods play a significant role. The variation equations are reduced to two basic formulas: to Lagrange's variational equation, and to Castigliano's variational formula. The variation Eq. (2.89) establishes the connection between the work of external and internal forces. This work is caused by the variation of given loads and required forces. The given loads, according to the meaning of the problem, should be the same both in the actual (real) state of the system and in all intermediate adjacent states of the system. The intermediate or adjacent states of the system are obtained by varying the variation of loads and δP_i should be taken equal to zero. As a consequence, from (2.89) another form of the Principle of Virtual Changes in the stress state will be obtained:

$$\delta U' = 0 \tag{2.92}$$

From (2.92) it follows that of all virtual changes in the internal forces those that lead to joint deformations must have additional work take a stationary value. A principle formulated in this way is called the Castigliano Variational Principle.

If the functional U' is expressed through generalized functions, the variation $\delta U'$ can be presented as follows:

$$\delta U'(P) = \sum_{i=1}^{n} \frac{\partial U'}{\partial P_i} \delta P_i \tag{2.93}$$

Expression (2.93) is substituted in (2.89) and obtained

$$\sum_{i=1}^{n} \Delta_i \delta P_i - \sum_{i=1}^{n} \frac{\partial U'}{\partial P_i} \delta P_i = 0 \tag{2.94}$$

or

$$\sum_{i=1}^{n} \left(\Delta_i - \frac{\partial U'}{\partial P_i} \right) \delta P_i = 0 \tag{2.95}$$

Since the values δP_i are arbitrary, then the equality (2.95) takes place when:

$$\frac{\partial U'}{\partial P_i} - \Delta_i = 0,$$

from where

$$\Delta_i = \frac{\partial U'}{\partial P_i} \tag{2.96}$$

Thus, a partial derivative of the additional energy by the generalized force is equal to the generalized displacement corresponding to that force. This is the conclusion of the formulation of Castigliano's theorem, which is valid for nonlinear systems.

Formula for a linearly deformed system is similar to formula (2.96). It can be obtained from expression (2.67) if in expression (2.67) we take a partial derivative of one of the generalized forces and take into account the dependence (2.66):

$$\frac{\partial U}{\partial P_j} = \delta_{j1} P_1 + \delta_{j2} P_2 + \cdots + \delta_{jj} P_j + \cdots + \delta_{j-1,n-1} P_{n-1} + \delta_{jn} P_n = \Delta_j \tag{2.97}$$

Thus, the derivative of the work of external forces or the potential energy of internal forces in any force acting on a system is equal to the displacement in the direction of that force. Here, it is appropriate to remind that the energy technique for determining displacements is based on Castigliano's theorem. To prove this theorem, let's compose an expression of potential energy for some system operating under a bending load. In the expression of potential energy we will consider only the influence of bending moments:

$$U = \sum \int_0^l \frac{M^2}{2EI} dx \tag{2.98}$$

The expression of the total moment in an arbitrary section of the structure from the system of forces can be represented in the following form:

$$M = \overline{M}_1 P_1 + \overline{M}_2 P_2 + \cdots + \overline{M}_i P_i + \cdots + \overline{M}_n P_n, \tag{2.99}$$

where $\overline{M}_1, \overline{M}_2, \ldots, \overline{M}_i, \ldots, \overline{M}_n$ are the values of bending moments caused by unit forces: $P_1 = 1$, $P_2 = 1, \ldots, P_i = 1, \ldots, P_n = 1$.

Let's take a partial derivative of the expression of potential energy U in the direction of the generalized force P_i:

$$\frac{\partial U}{\partial P_i} = \frac{\partial}{\partial P_i}\left(\sum \int_0^l \frac{M^2}{2EI} dx\right) = \sum \int_0^l \frac{M \frac{\partial M}{\partial P_i} dx}{EI}$$

because

$$\frac{\partial M}{\partial P_i} = \overline{M}_i \tag{2.100}$$

then

$$\frac{\partial U}{\partial P_i} = \sum \int_0^l \frac{\overline{M}_i M}{EI} dx. \tag{2.101}$$

The expression for determining the displacement is obtained on the right side of the formula.

$$\frac{\partial U}{\partial P_i} = \Delta_{ip}. \tag{2.102}$$

The sequence of the energy method for determining displacements using Castigliano's theorem is as follows: an appropriate force is applied to a system in the direction in which the displacement is sought; a full expression of the potential energy from the combined action of the applied force and load is composed; the expression of potential energy by the applied force is differentiated and a formula is: obtained by which the desired displacement is determined.

2.12.3 The Principle of Least Work

Let us now move on to the principle of least work. Imagine a statically indeterminate elastic system that is under the influence of the load (Fig. 2.13a). Let us turn this system into a statically determinate one by discarding some of the links imposed on it and replacing them with the corresponding reactions X_1, $X_2, \ldots, X_i, \ldots, X_{n-1}, X_n$ of the Fig. 2.13b. The reactions of the system are unknown quantities. Let us find the potential energy of the system at the lowest values of these unknowns.

Let us form an expression for the potential energy of deformation of a system under the action of a given load P and forces $X_1, X_2, \ldots, X_{n-1}, X_n$:

$$U = U(X_1, X_2, \ldots, X_{n-1}, X_n, P) \tag{2.103}$$

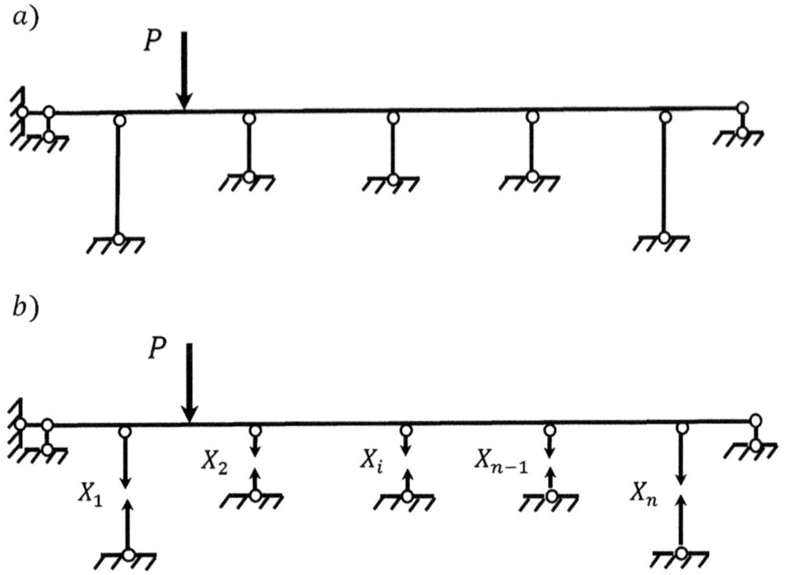

Fig. 2.13 (**a**) Statically indeterminate elastic system under the load, (**b**) Statically determinate system (primary system)

Based on Castigliano's theorem, we write that the displacements of the points of application of forces X_i in the direction of these forces are equal to zero. In this case, we obtain the following:

$$\frac{\partial U}{\partial X_1}=0, \quad \frac{\partial U}{\partial X_2}=0, \ \ldots, \ \frac{\partial U}{\partial X_i}=0, \ \ldots, \ \frac{\partial U}{\partial X_{n-1}}=0, \quad \frac{\partial U}{\partial X_n}=0 \tag{2.104}$$

As is known, the conditions of the form (2.104) determine the conditions for the extremum of the function U as a function of many variables X_i. In this case, the extremum of the function U is a minimum, since with an increase in any of the forces X_i , the potential energy increases. If we consider the value of the second derivative of the potential energy by one of the forces X_i using (2.66) and (2.67), then we can come to the same conclusion:

$$\frac{\partial^2 U}{\partial X_i^2}=\frac{\partial \Delta_i}{\partial X_i}=\delta_{11}>0. \tag{2.105}$$

It is also known that the positive sign of the second derivative indicates the minimum value of the function.

Thus, if the magnitude of forces X_i is arbitrarily changed without changing the given external load, then the value of the potential energy of the system will be greater than that which will be obtained with actual values of X_i. Therefore, only the actual values of X_i give the potential energy of deformation the lowest value and result in the least work that the external forces perform. This is the principle of least work or least action. The principle of least action is applied in the analysis of elastic systems. Condition (2.104) in an expanded form represent the canonical equations of the force method. The forces in the "extra, redundant" links of a statically indeterminate system are such that they correspond to a minimum of the potential energy.

2.13 Hamilton Principle

When solving problems of structural mechanics, which relate to the statics, dynamics, and stability of structures, and when studying the general laws of behavior of an elastic system under loading, a functional containing work of both internal and external forces is introduced. This functional is called the total potential energy of an elastic system and is determined in form of:

$$\ni = U - A \tag{2.106}$$

The total energy of an elastic system can be considered as the work performed by internal forces in a system transitioning from the deformed state to the initial no-deformed state. In this case, the internal forces perform the actual positive work U, and the external forces the possible negative work A, calculated as the total product of force and the path it has passed:

$$A = P\Delta$$

The total potential energy of the system is equal in its value and inverse in sign of the potential energy of deformation U, i.e.

$$\ni = -U$$

Let the elastic system be in motion under the action of a load applied to it. We will consider the transverse oscillations of the beam shown in Fig. 2.14. In accordance with the *Principle of D'Alembert* we will add forces of inertia $-\rho\frac{\partial^2 y}{\partial t^2}$ to active load, where ρ is a per unit length mass of the beam. Let us examine the elastic system in a state of equilibrium at any moment of time. Applying the principle of the virtual displacements to the system shown in the figure we will obtain:

$$\sum_{i=1}^{n} P_i\delta\Delta_i - \delta U - \int_0^l \rho\frac{\partial^2 y}{\partial t^2}\delta y dx = 0 \tag{2.107}$$

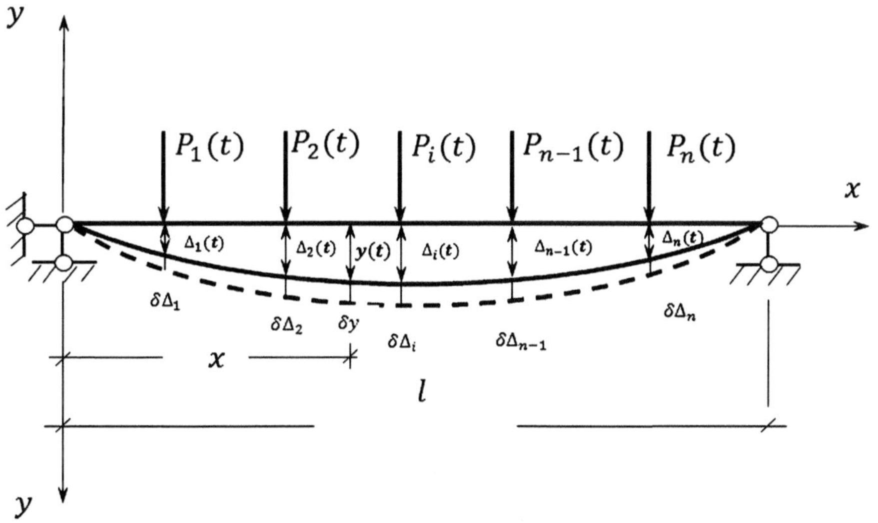

Fig. 2.14 Beam transverse (lateral) oscillations

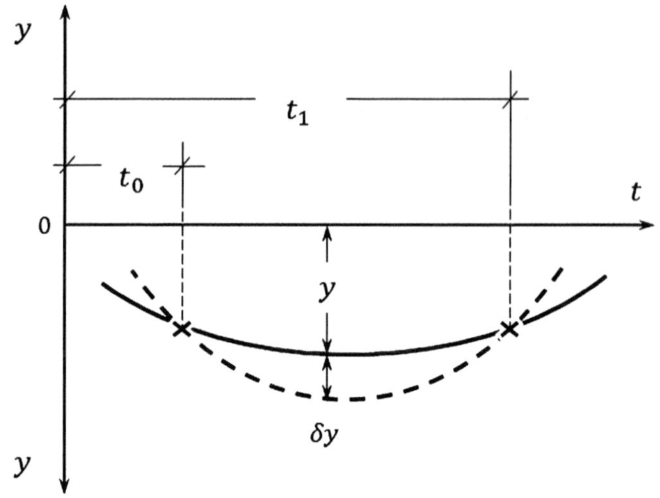

Fig. 2.15 Two trajectories of motion, which are close to each other

where the last member presented a variation of work of the inertial forces distributed along the length of the beam.

In expression (2.107), the forces P_i, displacements Δ_i, y , and potential energy U are the functions not only of the coordinate x but also of time t. Let's change in expression (2.107) the signs on the inverse and put the sign of variation in the first two members (terms) outside of parentheses. Then in the bracket in accordance with (2.106) the expression of total (full) potential energy of the system $\ni$ will remain, and we obtain:

$$\delta \ni + \int_0^l \rho \frac{\partial^2 y}{\partial t^2} \delta y dx = 0 \tag{2.108}$$

Let us now imagine that along with the real trajectory of motion, some trajectory close to it is present and intersecting with the actual trajectory at the moment of time t_0 and t_1 (Fig. 2.15). This means that we are variating the displacements of the system y, fixing (by obeying) their conditions:

$$\text{with } t = t_0 \text{ and } t = t_1 \quad \delta y = 0. \tag{2.109}$$

Let us integrate expression (2.108) over time within the limits from t_0 to t_1:

$$\int_{t_0}^{t_1} \delta \ni dt + \int_{t_0}^{t_1} \int_0^l \rho \frac{\partial^2 y}{\partial t^2} \delta y dx dt = 0 \tag{2.110}$$

Integrating by parts the second member of the obtained expression, we convert it to the form.

$$\int_{t_0}^{t_1} \int_0^l \rho \frac{\partial^2 y}{\partial t^2} \delta y dx dt = \int_0^l \rho \frac{\partial y}{\partial t} \cdot \delta y dx \Big|_{t_0}^{t_1} - \int_{t_0}^{t_1} \int_0^l \rho \frac{\partial y}{\partial t} \cdot \frac{\partial \delta y}{\partial t} dx dt \tag{2.111}$$

In accordance with condition (2.109), the first member of the right part of expression (2.111) becomes zero. As a result of it, we can write down:

$$\int_{t_0}^{t_1} \left[\delta \ni - \int_0^l \rho \frac{\partial y}{\partial t} \cdot \frac{\partial \delta y}{\partial t} dx \right] dt = 0 \tag{2.112}$$

It is easy to see that the integral in square brackets is the first variation of the kinetic energy of the system K. Actually, kinetic energy K,

$$K = \int_0^l \frac{\rho}{2} \left(\frac{\partial y}{\partial t} \right)^2 dx$$

and its variation will be:

$$\delta K = \int_0^l \rho \frac{\partial y}{\partial t} \cdot \frac{\partial \delta y}{\partial t} dx.$$

Thus expression (2.112) can be finally represented in the form of

$$\delta \int_{t_0}^{t_1} (\ni - K) dt = 0 \tag{2.113}$$

or

$$\delta \int_{t_0}^{t_1} (K - \ni) dt = 0.$$

Equality (2.113) expresses the principle of the least action or Hamilton's Principle, the meaning of which is that on the path of actual motion of the system in the time interval $t_0 - t_1$ the integral (2.113) has an extreme value. This principle is the basic criterion for solving the problems of dynamics of oscillations/vibrations of elastic systems.

2.14 The Principle of Reciprocity of Work

Let an arbitrary elastic system be in two different equilibrium states 1 and 2. In each of these states, a certain generalized static load acts on the system. Let us consider a beam which is loaded in the first case by force P_1, and in the second case by force P_2, Fig. 2.16a. The displacements (deflections and angles of rotation) of the system, we agree to denote Δ_{mn}, where the index m indicates the direction of displacement, and n indicates the reason that caused this displacement. Thus, Δ_{mn} is displacement in the direction of force m, from the forces n. Figure 2.16 shows displacements Δ_{11}, Δ_{12}, Δ_{21}, and Δ_{22} which are as follows: Δ_{11} is displacement in the direction of the force P_1, from the action of the force P_1; Δ_{12} is displacement in the direction of the force P_1, from the action of the force P_2; Δ_{21} is displacement in the direction of the force P_2, from the action of the force P_1, and Δ_{22} is displacement in the direction of the force P_2, from the action of the force P_2.

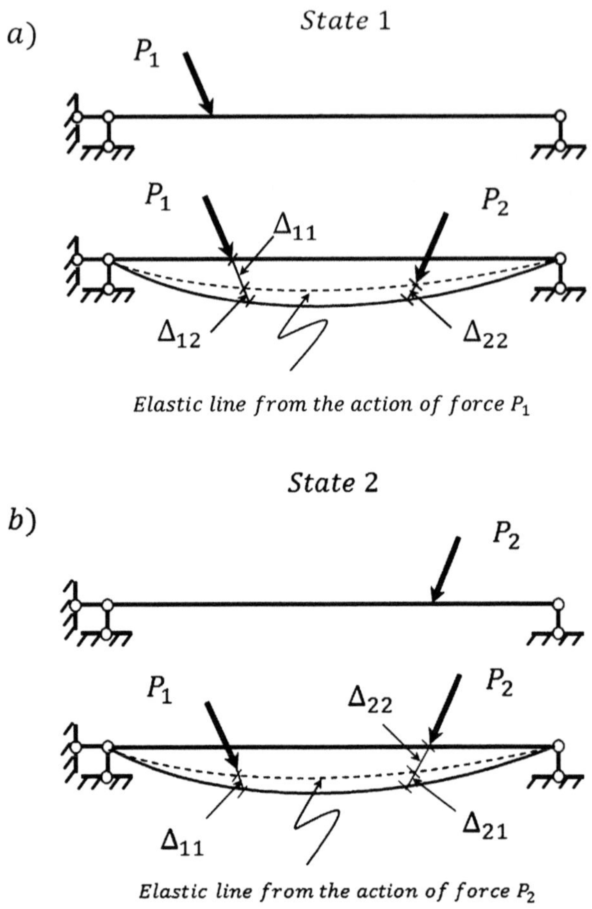

Fig. 2.16 Two states of equilibrium: (**a**) state 1 and (**b**) state 2

Let us imagine that first a statically increasing force P_1 is applied to the beam. As a result, the beam is deformed and the deformed state is shown by the dashed line in Fig. 2.16a. The force P_1 will perform the following work:

$$V_{11} = \frac{1}{2} P_1 \Delta_{11} \tag{2.114}$$

After the process of loading the beam with the first force is over, a static force P_2 will be applied to the deformed system. The second force causes additional deformation in the beam. In this case, the force P_2 will perform the following static work:

$$V_{22} = \frac{1}{2} P_2 \Delta_{22} \tag{2.115}$$

However, the force P_2 will do the work V_{22}, and at the same time will displace (shift) the point of application of force P_1. That is, it will involve force P_1 a second time in the work. In this case, the force P_1 will perform the virtual work. This virtual work is equal in magnitude to the product of the force P_1 and the path Δ_{12} traveled by it (displacement).

$$V_{12} = P_1 \Delta_{12} \tag{2.116}$$

Thus, the total work for the sequential loading of forces P_1 and P_2 will be:

$$V_{11} + V_{22} + V_{12} = \frac{1}{2} P_1 \Delta_{11} + \frac{1}{2} P_2 \Delta_{22} + P_1 \Delta_{12} \tag{2.117}$$

Let us now load the beam with static forces P_1 and P_2 in reverse sequence (Fig. 2.16b). Notice that the displacements Δ_{11} and Δ_{22} in the first and second cases of loading will be the same. Hence, the system under consideration is linearly deformable with small elastic deformations. By analogy with foregoing, for the total work of forces, we obtain the following expression:

$$V_{22} + V_{11} + V_{21} = \frac{1}{2} P_2 \Delta_{22} + \frac{1}{2} P_1 \Delta_{11} + P_2 \Delta_{21} \tag{2.118}$$

The total work of external forces in each of the considered cases of loading is equal to the potential energy of deformations U of the system. It is known that the potential energy of deformations of the system U does not depend on the order of loading of the elastic system, but is determined only by its initial and final states of the system. The initial and final states of the system in each of the considered loading cases are the same. Therefore, the total work in both cases equals to each other, hence it follows:

$$V_{12} = V_{21} \quad \text{or} \quad P_1 \Delta_{12} = P_2 \Delta_{21}. \tag{2.119}$$

Thus, the work of the forces of the first state on displacements in their directions, caused by the forces of the second state, is equal to the work of the forces of the second state on displacements in their directions, caused by the forces of the first state. This conclusion is the reciprocity work theorem or the *Betti Theorem*. The principle of reciprocity of work for linearly deformable systems is the source from which various practical methods of calculating problems in structural mechanics and elasticity theory are derived. In particular, the well-known theorems in structural mechanics on the reciprocity of unit displacements and reactions follow from the principle of reciprocity of work, which is widely used in calculations of statically indeterminate systems. If we assume that the generalized forces acting on the system (Fig. 2.16a) are equal to unity: $P_1 = P_2 = 1$, then from equality (2.119) we obtain:

$$\delta_{12} = \delta_{21}.$$

This expression is a reciprocity theorem for displacements.

If we consider two states of the system, which are characterized by a unit shift of any connections (Fig. 2.17), then from the theorem on the reciprocity of work it follows:

$$r_{12} = r_{21}$$

This expression is the reciprocity theorem of reactions.

If we consider two different states of the system, one of which is characterized by the application of a unit force $P_1 = 1$ to the system, and the other is characterized by a unit displacement of the beam support 2 (Fig. 2.18), then we obtain the following relation:

$$\delta'_{12} = -r'_{21}$$

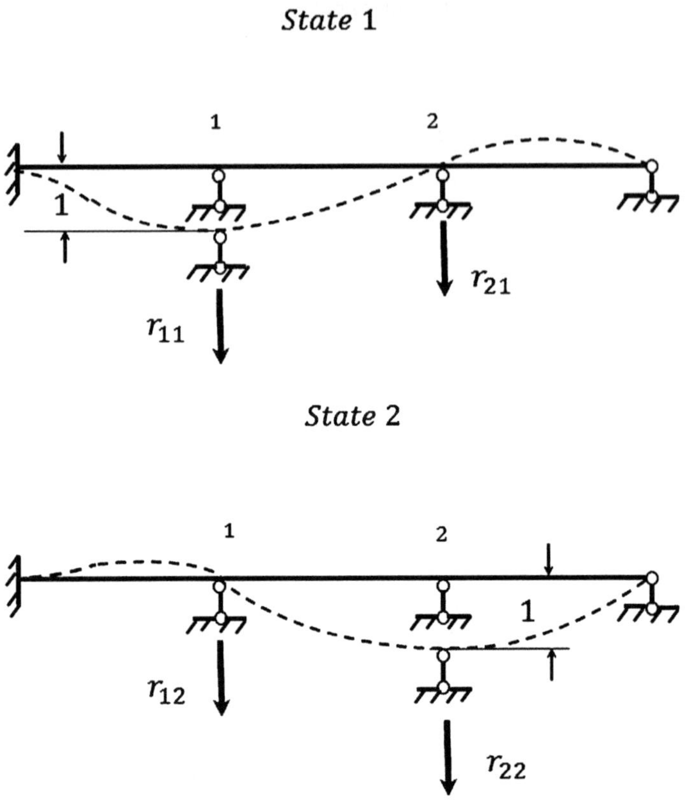

Fig. 2.17 Two states of link unit shift: state 1 and state 2

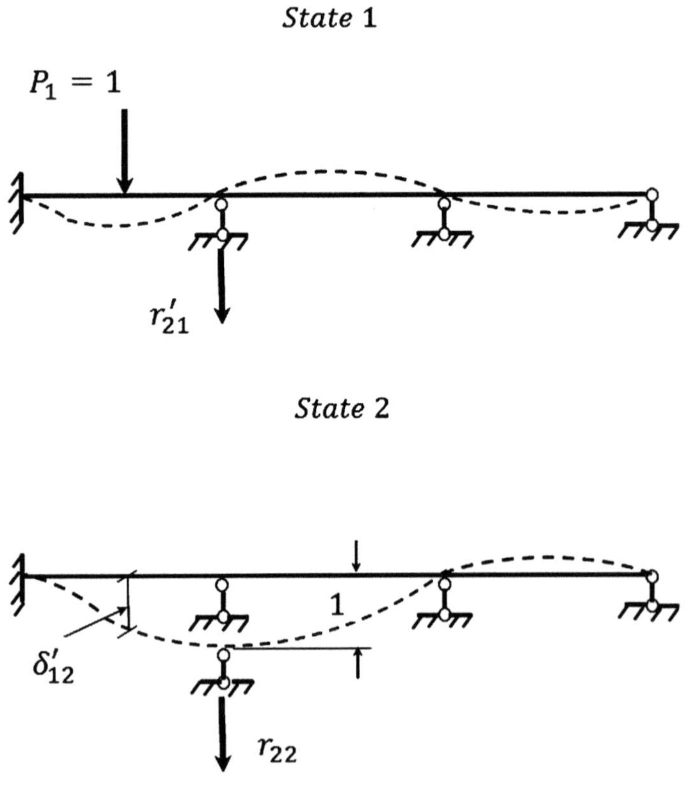

Fig. 2.18 Two different states: state 1 is characterized by the application of a unit force $P_1 = 1$, and state 2 is characterized by the link of a unit displacement

This expression is: a theorem on the reciprocity of unit displacements and reactions. The theorem on reciprocity of unit displacements and reciprocity of reactions is: called the *Rayleigh Theorem.*

From the given examples of linearly deformable systems, it follows that whatever the system under consideration, no matter how different the external influences on the system, in both states, the reciprocity of work must always and inevitably be carried out. In conclusion, we note that the principle of reciprocity of work is a consequence of the fact that there is a linear relationship between forces and displacements, and the system under consideration is conservative. That is, this is such a system in which, during the transition from one state to another, its total energy remains constant and does not depend on its intermediate states.

2.15 General Formula for Determining Displacements

Thus, the possible work is the work performed by external or internal forces on the displacements arising from other forces. The work of external or internal forces on possible displacements is called possible or virtual. The principle of possible displacements allows solving rather complex problems of determining the stress-strain state of structures.

Figure 2.19, a) shows a frame system that is under the influence of external forces and temperature. At the same time settlement of the frame support and its horizontal displacement is present.

Let us define the projection of the displacement of an arbitrary point k in an arbitrary direction i. Let us choose an auxiliary system to determine the desired displacement (Fig. 2.19b).

For this purpose, a unit force is applied at the point and in the direction in which the displacement is to be determined. The virtual system as a real system is in a state of equilibrium. The principle of virtual (possible) displacements is applicable to this system. That is, the work of all external and internal forces on infinitesimal displacements is equal to zero. As any infinitesimal displacements, the displacements of the k-th given system are selected, and the forces are taken from the auxiliary system i.

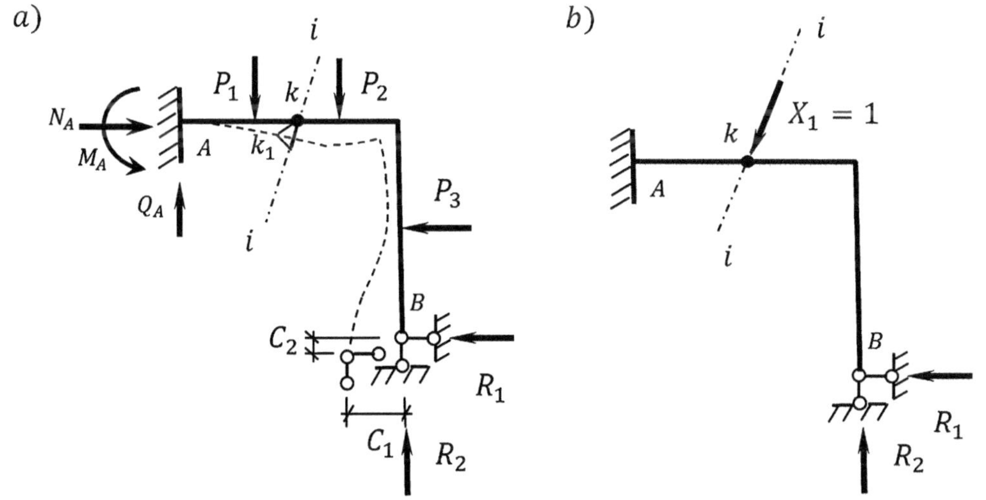

Fig. 2.19 (**a**) A frame system is under the influence of external forces, under the influence of temperature, as well as simultaneously from settlement; (**b**) an auxiliary frame system

$$1\cdot\Delta_{ik}+V+\sum R\cdot C=0, \tag{2.120}$$

from where

$$\Delta_{ik}=-V-\sum R\cdot C. \tag{2.121}$$

Let us consider several cases. From the frame system shown in Fig. 2.19a, we will cut out an infinitesimal element:

1. Assume that the frame element of length dx is under the action of normal forces. Then the linear deformation of this element will be:

$$\frac{N_k}{\mathrm{AE}}dx. \tag{2.122}$$

At this displacement, the internal normal force of the system $i-i$ will do the work (Fig. 2.19b). This work within the element dx will be:

$$-\frac{N_iN_k}{\mathrm{AE}}dx. \tag{2.123}$$

2. Suppose that the frame element of length dx is under the action of bending moments (Fig. 2.19a). Then the mutual displacements of the edges of the element, i.e., the angle of turning of this element will be:

$$\frac{M_k}{\mathrm{EI}}dx. \tag{2.124}$$

At this displacement, the internal moments of the system $i-i$ will do the work (Fig. 2.19b). This work within the element dx will be:

$$-\frac{M_iM_k}{\mathrm{EI}}dx. \tag{2.125}$$

3. Let us assume that the frame element of length dx is under the action of shear forces (Fig. 2.19a). Then the mutual displacements of the edges of element, i.e., the relative displacement of the edges of the element will be:

$$\mu\frac{Q_k}{\mathrm{AG}}dx. \tag{2.126}$$

At this displacement, the internal shear forces of the system $i-i$ will do the work (Fig. 2.19b). This work within the element dx will be:

$$-\mu\frac{Q_iQ_k}{\mathrm{AG}}dx. \tag{2.127}$$

4. Suppose that the frame element of length dx is under the uniform influence of the temperature. Then the axial deformation of the element can be represented in the following form: $\alpha t dx$, where t is the temperature corresponding to the axial deformation of the element dx and α is the coefficient of linear expansion of the material. At this displacement, the internal normal force of the system $i-i$ will do the work (Fig. 2.19b). This work within the element dx will be:

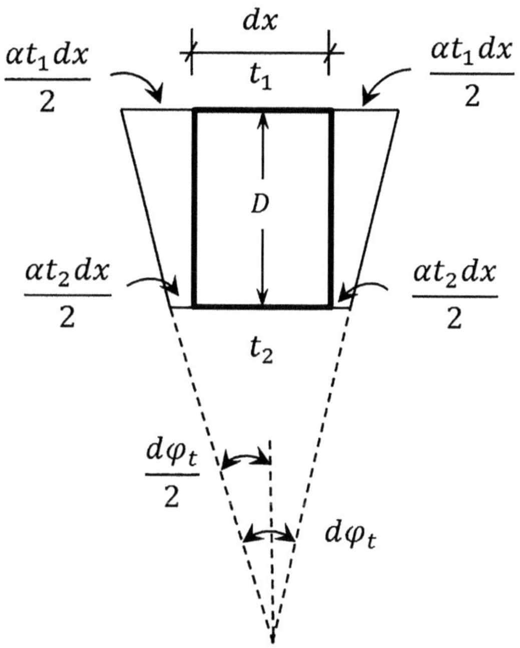

Fig. 2.20 Frame element of length ***dx*** is under the influence of uneven action of the temperature

$$-N_i \alpha t dx. \tag{2.128}$$

5. Assume that the frame element of length dx is under the influence of uneven action of the temperature (Fig. 2.20). The temperature on the upper surface of the frame element of length dx is designated as t_1, and the temperature on the lower surface is denoted as t_2 (Fig. 2.20). Let us suppose that $t_1 > t_2 > 0$, then it is clear from Fig. 2.20 that:

$$\frac{d\varphi_t}{2} = \frac{\frac{\alpha t_1 dx}{2} - \frac{\alpha t_2 dx}{2}}{D}; \qquad \frac{d\varphi_t}{2} = \frac{\alpha t' dx}{D} \tag{2.129}$$

Where $t' = t_1 - t_2$ is the temperature change.

At this displacement, the internal moments of the system $i - i$ will do the work (Fig. 2.19b). This work within the element dx will be:

$$-\frac{M_i \alpha t' dx}{D}. \tag{2.130}$$

Thus, the work of internal forces of the frame system $i - i$ (Fig. 2.19b) performed on the displacements of the frame system $k - k_1$ (Fig. 2.19a) will have the following form:

$$V = -\sum \int \frac{M_i M_k}{EI} dx - \sum \int \frac{N_i N_k}{AE} dx - \sum \int \mu \frac{Q_i Q_k}{AG} dx - \sum \int N_i \alpha t dx$$
$$-\sum \int \frac{M_i \alpha t'}{D} dx. \tag{2.131}$$

Now, let us substitute the values of expression (2.131) into (2.120). As a result, we obtain a general formula for determining displacements, i.e., Maxwell Mohr formula:

$$\Delta_{ik} = -\sum\int\frac{\mathrm{M_i}M_k}{\mathrm{EI}}dx+\sum\int\frac{\mathrm{N_i}N_k}{\mathrm{AE}}dx+\sum\int\mu\frac{\mathrm{Q_i}Q_k}{\mathrm{AG}}dx+\ \sum\int N_i\alpha t dx$$
$$+\sum\int\frac{M_i\alpha t'}{D}dx-\sum R\cdot C \tag{2.132}$$

The integrals in (2.132) extend along the entire length of the rods of the system. All rods of the system are located under the sum sign.

References

1. Timoshenko, S. P., & Gere, J. M. (1961). *Theory of elastic stability* (540p). McGraw-Hill.
2. Timoshenko, S. P. (1967). *Oscillations in engineering* (444p). Nauka.
3. Leontiev, N. N., & Sobolev, D. N. (1980). Variational principles of structural mechanics. In *Engineering problems of structural mechanics. Interuniversity collection of works* (pp. 114–139). MECI (Moscow Engineering and Construction Institute).
4. Rzhanitsin, A. R. (1984). *Structural mechanics* (398p). Bishaya Shkola.
5. Clough, R. W., & Penzien, J. (1993). *Dynamics of structures* (2nd ed., 738p). McGraw-Hill College.
6. Gelfand, I. M., & Fomin, S. V. (2000). *Calculus of variations* (240p) (R. A. Silverman, Trans.). Dover Publications Inc.
7. Slivker, V. (2007). Basic variational principles of structural mechanics. In *Foundations of engineering mechanics theory and applications* (700p). Springer-Verlag.
8. Anderson, J. C., & Naeim, F. (2012). *Basic structural dynamics* (280p). John Wiley & Sons, Inc.
9. Copra, A. K. (2016). *Dynamics of structures* (992p). Pearson.
10. Wunderlich, W., & Pilkey, W. D. (2019). *Mechanics of structures. Variational and computational methods* (2nd ed., 912p). CRC Press & Francis Group.
11. Kabe, A. M., & Sako, B. H. (2020). *Structural dynamics fundamentals and advanced applications* (Vol. I, 1st ed., 970p). Academic Press.
12. Katsikadelis, J. T. (2020). *Dynamic analysis of structures* (1st ed., 922p). Academic Press.

3 Energy Methods of Structural Dynamics

Abstract

This chapter presents energy methods of structural dynamics. Differential equations of motion can be obtained using energy principles and the calculus of variations, i.e., these are the same differential equations that are obtained using the static method. The main goal of energy principles is to find very effective approximate solutions to dynamic problems. Energy methods are associated with integration operations, allowing approximate solutions to have greater accuracy than using differentiation operations. Thus, the conditions for applying integration are better than the same for differentiation. Energy methods can also be considered as methods of approximating functions, since their application solves the problem of the best approximation of given functions in advance to the actual ones. Energy methods belong to the category of applied mathematical methods and can be used in many fields and problems of mechanics. In this chapter, we will discuss the following: derivation of equations of motion using D'Alembert Principle, static method, energy method, applicable problems and solutions, Rayleigh, Ritz, and Bubnov–Galerkin methods.

3.1 Derivation of Equations of Motion Using D'Alembert Principle

Two main methods are used to solve problems of the dynamics of structures: the static method and the energy method.

Static method The static method is based on the application of dynamic equilibrium equations. According to the D'Alembert Principle, the equations of equilibrium can be applied to moving bodies if inertia forces are added to the number of forces acting on the body. Thus, the equations of dynamic equilibrium differ from the equations of static equilibrium by accounting for the forces of inertia in the form of the product of masses or their moments of inertia and accelerations, that is, the second derivatives of linear or angular displacements over time.

The action of dynamic loads and inertial forces created by these loads are considered simultaneously. For each of the masses of a structure, a dynamic equilibrium equation of D'Alembert is composed:

L. G. Petrosian, *Dynamics of Structures*, https://doi.org/10.1007/978-3-031-63539-7_3

$$\begin{gathered}\sum X-m\ddot{x}=0,\quad \sum Y-m\ddot{y}=0,\quad \sum Z-m\ddot{z}=0,\\ \sum M_x-I^*_{mx}\ddot{\alpha}=0,\quad \sum M_y-I^*_{my}\ddot{\beta}=0,\quad \sum M_z-I^*_{mz}\ddot{\gamma}=0,\end{gathered}\tag{3.1}$$

where m is the mass (of a body, structure, any part of the structure, element, etc.) for which the dynamic equilibrium equation is composed; x, y, and z are coordinates of the center of gravity of the mass; α, β, and γ are the angles of rotation of the body relative to the coordinate axes; $\sum X$, $\sum Y$, and $\sum Z$ are the sums of projections of acting forces; $\sum M_x$, $\sum M_y$, and $\sum M_z$ are the sums of moments acting on the body of forces relative to the coordinate axes; I^*_{mx}, I^*_{my}, and I^*_{mz} are the moments of inertia of masses relative to the coordinate axes.

Among the numerous publications devoted to the application of energy calculation methods to solving problems of structure dynamics, let us highlight the works of the following authors: R. T. Harford [1], B.W. Young [2], H.S. Irving, C. I. Dyn [3], Khanh Chau Le, Lu Trong Khiem Nguyen [4], F. Guarrocino, A. Walker [5], H.L. Langhaar [6], and J.N. Reddy [7].

Energy method The energy method is based on the law of conservation of energy, according to which for a conservative system that oscillates without energy loss, i.e., the sum of kinetic (K) and potential (U) energies is a constant value:

$$K+U=\text{constant}\tag{3.2}$$

The potential energy of a system reaches its maximum value at the moment of the greatest deviation of the mass from the position of static equilibrium. At the same time, the velocity and with it the kinetic energy decreases to zero. At the moment of transition of the mass through the static equilibrium position, the potential energy of deformation (or the increment of potential energy as compared to that which corresponds to the position of static equilibrium) is equal to zero and the velocity of displacement reaches its maximum. Consequently, the kinetic energy also reaches its maximum.

The kinetic energy can be determined by the known formula:

$$K=\sum\frac{m_i v_i^2}{2}+\sum\int\frac{m(x)v^2}{2}dx,\tag{3.3}$$

where m_i is the concentrated mass; v_i is the velocity of the concentrated mass; $m(x)$ is the mass per unit length of the rod; v is the velocity of the distributed mass. The first summand (term) of formula (3.3) is the kinetic energy of concentrated masses. The second summand (term) is the kinetic energy of the distributed masses of the system.

The potential energy can be expressed through the work of external forces (V):

$$U=V=\frac{1}{2}\sum P_i\Delta_i\cos(\Delta_i,P_i)+\frac{1}{2}\sum q\Delta\cos(q,\Delta)dx,\tag{3.4}$$

or through the work of internal forces, which for a plane system will be equal to:

$$U=W=\frac{1}{2}\sum\int\frac{M^2}{EI}dx+\frac{1}{2}\sum\int\frac{N^2}{EA}dx+\frac{1}{2}\sum\mu\int\frac{Q^2}{GA}dx,\tag{3.5}$$

where P_i is the concentrated forces acting on the system; Δ_i is the displacement of the point of application of force P_i; $(P_i\Delta_i)$ is the angle between the direction of force P_i and displacement Δ_i ; q is the distributed load acting on the system; Δ is the displacement at the point of application of the distributed load; (Δ, q) is the angle between the direction of load and displacement; M, N and Q are, respectively, the bending moment, longitudinal force, and shear (transverse) force; E and G are, respectively, the elasticity module of material, and the shear deflection (deformation) module; A and I are, respectively, the area and moment of inertia of the cross-section; μ is the coefficient that takes into account the uneven distribution of shear stresses.

Based on the dependence (3.2) on the constancy of the sum of potential and kinetic energies, it follows that in a conservative system for free oscillations:

$$K_{\max} = U_{\max}. \tag{3.6}$$

To determine the frequencies of natural oscillations, let us define a possible form of oscillations. Each of the points of the system will make harmonic oscillations according to the law

$$y(t) = y_{\max} \sin \omega t, \tag{3.7}$$

and the velocity will be: $v = \dot{y}(t) = \omega y_{\max} \cos \omega t$. From whence it follows that $v_{\max} = \omega y_{\max}$. The expression of kinetic energy for a system with discrete masses has the following form:

$$K_{\max} = \frac{1}{2} \sum m_i v_{i\ \max}^2 = \frac{1}{2}\, \omega^2 \sum m_i\, v_{i\ \max}^2 \tag{3.8}$$

The expression of kinetic energy for a system with distributed mass, taking into account that

$$dK_{\max} = \frac{1}{2}\, v_{\max}^2 dm = \frac{1}{2} \omega^2 y_{\max}^2 dm \tag{3.9}$$

can be represented as:

$$K_{\max} = \frac{1}{2} \omega^2 \sum \int y^2 dm = \frac{1}{2} \omega^2 \sum m \int y_{\max}^2 dx. \tag{3.10}$$

Let us denote the kinetic energy $\overline{K}_{\max}$, at $\omega = 1$. If we accept that

$$\overline{K}_{\max} = \frac{1}{2} m \int y_{\max}^2 dx, \tag{3.11}$$

then

$$K_{\max} = \omega^2 \overline{K}_{\max} \tag{3.12}$$

As we already know, the potential energy of a system can be determined through the work of external forces P_i on displacements y_i:

$$U_{\max} = \frac{1}{2} \sum P_i y_{i\ \max} \tag{3.13}$$

or through the work of internal forces on the displacements of the bending elements:

$$U_{\max} = \frac{1}{2} \sum \int \frac{M^2}{EI} dx. \tag{3.14}$$

If the elastic line of the oscillation mode $y(x)$ of the system is given, then since $M = -EIy''$, the expression for potential energy is:

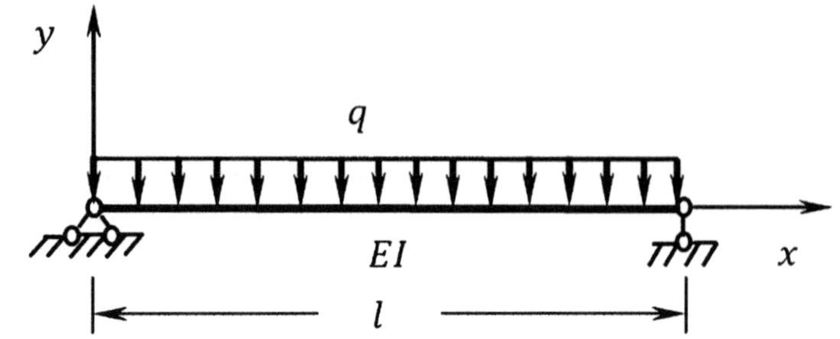

Fig. 3.1 Beam with hinge supports

$$U_{\max} = \frac{1}{2} \sum EI \int \left(y^{2}\right)^{2} dx \tag{3.15}$$

The frequency of free oscillations, taking into account formula (3.12), can be determined by the formula:

$$\omega^2 = \frac{U_{\max}}{K_{\max}}. \tag{3.16}$$

Problem 3.1

It is required to determine the first frequency of natural vibrations of a single-span beam hinged on both sides shown in Fig. 3.1. A beam with equally distributed mass m has a constant cross-section with stiffness EI. Here $y(x)$ is the elastic line of the beam from the distributed load q. It is assumed that the $y(x)$ is the function of the elastic beam line that satisfies the boundary conditions of the problem. Beam is shown in Fig. 3.1.

Solution

The beam is loaded with load q = constant. Deflection line at the origin from the left support $y(x) = \frac{q}{24EI}\left(l^3x - 2lx^3 + x^4\right)$. To simplify the problem let us assume that the distributed load $q = 24$, then we obtain:

(a) Deflection

$$y(x) = \frac{1}{EI}\left(l^3x - 2lx^3 + x^4\right),$$

(b) Bending Moment

$$M(x) = 12lx - 12x^2,$$

$$U_{\max} = \frac{1}{2}\int \frac{M^2}{EI}dx = \frac{1}{2}\int_0^l \frac{\left(12lx - 12x^2\right)}{EI}dx = \frac{12}{5}\frac{l^5}{EI}.$$

$$K_{\max} = \frac{1}{2}m\int \frac{y^2}{EI}dx = \frac{1}{2}m\int_0^l \frac{1}{(EI)^2}\left(l^3x - 2lx^3 - x^4\right)^2 dx = \frac{31}{1260}\frac{ml^9}{(EI)^2}.$$

The frequency of natural vibrations can be determined by the formula:

$$\omega^2 = \frac{U_{\max}}{K_{\max}} = \frac{12l^5 \cdot 1260(EI)^2}{5EI \cdot 31 \cdot l^9} = \frac{97.54EI}{l^4 m},$$

$$\omega = \frac{9.88}{l^2}\sqrt{\frac{EI}{m}}.$$

Now, let us set the elastic line of the beam as a function: $y(x) = a \sin \frac{\pi x}{l}$. This function satisfies the boundary conditions of the problem, on both supports of the beam, since at $x = 0 \quad y(0) = 0$, and $M(0) = 0$, and also at $x = l \quad y(l) = 0$, and $M(l) = 0$.

$$M(x) = -EIy''(x) = \frac{\pi^2}{l^2} EI \sin \frac{\pi x}{l},$$

$$U_{\max} = \frac{1}{2}\int \frac{M^2}{EI} dx = \frac{1}{2}\frac{\pi^4}{l^4} EI \int_0^l \sin^2 \frac{\pi x}{l} dx,$$

$$K_{\max} = \frac{1}{2} m \int \sin^2 \frac{\pi x}{l} dx,$$

$$\omega^2 = \frac{U_{\max}}{K_{\max}} = \frac{\pi^4}{l^4}\frac{EI}{m}; \quad \omega = \frac{\pi^2}{l^2}\sqrt{\frac{EI}{m}} = \frac{9.87}{l^2}\sqrt{\frac{EI}{m}}.$$

The first solution of the problem differs from the second one by only 0.1%.

Problem 3.2

It is required to find the first natural oscillation frequency of a cantilever beam of constant cross-section with a length l with constant stiffness EI with a uniformly distributed linear mass m as shown in Fig. 3.2.

Solution

Let us set the mode of oscillations/vibrations that will satisfy the boundary conditions of the cantilever beam. The boundary conditions satisfy the function $y(x) = a\left(1 - \cos \frac{\pi x}{2l}\right)$ at the origin of coordinates, i.e., in the point of built-in. For this curve $y(0) = 0$, $y'(0) = 0$, $y''(l) = 0$.

$$y' = a\frac{\pi}{2l} \sin \frac{\pi x}{2l}, \qquad y'' = a\frac{\pi^2}{4l^2} \cos \frac{\pi x}{2l},$$

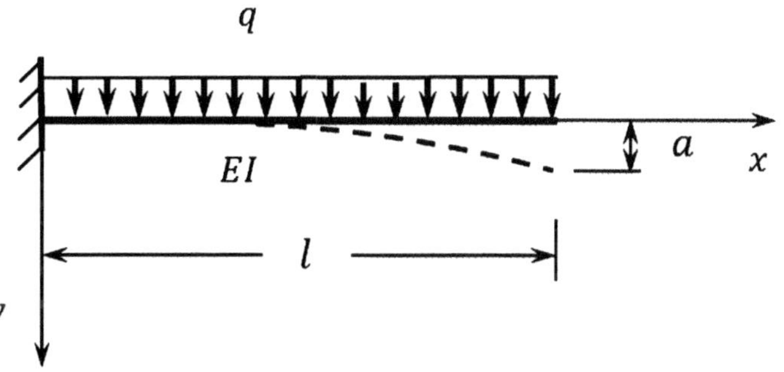

Fig. 3.2 Cantilever beam

$$U_{\max}=\frac{1}{2}EI\int_0^l (y'')^2dx=\frac{1}{2}EIa^2\frac{\pi^4}{16l^4}\int_0^l \cos^2\left(\frac{\pi x}{2l}\right)dx=\frac{1}{64}EIa^2\frac{\pi^4}{l^3},$$

$$K_{\max}=\frac{1}{2}m\int_0^l y^2dx=\frac{1}{2}ma^2\int_0^l\left(1-\cos\frac{\pi x}{2l}\right)^2dx=0.113ma^2l,$$

$$\omega^2=\frac{U_{\max}}{K_{\max}}=\frac{EIa^2\pi^4}{64l^3\cdot 0.113ma^2l}=13.469\frac{\pi^4}{l^4}\frac{EI}{m};\qquad \omega==\frac{3.67}{l^2}\sqrt{\frac{EI}{m}}.$$

The exact solution of the problem:

$$\omega=\frac{(1.875)^2}{l^2}\sqrt{\frac{EI}{m}}=\frac{3.52}{l^2}\sqrt{\frac{EI}{m}}$$

The error will be:

$$\frac{3.67-3.52}{3.67}\cdot 100\%=4\%$$

Problem 3.3

It is required to find the first natural oscillation frequency of a cantilever beam of constant cross-section with a length l with constant stiffness EI with a uniformly distributed linear mass m (see Fig. 3.2).

Solution

Let us set the mode of oscillations/vibrations that will satisfy the boundary conditions of the cantilever beam. The boundary conditions satisfy the function $y(x)=a\frac{x^2}{l^2}\cos\omega t$ at the origin of coordinates, i.e., in the point of built-in. For this curve $y(0)=0$, $y'(0)=0$, $y''(l)=0$. Let us write down the expression of kinetic energy for a given oscillation mode:

$$K=\int_0^l\frac{q}{2g}(y')^2dx=\int_0^l\frac{q}{2g}\left[a\frac{x^2}{l^2}-\omega\sin\omega t\right]^2dx=\frac{ql}{10g}a^2\omega^2\sin^2\omega t.$$

The maximum value of kinetic energy at $\omega=1$ will be

$$K_{\max}=\frac{qla^2}{10g}$$

Let us write down the expression of potential energy for a given oscillation mode:

$$U=\int_0^l\frac{EI}{2}\left(\frac{d^2y}{dx^2}\right)^2dx=\frac{EI}{2}\int_0^l\left(\frac{2a}{l^2}\cos\omega t\right)^2dx=\frac{2a^2EI}{l^3}\cos^2\omega t$$

The maximum value of potential energy will equal:

$$U_{\max}=2\frac{EI}{l^3}a^2$$

$$\omega^2=\frac{U_{\max}}{K_{\max}}=20\frac{EIg}{ql^4};\qquad \omega=\frac{4.47}{l^2}\sqrt{\frac{EI}{m}}.$$

Problem 3.4

It is required to find the first natural oscillation frequency of a cantilever beam of constant cross-section with a length l with constant EI with a uniformly distributed linear mass m (see Fig. 3.2).

Solution

Let us set the mode of oscillations/vibrations that will satisfy the boundary conditions of the cantilever beam. The boundary conditions satisfy the function

$$y(x) = \frac{4a}{l^4}\left(\frac{l^2x^2}{2} - \frac{lx^3}{3} + \frac{x^4}{12}\right)\cos\omega t$$

at the origin of coordinates, i.e., in the point of built-in. For this curve $y(0) = 0, y'(0) = 0, y''(l) = 0$. Let us write down the expression of kinetic energy for a given oscillation mode:

$$K = \int_0^l \frac{q}{2g}(y')^2 dx = \int_0^l \frac{q}{2g}\left[\frac{4a}{l^4}\left(\frac{l^2x^2}{2} - \frac{lx^3}{3} + \frac{x^4}{12}\right)\cdot(-\,\omega\sin\omega t)\right]^2 dx$$

$$= 0.1296\frac{qla^2}{g}\omega^2\sin^2\omega t$$

The maximum value of kinetic energy at $\omega = 1$ will be

$$K_{\max} = 0.1296\frac{qla^2}{g}$$

Let us write down the expression of potential energy for a given oscillation mode:

$$U = \int_0^l \frac{EI}{2}\left(\frac{d^2y}{dx^2}\right)^2 dx = \frac{EI}{2}\int_0^l\left[\frac{4a}{l^4}\left(l^2 - 2lx + x^2\right)\cos\omega t\right]^2 dx = \frac{8}{5}\frac{EIa^2}{l^3}\cos\omega t;$$

The maximum value of potential energy will be equal:

$$U_{\max} = 1.6\frac{EI}{3}a^2$$

$$\omega^2 = \frac{U_{\max}}{K_{\max}} = \frac{1.6EIg}{0.1296ql^4} = 12.36\frac{EIg}{ql^4}; \qquad \omega = \frac{3.517}{l^2}\sqrt{\frac{EI}{m}}.$$

As we mentioned above the exact solution of the problem:

$$\omega = \frac{(1.875)^2}{l^2}\sqrt{\frac{EI}{m}} = \frac{3.52}{l^2}\sqrt{\frac{EI}{m}}$$

Thus, if the elastic oscillation line is taken as a deflection line from a uniformly distributed load, we will get the closest result with respect to the exact solution of the problem.

Problem 3.5

It is required to find the first natural oscillation frequency of both ends built in beam of constant cross-section of length l with constant stiffness EI with a uniformly distributed linear mass m, shown in Fig. 3.3.

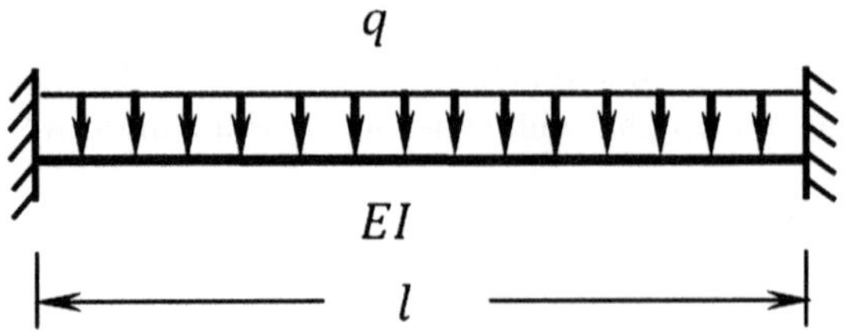

Fig. 3.3 Fixed-end beam

Solution

The equation of an elastic line from a uniformly distributed load q has the form:

$$y(x) = \frac{q}{24EI}\left(l^3x^2 - 2lx^3 + x^4\right).$$

The equation characterizing the bending moment in an arbitrary section of the beam is:

$$M(x) = \frac{q}{2}\left(lx - x^2 - \frac{l^2}{6}\right).$$

To simplify the problem, let us assume that the distributed load $q = 24$, then we obtain:

$$y(x) = \frac{1}{EI}\left(l^3x^2 - 2lx^3 + x^4\right);$$

$$M(x) = 12lx - 12x^4 - 2l^2;$$

$$U_{\max} = \frac{1}{2}\int_0^l \frac{M^2}{EI}dx = \frac{0.4l^5}{EI}, \qquad K_{\max} = \frac{1}{2}m\int_0^l y^2 dx = \frac{ml^9}{1260(EI)^2}.$$

$$\omega^2 = \frac{U_{\max}}{K_{\max}} = \frac{0.4l^5 \cdot 1260(EI)^2}{EIml^9} = \frac{504EI}{l^4m}, \qquad \omega = \frac{22.45}{l^2}\sqrt{\frac{EI}{m}}.$$

The exact solution of the problem:

$$\omega = \frac{4.73^2}{l^2}\sqrt{\frac{EI}{m}} = \frac{22.37}{l^2}\sqrt{\frac{EI}{m}}.$$

The error will be:

$$\frac{22.45 - 22.37}{22.45} \cdot 100\% = 0.36\%$$

Problem 3.6

It is required to determine the circular frequency of natural vibrations of the system shown in Fig. 3.4, using the energy method, (a) ignoring the weight of the beam; (b) accounting for the linear mass of the beam. The beam carries three concentrated masses: m_1, m_2, and m_3. The beam with stiffness $EI = \infty$ rests on a hinged fixed support and on two elastically pliable spring supports with stiffness r_{11}, and r_{22}, respectively. The shape of the vibrations/oscillations is shown in Fig. 3.4.

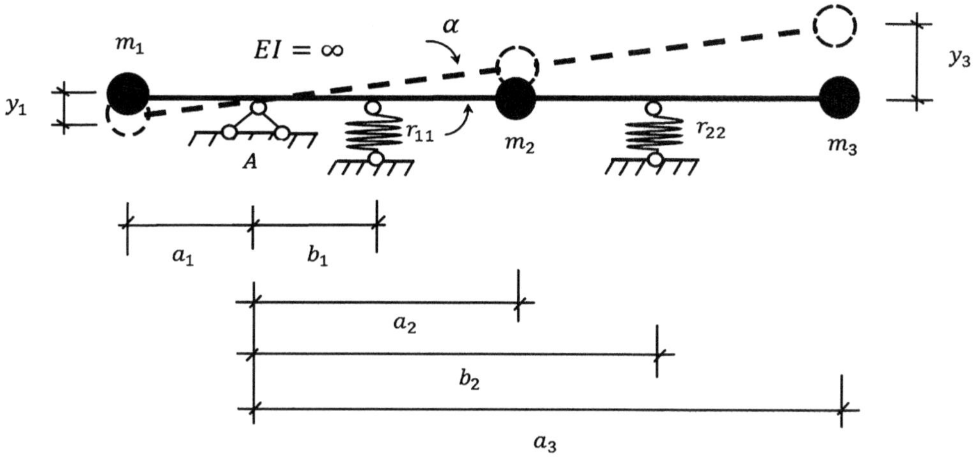

Fig. 3.4 The beam with stiffness $EI = \infty$

Solution

The displacement at any point at A distance x from a hinged fixed support can be expressed in the form:

$$y = \alpha x \sin(\omega t + \varphi_0); \qquad y_{\max} = \alpha x.$$

The expression for the velocity, respectively, is:

$$\dot{y} = \alpha x \omega \cos(\omega t + \varphi_0); \qquad \dot{y}_{\max} = \alpha x \omega.$$

The maximum value of the potential energy of the system will be:

$$U_{\max} = \frac{1}{2}\sum_{i=1}^{3} y_i^2 r_{ii} = \frac{1}{2}(\alpha b_1)^2 r_{11} + = \frac{1}{2}(\alpha b_2)^2 r_{22}.$$

The maximum value of the kinetic energy of the system without taking into account the linear mass of beam will be equal to:

$$K_{\max} = \frac{1}{2}\sum_{i=1}^{3} m_i(\dot{y}_i)^2 = \frac{1}{2}(\alpha a_1)^2\omega^2 m_1 + \frac{1}{2}(\alpha a_2)^2\omega^2 m_2 + \frac{1}{2}(\alpha a_3)^2\omega^2 m_3.$$

The expression for the circular frequency of oscillations of the system will be:

$$\omega^2 = \frac{b_1^2 r_{11} + b_2^2 r_{22}}{m_1 a_1^2 + m_2 a_2^2 + m_3 a_3^2}$$

The maximum value of the potential energy of the system remains unchanged when determining the frequency of natural oscillations, taking into account the mass per unit length. When taking into account the linear mass of the beam, the potential energy of an absolutely rigid beam with linear mass m_* should be added to the maximum value of the kinetic energy of the system. In this case, the expression for the kinetic energy of the system has the following form:

Fig. 3.5 Propped cantilever beam

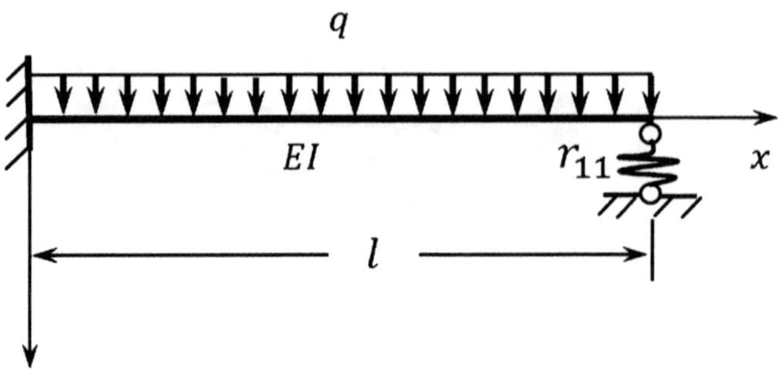

$$K_{m_*} = \frac{1}{2}\omega^2 \int_{-a_1}^{a_3} m_* (\alpha x)^2 dx = \frac{1}{6} m_* \alpha^2 \left(a_1^3 + a_3^3\right),$$

and then the circular frequency of natural vibrations of the system, taking into account the linear mass of the beam, will be equal to:

$$\omega^2 = \frac{b_1^2 r_{11} + b_2^2 r_{22}}{m_1 a_1^2 + m_2 a_2^2 + m_3 a_3^2 + \frac{1}{3} m_* \left(a_1^3 + a_3^3\right)}$$

Problem 3.7

It is required to determine and find the equation of natural oscillation frequencies for the system shown in Fig. 3.5, using the differential equation of transverse oscillations of the beam.

Solution

The differential equation of transverse oscillations of the beam has the following form:

$$\frac{\partial^2 y}{\partial t^2} + \frac{EI}{\overline{m}} \frac{\partial^4 y}{\partial x^4} = 0$$

Let us use the solution of the differential equation of transverse oscillations/vibrations of the beam to solve the problem:

$$\begin{aligned} y &= (C_1 \sin kx + C_2 \cos kx + C_3 \sinh kx + C_4 \cosh kx) \cdot (A\cos \omega t + B \sin \omega t) \\ &= X(x) T(t) \end{aligned}$$

For this problem: at

$$x = 0: \quad X = 0, \quad \frac{dX}{dx} = 0,$$

at

$$x = l: \quad EI \frac{d^3 X}{dx^3} = r_{11} X, \quad \frac{d^2 X}{dx^2} = 0.$$

Using boundary conditions gives four equations to determine the arbitrary constants C_i:

$$C_2 + C_4 = 0; \qquad C_1 + C_3 = 0;$$

$$-k^3C_1\cos kl + k^2C_2\sin kl + k^3C_3\cosh kl + k^3C_4\sinh kl$$

$$= \frac{r_{11}}{EI}(C_1\sin kl + C_2\cos kl + C_3\sinh kl + C_4\cosh kl);$$

$$-C_1k^2\sin kl - C_2k^2\cos kl + C_3k^2\sinh kl + C_4k^2\cosh kl = 0.$$

The equations obtained lead to two equations of the following form:

$$\left[-k^3(\cos kl + \cosh kl) + \frac{r_{11}}{EI}(\sinh kl - \sin kl)C_1\right]$$
$$+\left[k^2(\sin kl - \sinh kl) + \frac{r_{11}}{EI}(\cosh kl - \cos kl)\right]C_2 = 0 \qquad \text{(A)}$$
$$(\sin kl + \sinh kl)C_1 + (\cos kl + \cosh kl)C_2 = 0.$$

The equation for determining the frequency spectrum will be obtained if we compose the determinant of the Eq. (A) and equate it to zero.

$$\frac{r_{11}}{EI}l^3 = -\alpha^3\frac{1 + \cos\alpha\cosh\alpha}{\sin\alpha\cosh\alpha - \cos\alpha\sinh\alpha},$$

where $\alpha = kl$. At $r_{11} = \infty$, we obtain a hinged support; At $r_{11} = 0$ we obtain a free end.

3.2 Rayleigh Method

The Rayleigh Method is based on the equality of kinetic and potential energies of the system. According to the Rayleigh Method, when the system vibrates/oscillates, "suitable" amplitude functions are accepted, that is, functions that satisfy the geometric boundary conditions of a given problem. In the case of bending vibrations, sometimes the static deflection curve from the action of a given load is taken as a function of amplitude displacements. Moreover, in that case, the value of only one lower oscillation frequency of the system is determined. Let us show the application of the Rayleigh Method using the example of the cantilever beam shown in Fig. 3.6.

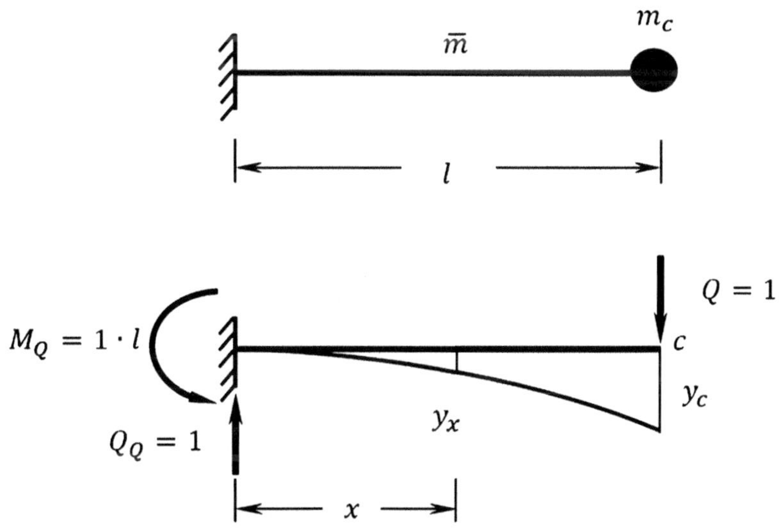

Fig. 3.6 Cantilever beam

Here $\overline{m}$ is the linear value of the distributed mass of the beam. The reduced mass of the beam is shown at the free end of the cantilever beam as a concentrated mass.

For a system with distributed mass, the displacement amplitude is taken as:

$$y(x,t) = y(x)\sin(\omega t + \lambda), \tag{3.17}$$

It assumes that the oscillations are harmonic in nature. Then, the expression for velocity will be:

$$v = y(x)\omega\cos(\omega t + \lambda). \tag{3.18}$$

Consequently, the kinetic energy of the system will take the following form:

$$K = \frac{1}{2}\int_0^l \overline{m}v^2 dx = \frac{1}{2}\omega^2\cos^2(\omega t + \lambda)\int_0^l \overline{m}y^2(x)dx. \tag{3.19}$$

The maximum value of kinetic energy in time takes the form:

$$K_{\max} = \frac{1}{2}\omega^2\int_0^l \overline{m}y^2(x)dx. \tag{3.20}$$

Let us write down the expression for potential energy:

$$U = \frac{1}{2}\sin^2(\omega t + \lambda)\int_0^l [y''(x)]^2 EIdx. \tag{3.21}$$

The maximum value of potential energy has the following form:

$$U_{\max} = \frac{1}{2}\int_0^l [y''(x)]^2 EIdx. \tag{3.22}$$

Following equality $K_{\max} = U_{\max}$ (3.6), the expressions for the maximum value of kinetic energy and the maximum value of potential energy will equate each other. As a result, we obtain the Rayleigh formula:

$$\omega^2 = \frac{\int_0^l [y''(x)]^2 EIdx}{\int_0^l \overline{m}y^2(x)dx.}. \tag{3.23}$$

Thus, according to the Rayleigh Method, without knowing $y(x)$, the displacement amplitude is set as approximately satisfying the boundary conditions of the given problem.

The scale of the accepted forms does not matter and can be any, that is, the displacements corresponding to any form can be increased or decreased an arbitrary number of times. This does not affect the final frequency value.

As noted earlier, in the case of bending vibrations of a system, the static deflection curve from the action of a given load is sometimes taken as a function of $y(x)$. Then the formula for the square of the natural frequency will take the form:

$$\omega^2 = g\frac{\int_0^l mvdx + \sum m_i v_i}{\int_0^l mv^2 dx + \sum m_i v_i^2}, \tag{3.24}$$

where g is the acceleration of gravity; m_i is the concentrated mass of the system; v_i is the velocity of the concentrated masses.

From the condition (3.6), the square of the circular frequency of natural oscillations can also be represented in the following form:

$$\omega^2 = \frac{2U_{\max}}{\sum_{i=1}^{n} m_i[y(x_i)]^2 + \int_l m(x)[y(x)]^2 dx}. \tag{3.25}$$

This formula gives slightly overestimated values of the natural frequency ω^2.

Problem 3.8

It is required to determine the oscillation frequency of the cantilever beam shown in Fig. 3.6, using the Rayleigh Method.

Solution

For the ordinate of function $y(x)$, the deflection of the beam from a concentrated force has the form:

$$y = \frac{1}{EI} \cdot \frac{1}{2}\left(lx^2 - \frac{x^3}{3}\right),$$

While,

$$y'' = \frac{1}{EI}(l - x)$$

The first equation describes a cubic parabola. The second equation corresponds to the diagram of the bending moment M from the force $Q = 1$. Let us substitute first and second expressions into (3.23). As a result, we obtain:

$$\omega^2 = 12.73\frac{EI}{l^4\overline{m}}; \qquad \omega = \frac{3.57}{l^2}\sqrt{\frac{EI}{\overline{m}}}.$$

The exact frequency value is:

$$\omega = \frac{3.515}{l^2}\sqrt{\frac{EI}{\overline{m}}}$$

The percentage of deviation from the exact calculation is 1.57%. (The percentage of error is 1.57%)

3.3 Ritz Method

The Ritz Method is a further development of the Rayleigh Method. The Ritz Method makes it possible to calculate both the first fundamental frequency of natural oscillations of a system and the frequencies of higher tones of oscillations/vibration. According to the Ritz Method, $Y(x)$ is given as a truncated series. The members of this series must satisfy the boundary conditions of the problem. To determine the coefficients a_i ($i = 1, 2, \ldots, n$) from the minimum condition of expression (3.23), the following equations are compiled:

$$\frac{\partial}{\partial a_i}\left(\frac{U}{K}\right) = 0; \quad \text{or} \quad K\frac{\partial U}{\partial a_i} - U\frac{\partial K}{\partial a_i} = 0 \tag{3.26}$$

Let us substitute here $\omega^2 = \frac{U}{K}$, according to (3.23) we obtain:

$$\left(K\frac{\partial U}{\partial a_i} - \omega^2\frac{\partial K}{\partial a_i}\right) = 0, \qquad (i = 1, 2, \ldots, n) \tag{3.27}$$

Taking into account that $K \neq 0$.

$$\frac{\partial}{\partial a_i}\left(U - \omega^2 K\right) = 0; \tag{3.28}$$

Thus, a system of homogeneous equations with respect to a_i have been obtained.

Typically, a function that approximates the shape of an elastic line during vibrations is taken in the form:

$$Y(x) = \sum_{i=1}^{n} a_i\varphi_i(x) \tag{3.29}$$

where $a_1, a_2, \ldots, a_n$ are the coefficients that need to be determined. These coefficients are chosen from the condition of the maximum of the right side of the expression (3.25), namely:

$$\frac{\partial}{\partial a_i} \cdot \frac{U_{\max}}{\sum\limits_{i=1}^{n} m_i[y(x_i)]^2 + \int_l m(x)[y(x)]^2 dx} = 0 \tag{3.30}$$

$$(i = 1, 2, \ldots, n)$$

The functions $\varphi_1(x)$, $\varphi_2(x)$, …, $\varphi_n(x)$ should be linearly independent, satisfy the boundary conditions of the problem, and correspond to possible forms of oscillations/vibrations.

The maximum value of potential energy (3.22) is substituted into (3.26). Then, the maximum condition for transverse oscillation/vibrations of a beam with concentrated and distributed masses can be written in the form:

$$\frac{\partial}{\partial a_i} \cdot \frac{\int_0^l [y''(x)]^2 EI dx}{\sum\limits_{i=1}^{n} m_i[y(x_i)]^2 + \int_l m(x)[y(x)]^2 dx} = 0 \tag{3.31}$$

$$(i = 1, 2, \ldots, n)$$

If, according to the Rayleigh Method, to calculate the fundamental frequency of the shape of static deflection from an action of a given load is chosen as the minimizing modes, then, according to the Ritz Method, the minimizing modes are constructed as linear combinations of a certain number of modes. These modes are called basic and must be chosen so as to be close, respectively to the first, second, third, and so on natural modes of oscillation/vibration.

For the first approximation, one term of the sum (3.29) is usually given, namely, $\varphi_i(x)$ with the coefficient a_i. Such an approximation makes it possible to calculate only the first fundamental frequency of natural oscillations. For the second approximation, two terms are given $Y(x) = a_1\varphi_1(x) + a_2\varphi_2(x)$ of the sum (3.29). As a result, two frequencies of natural oscillations/vibration

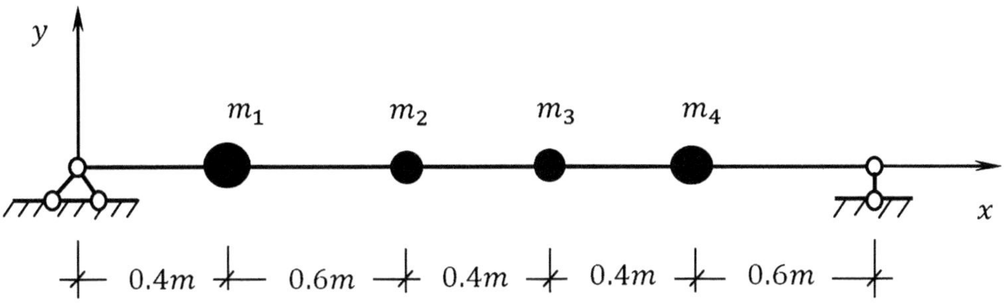

Fig. 3.7 A weightless beam with four discrete masses m_1, m_2, m_3, and m_4

are obtained. The first fundamental frequency is calculated with the accuracy of the second approximation, and the second frequency is calculated with the accuracy of the first approximation.

Sufficiently good results are obtained if beam functions are used as functions of $\varphi_i(x)$. For transverse bending vibrations of rods and beams supported at the ends, the basic forms are used (Fig. 3.7).

As an example, let us choose functions that are close to the basic forms of natural oscillations:

$$\sin\frac{\pi x}{l},\ \sin\frac{2\pi x}{l},\ \sin\frac{3\pi x}{l},\cdots,$$

where l is the length of the beam or rod; x is the distance from the left end to the point where the concentrated mass of the system is located. The first function $\sin\frac{\pi x}{l}$ does not change its sign along the length of the beam. The second function $\sin\frac{2\pi x}{l}$ changes its sign once along the length of the beam. The third function $\sin\frac{3\pi x}{l}$ changes its sign twice, and so on. The selected beam functions satisfy the boundary conditions of the problem, i.e., the fixing conditions for $x = 0$ and $x = l$, since they vanish together (they are zeroed out together) with the second derivatives along the x axis. These functions are also mutually orthogonal, since, with integers when h and k are not equal to each other:

$$\int_0^l \sin\frac{h\pi x}{l}\sin\frac{k\pi x}{l}dx = 0 \qquad (h = 1, 2, 3, \ldots, s;\quad k = 1, 2, 3, \ldots, n; h \neq k)$$

Potential energy of a rod or beam at transverse oscillations near the position of stable equilibrium is determined by the above formula (3.22)

For other cases of beam support, i.e., other boundary conditions, vibration modes/shapes close to their corresponding forms are constructed by means of the expression:

$$Y(x) = A\sinh\frac{k\pi x}{l} + B\sin\frac{k\pi x}{l} + C\cosh\frac{k\pi x}{l} + D\cos\frac{k\pi x}{l},$$

where the constants A, B, C, and D must be preliminarily found from the supporting conditions of the beam (the beam-support conditions). The application of such functions leads to sufficiently accurate values of the first and second frequencies and to very good approximations for the frequencies of the higher tones of oscillations/vibrations.

The frequencies of natural oscillations calculated by the Ritz Method exceed the exact values of the frequencies, since replacing the true form of oscillations/vibrations with a suitable form is equivalent to superimposing additional links to the system. These links increase the rigidity of the system.

Problem 3.9

It is required to determine the first and second frequency of transverse bending oscillations/vibrations of the beam shown in Fig. 3.7. A hinged supported beam of length $l = 240$ cm carries four concentrated masses: $m_1 = 100\frac{\text{k}\Gamma\cdot\text{s}^2}{\text{cm}}$; $m_2 = 60\frac{\text{k}\Gamma\cdot\text{s}^2}{\text{cm}}$; $m_3 = 60\frac{\text{k}\Gamma\cdot\text{s}^2}{\text{cm}}$, and $m_4 = 80\frac{\text{k}\Gamma\cdot\text{s}^2}{\text{cm}}$. The moment of inertia of the cross-section of the beam is assumed to be equal: $I = 100\ \text{cm}^4$. The modulus of elasticity of the steel beam is equal to: $E = 2.1\cdot\frac{10^6\ \text{k}\Gamma}{\text{cm}^2}$

Solution

Taking into account the boundary conditions of the problem, we will try to approximate the mode of the elastic line of the beam during vibrations by the following expression:

$$y(x) = a_1 \sin\frac{\pi x}{l} + a_2 \sin\frac{2\pi x}{l}$$

Let us calculate the values of function $y(x)$ for the points where the masses of the beam are concentrated, i.e., for $x_1 = 40$ cm, $x_2 = 100$ cm, $x_3 = 140$ cm, $x_4 = 180$ cm:

$$y(x_1) = 0.5a_1 + 0.8660a_2;$$

$$y(x_2) = 0.9659a_1 + 0.8660a_2;$$

$$y(x_3) = 0.9659a_1 - 0.5a_2;$$

$$y(x_4) = 0.7071a_1 - a_2.$$

The expression of the maximum kinetic energy value is as follows:

$$K_{\max} = \frac{\omega^2}{2}\left[100y^2(x_1) + 60y^2(x_2) + 60y^2(x_3) + 80y^2(x_4)\right]$$
$$= \omega^2\left[90.29a_1^2 + 94.39a_2^2 - 13.54a_1a_2\right]\cdot 10^{-3}$$

The potential energy is calculated by formula (3.22):

$$U_{\max} = \frac{EI\pi^4}{4l^3}\left(a_1^2 + 16a_2^2\right).$$

Let us designate that $\omega^2 = \frac{EI\pi^4}{4l^3}$, then Eq. (3.27) can be written in the form:

$$\omega^2 a_1 = (90.29a_1 - 6.77a_2)\cdot 10^{-3}$$

$$\omega^2 a_2 = (-6.77a_1 - 94.39a_2)\cdot 10^{-3}$$

The frequency equation takes the following form:

$$\begin{vmatrix} 90.29\cdot 10^{-3} - \omega^2 & -6.77\cdot 10^{-3} \\ -6.77\cdot 10^{-3} & 94.39\cdot 10^{-3} - \omega^2 \end{vmatrix} = 0,$$

or

$$\omega^4 - 96.19\cdot 10^{-3}\omega^2 + 529.79\cdot 10^{-6} = 0.$$

From this equation, we find:

$$\omega_1 = 5.87 \cdot 10^{-3}; \quad \omega_2 = 90.31 \cdot 10^{-3}.$$

3.4 Bubnov–Galerkin Method

The Bubnov–Galerkin Method is a general direct variational method for solving differential equations. This method follows from the Principle of Virtual Displacements, and its theoretical foundations are presented in the mathematical literature and in the theory of elasticity.

Let us assume that it is necessary to solve the equation:

$$Ly(x) - f = 0, \tag{3.32}$$

where L is a linear operator. The solution of this equation must satisfy the boundary conditions on the contour.

According to the Bubnov–Galerkin Method, a complete sequence of coordinate functions φ_1, φ_2, …, φ_n is selected, each of which satisfies boundary conditions of the problem The solution of Eq. (3.1) is sought in the form:

$$Y(x) = \sum_{i=1}^{n} a_i \varphi_i(x), \tag{3.33}$$

where a_i are constants.

After substituting $Y(x)$ into Eq. (3.32) and multiplying by $\varphi_i (i = 1, 2, \ldots, n)$, we integrate over the domain D. As a result, we obtain a system of linear algebraic equations:

$$\int_D \left[L\left(\sum a_i \varphi_i \right) - f \right] \varphi_i dD = 0, \tag{3.34}$$

with respect to constants $a_1, a_2, \ldots a_n$, from which we find the values of a_i.

The essence of the Bubnov–Galerkin Method will be shown using the example of determining the frequency of bending vibrations. As well as in the Ritz Method in the expression $y(x, t) = y(x)T(t)$, we set $y(x)$ in the form of a truncated series with n coefficients a_i:

$$y(x) = a_1 f_1(x) + a_2 f_2(x) + \ldots + a_i f_i(x) + \ldots + a_n f_n(x). \tag{3.35}$$

Each term of the series satisfies the boundary conditions and expresses the virtual displacement. The left part of the differential equation:

$$\frac{d^2}{dx^2}\left(EI \frac{d^2 y(x)}{dx^2} \right) - m\omega^2 y(x) = 0, \tag{3.36}$$

can be considered as the sum of internal and inertial forces per unit length of the beam with the accuracy of a multiplier $T(t) = A \sin(\omega t + \lambda)$. The expression of the virtual work in the state of equilibrium is obtained by multiplying it by $a_i f_i(x) dx$ and integrating it within the length of the beam. This expression according to the beginning of possible virtual displacements is equal to zero:

$$\int_l \left\{ \frac{d^2}{dx^2} \left[EI \frac{d^2}{dx^2} \left(\sum_{i=1}^{n} a_i f_i(x) \right) \right] - m\omega^2 \sum_{i=1}^{n} a_i f_i(x) \right\} f_i(x) dx = 0. \tag{3.37}$$

The equation can also be seen as a condition of orthogonality of the left part of differential equation (3.36) to all functions of (3.35).

Thus, from the condition that the determinant of the system of homogeneous equation (3.37) is equal to zero, it is possible to obtain equations for finding the frequencies of natural oscillations ω of the system. According to the method of Bubnov–Galerkin the approximating functions $f_i(x)$ should satisfy both kinematic and force boundary conditions.

Thus, a nontrivial solution of a system of homogeneous algebraic equations leads to a characteristic equation of the n-th degree for calculating the natural oscillation frequencies of the system. When calculating the first fundamental frequency with sufficient accuracy for practice, only one term in the series (3.35) can be limited. Two terms of series (3.35) can be limited to verify the result obtained.

To determine free vibrations of a cantilever beam, it is necessary that the functions $f_i(x)$, in addition to the kinematic boundary conditions $f_i(0) = 0$, and $f_i'(0)=0$ also satisfy the force boundary conditions:

$$\frac{d^2 f_i}{dx^2}=0; \quad \frac{d^3 f_i}{dx^3} = -m\omega^2 f_i, \tag{3.38}$$

for $x = l$.

When using the Ritz Method, it is possible for $x = l$, to satisfy only the kinematic boundary conditions:

$$\frac{d^2 f_i}{dx^2}=0; \quad \frac{d^3 f_i}{dx^3}=0. \tag{3.39}$$

For example, for a cantilever beam with a concentrated mass m at the end, the mass (weight of the load) is taken into account in the expression:

$$\Phi = \frac{1}{2}\int_0^l \left[\frac{EI}{2}\left(\frac{d^2 y(x)}{dx^2}\right)^2 - \frac{m}{2}\omega^2 y^2(x)\right] dx - \frac{m}{2} y^2(l)\omega^2. \tag{3.40}$$

Thus, the Bubnov–Galerkin Method, like the Ritz Method, is universal and can be applied to solving various boundary value problems.

All the limitations in the choice of approximation same functions made in the description of the Ritz Method must be taken into account in the Bubnov–Galerkin Method as well, when choosing a minimizing form, i.e., the function when choosing a minimizing form, i.e., the function of the oscillations/vibrations modes.

Problem 3.10

It is required to find the fundamental frequency of transverse vibrations of the beam of variable stiffness shown in Fig. 3.8.

The height of the beam h and its mass m change according to the harmonic law:

$$h = h_0\left(1 + \sin\frac{\pi x}{l}\right), \quad m = m_0\left(1 + \sin\frac{\pi x}{l}\right)$$

The moment of inertia of the cross-section of the beam varies according to the harmonic law:

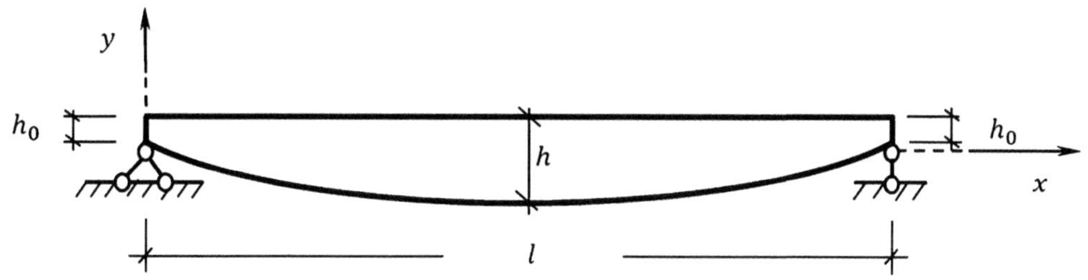

Fig. 3.8 Scheme of the beam

$$I = I_0\left(1 + \sin\frac{\pi x}{l}\right)^3,$$

where I, and h_0 are, respectively, the moment of inertia and the minimum height of the beam on its supports; m_0 is the linear mass of a beam of constant stiffness with height h_0.

Solution

Using the Bubnov–Galerkin Method, we choose as the basis beam functions:

$$y_1 = \sin\frac{\pi x}{l},$$

As a first approximation, let us limit ourselves to one term of the series. The chosen function must satisfy the four boundary conditions of the problem, i.e., the conditions of hinged support of the beam:

$$y_1(0) = y''(0) = 0$$

$$y_1(l) = y''(l) = 0$$

This function is substituted into expression (3.37), having previously denoted by K_{11} and U_{11}, we get:

$$K_{11} = m_0\int_0^l\left(1 + \sin\frac{\pi x}{l}\right)\sin^2\frac{\pi x}{l}dx = 0.356m_0 l,$$

$$U_{11} = -\frac{EI_0\pi^2}{l^2}\int_0^l\left(1 + \sin\frac{\pi x}{l}\right)^3\sin\frac{\pi x}{l}dx = 1.626\frac{EI_0\pi^4}{l^3},$$

from where:

$$\omega = \frac{2.16\pi^2}{l^2}\sqrt{\frac{EI_0}{m_0}}$$

As basic approximating functions for calculating the second approximation of the first frequency, we take the following functions:

$$y_1 = \sin\frac{\pi x}{l}, \qquad y_2 = \sin\frac{3\pi x}{l}.$$

The minimizing form of oscillations with two terms of the series:

$$y = a_1 \sin\frac{\pi x}{l} + a_2 \sin\frac{3\pi x}{l}.$$

is substituted into expression (3.37) and we calculate the values for K_{11} and U_{11}:

$$K_{11} = 0.356 m_0 l;\, K_{12} = K_{21} = -0.085 m_0 l;\, K_{22} = 0.414 m_0 l;$$

$$U_{11} = 1.62\frac{EI_0}{l^3}\pi^4; \quad U_{12} = U_{21} = -6.98\frac{EI_0}{l^3}\pi^4; \quad U_{22} = 98.92\frac{EI_0}{l^3}\pi^4.$$

The frequency equation will have the following form:

$$\begin{vmatrix} K_{11}\omega^2 - U_{11} & K_{11}\omega^2 - U_{12} \\ K_{21}\omega^2 - U_{21} & K_{22}\omega^2 - U_{22} \end{vmatrix} = 0,$$

After performing some transformations and substituting the values of K_{11} and U_{11}, we obtain the following equation:

$$\begin{vmatrix} 0.924 - 3.238\lambda & 0.085 - 6.98\lambda \\ 0.085 - 6.98\lambda & 0.828 - 197.83\lambda \end{vmatrix} = 0,$$

Where:

$$\lambda = \frac{EI_0\pi^4}{m_0 l^2 \omega^2};$$

from here:

$$\omega = \frac{1.781\pi^2}{l^2}\sqrt{\frac{EI_0}{m_0}}.$$

In the third approximation, the value of the frequency of the fundamental tone of oscillations:

$$\omega = \frac{1.779\pi^2}{l^2}\sqrt{\frac{EI_0}{m_0}}$$

is sufficient from a practical point of view.

References

1. Harford, R. T. (1979). *Energy methods in stress analysis: With an introduction to finite element techniques* (432p). Publisher Ellis Horwood Ltd.
2. Young, B. W. (1981). *Energy methods of structural analysis. Theory, worked examples and problems* (168p, 476p). Red Globe Press.
3. Shames, I. H., & Clive, I. (2006). *Dyn energy and finite element method in structural mechanics* (750p). New Age International Publishers.
4. Le, K. C., & Nguyen, L. T. K. (2014). *Energy methods in dynamics. (Interaction of mechanics and mathematics)* (413p). Springer.
5. Guarrocino, F., & Walker, A. (2015). *Energy methods in structural mechanics: A comprehensive introduction to matrix and finite element methods of analysis* (444p). Thomas Telford University of Napoli.
6. Langhaar, H. L. (2016). *Energy methods in applied mechanics* (1st ed., 368p). Dover Publications.
7. Reddy, J. N. (2017). *Variational methods in applied mechanics* (3rd ed., 712p). Wiley and Sons.

4 Free Vibrations of a System with Single Degree of Freedom (SDOF)

Abstract

This chapter presents the study of free vibrations of a system with single degree of freedom (SDOF) and the following topics are discussed: analysis of free vibration, components of the basic dynamic system such as degrees of freedom, mass, elastic properties of the system, equation of motion for (SDOF), damping that including critically damped system, over-critically damped system, under-critically damped system, no damped system, as well as the total energy of the vibrating mechanical system and applicable problems and solutions.

4.1 Analysis of Free Vibration and Components of the Basic Dynamic System

A dynamic response involving deformations is usually oscillatory in nature. Free oscillation takes place when a system oscillates/vibrates under the action of forces inherent to the system itself due to an initial disturbance and absence of externally applied forces. The system subsequently oscillates about its equilibrium position. Such unforced oscillations are called free or natural because they are not driven by an external force.

As is known, time is introduced as a new coordinate when the action of dynamic loads is calculated. Therefore, the task of dynamics is to determine the stress-strain state in time. The inertial forces that depend on accelerations are an essential feature in dynamic calculations. The next circumstance that distinguishes dynamics from statics is the need to take into account resistance forces. The nature of the resistance forces is complex and diverse. Therefore, taking into account the forces of resistance is more complicated than taking into account the forces of inertia. It must be remembered that the forces of resistance are always present. However, in some cases, they can be neglected or approximated, depending on the conditions of a given problem.

The simplest case of oscillatory motion is the oscillation of a conservative linear system with a single degree of freedom. The free oscillations/vibrations of a system with a single degree of freedom are considered below in the example of a cantilever beam supported only at one end (built-in, or fixed support) and with a lumped mass at the free end (Fig. 4.1). We will consider the motion of a cantilever beam as a function of time with stiffness k in the lateral direction.

During free oscillations of the cantilever beam at any moment in time, the following forces: restoring force R, force of resistance or damping force D, and force of inertia I will act on the mass

L. G. Petrosian, *Dynamics of Structures*, https://doi.org/10.1007/978-3-031-63539-7_4

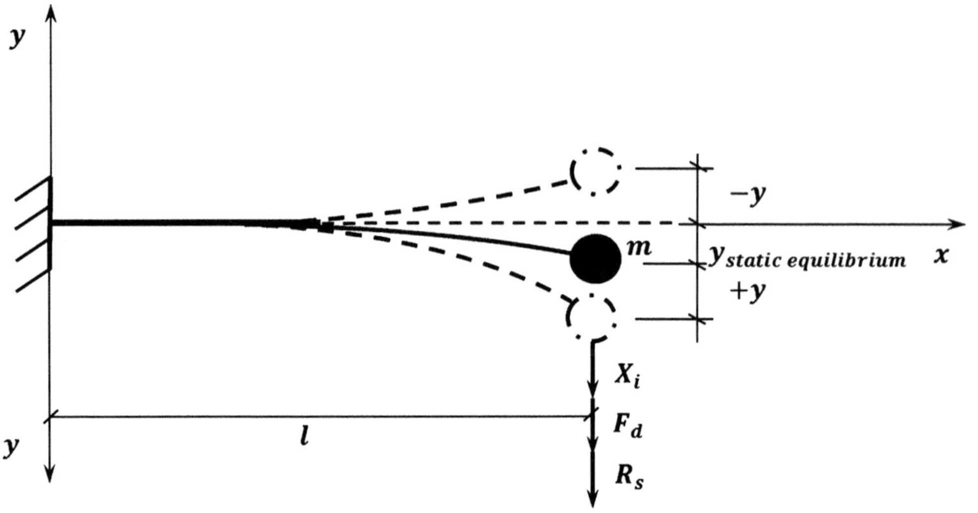

Fig. 4.1 Cantilever beam is a system with SDOF

which is deviated from the position of a static equilibrium by value y. Let us consider each of these forces, assuming positive forces, displacements, speeds, and accelerations directed downward.

Restoring force Restoring force is the force of the elastic reaction of a system and arises out of the deviation of mass m from the position of a static equilibrium. This force tends to return the mass to the position of static equilibrium. The restoring force is proportional to the deviation y of the point in which the mass is concentrated, i.e.:

$$R_s = -ky \tag{4.1}$$

The coefficient of proportionality or coefficient of rigidity, k, is the reaction of the beam at the point when the displacement of this point equal to one unit of displacement. Coefficient of proportionality of a system is the value that depends on the elastic and geometrical characteristics of the system and can be found from the general expression for displacement of one unit.

$$\bar{y} = \delta k = 1 \text{ or } k = 1/\delta$$

where δ is the displacement of the considered point, caused by a unit of force. So, for example, for the considered cantilever beam (Fig. 4.2) the rigidity of which is equal to EI, at the action of force $P = 1$ we have:

$$\delta = 1 \cdot \frac{l^3}{3EI} \text{ and } k = (3EI)/l^3$$

For the simple beam, rested on two hinged supports at both ends (Fig. 4.3), we have the dependence between the deformation at point $x = l_1$ and force P in the form of:

$$y = \frac{Pl_1^2(l - l_1)^2}{3lEI}; \quad y = P/k,$$

where:

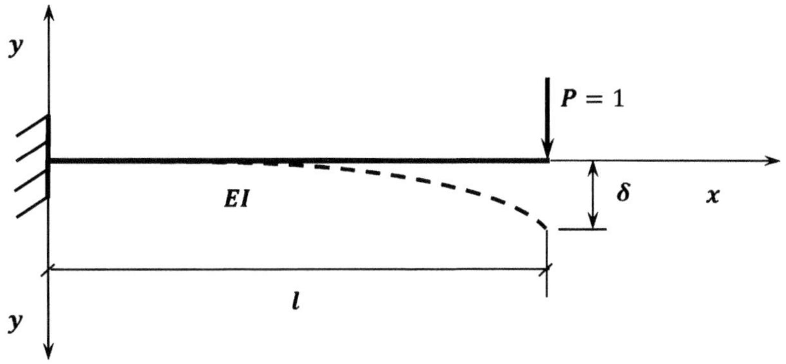

Fig. 4.2 Cantilever beam

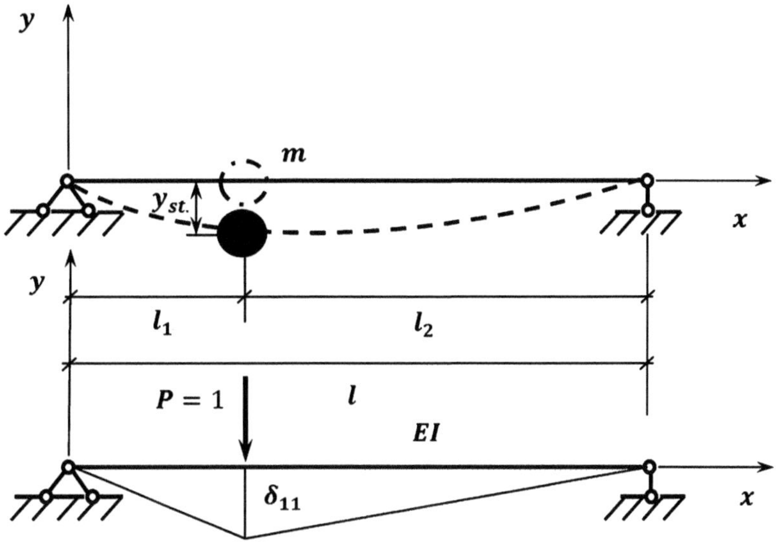

Fig. 4.3 Simple beam is as a system with SDOF

$$k = \frac{3lEI}{l_1^2 (l - l_1)^2}.$$

The next force that we need to consider is damping. A structure is characterized by its mass, stiffness, and damping. Resistance force F_d arises from the resistance of the environment, i.e., air or water, as well as from the viscosity or partial elasticity of the material of the structure, and from the friction in the supporting system. With all these inelastic resistances, energy absorption takes place. This absorbing force, which is also called a damping force, works to counteract any oscillations and vibrations. There are several ways to define a damping force. However, damping should be determined through experiments.

To simplify the mathematical side of the task, it is assumed that the resistance force or damping force F_d is proportional to the velocity of the oscillations $v = \dot{y}$. The force of resistance is applied to the mass and directed to the side opposite to its movement.

$$F_d = -c\dot{y} \tag{4.2}$$

where c is coefficient of proportionality or damping coefficient.

If the mass is moving downward, then the velocity $\dot{y}$ will be positive and F_d will be negative and will be acting to pull the mass back up and vice versa, so that it has a sign opposite to that of v. Damping or energy dissipation is a complex phenomenon for which numerous analytical models exist to describe its effects.

The highest velocity of motion of the beam reaches the position of the line of static equilibrium during oscillations of the system. The velocity of motion of the beam is equal to zero at the place of the greatest deviation of the system from the line of static equilibrium.

Inertial force The inertial force X_i, in accordance with the Principle of D'Alembert, is equal to the product of mass m to the acceleration that is to the second derivative of the displacement y over the time $t(X_i = I_i = m\ddot{y})$. The force of inertia is directed toward the opposite acceleration.

The general form of the differential equation describing the free oscillations follows directly from Lagrange's equation (2.83) or from simply balancing the forces on the mass.

Equation of dynamic equilibrium The equation of dynamic equilibrium of all forces acting on mass m has the following form:

$$\sum Y = X_i + F_d + R_s \tag{4.3}$$

Also, as shown in Fig. 4.1, we will measure all displacements of the mass of the cantilever beam from its equilibrium position. Therefore, the $y = 0$ position will correspond to the center of gravity for the mass as it hangs on the free end of the cantilever beam. According to the equation of dynamic equilibrium of all forces acting on the mass, let us substitute the corresponding expressions of the forces X_i, F_d, and R_s with opposite signs to obtain the ordinary homogeneous linear differential equation of the second kind:

$$m\ddot{y} + c\dot{y} + ky = 0 \tag{4.4}$$

where the term $m\ddot{y}$ represents the inertia force due to acceleration $\ddot{y}$ of mass m; the term $c\dot{y}$ represents the viscous damping force equal to the product between the damping coefficient c and velocity $\dot{y}$; the term ky represents the reactions produced by the system with stiffness k due to the displacement y. The mass m, k, and stiffness ($k > 0$) are greater than zero for a physical/mechanical system. A dot ($\cdot$) over a variable indicates differentiation with respect to time. Let us normalize Eq. (4.4) by division by m, and let:

$$2\zeta = \frac{c}{m}; \qquad \omega_0^2 = \frac{k}{m} \tag{4.5}$$

so that the equation becomes:

$$\ddot{y} + 2\zeta\omega_0\dot{y} + {\omega_0}^2 y = 0 \tag{4.6}$$

using the natural angular frequency ω_0, and the damping ratio ζ,

$$\omega_0 = \sqrt{\frac{k}{m}}; \qquad \zeta = \frac{c}{2\sqrt{km}}. \tag{4.7}$$

The characteristics of the free vibration behavior depend on the magnitude of the damping ratio. The natural frequency is determined by the stiffness and mass of the system.

For Eq. (4.4), the initial conditions are:

$$y|_{t=t_0} = y_0, \qquad \left.\frac{dy}{dt}\right|_{t=t_0} = v_0,$$

It is clear from the physical considerations of oscillations that the initial displacement y_0 and initial velocity v_0 of the oscillating mass of the system completely determine the oscillation process. If y_1 and y_2 are two partial solutions of Eq. (4.4), then their sum $y = y_1 + y_2$ is also a solution of the same equation. It follows from the above that if two partial solutions of y_1 and y_2 of Eq. (4.4) are found, then their linear combination

$$y = C_1 y_1 + C_2 y_2 \tag{4.8}$$

where C_1 and C_2 are arbitrary constants also is a solution of the same equation. The general solution of the differential equation of the second order is obtained as a result of two integrations and therefore contains two arbitrary constants. Hence expression (4.8) is the general solution of Eq. (4.4). Of course, y_1 and y_2 should not be proportional, because if $y_2 = ky_1$, then

$$C_1 y_1(t) + C_2 y_2(t) = (C_1 + C_2 k) y_1(t) = C_1 y_1(t)$$

there is only one arbitrary constant.

Based on the property of the exponent's proportionality to its derivatives, we search for a solution to Eq. (4.4) by using Euler in the form

$$y = e^{pt} \tag{4.9}$$

where p is some constant to be determined. The solution of Eq. (4.6) is of exponential type and is found by substitution of the partial solution $y = e^{pt}$ into the homogeneous equation of motion, yielding the characteristic equation. Let us substitute the following expression:

$$\frac{dy}{dt} = pe^{pt}, \qquad \frac{d^2y}{dt^2} = p^2 e^{pt} \tag{4.10}$$

into (4.4) and reduce by e^{pt}. We obtain a quadratic characteristic equation to determine p.

$$mp^2 + cp + k = 0 \tag{4.11}$$

As known from algebra when solving a quadratic equation, different cases can be presented depending on the sign of discriminant.

$$D = c^2 - 4mk \tag{4.12}$$

Case 1. Critically Damped System When $c = c_c$. If $c = 2\sqrt{mk}$, the discriminant is zero. This special value of damping when the discriminant of equation © is equal to zero is called a critical damping coefficient and is denoted by c_c.

$$c_c = 2\sqrt{mk} \tag{4.13}$$

The ratio of actual damping rate to the critical damping coefficient is called the damping ratio ζ,

$$\zeta = \frac{\text{Damping Coefficient}}{\text{Critical Damping Coefficient}} = \frac{c}{c_c} = \frac{c}{2\sqrt{mk}} \tag{4.14}$$

In this case, the two roots are real:

$$p_{1,2} = -a$$

And the solution at any time t will be:

$$y = C_1 e^{-at} + C_2 t e^{-at} = e^{-at}(C_1 + C_2 t) \tag{4.15}$$

here e is the base of natural logarithms $e = 2.72$.

Notice that as $t \to \infty$ the displacement will approach zero. The motion is aperiodic and critically damped. In this case, $\zeta = 1$ and the system returns to its equilibrium position without oscillating.

The characteristics of the free oscillation behavior depend on the magnitude of the damping ratio ζ.

Case 2. Overcritically Damped System When $c > c_c$ The system is overcritically damped. If forces of friction are great, then motion is aperiodic and overcritical damped; in other words, if $c^2 > 4mk$, then the Eq. (4.11) has real roots:

$$p_{1,2} = \frac{-c \pm \sqrt{c^2 - 4mk}}{2m} \tag{4.16}$$

Let us designate $p_1 = -a$, $p_2 = -b$, as both are negative and unequal. Then from Eqs. (4.8) and (4.9), we obtain the general solution of Eq. (4.4) in the form:

$$y = C_1 e^{-at} + C_2 e^{-bt} \tag{4.17}$$

where C_1 and C_2 are some constants. Thus, with a large friction the deviation of the system from the equilibrium position with the increasing t approaches zero according to the exponential law, without oscillations. The real constants C_1 and C_2 are determined from the initial conditions: initial displacement y_0 and initial veleocity v_0. The system does not oscillate and returns to its equilibrium position, as in the previous case, but at a slower rate.

Case 3. Undercritically Damped System When $0 < c < c_c$, the system is under-damped. If the damping rate is positive, but less than the critical damping coefficient, the mechanical system will oscillate freely from the same initial displacement/deviation y_0 and initial velocity v_0.

Thus, at the low friction when $c^2 < 4mk$, Eq. (4.11) has conjugate complex roots.

$$p_{1,2} = -\frac{c}{2m} \pm i\sqrt{\frac{k}{m} - \frac{c^2}{4m^2}} = \gamma \pm i\omega, \tag{4.18}$$

Here it is designated $\gamma = \frac{c}{2m}$, $\omega = \sqrt{\frac{k}{m} - \frac{c^2}{4m^2}}$. Bearing in mind (4.9), we obtain a general solution (4.8).

$$y = C_1 e^{-\gamma t + i\omega t} + C_2 e^{-\gamma t - i\omega t} = e^{-\gamma t}\left(C_1 e^{i\omega t} + C_1 e^{-i\omega t}\right). \tag{4.19}$$

The multipliers $e^{i\omega t}$ and $e^{-i\omega t}$ are periodic functions. Their period is equal: $T = \frac{2\pi}{\omega}$

Therefore, ω is a circular frequency of oscillations. Expression (4.19) will be the solution of Eq. (4.4) for any constants C_1 and C_2. In order to obtain a real solution, we will take any of $C_1 = \frac{1}{2} r e^{i\varphi}$ and assume that $C_2 = \frac{1}{2} r e^{-i\varphi} = C_1^*$, then we obtain:

$$y = e^{-\gamma t}\left(\frac{1}{2} r e^{i\varphi} e^{i\omega t} + \frac{1}{2} r e^{-i\varphi} e^{-i\omega t}\right) = \frac{1}{2} r e^{-\gamma t}\left[e^{i(\omega t+\varphi)} + e^{-i(\omega t+\varphi)}\right]. \tag{4.20}$$

Let's use Euler's formula: $e^{i\alpha} = \cos\alpha + i\sin\alpha$. We obtain:

$$y = \frac{1}{2} r e^{-\gamma t}[\cos(\omega t+\varphi)i + \sin(\omega t+\varphi) + \cos(\omega t+\varphi) - i\sin(\omega t+\varphi)] = r e^{-\gamma t}\cos(\omega t+\varphi).$$

The solution $y = re^{-\gamma t}\cos(\omega t+\varphi)$ is real, and it contains two arbitrary constants r and φ. This solution can be re-written in the following form, by opening the cosine of the sum:

$$\begin{aligned} y &= re^{-\gamma t}(\cos\omega t\cos\varphi - \sin\omega t\sin\varphi) = (r\cos\varphi)e^{-\gamma t}\cos\omega t + (-r\sin\varphi)e^{-\gamma t}\sin\omega t \\ &= C_1 e^{-\gamma t}\cos\omega t + C_2 e^{-\gamma t}\sin\omega t, \end{aligned} \tag{4.21}$$

where through C_1 and C_2, the corresponding multipliers in brackets are designated (note: C_1 and C_2 are not equal to C_1 and C_2 in formula (4.19) where they are determined by the initial conditions). The motion is oscillatory and is under-critically damped.

Let us re-write the Eq. (4.16), using critical damping coefficient (4.13) and the damping ratio (4.14):

$$p_{1,2} = -\zeta\omega_n \pm i\omega_n\sqrt{1-\zeta^2} \tag{4.22}$$

Defining the damped natural frequency as:

$$\omega_d = \omega_n\sqrt{1-\zeta^2} \tag{4.23}$$

Equation (4.22) can be rewritten in the following form:

$$p_{1,2} = \zeta\omega_n \pm i\omega_d \tag{4.24}$$

With these new damping variables ω_n, and ω_d we can re-write the solution for differential equation of motion (4.21) for a damped unforced system as:

$$y = e^{-\zeta\omega_n t}(a\cos\omega_d t + b\sin\omega_d t) \tag{4.25}$$

where a and b are constants to be determined by the initial conditions: $y(0) = y_0;\ \dot{y}(0) = v_0$.

$$y = e^{-\zeta\omega_n t}\left(y_0\cos\omega_d t + \frac{v_0 + \zeta\omega_n y_0}{\omega_d}\sin\omega_d t\right) \tag{4.26}$$

or

$$y = y_0 e^{-\zeta\omega_n t}\left(\cos\omega_d t + \frac{\zeta\omega_n}{\omega_d}\sin\omega_d t\right) + \frac{v_0}{\omega_d}e^{-\zeta\omega_n t}\sin\omega_d t \tag{4.27}$$

Civil and industrial structures usually have low damping ratios $\zeta \le 5\%$. Thus, the dumped natural frequency, ω_d, is typically close to the natural frequency, ω_n (Fig. 4.4).

For civil engineers, structure engineers, and researchers, the most important case is Case #3: an under-damped system (when $0 < c < c_c$), because structures of interest are buildings, bridges, damps, tunnels, structures on an elastic foundation, etc. All above-mentioned engineering structures fall into one category, as typically, their damping ratio is less than 0.1. Therefore, we have little reason to study the dynamics of critically damped systems $c = c_c$ or over-damped system $c > c_c$. Thus, we will conduct comprehensive studies of under-damped system.

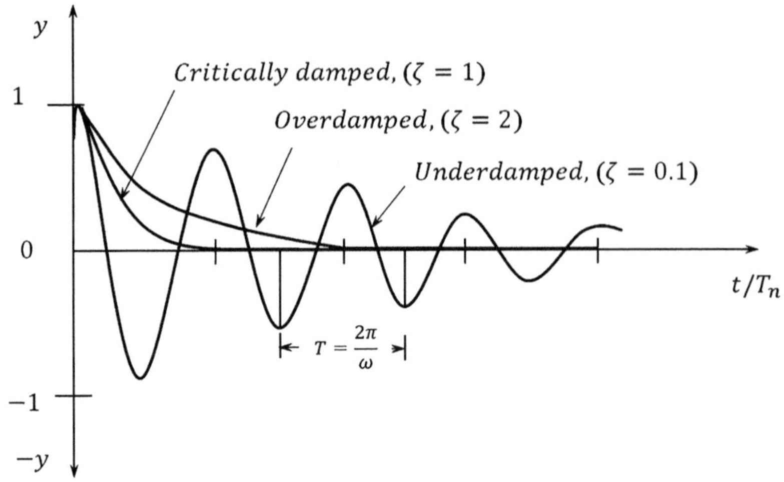

Fig. 4.4 Free oscillation of critically damped over-damped and under-damped system [1]

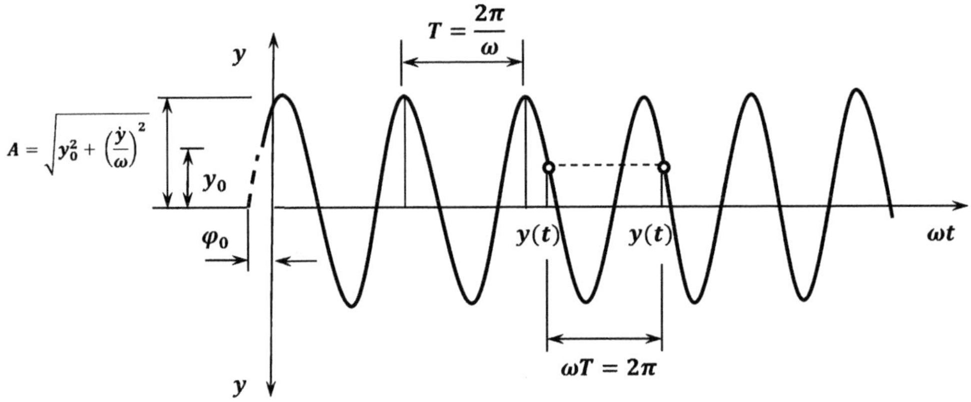

Fig. 4.5 Displacement $y(t)$, velocity $\dot{y}(t)$ as function of time t, and A- Constant amplitude oscillation

Case 4. No Damped System When $c = 0$, there is no damping. In this case, the characteristic equation (4.11) takes the form:

$$mp^2 + k = 0, \tag{4.28}$$

Wherefrom:

$$p_{1,2} = \pm i\sqrt{\frac{k}{m}}. \tag{4.29}$$

This is called the natural frequency of the undamped mechanical system, which oscillates freely relative to the deflection line of a static equilibrium. See Fig. 4.5. The solution of differential equation is:

$$y = C_1 e^{i\omega t} + C_2 e^{-i\omega t} = r\cos(\omega t + \varphi), \quad \text{where } \omega = \sqrt{\frac{k}{m}} \tag{4.30}$$

which is a simple cosine wave, and the motion is called the simple harmonic motion at their natural angular frequency ω rad/s. The amplitudes depend on the initial displacement y_0 and the initial velocity v_0. We can rewrite solution (4.30) in the following form:

$$y = y_0 \cos(\omega_0 t) + \frac{v_0}{\omega_0} \sin(\omega_0 t) \tag{4.31}$$

Assume that where $C_1 = A \sin \varphi_0$ and $C_2 = A \cos \varphi_0$, where A and φ_0 are some new constants. Then, using the sine formula of the sum of two angles, instead of (4.31) we can write:

$$y = A \sin(\omega t + \varphi_0), \tag{4.32}$$

Taking into account the values of C_1 and C_2, we obtain the constants A and φ_0 in the form:

$$A = \sqrt{C_1^2 + C_2^2} = \sqrt{y_0^2 + \left(\frac{\dot{y}}{\omega}\right)^2} \tag{4.33}$$

$$\tan \varphi_0 = C_1/C_2 = \omega y_0/\dot{y}_0$$

Wherefrom:

$$\varphi_0 = \text{arc } tg \frac{y_0 \omega}{\dot{y}_0} \tag{4.34}$$

φ_0 is the initial phase of free oscillations and characterizes the deviation of mass at the beginning of the motion.

Figure 4.5 shows a graph of mass motion for this case. The period of oscillation is the time of one full cycle of oscillation and has units of seconds.

$$T = \frac{2\pi}{\omega} \tag{4.35}$$

Therefore, ω is the number of oscillations in 2π s. This value is called the angular frequency. It is one of the main characteristics of the system's oscillations and is calculated by the formula:

$$\omega = \frac{1}{\sqrt{m\delta}} \tag{4.36}$$

The angular frequency is determined by:

$$\omega = \frac{2\pi}{T}, \tag{4.37}$$

and has units of radian per second.

The natural frequency (standard frequency) f is the number of oscillations per time unit, whereby:

$$f = \frac{1}{T} = \frac{\omega}{2\pi} \tag{4.38}$$

and has units of cycles per second, or Herts.

The oscillatory motion of an elastic system with single degree of freedom (SDOF) in the absence of a disturbing force is called a harmonic oscillation and is represented in the coordinates y by a sinusoidal curve as in Fig. 4.6.

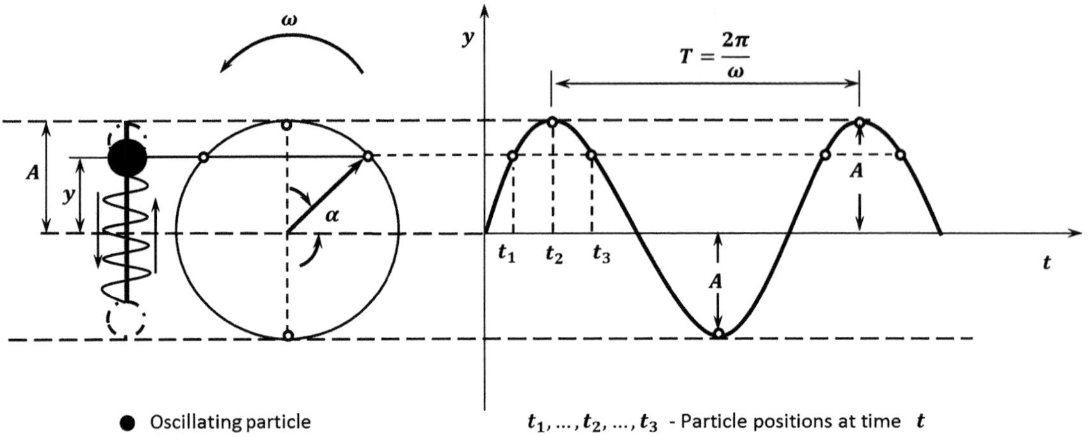

Fig. 4.6 Graph plot of mass'(particle's) position as time progresses

4.2 Total Energy of an Oscillating/Vibrating Mechanical System

Let us follow the behavior of the total energy of the oscillating mechanical system. The expression of the total energy has the form:

$$E_{\text{total}} = E_{\text{kinetic}} + E_{\text{elastic}} = \frac{mv^2}{2} + \frac{ky^2}{2} = \frac{1}{2}\left[m\left(\frac{dy}{dt}\right)^2 + ky^2\right] \tag{4.39}$$

while in the right part of the equation, the first summand is equal to the kinetic energy and the second summand is equal to the potential energy of the oscillating mechanical system. To solve, let us substitute here formula (4.30), we obtain:

$$E_{\text{total}} = \frac{1}{2}m\omega^2 r^2 \sin^2(\omega t + \varphi) + \frac{1}{2}kr^2 \cos^2(\omega t + \varphi) \equiv \frac{1}{2}kr^2 \tag{4.40}$$

Thus, the mechanical energy of undamped free vibrations is constant. If friction is present ($c \neq 0$), the full energy of the system decreases and dissipates. In a way, by differentiating formula (4.39) and using the differential equation (4.4), we will find

$$\frac{E_{\text{total}}}{dt} = \frac{1}{2}\left[2m\frac{dy}{dt}\frac{d^2y}{dt^2} + 2ky\frac{dy}{dt}\right] = \frac{dy}{dt}\left(-c\frac{dy}{dt} - ky\right) + ky\frac{dy}{dt} = -c\left(\frac{dy}{dt}\right)^2 \tag{4.41}$$

that the derivative is negative, hence the energy decreases. The system is under damping.

The graph of free-damping oscillations is shown in Fig. 4.7.

The solution of the differential equation (4.4) could be rewritten in the following form as well:

$$y = a_0 e^{-\frac{ct}{2m}} \sin(\omega t + \varphi_0) \tag{4.42}$$

Here y_0 is the initial displacement of mass:

$$y_0 = a_0 \sin \varphi_0,$$

and φ_0 is the initial phase of free oscillations:

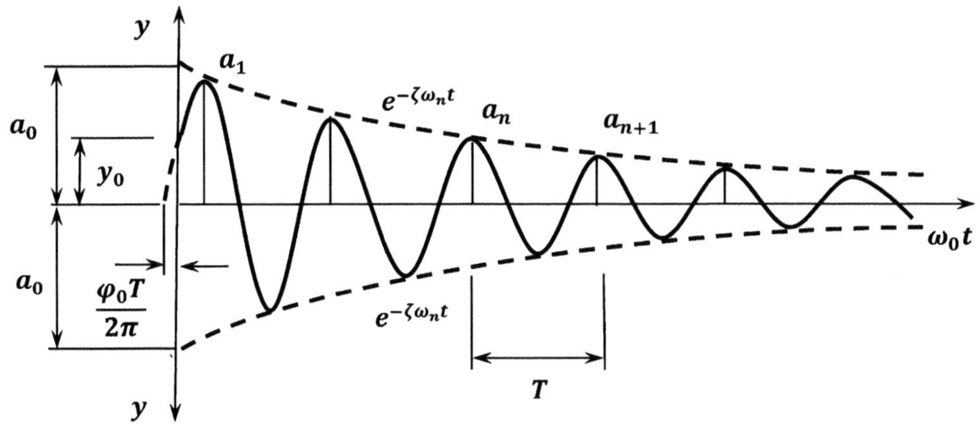

Fig. 4.7 Graph of free damping oscillations

$$\varphi_0 = \tan^{-1}\frac{y_0\omega}{v_0 + y_0\frac{c}{m}} = \sin^{-1}\frac{y_0}{a_0} \tag{4.43}$$

and a_0 is the initial amplitude of free oscillations:

$$a_0 = \sqrt{y_0^2 + \left(\frac{v_0 + y_0\frac{c}{2m}}{\omega}\right)^2} \tag{4.44}$$

where v_0 is the initial velocity of mass oscillations. The angular frequency is determined by the following formula:

$$\omega = \sqrt{\frac{k}{m} - \left(\frac{c}{2m}\right)^2} \tag{4.45}$$

The dynamic response of damped systems decays over time. Let's find a natural logarithm of the ratio of two consecutive single-valued amplitudes of free oscillations:

$$\delta = \ln\frac{a_n}{a_{n+1}} = \ln\frac{e^{-\frac{ct_n}{2m}}}{e^{-\frac{ct_{n+1}}{2m}}} = \ln e^{\frac{c}{2m}(t_{n+1}-t_n)} = \ln e^{\frac{ct}{2m}} = \frac{ct}{2m} = nT \tag{4.46}$$

or

$$\delta = \ln\frac{a_n}{a_{n+1}} = \ln e^{\zeta\omega_0 T} = \frac{2\pi\zeta}{\sqrt{1-\zeta^2}} \tag{4.47}$$

The value of the natural logarithm characterizes the rate of damping of oscillations and is called the logarithmic decrement of damping oscillations. The value $n = \frac{c}{2m}$ is called the coefficient of damping.

In Eq. (4.4), it has been assumed that the resistance force F_d is proportional to the velocity of the oscillations $v = \dot{y}$. Such dependence on the diagram of force-deformation corresponds to an elliptical hysteresis loop. Dependence between force P and displacements y in the damped systems is nonlinear. Therefore, during each cycle of oscillations on the diagram $P - y$ the closed ellipse-shaped curve is obtained, called the hysteresis loop (Fig. 4.8). With harmonic oscillations, it is possible to accept that the loop of a hysteresis has the ellipse form with the center at the beginning of coordinates.

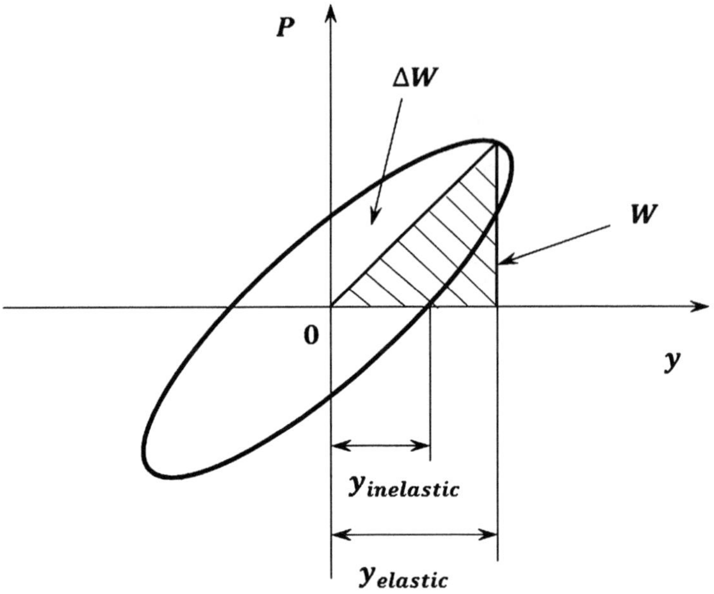

Fig. 4.8 The graph shows the ratio of the work of inelastic resistance per cycle to the work of elastic forces per quarter cycle

The ratio of the work of inelastic resistances per cycle W to the work of ΔW elastic forces per quarter of a cycle, i.e., on the way from an unloaded state to an extremely deformed state of a system, is called the absorption coefficient.

In Fig. 4.8, the work ΔW is represented by the area of an ellipse, and the work W is represented by the area of the hatched triangle. Thus, the absorption coefficient is equal to the ratio of the dissipated energy for the i period $W_i - W_{i+1}$ of oscillations to the potential energy W_i at the beginning of the oscillation cycle.

$$\psi = \frac{\Delta W}{W} = -\int_{t}^{t+T} \frac{\Delta W}{W} = 2\int_{t}^{t+T} \frac{dy}{y} = 2\ln\frac{a_n}{a_{n+1}} = 2\delta. \tag{4.48}$$

The values of the coefficient of inelastic resistance of materials in the structure γ are usually entered into the calculation. The coefficient of inelastic resistance of materials is the ratio of the amplitude of inelastic and elastic deformation, that is:

$$\gamma = \frac{y_{\text{inelastic}}}{y_{\text{elastic}}} = \frac{\psi}{2\pi} = \frac{\delta}{\pi} = \frac{cT}{2\pi m} = \frac{c}{m\omega}. \tag{4.49}$$

From formula (4.49), it is possible to express the initially accepted coefficient of proportionality or damping coefficient c through other characteristics of damping. Typically, frequency-independent models of internal friction are used to calculate structures. Damping parameters, in particular, the loss factor (coefficient) $\gamma = \frac{\delta}{\pi}$ in frequency-independent models of internal friction, do not depend on the oscillation frequency.

$$c = \frac{2m\delta}{T} = \frac{m\psi}{T} = m\gamma\omega = 2mn = \gamma\sqrt{km} \tag{4.50}$$

The number of oscillations per minute is called the technical frequency and is determined by the following formulas:

$$n = 60f = \frac{60}{T} = \frac{30\omega}{\pi} = \frac{30}{\pi}\sqrt{\frac{981}{y_{st}}} \approx \frac{300}{\sqrt{y_{st}}} \tag{4.51}$$

These formulas are used in calculations in cases where the design scheme can be represented as a system with single degree of freedom.

Numerous experiments have shown that the shape of a hysteresis curve (for a case of an elliptic curve) in stationary harmonic oscillations is practically independent of frequency. For many construction materials in a certain range of frequencies and amplitudes, the characteristics of damping can be kept constant. However, according to the Theory of Viscous Friction, the damping is frequency-dependent and the parameters of damping δ or γ can be its quantitative characteristics only at fixed frequencies. This contradiction is a disadvantage of the Voigt Hypothesis in the problems of accounting of internal friction forces. Therefore, the Voigt Hypothesis should be used only for illustrative purposes. For this reason, various methods of correcting the Voigt's hypothesis are used in practical calculations of structures and research works, which provide frequency independence of parameters of damping [2, 3]. The most complete reflection of this issue is found in the model of the frequency-independent elastic-viscous resistance. This model allows for such adjustments not only for a system with single degree of freedom (SDOF) but also for arbitrary linear systems with discrete and continuous parameters [4].

The given formulas are used in the calculations in such cases when the dynamic design model (DDM) of actual structure can be considered as a system with single-degree-of-freedom (SDOF). Let us consider several examples of how to determine the frequency of oscillations.

Problem 4.1

When determining the frequencies of vertical oscillations of a foundation with a mass of m (the weight of foundation: column footing $Q = 100T$.) and an area of $A = 10m^2$ (Fig. 4.9) lying on an elastic base, we can accept this system as a system with single degree of freedom.

Solution

The circular frequency of vibrations will be determined by the formula:

Fig. 4.9 Column footing

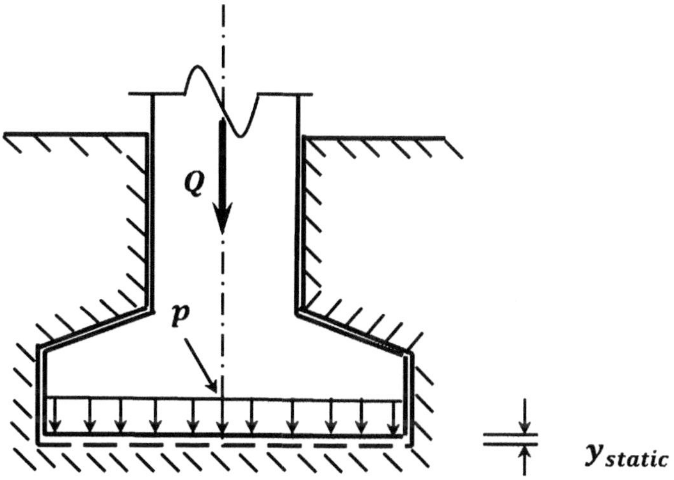

$$\omega = \sqrt{\frac{A\,k_{\text{comp}}}{m}} = \sqrt{\frac{A\,k_{\text{comp}}g}{Q}} = \sqrt{\frac{10m^2 \cdot \frac{2.5\ \text{k}\Gamma}{\text{cm}^3} \cdot \frac{9.81\ m}{\text{s}^2}}{100T}} = 49.52\ s^{-1}$$

where k_{comp} is the coefficient of elastic compression of the ground, have units of kΓ/cm^3. Let us assign:

$$k_{\text{comp}} = 2.5\ \text{k}\Gamma/\text{cm}^3$$

The pressure on the ground is:

$$p = \frac{Q}{A} = \frac{100}{10} = \frac{10T}{m^2} = 1\ \text{k}\Gamma/\text{cm}^2$$

at such pressure, the elastic static sediment of the foundation will be equal:

$$y_{\text{st}} = \frac{p}{k} = \frac{1}{2,5} = 0.4\ \text{cm}$$

The technical frequency of free vibrations/oscillations

$$n = \frac{300}{\sqrt{y_{\text{st}}}} = \frac{300}{\sqrt{0.4}} = 476\ \text{min}^{-1}$$

Problem 4.2
It is required to determine the frequency of longitudinal oscillations of a post with a heavy load at the end can be determined (Fig. 4.10).

Solution
If the weight of a post can be neglected, then the static displacement of a load can be determined by the formula:

$$y_{\text{st}} = \frac{Qh}{EA} \tag{A}$$

where

E is the elastic modulus of the post's material (kΓ/cm^2);
A is the cross-section area;
h is the height of the post.

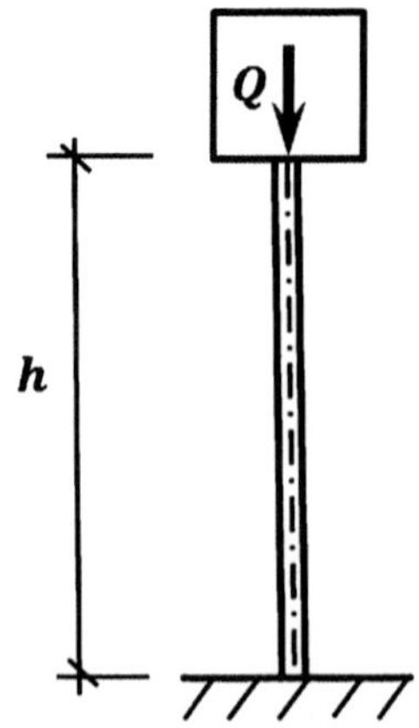

Fig. 4.10 Post with a heavy load

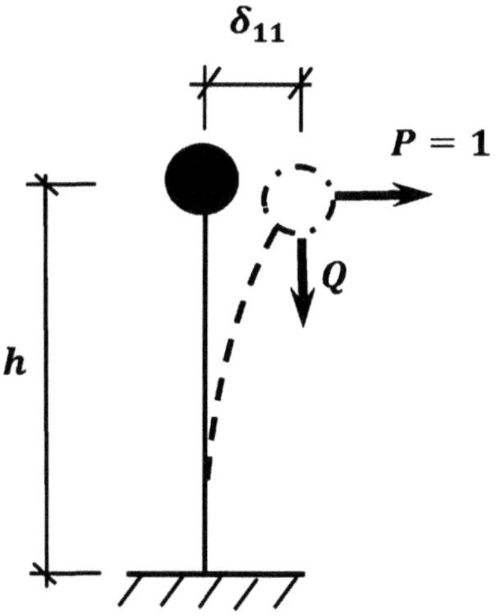

Fig. 4.11 The DDM of weightless post with one concentrate mass

The circular frequency of natural oscillations for this case will be represented by the following formula:

$$\omega = \sqrt{\frac{EA}{mh}} \tag{B}$$

where $m = Q/g$ – is the mass of load.

In general, the mass on a post can make horizontal oscillations at which the post will be in conditions of lateral and longitudinal bending, Fig. 4.11. This means that δ_{11} must be determined taking into account the compression of the post by the weight of the mass Q (Fig. 4.11). Using the approximation formula of longitudinal-lateral bending, we obtain:

$$\delta_{11} = \frac{h^3}{3EI} : \left(1 - \frac{Q}{P_e}\right) = h^3 : 3EI\left(1 - \frac{mg}{P_e}\right).$$

The frequency of free oscillations according to the formula:

$$\omega = \sqrt{\frac{1}{m\delta_{11}}} = \sqrt{\frac{g}{y_{\text{st}}}},$$

takes the form:

$$\omega = \sqrt{\frac{3EI}{mh^3}\left(1 - \frac{mg}{P_e}\right)}$$

where

$$P_e = \frac{\pi^2 EI}{4h^2}$$

is the Euler force.

Here are additional examples of how to determine the frequency of free oscillations.

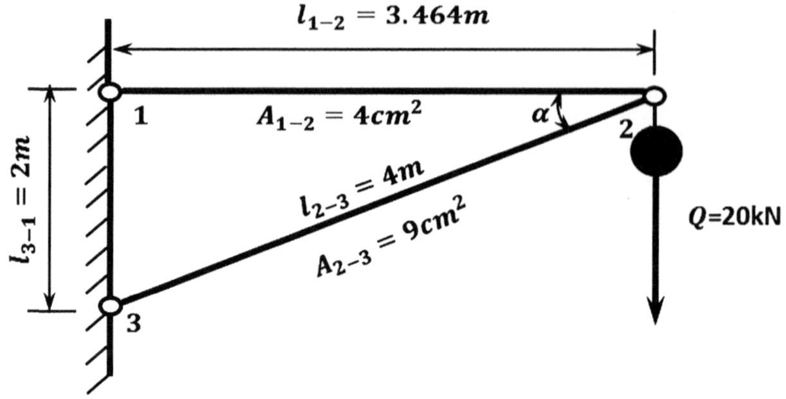

Fig. 4.12 The mass of the triangular truss is concentrated in one of its nodes

Problem 4.3

It is required to determine the technical frequency of a steel ancon truss loaded with a concentrated force $Q = 20$ kN. The member weight of an ancon truss can be neglected (Fig. 4.12; the elastic module of bracket's members $E = 210$ GPa; $A_{1-2} = 4$ cm^2; $A_{2-3} = 9$ cm^2).

Solution

For systems working on tension and compression:

$$y_{st} = \sum \frac{N_i N_k}{AE} l.$$

where

N_i is the force in the elements of ancon truss from the unit force acting in the direction of the desired displacement; and

N_k is the force in the elements of ancon truss from a given load.

Taking into account that:

$$N_k = N_i Q,$$

Let us define the static deflection of the system:

$$y_{st} = \frac{Q}{E} \sum \frac{N_i^2}{A} = \frac{20 \cdot 10^3}{210 \cdot 10^9} \left(1.732^2 \frac{3.464}{4 \cdot 10^{-4}} + 2^2 \frac{4}{9 \cdot 10^{-4}} \right) = 0.4 \text{ cm}.$$

The technical frequency of free oscillations:

$$n = \frac{300}{\sqrt{y_{st}}} = \frac{300}{\sqrt{0.4}} = 474.34 \text{ m}^{-1}.$$

Problem 4.4

It is required to determine how the frequency of free oscillations of a beam varies, depending on the location of the mass in the span of the beam. The beam weight is disregarded (Fig. 4.13).

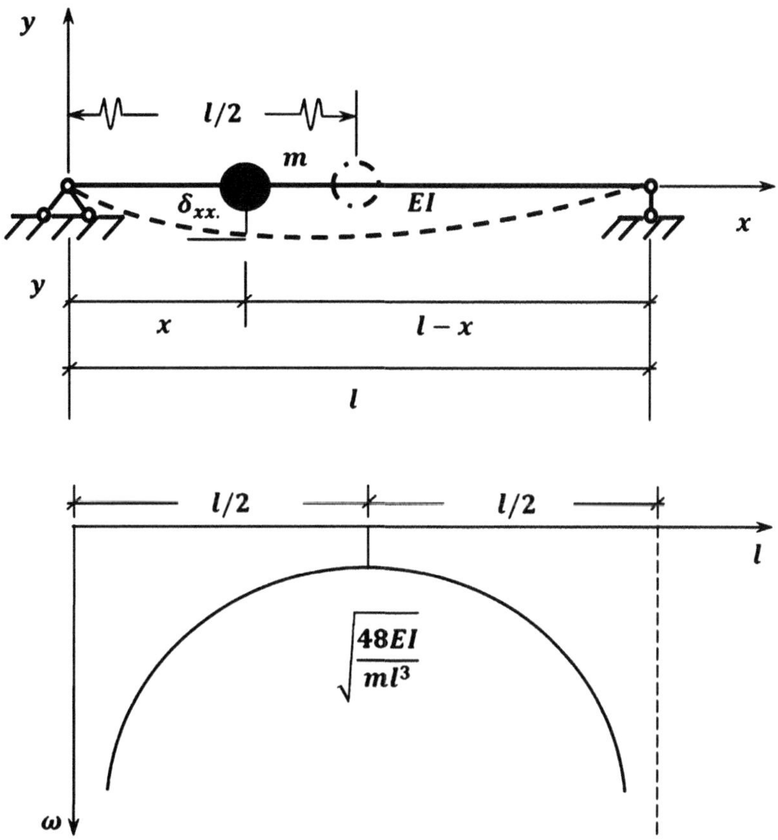

Fig. 4.13 Simple beam is as a system with SDOF

Solution

Let us determine the deflection of the beam at the point x from the unit force acting at the same point:

$$\delta_{xx} = \frac{x^2(l-x)^2}{3EI \cdot l}; \quad k = \frac{1}{\delta_{xx}} = \frac{3EI \cdot l}{x^2(l-x)^2}.$$

The expression of oscillation will take the following form:

$$\omega = \sqrt{\frac{k}{m}} = \frac{1}{x(l-x)}\sqrt{\frac{3EI \cdot l}{m}},$$

when $x = 0$ and $x = l$; $\omega \to \infty$;

when $= l/2$

$$\omega = \frac{4}{l^2}\sqrt{\frac{3EI \cdot l}{m}} = \sqrt{\frac{48EI}{ml^3}}.$$

Thus, when the concentrated mass approaches the beam supports, the oscillation frequency increases unlimitedly. The lowest value of frequency is obtained in the case when the concentrated mass is in the middle of the beam span. This corresponds to the fundamental tone of oscillations as well as the first basic mode of the oscillating beam with single degree of freedom.

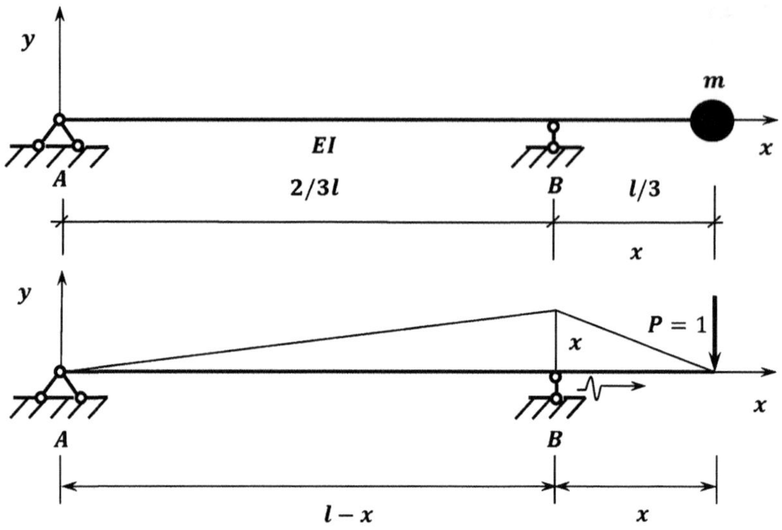

Fig. 4.14 Overhanging beam that overhangs the right support and carries concentrated mass at the end

Problem 4.5
A beam on the two supports has an overhang with a concentrated mass *m* at the end (Fig. 4.14). The right support of the beam "B" is required to move to the right, so that the frequency of vibration of the mass m is increased by two, three, and five times. The length of the overhang is equal to *l*/3.

Solution
The given system has single degree of freedom, and the frequency of free vibrations is determined by the following expression:

$$\omega = \sqrt{\frac{1}{m\delta_{11}}}.$$

The displacement of the end of the overhang from a unit force in an arbitrary position of the right support will be:

$$\delta_{11} = \frac{x(l-x)}{2} \cdot \frac{2}{3}x \cdot \frac{1}{EI} + \frac{x \cdot x}{2} \cdot \frac{2}{3}x \cdot \frac{1}{EI} = \frac{x^2 l}{3EI}.$$

When $= l/3$, the displacement:

$$\delta_{11} = \frac{l^3}{27EI},$$

the frequency of oscillation has the following value:

$$\omega_{x=l/3} = \sqrt{\frac{27EI}{ml^3}}.$$

Let's write down the first condition of the problem:

$$\omega_x = 2\omega_{x=l/3}$$

$$\sqrt{\frac{3EI}{mx^2 l}} = 2\sqrt{\frac{27EI}{ml^3}}.$$

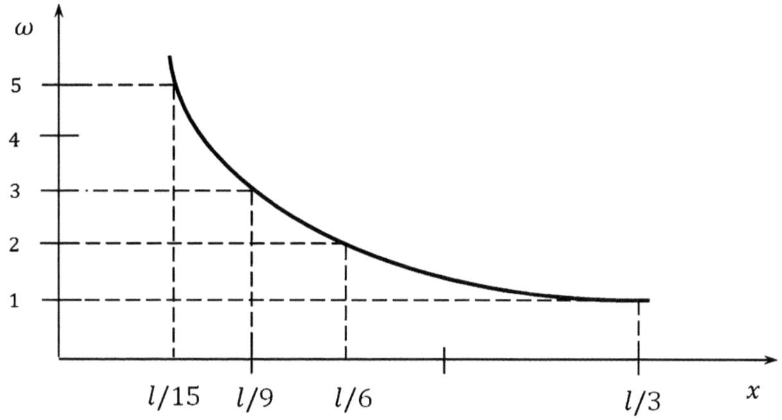

Fig. 4.15 Dependence of the frequency of free oscillations on the length of overhang

The obtained solution for the first condition of problem is: $x = l/6$

Thus, when moving the right support to the right by $l/6$ bay, the frequency of oscillations of mass of m has to increase twice. Let's satisfy the second condition of the problem:

$$\omega_x = 3\omega_{x=l/3},$$

we obtain:

$$\sqrt{\frac{3EI}{mx^2 l}} = 3\sqrt{\frac{27EI}{ml^3}}.$$

The obtaining solution for the second condition of problem is: $x = l/9$. Thus, when moving the right support to the right on $l/9$ bay the frequency of oscillations of mass m has to increase three times.

And finally, we will write down the third condition of the problem:

$$\omega_x = 5\omega_{x=l/3},$$

we obtain:

$$\sqrt{\frac{3EI}{mx^2 l}} = 5\sqrt{\frac{27EI}{ml^3}}.$$

The obtained solution for the third condition of problem is $x = l/15$. The frequency of oscillations of mass m has to increase five times. Figure 4.15 shows the dependence of the frequency of free vibrations of a beam on the length of the overhang with a concentrated mass m at the end.

References

1. Copra, A. K. (2016). *Dynamics of Structures* (992 p). Pearson.
2. Tseitlin, A. I., & Guseva, N. I. (1979). *Statistical methods of calculation of construction to the group of dynamic impacts* (175p). Stroyisdat.
3. Petrosian, L. G. (1986). The calculation of the internal friction in the one-dimensional dynamic systems. In *Structure mechanics* (pp. 40–44). YPTU.
4. Tseitlin, A. I., Kusainov, A. A., & Petrosian, L. G. (1986). *The free oscillations of dissipative systems* (Vol. 9, pp. 50–56). AS Kazakh SSR.

5 Forced Vibrations of a System with Single Degree of Freedom (SDOF)

Abstract

This chapter presents forced vibrations of a system with SDOF. Analysis of forced vibrations, differential equation of forced vibrations, dynamic coefficient, beat phenomenon, problem, and solutions are discussed. The chapter also deals with forced harmonic vibrations of a damped system with SDOF. Analysis of forced harmonic vibrations, differential equation of motion, resonance phenomenon and definition, and applicable problems and solutions are presented.

5.1 Analysis of Forced Vibrations

Forced oscillations occur in a mechanical system as a result of excitation of external forces. Among numerous publications of the last two decades devoted to the vibrations/oscillations of structures, let me note the following major works: R.W. Clough., J. Penzien [1], J.C. Anderson, F. Naeim [2], A.K. Copra [3], H.R. Busby, G.H. Staab, [4], R.R. Jr. Craig, A.J. Kurdila [5], J.T. Katsikadelis [6], and others. In this section, we will limit ourselves to studying the properties of the forced oscillations of the linear oscillator under the action of an external periodic force, changing according to a harmonic law $P(t) = P \sin \theta t$ (Fig. 5.1).

Differential Equation of Forced Oscillations Differential equations of forced oscillations without taking into account the forces of resistance have the form of:

$$m\ddot{y} + ky = P(t), \tag{5.1}$$

or

$$\ddot{y} + \omega^2 y = \frac{P}{m} \sin \theta t. \tag{5.2}$$

where

P is the amplitude of the excitation force.
θ is the frequency of the excitation force.

L. G. Petrosian, *Dynamics of Structures*, https://doi.org/10.1007/978-3-031-63539-7_5

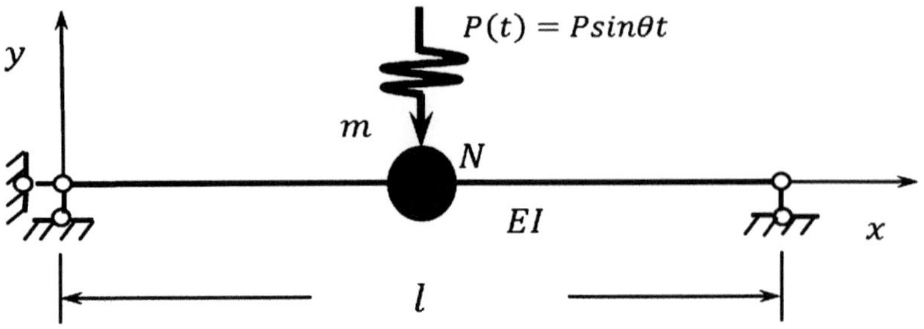

Fig. 5.1 Forced oscillations of a system with SDOF

The general solution of this equation, according to the common rules of finding solutions of equations with the right part, is the sum of the general solution of a homogeneous equation, as shown below:

$$\ddot{y} + \omega^2 y = 0 \tag{5.3}$$

and the particular solution of the inhomogeneous equation (5.2):

$$y = C_1 \sin \omega t + C_2 \cos \omega t + \frac{P \sin \theta t}{m(\omega^2 - \theta^2)} \tag{5.4}$$

Here, C_1 and C_2 are arbitrary constants, determined from the initial conditions.

The first two summands of this equation define the free undamped oscillations, and the third summand defines forced oscillations. The second term is of interest in practical calculations since the real systems have damping over time, i.e., natural oscillations of a system will be damped and only forced oscillations with frequency θ will remain. The solution of the equation of forced oscillations can be transformed as follows:

$$y = \frac{P \sin \theta t}{\omega^2 \cdot m\left(1 - \frac{\theta^2}{\omega^2}\right)} = \frac{P \sin \theta t}{k\left(1 - \frac{\theta^2}{\omega^2}\right)} = \frac{P \cdot \delta \sin \theta t}{1 - \frac{\theta^2}{\omega^2}} = \mu y_{st} \sin \theta t \tag{5.5}$$

where $y_{st}$is the static deflection of the system from the force P.

Dynamic Coefficient It follows from formula (5.5) that when the frequency of the disturbing force approaches the frequency of natural oscillations ω, the amplitudes of displacements y begin to increase rapidly and approach infinity. The coefficient μ

$$\mu = \frac{1}{1 - \frac{\theta^2}{\omega^2}} \tag{5.6}$$

is called the dynamic coefficient, which shows how often the dynamic effect of a disturbing force exceeds the static effect of its amplitude. In other words, it is the factor by which static deflection or displacement, produced in a beam (Fig. 5.1) by a force of the elastic reaction of the system ($R_s = -ky$) that arises at the deviation of mass m from the position of a static equilibrium, must be multiplied by the dynamic coefficient in order to obtain the amplitude of oscillations/vibrations resulting from the same force when the force acts dynamically. When the frequency of forced oscillations becomes equal to the frequency of natural oscillations, then the phenomenon of resonance occurs, and the amplitudes of oscillations approach infinity. Figure 5.2 shows a graph of changes in magnitude of the dynamic coefficient, depending on the ratio of the frequency of the disturbing force to the frequency of free oscillations, taken as an absolute value.

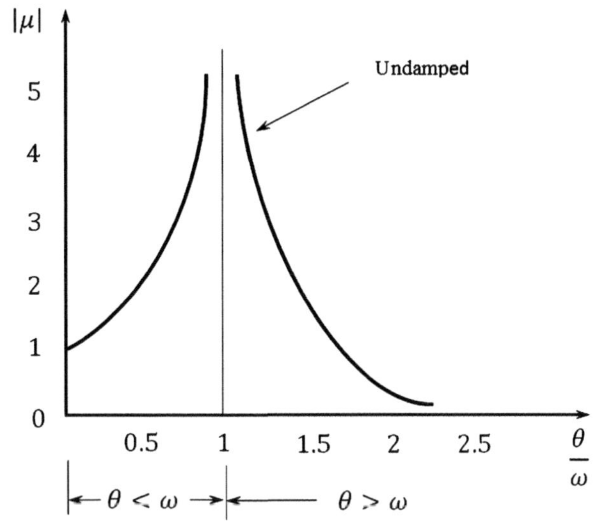

Fig. 5.2 Dynamic coefficient or dynamic amplification factor or deformation response factor (Frequency ratio θ/ω)

From (5.5), it follows that forced oscillations occur with the frequency of a disturbing force. Forced oscillations, as opposed to free oscillations, do not depend on the initial conditions of oscillations. Therefore, in order to change the amplitude of forced oscillations, it is necessary to change the scheme, geometry of the construction, its rigidity, mass distribution, etc. In the case of free oscillations, it is sufficient to change the initial deviation or initial velocity of oscillations.

At $\theta = 0$, the coefficient $\mu = 1$, and at $\theta \ll \omega$ the value of μ is close to 1. When $\theta \to \infty$, than $\mu \to 0$.

At $\theta/\omega < 1$, the dynamic coefficient is positive, and forced oscillations occur in the same phase as the oscillations of the disturbing force.

At $\theta/\omega > 1$, the dynamic coefficient becomes negative. This is explained by the fact that the shift of a phase by π takes place, i.e., the law of change of an external force does not coincide with the law of motion of the mass of the system m; when the force is directed downwards, the system moves to the top, and vice versa.

At $\theta/\omega \gg 1$, the forced oscillations of a system will be so small that the mass m will practically stop oscillating.

When the frequencies are equal $\omega = \theta$, i.e., at resonance, formula (5.6) leads to the value $\mu = \infty$. The unlimited increase in the amplitudes of oscillations is explained by the fact that the absorption of energy during such oscillations is not taken into account. However, in that case, formula (5.6), obtained without taking into account the natural oscillations, is not accurate enough. A more accurate solution can be obtained by adding the general solution of a homogeneous equation (5.3) to the partial solution of the inhomogeneous equation (5.2), i.e., consider expression (5.4):

Let at $t = 0$ $y(0) = y_0$ and $\dot{y}(0) = \dot{y}_0$. Then the constants C_1 and C_2 from (5.4) will be

$$C_1 = \frac{\dot{y}_0}{\omega} - \frac{P\theta}{m\omega\left(\omega^2 - \theta^2\right)}; \qquad C_2 = y_0 \tag{5.7}$$

and

$$y = y_0 \cos \omega t + \frac{\dot{y}_0}{\omega} \sin \omega t - \frac{P\theta \sin \theta t}{m\omega\left(\omega^2 - \theta^2\right)} + \frac{P \sin \theta t}{m\left(\omega^2 - \theta^2\right)} \tag{5.8}$$

The first two summands of formula (5.8) describe free oscillations of a system in the absence of a disturbing force. The third component describes harmonic oscillations with the natural frequency ω, but with an amplitude depending on the disturbing force. This oscillation also refers to free oscillations and always accompanies forced oscillations of a system under any initial conditions on which it does not depend. This oscillation is called free accompanying oscillation. The fourth term characterizes the number of forced oscillations occurring with the frequency of the disturbing force θ.

For example, if at the beginning of process of oscillations, a system was in the position of equilibrium, i.e., if $y(0) = y_0 = 0$, $\dot{y}(0) = \dot{y}_0 = 0$, then the solution has the following form:

$$y = \frac{P\sin\theta t}{m\left(\omega^2-\theta^2\right)} - \frac{P\theta\sin\omega t}{m\omega\left(\omega^2-\theta^2\right)} = \frac{P(\omega\sin\theta t - \theta\sin\omega t)}{m\omega\left(\omega^2-\theta^2\right)} \tag{5.9}$$

Particularly interesting is the case when the frequency of the disturbing force θ coincides with the frequency of the natural oscillations ω. Formula (5.9) cannot be used because the numerator and denominator of this formula are turned into zero. Let us disclose an uncertainty according to Lopital's rule. The fraction in (5.9) is replaced with a limit at $\theta = \omega$ of the ratio of derivatives with respect to the numerator and denominator:

$$\lim_{\theta\to\omega} P\left[\frac{\omega t\cos\theta t - \sin\omega t}{-2m\omega\theta}\right] = -\frac{Pt}{2m\omega}\cos\omega t + \frac{P}{2m\omega^2}\sin\omega t \tag{5.10}$$

Thus, the common integral (5.8) takes the form:

$$y = y_0\cos\omega t + \frac{\dot{y}_0}{\omega}\sin\omega t + \frac{P}{2m\omega^2}\sin\omega t - \frac{Pt}{2m\omega}\cos\omega t \tag{5.11}$$

Here, forced oscillations are represented by the nonperiodic fourth member of (5.11). In contrast to all the cases shown above, i.e., when the swing of oscillations was limited, here the swing of oscillations grows infinitely due to the multiplier t ($\zeta = 0; \theta = \omega; y_0 = \dot{y}_0 = 0$) (Fig. 5.3).

Of course, this should be considered only as a mathematical idealization of real processes; when the amplitude of oscillations becomes too large, then the assumptions under which the formula is derived are no longer valid, and in addition, the resulting forces become so large that they destroy the oscillating structure.

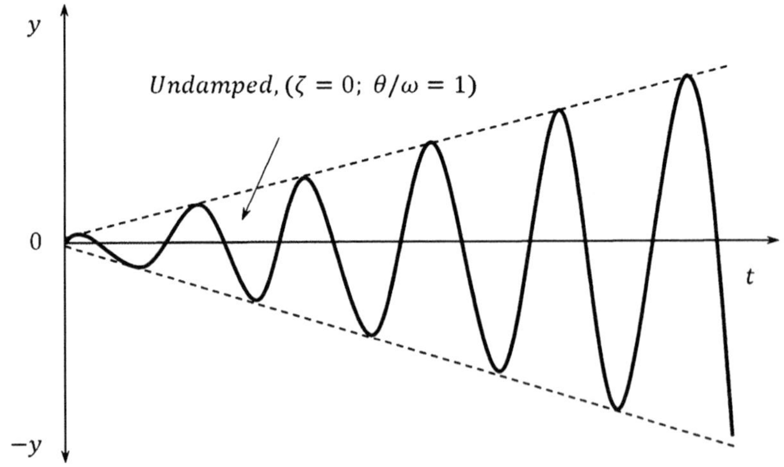

Fig. 5.3 Displacement response ratio of an undamped SDOF system

$$y = y_{st}\frac{\omega t}{2}\cos\omega t \tag{5.12}$$

Formula (5.12) shows that even in an ideal elastic system the infinite value of the amplitude is reached only after an infinitely wide interval of time of swinging. It should be noted that the amplitude of oscillations in resonance in the example above was obtained without accounting for the energy of dissipation, even though at such resonance the influence of damping is especially big. Thus in reality, due to the presence of dissipative forces, the amplitude of the oscillations will not unlimitedly grow. In addition, in the area of large deformations in resonant and near-resonant modes of oscillation, the solution of linear oscillation theory may not be valid or tenable.

The amplitude of forced oscillations from (5.4) can be rewritten in the form:

$$A = \frac{P\sin\theta t}{m\left(\omega^2-\theta^2\right)};\quad \text{when}\quad \theta<\omega$$

and (5.13)

$$A = \frac{P\sin\theta t}{m\left(\theta^2-\omega^2\right)};\quad \text{when}\quad \theta>\omega$$

The graph of the amplitude of forced oscillations depending on the ratio of the frequency of forced oscillations to the frequency of the free natural oscillations θ/ω of the system is shown in Fig. 5.2. The phase of forced oscillations changes abruptly by the value of π, when passing through the resonance, due to $\theta < \omega$

$$y = -\frac{P\sin\theta t}{m\left(\omega^2-\theta^2\right)} = \frac{P}{m\left(\omega^2-\theta^2\right)}\sin(\theta t+\pi) = A\sin(\theta t+\pi) \tag{5.14}$$

When $\theta < \omega$, the motion of the system occurs in the direction opposite to the direction of the force P.

Let us consider the case when the frequency of the disturbing force θ is very close to the natural frequency or frequency of free oscillations ω that satisfies the following condition:

$$|\theta-\omega| \ll \theta+\omega.$$

At $t = 0$, i.e., when the mass of the system is at the equilibrium line in an undeflected state, from the general solution (5.4) we have $C_2 = 0$, leaving us with:

$$y = C_1\sin\omega t + \frac{P\sin\theta t}{m\left(\omega^2-\theta^2\right)} \tag{5.15}$$

and

$$\dot{y} = v = C_1\omega\cos\omega t + \frac{P\theta\cos\theta t}{m\left(\omega^2-\theta^2\right)} \tag{5.16}$$

Substituting $\dot{y} = v = 0$, and $t = 0$, in formula (5.16), we obtain the arbitrary constant C_1:

$$C_1\omega + \frac{P\theta}{m\left(\omega^2-\theta^2\right)} = 0;\qquad C_1 = -\frac{P\theta}{m\left(\omega^2-\theta^2\right)\omega}.$$

Hence, substituting into (5.15), we find for the required solution:

$$y = \frac{P}{m(\omega^2 - \theta^2)}\left(-\frac{\theta}{\omega}\sin\omega t + \sin\omega t\right) \tag{5.17}$$

Let us set $\theta/\omega = 1$ in expression (5.17), obtaining:

$$y = \frac{P}{m}\frac{\sin\omega t - \sin\theta t}{\theta^2 - \omega^2}$$

If we convert the difference of the sine terms into a product, we obtain:

$$y = -\frac{P}{m}\frac{2\cos\left(\frac{\theta+\omega}{2}t\right)\sin\left(\frac{\theta-\omega}{2}t\right)}{(\theta+\omega)(\theta-\omega)} \tag{5.18}$$

If we now denote the small quantity $\theta - \omega = 2\varepsilon$ and note that $\theta + \omega \approx 2\theta$, we can rewrite expression (5.18) in the following form:

$$y \approx -\frac{P}{m}\frac{\sin\varepsilon t}{2\theta\varepsilon}\cos\theta t \tag{5.19}$$

Since ε is a small value, the function $\sin\varepsilon t$ changes slowly and its period $2\pi/\varepsilon$ is large. Hence, the form of expression (5.19) shows that y can be regarded as essentially a periodic function $\cos\theta t$ of frequency θ, with the slowly varying amplitude from 0 to $y = A_+ = A_-$.

$$A_{\pm} \approx -\frac{P}{m}\frac{\sin\varepsilon t}{2\theta\varepsilon} \tag{5.20}$$

and with period:

$$T = \frac{\pi}{\varepsilon} = \frac{2\pi}{|\theta - \omega|} \tag{5.21}$$

Beat Phenomenon This type of oscillation is called beats which occur whenever the frequency of forced oscillations is closed to the natural frequency of a system. Figure 5.4 shows the general nature of this type of oscillations when frequency of forced oscillations θ is nearly, but not quite, equal to the natural frequency of the system ω. The period of beats increases when the frequency of disturbing force θ approaches the frequency of natural oscillations ω, i.e., when the frequencies approach the conditions of resonance.

Thus, the swing of oscillation will periodically increase and decrease during the phenomenon of beats with the maximum amplitude value corresponding to formula (5.13).

For a limited case $\omega = \theta$, we can present εt instead of $\sin\varepsilon t$ in expression (5.19) and obtain:

$$y = -\frac{Pt}{2m\theta}\cos\theta t \tag{5.22}$$

The amplitude of the oscillations of motion, described by Eq. (5.22), increases unlimitedly with time, as it is shown in Fig. 5.3.

The phenomenon of beats in practice can be observed, for example, when installing several machines with close frequencies of disturbance in the same building.

Sometimes the same system is involved in several oscillations. For example, when two harmonic forces apply to the same system with close frequencies, a special phenomenon of oscillations occurs. During the beats, the swing of oscillations slowly increases and decreases periodically. When the

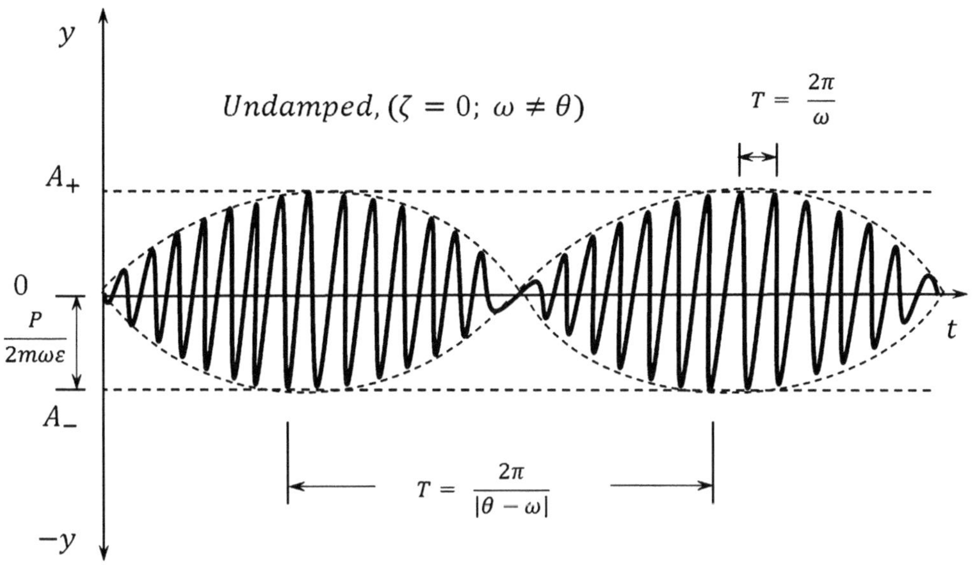

Fig. 5.4 The phenomenon of beats

oscillations are added, for example, the oscillations $y_1 = A \sin \theta_1 t$ and $y_2 = A \sin \theta_2 t$ we get the oscillation:

$$y = y_1 + y_2 = A(\sin \theta_1 t + \sin \theta_2 t) = 2A \sin \frac{\theta_1 + \theta_2}{2} t \cos \frac{\theta_1 - \theta_2}{2} t, \tag{5.23}$$

occurring with a frequency:

$$\theta = \frac{\theta_1 + \theta_2}{2}.$$

The oscillation amplitude changes from the minimum value:

$$A_{\min} = \left| \frac{P_1}{k\left|1 - \frac{\theta_1^2}{\omega^2}\right|} - \frac{P_2}{k\left|1 - \frac{\theta_2^2}{\omega^2}\right|} \right|, \tag{5.24}$$

to the maximal value:

$$A_{\max} = \left| \frac{P_1}{k\left|1 - \frac{\theta_1^2}{\omega^2}\right|} + \frac{P_2}{k\left|1 - \frac{\theta_2^2}{\omega^2}\right|} \right|. \tag{5.25}$$

The beat period is the time between the moments of reaching successive peaks of amplitude and are:

$$T = \frac{2\pi}{|\theta_1 - \theta_2|}. \tag{5.26}$$

Since the frequencies θ_1 and θ_2 are close in value to each other, the multiplier $\cos \frac{\theta_1 - \theta_2}{2} t$ changes very slowly, and the multiplier $\sin \frac{\theta_1 + \theta_2}{2} t$ has almost the same frequency as the summed oscillations. When the amplitudes of oscillations are different, then the phase of the summed oscillations also

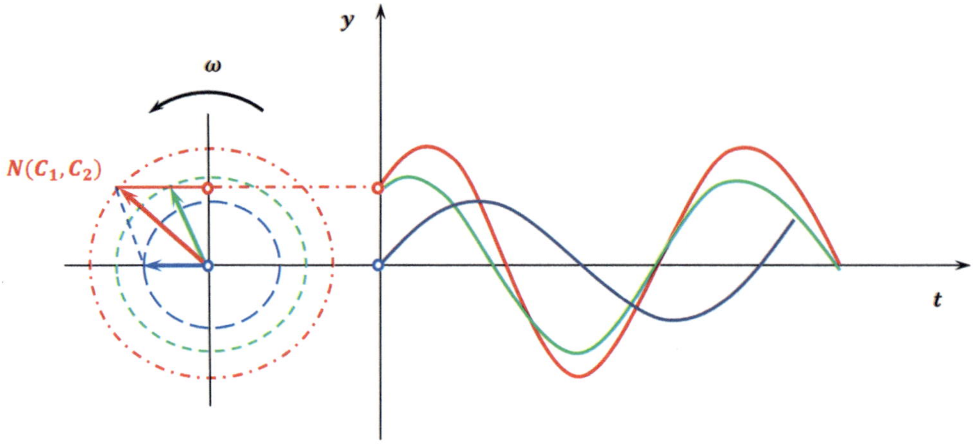

Fig. 5.5 Vector connecting the origin of the coordinates with the point $N(C_1, C_2)$ over a period of time

changes slowly. Sometimes we cannot perceive the components of oscillations due to the fact that their frequency is very high; however, we notice a slow change in the amplitude of the summed oscillatory process.

Let us consider a case when the frequencies of the summable oscillations are the same. In this case, the resulting oscillation is obtained by the same frequency. The total oscillation can be represented as $y_1 = A_1 \sin \theta t + A_2 \cos \theta t$, and $y_2 = B_1 \sin \theta t + B_2 \cos \theta t$, when we add them together, we obtain:

$$y = y_1 + y_2 = (A_1 + B_1) \sin \theta t + (A_2 + B_2) \cos \theta t \tag{5.27}$$

i.e., the oscillation of the same frequency θ.

The described method of adding oscillations is not convenient, because it is difficult to determine what will the amplitude of the total oscillation be. Here, the geometric interpretation is more convenient that the mathematical formula: $y = C_1 \sin \theta t + C_2 \cos \theta t$. The image of the oscillation is in the form of a vector connecting the origin of the coordinates with the point $N(C_1, C_2)$, as shown in Fig. 5.5. Expression (5.27) shows that when the oscillations are added, the coordinates of the vectors are added as well, i.e., the vectors are added according to the parallelogram rule (Fig. 5.5; Ref. to Phase difference. Sine & cosine waves [Cyberphysics.Co.UK]).

If both summed oscillations have the same phase, then they are represented by vectors of the same direction. When such vectors are added, their lengths are added. Thus, the amplitudes of the oscillations are added when the oscillations have the same frequency and phase. If phases of summed oscillations differ from each other by π, then their amplitudes are subtracted. In this case, the vibrations dampen each other in general. When phases of oscillations differ by some angle, the amplitude of the total oscillation is equal to the length of a diagonal of the parallelogram (Ref. to Phase difference. Sine & cosine waves [Cyberphysics.Co.UK]) (Fig. 5.6).

Problem 5.1

A weightless beam with one concentrated mass in the middle of the span of the beam is shown in Fig. 5.1. It is required to determine the amplitude of forced oscillations for SDOF system under the action of an external periodic force, changing according to a harmonic law $P(t) = P \sin \theta t$

Solution

The amplitude of forced oscillations can be determined by the following formula:

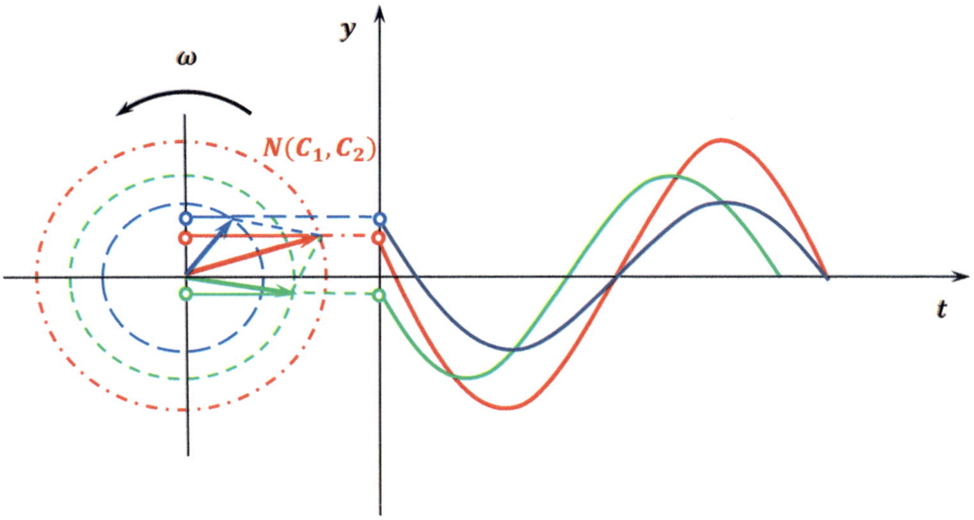

Fig. 5.6 Vector connecting the origin of the coordinates with the point $N(C_1, C_2)$, over a period of time

$$A = \frac{P}{m}\frac{1}{\omega^2-\theta^2} = \frac{P}{\delta_{11}}\frac{1}{\frac{m}{\delta_{11}}\left(\omega^2-\theta^2\right)} = y_{st}\frac{1}{1-\left(\frac{\theta}{\omega}\right)^2}.$$

where

$$\delta_{11} = \frac{48EI}{l^3}; \qquad y_{st} = \frac{Pl^3}{48EI};$$

$$\omega = \sqrt{\frac{1}{m\delta_{11}}} = \frac{4}{l}\sqrt{\frac{3EI}{ml}}.$$

Problem 5.2

An asynchronous electric motor weighting $G = 1500$ $k\Gamma$ is mounted at the end of a T-shape rolled-steel cantilever beam, with the following stiffness characteristics: (Profile # 16; $I = 873$ cm^4—moment of inertia; $W = 109$ cm^3—the section modulus; $E = 2.1 \cdot 10^6$kg/cm^2—module of elasticity; $n = 800$ rev/min the number of revolutions per minute; $P = 200$ kg—the weight of unbalanced parts of electric motor; $e = 0.3$ cm—the eccentricity). It is required to determine the dynamic deflection and stress, neglecting the beam's weight (Fig. 5.7).

Solution

The frequency of forced vertical oscillations can be determined by the formula:

$$\theta = \frac{2\pi n}{60} = \frac{6.28 \cdot 800}{60} = 83.73\ \text{s}^{-1}$$

The centrifugal force:

$$P_{\text{centrifugal}} = \frac{mv^2}{e} = \frac{m(\theta e)^2}{e} = m\theta^2 e = \frac{200}{9.81}83.73^2 \cdot 0.3 = 428.8\ \text{kg}$$

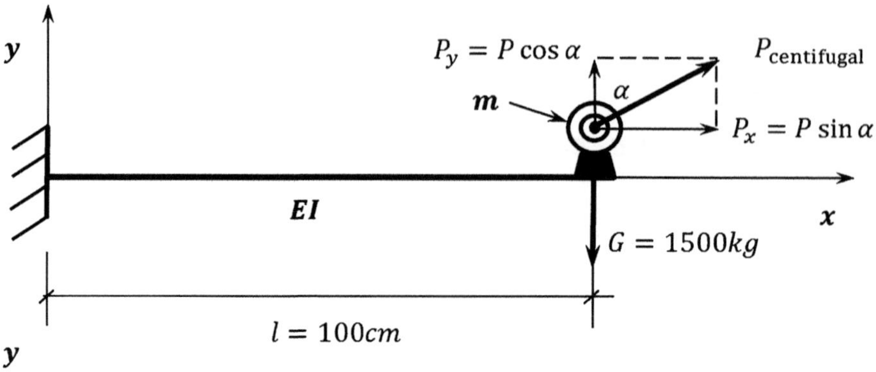

Fig. 5.7 Forced vibration of the cantilever beam

The disturbing force:

$$P = 428.8\sin\theta t.$$

The static deflection from the weight of asynchronous electric motor:

$$\delta_{st} = \frac{1500\cdot 100^3}{3\cdot 2.1\cdot 10^6\cdot 873} = 0.27\text{ cm}$$

The natural frequency of oscillation of cantilever beam given as a SDOF system:

$$\omega = \sqrt{\frac{g}{\delta_{st}}} = \sqrt{\frac{981}{0.27}} = 68.28\text{ s}^{-1}$$

The ratio of frequency of forced oscillations to the natural frequency of oscillations of cantilever beam indicates no resonance.

$$\frac{\theta}{\omega} = \frac{83.73}{60.28} = 1.38$$

Let us determine the amplitude of forced oscillations:

$$A = \frac{P}{m}\frac{1}{\omega^2 - \theta^2} = \frac{428.8}{\frac{1500}{981}}\frac{1}{\left(83.73^2 - 60.28^2\right)} = 0.08\text{ cm}$$

Adding this value to the static displacement of the beam δ_{st} we obtain the deflection of the end of cantilever beam:

$$A + \delta_{st} = 0.27 + 0.08 = 0.35\text{ cm}$$

Now, let us calculate the additional force X causing the deflection of the beam:

$$\frac{X\cdot 100^3}{3\cdot 2.1\cdot 10^6\cdot 873} = 0.08\text{ cm};\qquad X = 440\text{ kg}$$

The greatest value of the bending moment at the built-in end where $x = l = 100\text{ cm} = 1\text{ m}$

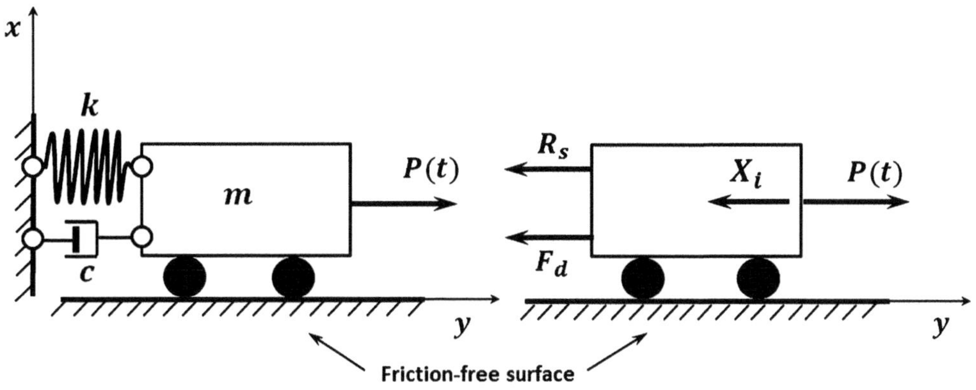

Fig. 5.8 Model of a damped system with SDOF

$$M = (1500 + 440) \cdot 1 = 1940 \text{ kgm}.$$

The maximum stresses:

$$\sigma_{\max} = \frac{M_{\max}}{W} = \frac{194000}{109} = 1780 \text{ kg/cm}^2$$

5.2 Forced Harmonic Vibrations of a Damped System with Single-Degree-of-Freedom (SDOF)

5.2.1 Analysis of Forced Harmonic Vibrations

Above we assumed that there was no internal friction in the system. The simplest model of a real system with a single degree of freedom can be chosen as the model as shown in Fig. 5.8. This mechanical model of a rigid body of mass is constrained to move in only one translational direction, whose position is completely defined by the single displacement coordinate y.

Differential Equation of Motion In the direction of motion under the action of an external periodic force changing according to a harmonic law $P(t) = P \sin \theta t$, we have the following: the force of inertia $Y_i = -m\ddot{y}$; the strength of spring elasticity $R_s = -ky$; the force of internal friction or damping force $F_d = -c\dot{y}$. In the presence of damping, below we will consider a critically under-damped case, i.e. (when $0 < c < c_c$), which is the most practically applicable case in all engineering structures. The differential equation of motion for forced oscillations under a harmonic load takes the form:

$$m\ddot{y} + c\dot{y} + ky = P \sin \theta t \tag{5.28}$$

or

$$\ddot{y} + \frac{c}{m}\dot{y} + \omega^2 y = \frac{P}{m} \sin \theta t \tag{5.29}$$

The general integral of Eq. (5.29) consists of the general solution of the corresponding homogeneous equation (4.4), and a particular solution of Eq. (5.29):

$$y = e^{-nt}(C_1 \sin \omega t + C_2 \cos \omega t) + \frac{P}{\sqrt{(\omega^2 - \theta^2)^2 + 4n^2\theta^2}} \sin(\theta t - \varepsilon) \tag{5.30}$$

where ε is a shift in the phase of forced oscillations with respect to the phase of the disturbing force, which is observed in the presence of internal friction in the system.

$$\varepsilon = \tan^{-1}(C_2/C_1) = \tan^{-1}\frac{c\theta}{\omega^2 - m\theta^2} \tag{5.31}$$

If at the initial moment of oscillations of the system $t = 0$, $y(0) = y_0$, and $\dot{y}(0) = \dot{y}_0$, then the solution takes the following form:

$$\begin{aligned} y &= e^{-nt}\left(y_0 \cos \omega t + \frac{\dot{y}_0 + n y_0}{\omega} \sin \omega t\right) + \\ &+ Ae^{-nt}\left(\sin \varepsilon \cos \omega t + \frac{n \sin \varepsilon - \theta \cos \varepsilon}{\omega} \sin \omega t\right) + A \sin(\theta t - \varepsilon) \end{aligned} \tag{5.32}$$

The first two summands of the general solution (5.32) correspond to free and free accompanying oscillations; the third component defines the forced harmonic oscillations occurring with the frequency θ. The internal friction in this case has a dual effect. First, the frequency of oscillations decreases, meaning instead of ω it becomes equal only: $\sqrt{\omega^2 - n^2}$. Second, because of the appeared multiplier e^{-nt}, the amplitude of oscillations decreases over time. At the same time, the amplitude of oscillations decreases in $e^{\frac{2\pi n}{\omega}}$ times, for the period of time $\left[0; \frac{2\pi}{\omega}\right]$. Thus, after some time passes, the first two summands of Eq. (5.32) decrease and eventually will become negligible and only forced oscillations will remain. The system will oscillate in the steady-state mode. Let us note that the amplitude of forced oscillations does not depend on the initial conditions and time. The amplitude of established oscillations will be determined by the formula:

$$A = \frac{P}{m\sqrt{(\omega^2 - \theta^2)^2 + 4n^2\theta^2}} \tag{5.33}$$

The ratio of the largest amplitude of oscillations or dynamic deflection to static displacement from the action of the force P is called dynamic coefficient. The dynamic coefficient characterizes the dynamic effect caused by a disturbing force. Taking into account that:

$$y_{st} = P\delta = \frac{P}{m\omega^2} \tag{5.34}$$

from (5.33) we will find the value of the dynamic coefficient:

$$\mu = \frac{A}{y_{st}} = \frac{1}{\sqrt{\left(1 - \frac{\theta^2}{\omega^2}\right)^2 + \left(\frac{c\theta}{m\omega^2}\right)^2}} = \frac{1}{\sqrt{\left(1 - \frac{\theta^2}{\omega^2}\right)^2 + \left(\gamma\frac{\theta}{\omega}\right)^2}} \tag{5.35}$$

Resonance Phenomenon and Definition Figure 5.9 shows the dependence of the dynamic coefficient on the frequency of the external excitation force for some values of the loss coefficient—γ. The amplitude of the forced oscillations and the dynamic coefficients increase significantly, when the frequency of forced oscillations θ coincides with the frequency of the system's natural oscillations. Due to damping, the dynamic coefficients do not turn to infinity anymore.

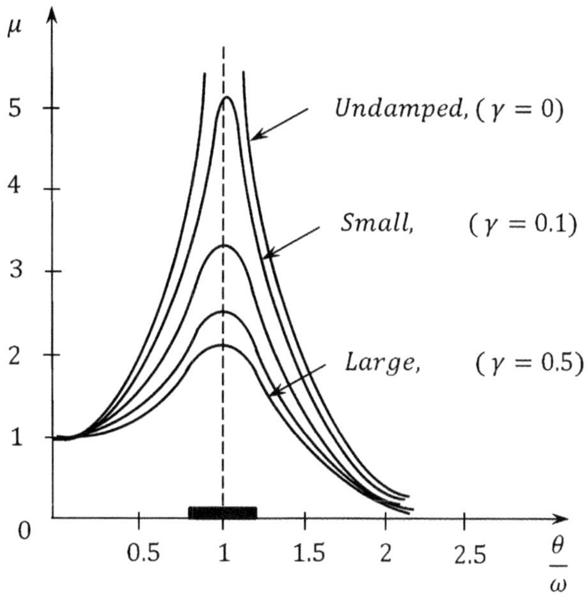

Fig. 5.9 Resonance curves with increasing levels of damping

When the frequencies coincide $\theta = \omega$ from formula (5.35), we have:

$$\mu = \frac{1}{\gamma} \tag{5.36}$$

Let's determine the greatest value of μ, for that we calculate the minimum value of:

$$\frac{1}{\mu^2} = \left(1 - \frac{\theta^2}{\omega^2}\right)^2 + \gamma^2 \frac{\theta^2}{\omega^2} \tag{5.37}$$

Or,

$$\frac{1}{\mu^2} = \left(1 - u^2\right)^2 + \gamma^2 u^2 \tag{5.38}$$

where

$$u^2 = \frac{\theta^2}{\omega^2}$$

The minimum for $\frac{1}{\mu^2}$, equal to $\gamma^2 - \frac{\gamma^4}{4}$, we obtain at $u^2 = 1 - \frac{\gamma^2}{2}$. Then

$$\mu_{\max} = \frac{1}{\sqrt{\gamma^2 - \frac{\gamma^4}{4}}} = \frac{1}{\sqrt{1 - \frac{\gamma^2}{4}}} \tag{5.39}$$

At $\gamma = 0$, i.e., without taking into account the damping, from (5.35) we obtain a familiar formula for determining the dynamic coefficient (5.6). The graph of the phase shift change ε of forced oscillations depending on the frequency θ is shown in Fig. 5.10. At the resonance, the amplitude of forced oscillations is inversely proportional to the logarithmic decrement, and the shift of the phase of

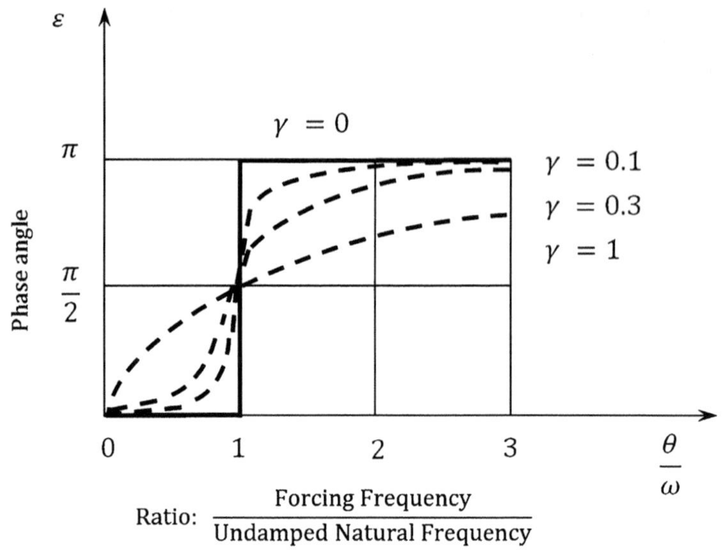

Fig. 5.10 Phase angle for a damped system excited by a harmonic force

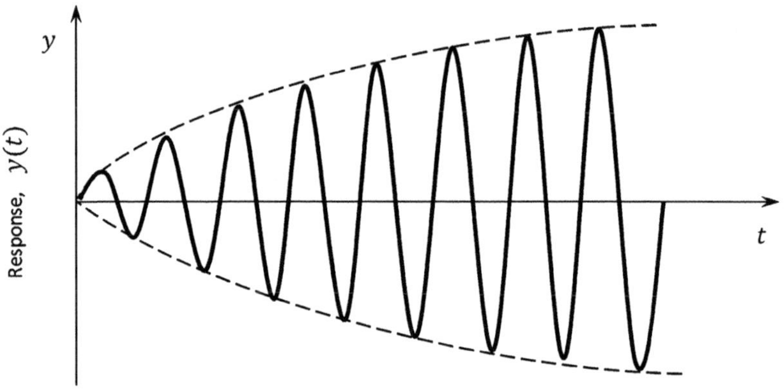

Fig. 5.11 Response of a damped system to a harmonic force of frequency $\theta = \omega$. $y(0) = \dot{y}(0) = 0$

oscillations with respect to the oscillations of the disturbing force is: $\varepsilon = -\pi/2$, i.e., the displacements reach their maximum at the moments of time when the values of this force are equal to zero.

At the same time, we see that the phase angle varies within the limits $0 < \varepsilon < 180^\circ$, and especially quickly this change occurs in the near resonance and resonance zones. At $\gamma = 0$, i.e., when the forces of internal friction disappear, the phase shift occurs abruptly. In practical calculations of engineering structures, working in modes far enough from resonant, the amplitudes of forced oscillations at a relatively weak internal friction are slightly different from the corresponding amplitudes of forced oscillations of an undamped system. In this case, the amplitudes of forced oscillations are determined by the following formula:

$$A = \frac{P}{\omega^2 - \theta^2} \tag{5.40}$$

Figure 5.11 shows the effect of damping on the growth limits of amplitudes of resonance of oscillations. The number of cycles (about 3–6) for which the amplitude of displacements of the damped system almost reaches the maximum value depends on the damping parameters.

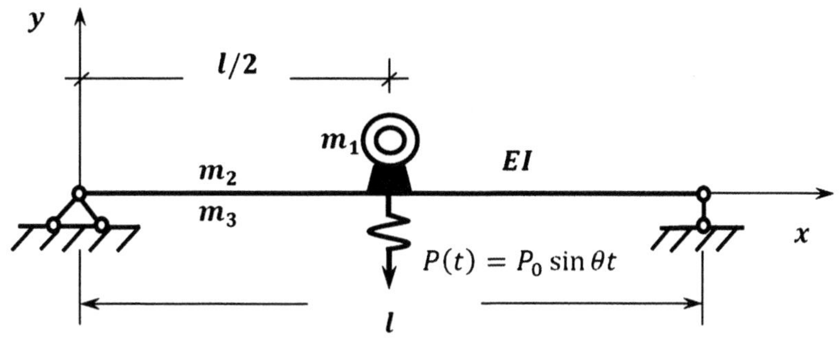

Fig. 5.12 Forced vibration of the hinge-supported beam

Note: In the case of forced oscillations, it is preferable to use the term "internal friction" or "damping," since the attenuation of oscillations occurs at (or during) free oscillations.

Problem 5.3

It is required to determine the dynamic coefficient, the amplitude of forced vibrations, and the highest normal stresses in a beam shown in Fig. 5.12. The concentrated mass of asynchronous electric motor $m_1 = 2500$ kg is mounted on the middle of the span of a T-shape rolled-steel simple beam. (Profile #24; $I = 3460$ cm^4—moment of inertia; $W = 289$ cm^3—the section modulus; $E = 2.1 \cdot 10^6$ kg/cm^2—module of elasticity; $m_3 = 27.3$ kg/m; $l = 3$ m). Mass of the distributed load: $m_2 = 300$ kg/m. The electrical engine is making 300 revolutions per minute and developing the vertical component of the centrifugal force $P(t) = P_0 \sin \theta t$. The amplitude of the disturbing force $P_0 = 6000$ N.

Solution

A uniform mass in half the size will focus in the middle of the span, i.e., we will replace the given system with the system presented in Fig. 5.12.

The reduced mass of the oscillatory system will take the form:

$$m = m_1 + \frac{(m_2 + m_3)}{2} l = 2500 + \frac{(300 + 27.3)}{2} 3 = 2990.95 \frac{\text{kg} \cdot \text{s}^2}{\text{m}}$$

We will determine the gravity of the reduced mass:

$$Q = \text{mg} = 2990.95 \cdot 9.81 = 29341.2\ N$$

The static deflection of the beam caused by this force:

$$y_{\text{st}} = \frac{Ql^3}{48EI} = \frac{29341.2 \cdot 3^3}{48 \cdot 210 \cdot 10^9 \cdot 3460 \cdot 10^{-8}} = 0.0023 \text{ m}$$

Angular frequency of natural oscillations:

$$\omega = \sqrt{\frac{g}{y_{\text{st}}}} = \sqrt{\frac{9.81}{0.0023}} = 65.31 \text{ s}^{-1}$$

The frequency of forced oscillations is equal to the frequency of disturbing force:

$$\theta = 600\frac{2\pi}{60} = 62.80\ \mathrm{s}^{-1}$$

The coefficient of losses γ for a rolled-steel beam under the influence of inertia force is assumed to be equal to: $\gamma = 0.025$; since, in our case, the amplitude of the vertical component of the centrifugal force is more than 1000 N., if the amplitude of the inertial force is less than 1000 N, then the coefficient of losses γ for a steel beam will be equal to $\gamma = 0.01$. The dynamic coefficient is determined by formula (5.35):

$$\mu = \frac{1}{\sqrt{\left(1-\frac{\theta^2}{\omega^2}\right)^2 + \left(\frac{c\theta}{m\omega^2}\right)^2}} = \frac{1}{\sqrt{\left(1-\frac{62.80^2}{65.31^2}\right)^2 + \left(0.025\,\frac{62.80}{65.31}\right)^2}} = 3.62$$

As we see, far away from the resonance, the damping accounting does not actually change the value of the dynamic coefficient. When $\theta = \omega$, i.e., at resonance:

$$\mu = \frac{1}{\gamma} = \frac{1}{0.025} = 40$$

Let us determine the bending moment and static stresses due to dynamic forces:

$$M_{\mathrm{st}} = \frac{2500 \cdot 9.81 \cdot 3}{4} + \frac{327.3 \cdot 9.81 \cdot 3^2}{8} = 22005.9\ \mathrm{N \cdot m};$$

$$\sigma_{\mathrm{st}} = \frac{M_{\mathrm{st}}}{W} = \frac{22005.9}{\left(289 \cdot 10^{-6}\right)1000} = 76145\ \mathrm{kPa}$$

Determining the amplitude of the bending moment and the stress from the dynamic load:

$$M_{\mathrm{dyn}} = \mu\frac{Pl}{4} = 3.62\frac{6000 \cdot 3}{4} = 16290\ \mathrm{N \cdot m}$$

$$\sigma_{\mathrm{dyn}} = \frac{M_{\mathrm{dyn}}}{W} = \frac{16290}{\left(289 \cdot 10^{-6}\right)1000} = 56367\ \mathrm{kPa}$$

Total stresses:

$$\sigma = \sigma_{\mathrm{st}} + \sigma_{\mathrm{dyn}} = 76145 + 56367 = 132512\ \mathrm{kPa}.$$

To determine the static deflection from force $P = 6000$ N, using the deflection in the middle of the span of the beam from force $P = 29341.2$ N, we obtain:

$$y^*_{\mathrm{st}} = 0.0023\frac{6000}{29341.2} = 0.00047\ \mathrm{m}$$

Now, let us determine the amplitude of the dynamic deflection:

$$y_{\mathrm{dyn}} = \mu \cdot y^*_{st} = 3.62 \cdot 0.00047 = 0.002.$$

The maximum deflection value of the beam under load will be equal to the sum of the static deflection from force Q and the amplitude of the dynamic deflection from force P.

$$y_{\max} = 0.0023 + 0.002 = 0.0043\ \mathrm{m}$$

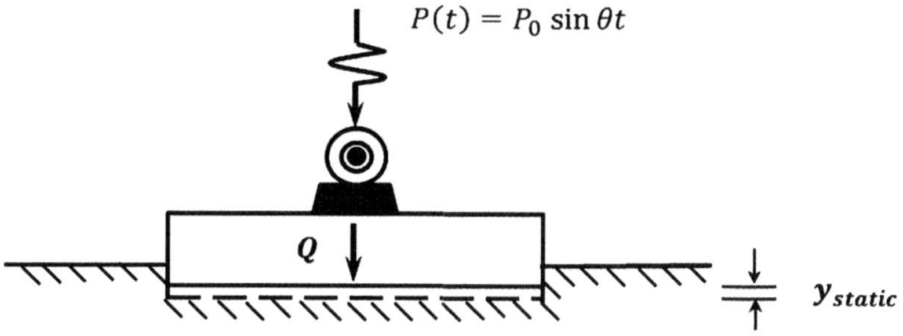

Fig. 5.13 Foundation plate under forced vibration

Problem 5.4

Let us consider forced oscillations of a foundation plate with a weight of $Q = 400$ kN, with area of the base $A = 5$ m^2, lying on an elastic base under the dynamic load (Fig. 5.13). The electric motor is mounted on the foundation plate. The electrical engine is making 250 revolutions per minute and developing the vertical component of the harmonic centrifugal force $P(t) = 25 \sin \theta t$. It is required to determine the stresses of foundation base with and without damping, if the coefficient of inelastic resistance of materials or coefficient of losses $\gamma = 0.2$, and coefficient of bed $K = 5600$ kN/m^3.

Solution

Total stress at the base of the foundation consists of the sum of static and dynamic stresses:

$$\sigma = \sigma_{\text{st}} + \sigma_{\text{dyn}};$$

$$\sigma_{\text{st}} = \frac{Q}{A} = \frac{400\ \text{kN}}{5\ \text{m}^2} = 80\ \text{kPa};$$

$$\sigma_{\text{dyn}} = \mu \frac{P_0}{A} = \mu \frac{25\ \text{kN}}{5\ \text{m}^2} = 5\mu(\text{kPa}).$$

Let us rewrite formula (5.35) for this particular problem. The following formula to determine the dynamic coefficient with damping will take the form:

$$\mu = \frac{1}{\sqrt{\left(1 - \frac{n_f{}^2}{n_b{}^2}\right)^2 + \left(\gamma \frac{n_f}{n_b}\right)^2}}$$

The dynamic coefficient without damping will take the form:

$$\mu = \frac{1}{1 - \left(\frac{n_f}{n_b}\right)^2}$$

Here, n_f is the technical frequency of free oscillation of the engine $n_f = 250$ rev/min; n_b is the technical frequency of free oscillations of the foundation base, which is unknown and determined by the formula:

$$n_b = \frac{300}{\sqrt{y_{st}}};$$

Here, y_{st} is the static displacement (settlement) of the foundation plate under the influence of its own weight:

$$y_{\mathrm{st}} = \frac{\sigma_{\mathrm{st}}}{K} = \frac{80\ \mathrm{kPa}}{5600\ \mathrm{kN/m^3}} = 0.014\ \mathrm{m} = 1.4\ \mathrm{cm}$$

Now let us determine the technical frequency of natural oscillations of the foundation:

$$n_b = \frac{300}{\sqrt{1.4}} = 254\ \mathrm{rev/\ min}$$

Then we proceed to determine the dynamic coefficient:

(a) Accounting for damping:

$$\mu = \left(\sqrt{\left(1 - \frac{{n_f}^2}{{n_b}^2}\right)^2 + \left(\gamma \frac{n_f}{n_b}\right)^2}\right)^{-1} = \left(\sqrt{\left(1 - \frac{250^2}{254^2}\right)^2 + 0.2^2\left(\frac{250}{254}\right)^2}\right)^{-1} = 5$$

(b) Ignoring damping:

$$\mu = \frac{1}{1 - \left(\frac{250}{254}\right)^2} = 32.3$$

Finally, let us determine the total stress at the base of the foundation:

(a) Damping is considered:

$$\sigma = \sigma_{\mathrm{st}} + \sigma_{\mathrm{dyn}} = 80 + 5 \cdot 5 = 105\ \mathrm{kPa}$$

(b) Damping is ignored:

$$\sigma = \sigma_{\mathrm{st}} + \sigma_{\mathrm{dyn}} = 80 + 5 \cdot 32.3 = 241.5\ \mathrm{kPa}$$

As we see the accounting for damping reduces the dynamic stresses.

Problem 5.5

It is required to determine the frequency of natural oscillations, dynamic stresses, and dynamic deflection in a continuous beam, as shown in Fig. 5.14. The electric motor is mounted in the middle of the second span of the T-shaped rolled- steel continuous beam. The weight of the engine is $Q = 60$ kN. The electrical engine is making 400 revolutions per minute and developing the vertical component of the centrifugal force $P(t) = P_0 \sin \theta t$. The amplitude of the disturbing force $P_0 = 20$ kN (The beam profile #50; $I = 39290\ \mathrm{cm}^4$—moment of inertia; $W = 1572\ \mathrm{cm}^3$—the section modulus; $E = 2.1 \cdot 10^6 \mathrm{kg/cm}^2$—module of elasticity; $l_1 = l_2 = l_3 = 8$ m). The weight of the beam is neglected.

Solution

The angular frequency of forced oscillations is equal to the frequency of disturbing force:

$$\theta = \frac{2\pi n_1}{60} = 400\frac{2\pi}{60} = 62.80\ \mathrm{s}^{-1} = 41.87\ \mathrm{s}^{-1}$$

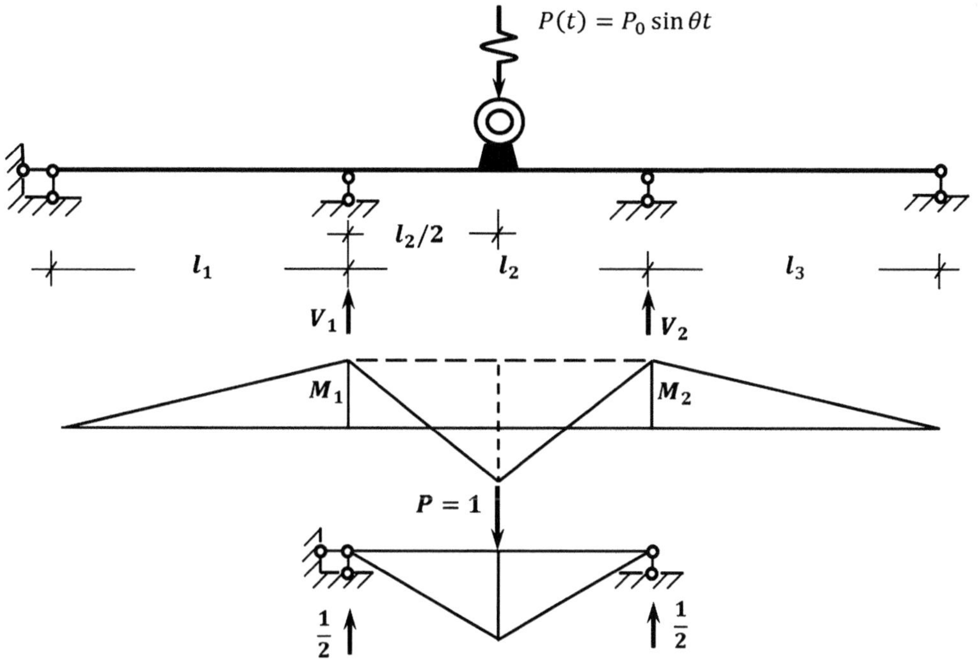

Fig. 5.14 Continuous beam under forced vibration

The technical frequency of free oscillations will be determined by the following formula:

$$n = \frac{300}{\sqrt{y_{st}}};$$

To determine the static deflection, we will perform a static calculation of the continuous beam.

The diagram of bending moments may be found by superimposing the diagram of positive moment for a simple beam, subjected to a concentrated load at the mid-span to the diagram of moments for a continuous beam. The maximum ordinate of bending moments is at the center of mid-span and equals $\frac{Pl}{4}$. The diagram of negative moments consists of a rectangle, the constant ordinate of which is the unknown including moment M_1 or M_2; this superimposed diagram is shown in Fig. 5.14. Applying the Theorem of Three Moments we obtain:

$$2M_1(l_1 + l_2) + M_2 l_2 = 2M_1(8 + 8) + M_2 8 = -6R_A^f$$

From the conditions of symmetry and equilibrium, we have:

$$M_1 = M_2 = M, V_1 = V_2 = 1/2Q$$

Therefore, consequently:

$$2M(8 + 8) + M8 = -6R_A^f$$

Now let us determine the fictitious reactions of the beam:

$$R^f = R_A^f = R_B^f = \frac{Ql_2}{4}\frac{l_2}{2}\frac{1}{2} = \frac{l^2}{16} = 4$$

The unknown end moments:

$$20M = -3R^f = -3 \cdot 4 = -12$$

$$M = -6\ \text{kN} \cdot \text{m}$$

Alternatively, M may be found from the elastic–curve equation as follows: difference in slopes of the elastic curve at the end and center of the beam is equal to zero, i.e., the tangent to the elastic curve is horizontal at both points. Hence, according to the integral expression of the elastic–curve equation, the deflection:

$$\Delta = \int_0^{\frac{l}{2}} \frac{M}{EI} dx = 0$$

This equation may be solved by using direct integration, or it may be solved semi-graphically in accordance with the moment-area method. By integration, the solution is:

$$\Delta = \int_0^{\frac{l}{2}} \frac{M}{EI} dx = \frac{1}{EI} \int_0^{\frac{l}{2}} \left(M_1 + \frac{Q}{2}x\right) dx = 0$$

$$\left[M_1 x + \frac{Q}{4}x^2\right]_0^{\frac{l}{2}} = 0; \qquad M_1 = -\frac{1}{8}Ql = -6\ \text{kN} \cdot \text{m}$$

We can apply the moment-area method for determining the maximum deflection, i.e., "the value of static displacement - y" of the beam under the electric motor, mounted in the middle of the second span.

$$y_{st} = \frac{1}{EI}\left(\frac{Ql}{4} \cdot \frac{l}{2} \cdot \frac{1}{2} \cdot \frac{2}{3} \cdot \frac{l}{4} \cdot 2 - \frac{l}{4} \cdot l \cdot \frac{1}{2} \cdot M\right) = \frac{Ql^3}{48EI} - \frac{Ml^2}{8EI} =$$

$$= \frac{6000 \cdot 800^3}{48 \cdot 2.1 \cdot 10^6 \cdot 39290} - \frac{6 \cdot 0.6 \cdot 800^2 \cdot 10^5}{8 \cdot 2.1 \cdot 10^6 \cdot 39290} = 0.776 - 0.349 = 0.427\ \text{cm}$$

Thus, the technical frequency of free oscillations:

$$n = \frac{300}{\sqrt{y_{st}}} = \frac{300}{\sqrt{0.427}} = 460\ \text{min}^{-1}$$

Now let us determine the dynamic coefficient without taking into account damping. In our case, from the given condition of the problem: $n > n_1$:

$$\mu = \frac{1}{1 - \left(\frac{\theta}{\omega}\right)^2} = \frac{1}{1 - \left(\frac{n_1}{n}\right)^2} = \frac{1}{1 - \left(\frac{400}{460}\right)^2} = 4.08$$

The static deflection of the beam from the force $P_0 = 20$ kN will be found using the result obtained from the force $Q = 60$ kN:

$$y^*_{st} = 0.427 \frac{P_0}{Q} = 0.427 \frac{20}{60} = 0.142\ \text{cm}$$

The amplitude of forced oscillations will equal to dynamic deflection:

$$y_{\text{dyn}} = \mu \cdot y^{*}{}_{\text{st}} = 4.08 \cdot 0.142 = 0.579 \text{ cm}$$

The maximum deflection value of the beam under load will be equal to the sum of the static deflection from force Q and the amplitude of the dynamic deflection from force P_0.

$$y_{\max} = y_{\text{st}} + y_{\text{dyn}} = 0.427 + 0.579 = 1.006 \approx 1 \text{ cm}$$

Let us determine the maximum value of bending moment taking into account the dynamic influence of the force P_0:

$$\begin{aligned} M_{\max} &= \left[\left(\frac{Ql}{4} - MQ\right) + \left(\mu\frac{P_0 l}{4} - \mu M P_0\right)\right] = \\ &= \left[\left(\frac{6000 \cdot 800}{4} - 6 \cdot 0.6 \cdot 10^5\right) + \left(4.08\frac{2 \cdot 10^3 \cdot 800}{4} - 2 \cdot 0.6 \cdot 4.08 \cdot 10^5\right)\right] = \\ &= 1982000 \text{ kg} \cdot \text{cm} \end{aligned}$$

Finally, let's determine the total stress:

$$\sigma_{\max} = \frac{M_{\max}}{W} = \frac{1982000}{1572} = 1260 \text{ kg/cm}^2;$$

For the case when the electric motor is turned off:

$$M_{\max} = \frac{Ql}{4} - MQ = \frac{6000 \cdot 800}{4} - 6 \cdot 0.6 \cdot 10^5 = 840000 \text{ kg} \cdot \text{cm}$$

The corresponding stresses for the case when the electric motor is turned off:

$$\sigma_{\max} = \frac{840000}{1572} = 534 \text{ kg/cm}^2$$

Thus, the dynamic load significantly increases the stresses in the structure compared to the static load, even when the structure is not in the resonance zone.

References

1. Clough, R. W., & Penzien, J. (1993). *Dynamics of Structures* (2nd ed., 738p). McGraw-Hill College.
2. Anderson, J. C., & Naeim, F. (2012). *Basic Structural Dynamics* (280p). Wiley.
3. Copra, A. K. (2016). *Dynamics of Structures* (992 p). Pearson.
4. Busby, H. R., & Staab, G. H. (2017). *Structural Dynamics: Concepts and Applications* (568p). CRC Press.
5. Craig, R. R. J. R., & Kurdila, A. J. I. (2011). *Fundamentals of Structural Dynamics* (2nd ed., 752p). Wiley.
6. Katsikadelis, J. T. (2020). *Dynamic Analysis of Structures* (1st ed., 922p). Academic.

6 Motion of an Elastic System with SDOF Under the Action of a Force Varying According to Arbitrary Law

Abstract

This chapter presents the motion of an elastic system with SDOF under the action of a force varying by arbitrary law. In this chapter, we will discuss the following: the reaction of a system with SDOF under impulsive and suddenly applied loads, the action of a sudden applied force of constant value, the action of a force growth according to linear law, the action of a suddenly applied force of constant value and suddenly terminated its action, short-term impulse of force, the action of periodic instantaneous impulses, and applicable problems and solutions.

6.1 Differential Equations of Motion

As is known, the influence function or the Green's function can be applied to solve linear second-order inhomogeneous (heterogeneous) equations with constant coefficients (Fig. 6.1).

The equation describes the motion of a mechanical system under an external force dependent only on time.

$$m\ddot{y} + c\dot{y} + ky = P(t) \tag{6.1}$$

were, the term $m\ddot{y}$ represents the inertia force due to acceleration $\ddot{y}$ of mass m; the term $c\dot{y}$ represents the viscous damping force equal to the product between the damping coefficient c and velocity $\dot{y}$; the term ky represents the reactions produced by the system with stiffness k due to the displacement y. The equation of motion, taking into account the value of the loss coefficient (4.49), can be represented in the following form:

$$m\ddot{y} + \gamma\sqrt{km}\dot{y} + ky = P(t) \tag{6.2}$$

Here $\gamma\sqrt{km}\dot{y}$ is the internal friction force.

The mass m, k, and stiffness (rigidity) ($k > 0$) are greater than zero for a physical/mechanical system. A dot ($\cdot$) over a variable indicates differentiation with respect to time.

The solution will be sought under the initial conditions set at 0:

L. G. Petrosian, *Dynamics of Structures*, https://doi.org/10.1007/978-3-031-63539-7_6

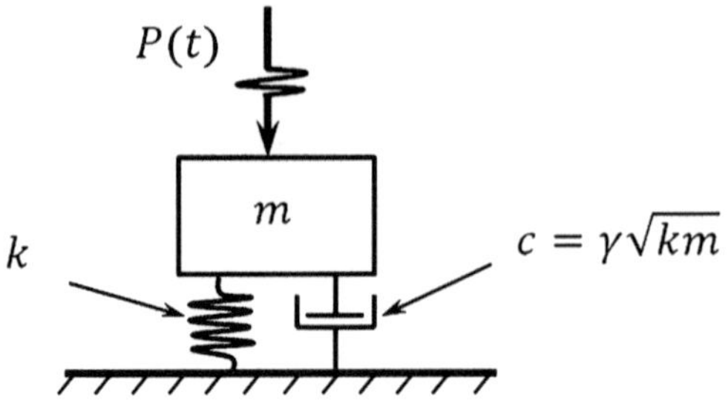

Fig. 6.1 Dynamic model of a system with single degree of freedom

$$y|_{t=t_0}=0, \qquad v=\left.\frac{dy}{dt}\right|_{t=t_0}=0 \tag{6.3}$$

To build the Green's function, we will solve the following equation:

$$m\ddot{y}+c\dot{y}+ky=\delta(t-\tau) \tag{6.4}$$

under the initial conditions (6.3). Here δ is the delta function. Assuming that $t>t_0$, we obtain that $y\equiv 0$ at $t_0<t<\tau$. Let us integrate the Eq. (6.4) from $t=\tau-0$ to $t=\tau+0$, we obtain the conditions:

$$\left.\frac{dy}{dt}\right|_{\tau+0}=\frac{1}{m} \tag{6.5}$$

and

$$y|_{\tau+0}=y|_{\tau-0}=0 \tag{6.6}$$

the integrals from the second and third terms in (6.4) are equal to zero.

So, we can see that at $t=\tau$ the derivative $\frac{dy}{dt}$ has a finite jump, therefore, the function y itself at $t=\tau$ does not have a jump, i.e., (6.6). Thus, at $t>\tau$ it is required to find a solution of homogeneous equation (4.4):

$$m\ddot{y}+c\dot{y}+ky=0 \tag{6.7}$$

under the initial conditions (6.5) and (6.6).

Based on the general solution of the Eq. (6.7):

$$y=C_1e^{p_1t}+C_2e^{p_2t} \tag{6.8}$$

where p_1 and p_2 are the roots of the characteristic equation (4.11), we obtain the desired solution:

$$\begin{aligned}y&=\frac{e^{-p_1\tau}}{m(p_1-p_2)}e^{p_1t}+\frac{e^{-p_2\tau}}{m(p_2-p_1)}e^{p_2t}\\&=\frac{1}{m(p_1-p_2)}\left[e^{p_1(t-\tau)}-e^{p_2(t-\tau)}\right]\end{aligned} \tag{6.9}$$

The presented solution is valid for a weak and/or strong damping under accepted assumptions. See Chap. 4, Sect. 4.1. So, let us write down the function of influence, i.e., the Green's function:

$$G(t;\tau)=\begin{cases}0 & (t_0<t<\tau)\\ \dfrac{1}{m(p_1-p_2)}\left[e^{p_1(t-\tau)}-e^{p_2(t-\tau)}\right] & (\tau<t<\infty)\end{cases} \tag{6.10}$$

The obtained function is continuous, however, at $t = \tau$ the function has a jump. Hence. we obtain the solution of Eq. (6.1) under the initial conditions equal to zero (6.3)

$$y(t) = \frac{1}{m(p_1 - p_2)} \int_{t_0}^{t} \left[e^{p_1(t-\tau)} - e^{p_2(t-\tau)} \right] P(\tau) d\tau \tag{6.11}$$

Now, let us consider the action of a disturbing force, the law of change of which is known over time. We write the differential equation of motion of an elastic system with single degree of freedom without damping under the action of external force *P*(*t*), i.e., Eq. (6.1) is replaced by:

$$m\ddot{y} + ky = P(t) \tag{6.12}$$

In this case, the characteristic equation has roots, $p_{1,\,2} = \pm i\omega_0$, where

$$\omega_0 = \sqrt{\frac{k}{m}} \tag{6.13}$$

ω_0 is the frequency of free oscillations, i.e., oscillations without external force. Formula (6.11), transformed by Euler's formula, gives a solution under the initial conditions equal to zero (6.3)

$$\begin{aligned} y &= \frac{1}{m \cdot 2i\omega_0} \int_{t_0}^{t} \left[e^{i\omega_0(t-\tau)} - e^{-i\omega_0(t-\tau)} \right] P(\tau) d\tau = \\ &= \frac{1}{m\omega_0} \int_{t_0}^{t} \sin \omega_0(t-\tau) \; P(\tau) d\tau \end{aligned} \tag{6.14}$$

Let us use the following formula:

$$\sin \omega_0(t-\tau) = \sin \omega_0 t \cdot \cos \omega_0 \tau - = \sin \omega_0 \tau \cdot \cos \omega_0 t,$$

and present the solution as:

$$y(t) = \frac{1}{m\omega_0} \sin \omega_0 t \cdot \int_{t_0}^{t} P(\tau) \cos \omega_0 \tau \; d\tau - \frac{1}{m\omega_0} \cos \omega_0 t \cdot \int_{t_0}^{t} P(\tau) \sin \omega_0 \tau \; d\tau$$

If the action of the force *P*(*t*) ceases after some time t_1, the integrals in the formula cease to depend on time at $t > t_1$ and the solution takes the form:

$$y(t) = a \cos \omega_0 t + b \sin \omega_0 t,$$

If $t > t_1$, where,

$$a = -\frac{1}{m\omega_0} \int_{t_0}^{t_1} P(\tau) \sin \omega_0 \tau \; d\tau, \qquad b = \frac{1}{m\omega_0} \int_{t_0}^{t_1} P(\tau) \cos \omega_0 \tau \; d\tau$$

Thus, if initially a system or body was at rest, then an external force *P*(*t*) acted on it for some time, then at the end of the force action, the system will oscillate/vibrate freely with frequency ω_0.

Formula (6.14) can be written differently by introducing the complex deviation of the system or body from the equilibrium position:

$$w(t) = \frac{1}{m\omega_0} \int_{t_0}^{t} \mathrm{e}^{\mathrm{i}\omega_0(\mathrm{t}-\tau)} \; P(\tau) d\tau = \frac{1}{m\omega_0} \int_{t_0}^{t} \mathrm{e}^{-\mathrm{i}\omega_0\tau} \; P(\tau) d\tau \cdot e^{i\omega_0 t},$$

The real deviation *y*(*t*) serves as the imaginary part. If t_0 is the initial moment of action of the force, then the formula can be represented as:

$$w(t)=\frac{1}{m\omega_0}\int_{-\infty}^{t} \mathrm{e}^{-\mathrm{i}\omega_0\tau}\ P(\tau)d\tau\cdot e^{i\omega_0 t},$$

because at this transformation a part equal to zero is added to the integral. If the force acts only up to a certain moment t_1 of time, then, starting from this moment, we can write:

$$w(t)=\frac{1}{m\omega_0}\int_{-\infty}^{\infty} \mathrm{e}^{-\mathrm{i}\omega_0\tau} P(\tau)d\tau\cdot e^{i\omega_0 t}.$$

This integral is extended over the interval from $t=t_0$ to $t=t_1$. Thus, we have harmonic oscillations/ vibrations with the frequency ω_0 and the corresponding amplitude:

$$\frac{1}{m\omega_0}\left|\int_{-\infty}^{\infty} \mathrm{e}^{-\mathrm{i}\omega_0\tau} P(\tau)d\tau\right|.$$

The general integral of Eq. (6.12) has the following form:

$$y=y_0\cos\ \omega t+\frac{\dot{y}_0}{\omega}\ \sin\omega t+\frac{1}{m\omega}\int_0^t P(\tau)\sin\omega(t-\tau)d\tau \tag{6.15}$$

here y_0 and $\dot{y}_0$, respectively, the displacement and velocity at the initial time $t=0$; τ is the current coordinate of time. The initial conditions at $t=0$, $y_0(0)=0; \dot{y}_0(0)=0$ draw in zero the first two members of Eq. (6.15). Only the displacements caused by a disturbing (external) force remain.

$$y=\frac{1}{m\omega}\int_0^t P(\tau)\sin\omega(t-\tau)d\tau \tag{6.16}$$

Thus, we obtain a solution for the forced steady-state oscillations of a system with SDOF. Solution (6.16) is known as Duhamel's integral or the convolution integral, which is applicable to linear systems. At the same time, the particular solution (6.16) has the property that at $t=0$, it and its first derivative—over t turn to zero. At the action of a harmonic external force $P(\tau)=P\sin\theta\tau$, we obtain:

$$y=\frac{P}{m\omega}\int_0^t \sin\theta\tau\sin\omega(t-\tau)d\tau=\frac{P}{m\left(\omega^2-\theta^2\right)}\sin\theta t-\frac{\theta}{\omega}\sin\omega t \tag{6.17}$$

or at the stage of steady-state oscillations, we obtain the following expression:

$$y=\frac{P\sin\theta t}{m\cdot\omega^2\left(1-\frac{\theta^2}{\omega^2}\right)} \tag{6.18}$$

The general solution (6.15) follows a particular solution for a case of action of the disturbing force of a constant value and a force that is suddenly applied and suddenly ceases its action.

In general, a partial solution of Eq. (6.16) can be obtained by the method of variation of arbitrary Lagrange constants; however, the validity of this solution can be proved by substituting into the original differential equation (6.12). Let us differentiate (6.16):

$$
\begin{aligned}
\dot{y} &= \frac{1}{m}\int_0^t P(\tau)\ \cos\omega(t-\tau)d\tau \\
\ddot{y} &= -\ \frac{\omega}{m}\int_0^t P(\tau)\sin\omega(t-\tau)d\tau + \frac{P(t)}{m}
\end{aligned}
\tag{6.19}
$$

and, after substituting (6.19) into (6.12), we obtain:

$$
-\omega\int_0^t P(\tau)\ \sin\omega(t-\tau)d\tau + P(t) + \frac{k}{m\cdot\omega}\ \int_0^t P(\tau)\ \sin\omega(t-\tau)d\tau = P(t) \tag{6.20}
$$

The angular frequency is determined by the following formula:

$$
\omega = \sqrt{\frac{k}{m} - \left(\frac{c}{2m}\right)^2} \tag{6.21}
$$

In addition to the above, let us consider the case when a harmonic external load acts on an oscillatory system without friction under the initial conditions equal to zero. Then, the solution of the problem is found by formula (6.14), and assuming for simplicity that $t_0 = 0$, we obtain:

$$
\begin{aligned}
y &= \frac{1}{m\omega_0}\int_0^t \sin\omega_0(t-\tau)P\sin\omega(\tau)d\tau \\
&= \frac{P}{m\omega_0}\int_0^t \frac{1}{2}[\cos(\omega\tau-\omega_0(t-\tau)) - \cos(\omega\tau+\omega_0(t-\tau))]d\tau \\
&= \frac{P}{2m\omega_0}\left[\frac{\sin(\omega\tau-\omega_0(t-\tau)}{\omega+\omega_0} - \frac{\sin(\omega\tau+\omega_0(t-\tau)}{\omega-\omega_0}\right]_{\tau=0}^{t} \\
&= \frac{P}{2m\omega_0}\left[\frac{\sin\omega t}{\omega+\omega_0} - \frac{\sin\omega t}{\omega-\omega_0} - \frac{\sin(-\omega_0 t)}{\omega+\omega_0} + \frac{\sin\omega_0 t}{\omega-\omega_0}\right] \\
&= \frac{P}{m\omega_0(\omega+\omega_0)(\omega-\omega_0)}(\omega\sin\omega_0 t - \omega_0\sin\omega t).
\end{aligned}
$$

Let us assume that the frequency of an external force is close to the natural frequency of oscillations/vibrations of the system itself. Then let us transform the right part of the obtained formula:

$$
\begin{aligned}
y &= \frac{P}{m\omega_0(\omega+\omega_0)(\omega-\omega_0)}[\omega_0\sin\omega_0 t - \omega_0\sin\omega t + (\omega-\omega_0)\sin\omega_0 t\,] \\
&= \frac{P}{m(\omega+\omega_0)(\omega-\omega_0)}(\sin\omega_0 t - \sin\omega t) + \frac{P}{m\omega_0(\omega+\omega_0)}\sin\omega_0 t \\
&= \frac{2P}{m(\omega+\omega_0)(\omega-\omega_0)}\sin\frac{\omega_0-\omega}{2}t\cos\frac{\omega_0+\omega}{2}t + \frac{P}{m\omega_0(\omega+\omega_0)}\sin\omega_0 t
\end{aligned}
$$

Following the assumption, we can replace $\omega_0 + \omega$ with $2\omega_0$, and then obtain:

$$
y \approx \frac{-P}{m\omega_0(\omega-\omega_0)}\sin\frac{\omega-\omega_0}{2}t\cos\omega_0 t + \frac{P}{2m\omega_0^2}\sin\omega_0 t
$$

Here, the first summand is of the greatest interest and can be represented in the following form:

$$
\mathrm{Y(t)}\cos\omega_0 t
$$

where,

$$Y(t) = \frac{-P}{m\omega_0(\omega - \omega_0)} \sin \frac{\omega - \omega_0}{2} t$$

i.e. as the equation of harmonic vibrations with frequency ω_0 and a slowly changing amplitude. The amplitude varies from zero to

$$Y_0 = \max |Y(t)| = \frac{P}{m\omega_0(\omega - \omega_0)} \tag{6.22}$$

with period

$$T = \frac{2\pi}{|\omega - \omega_0|}. \tag{6.23}$$

The reader was familiarized with this type of oscillations with beats phenomenon, the vibration graph of which is shown in Fig. 5.4 of Chap. 5. They are obtained as a result of interference of forced and free oscillations/vibrations due to their proximity in frequency and amplitude. If the vibrating system has low damping, then the vibration process will begin with beats, then the free vibrations after some time will disappear and only forced vibrations will remain.

Let us consider the case when there is no friction in the vibrating system, and the frequency of forced vibrations coincides with the frequency of free vibrations of the system. Here, we will use formula (6.14), where for simplicity we take $t_0 = 0$:

$$\begin{aligned} y &= \frac{1}{m} \int_0^t \sin \omega_0 (t - \tau) \cdot P \sin \omega_0 \tau d\tau \\ &= \frac{P}{2m} \int_0^t [\cos \omega_0 (t - 2\tau) - \cos \omega_0 t] \, d\tau \\ &= \frac{P}{2m} \left[\frac{\sin \omega_0 (t - 2\tau)}{-2\omega_0} - \tau \cos \omega_0 t \right]_{\tau=0}^{t} \\ &= \frac{P}{2m\omega_0} \sin \omega_0 t - \frac{P}{2m} t \cos \omega_0 t. \end{aligned} \tag{6.24}$$

The first resulting summand represents free harmonic vibrations. The second summand represents the vibration, the amplitude of which tends to infinity over time according to the linear law. This is a very important phenomenon of resonance.

6.2 Reaction of a System with a Single Degree of Freedom (SDOF) Under Impulsive and Suddenly Applied Loads

Let us consider a special class of dynamic excitation: impulsive loads or shock loads. In most cases, impulsive loads by the nature of the application are close to concentrated load. The concentrated load that acts for a very short time but with a time integral that is finite is called an impulsive force or an impulsive load. Since force is a vector quantity, impulse is also a vector in the same direction. An impulsive load of a single action is a short-term load with constant direction and one predominant maximum value during a continuous action. If a load during the action has several comparable maximums, then its action can be considered as a sequence of actions of a number of independent impulses. The value of the maximum response of a system to an action of impulsive load depends on damping considerably less than for periodic, harmonic, or general vibrational loads. Significant contributions to the solution of general problems of vibrations of structures under impulsive loads were made by R.W. Clough, J. Penzien [1], Y.G. Panovko, N.N. Gubanova [2], E.S. Sorokin [3, 5], A.I. Tseitlin [4], A.K. Copra [6], A.M. Kabe., B.H. Sako [7], J.T. Katsikadelis [8], G.K Clein [9], L.M. J.W. Tedesco, W.G. McDougal, C. Allen Ross [10], R.R. Craig, A.J. Kurdila [11], P.L. Gatti [12], and others.

6.2.1 The Action of a Sudden Applied Force of Constant Value (Magnitude)

Let us consider the following: at the initial moment of time $t = 0$ a constant force $P(t) = P_0$ which is a Heaviside step function is applied to an elastic system in a state of rest, i.e., $y_0(0) = 0, \quad \dot{y}_0(0) = 0$. The equation of dynamic equilibrium without damping at $t > 0$ has the form:

$$\ddot{y} + \omega^2 y = \frac{P_0}{m} \tag{6.25}$$

The general solution of this equation, including free and forced oscillations of the system, has the following form:

$$\begin{aligned} y(t) &= C_1 \sin \omega t + C_2 \cos \omega t + \frac{P_0}{m\omega^2} \\ y(t) &= A \sin(\omega t + \varphi_0) + \frac{P_0}{m\omega^2}(1 - \cos \omega t) \end{aligned} \tag{6.26}$$

From the initial conditions, we obtain:

$$C_2 = -\frac{P_0}{m \cdot \omega^2}; \qquad C_1 = 0.$$

Then the final solution (6.26) will take the form:

$$y(t) = \frac{P_0}{m\omega^2}(1 - \cos \ \omega t) = y_{st} \ (1 - \cos \ \omega t) \tag{6.27}$$

Solution (6.26) can also be obtained from (6.16):

$$\begin{aligned} y &= \frac{1}{m\omega^2} \int_0^t P(\tau) \sin \omega(t-\tau) d\tau = \frac{P_0}{m\omega} \int_0^t \sin \omega(t-\tau) \ d\tau = \\ &= \frac{P_0}{m\omega^2} [\cos \omega(t-\tau)]_0^t = y_{st}(1 - \cos \omega t) \end{aligned} \tag{6.28}$$

The graph of this motion is shown in Fig. 6.2 [9]. The maximum value of the dynamic coefficient at forced oscillations will be:

$$\mu = y_{max}/y_{st} = 2$$

when $\omega t = \pi, \ 3\pi, \ \ldots$.

The displacement of the system varies from zero to $2y_{st}$—a double static displacement from the force P_0. Thus, when the load is applied instantaneously, the system's design parameters of oscillations are doubled compared to the static (i.e., rather slow) loading of the same load. Expression (6.27) can be given the following form:

$$y = 2y_{st} \sin^2 \frac{\omega t}{2} = 2y_{st} \sin^2 \frac{\pi t}{T} \tag{6.29}$$

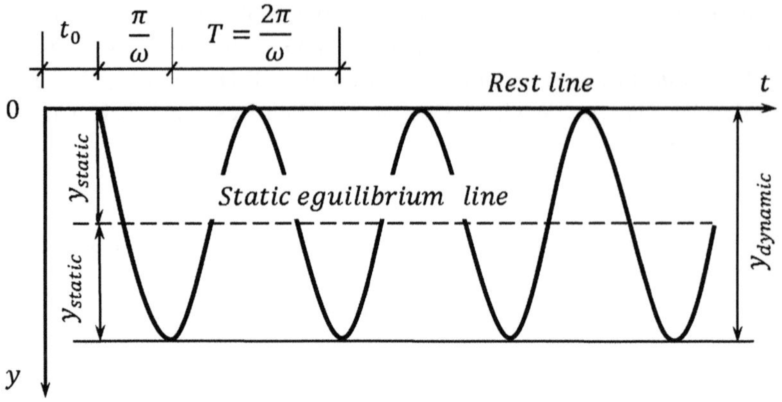

Fig. 6.2 A graph of oscillatory motion of the point of application of force P

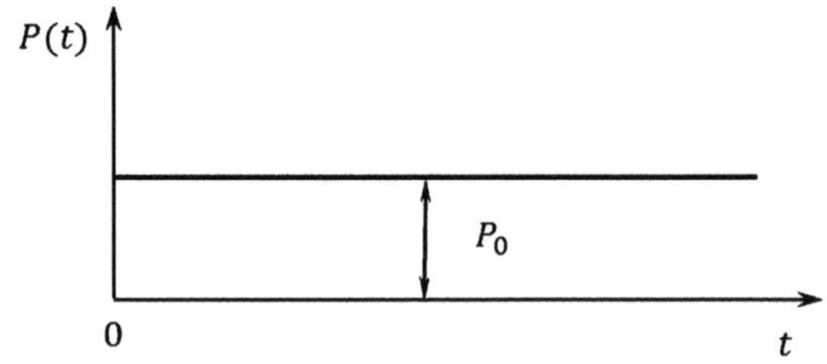

Fig. 6.3 A graph of a force action of a certain value suddenly applied to the system while maintaining its constant action

Problem 6.1
It is required to determine the dynamic coefficient under the action of an instantaneous applied load and then remaining constant at $t \leq 0,\ P(\tau) = 0$, at $t \geq 0,\ P(\tau) = P_0 =$ constant (Fig. 6.3).

Solution

1. To determine the dynamic coefficient without damping:

$$\begin{aligned} y &= \delta_{11}\,\omega \int_0^t P(\tau)\sin\omega(t-\tau)d\tau = \delta_{11}\,\omega P_0 \int_0^t \sin\omega(t-\tau)\,d\tau \\ &= \delta_{11}\,\omega P_0 \frac{1}{\omega}\cos\omega(t-\tau)\Big|_0^t = \\ &= y_{st}(1-\cos\omega t), \end{aligned}$$

   then

$$y = y_{st}(1-\cos\omega t) = 2y_{st}\sin^2\frac{\omega t}{2}.$$

   The oscillation graph is shown in Fig. 6.2. The dynamic coefficient $\mu = 2$.
2. To determine the dynamic coefficient with damping, let us use the Duhamel's integral (6.16) to determine the displacement function y.

$$y = \frac{1}{\omega m}P_0 \int_0^t e^{-\alpha(t-\tau)}\sin\omega(t-\tau)d\tau. \tag{A}$$

   Let us calculate the integral using the formula (I.S. Gradstain, and I.M. Ridjick) [13]

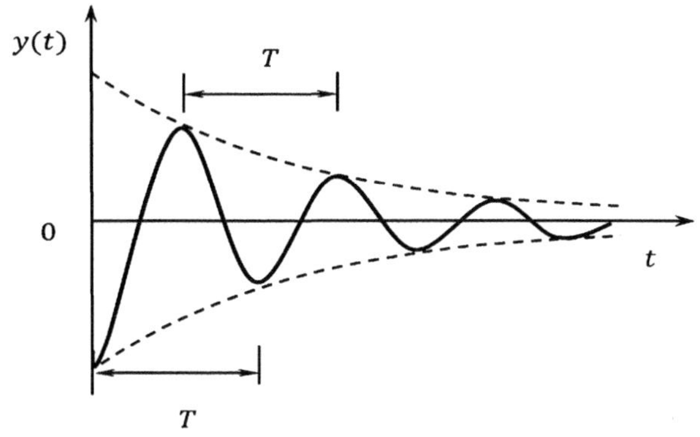

Fig. 6.4 A graph of damped oscillations

$$\int e^{ax}\sin bx dx = \frac{e^{ax}}{a^2+b^2}(a\sin bx - b\cos bx)$$

$$\int_0^t e^{-\alpha(t-\tau)}\sin\omega(t-\tau)d\tau = \frac{1}{\omega^2+\alpha^2}\left[\omega - e^{-\alpha t}(\alpha\sin\omega t + \omega\cos\omega t)\right]$$

Let us substitute the calculated integral into (A). The following expression will be obtained after transformations:

$$y = y_{st}\left[1 - e^{-\alpha t}\left(\frac{\alpha}{\omega}\sin\omega t + \cos\omega t\right)\right] \tag{B}$$

The graph of oscillations is shown in Fig. 6.4.

It should be noted that if in (B) we take $\alpha = 0$, then we obtain (6.29). To determine the maximum displacements, let us find the derivative of $\dot{y}$ for (B). Whence we find that the maximum displacements will be at $t_1 = 0.5T;\ 1.5T;\ 2.5T$. The minimum displacements will be at $t_2 = T,\ 2T,\ 3T,\ \ldots$
Substituting the found values of t_1 and t_2 in (B), we obtain:

$$y_{\max} = y_{st}\left[1 - e^{-\alpha t_1}\right]; \qquad y_{\min} = y_{st}\left[1 - e^{-\alpha t_2}\right]$$

The first maximum displacement will be at $t_1 = 0.5T = 0.5\frac{2\pi}{\omega} = \pi/\omega$

$$y_{\max} = y_{st}\left[1 + e^{-\frac{\pi\alpha}{\omega}}\right] = y_{st}\left(1 + e^{-0.5\varepsilon\pi}\right) = \mu y_{st}.$$

Then the dynamical coefficient will be:.

$$\mu = 1 + e^{\frac{-\pi\alpha}{\omega}} = 1 + e^{-0.5\varepsilon\pi}$$

Numerical values of the dynamic coefficient in the case of vibrations/oscillations of the structure in air will be: $\varepsilon = 004;\quad \mu = 1.94$, where ε is the coefficient of inelastic resistance for metal structures. The dynamic coefficient in the case of vibrations/oscillations of the structure in a damping medium will be, respectively: $\alpha = 0.5\omega;\quad \mu = 1.208$

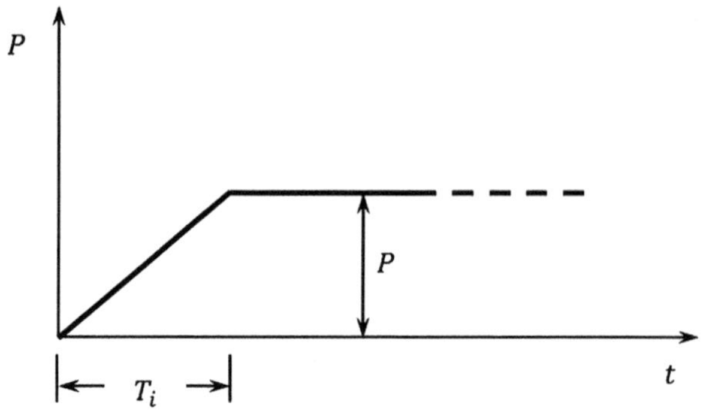

Fig. 6.5 Action of a force that increases linearly to a certain value and retains its maximum value

6.2.2 The Action of a Force Increasing According to the Linear Law

Let us consider the action of a force that increases in accordance with the linear law to the value of P and then remains constant, Fig. 6.5. The law of increasing force is as follows:

$$P(t) = \frac{P}{T_i}\ t, \tag{6.30}$$

where T_i is the time of load increase.

At $t \geq T_i$ the force becomes constant with the value P, and expression (6.16) takes the form:

$$y(t) = \frac{P}{m\omega T_i} \int_0^t \tau \sin \omega(t-\tau)\ d\tau \tag{6.31}$$

By integration in parts (6.31), we obtain:

$$y(t) = y_{st}\left(\frac{t}{T_i} - \frac{1}{2\pi}\frac{T_1}{T_i}\sin 2\pi \frac{t}{T_1}\right) \tag{6.32}$$

where T_1 is the period of system's natural oscillations.

At the moment of transition, that is, when $t = T_i$ we can calculate $y(T_i)$ and $\dot{y}(T_i)$. Furthermore, using (6.15) with new initial conditions, we obtain a solution for $t > T_i$. Figure 6.6 illustrates two solutions for $T_i > T_1$. In the case when T_1 is a multiple of T_i, the oscillations disappear at $t \geq T_i$.

Figure 6.6 clearly demonstrates the meaning of the expression: the action of a statically applied force. When the ratio of the period of time T_i of rising force to the period of natural oscillations T_1, the oscillations become negligible.

The dynamic coefficient after the increase of the load takes the following form:

$$\mu = 1 - \frac{\sin \omega t}{\omega T_i} + \frac{\sin \omega(t - T_i)}{\omega T_i}, \tag{6.33}$$

The value of the dynamic coefficient depends on the rate of increase of the load. For $T_i \rightarrow 0$, from (6.33), we obtain:

$$\mu = 1 - \cos \omega t \tag{6.34}$$

When $T_i < T_1$, there is a “dynamic effect,” i.e., the dynamic coefficient increases with a decrease in the period of time of load increase T_i.

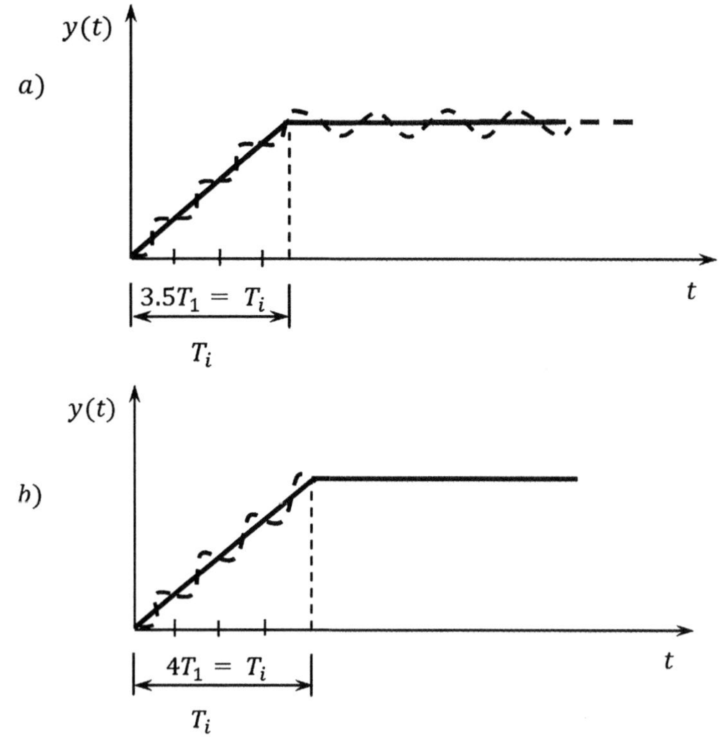

Fig. 6.6 (**a**, **b**) Two solutions for $T_i > T_1$

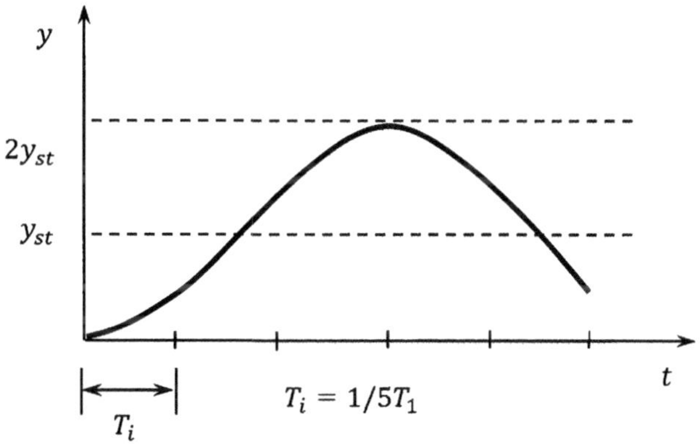

Fig. 6.7 A graph of oscillatory motion

Figure 6.7 shows a graph of oscillatory motion for the case $T_i = 1/5T_1$. The highest value of the dynamic coefficient reaches 1.9. The dynamic coefficient in the case of increasing of the load according to the linear law does not depend on the rate of load increase and over time tends to be one.

The dynamic coefficient is determined by the formula:

$$\mu = 1 - \frac{\sin \omega t}{\omega t} = 1 - \frac{\sin 2\pi \frac{t}{T_1}}{2\pi \frac{t}{T_1}}, \tag{6.35}$$

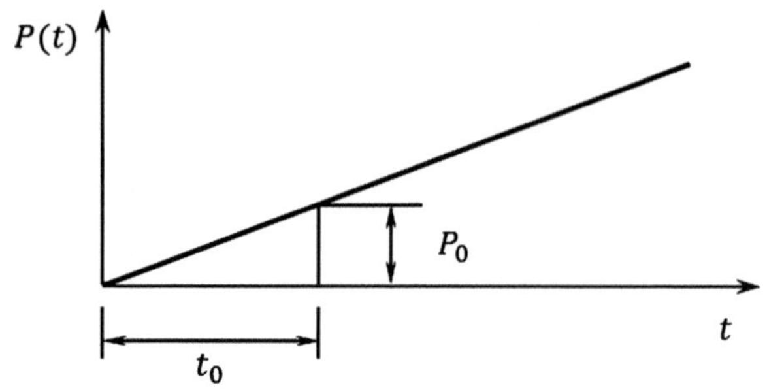

Fig. 6.8 Action of a force that increases linearly

Problem 6.2
It is required to perform a dynamic calculation on the load, which changes according to the linear law (Fig. 6.8).

At $t \leq 0$, $P(t) = 0$, and at $t > 0$, $P(t) = P_0 t/t_0$.

Solution
Let us use the Duhamel's integral (6.16), then we get:

$$y = \delta_{11}\,\omega \int_0^t P(\tau)\sin\omega(t-\tau)d\tau = \delta_{11}\,\omega \frac{P_0}{t_0}\int_0^t \tau \sin\,\omega(t-\tau)d\tau \tag{C}$$

The integral

$$\int_0^t \tau \sin\ \omega(t-\tau)d\tau \tag{D}$$

is calculated using integration by parts ($\int udv = uv - \int vdu$). As a result, we will obtain the following expression:

$$\int_0^t \tau \sin\,\omega(t-\tau)d\tau = \frac{t}{\omega}\left(1 - \frac{\sin\omega t}{\omega t}\right) \tag{E}$$

When calculating the integral (D), it was assumed that: $u = \tau,\ dv = \sin\,\omega(t-\tau)d\tau$. The expression (E) is substituted in (C), we get:

$$y = \delta_{11}\,\omega \frac{P_0}{t_0}\,\frac{t}{\omega}\left(1 - \frac{\sin\omega t}{\omega t}\right) = \delta_{11}\,P(t)\left(1 - \frac{\sin\omega t}{\omega t}\right). \tag{F}$$

For each moment of time, the dynamical coefficient μ is equal to:

$$\mu = 1 - \frac{\sin\omega t}{\omega t} \tag{G}$$

The graph of displacements which is constructed by formula (F) is shown in Fig. 6.9a.

The graph of the dynamic coefficient, which is based on the formula (G), is shown in Fig. 6.9b.

The maximum values of μ for each of the time intervals will be: at $t = \frac{3}{4}\mathrm{T} - 1.21$, at $t = \frac{7}{4}\mathrm{T} - 1.09$, at $t = \frac{11}{4}\mathrm{T} - 1.058$, at $t = \frac{15}{4}\mathrm{T} - 1.043$, at $t = \frac{19}{4}\mathrm{T} - 1.033$, at $t = \frac{39}{4}\mathrm{T} - 1.016$.

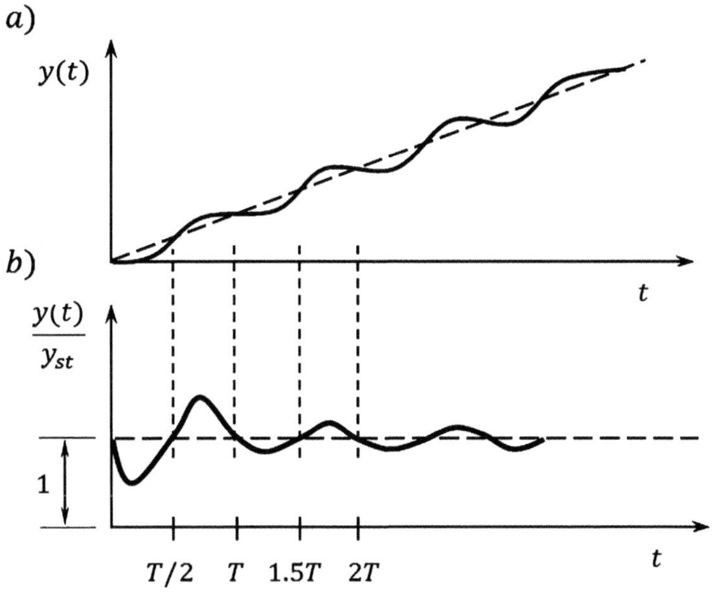

Fig. 6.9 (**a**) A graph of displacements; (**b**) a graph of dynamic coefficient

Table 6.1 The values of the dynamic coefficient μ

T_p/T	0	0.2	0.4	0.5	0.75	1	1.5	2
μ	0	0.66	1.05	1.2	1.42	1.55	1.69	1.76

6.2.3 The Action of a Suddenly Applied Force of Constant Value and a Suddenly Terminated Action

It is assumed that the force P is applied suddenly at time t_0 and remains a constant value during a period T_p. For a period $t_0 < t < t_0 + T_p$, the previous solution is valid. During the period $t > t_0 + T_p$, i.e., after the termination of the force P, the system will make free oscillations in accordance with the first term of expression (6.26). At the same time, the initial conditions of this movement $t = t_0 + T_p$ coincide with the conditions of movement at the end of the previous period. Thus, the initial conditions of this motion for this case allow us to determine the amplitude of oscillations and the dynamic coefficient of force P.

$$\mu = \; = \pm \; 2\sin\frac{\omega T_p}{2} = \pm \; 2\sin\frac{\pi T_p}{T} \tag{6.36}$$

The value of this coefficient depends on the ratio between the period T_p of the load and the period T of free oscillations of the system. Table 6.1 shows the values of the dynamic coefficient μ depending on the value of T_p/T. R W. Clough, J. Penzien [1].

Regardless of the law of change of the disturbing force $P = P(t)$, if the duration T_p of its action is very small compared to the period of free oscillations of the system, then at $t > T_p$

$$y \approx \frac{S}{m\omega}\sin\omega t \tag{6.37}$$

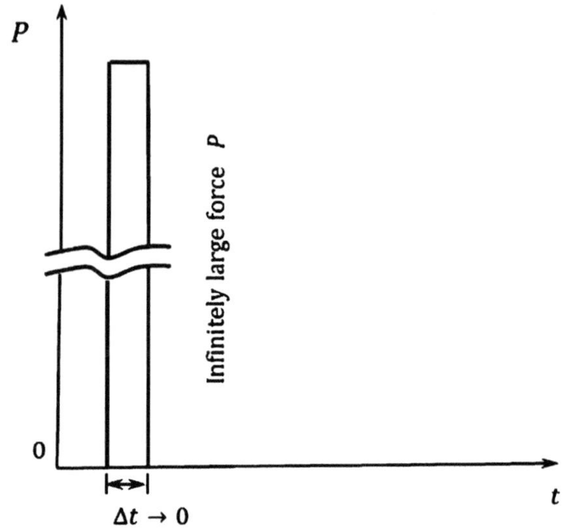

Fig. 6.10 The impulse (momentum) of force

By multiplying and dividing the right part of the expression for the dynamic coefficient by $\frac{\omega T_p}{2}$ we obtain:

$$\mu = \pm \omega T_p \frac{\sin \frac{\omega T_p}{2}}{\frac{\omega T_p}{2}} \tag{6.38}$$

6.2.4 Short-Term Impulse (Momentum) of Force

Duration of a short-term load can vary within very wide limits. In order to take into account the effects of a short-term high-intensity load (Fig. 6.10) we should consider the following: If the impulse (momentum) of the concentrated force $P(t)$ has a value of S, and acts for a small period of time (period) T_p^0, then the average value of the force for this period will be

$$P = \frac{S}{T_p^0} \tag{6.39}$$

Let us substitute the value of the force P in formula (6.29), we obtain:

$$y = \frac{2S}{\omega^2 m T_p^0} \sin^2 \frac{\omega t}{2}. \tag{6.40}$$

At $t > T_p$, i.e., after the termination of the force P, at any time the displacement is determined by the formula:

$$\begin{aligned} y &= \frac{1}{m\omega^2} \int_0^t P(\tau) \sin \omega(t-\tau) d\tau = \frac{1}{m\omega^2} \int_0^{T_p} P \sin \omega(t-\tau) d\tau + \\ &+ \int_{T_p}^t 0 \cdot \sin \omega(t-\tau) d\tau = \frac{P}{m\omega^2} \left[\cos \omega(t-T_p) - \cos \omega t\right] \\ &= \frac{2P}{m\omega^2} \sin \frac{\omega T_p}{2} \sin \omega \left(t - \frac{T_p}{2}\right) = 2y_{st} \sin \frac{\omega T_p}{2} \sin \omega \left(t - \frac{T_p}{2}\right) \end{aligned} \tag{6.41}$$

This solution describes free oscillations of the system with frequency ω and amplitude:

$$y = A = 2y_{st}\sin\frac{\omega T_p}{2} = 2y_{st}\sin\frac{\pi T_p}{T}. \tag{6.42}$$

Note that for $t = T_p$, i.e., for the moment of termination of the force, formula (6.41) coincides with expression (6.27).

If $\frac{\omega T_p}{2} = \frac{\pi}{6}$, i.e., the duration of action of the force is equal to 1/6 of the period of free oscillations, then the effect of the force is equal to the static effect.

The dynamic effect of the force begins to exceed its static effect when the load is applied for more than 1/6 of duration of the natural oscillation period. The greatest dynamic effect takes place when the duration of action of the force $T_p = \pi/\omega$, is equal to half of the period. The greatest displacement of the system is achieved after termination (cessation) of the action of the force at $t > T_p$.

$$y_{max} = \frac{2P}{m\omega^2}\sin\frac{\omega T_p}{2} = 2y_{st}\sin\frac{\omega T_p}{2} = 2\sin\frac{\pi T_p}{T} \tag{6.43}$$

hence, the dynamic coefficient is:

$$\mu = 2\sin\frac{\omega T_p}{T} = 2\sin\frac{\pi T_p}{T} \tag{6.44}$$

The impulse is the product of an infinitely large force P and an infinitely small time of its action Δt, and in this case, the magnitude of the impulse itself is finite. The momentum is equal to the amount of motion of the mass at the initial moment of time $m\dot{y}(0)$.

In formula (6.36), to determine the dynamic coefficient we take its average value P, instead of the force $P(t)$. Both sides of expression (6.44) are multiplied by P and, in addition, the right side is multiplied and simultaneously divided by $\frac{\omega T_p^0}{2}$. Then on the left side, we obtain a static force equivalent to the given impulse in its action [9]:

$$\begin{aligned} P_{\text{equivalent}} &= \mu P = \pm\ 2P\ \sin\frac{\omega T_p^0}{2}\cdot\frac{\sin\frac{\omega T_p^0}{2}}{\frac{\omega T_p^0}{2}} = \pm P\omega T_p^0\frac{\sin\frac{\omega T_p^0}{2}}{\frac{\omega T_p^0}{2}} \\ &= \pm\omega S\frac{\sin\frac{\omega T_p^0}{2}}{\frac{\omega T_p^0}{2}} \end{aligned} \tag{6.45}$$

Since for any angle α, the ratio $\frac{\sin\alpha}{\alpha} < 1$ and this ratio tends to one when $\alpha \to 0$ tends to zero, then the most dangerous of all shock impulses that have a given the value S is an instantaneous impulse for which.

$$P_{\text{equivalent}} = \pm\omega S \tag{6.46}$$

6.2.5 The Action of Periodic Instantaneous Impulses

In some cases, the impulsive load can be multiple, i.e., its action may be represented in the form of sequential application of loads at equal intervals as shown in Fig. 6.11.

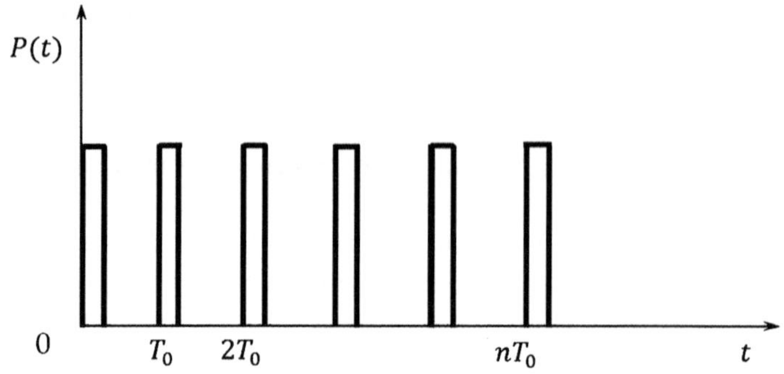

Fig. 6.11 Periodic instantaneous impulses

Let us consider the problem of action of periodic equal instantaneous impulses S_0 on an oscillatory elastic system. It is known that the most dangerous of all impulses is an instantaneous impulse. The duration of the action of such impulse is assumed to be zero, but the impulse S_0 is considered to be finite. The oscillations caused by the instantaneous impulse are determined by the initial conditions of the problem posed:

$$y(0)=0, \qquad \dot{y}(0)=\frac{S_0}{m} \tag{6.47}$$

The dynamic displacement of the system

$$y=\frac{S_0}{m\omega}\sin\omega t \tag{6.48}$$

occurs with its own frequency ω. The largest deviation will be:

$$y_{\max}=\frac{S_0}{m\omega}=\frac{S_0}{\sqrt{km}} \tag{6.49}$$

Let T be the period of application of impulses, then $t = 0,\ T_0,\ 2T_0,\ \ldots,$ are the moments of application of the first, second, and so on impulses.

Let us consider the most general case of an action on an oscillatory system with single degree of freedom of an arbitrary disturbing force *P*(*t*). As indicated, the equation of motion:

$$\ddot{y}+\omega^2 y=\frac{1}{m}P(t) \tag{6.50}$$

with zero initial conditions has the following solution:

$$y=\frac{1}{m\omega}\int_0^t P(\tau)\sin\omega(t-\tau)d\tau \tag{6.51}$$

or

$$y=A\ \sin\omega t+B\ \cos\ \omega t, \tag{6.52}$$

where

$$A=\frac{1}{m\omega}\int_0^t P(\tau)\cos\ \omega\tau\ d\tau \tag{6.53}$$

$$B = -\frac{1}{m\omega}\int_0^t P(\tau)\sin\omega\tau d\tau \tag{6.54}$$

If at the moments of time $t = 0,\ T_0,\ 2T_0, \ldots$ the same impulses S_0 are applied to the system, then for the time interval $[n\ T_0,\quad (n+1)\ T_0]$ using expressions (6.53) and (6.54), we find:

$$A = \frac{S_0}{m\omega}[1 + \cos\omega T_0 + \cos 2\omega T_0 + \cdots + \cos nT_0] \tag{6.55}$$

$$B = -\frac{S_0}{m\omega}[\sin\omega T_0 + \sin 2\omega T_0 + \cdots + \sin nT_0] \tag{6.56}$$

The sum of expression (6.54) is multiplied by $2\sin\frac{\omega T_0}{2}$

$$2\sin\frac{\omega T_0}{2}\sin\omega T_0 = \cos\frac{\omega T_0}{2} - \cos\frac{3\omega T_0}{2},$$
$$2\sin\frac{\omega T_0}{2}\sin 2\omega T_0 = \cos\frac{3\omega T_0}{2} - \cos\frac{5\omega T_0}{2},$$
$$\cdots\cdots\cdots\cdots\cdots\cdots\cdots\cdots\cdots\cdots$$
$$2\sin\frac{\omega T_0}{2}\sin n\omega T_0 = \cos\frac{2n-1}{2}\omega T_0 - \cos\frac{2n+1}{2}\omega T_0,$$

and sum all these equations. As a result, the expression in the square brackets (6.56) will take the form:

$$\frac{\cos\frac{\omega T_0}{2} - \cos\frac{2n+1}{2}\omega T_0}{2\sin\frac{\omega T_0}{2}} = \frac{\sin\frac{n+1}{2}\omega T_0 \cdot \sin\frac{n}{2}\omega T_0}{\sin\frac{\omega T_0}{2}}.$$

Let us similarly transform the sum in the square brackets of formula (6.55):

$$\frac{\sin\frac{2n+1}{2}\omega T_0 + \sin\frac{\omega T_0}{2}}{2\sin\frac{\omega T_0}{2}}.$$

Now the expressions for A and B will take the form:

$$A = \frac{S_0}{m\omega}\cdot\frac{\sin\frac{2n+1}{2}\omega T_0 + \sin\frac{\omega T_0}{2}}{2\sin\frac{\omega T_0}{2}}; \tag{6.57}$$

$$B = -\frac{S_0}{m\omega}\cdot\frac{\sin\frac{n+1}{2}\omega T_0 \cdot \sin\frac{n}{2}\omega T_0}{\sin\frac{\omega T_0}{2}}. \tag{6.58}$$

The displacements of the system on any interval are determined through the constants A and B using formulas (6.57) and (6.58) and then the solution in (6.32) can be used.

Note, however, that the resulting solution is not periodic and varies from one period to another. Therefore, for a complete description of the motion, it is necessary to repeat the calculations for each new period. Other options for solving this problem are given in the book Y.G. Panovko [2].

The equations of motion (6.2) for a system of a single degree of freedom and viscous elastic properties, described by the frequency-independent model of internal friction (E.S. Sorokin [3], and A.I. Tseitlin [4] Fig. 6.1), are presented in the form:

$$m\ddot{y} + \gamma\sqrt{km}\,\dot{y} + ky = P(t) \tag{6.59}$$

Here: $m\ddot{y}$ is the force of inertia; $\gamma\sqrt{km}\dot{y}$ is the internal friction force; ky is the elastic force of the system.

Let us reduce Eq. (6.59) to the characteristic form:

$$\ddot{y} + \gamma\omega_0\dot{y} + \omega_0^2 y = m^{-1}P(t) \tag{6.60}$$

where the following notation is introduced: $\omega_0 = \sqrt{\frac{k}{m}}$.

The solution of Eq. (6.60) for free vibrations, i.e., when $P(t) = 0$ has the following form:

$$y(t) = Ae^{\left(-\frac{\gamma\omega_0}{2}+i\omega\right)t} + Be^{\left(-\frac{\gamma\omega_0}{2}+i\omega\right)t}, \tag{6.61}$$

where

$$\omega = \omega_0\sqrt{1-\frac{\gamma^2}{4}} \tag{6.62}$$

Assuming further for a single instantaneous impulse under the initial conditions equal to zero

$$y|_{t=t_0} = Y_0 = 0; \qquad v = \frac{dy}{dt}\bigg|_{t=t_0} = \dot{Y}_0 = \frac{S_0}{m} = \frac{1}{m}. \tag{6.63}$$

Here S_0 is an initial single instantaneous impulse, and Y_0 is un initial displacement

Let us find the response of the system, i.e., its displacement, which is called the impulse transition function:

$$y(t) = H(t) = \frac{1}{m\omega}\, e^{-\frac{\gamma\omega_0 t}{2}} \sin \omega t. \tag{6.64}$$

The graph of the impulse transition function describes a damped process with frequency ω and logarithmic decrement of oscillations /vibrations:

$$\delta = \ln \frac{H\left(\frac{n\pi}{2\omega}\right)}{H\left[\frac{(n+1)\pi}{2\omega}\right]} = \frac{\pi\gamma}{\sqrt{1-\frac{\gamma}{4}}} \approx \pi\gamma. \tag{6.65}$$

The values of the loss coefficients for various building materials are given in Table 6.2. It follows from Table 6.2, that t $\gamma \leq 1$ and, therefore, the value of $\gamma^2/4$ from expression (6.62) can be neglected compared to unity, i.e., we can assume that $\delta = \pi\gamma$ and $\omega = \omega_0$. Since ω_0 is the natural frequency of an oscillating/vibrating system without internal friction at $\gamma = 0$, then a system with single degree of freedom with weak damping performs natural oscillations with a frequency that corresponds to an undamped system. It follows from the formula:

$$\omega_0 = \sqrt{\frac{k}{m}}, \tag{6.66}$$

that the frequency of natural oscillations/vibrations is higher the greater the rigidity of the system or the smaller its mass, and vice versa.

Values of the loss coefficients for various building materials are given in Table 6.2 [14]

Maximum displacement of the system according to (6.64) is obtained in the first quarter of the period and it is equal to:

Table 6.2 Values of loss coefficients for different building materials

Type of structures	Loss coefficient/factor	
	Vibration level	Strength
Reinforced concrete structures	0.05	0.1
Prestressed reinforced concrete structures	0.025	0.05
Metal structures	0.01	0.025
Large-span metallic coverings	0.007	0.02
Aluminum structures	0.007	0.025
Brick masonry	0.025	0.04
Wooden structures	0.03	0.05
One-story & multi-story buildings	0.05	0.1
Metal towers, and masts	0.02	0.04

$$H_{\max} = \frac{1}{m\omega} e^{-\frac{\pi\omega_0}{4\omega}} \approx \frac{1}{m\omega_0}. \tag{6.67}$$

Maximum displacement of the system under the action of an instantaneous impulse of magnitude S_0 will be:

$$y_{\max} = \frac{S_0}{m\omega_0} = \frac{S_0}{\sqrt{km}} \tag{6.68}$$

If an instantaneous impulse of magnitude S_0 acts on the system, it follows from expression (6.68) that the displacement of the system under the action of impulsive or impact load can be reduced by increasing the stiffness and mass of the system. According to formula (6.68), the maximum elastic force of the system under the action of impulse is equal to:

$$K_{\max} = ky_{\max} = \frac{S_0 k}{\sqrt{km}} = S_0\omega_0 \tag{6.69}$$

Here ω_0 is the natural frequency of the system without internal friction, i.e., at $\gamma = 0$. It follows from formula (6.69) that when calculating the strength, it should be borne in mind that an increase in the rigidity of the structure leads to an increase in forces. An increase in the mass of the structure and a decrease in its stiffness for impulsive and impact loads leads to a decrease in the forces and, consequently, stresses in the structural elements. Any dynamic load can be represented as a sequence of instantaneous impulses.

$$P(t) = \int_0^t P(\tau)\,\delta(t-\tau)d\tau \tag{6.70}$$

where $\delta(t - \tau)$ is a delta function that describes the instantaneous momentum. The general solution of the equation of motion of the system (6.59) under initial conditions equal to zero as the following form:

$$y(t) = \frac{1}{m\omega} \int_0^t P(\tau)\,e^{\frac{-\gamma\omega_0(t-\tau)}{2}} \sin\omega(t-\tau)d\tau \tag{6.71}$$

At $\gamma = 2$ the frequency (6.62) equals to zero and the system does not oscillate/vibrate. This value of the loss coefficient and the corresponding value of the relative damping parameter are called critical $\mu = \gamma/2 = 1$.

Another important characteristic of the system is the transfer function. The transfer function is defined as the solution of the equation of motion (6.59) under the influence of a unit of harmonic force:

$$P(t) = e^{i\omega t} \tag{6.72}$$

Let us imagine that the oscillations occur in accordance with

$$y(t) = \Phi(\omega) \cdot e^{i\omega t}, \tag{6.73}$$

where $\Phi(\omega)$ is the transfer function, we obtain:

$$\Phi(\omega) = A(\omega)e^{-\varphi(\omega)} = \frac{1}{m\omega_0^2\sqrt{(1-\xi^2)+\gamma^2\xi^2}} e^{-i\frac{\gamma\xi}{1-\xi^2}}. \tag{6.74}$$

Here $\xi = \omega/\omega_0$, $A(\omega)$ is the amplitude–frequency characteristic of the system, $\varphi(\omega)$ is the phase-frequency characteristic of the system. The graph of the amplitude response of a system with single degree of freedom is shown in Fig. 6.12.

Analysis of the graph shows that the amplitude of harmonic oscillations/vibrations of a system with a single degree of freedom increases sharply only in the resonance zone. In the pre-resonance zone, the amplitude differs little from the static displacement of the system, which is caused by a unit force. The amplitude of oscillations/vibrations in the trans-resonant domain decreases rapidly with increasing frequency. The oscillation phase changes sign at the resonant frequency.

$$\varphi = \frac{\gamma\xi}{1-\xi^2} \tag{6.75}$$

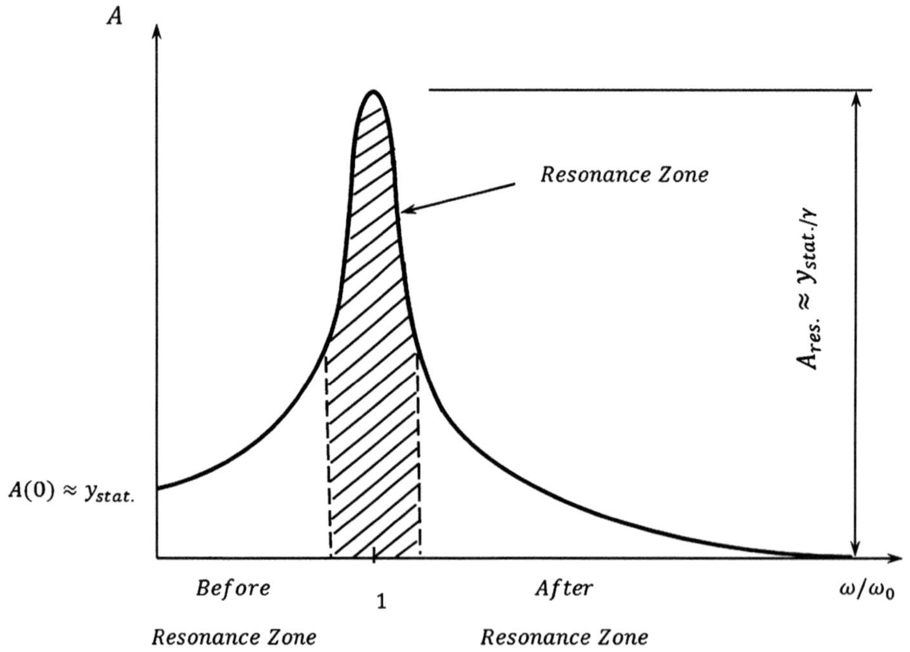

Fig. 6.12 Amplitude characteristics of a system with a single degree of freedom

The oscillations/vibrations occur practically in the same phase with the excitation force until resonance occurs. In the trans-resonant zone, displacement and external force have opposite signs. Maximum value of the amplitude

$$A_{\text{res.}} = \frac{1}{k\gamma} = \frac{y_{\text{stat.}}}{\gamma} \tag{6.76}$$

corresponds to the resonant frequency.

$$\omega_{\text{res}} = \omega_0\sqrt{1 - \frac{\gamma^2}{4}} \approx \omega_0 \tag{6.77}$$

Hence it follows that the internal friction parameter, i.e., the loss coefficient is one of the most important characteristics of the system. The loss coefficient, together with the static displacement completely, determines the amplitude of displacements in the most dangerous resonance region of vibrations /oscillations.

The general solution of Eq. (6.59) for any load $P(t)$ can be found by first determining the transfer function of the system.

From (6.74), it follows that the transfer function of a conservative system at $\gamma = 0$ has the form:

$$\Phi_k(\omega) = \frac{1}{k\left(1 - \xi^2\right)} = \frac{1}{m\left(\omega_0^2 - \omega^2\right)}. \tag{6.78}$$

Expression (6.74) must be replaced by Eq. (6.78) in order to move from the transfer function of the conservative system to the transfer function of the dissipative system: The amplitude of the oscillations of a dissipative system is as follows:

$$A_k = \frac{1}{m\sqrt{\left(\omega_0^2 - \omega^2\right)^2 + \gamma^2\omega^2}} \tag{6.79}$$

Problem 6.3

Force P is suddenly applied to a beam with mass m, and after a some time τ, the same force P is added (Fig. 6.13). It is required to describe the process of beam vibrations after the application of the second force at $\tau = T$ and $\tau = T/2$ (T is the period of free oscillations).

Solution

Let us write the Duhamel integral for the given dynamic load $P(t) = P$ at $0 \leq t \leq \tau$:

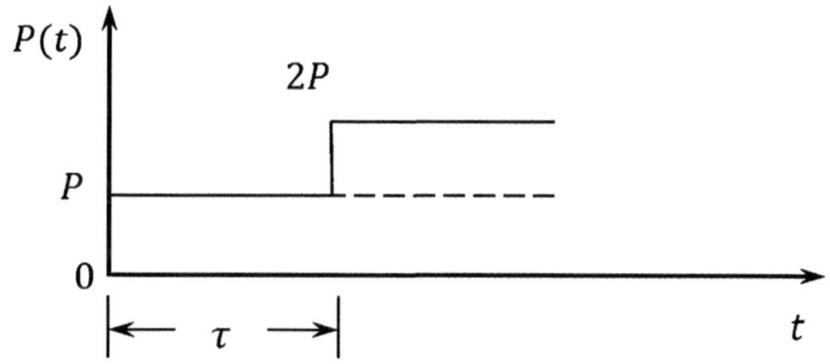

Fig. 6.13 A force action of certain value applied to the system suddenly and then the same force added again after a period of time

$$y_{\text{dyn}} = P\delta_{11}\omega \int_0^t \sin \omega(t-\tau)d\tau.$$

Let us calculate the integral and find y_{dyn}.

$$y_{\text{dyn}} = y_{\text{st}}(1 - \cos \omega t)$$

When $<\tau$.

$$y_1(t) = y_{\text{st}}(1 - \cos \omega t)$$

When $>\tau$, the vibration/oscillations from the second force are superimposed on the oscillations from the first force:

$$\begin{aligned} y_2(t) &= y_{\text{st}}(1 - \cos \omega t) + y_{\text{st}}[1 - \cos \omega(t-\tau)] \\ &= y_{\text{st}}[2 - \cos \omega(t-\tau) - \cos \omega t], \end{aligned}$$

or

$$y_2(t) = [2 - \cos \omega t \cos \omega\tau - \sin \omega t \sin \omega\tau - \cos \omega t].$$

1. At $\tau = T = 2\pi/\omega$; $y_2(t) = 2\ y_{\text{st}}(1 - \cos \omega t)$.
2. At $\tau = T/2 = \pi/\omega$; $y_2(t) = 2\ y_{\text{st}}$, i.e., there are no oscillations in this case.

Problem 6.4

It is required to investigate the effect of damping in a system with single degree of freedom under the action of a dynamic load. According to the condition of the problem, the dynamic load acts suddenly and remains constant for a long period of time: $P(t) = P$ at $t > 0$. Figure 6.14.

Solution

When damping is taken into account, the displacement is determined by the Duhamel integral. The expression for the acting load is substituted into the Duhamel integral. As a result, we obtain:

$$y(t) = \frac{P}{m\overline{\omega}} \int_0^t \mathrm{e}^{-n(t-\tau)} \sin \overline{\omega}(t-\tau)d\tau = y_{st}\left[1 - \mathrm{e}^{-nt}\left(\cos \omega t + \frac{n}{\overline{\omega}} \sin \overline{\omega} t\right)\right],$$

where n is damping coefficient; ω is the frequency of natural oscillations with damping $\overline{\omega} = \sqrt{\omega^2 - n^2}$.

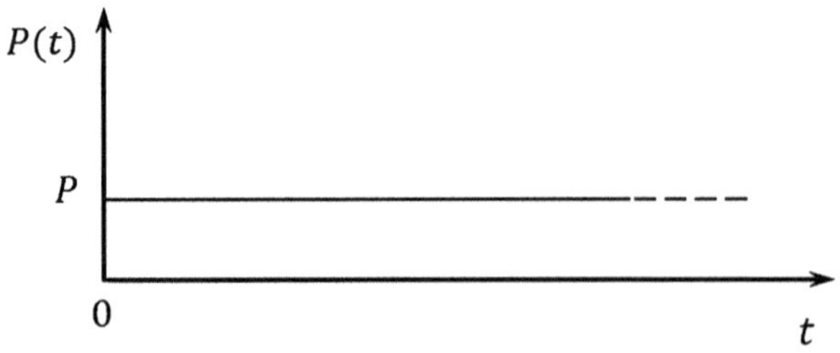

Fig. 6.14 A force action of a certain value applied to the system suddenly and remaining constant for a long period

Note that the expression for displacements in the absence of damping has the following form:

$$y(t) = y_{st}(1 - \cos \omega t).$$

Since ≪1, it can be assumed that the frequencies with and without damping are equal, i.e., $\overline{\omega} = \omega$. Then let us assume that $n = \gamma\omega/2$, where γ is the coefficient of inelastic resistance of material. Then the expression for displacements, with damping, will take the form:

$$y(t) = y_{st}\left[1 - \mathrm{e}^{-nt}\left(\cos \omega t + \frac{\gamma}{2} \sin \omega t\right)\right].$$

Assume that the duration of the load time is equal to half of the vibration/oscillation period:

$$t = \frac{T}{2} = \frac{\pi}{\omega}.$$

Then for the case without damping from (6.27), we find:

$$y(t) = y_{\mathrm{st}}(1 - \cos \pi) = 2y_{\mathrm{st}}$$

Displacements with damping:

1. If $\gamma = 0.05$; then $y = y_{\mathrm{st}}(1 - e^{-0.0763} \cos \pi) = 1.926y_{\mathrm{st}}$
2. If $\gamma = 0.1$; then $y = y_{\mathrm{st}}(1 - e^{-0.157} \cos \pi) = 1.854y_{\mathrm{st}}$

Problem 6.5

Instant impulse S is suddenly applied to a truss with the mass m, concentrated in the middle of the span (Fig. 6.15a). It is required to find the maximum displacement of the mass of the truss (Fig. 6.15a). It is given that the height of the truss f is equal to half the span l of the truss.

Solution

The maximum displacement of the mass: $y = S\omega\delta_{11}$. The frequencies of free vibrations are determined from the expression $\omega = \left(\sqrt{m \cdot \delta_{11}}\right)^{-1}$, therefore:

$$y_{\max} = S\sqrt{\frac{\delta_{11}}{m}}$$

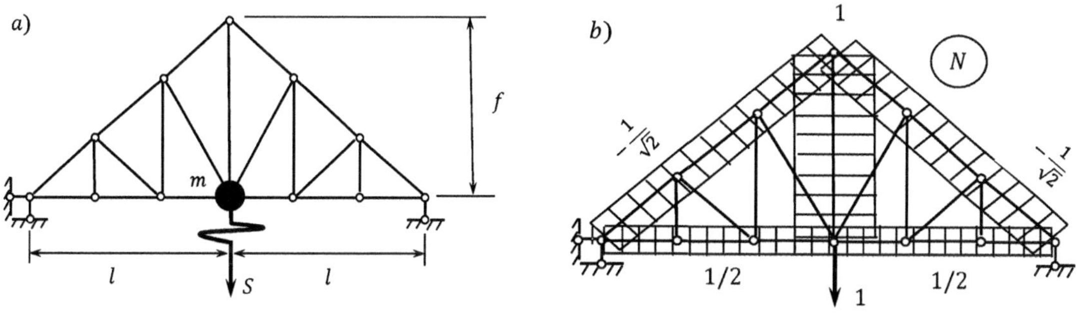

Fig. 6.15 (**a**) Double Hove truss with lumped mass in the middle of the span; (**b**) Longitudinal forces

At the point where the concentrated mass is located, we apply a unit force. We are building a diagram of longitudinal forces to find a unit displacement in the middle of the span of the truss. Figure 6.15b.

$$\delta_{11} = \sum_i \frac{N_i^2}{E_i A_i}$$

Let us calculate the unit displacement:

$$\delta_{11} = \frac{1}{EA}\left(1 \cdot 1 \cdot 1 + 2\sqrt{2} \cdot l \frac{1}{\sqrt{2}} + 2 \cdot l \frac{1}{2} \cdot \frac{1}{2}\right) = \frac{1.5 + \sqrt{2}}{EA}$$

The expression for the maximum displacement will be as follows:

$$y_{\max} = S\sqrt{\frac{1.5 + \sqrt{2}}{EA}}$$

References

1. Clough, R. W., & Penzien, J. (1993). *Dynamics of structures* (2nd ed., 738p). McGraw-Hill College.
2. Panovko, Y. G., & Gubanova, N. N. (1964). *Stability and oscillations of elastic systems* (335p). Fizmatgiz.
3. Sorokin E. S. (1983). *On the applied method of dynamic calculation of liner elastic dissipative systems*. MIET. Issues.720, pp. 51–60.
4. Tseitlin, A. I. (1978). About the linear models of the frequency-independent internal friction. *Izvestia AS USSR, MTT, 3*, 18–28.
5. Sorokin, E. S. (1977). Calculation of structures under the action of impulsive loads. In *Dynamics of structures* (pp. 93–123). Stroyizdat.
6. Copra, A. K. (2016). *Dynamics of Structures* (992 p). Pearson.
7. Kabe, A. M., & Sako, B. H. (2020, June 17). *Structural dynamics fundamentals and advanced applications* (Vol. I, 1st ed., 970p).
8. Katsikadelis, J. T. (2020, June 23). *Dynamic analysis of structures* (1st ed., 922p). Academic.
9. Clein, G. K. (1972). Fundamentals of the dynamics of rod systems. In *Manual for the course of structural mechanics* (pp. 124–219).
10. Tedesco, J. W., McDougal, W. G., & Allen, R. C. (1998). *Structural dynamics theory and applications* (1st ed., 832p). Pearson.
11. Craig, R. R., & Kurdila, A. J. (2006). *Fundamentals of structural dynamics* (735p). Wiley.
12. Gatti, P. L. (2014). *Applied structural and mechanical vibrations: Theory and methods* (668p). CRC Press.
13. Gradstein, I. S., & Ridjick, L. M. (1991). *Tables of integrals, sums, series and products* (1100p). Fizmatgiz.
14. Vainer, D., & Tseitlin, A. I. (1994). *Vibration damage in industrial construction* (336p). Swedish Council for Research in Construction Buildings Protection Center.

7 Impact Loading

Abstract

This chapter is devoted to impact loading. An impact occurs when two or more bodies collide. The impact could be caused by a weight falling on a designated object. An important characteristic of an impact load is the generation of relatively large forces at points of contact for relatively short periods of time. An impact load is defined as the force provided by a sudden load on the designated object. Analysis of impact, dynamic deflection, numbers of problems, and solutions are discussed.

7.1 Analysis of Impact Loading

The action on an elastic system of a sudden load, which before its impact had a certain speed or acceleration, is called an impact. When a moving body hits an elastic system, the kinetic energy of that body's (load's) motion is transferred to the elastic system. Impact loading analysis has fairly wide applications in civil and mechanical engineering, including structural design, foundational design, etc.

A great contribution to the development of the theory and analysis of impact problems was made by the following scientists and researchers: S.P. Timoshenko [1]. Timoshenko's theory of transverse impact. (This theory takes into account local deformations of the body striking a beam, and combines the essential provisions of Saint-Venant and Hertz), S.P. Timoshenko, D.H. Young, and W. Weaver [2], J.A. Simmons, F. Hauser, and J. E. Dorn [3], W. Goldsmith [4], N. Cristescu [5], E.S. Sorokin [6], N.A. Kilchevcki [7], G.K Clain [8], Norman Jones [9], and others.

Impact load is calculated as the total kinetic energy of dissipation divided by local deformation. The kinetic energy of an impact load, as in the example of the falling weight, will be transferred to the elastic system and stored as deformation/compression. As a result, two equal interaction forces between the load and the elastic system arise and are accompanied by the deformation of the elastic system. Each of these forces is called the impact force. We assume that the moving body is absolutely rigid, and the elastic system or structures, i.e., the struck object does not receive local deformations either. Under the foregoing assumption, the impact phenomenon is considered instantaneous. The speed of the falling load v_1 at the initial moment of contact with the structure instantly decreases to the speed v_0. This speed is the total initial speed of movement of the load and structure (total mass: $m_1 + m_2$). *Note* that the velocity v_1 corresponds to the height of the fall h: $v_1 = \sqrt{2gh}$ or height of fall $h + \mu y_{st}$: $v = \sqrt{2g(h + \mu y_{st})}$. At the time of impact, the mass of the load is instantly added to the structure. From this moment, the total speed of their combined movement (masses of the load and

L. G. Petrosian, *Dynamics of Structures*, https://doi.org/10.1007/978-3-031-63539-7_7

structure) equals to v_0. The velocity v_0 corresponds to the moment of time $t = 0$. We assume that the structure was at rest before the impact. The initial velocity of motion v_0 of an elastic system is derived from the law of conservation of the amount of motion. The mass and speed of the load before the impact will be denoted by m_1 and v_1, the reduced mass of the elastic system at the impact point is m_2.

To determine v_0, we use the following equality:

$$m_1 v_1 = (m_1 + m_2) v_0 \tag{7.1}$$

The amount of motion before the impact is equal to the amount of motion after the impact. From where:

$$v_0 = \frac{m_1}{m_1 + m_2} v_1 \tag{7.2}$$

where $m_1 + m_2 = m$ is the total mass of the falling load and the structure. Therefore, the initial velocity v_0 can be expressed through an instantaneously applied impulse:

$$v_0 = \frac{m_1 v_1}{m} = \frac{S}{m} \tag{7.3}$$

Let us consider the case of a free fall of a load $Q = mg$ on a cantilever beam using the example shown in Fig. 7.1. As we know the equal interaction forces between the falling weight and the beam have arisen and are accompanied by a deformation of the beam.

An impact force has a well-defined duration or period of action T_p^0. The period of action of an impact load is usually measured in hundredths or thousandths of a second and during this period varies in magnitude. An approximate graph of an impact force is shown in Fig. 7.2.

An impact force is characterized by its largest value P, period T_p^0, and momentum, equal to the area of the impact diagram:

$$S = \int_0^{T_p^0} P(t) dt \tag{7.4}$$

As we know, at the same time, the impulse of the impact force is equal to the momentum of the impacting mass m:

$$S = mv, \tag{7.5}$$

where v is the speed of the moving mass m.

If the force during the period T_p remains constant and equal to P, then the impulse of such force will be:

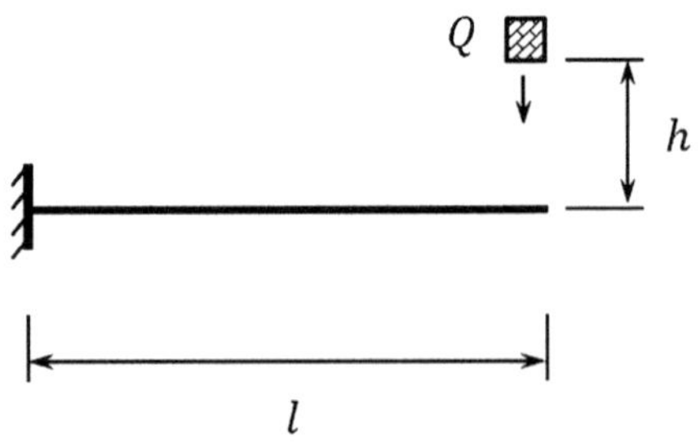

Fig 7.1 Cantilever beam under an impact load

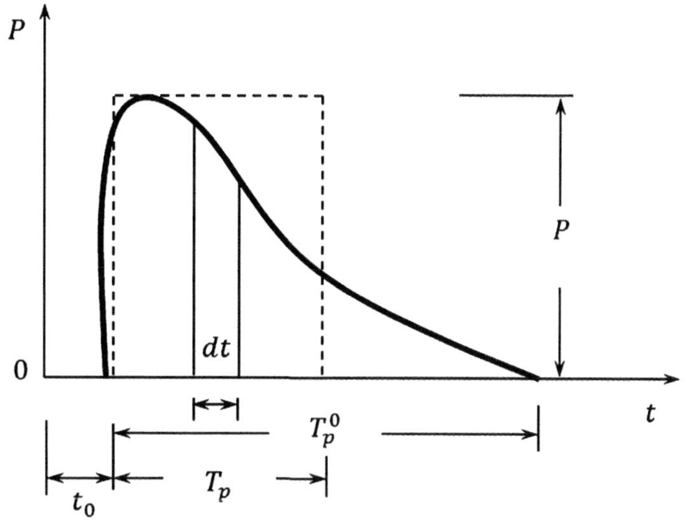

Fig. 7.2 Approximate graph of an impact force

$$S = PT_p, \tag{7.6}$$

The impact force, period of its action, and the law of time variation acting on an elastic system depend on the values of the load's mass, the speed of its movement, elastic properties of the system (i.e., the structure itself), and system's surface at the impact site. The curve of the force of impact can be approximated by an equal-sized rectangle, e.g., on the basis of the equality of the greatest ordinates or on the basis of the equality of periods $T_p = T_p^0$ (Fig. 7.2).

Let us consider formula (6.45). From the formula it follows that in order to find a static force that is equivalent to the force of an inelastic impact, it is necessary to know the magnitude of the impulse of the impact load, the period of its action, as well as the circular frequency of natural oscillations of the system.

Suppose that the impulse is instantaneous, then the equivalent static force will be expressed by formula (6.46). The dynamic coefficient of the falling weight $Q = mg$ will be equal to:

$$\mu = \frac{P_{\text{equivalent}}}{Q} = \frac{\omega S}{Q} = \frac{\sqrt{\frac{g}{y_{st}}}\, mv}{Q} = \frac{\sqrt{\frac{g}{y_{st}}}\, \frac{Q}{g} v}{Q} = \frac{v}{\sqrt{g y_{st}}} \tag{7.7}$$

From formula (7.7), it follows that the effect of an impact load depends on the size of the impulse and also on the frequency of natural oscillations of the system itself. The more rigid the structure, the greater the dynamic effect, and the greater the dynamic coefficient of impact under the same impulse.

Dynamic Deflection The expression for the velocity of a free fall from a height, accounting for the dynamic deflection of the beam is: $h + \mu y_{st}$, $v = \sqrt{2g(h + \mu y_{st})}$, which we substitute in formula (7.7). Furthermore, solving the quadratic equation, we obtain the expression for the dynamic coefficient:

$$\mu = 1 + \sqrt{1 + \frac{2h}{y_{st}}}. \tag{7.8}$$

In the case where the drop height h is much greater than the static deflection y_{st} of a system, the units that are in formula (7.8) can be neglected. Then the approximate formula of the dynamic coefficient will take the form:

$$\mu = \sqrt{\frac{2h}{y_{st}}}. \tag{7.9}$$

The static deflection of the system, taking into account the concentrated mass at the impact site from the weights of both masses, will have the following form:

$$y_{st}^{*} = y_{st}\frac{m + m_1}{m} = y_{st}\left(1 + \frac{m_1}{m}\right). \tag{7.10}$$

Instead of the static deflection y_{st}, we substitute the value of y_{st}^{*} into formula (7.9), and obtain:

$$\mu = \sqrt{\frac{2h}{y_{st}} \cdot \frac{1}{1 + \frac{m_1}{m}}} \tag{7.11}$$

A uniformly distributed mass can be reduced to concentrated at the impact site by multiplying it by some coefficient β, which depends on the boundary conditions of the problem, i.e., the conditions for end support of the beam, as in this case.

The magnitude of the impulse at the direct central impact of the body on the structure can also be determined by the formula:

$$S = mv_0(1 + k_0), \tag{7.12}$$

where m is the mass of the striking body or falling weight; v_0 is the velocity of that body, weight, or load at the beginning of the impact; and k_0 is the recovery factor for the impact. This coefficient depends on the properties of the colliding bodies (see Table 7.1. [6]).

Here, let me remind the reader that the impact is considered direct and central when the velocity of the striking body, weight, or load is normal to the center of mass of the system and coincides in direction with the displacement of the system. For example, an approximate solution to a problem can be obtained in the case when the mass of a beam is negligible compared to the mass of the falling body, weight, or load. In this case, we assume that the bending curve of the beam during the impact has a shape corresponding to the static bending curve.

Table 7.1 Approximate values of the coefficient k_0

Material of the impact surface of the structure that perceives the strike	Material and shape of the striking body				
	Hard materials (steels and alloys).		Copper, aluminum, wood, hard plastics, stone, concrete, etc.		Soft plastic materials (asphalt, clays, resins, oils, etc.
	Ball	Parallelepiped	Ball	Parallelepiped	Any shape
Steel	0.60	0.35	0.40	0.25	0
Wood	0.55	0.30	0.40	0.20	0
Stone	0.40	0.20	0.30	0.15	0
Concrete	0.35	0.15	0.25	0.10	0
Xylonite	0.20	0.10	0.10	0.05	0
Asphalt	0	0	0	0	0

Then the greatest deflection and the greatest stresses are determined from the energy of the system. In our example, there is a relationship between the deflection at the end of the cantilever beam and the acting force:

$$y = \frac{Pl^3}{3\mathrm{EI}} \tag{7.13}$$

The potential energy of deformation has the following form:

$$V = \frac{Py}{2} = \frac{3\mathrm{EI}}{2l^3}y^2 \tag{7.14}$$

If the load Q falls from a height h, then the work performed by this load is equal to:

$$W = Q(h + y_{\mathrm{dyn}}) \tag{7.15}$$

and the dynamic deflection y_{dyn} is determined from the equation:

$$Q(h + y_{\mathrm{dyn}}) = \frac{3\mathrm{EI}}{2l^3}y_{\mathrm{dyn}}^2, \tag{7.16}$$

from where:

$$y_{\mathrm{dyn}} = y_{\mathrm{st}} + \sqrt{y_{\mathrm{st}}^2 + 2hy_{\mathrm{st}}} \tag{7.17}$$

where

$$y_{st} = \frac{Ql^3}{3\mathrm{EI}}$$

represents the static deflection of the beam from the action of the load Q. We neglected the mass of the beam when determining the dynamic deflection. We assumed that the kinetic energy of the falling load Q has completely converted into the potential energy of the beam deformation. Actually, some part of the kinetic energy will be lost on impact. To obtain a more accurate solution, it is necessary to take into account the mass of the beam being impacted.

In either of the cases when a direct central impact of a body which has mass Q/g and velocity $v_{\max}$ on a system with mass m, which is in a state of rest, and/or if the deformation of the contact system with mass Q_1/g which is in a state of rest, and/or if the deformation of the contact point is in elastic deformation, the final total velocity v after the impact can be determined from the equation:

$$\frac{Q}{g}v_{\max} = \frac{Q + Q_1}{g}v \tag{7.18}$$

from where:

$$v = v_{\max}\frac{Q}{Q + Q_1} \tag{7.19}$$

At the moment of impact, the velocities v of the striking body Q and the beam Q_1 are the same only at the point of contact. Therefore, when determining the velocity v, not the entire mass of the beam but some part of it, i.e., the reduced mass, should be involved. The value of this reduced mass depends on

Table 7.2 Coefficients of reduction for uniformly distributed mass

#/#	Boundary conditions of simple beam	β
1	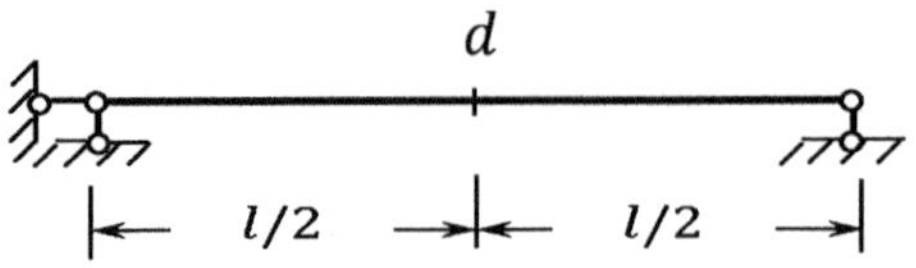	0.493
2	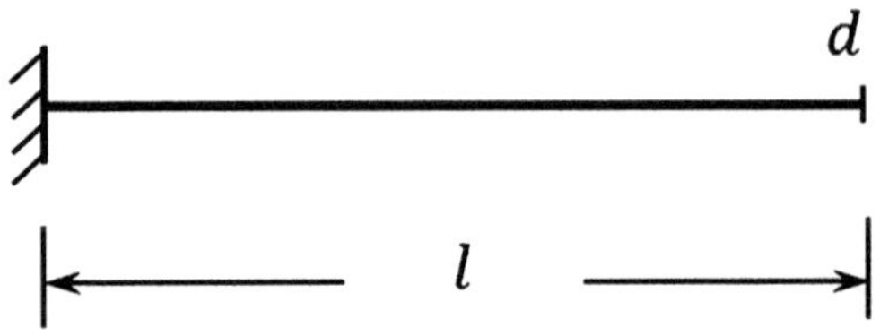	0.237
3	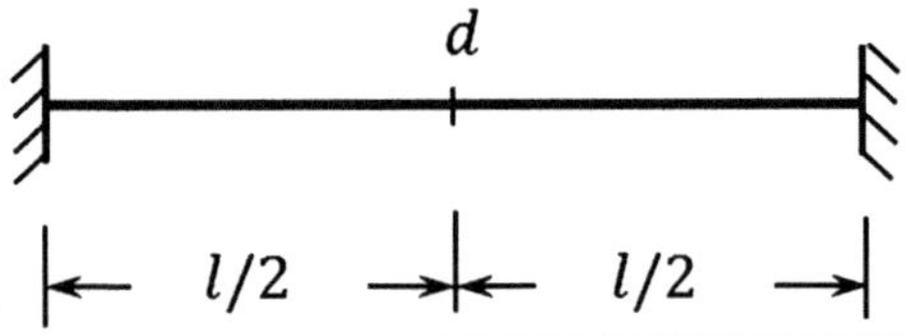	0.375
4	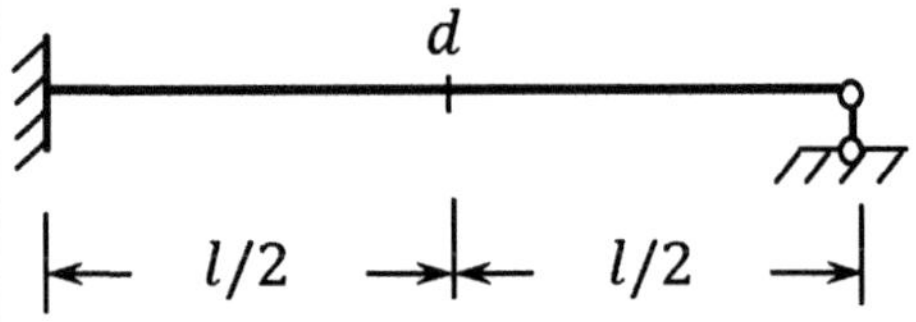	0.460

the shape of the bending curve; hence it depends on the support conditions of the ends of the beam (Table 7.2. [8]) and can be determined assuming that the bending curve is the same as the static curve.

Since in our case the load acts on the end of the cantilever beam, the equation of its elastic line will have the form:

$$y = \frac{Q}{6\mathrm{EI}} x(3l - x) \tag{7.20}$$

To determine the kinetic energy of the system, let us assume that the velocity v_x of the beam element which is located at a distance x from the end of the cantilever beam, Fig. 7.3, is proportional to the displacement of this section y_x from the static load, which is applied as a force P at the point of impact. This proportionality condition can be expressed by the following equality:

$$\frac{v_x}{v_{\max}} = \frac{y_x}{y_{\max}} \tag{7.21}$$

where $v_{\max}$, and $y_{\max}$, respectively, are the velocity and deflection of the point of impact of the cantilever beam.

Assume that the impact point is located at the end of the cantilever beam. Let us write the equation of deflection (7.20) of the cantilever beam in the following form:

$$y_x = \frac{Ql^3}{3EI}\left(\frac{3x^2}{2l^2} - \frac{x^3}{2l^3}\right) = y_{\max}\left(\frac{3x^2}{2l^2} - \frac{x^3}{2l^3}\right) \tag{7.22}$$

Therefore,

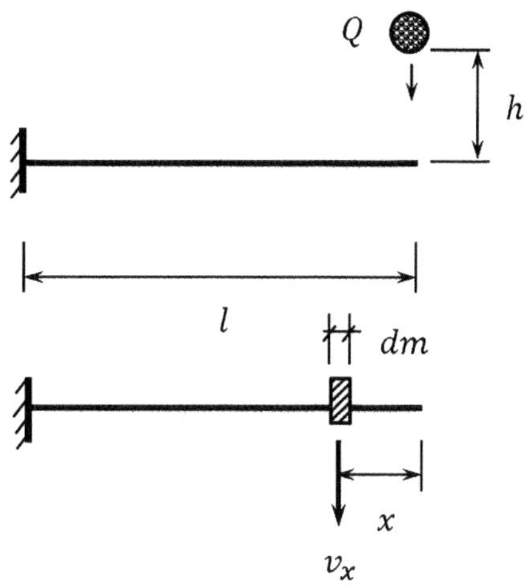

Fig. 7.3 Cantilever beam under impact load

$$v_x = v_{\max}\left(\frac{3x^2}{2l^2} - \frac{x^3}{2l^3}\right), \quad (0 \leq x \leq l). \tag{7.23}$$

The kinetic energy of the system is determined by the formula:

$$K = \int_0^l \frac{v_x^2}{2} dm = \int_0^l A\rho \frac{v_x^2}{2} dx = \int_0^l A\rho v_{\max}^2 \left(\frac{3x^2}{2l^2} - \frac{x^3}{2l^3}\right)^2 dx = \frac{33}{140}\frac{Qv_{\max}^2}{2g} \tag{7.24}$$

where ρ is the density of the material of the beam; A is the cross-sectional area of the beam, and g is the acceleration of gravity. Then:

$$v = v_{\max}\frac{Q}{Q + \frac{33}{140}Q_1} \tag{7.25}$$

where $\frac{33}{140}Q_1$ is the reduced weight (mass) of the beam. The kinetic energy of the system is equal to:

$$\frac{\left(Q + \frac{33}{140}Q_1\right)}{2g}v^2 = \frac{Qv_{\max}^2}{2g}\frac{1}{1 + \frac{33}{140}\frac{Q_1}{Q}}$$

In expression (7.20), substitute $\frac{Qv_{\max}^2}{2g}$ instead of Qh, and thereby take into account the influence of the mass of the beam. Then the dynamic deflection will be equal to:

$$y_{\text{dyn}} = y_{\text{st}} + \sqrt{y_{\text{st}}^2 + 2hy_{\text{st}}\frac{1}{1 + \frac{33}{140}\frac{Q_1}{Q}}} \tag{7.26}$$

Now let us consider a hinge-supported beam with a constant cross-section, in the middle of which the impact load Q falls (Fig. 7.4).

Let us write down the equation of the elastic line of the beam:

$$y_x = \frac{Q}{48\text{EI}}x(3l^2 - 4x^2) = \frac{Ql^3}{48\text{EI}}\left(\frac{3x}{l} - 4\frac{x^3}{l^3}\right) = v_{\max}\left(\frac{3x}{l} - 4\frac{x^3}{l^3}\right) \tag{7.27}$$

Then:

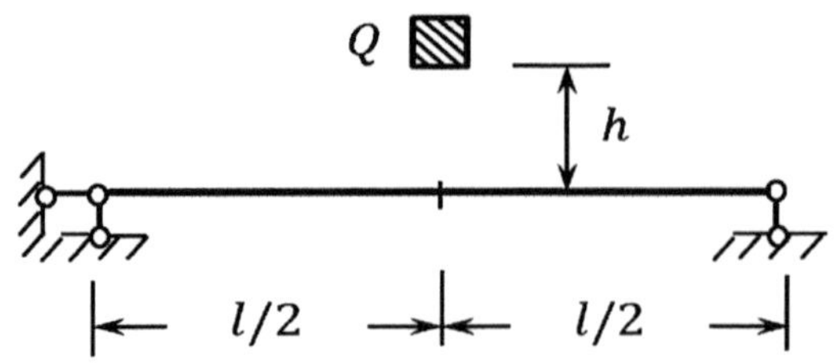

Fig. 7.4 Hinge supported beam under an impact load

$$v_x = v_{\max}\left(\frac{3x}{l} - 4\frac{x^3}{l^3}\right) \quad (0 \le x \le l/2) \tag{7.28}$$

The kinetic energy of the system will be determined by the equality:

$$K = \int_0^l \frac{v_x^2}{2} dm = \int_0^l A\rho \frac{v_x^2}{2} dx = 2\int_0^{\frac{l}{2}} \frac{1}{2} A\rho v_{\max}^2 \left(\frac{3x}{l} - 4\frac{x^3}{l^3}\right)^2 dx = \frac{17}{35}\frac{Qv_{\max}^2}{2g} \tag{7.29}$$

Let us move on to determining the kinetic energy for the hinge-supported beam, which has a reduced mass in the middle of the span. Taking into account that the speed of this mass is equal to $v_{\max}$, we obtain:

$$K = \frac{m_1}{2} v_{\max}^2 = \frac{\beta Q_1}{g} \frac{v_{\max}^2}{2}. \tag{7.30}$$

Equating the two expressions obtained above, we obtain: $\beta = 17/35$.

Then we transform the expression of dynamic deflection. After that, we substitute the obtained coefficient β value into formula (7.30). The value of dynamic coefficient for the beam when the falling weight hits the point in the middle of the span depends on the conditions of the beam support. Here Q_1 is the weight of the beam, and Q is the weight of the falling weight.

Table 7.2 [8] shows the coefficients of reduction that bring the uniformly distributed mass of the beam to the point of impact. The place or point of impact on the beam is indicated by the letter "d."

After determining the highest value of the dynamic coefficient, the dynamic stress is determined by the formula:

$$\sigma_{\text{dyn}} = \mu\sigma_{\text{st}}. \tag{7.31}$$

Of course, the dynamic stress from the falling load is added to the static stress from the structure's own weight. Note that taking into account the mass of the structure is essential.

From formulas (7.9) and (7.11), it can be seen that when y_{st} increases, the dynamic coefficient μ decreases. At $h = 0$, which corresponds to the sudden action of force, the dynamic coefficient is equal to two $\mu = 2$.

From formulas (7.9) and (7.11), it can be seen when y_{st} increases, the dynamic coefficient μ decreases. At $h = 0$, which corresponds to the sudden action of force, the dynamic coefficient is equal to two $\mu = 2$.

The value of the dynamic coefficient increases rapidly as the height h of the load falling on a structure increases. Formula (7.11) can be used in both lateral and longitudinal impacts.

Problem 7.1

It is required to determine the greatest deflection and the highest normal stress of a load falling on a rolled – steel beam from the height $h = 1$ m, as shown in Fig. 7.1. The length of the beam is 3 m; actual weight of the beam is $Q_1 = 370.33$ kg; Depth of beam: 18 in $= 45.72$ cm; $I = 1256.5\ \text{in}^4 = 52299.5\ \text{cm}^4$; $E = 2.1 \cdot 10^6 \text{kg/cm}^2$.

Solution

The static deflection of a beam from a load of weight $Q = 100$ k:

$$y_{st} = \frac{Ql^3}{3EI} = \frac{100 \cdot 300^3}{3 \cdot 2.1 \cdot 10^6 \cdot 52299.5} = 0.0082\ \text{cm}.$$

The actual weight of the beam, reduced to the point of impact, is taken to be one-quarter of the actual weight. See Table 7.1 $\beta = 0.237$:

$$Q_1 = 0.237 \cdot 370.33\ \text{kg} = 87.77\ \text{kg}$$

The dynamic impact coefficient is determined by formula (7.11):

$$\mu = \sqrt{\frac{2h}{y_{st}} \cdot \frac{1}{1 + \frac{m_1}{m}}} = \sqrt{\frac{2 \cdot 100}{0.0082\left(1 + \frac{87.77}{100}\right)}} = 113.96 \approx 114$$

Now, let us determine the impact force:

$$P = \mu \cdot Q = 114 \cdot 100 = 11400\ \text{kg}.$$

Since the height of the load drop is significantly greater than the static deflection of the beam, we use the following formula to calculate the dynamic deflection:

$$\delta_{dyn} = \sqrt{\frac{2hy_{st}}{1 + \frac{Q_1}{Q}}} = \sqrt{\frac{2 \cdot 100 \cdot 0.0082}{1 + \frac{87.77}{100}}} = 0.935$$

Let us check the dynamic deflection value obtained by using the previous formula:

$$\delta_{dyn} = \mu \cdot y_{st} = 114 \cdot 0.0082 = 0.9348 \approx 0.935$$

The greatest stress on impact will be in the building-in section of the cantilever beam:

$$\sigma_{dyn} = \mu \cdot \sigma_{st} = 114 \cdot \frac{100 \cdot 300}{52299.5} \cdot 22.86 = 1495\ \text{kg/cm}^2$$

Problem 7.2

A load weighing 300 kg from height $h = 20$ cm falls on a steel beam in the middle of its span (Fig. 7.5). It is required to determine the impact force if the modulus of elasticity of a rolled-steel beam is: $E = 2.1 \cdot 10^6\ \text{kg/cm}^2$; depth of the beam is: 15 in $= 38.1$ cm, the moment of inertia of the cross-section of the beam is: $I = 28603.42\ \text{cm}^4$, the self-weight of the beam is $Q_1 = 816.32$ kg., and the length of the beam is 8 m.

Solution

Static deflection of a beam from a load of weight $Q = 300$ kg.

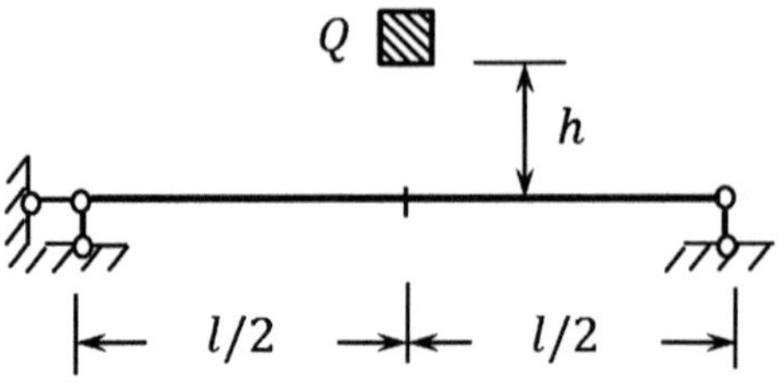

Fig. 7.5 Hinge-supported beam under an impact load

$$y_{st} = \frac{Ql^3}{48EI} = \frac{300 \cdot 800^3}{48 \cdot 2.1 \cdot 10^6 \cdot 28603.42} = 0.053 \text{ cm}.$$

The actual weight of the beam, reduced to the point of impact, is taken to be equal to half of the actual weight. See Table 7.1. $\beta = 0.493 \approx 0.5$:

$$Q_1 = 0.493 \cdot 816.32 \text{ kg} = 402.45 \text{ kg}$$

The dynamic impact coefficient is determined by formula (7.11). In this case, the ratio of the masses is replaced by the ratio of the corresponding weights:

$$\mu = \sqrt{\frac{2h}{y_{st}} \cdot \frac{1}{1 + \frac{Q_1}{Q}}} = \sqrt{\frac{2 \cdot 20}{0.053\left(1 + \frac{402.45}{300}\right)}} = 17.95 \approx 18$$

Now, let us determine the impact force:

$$P = \mu \cdot Q = 18 \cdot 300 = 5400 \text{ kg}.$$

If the weight of the beam is neglected, then the magnitude of the dynamic coefficient and the impact force will be much greater:

$$\mu = \sqrt{\frac{2h}{y_{st}}} = \sqrt{\frac{2 \cdot 20}{0.053}} = 27.47 \approx 27.5$$

The impact force will accordingly be:

$$P = \mu \cdot Q = 27.5 \cdot 300 = 8250 \text{ kg}.$$

Thus, the force of impact of the weight, when falling even from a low height, is much greater than the weight of this load. Moreover, the impact force increases as the rigidity of the structure increases.

Problem 7.3

It is required to determine the force of the impact of an ice floe, weighing 2000 kg, on the pile at the level of the water surface (Fig. 7.6). The reinforced concrete shell pile is driven into the river bottom at the depth of $h = 4$ m. The ground of the riverbed is dense, and, therefore, the shell pile is conditionally considered to be constrained into the ground at the level of the river bottom. The internal diameter of the pile is: $d = 2$ m, the wall thickness of the pile is: $\delta = 16$ cm. The reinforced concrete pile's modulus of elasticity is assumed to be E = 3,400,000 kg/cm^2. The speed of the river flow is: =1 m/s. Additionally, we are asked to determine the stresses in the shell pile that arise from the impact of the ice floe.

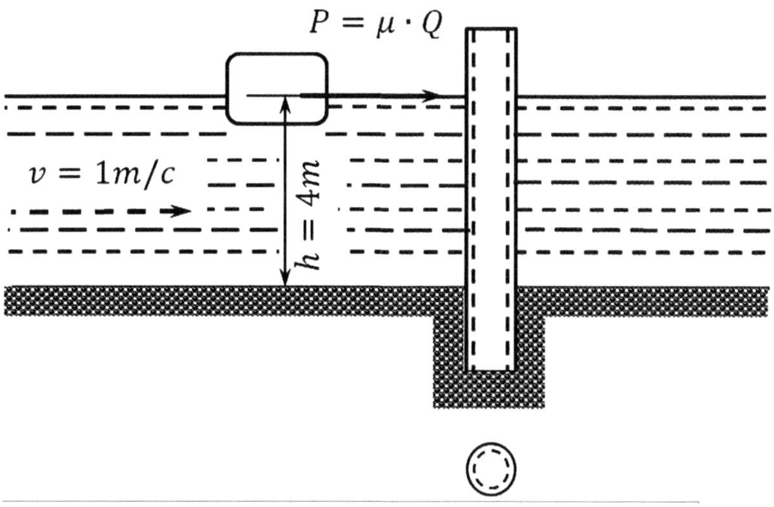

Fig. 7.6 A dynamic force action of an ice floe on the pile at the level of water surface

Solution
The moment of inertia of the annular section at the radius of the median surface $r = 1.08\ m$.

$$I = \pi r^3 \delta = 3.14 \cdot 1.08^3 \cdot 0.16 = 0.63\ \text{m}^4 = 6.3 \cdot 10^7 \text{cm}^4.$$

Shell pile's actual weight that is brought to the point of impact is taken equal to one-quarter of the total weight of its ground part (See Table 7.1). $\beta = 0.237 \approx 0.25$, that is:

$$Q_1 = 0.237 \cdot 2\pi r \delta h \gamma_{\frac{r}{c}} = 0.237 \cdot 2 \cdot 3.14 \cdot 1.08 \cdot 0.16 \cdot 4 \cdot 2.5 = 2.57\ T = 25.7\ \text{kN}.$$

The static deflection of the pile from the force Q equals to the weight of the ice floe:

$$y_{st} = \frac{Qh^3}{3EI} = \frac{2000 \cdot 400^3}{3 \cdot 3.4 \cdot 10^5 \cdot 6.3 \cdot 10^7} = 0.0020\ \text{cm}.$$

The dynamic coefficient of an ice floe impact on the pile is determined by formula (7.11):

$$\mu = \frac{v}{\sqrt{g y_{st}\left(1 + \frac{Q_1}{Q}\right)}} = \frac{100}{981 \cdot 0.0020\left(1 + \frac{2.57}{2}\right)} = 22.3$$

The impact force will accordingly be:

$$P = \mu \cdot Q = 22.3 \cdot 2.0 = 44.61T = 446.1\ \text{kN}$$

The greatest bending moment in the pile at the level of the river bottom will be:

$$M = M_{st} \cdot \mu = 20 \cdot 4 \cdot 22.3 = 1784\ \text{kN} \cdot \text{m}$$

The highest normal stress in the pile is determined using the following formula:

$$\sigma = \frac{M}{W} = \frac{1784 \cdot 2.16}{0.63} = 61.2\ \text{MPa}$$

Table 7.3 shows various impact points on beams and plates/slabs with different boundary conditions of support, as well as expressions for determining dynamic coefficients and dynamic deflections. The formulas given in the table can be successfully applied in engineering practice.

Table 7.3 Expressions for determining dynamic coefficients and dynamic deflections depending on the boundary conditions of supports for beams and plates

##	Point of impact on the beam and on the plate	Dynamic coefficient and dynamic deflection
1.	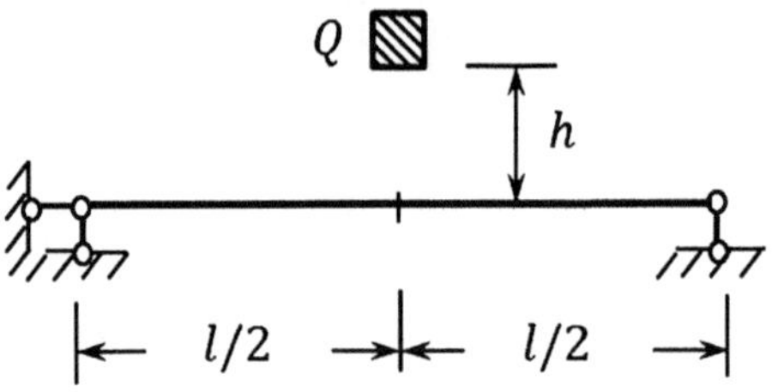	$\mu = 1 + \sqrt{1 + \frac{48EI}{Ql^3} \cdot \frac{v^2}{g}}$ $y_{dyn} = y_{st} + \sqrt{y_{st}^2 + \frac{v^2}{g} y_{st} \frac{1}{1+\frac{17Q_1}{35Q}}}$
2.	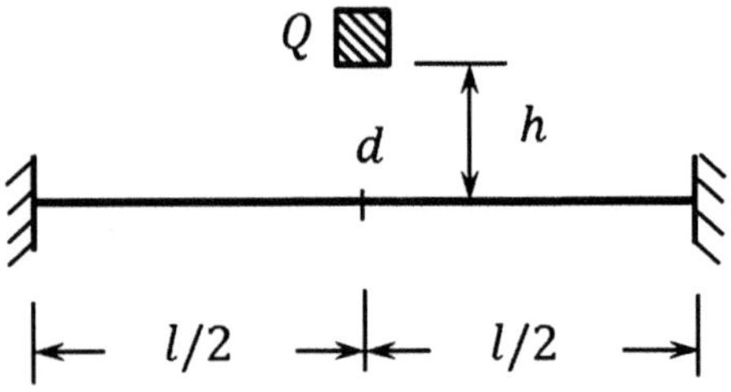	$\mu = 1 + \sqrt{1 + \frac{192EI}{Ql^3} \cdot \frac{v^2}{g}}$ $y_{dyn} = y_{st} + \sqrt{y_{st}^2 + 2hy_{st}}$
3.	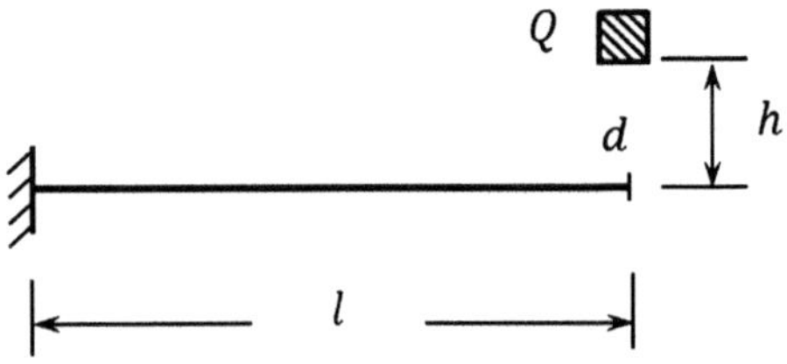	$\mu = 1 + \sqrt{1 + \frac{3EI}{Ql^3} \cdot \frac{v^2}{g}}$ $y_{dyn} = y_{st} + \sqrt{y_{st}^2 + \frac{v^2}{g} y_{st} \frac{1}{1+\frac{33Q_1}{140Q}}}$
4.	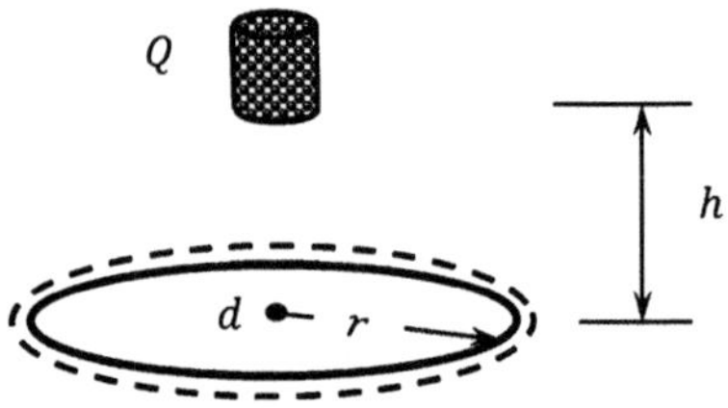	The slab (plate) is supported along the contour and free in tangential directions. $\mu = 1 + \sqrt{1 + \frac{4\pi Et^3}{3(1-\nu)(3+\nu)r^2Q} \cdot \frac{v^2}{g}}$ Where: t is the slab (plate) thickness; v is the Poisson's coefficient.
5.	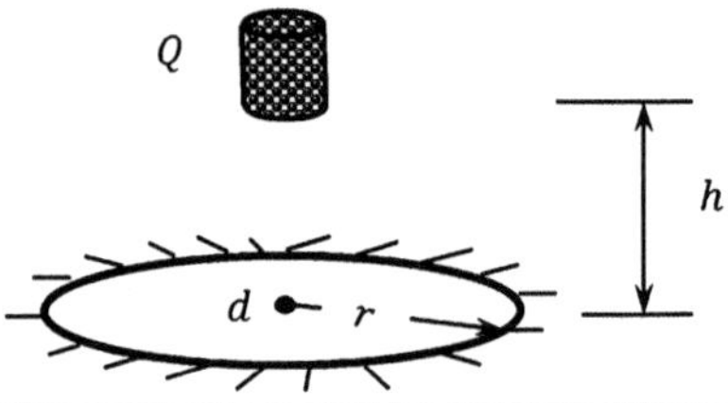	The slab (plate) is built in along the contour. $\mu = 1 + \sqrt{1 + \frac{4\pi Et^3}{3(1-\nu^2)r^2Q} \cdot \frac{v^2}{g}}$
6.	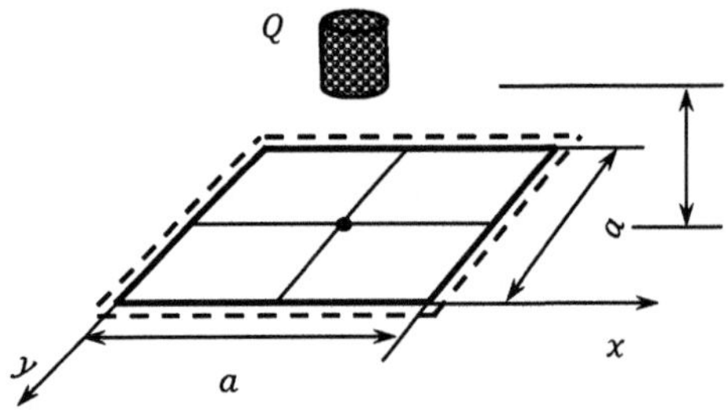	$\mu = 1 + \sqrt{1 + \frac{\pi^2 Et^3(\pi^3+8\nu^2)}{192(1-\nu^2)a^2Q\left(1+k\frac{Q_1}{Q}\right)} \cdot \frac{v^2}{g}}$ Where: $Q_1 = 4\gamma a^2 t$; $k = 0.45$, k - is the coefficient of weight reduction of the entire plate to the point of impact.

Problem 7.4
It is required to determine the displacement at a certain point of the system on which the load Q falls from the height h. According to the condition of the problem, the system can make linear bending oscillations/vibrations.

Solution
Let us separate the point of impact on which the load falls and consider its equilibrium from the action of the falling load Q, from the internal resistance force R_s and from the inertial force I. The inertial force I, in accordance with the Principle of D'Alembert (Chap. 4, Sect. 4.1) can be determined by formula:

$$I = -\frac{Q}{g}\frac{d^2y}{dt^2}; \qquad \text{or} \qquad I = -m\ddot{y}$$

The force of inertia is directed toward the opposite acceleration.

The equation of dynamic equilibrium of all forces acting on the point of impact has the following form:

$$\sum Y = 0; \qquad I - R_s + Q = 0 \tag{a}$$

Let us substitute the corresponding expressions of the forces I, R_s and Q with opposite signs to obtain the ordinary homogeneous linear differential equation:

$$-m\ddot{y} - R_S + Q = 0 \tag{b}$$

The restoring force is proportional to the deviation y of the point in which the mass is concentrated, i.e.:

$$R_s = -ky$$

The coefficient of proportionality or coefficient of rigidity, k, is the reaction of the beam at the point of impact on which the load falls, when the displacement of this point is equal to one unit of displacement. Coefficient of proportionality of a system is the value that depends on the elastic and geometrical characteristics of the system and can be found from the general expression for displacement of one unit.

$$\bar{y} = \delta k = 1 \text{ or } k = 1/\delta$$

where δ is the displacement of the considered point, caused by a unit of force. Thus, taking into account the foregoing the above, the following equation is obtained:

$$\ddot{y} + \omega^2 y = g \tag{c}$$

where $\omega^2 = \frac{k}{m}$

The general solution of an ordinary differential equation with the right-hand side has the following form:

$$y = C_1 \cos \omega t + C_2 \sin \omega t + \frac{g}{\omega^2} \tag{d}$$

where $\omega^2 = \frac{k}{m}$, $\quad y_{st.} = \frac{g}{\omega^2} = \frac{gkQ}{g} = Q\bar{y}$. $\bar{y}$ is the static displacement of the point of impact at $R_s = 1$

Taking that into account, we have:

$$y = y_{st.} + C_1 \cos \omega t + C_2 \sin \omega t. \tag{e}$$

Thus, the load Q, after falling on the system will oscillate together with the system relative to its equilibrium position. In this case $y = y_{st.}$ The constants C_1 and C_2 are determined from the initial conditions of the impact. At $t = 0 : y(0) = 0; \dot{y} = v_0; \ \ v_0 = \sqrt{2gh}$. The initial conditions of the problem are used in solution (e). As a result, to determine the constants C_1 and C_2 we obtain the following equations:

$$y_{st.} + C_1 = 0; \omega c = \sqrt{2gh},$$

whence

$$C_1 = -y_{st.}; \qquad C_2 = \frac{\sqrt{2gh}}{\omega}. \tag{f}$$

Let us substitute the values of constants C_1 and C_2, that is, the expression (f) into (e),

$$y = y_{st.}(1 - \cos \omega t) + \frac{\sqrt{2gh}}{\omega} \sin \omega t \tag{g}$$

as a result, we obtain an equation with which we can determine the displacement at the point of impact. Thus, a weightless system oscillates under impact as a system with single degree of freedom if the impact occurs at one point and if separate linear or torsional deformations.

References

1. Timoshenko, S. P. (1967). *Oscillations in Engineering* (444p). Nauka.
2. Timoshenko, S., Young, D. H., & Weaver, W. (1974). *Vibration Problems in Engineering* (4th ed., pp. 432–434). Wiley.
3. Simmons, J. A., Hauser, F., & Dorn, J. E. (1960). *Mathematical Theories of Plastic Deformation under Impulsive Loading* (pp. 177–229). University of California Publications in Engineering.
4. Goldsmith, W. (2001). *Impact. The theory and Physical Behavior of Colliding Solids* (390p). Dover Publications. Inc..
5. Cristescu, N. (1967). *Dynamic Plasticity. Series in Applied Mathematics and Mechanics* (Vol. 4, 620p). North-Holland Publish Company.
6. Sorokin, E. S. (1971). *To the Theory of Internal Friction with the Oscillations of Elastic Systems* (131p). Gosstroyizdat.
7. Kilchevcki, N. A. (1976). *Dynamic Contact Compression of Solid Bodies. Impact* (319p). Naukova Dumka.
8. Clain, G. K. (1982). Fundamentals of dynamics of rod system. In *Structure Mechanics Cource* (320p). Vishaya Shkola.
9. Jones, N. (2012). *Structural Impact* (2nd ed., 600p). University of Liverpool. Cambridge University Press.

Multi-Degree-of-Freedom System

8

Abstract

This chapter is devoted to discussion of free vibration of systems with multi-degrees of freedom. Natural transverse vibrations of beams and frames are considered as systems with a *n*th number of concentrated masses. When composing equations and choosing a computational dynamic model, classical methods of calculating rod systems are used: the method of forces and the method of displacements. The chapter includes extensive practical examples and problems.

8.1 Free Vibrations of Systems with Multi-Degrees of Freedom or Systems With a Finite Number of Degrees of Freedom

Let us consider the construction of motion equations for a discrete system that makes free oscillations on the example of a weightless beam with point masses (Fig. 8.1). The differential equations of motion can be obtained using the force Method known from structural mechanics.

$$y_i = \sum_{k=1}^{n} \delta_{ik} I_k \qquad (i = 1, 2, \ldots, n) \tag{8.1}$$

where δ_{ik} is a unit displacement; $I_k$is the force of inertia of concentrated mass k.

The displacement of the center of each mass of the system under the influence of inertial forces I_1, I_2, . . ., I_n will take the form:

$$\begin{aligned} y_1 &= \delta_{11} I_1 + \delta_{12} I_2 + \ldots + \delta_{1n} I_n \\ y_2 &= \delta_{21} I_1 + \delta_{22} I_2 + \ldots + \delta_{2n} I_n \\ &\ldots\ldots\ldots\ldots\ldots\ldots\ldots\ldots \\ y_n &= \delta_{n1} I_1 + \delta_{n2} I_2 + \ldots + \delta_{nn} I_n \end{aligned} \tag{8.2}$$

Let us substitute the values of inertial forces $I_1 = -m_1 \ddot{y}_1, I_2 = -m_2 \ddot{y}_2, \ldots, I_n = -m_n \ddot{y}_n$ we obtain:

L. G. Petrosian, *Dynamics of Structures*, https://doi.org/10.1007/978-3-031-63539-7_8

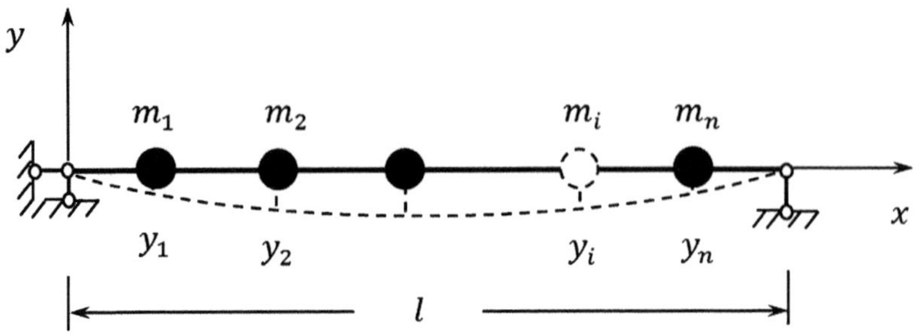

Fig. 8.1 A weightless beam with m_n concentrated masses

$$\begin{aligned} y_1 &= -\delta_{11}m_1\ddot{y}_1 - \delta_{12}m_2\ddot{y}_2 - \ \ldots\ -\delta_{1n}m_n\ddot{y}_n \\ y_2 &= -\delta_{21}m_1\ddot{y}_1 - \delta_{22}m_2\ddot{y}_2 - \ \ldots\ -\delta_{2n}m_n\ddot{y}_n \\ &\ldots\ldots\ldots\ldots\ldots\ldots\ldots\ldots \\ y_n &= -\delta_{n1}m_1\ddot{y}_1 - \delta_{n2}m_2\ddot{y}_2 - \ \ldots\ -\delta_{nn}m_n\ddot{y}_n \end{aligned} \tag{8.3}$$

The system of differential Eq. (8.3) allows n particular solutions.

$$\begin{aligned} y_1 &= A_1 \sin(\omega t + \varphi_0) \\ y_2 &= A_2 \sin(\omega t + \varphi_0) \\ &\ldots\ldots\ldots\ldots \\ y_n &= A_n \sin(\omega t + \varphi_0) \end{aligned} \tag{8.4}$$

which describe harmonic oscillations with frequency ω. Here $y_1, y_2, \ldots, y_n$ represent deviations of system masses; $A_1, A_2, \ldots, A_n$ are amplitudes of the deflected masses of the system; φ_0 is the initial phase of oscillations. The expressions for displacements and their second derivatives in time are:

$$\begin{aligned} \ddot{y}_1 &= -\omega^2 A_1 \sin(\omega t + \varphi_0) \\ \ddot{y}_2 &= -\omega^2 A_2 \sin(\omega t + \varphi_0) \\ &\ldots\ldots\ldots\ldots \\ \ddot{y}_n &= -\omega^2 A_n \sin(\omega t + \varphi_0) \end{aligned} \tag{8.5}$$

substituting into system of Eq. (8.3), after reduction by $\sin(\omega t + \varphi_0)$, we arrive at the following relationships:

$$\begin{aligned} A_1 &= \delta_{11}m_1A_1\omega^2 + \delta_{12}m_2A_2\omega^2 + \ldots + \delta_{1n}m_nA_n\omega^2 \\ A_2 &= \delta_{21}m_1A_1\omega^2 + \delta_{22}m_2A_2\omega^2 + \ldots + \delta_{2n}m_nA_n\omega^2 \\ &\ldots\ldots\ldots\ldots\ldots\ldots\ldots\ldots \\ A_n &= \delta_{n1}m_1A_1\omega^2 + \delta_{n2}m_2A_2\omega^2 + \ldots + \delta_{nn}m_nA_n\omega^2 \end{aligned}$$

or

$$\begin{aligned} &\left(\delta_{11}m_1 - \frac{1}{\omega^2}\right)A_1 + \delta_{12}m_2A_2 + \ldots + \delta_{1n}m_nA_n = 0, \\ &\delta_{21}m_1A_1 + \left(\delta_{22}m_2 - \frac{1}{\omega^2}\right)A_2 + \ldots + \delta_{2n}m_nA_n = 0, \\ &\delta_{n1}m_1A_1 + \delta_{n2}m_2A_2 + \ldots + \left(\delta_{nn}m_n - \frac{1}{\omega^2}\right)A_n = 0. \end{aligned} \tag{8.6}$$

It is obvious that the trivial solution of this system of algebraic equations corresponds to the case when the system is at rest. Solutions other than zero will be obtained if we equate to zero the determinant which is composed of coefficients at these amplitudes.

$$D = \begin{vmatrix} \delta_{11}m_1 - \dfrac{1}{\omega^2} & \delta_{12}m_2 \ldots \delta_{1n}m_n \\ \delta_{21}m_1 & \delta_{22}m_2 - \dfrac{1}{\omega^2} \ldots \delta_{2n}m_n \\ \cdots\cdots & \cdots\cdots \\ \delta_{n1}m_1 & \delta_{n2}m_2 \ldots \delta_{nn}m_n - \dfrac{1}{\omega^2} \end{vmatrix} = 0 \tag{8.7}$$

Thus, an equation of frequencies is obtained, which is an algebraic equation of the nth degree with respect to ω^2.

$$\lambda^n - B_1\lambda^{n-1} + B_2\lambda^{n-2} - \ \ldots \ -B_{n-1}\lambda + (-1)^n B_n = 0 \tag{8.8}$$

here λ is a dimensionless number. For convenience of calculations, the displacements δ_{ik} and masses m_i can be represented in dimensionless form. Then dimensionless displacements and masses will be: $\delta_{ik} = \delta_{ik}/\delta_0$; $m_i = m_i/m_0$. Thus the dimensionless number is represented as:

$$\lambda = 1/\omega^2\delta_0 m_0$$

In mechanics, the equations of frequencies are called the secular (century) equation. This name arose in connection with the problems of celestial mechanics about periodic deviations of planets from their orbits. If this determinant is expanded, then we obtain a polynomial of degree n with respect to ω^2. In statically stable systems, the determinant

$$D = \begin{vmatrix} \delta_{11} & \delta_{12} \ldots \delta_{1n} \\ \delta_{21} & \delta_{22} \ldots \delta_{2n} \\ \cdots & \cdots \\ \delta_{n1} & \delta_{n2} \ldots \delta_{nn} \end{vmatrix} = 0$$

and its primary (major) minors are positive. As the diagonal terms increase, the determinant increases. Therefore, to satisfy the equality (8.8), we must take $\omega^2 > 0$.

Then the squares of the frequencies will be positive, and the frequencies themselves are valid.

All the roots of this equation are positive and each of them corresponds to its own frequency value: $\omega_1 = \frac{1}{\sqrt{\lambda_1}}$, $\omega_2 = \frac{1}{\sqrt{\lambda_2}}$, $\ldots, \omega_n = 1/\sqrt{\lambda_n}$.

The totality of all the frequencies of the system's natural oscillations makes up its frequency spectrum. The spectrum of natural oscillation frequencies is an important dynamic characteristic of the structure. The lowest frequency is called the fundamental frequency, or the fundamental tone of oscillations, and the other frequencies are called overtones. For engineering problems, it is usually sufficient to find the lowest frequency. The lowest frequency is the most dangerous because resonance at the lowest frequency leads to the greatest dynamic effect. In addition, even if the frequency of the disturbing force exceeds the lowest frequency of natural oscillations of the system, the resonance will still occur in the starting and stopping modes of machines and mechanisms.

When a second- or third-order determinant is included in the frequency equation, its solution is not particularly difficult. For higher-order determinants, determining frequencies is a very difficult task. For large values of n, frequencies are determined only with the help of a computer. Therefore, all the transformations that simplify the form of the determinant in the secular equation and facilitate its disclosure acquire special value. The most effective methods for solving secular equations, which require a relatively little time, are given in the work of V.V. Voevodin [1], J. H. Wilkinson, C. Reinsch [2], G.H. Golub [4], D.K. Fadeev, V.N. Fadeeva [5], J.H. Wilkins [6], and A. Melman A [7].

Each mode of natural oscillations corresponds to its own frequency, and the number of modes of natural oscillations of an elastic system is equal to the number of its degrees of freedom. Each value of ω_i^2 corresponds to a system of ratios between the amplitudes

$$\frac{A_{2k}}{A_{1k}}; \frac{A_{3k}}{A_{1k}}; \ldots; \frac{A_{nk}}{A_{1k}},$$

which determine the natural mode of system oscillations. The system of ratios between the amplitudes is also called the primary (major), or normal coordinates of the system. These ratios are obtained from the differential equations of motion if particular solutions (8.4) are substituted in them.

The differential equations of motion (8.1) are the canonical equations of the method of forces. The coefficients are the displacement of the system: δ_{ik} is the displacement of the point of application of the i-th concentrated mass in the direction of motion of the i-th inertial force from the unit force, which acts at the point of application of the k-th concentrated mass. This fact allows us to point out the case of simplification of the canonical equations known from the statics of structures.

If the directions of the displacements $y_1, y_2, \ldots, y_n$ are chosen in such a way, that the secondary (minor) displacements δ_{ik} are zero, then the frequency equations, that is, the system of equations will break up into separate equations. These equations will contain only the primary (major), displacements. In this case, the displacements $y_1, y_2, \ldots, y_n$ that is the primary (major) coordinates, correspond to the primary (major) oscillation modes. Each of the main modes has its own specific frequency, which is determined by the formula:

$$\omega_i = \frac{1}{\sqrt{\delta_{ii} m_i}} \tag{8.9}$$

as for systems with a single degree of freedom.

The main directions of displacements in practice can be found in rare cases, just as in the method of forces only in rare cases it is possible to obtain a complete separation of unknowns in canonical equations. So, for example, for systems with two degrees of freedom, the choice of the primary (major) coordinates is always possible. For systems with more than three degrees of freedom, in general, the choice of the primary (major) coordinates is very difficult.

In the case when the number of degrees of freedom is equal to two, the frequency equation takes the form:

$$\begin{vmatrix} \delta_{11} m_1 - \dfrac{1}{\omega^2} & \delta_{12} m_2 \\ \delta_{21} m_1 & \delta_{22} m_2 - \dfrac{1}{\omega^2} \end{vmatrix} = 0 \tag{8.10}$$

or in expanded form:

$$\left(\delta_{11} m_1 \omega^2 - 1\right)\left(\delta_{22} m_2 \omega^2 - 1\right) - \delta_{12}^2 m_1 m_2 \omega^4 = 0 \tag{8.11}$$

Due to the linearity of the basic equations of motion, the general solution of the system of differential equations is the sum of particular solutions (8.4), and for a system with two degrees of freedom we have:

$$\begin{aligned} y_1 &= A_1 \sin(\omega_1 t + \varphi_1) + B_1 \sin(\omega_2 t + \varphi_2) \\ y_2 &= A_2 \sin(\omega_1 t + \varphi_1) + B_2 \sin(\omega_2 t + \varphi_2) \end{aligned} \tag{8.12}$$

where ω_1 and ω_2 are solutions of the secular equation (8.10):

$$(m_1 \neq m_2)$$

$$\begin{aligned} \omega_{1,2}^2 = &\frac{1}{2(\delta_{11}\,\delta_{22} - \delta_{12}^2)m_1 m_2} \left[(\delta_{11}m_1 + \delta_{22}m_2) \pm \right. \\ &\left. \pm \sqrt{(\delta_{11}m_1 + \delta_{22}m_2)^2 - 4(\delta_{11}\,\delta_{22} - \delta_{12}^2)m_1 m_2}\right] \end{aligned} \tag{8.13}$$

$$(m_1 = m_2 = m)$$

$$\omega_{1,2} = \frac{\sqrt{2}}{\sqrt{m\left[\delta_{11} + \delta_{22} \pm \sqrt{(\delta_{11} - \delta_{22})^2 + 4\delta_{12}^2}\right]}} \tag{8.14}$$

Since each value of ω_i corresponds to its own ratio between the amplitudes, then with respect to a system with two degrees of freedom the following results are obtained:

$$\frac{A_2}{A_1} = -\frac{\delta_{11}m_1\omega^2 - 1}{\delta_{12}m_2\omega^2} = -\frac{\delta_{21}m_1\omega^2}{\delta_{22}m_2\omega^2 - 1}. \tag{8.15}$$

Moreover, the values of ω obtained from the equation of frequencies ensure equality between the second and third terms (8.15). Let us introduce the following notations:

$$\nu_1 = -\frac{\delta_{11}m_1\omega_1^2 - 1}{\delta_{12}m_2\omega_1^2}; \qquad \nu_2 = -\frac{\delta_{11}m_1\omega_2^2 - 1}{\delta_{12}m_2\omega_2^2} \tag{8.16}$$

Then $A_2 = \nu_1 A_1$, $B_2 = \nu_2 B_1$, and Eq. (8.12) will take the form:

$$\begin{aligned} y_1 &= A_1 \sin(\omega_1 t + \varphi_1) + B_1 \sin(\omega_2 t + \varphi_2); \\ y_2 &= A_1\nu_1 \sin(\omega_1 t + \varphi_1) + B_1\nu_2 \sin(\omega_2 t + \varphi_2). \end{aligned} \tag{8.17}$$

Expression (8.17) contains four arbitrary constants and is a general solution. To determine the constants, it is necessary to have four conditions, such as the values of displacements and velocities at the initial moment $t = 0$

$$\begin{aligned} &y_1 = \eta_1, \quad y_2 = \eta_2 \\ &\frac{dy_1}{dt} = v_1, \quad \frac{dy_2}{dt} = v_2 \end{aligned}$$

or from (8.17) we obtain:

$$\begin{aligned} \eta_1 &= A_1 \sin\varphi_1 + B_1 \sin\varphi_2, \\ \eta_2 &= A_1\nu_1 \sin\varphi_1 + B_1\nu_2 \sin\varphi_2, \\ v_1 &= \omega_1 A_1 \cos\varphi_1 + \omega_2 B_1 \cos\varphi_2, \\ v_2 &= \nu_1\omega_1 A_1 \cos\varphi_1 + v_2\omega_2 B_1 \cos\varphi_2. \end{aligned} \tag{8.18}$$

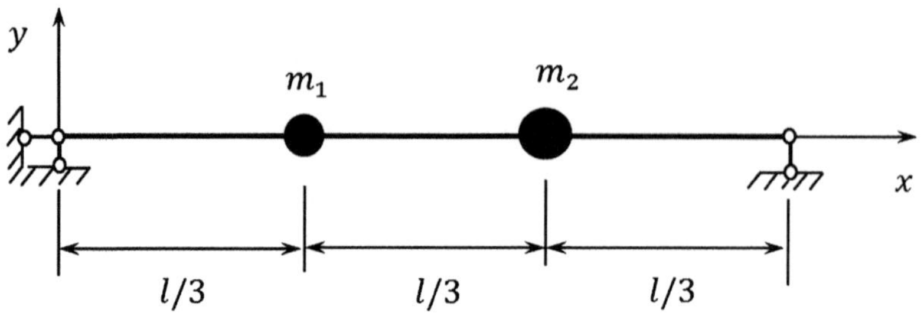

Fig. 8.2 A weightless beam with two discrete masses m_1, and m_2

Solution (8.18) allows us to conclude that free oscillations of real systems can occur both in one of their eigenmodes (natural modes), and in a mode (form) that consists of a set of several eigenmodes (natural modes).

Problem 8.1

Determine the frequencies of natural oscillations of a weightless beam with two concentrated masses m_1, and m_2 (located in the third of the span. Figure 8.2). From the condition of the problem, mass m_2 is twice as large as mass m_1.

Solution

Since the number of degrees of freedom is equal to two, then the equation of frequency will take the form:

$$\begin{vmatrix} \delta_{11}m_1 - \frac{1}{\omega^2} & \delta_{12}m_2 \\ \delta_{21}m_1 & \delta_{22}m_2 - \frac{1}{\omega^2} \end{vmatrix} = 0$$

Following the condition of the problem let's introduce the following designation:

$$m_1 = m_0; \;\; m_2 = 2m_0.$$

To determine the displacements, let's construct the unit force diagrams in Fig. 8.3.

The displacements from the unit forces are calculated by multiplying the diagrams of bending moments constructed from them.

$$\delta_{11} = \delta_{22} = \frac{1}{EI}\left(\frac{2}{9}l \cdot \frac{l}{3} \cdot \frac{1}{2} \cdot \frac{2}{3} \cdot \frac{2}{9}l + \frac{2}{9}l \cdot \frac{2}{3}l \cdot \frac{1}{2} \cdot \frac{2}{3} \cdot \frac{2}{9}l\right) = \frac{4l^3}{243EI},$$

$$\delta_{12} = \delta_{21} = \frac{1}{EI}\left[\frac{2}{9}l \cdot \frac{l}{3} \cdot \frac{1}{2} \cdot \frac{2}{3} \cdot \frac{2}{9} \cdot l \cdot 2 + \frac{2}{9}l \cdot \frac{l}{3} \cdot \frac{1}{2}\left(\frac{2}{3} \cdot \frac{l}{9} + \frac{1}{3} \cdot \frac{2}{9}l\right) + \right.$$
$$\left. + \frac{l}{9} \cdot \frac{l}{3} \cdot \frac{1}{2}\left(\frac{2}{3} \cdot \frac{2}{9}l + \frac{1}{3} \cdot \frac{l}{9}\right)\right] = \frac{7l^3}{486EI}.$$

Therefore,

$$\begin{vmatrix} \frac{8l^3}{486EI}m_0 - \frac{1}{\omega^2} & \frac{7l^3}{486EI}2m_0 \\ \frac{7l^3}{486EI}m_0 & \frac{8l^3}{486EI}2m_0 - \frac{1}{\omega^2} \end{vmatrix} = 0$$

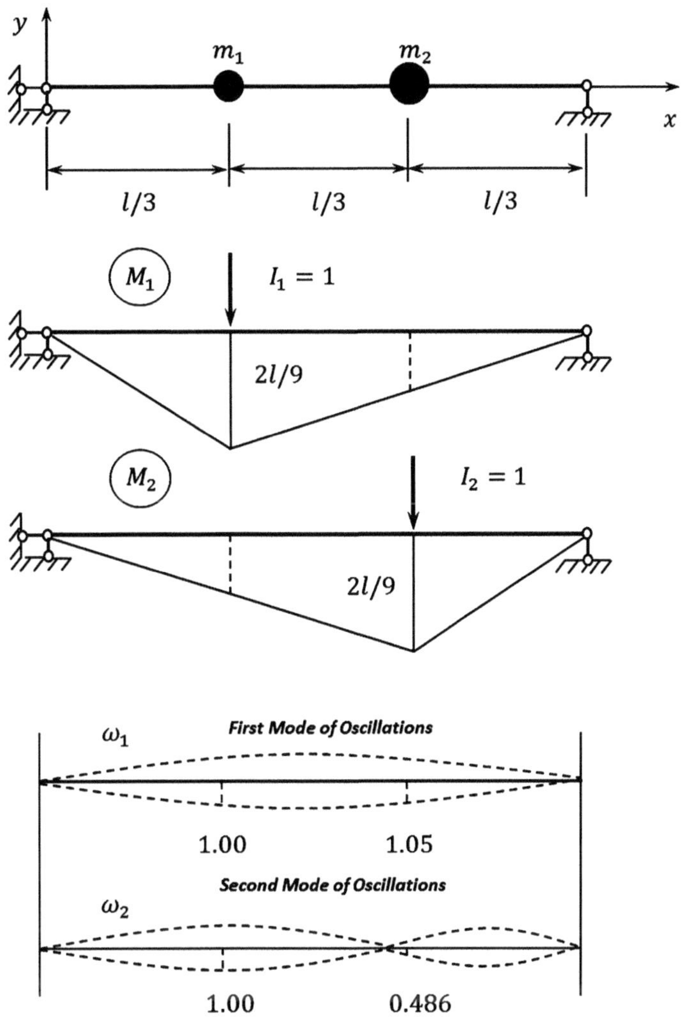

Fig. 8.3 The unit force diagrams and main modes of oscillations/vibrations

The square equation with respect to $\frac{1}{\omega^2}$ is obtained by expanding the determinant:

$$\left(\frac{8l^3}{486EI}m_0 - \frac{1}{\omega^2}\right)\left(\frac{8l^3}{486EI}2m_0 - \frac{1}{\omega^2}\right) - \frac{7l^3}{486EI}m_0 \cdot \frac{7l^3}{486EI}2m_0 = 0$$

Let's introduce the following designation:

$$\frac{l^3}{486EI}m_0 = k$$

then

$$\left(8k\omega^2 - 1\right)\left(16k\omega^2 - 1\right) - 7k \cdot 14k \cdot \omega^4 = 0$$

from which

$$\omega_1^{-2} = 22.68k; \qquad \omega_2^{-2} = 1.32k$$

Therefore,

$$\omega_1 = \frac{1}{\sqrt{22.68k}} = 4.6\sqrt{\frac{EI}{m_0 l^3}}; \qquad \omega_2 = \frac{1}{\sqrt{1.32k}} = 19.2\sqrt{\frac{EI}{m_0 l^3}}$$

Since each frequency corresponds to its own mode of oscillations (8.6), Eq. (8.6) will take the following form:

$$\left(\delta_{11}m_1 - \frac{1}{\omega^2}\right)A_1 + \delta_{12}m_2A_2 = 0$$

$$\delta_{21}m_1A_1 + \left(\delta_{22}m_2 - \frac{1}{\omega^2}\right)A_2 = 0$$

or

$$\left(8k - \omega^{-2}\right)A_1 + 14kA_2 = 0$$

$$7kA_1 + \left(16k - \omega^{-2}\right)A_2 = 0$$

Substituting the value $\omega_1^{-2} = 22.68k$ into one of the equations, we obtain:

$$A_2 = 1.05A_1,$$

and if we substitute the value $\omega_2^{-2} = 1.32k$, we obtain:

$$A_2 = -0.468A_1.$$

Based on the data obtained, we will construct the oscillation modes for the beam that correspond to the frequencies: ω_1 and ω_2. See Fig. 8.3

Let's simplify the problem under consideration and determine frequencies of natural oscillations of a weightless beam with two equal concentrated masses, Figure 8.4.

Let's substitute the values δ_{ik} into the frequency equation. The equation of frequencies for beam with two equal concentrated masses will take the following form:

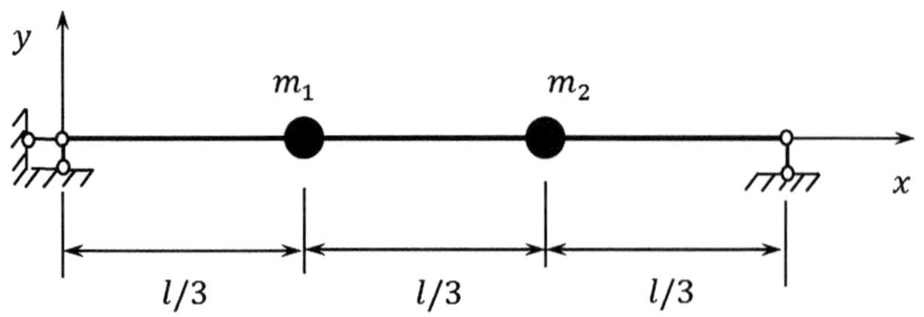

Fig. 8.4 A weightless beam with two equal discrete masses

$$\begin{vmatrix} \frac{4l^3}{243EI}m - \frac{1}{\omega^2} & \frac{7l^3}{486EI}m \\ \frac{7l^3}{486EI}m & \frac{4l^3}{243EI}m - \frac{1}{\omega^2} \end{vmatrix} = 0$$

The square equation with respect to $\frac{1}{\omega^2}$ is obtained by expanding the determinant:

$$\left(\frac{1}{\omega^2}\right)^2 - \frac{16l^3m}{486EI} \cdot \frac{1}{\omega^2} + \frac{15l^6m^2}{(486EI)^2} = 0,$$

where

$$\omega_1 = 5.692\sqrt{\frac{EI}{ml^3}}, \qquad \omega_2 = 22.045\sqrt{\frac{EI}{ml^3}}.$$

Let's find the characteristics of the primary (major) modes of oscillation:

$$\nu_1 = -\frac{\delta_{11}m_1\omega_1^2 - 1}{\delta_{12}m_2\omega_1^2} = -\frac{8}{7} + \frac{486}{7} \cdot \frac{EI}{ml^3} \cdot \frac{5ml^3}{162EI} = 1,$$

$$\nu_2 = -\frac{\delta_{11}m_1\omega_2^2 - 1}{\delta_{12}m_2\omega_2^2} = -\frac{8}{7} + \frac{486}{7} \cdot \frac{EI}{ml^3} \cdot \frac{ml^3}{486EI} = -1.$$

Therefore, the general solution (8.12) will take the form:

$$y_1 = A\sin(\omega_1 t + \varphi_1) + B\sin(\omega_2 t + \varphi_2)$$

$$y_2 = A\sin(\omega_1 t + \varphi_1) - B\sin(\omega_2 t + \varphi_2)$$

Thus, the concentrated masses of the system under consideration oscillate in the same phase with a lower frequency and in the opposite phase with a higher frequency. From this, it follows that the shape of the bending of the beam at the oscillations of the lowest (first) mode will be symmetric, which coincides with the action of the weight of both masses. The shape of the bending of the beam at the oscillations that correspond to the second mode will be an inversely symmetric (Fig. 8.3).

When determining the frequencies of free oscillations in a frame system, usually a uniformly distributed mass along the length of each bar of the frame is concentrated at the rod midpoint. The deformed shape of the frame axes is considered for determining the number of degrees of freedom of the system. The number of degrees of freedom of the system is determined by the number of possible displacements of the centers of mass of the frame. In some cases, the task can be simplified. For example, with the inversely symmetrical shape of oscillations of the frames, vertical displacements in the horizontal elements of the frame can be neglected, and only their horizontal displacements can be considered. Let us demonstrate the above by an example of determining the frequency of free vibrations of frames.

Problem 8.2

It is required to find the lowest circular frequency of natural oscillations of the frame shown in Fig. 8.5a. The rigidity *EI* and linear mass *m* of horizontal and vertical frame elements are assumed to be constant.

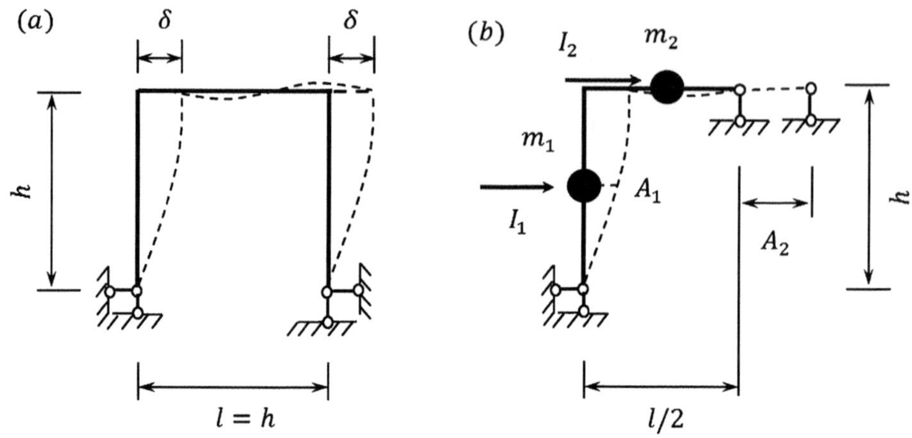

Fig. 8.5 Dynamic design model (DDM): (**a**) Frame structure; (**b**) Weightless frame with discrete masses

Solution

Consider the inversely symmetric shape of the frame oscillations. This mode of oscillation gives the lowest circular frequency of natural oscillations. Here, we will try to use the inversely symmetric form (deflection) of frame oscillations. We will cut the given frame along the axis of symmetry and get a new frame scheme with a horizontally movable support Fig. 8.5b. Here it is possible to ignore the longitudinal deformations of the frame elements, as well as the vertical movements of the crossbar points. Thus, we obtain a system with two degrees of freedom: mass $m_1 = mh$ has the amplitude of horizontal oscillations A_1, and mass $m_2 = ml/2$ has the amplitude A_2. For inertia forces, let us write the following expressions:

$$I_1 = m_1\omega^2 A_1; \qquad I_2 = m_2\omega^2 A_2.$$

To simplify the problem, we will introduce the following notation $x = mh\omega^2$, we will obtain:

$$I_1 = A_1 x; \qquad I_2 = 0.5A_2 x.$$

Following the calculated dynamic scheme of the problem, the frame has two degrees of freedom. Figure 8.5b. Let's write down equations based on Eq. (8.2):

$$A_1 = A_1 x\delta_{11} + 0.5A_2 x\delta_{12};$$

$$A_2 = A_1 x\delta_{21} + 0.5A_2 x\delta_{22}.$$

The determinant of the system of coefficients at A_1 and A_2 to equate to zero:

$$\begin{vmatrix} (x\delta_{11} - 1\,) & 0.5x\delta_{12} \\ x\delta_{21} & (0.5x\delta_{22} - 1) \end{vmatrix} = 0,$$

or in expanded form we obtain the following equation:

$$0.5x^2\left(\delta_{11}\delta_{22} - \delta_{12}^2\right) - x(\delta_{11} + 0.5\delta_{22}) + 1 = 0.$$

Further using the rule of bending moment diagram multiplication, we find displacements:

$$\delta_{11} = \frac{5}{24}\frac{h^3}{EI}; \quad \delta_{22} = \frac{h^3}{2EI}; \quad \delta_{12} = \frac{5}{16}\frac{h^3}{EI}.$$

The values of the displacements are substituted into the previous equation, which is written in the expanded form:

$$5u^2 - 704u + 1536 = 0,$$

where the following notation is introduced:

$$u = xh^3/EI.$$

Having solved this equation, we find:

$$u_1 = 2.218; \quad u_2 = 138.52.$$

The lowest circular frequency of oscillations:

$$\omega_1 = \frac{1.489}{h^2}\sqrt{\frac{EI}{m}}.$$

Problem 8.3
It is required to determine the frequencies of free oscillations of a continuous beam with two concentrated masses using the method of forces and the method of displacement (Fig. 8.6). The rigidity EI and linear mass m of a continuous beam are assumed to be constant. By the condition of the problem, the mass $m_1 = m_2 = m$.

Solution
Since the given system and the masses located on it are symmetrical, the simplifications associated with this can be used when choosing a design of a dynamic scheme.

Let's start by plotting M_1 and M_2 diagrams from unit group forces (Fig. 8.7a, b). We find the displacements of a continuous beam by multiplying the corresponding diagrams of bending moments:

$$\delta_{11} = \frac{l^3}{24EI}; \quad \delta_{22} = \frac{14l^3}{768EI}.$$

The frequency equation for inversely symmetric oscillations (Fig. 8.7c):

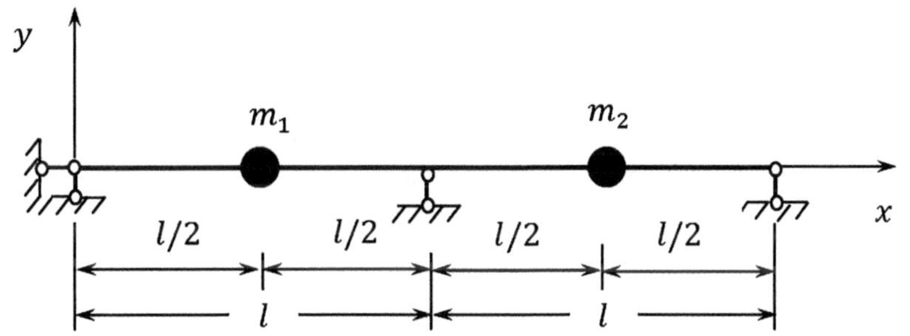

Fig. 8.6 A weightless continuous beam with discrete masses

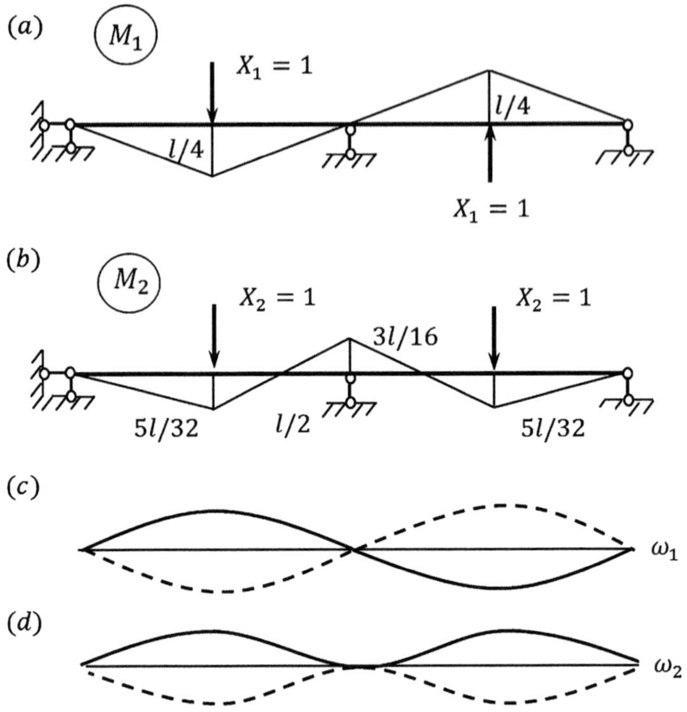

Fig. 8.7 (**a**, **b**) The bending moment diagrams from unit group forces. (**c**, **d**) Main modes of oscillations/vibrations corresponding to the first and second frequency of vibrations/oscillations

$$\delta_{11}\frac{m}{2} - \frac{1}{\omega_1^2} = 0,$$

from here:

$$\omega_1 = \sqrt{\frac{2}{\delta_{11}m}} = \sqrt{\frac{48EI}{l^3 m}} = 6.94\sqrt{\frac{EI}{l^3 m}}.$$

The frequency equation for symmetrical oscillations (Fig. 8.7d):

$$\delta_{22}\frac{m}{2} - \frac{1}{\omega_2^2} = 0,$$

from here

$$\omega_2 = \sqrt{\frac{2}{\delta_{22}m}} = \sqrt{\frac{768EI}{7l^3 m}} = 10.45\sqrt{\frac{EI}{l^3 m}}.$$

Now, let's move onto determining the oscillation frequencies using the displacement method.

The essence of the displacement method is that inertia forces are expressed as functions of displacement. For example, the inertia forces of the masses m_1, m_2,. . ., m_i, . . .,m_n for the system with a finite number of degrees of freedom shown in Fig. 8.8.

can be represented by the following system of differential equations:

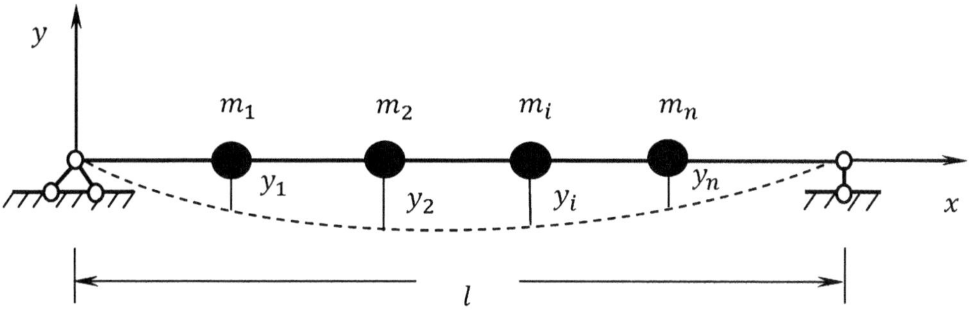

Fig. 8.8 A weightless beam with m_n concentrated masses

$$\begin{aligned} -m_1\ddot{y}_1 &= y_1 r_{11} + y_2 r_{12} + \ldots + y_i r_{1i} + \ldots + y_n r_{1n} \\ -m_2\ddot{y}_2 &= y_1 r_{21} + y_2 r_{22} + \ldots + y_i r_{2i} + \ldots + y_n r_{2n} \\ -m_i\ddot{y}_i &= y_1 r_{i1} + y_2 r_{i2} + \ldots + y_i r_{ii} + \ldots + y_n r_{in} \\ -m_n\ddot{y}_n &= y_1 r_{n1} + y_2 r_{n2} + \ldots + y_i r_{ni} + \ldots + y_n r_{nn} \end{aligned} \tag{8.19}$$

The reactions r_{11}, r_{22}, r_{ii}, r_{nn}, and $r_{12} = r_{21}$, $r_{1n} = r_{n1}$, $r_{1i} = r_{i1}$, $r_{2i} = r_{i2}$, $r_{2n} = r_{n2}$, $r_{in} = r_{ni}$ are calculated as usual from unit displacements. These displacements are applied in sections where the concentrated masses of the system are located. Expressions (8.4) and (8.5) remain valid for displacements and their second derivatives. These expressions are substituted into Eq. (8.19), as the result of which we obtain the following system of homogeneous equations:

$$\begin{aligned} &\left(r_{11} - m_1\omega^2\right)A_1 + r_{12}A_2 + \ldots + r_{1i}A_i + \ldots r_{1n}A_n = 0 \\ &r_{21}A_1 + \left(r_{22} - m_2\omega^2\right)A_2 + \ldots + r_{2i}A_i + \ldots + r_{2n}A_n = 0 \\ &r_{i1}A_1 + r_{i2}A_2 + \ldots + \left(r_{ii} - m_i\omega^2\right)A_i + \ldots + r_{in}A_n = 0 \\ &r_{n1}A_1 + r_{n2}A_2 + \ldots + r_{ni}A_i + \ldots + \left(r_{nn} - m_n\omega^2\right)A_n = 0 \end{aligned} \tag{8.20}$$

Then, let us compose the determinant from the coefficients of this equation. The frequencies of natural vibrations of the system can be found by solving the determinant. The frequencies of natural vibrations of the system with any number of degrees of freedom n can be determined by the same principle. The determinant is written in the following form:

$$D = \begin{vmatrix} \left(r_{11} - m_1\omega^2\right) & r_{12} \;\ldots & r_{1i} \;\ldots & r_{1n} \\ r_{21} & \left(r_{22} - m_2\omega^2\right) & \ldots r_{2i} \ldots & r_{2n} \\ r_{i1} & r_{i2} \ldots & \left(r_{ii} - m_i\omega^2\right) \;\ldots & r_{in} \\ r_{n1} & r_{n2} \ldots & r_{ni} \ldots & \left(r_{nn} - m_n\omega^2\right) \end{vmatrix} = 0 \tag{8.21}$$

A beam that carries concentrated masses does not always coincide with one of the cases of standard tables for determining displacements when calculating beams and frames for a static load. In such cases, the reactions of the system must be calculated through displacements. Therefore, this method turns out to be more laborious than the previous one, the force method.

Problem 8.4

Determine the frequency of free vibrations of the continuous beam considered earlier and shown in Fig. 8.6. by the method of displacements.

Solution

Reactions caused by unit displacements can be found based on previously found displacements from unit forces.

$$r_{11} = \frac{1}{\delta_{11}} = \frac{24EI}{l^3}; \qquad r_{22} = \frac{1}{\delta_{22}} = \frac{384EI}{7l^3}$$

The frequency equation for inversely symmetrical and symmetrical vibrations:

$$r_{11} - \frac{m}{2}\omega_1^2 = 0,$$

$$r_{22} - \frac{m}{2}\omega_2^2 = 0.$$

From here the corresponding frequencies can be obtained

$$\omega_1 = \sqrt{\frac{2r_{11}}{m}} = 6.94\sqrt{\frac{EI}{l^3 m}}.$$

$$\omega_2 = \sqrt{\frac{2r_{22}}{m}} = 10.45\sqrt{\frac{EI}{l^3 m}}.$$

8.2 An Approximate Estimate of the Frequency of the Fundamental Tone of Vibrations

The determination of all the frequencies of the system's free vibrations/oscillations is often superfluous (redundant), and in such problems the determination of only the lowest (first) frequency of vibrations is sufficient. As we know, this may be the case when the frequency of the disturbing load is lower than the first oscillation/vibration frequency of the structure, and therefore, resonance with higher frequencies is already excluded. In this regard, S.A. Bernshtein's method deserves attention. The frequencies that are determined (found) by this method depend on the elastic properties, displacements, and loads of the structure. This method, being essentially a sequential approximation method, allows at any stage of the calculation to evaluate discrepancies with the exact result. To determine the boundaries, the limits between which the frequency of the fundamental tone of oscillations/vibrations can be used the formula of S. A. Bernshtein [3]. These limits are called two-sided estimates of the frequency ω_1 and can be written in the following form:

$$\frac{1}{\sqrt[4]{B_2}} < \omega_1 < \frac{1}{\sqrt{\frac{B_1}{2}\left(1 + \sqrt{\frac{2B_2}{B_1^2} - 1}\right)}} \tag{8.22}$$

Where

$$B_1 = \int_0^l m(x)\delta(x.x)dx + \sum_i m_i\delta_{ii}; \tag{8.23}$$

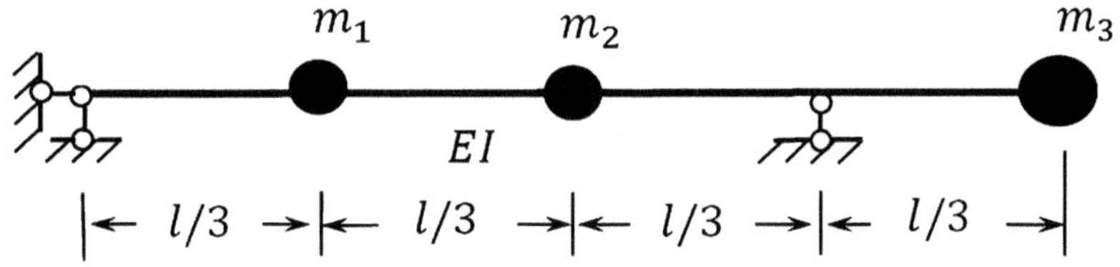

Fig. 8.9 A weightless beam with three discrete masses

$$B_2 = \int_0^l \int_0^l m(x)m(s)\delta(x,s)dxds + \sum_i \sum_k m_i m_k \delta(x_i, x_k). \tag{8.24}$$

For a system with lumped masses only we have

$$B_1 = \sum_{i=1}^{i=n} m_i \delta_{ii} \tag{8.25}$$

$$B_2 = \sum_{i=1}^{i=n} m_i^2 \delta_{ii}^2 + 2\sum_{i<k} m_i m_k \delta_{ik}^2 \tag{8.26}$$

Here δ_{ii} and δ_{ik} are the main and, respectively, secondary displacements of the system from the action of unit forces at the points of application of concentrated masses m_i, and m_k.

Problem 8.5

It is required to determine the lowest frequency of free vibrations of the beam shown in Fig. 8.9.

By the condition of problem, the mass m_3 is three times the mass of m_1. Following the condition of the problem let us introduce the following designation: $m_1 = m_2 = m_0$; $m_3 = 3m_0$.

To determine the displacements, which are included in formula (A), let's build a diagram of the unit forces (Fig. 8.10). The unit forces are applied at the points of concentrated masses of the system.

Let's find the displacements using the unit diagrams shown in Fig. 8.10.

$$\delta_{11} = \delta_{22} = \frac{1}{EI}\left(\frac{2}{9}l\frac{1}{3}l\frac{1}{2}\frac{2}{3}\frac{2}{9}l + \frac{2}{9}l\frac{2}{3}l\frac{1}{2}\frac{2}{3}\frac{2}{9}l\right) = \frac{4l^3}{243EI} = \frac{8l^3}{486EI}$$

$$\delta_{33} = \frac{1}{EI}\left(\frac{l}{3}l\frac{1}{2}\frac{2}{3}\frac{l}{3} + \frac{l}{3}\frac{l}{3}\frac{1}{2}\frac{2}{3}\frac{l}{3}\right) = \frac{4l^3}{81EI} = \frac{24l^3}{486EI}$$

$$\delta_{12} = \delta_{21} = \frac{1}{EI}\left[\frac{2}{9}l\frac{1}{3}l\frac{1}{2}\frac{2}{3}\frac{1}{3}l\cdot 2 + \frac{2}{9}l\frac{1}{3}l\frac{1}{2}\left(\frac{2}{3}\frac{1}{9}l + \frac{1}{3}\frac{2}{9}l\right) + \right.$$
$$\left. + \frac{1}{9}l\frac{1}{3}l\frac{1}{2}\left(\frac{2}{3}\frac{2}{9}l + \frac{1}{3}\frac{1}{9}l\right)\right] = \frac{7l^3}{486EI}$$

The remaining side movements that are included in formula (A) are determined in the same way:

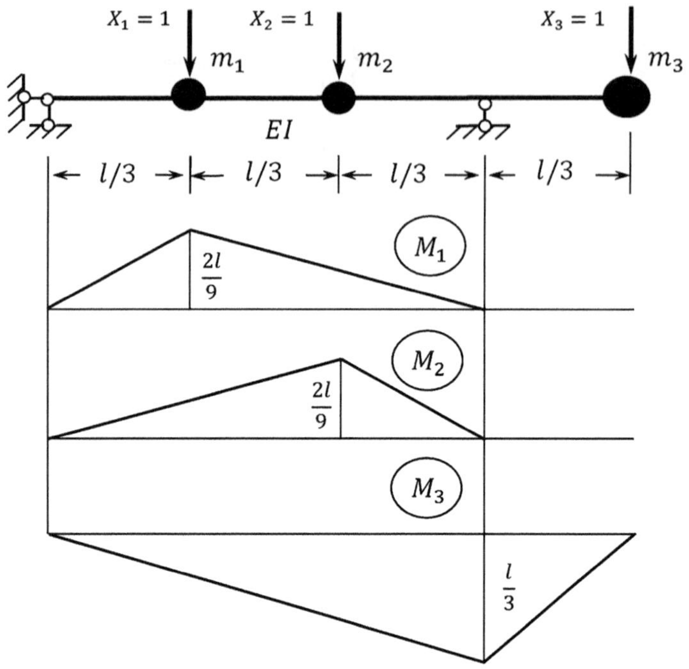

Fig. 8.10 The bending moment diagrams from unit forces applied at the points of concentrated masses of the beam system

$$\delta_{13} = \delta_{31} = -\frac{8l^3}{486EI}; \delta_{23} = \delta_{32} = -\frac{10l^3}{486EI}.$$

Let's write down the values of the parameters included in the formula for determining the fundamental frequency of free oscillations of the problem under consideration:

$$B_1 = \sum_{i=1}^{i=n} m_i\delta_{ii} = m_1\delta_{11} + m_2\delta_{22} + m_3\delta_{33} = \frac{8l^3 m_1}{486EI} + \frac{8l^3 m_2}{486EI} + \frac{24l^3 m_3}{486EI} =$$

$$= \frac{8l^3 m_0}{486EI} + \frac{8l^3 m_0}{486EI} + \frac{24l^3 \cdot 3m_0}{486EI} = \frac{88l^3 m_0}{486EI}$$

$$B_2 = \sum_{i=1}^{i=n} m_i^2\delta_{ii}^2 + 2\sum_{i<k} m_i m_k \delta_{ik}^2 = m_1^2\delta_{11}^2 + m_2^2\delta_{22}^2 + m_3^2\delta_{33}^2 +$$

$$+2\left(m_1 m_2 \delta_{12}^2 + m_2 m_3 \delta_{23}^2 + m_1 m_3 \delta_{13}^2\right) = \left(\frac{8l^3 m_0}{486EI}\right)^2 + \left(\frac{8l^3 m_0}{486EI}\right)^2 +$$

$$+\left(\frac{24l^3 \cdot 3m_0}{486EI}\right)^2 + 2\left[\left(\frac{7l^3}{486EI}\right)^2 m_0 m_0 + \left(\frac{10l^3}{486EI}\right)^2 m_0 3m_0 +\right.$$

$$\left.+\left(\frac{8l^3}{486EI}\right)^2 m_0 3m_0\right] = 6394\left(\frac{l^3 m_0}{486EI}\right)^2$$

The desired frequency of free oscillations is within the following limits:

$$\omega_1 > \frac{1}{\sqrt[4]{B_2}} > \frac{1}{\sqrt[4]{6394\left(\frac{l^3 m_0}{486EI}\right)^2}} = 2.45\sqrt{\frac{EI}{m_0 l^3}}$$

$$\omega_1 < \frac{1}{\sqrt{\frac{B_1}{2}\left(1+\sqrt{\frac{2B_2}{B_1^2}}-1\right)}} < \frac{1}{\sqrt{\frac{88 l^3 m_0}{486EI\cdot 2}\left(1+\sqrt{\frac{2\cdot 6394}{88^2}}-1\right)}} = 2.47\sqrt{\frac{EI}{m_0 l^3}}$$

The average frequency of the fundamental frequency of free oscillations will be:

$$\omega_1 = 2.46\sqrt{\frac{EI}{m_0 l^3}}$$

References

1. Voevodin, V. V. (1966). *Numerical methods of algebra. Theory and algorithms* (246p). Science.
2. Wilkinson, J. H., & Reinsch, C. (1971). *Linear algebra, handbook for automatic computation* (Vol. II). Springer.
3. Bernshtein, S. A. (1960). *Basics of the dynamics of structures* (160p). Gosstroyizdat.
4. Golub, G. H. (1973). Some modified matrix eigenvalue problems. *SIAM Review, 15*, 318–334.
5. Fadeev, D. K., & Fadeeva, V. N. (1980). *Computational methods of linear algebra* (665p). Fizmatgiz.
6. Wilkins, J. H. (1988). *The algebraic Eigenvalue problem* (680p). Oxford University Press Academic.
7. Melman, A. (1995). Numerical solution of a secular equation. *Numerische Mathematik, 69*, 483–493.

Forced Vibration of a System with Multi-Degree of Freedom

9

Abstract

This chapter presents forced vibrations of a system with multi-degree of freedom. An elastic system with concentrated (lumped) masses acting on vibrational forces is considered. Bending moments, shear, and longitudinal forces that vary over time when a harmonic vibrational load on an elastic system with many degrees of freedom is discussed. The chapter includes extensive practical examples and problems with solutions.

9.1 Equations of Motion Under Forced Vibrations

Let us apply the inertial load method (D'Alembert principle) to a system with *n* degrees of freedom. As an example, consider a beam that carries *n* concentrated masses, Fig. 9.1.

The bending moments, transverse, and longitudinal forces under the action on an elastic system with many degrees of freedom of a vibrating harmonic load change over time. The largest values of the oscillation amplitudes depend on the frequency of the disturbing (perturbing) force or on the frequencies of the disturbing forces acting on the system. If all disturbing forces acting on the system have the same frequency θ and change in the same phase, then the forces of inertia, bending moments, transverse, and longitudinal forces reach the highest (maximum) values at the same time. For systems with many degrees of freedom, several resonance cases are possible. The resonance may occur when the frequency coincides with certain frequencies of free vibrations/oscillations of the system.

The task of dynamic calculation includes determining the amplitudes of internal forces and stresses, as well as checking the system for resonance. Let us consider the elastic system shown in Fig. 9.1, with point masses, on which vibration forces act $P_i(t) = P_i \sin \theta t$. $(i = 1, \quad 2, \quad \ldots, \quad n)$ [1, 2].

Using the force method and based on the D'Alembert principle,

$$I_i = -m_i \frac{d^2 y_i}{dt^2}$$

we introduce the forces of inertia and the displacement of each mass m_i at an arbitrary point in time t is expressed as follows:

$$y_i = \delta_{i1} I_1 + \delta_{i2} I_2 + \cdots + \delta_{ii} I_i \ldots + \cdots + \delta_{in} I_n + \Delta_{ip} \tag{9.1}$$

L. G. Petrosian, *Dynamics of Structures*, https://doi.org/10.1007/978-3-031-63539-7_9

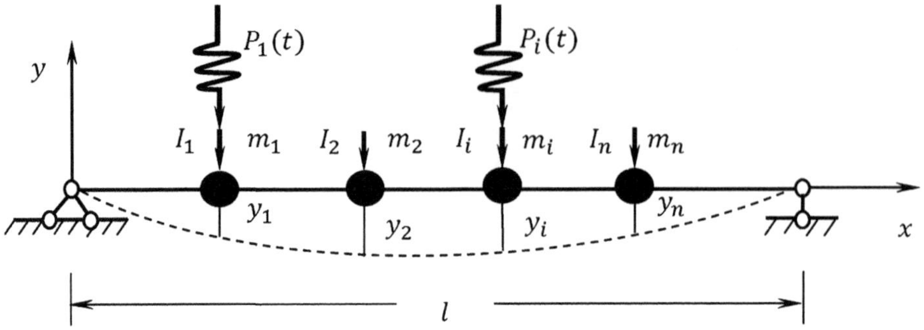

Fig. 9.1 Beam with lumped masses acting on vibrational forces

or

$$y_i = -\sum_{k=1}^{n} \delta_{ik} m_k \frac{d^2 y_k}{dt^2} + \Delta_{ip} \qquad i = 1,\ 2,\ \ldots,\ n \tag{9.2}$$

Thus, the differential equations (9.1) and (9.2) describe the oscillatory motions of the system under consideration.

Where $I_1,\ I_2,\ \ldots,\ I_i,\ \ldots,\ I_n$ are inertial forces of the corresponding masses; $\delta_{i1},\ \delta_{i2},\ \ldots,\ \delta_{ii},\ \ldots,\ \delta_{in}$ are displacements in the direction of the inertial forceI_i, which are caused by the unit forces I_1, $I_2,\ \ldots,\ I_i,\ \ldots,\ I_n$. The unit forces are applied at the locations of the corresponding masses. Δ_{ip} is the displacement of point i from the amplitude values of the vibration loads. The diagram from loads (diagram of M_p) in determining these displacements is integrated or multiplied with the corresponding unit diagrams.

The displacement of mass m_i and its acceleration in harmonic forced oscillations with a frequency θ are expressed as follows:

$$y_i = a_i \sin \theta t; \qquad \ddot{y}_i = -a_i \theta^2 \sin \theta t = y_i \theta^2 \tag{9.3}$$

The force of inertia of the mass m_i can be represented as:

$$I_i = -m_i \ddot{y}_i = m_i y_i \theta^2 \tag{9.4}$$

Hence it follows that

$$y_i = \frac{I_i}{m_i \theta^2} \tag{9.5}$$

In the case of free oscillations $\Delta_{ip} = 0, \quad i = 1, 2, \ldots n$, we obtain:

$$y_i = -\sum_{k=1}^{n} \delta_{ik}\ m_k \frac{d^2 y_k}{dt^2} \qquad i = 1, 2, \ldots, n \tag{9.6}$$

Thus, Eq. (9.6) is a homogeneous system of differential equations. The particular solutions of which have the following form:

$$y_i = a_i \sin(\omega t + \varphi_0), \qquad i = 1, 2, \ldots, n$$

The partial solution of the differential equation is substituted in Eq. (9.6) as a result, we obtain a system of linear homogeneous equations with respect to a_i, with free oscillations/vibrations. From the condition of nonzero values of a_i, the determinant composed of coefficients at the amplitudes of free oscillations of the system should be set to zero:

$$\begin{vmatrix} \delta_{11}-\dfrac{1}{m_1\omega^2} & \delta_{12}\ldots\delta_{1i}\ldots\delta_{1n} \\ \delta_{21} & \delta_{22}-\dfrac{1}{m_2\omega^2}\ldots\delta_{2i}\ldots\delta_{2n} \\ \ldots & \ldots \\ \delta_{i1} & \delta_{i2}\ldots\delta_{ii}-\dfrac{1}{m_i\omega^2}\ldots\delta_{in} \\ \ldots & \ldots \\ \delta_{n1} & \delta_{n2}\ldots\delta_{ni}\ldots\ \delta_{nn}-\dfrac{1}{m_n\omega^2} \end{vmatrix}=0$$

This condition in an expanded form gives an equation of the n-th degree for determining n values of ω^2. The corresponding frequency values represent the frequency spectrum of the system, i.e. the set of its free frequencies. The number of frequencies of the system is equal to the number of its degrees of freedom. The frequencies of free oscillations are distributed as follows: $\omega_1<\omega_2<\cdots<\omega_i<\cdots<\omega_n$.

Expression (9.5) is substituted into the original canonical equation, we get:

$$\delta_{i1}I_1+\delta_{i2}I_2+\cdots+\delta^*_{ii}I_i\ldots+\cdots+\delta_{in}I_n+\Delta_{ip}=0, \tag{9.7}$$

here

$$\delta^*_{ii}=\delta_{ii}-\frac{1}{m_i\theta^2};\qquad \Delta_{ip}=\sum\int\frac{\overline{M}_iM_p}{EI}dx \tag{9.8}$$

Any equations describing the forced oscillations of the system under the action of a vibrational periodic load are composed in the same way. As a result, the system of equations, which allows us to determine the largest (maximum) amplitude values of the forces of inertia will be obtained. For this, expression (9.5) is substituted into (9.1). The system of canonical equations with respect to inertia forces will have the following form:

$$\begin{aligned} &\left(\delta_{11}-\frac{1}{m_1\theta^2}\right)I_1+\delta_{12}I_2+\cdots+\delta_{1i}I_i+\cdots+\delta_{1n}I_n+\Delta_{1p}=0\\ &\delta_{21}I_1+\left(\delta_{22}-\frac{1}{m_2\theta^2}\right)I_2+\cdots+\delta_{2i}I_i+\cdots+\delta_{2n}I_n+\Delta_{2p}=0\\ &\cdots\cdots\cdots\cdots\cdots\cdots\cdots\cdots\cdots\cdots\\ &\delta_{i1}I_1+\delta_{i2}I_2+\cdots+\left(\delta_{ii}-\frac{1}{m_i\theta^2}\right)I_i+\cdots+\delta_{in}I_n+\Delta_{ip}=0\\ &\cdots\cdots\cdots\cdots\cdots\cdots\cdots\cdots\cdots\cdots\\ &\delta_{n1}I_1+\delta_{n2}I_2+\cdots+\delta_{ni}I_i+\cdots+\left(\delta_{nn}-\frac{1}{m_n\theta^2}\right)I_n+\Delta_{np}=0 \end{aligned} \tag{9.9}$$

Thus, the solution of the system of Eq. (9.9) provides the final values of the maximum inertia forces I_i for any given value of θ.

For the case when the frequency of the disturbing (perturbing) force coincides with any of the natural oscillation frequencies ω_i, we have a resonance phenomenon: i.e., an unlimited increase in oscillation amplitudes.

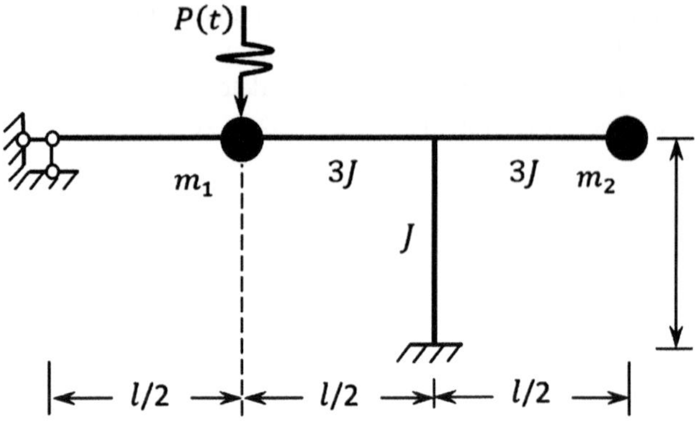

Fig. 9.2 Frame structure under the action of a periodic load

The dynamic diagram of bending moments is constructed after the inertia forces are determined from the solution of Eq. (9.9). On the basis of the principle of the independent action of forces, the dynamic diagram of bending moments is constructed by adding unit diagrams of bending moment with the diagram of bending moments from the vibration load.

$$M = \overline{M}_1 I_1 + \overline{M}_2 I_2 + \cdots + \overline{M}_i I_i + \cdots + \overline{M}_n I_n + M_p \tag{9.10}$$

Previously, unit diagrams of bending moment are multiplied by the found values of the corresponding inertial forces.

In dynamic calculation, the use of symmetry of the system is only possible in a symmetrical mass arrangement.

With an asymmetric vibration load, the decomposition of inertia forces into symmetrical and inversely symmetrical groups of forces leads to the breakdown of the system of canonical equations into two independent systems.

Problem 9.1

An example of calculation of the frame system under the action of a periodic dynamic load $P(t) = P_0 \sin \theta t$. Let us consider the frame structure (Fig. 9.2) under dynamic load. The load changes according to the law: $P_0 \sin \theta t$.

The following dynamic and geometric characteristics of the system are given:

$$P(t) = P_0 \sin \theta t;$$
$$\theta = 0.8\omega_{\min};$$
$$P_0 = 9\,\text{kN};\, G = 20\,\text{kN};$$
$$l = 8\,\text{m};\, = h = 4\,\text{m}.$$

It is required to determine:

1. ω_1, ω_2 are the frequency of free oscillations of the system;
2. The greatest values of inertial forces I_1, I_2; and
3. Build diagrams M, Q, and N from the weight (G) of concentrated masses and the joint effect of inertial forces of dynamic load.

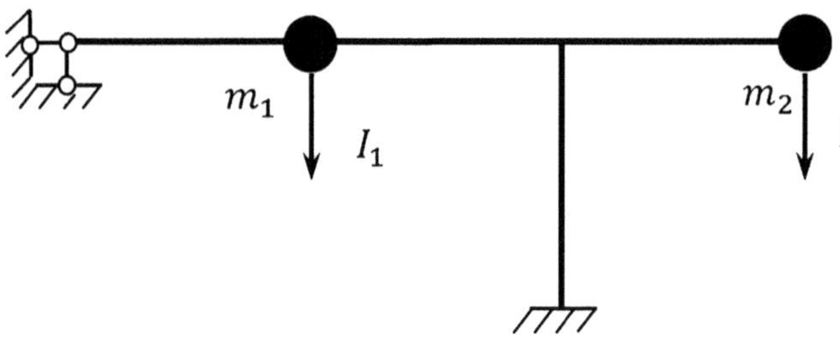

Fig. 9.3 Dynamic design model (DDM) for weightless frame with two lumped masses under periodic vibration load

Solution

The system in question has two degrees of freedom and therefore the free oscillations (vibrations) are characterized by two frequencies. The inertia forces arising from the dynamic load are shown in Fig. 9.3.

Determination of the free oscillations.

For a system with two degrees of freedom, the equation has the following form:

$$\begin{vmatrix} \delta_{11} - \dfrac{1}{m_1 \cdot \omega^2} & \delta_{12} \\ \delta_{21} & \delta_{22} - \dfrac{1}{m_2 \cdot \omega^2} \end{vmatrix} = 0 \tag{A}$$

In the problem considered: $m_1 = m_2 = m = G/g = 20/9.81 = 2.039$,

$$\delta_{jk} = \sum \int \frac{\overline{M}_j \overline{M}_k}{\mathrm{EI}} dx, \quad j, k = 1, 2.$$

Thus, to determine δ_{jk} we can build diagrams of moments $\overline{M}_1, \overline{M}_2$, respectively, from $I_1 = 1$, and from $I_2 = 1$ unit forces applied in the points of concentrated masses:

Building a diagram: $\overline{M}_1$

By using bars to be inextensible, it becomes possible to make a distinction between frames with displaceable and non-displaceable joints. Frames, in which none of the joints can experience sideways, will be referred to as non-displaceable frame, all others, displaceable.

We apply the method of displacements for the frame in question in the following sequence.

Determining the number of unknowns:

Using the displacement method for the frame shown in Fig. 9.4.

In this case, we have non-displaceable frame.

$$n = n_{\text{nodes/joints}} + n_{\text{sideways}} = 1 + 0 = 1$$

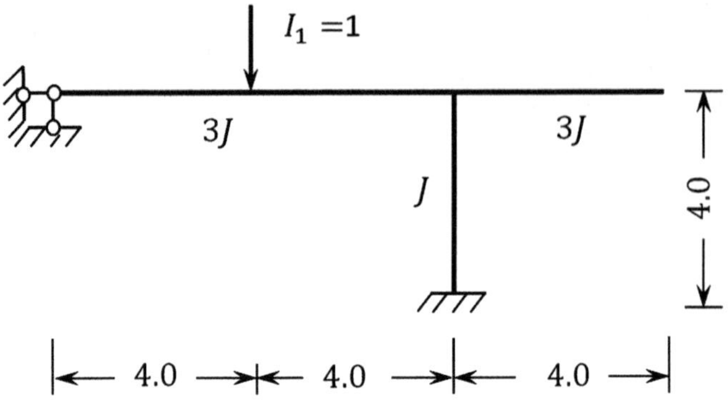

Fig. 9.4 Frame structure under the action of inertial force I_1

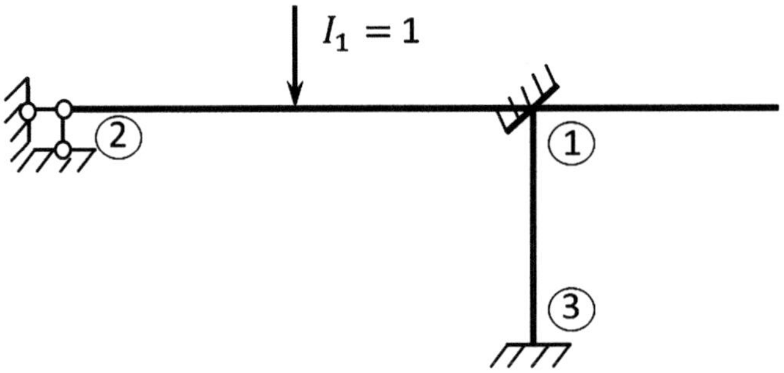

Fig. 9.5 Primary system for the method of displacements

Selection of the primary system for the method of displacements:

Canonical equation of the chosen method:

$$r_{11}Z_1 + R_{1I_1} = 0,$$

Determining the linear stiffness of horizontal and vertical elements of the frame (Fig. 9.5):

$$i_{1-2} = 3EI/l = 3/8 = 0.375,$$
$$i_{1-3} = EI/h = 1/4 = 0.25\ .$$

Plotting the diagrams of moments from a unit force and from an external load (Fig. 9.6):

$$r_{11} = 3i_{12} + 4i_{13} = 3 \cdot 0.375 + 4 \cdot 0.25 = 2.125,$$
$$R_{1I_1} = \frac{3}{16}l = 1.5,$$
$$Z_1 = -\frac{R_{1I_1}}{r_{11}} = -\frac{1.5}{2.125} = -0.706.$$
$$\overline{M}_1 = M_1 \cdot Z_1 + M_{I_1}$$

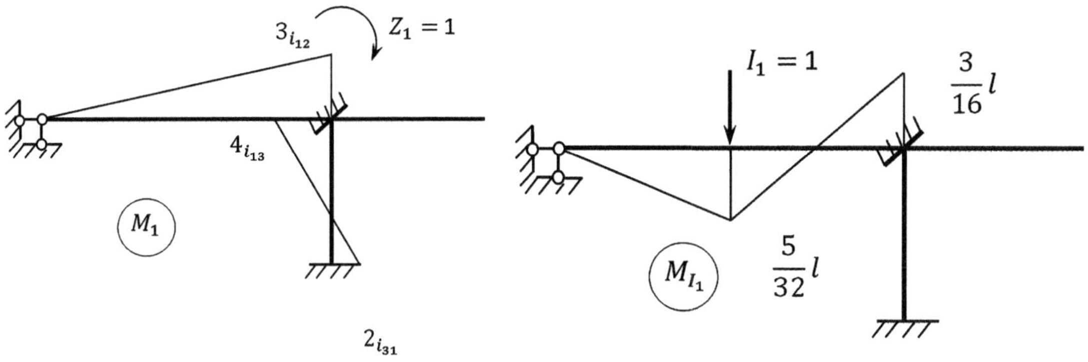

Fig. 9.6 Bending moment diagrams from a unit force and from an external load

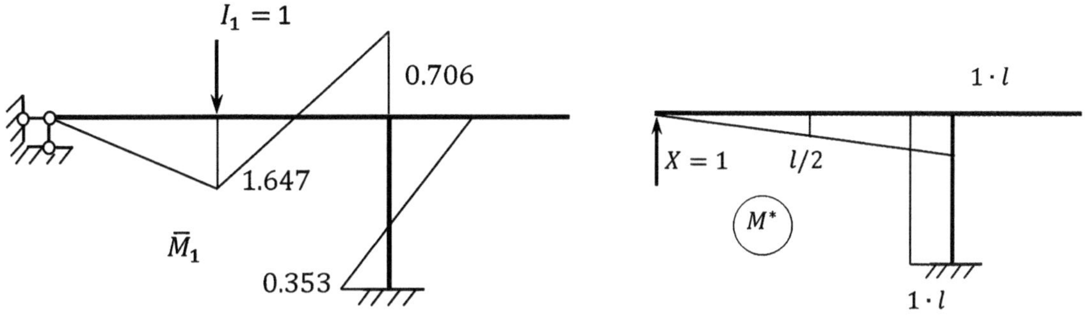

Fig. 9.7 Bending moment diagrams from a unit force of inertia I_1 and an additional diagram of moments from a unit force acting in the vertical support link of a given frame

Note: *Here, the bending moment diagrams have been shown from the side of the stretched zones of the elements of the frame* (Fig. 9.7).

To check the diagram of moments, let's construct an additional diagram of moments from a unit force acting in the vertical support link of the given frame.

$$\sum \int \frac{\overline{M}_1 M}{EI} dx = 0,$$

$$\left(-\frac{0.706 \cdot l}{EI} \cdot \frac{2}{3} l + \frac{1}{2} \cdot \frac{l}{4} \cdot l \cdot \frac{l}{2} \right) \frac{1}{3EI} - \frac{0.706 - 0.353}{2} h \cdot l \frac{1}{EI} = 0, \quad 0 = 0.$$

Construction (plotting) of moments diagrams:

In this case, it is also advisable to use the displacement method, while the values of the moments in the points remain the same, only the diagram of moments from the external load changes. The primary system and diagrams of moments from external load are shown in Fig. 9.8.

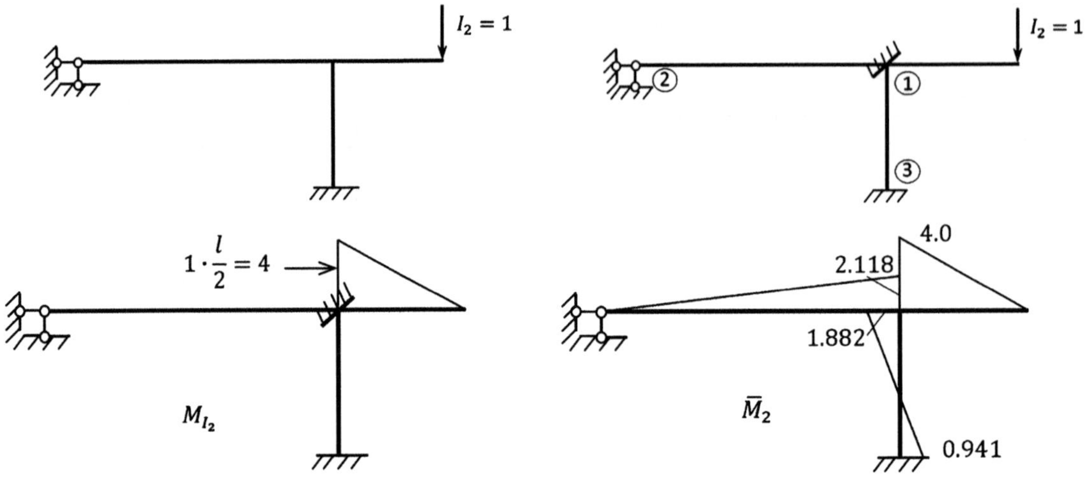

Fig. 9.8 Primary system and diagrams of moments from external load

$$r_{11}Z_1' + R_{1I_2} = 0, \qquad r_{11} = 2.125,$$
$$R_{1I_2} = -1 \cdot \frac{l}{2} = -4, \qquad Z_1' = -\frac{-4}{2.125} = 1.882,$$
$$\overline{M}_2 = M_1 Z_1' + M_{I_2}.$$

Checking the diagram of moments

$$\sum \int \frac{\overline{M}_2\, M^*}{\text{EI}}\, dx = 0,$$
$$-\frac{2.118 \cdot l}{2} \cdot \frac{2}{3} l \frac{1}{\text{EI}} + \frac{1.882 - 0.941}{2} \cdot h \cdot l \frac{1}{\text{EI}} = 0, \quad 0 = 0.$$

Determining the displacements: $\delta_{11}, \delta_{12}, \delta_{22}$

$$\delta_{11} = \sum \int \frac{M_1^2}{\text{EI}} dx \quad \text{or} \quad \delta_{11} = \sum \int \frac{\overline{M}_1 M_1^o}{\text{EI}} dx,$$

The diagram of bending moments $\overline{M}_1^o$ for the chosen primary system of the method of forces and the diagram of bending moments from the unit inertial force $I_1 = 1$, are shown in Fig. 9.9:

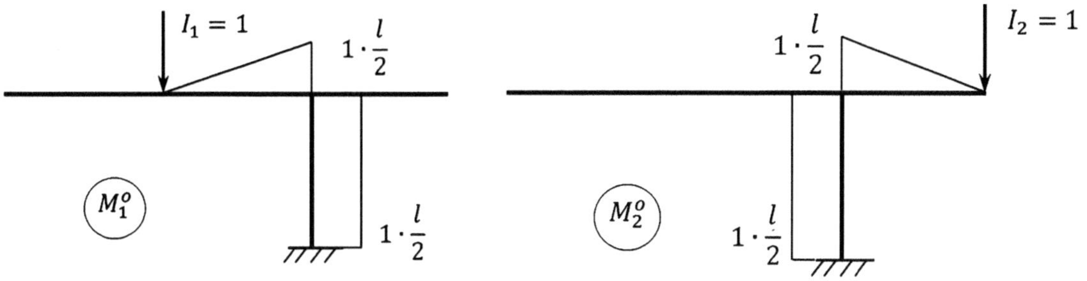

Fig. 9.9 Bending moment diagram for a primary system of the force method and diagram of bending moments due to a unit inertial force I_1

$$\delta_{11} = \sum \int \frac{\overline{M}_1 M_1^o}{\mathrm{EI}} dx = \frac{1}{2}\frac{l}{2} \cdot \frac{l}{2}\left(\frac{2}{3} \cdot 0.706 - \frac{1}{3} \cdot 1.647\right)\frac{1}{3\mathrm{EI}} + $$
$$+ \frac{0.706 - 0.353}{2} \cdot h \cdot \frac{l}{2} \cdot \frac{1}{\mathrm{EI}} = \frac{2.615}{\mathrm{EI}},$$
$$\delta_{12} = \delta_{21} = \sum \int \frac{\overline{M}_1 \overline{M}_2}{\mathrm{EI}} dx,$$
$$\delta_{12} = \sum \int \frac{\overline{M}_1 M_2^o}{\mathrm{EI}} dx = -\frac{0.706 - 0.353}{2} \cdot h \cdot \frac{l}{2} \cdot \frac{1}{\mathrm{EI}} = -\frac{2.824}{\mathrm{EI}},$$
$$\delta_{22} = \sum \int \frac{\overline{M}_2^2}{\mathrm{EI}} dx = \sum \int \frac{\overline{M}_2 M_2^0}{\mathrm{EI}} dx = \frac{1}{2} \cdot \frac{l}{2} \cdot \frac{l}{2} \cdot \frac{2}{3} \cdot 4 \cdot \frac{1}{3\mathrm{EI}} + $$
$$+ \frac{1.882 - 0.941}{2} \cdot h \cdot \frac{l}{2} \cdot \frac{1}{\mathrm{EI}} = \frac{14.639}{\mathrm{EI}}$$

Determining the frequencies ω_1, and ω_2:

For the case when $m_1 = m_2 = m$ from the Eq. (A) we get:

$$\omega = \frac{\sqrt{2}}{\sqrt{m\left[\delta_{11} + \delta_{22} \pm \sqrt{(\delta_{11} - \delta_{22})^2 + 4\delta_{12}^2}\right]}} \tag{B}$$

The roots of the Eq. (A) with a negative value have no physical meaning. Let us substitute the displacements δ_{11}, δ_{22} and δ_{12} values in (B)

$$\omega = \frac{1.414\sqrt{\mathrm{EI}}}{\sqrt{2.039\left(2.615 + 14.639 \pm \sqrt{(2.615 - 14.639)^2 + 4(2.824)^2}\right)}}$$
$$= \frac{1.414\sqrt{\mathrm{EI}}}{\sqrt{35.181 \pm 27.086}}.$$

Thus, the frequencies of free oscillations of a frame system with two degrees of freedom have the following values:

$$\omega_1 = 0.179\sqrt{\mathrm{EI}}\ \mathrm{s}^{-1}; \qquad \omega_2 = 0.497\sqrt{\mathrm{EI}}\ \mathrm{s}^{-1} \tag{C}$$

Determination of the values of inertial forces:

When a dynamic harmonic load $P(t) = P_o \sin \theta t$ acts on a structure, inertial forces are determined from the following system of equations:

$$\begin{cases} \left(\delta_{11} - \dfrac{1}{m_1 \cdot \theta^2}\right) I_1 + \delta_{12} I_2 + \Delta_{1p} = 0 \\ \delta_{21} I_1 + \left(\delta_{22} - \dfrac{1}{m_2 \cdot \theta^2}\right) I_2 + \Delta_{2p} = 0 \end{cases} \tag{D}$$

Let us write down the obtained values of displacements, masses, and frequencies of oscillations:

$$\delta_{11} = \frac{2.615}{\mathrm{EI}}, \quad \delta_{22} = \frac{14.639}{\mathrm{EI}}, \quad \delta_{12} = \delta_{21} = -\frac{2.824}{\mathrm{EI}}$$

$$m_1 = m_2 = m = G/g = 20/9.81 = 2.039,$$

$$\theta = 0.8\omega_{\min} = 0.8\omega_1 = 0.8 \cdot 0.179\sqrt{\mathrm{EI}} = 0.143\sqrt{\mathrm{EI}}$$

$$\theta^2 = \left(0.143\sqrt{\mathrm{EI}}\right)^2 = 0.0205\,\mathrm{EI},$$

$$\frac{1}{m \cdot \theta^2} = \frac{1}{2.039 \cdot 0.0205\ \mathrm{EI}} = \frac{23.924}{\mathrm{EI}}$$

$$\Delta_{1p} = \delta_{11} P_o = \frac{2.615}{\mathrm{EI}} \cdot 9 = \frac{23.535}{\mathrm{EI}}$$

$$\Delta_{2p} = \delta_{21} P_o = -\frac{2.824}{\mathrm{EI}} \cdot 9 = \frac{25.416}{\mathrm{EI}}$$

Substituting the above values of displacements, frequencies, and masses of the system into (D), we obtain:

$$\begin{cases} 21.309 I_1 + 2.824 I_2 + 23.535 = 0 \\ 2.824 I_1 + 9.285 I_2 - 25.416 = 0 \end{cases}$$

$$I_1 = 1.529\ \mathrm{kN}; \quad I_2 = -3.302\ \mathrm{kN}$$

Construction of a diagram of moments from the weight of concentrated masses and from the joint action of dynamic load (Fig. 9.10):

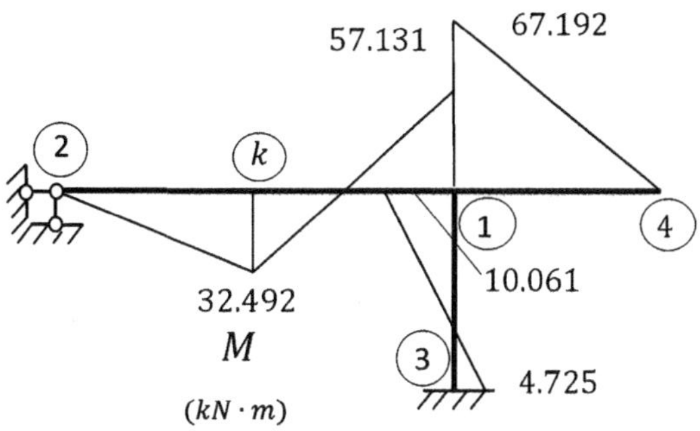

Fig. 9.10 Dynamic diagram of bending moments

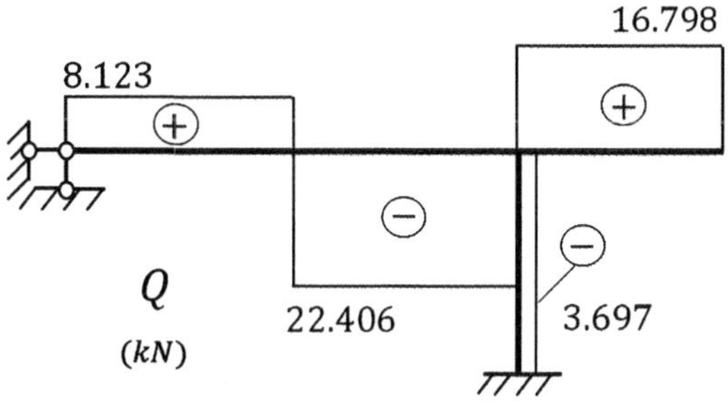

Fig. 9.11 Diagram of shear forces

$$M = G \cdot \overline{M}_1 + G \cdot \overline{M}_2 + I_1 \cdot \overline{M}_1 + I_2 \cdot \overline{M}_2 + P_o \cdot \overline{M}_1 = (G + I_1 + P_o)\overline{M}_1 + (G + I_2)$$

$$\overline{M}_2 = 30.529\overline{M}_1 + 16.798\overline{M}_2,$$

$$M_{12} = 30.529 \cdot 0.706 + 16.798 \cdot 2.118 = 57.131,$$

$$M_{1k} = 30529(-1.647) + 16.798 \cdot 1.059 = -32.492,$$

$$M_{13} = 30.529 \cdot (-0.706) + 16.798 \cdot 1.882 = 10.061,$$

$$M_{31} = 30.529 \cdot 0.363 + 16.798 \cdot (-0.941) = -4.725\,.$$

Construction of a shear force diagram (Fig. 9.11):

Since the moment diagram consists only of line intervals, we will use the following formula:

$$Q = \frac{dM}{dx} = \tan\alpha,$$

$$Q_{13} = Q_{31} = -\frac{10.061 + 4.725}{h} = -3.697,$$

$$Q_{2k} = \frac{32.492}{4.0} = 8.123,$$

$$Q_{k1} = Q_{1k} = -\frac{32.492 + 57.131}{4.0} = -22.406,$$

$$Q_{14} = Q_{41} = \frac{67.192}{4.0} = 16.798.$$

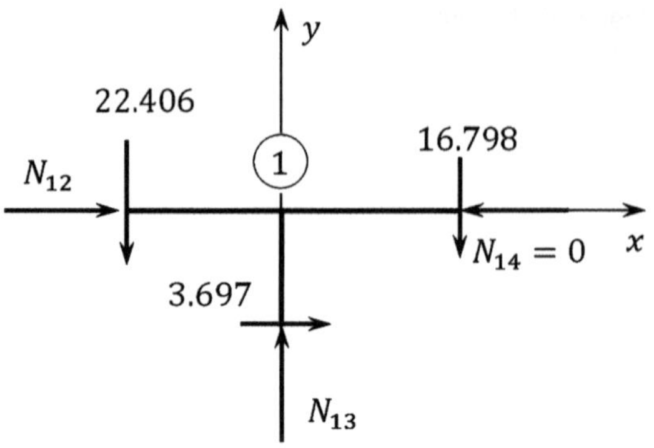

Fig. 9.12 Equilibrium of frame nodes

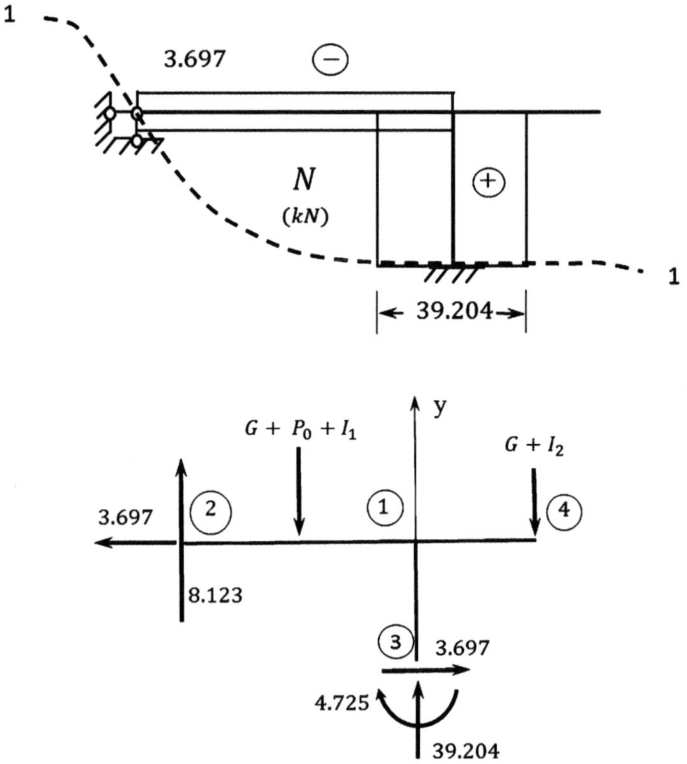

Fig. 9.13 Diagram of normal forces and equilibrium of frame nodes

Building a diagram of normal forces (Fig. 9.13):

Consider the equilibrium of the frame node 1 (Fig. 9.12). It is obvious, that $N_{14} = 0$ (Fig. 9.13).

$$\sum X = 0, \qquad N_{12} + 3.697 = 0, \qquad N_{12} = -3.697$$

$$\sum Y = 0, \quad N_{13} - 22.406 - 16.798 = 0, \qquad N_{13} = -39.204$$

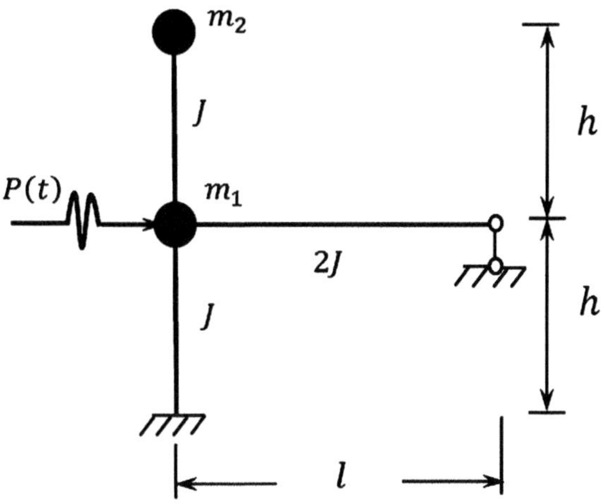

Fig. 9.14 Dynamic design model (DDM) for weightless frame with two lumped masses under periodic vibration load

Checking the diagram of shear forces and normal (axial) forces based on section method 1- 1 (Fig. 9.12).

$$\sum X=0; \quad 3.697-3.697=0,$$
$$\sum Y=0; \quad 8.123+39.204-(G+P_0+I_1+G+I_2)=0$$
$$47.327-(20\cdot 2+g+1.529-3.202)=0; \quad 47.327-47.327=0.$$

Problem 9.2
Let us consider the frame structure (Fig. 9.14) under dynamic load. The load changes according to the law: $P_0 \sin \theta t$.

The following dynamic and geometric characteristics of the system are given:

$$P(t)=P_0 \sin \theta t;$$
$$\theta=0.8\omega_{\min};$$
$$P_0=8\,\text{kN}; G=30\,\text{kN};$$
$$l=6\,\text{m}; =h=4\,\text{m}.$$

It is required to determine:

1. ω_1, ω_2—the frequency of free oscillations of the system.
2. The greatest values of inertial forces I_1, I_2
3. Build diagrams *M*, *Q*, and *N* from the weight (*G*) of concentrated masses and the joint effect of inertial forces of dynamic load.

Solution
The system in question has two degrees of freedom and therefore the free oscillations (vibrations) are characterized by two frequencies. The inertia forces arising from the dynamic load are shown in Fig. 9.15.

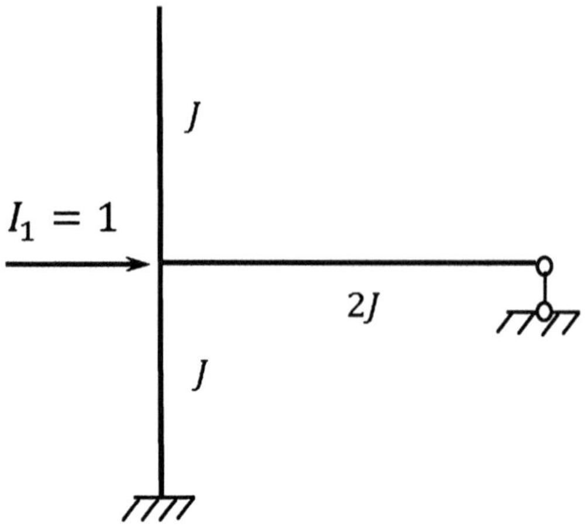

Fig. 9.15 Frame structure under the action of inertial force I_1

Determination of the free oscillations of the problem.

For a system with two degrees of freedom, the equation has the following form:

$$\begin{vmatrix} \delta_{11} - \dfrac{1}{m_1 \cdot \omega^2} & \delta_{12} \\ \delta_{21} & \delta_{22} - \dfrac{1}{m_2 \cdot \omega^2} \end{vmatrix} = 0 \qquad (A')$$

In the problem considered: $m_1 = m_2 = m = G/g = 30/9.81 = 3.058$,

$$\delta_{jk} = \sum \int \frac{\overline{M}_j \overline{M}_k}{\mathrm{EI}} dx, \quad j, k = 1, 2.$$

The solution of this problem is similar to the solution of the previous problem, however, in this case, it is more expedient to use the method of forces.

Note: *Here, the integration of the diagrams of bending moments is replaced with multiplication of the corresponding areas of the bending moment diagrams by the ordinates that correspond to the centers of gravity of the multiplied diagram. For this reason, it is more convenient to build a diagram of bending moments along the stretched zone of the frame elements.*

Building a diagram: $\overline{M}_1$

Determining the number of unknowns using the force method: $n = 1$
Selection of the primary system for the method of forces (Fig. 9.16a) *:*
Canonical equation of the chosen method:

$$\overline{\delta}_{11} X_1 + \Delta_{1I_1} = 0$$

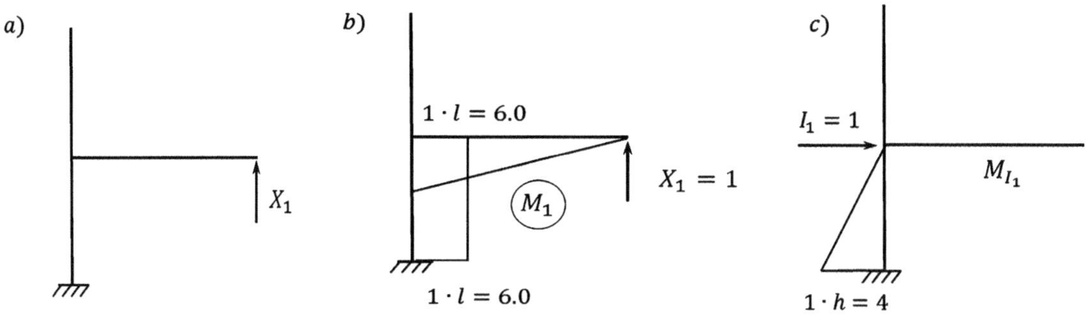

Fig. 9.16 (**a**) Primary system for the method of forces; (**b**) Bending moment Diagrams due to a unit force and an external load

Fig. 9.17 Bending moment diagrams due to a unit force and an external load

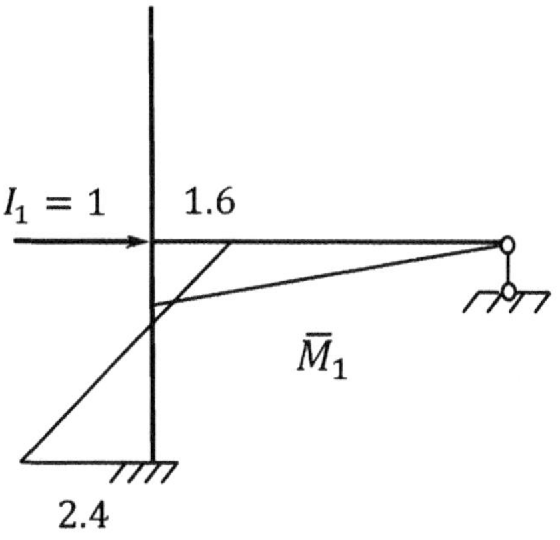

Plotting a diagram of moments from a unit force and from an external load (Fig. 9.16b, c):

$$\mathrm{EI}\bar{\delta}_{11} = \frac{l \cdot l}{2} \cdot \frac{2}{3} l \cdot \frac{1}{2} + l \cdot h \cdot l = \frac{l^3}{6} + l^3 h = 180.00$$

$$\mathrm{EI}\Delta_{1I_1} = -\frac{h \cdot h}{2} \cdot l \cdot \frac{1}{1} = -\frac{h^2 l}{2} = -48.00,$$

$$X_1 = -\frac{\Delta_{1I_1}}{\bar{\delta}_{11}} = -\frac{-48}{180} = 0.267,$$

Plotting a diagram of moments from a unit force and from an external load (Fig. 9.17):

$$\overline{M}_1 = M_1 \cdot X_1 + M_{I_1} \ .$$

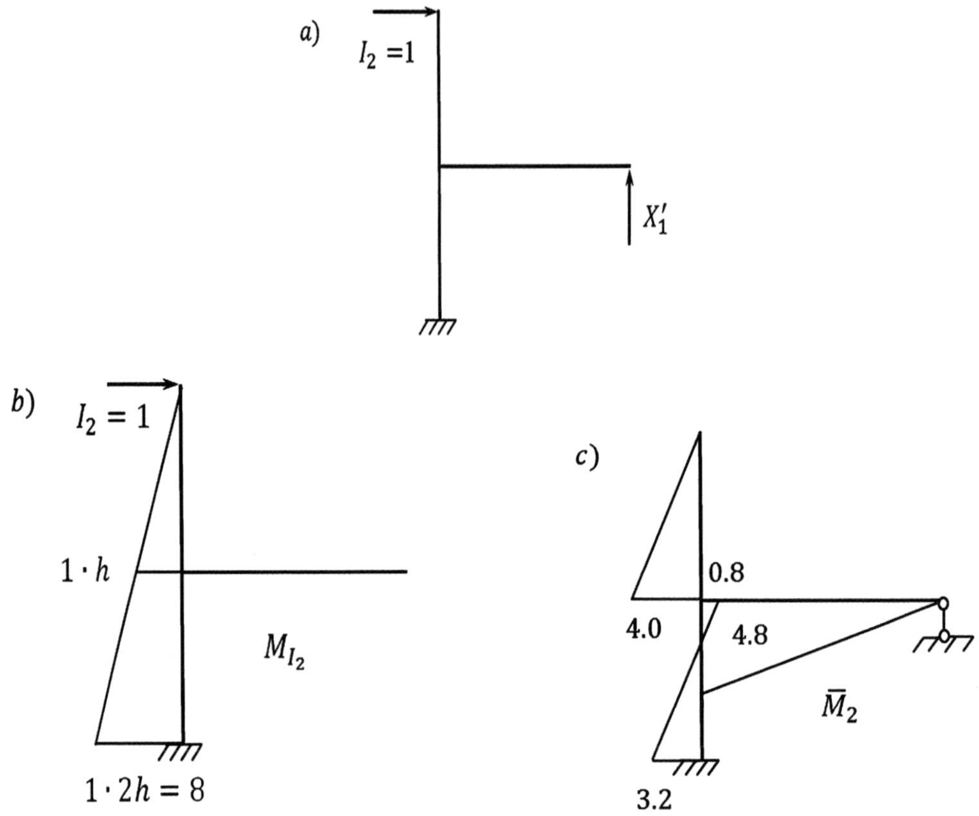

Fig. 9.18 Bending moment diagrams due to a unit force and diagram of moment $\overline{M}_2$

Checking the diagram of moments

$$\sum \int \frac{\overline{M}_1 M_1}{\mathrm{EI}} dx = 0,$$

$$\frac{1.6 \cdot 6}{2} \cdot \frac{2}{3} l \cdot \frac{1}{2\mathrm{EI}} - \frac{2.4 - 1.6}{2} \cdot h \cdot l \frac{1}{\mathrm{EI}} = 0;$$

$$0 = 0.$$

Plotting the diagrams of moment (Fig. 9.18)*:* $\overline{M}_2$

In this case, it is once again more efficient to use the method of forces.

$$\mathrm{EI}\Delta_{1I_2} = -\frac{h + 2h}{2} \cdot h \cdot l = \frac{4 + 8}{2} \cdot 4 \cdot 6 = -144.00,$$

$$X_1' = -\frac{\Delta_{1I_2}}{\overline{\delta}_{11}} = -\frac{-144}{180} = 0.80,$$

$$\overline{M}_2 = M_1 \cdot X_1 + M_{I_2} \ .$$

Checking the diagram of moments $\overline{M}_2$

$$\sum\int\frac{\overline{M}_2\,M_1}{EI}\,dx=0,$$

$$\frac{4.8\cdot l}{2}\cdot\frac{2}{3}l\cdot\frac{1}{2\mathrm{EI}}-\frac{3.2-0.8}{2}\cdot h\cdot l\frac{1}{\mathrm{EI}}=(28.8-28.8)\frac{1}{\mathrm{EI}}=0;$$

Determining the displacements: δ_{11}, δ_{12}, *and* δ_{22}.

$$\delta_{11}=\sum\int\frac{\overline{M}_1^{\,2}}{\mathrm{EI}}\,dx=\sum\int\frac{\overline{M}_1\,M_{I_1}}{\mathrm{EI}}\,dx=\frac{4\cdot 4}{2}\left(\frac{2}{3}\cdot 2.4-\frac{1}{3}\cdot 1.6\right)\frac{1}{\mathrm{EI}}=\frac{8.533}{\mathrm{EI}},$$

$$\delta_{22}=\sum\int\frac{\overline{M}_2^{\,2}}{\mathrm{EI}}\,dx=\sum\int\frac{\overline{M}_2\,M_2}{\mathrm{EI}}\,dx=\frac{4\cdot 4}{2}\cdot\frac{2}{3}\cdot 4\frac{1}{\mathrm{EI}}+$$

$$+\left[\frac{3.2\cdot 4}{2}\left(\frac{2}{3}\cdot 8+\frac{1}{3}\cdot 4\right)-\frac{0.8\cdot 4}{2}\left(\frac{2}{3}\cdot 4+\frac{1}{3}\cdot 8\right)\right]\frac{1}{\mathrm{EI}}=\frac{55.467}{\mathrm{EI}},$$

$$\delta_{12}=\delta_{21}=\sum\int\frac{\overline{M}_1\,\overline{M}_2}{\mathrm{EI}}\,dx=\sum\int\frac{M_{I_1}\overline{M}_2}{\mathrm{EI}}\,dx=\frac{4\cdot 4}{2}\left(\frac{2}{3}\cdot 3.2-\frac{1}{3}\cdot 0.8\right)\frac{1}{\mathrm{EI}}$$

$$=\frac{14.933}{\mathrm{EI}}.$$

Determining the frequencies ω_1, *and* ω_2*:*

For the case when $m_1=m_2=m=G/g=30/9.81=3.06$ from the Eq. (A′), we get:

$$\omega=\frac{\sqrt{2}}{\sqrt{m\left[\delta_{11}+\delta_{22}\pm\sqrt{(\delta_{11}-\delta_{22})^2+4\delta_{12}^2}\right]}},\qquad\text{(B}'\text{)}$$

the displacements δ_{11}, δ_{22}, and δ_{12} values in (B)

$$\omega=\frac{1.414\sqrt{\mathrm{EI}}}{\sqrt{3.06\left(8.533+55.467\pm\sqrt{(8.533-55.467)^2+4(14.933)^2}\right)}}=$$

$$=\frac{1.414\sqrt{\mathrm{EI}}}{\sqrt{3.06(64.00\pm 55.631)}}=\frac{1.414\sqrt{\mathrm{EI}}}{\sqrt{195.84\pm 170.231}},$$

Thus, the frequencies of free oscillations are as follows:

$$\omega_1=0.074\sqrt{\mathrm{EI}}\ \mathrm{s}^{-1};\qquad \omega_2=0.279\sqrt{\mathrm{EI}}\ \mathrm{s}^{-1}$$

Number the frequencies of free oscillations expressed through $\sqrt{\mathrm{EI}}$ from the following condition $\omega_1<\omega_2$.

The inertial forces are determined from the following system of equations:

$$\begin{cases} \left(\delta_{11} - \dfrac{1}{m_1 \cdot \theta^2}\right) I_1 + \delta_{12} I_2 \ + \Delta_{1p} = 0 \\ \delta_{21} I_1 + \left(\delta_{22} - \dfrac{1}{m_2 \cdot \theta^2}\right) I_2 + \Delta_{2p} = 0 \end{cases} \qquad (\mathrm{D}')$$

Let us write down the obtained values of displacements, masses, and frequencies of oscillations:

$$\delta_{11} = \frac{8.533}{\mathrm{EI}}, \quad \delta_{22} = \frac{55.467}{\mathrm{EI}}, \quad \delta_{12} = \delta_{21} = -\frac{14.933}{\mathrm{EI}}$$

$$m_1 = m_2 = m = 3.06,$$

$$\theta = 0.8\omega_{\min} = 0.8\omega_1 = 0.8 \cdot 0.074\sqrt{\mathrm{EI}} = 0.0592\sqrt{\mathrm{EI}}$$

$$\theta^2 = \left(0.0592\sqrt{\mathrm{EI}}\right)^2 = 0.0035\mathrm{EI},$$

$$\frac{1}{m \cdot \theta^2} = \frac{1}{3.06 \cdot 0.0035\ \mathrm{EI}} = \frac{93.371}{\mathrm{EI}},$$

$$\Delta_{1p} = \delta_{11} P_o = \frac{8.53}{\mathrm{EI}} \cdot 8 = \frac{68.24}{\mathrm{EI}}$$

$$\Delta_{2p} = \delta_{21} P_o = \frac{14.933}{\mathrm{EI}} \cdot 8 = \frac{119.464}{\mathrm{EI}}$$

Substituting the above values of displacements, frequencies, and masses of the system into (D′), we obtain:

$$\begin{cases} 84.838 \cdot I_1 - 14.933 \cdot I_2 - 6.824 = 0 \\ 14.933 \cdot I_1 - 37.904 \cdot I_2 + 11.946 = 0 \end{cases}$$

$$I_1 = 1.4604\ \mathrm{kN}; I_2 = -3.72705\ \mathrm{kN}$$

In this problem, the bending of the frame elements from the weight of the concentrated masses does not occur ($\sum M(G) = 0$), (Fig. 9.19).

Plotting a diagram of moments from the joint action of dynamic load (Fig. 9.20):

$$M = (P_o + I_1) \cdot \overline{M}_1 + I_2 \cdot \overline{M}_2 = 9.46\overline{M}_1 + 3.727\overline{M}_2.$$

Construction of a shear force diagram.

Since the moment diagram consists only of line segments, we will use the following formula:

$$Q = \frac{dM}{dx} = \tan\alpha,$$

The diagram of normal forces shown in Fig. 9.21b is based on the diagram of shear forces shown in Fig. 9.21a. (See building a diagram of normal forces in the previous problem.)

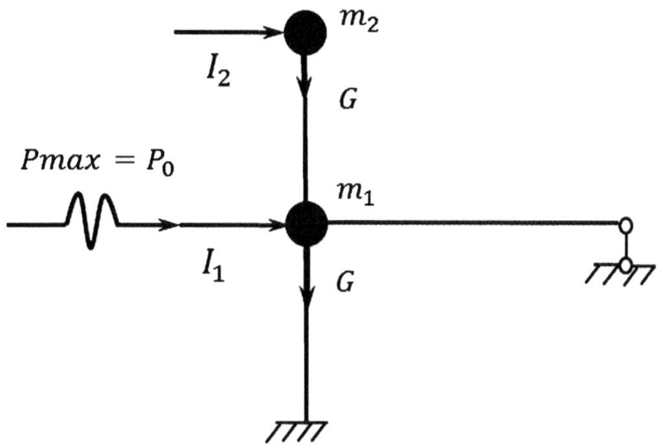

Fig. 9.19 Dynamic design model (DDM) for weightless frame with two lumped masses under a periodic vibration load

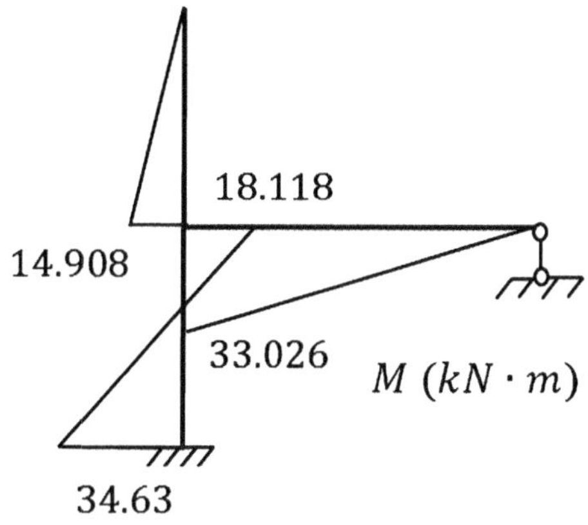

Fig. 9.20 Dynamic diagram of bending moments

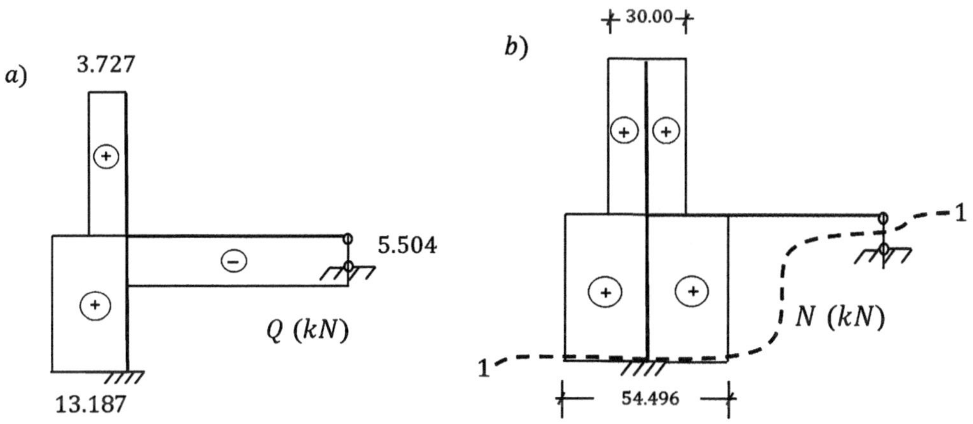

Fig. 9.21 (**a**) Diagram of shear forces and (**b**) diagram of normal forces

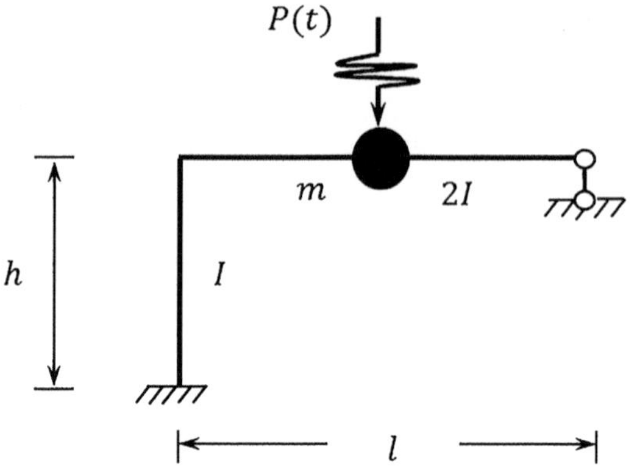

Fig. 9.22 Frame structure under the action of a periodic load

Checking the diagram of shear forces and normal (axial) forces based on section method. See previous problem: *Construction of a shear force diagram.*

$$\sum X = 0, \quad I_1 + P_0 + I_2 - 13.187 = 0$$
$$3.727 + 8 + 1.46 - 13.187 = 0, \quad 0 = 0$$
$$\sum Y = 0, \quad 54.496 + 5.504 - 30 - 30 = 0, \quad 0 = 0$$

Problem 9.3

As an example of calculation of the frame under the action of a periodic dynamic load, let us consider the frame structure shown in Fig. 9.22. The load changes according to the law $P(t) = P_0 \sin \theta t$. According to the condition of the problem, the mass of the frame girder is concentrated in the middle of the span.

The dynamic and geometric characteristics of the system are as follows:

$$P(t) = P_0 \sin \theta t;$$

$$\theta = 0.8\omega_{\min};$$

$$P_0 = 15 \text{ kN}; \quad G = 40 \text{ kN};$$

$$l = 5m; \; = h = 3m$$

The following should be determined:

1. ω_1, ω_2—the frequency of free oscillations of the system
2. The maximum values of inertial forces I_1, I_2.
3. Diagrams M, Q,and N from the weight (G) of concentrated mass and the joint effect of inertial forces of dynamic load

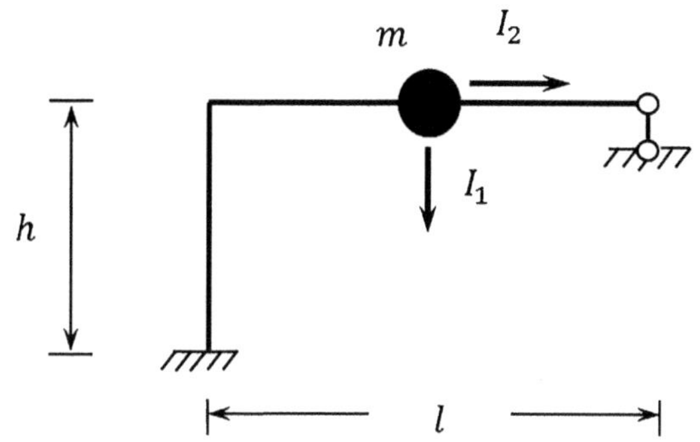

Fig. 9.23 DDM for weightless frame with one lumped mass under periodic vibration load

Solution

The system in question has two degrees of freedom and, therefore, the free oscillations (vibrations) are characterized by two frequencies. The inertia forces arising from the dynamic load are shown in Fig. 9.23.

Determination of Free Oscillations of the Problem

A system with two degrees of freedom is given as follows:

$$\begin{vmatrix} \delta_{11} - \dfrac{1}{m_1 \cdot \omega^2} & \delta_{12} \\ \delta_{21} & \delta_{22} - \dfrac{1}{m_2 \cdot \omega^2} \end{vmatrix} = 0 \quad (A'')$$

In the problem considered, $m = G/g = 40/9.81 = 4.077\ \text{kN} \cdot \text{s}^2/\text{m}$

$$\delta_{jk} = \sum \int \frac{\overline{M}_j \overline{M}_k}{\text{EI}} dx, \quad j, k = 1, 2.$$

In this case, it is more expedient to use the method of forces.

Note *Here, the integration of the diagrams of bending moments is replaced with multiplication of the corresponding areas of the bending moment diagrams by the ordinates that correspond to the centers of gravity of the multiplied diagram. For this reason, it is more convenient to build a diagram of bending moments along the stretched zone of the frame elements.*

Determining the number of unknowns: $n = 1$.
Selection of the primary system for the method of forces (Fig. 9.24b).

Canonical equation of the chosen method:

$$\overline{\delta}_{11} \cdot X_1 + \Delta_{1I_1} = 0$$

Thus, to determine δ_{11} and δ_{22}, we can build diagrams of moments $\overline{M}_1$, $\overline{M}_2$, respectively, from $I_1 = 1$ and from $I_2 = 1$ unit forces applied in the point of concentrated mass.

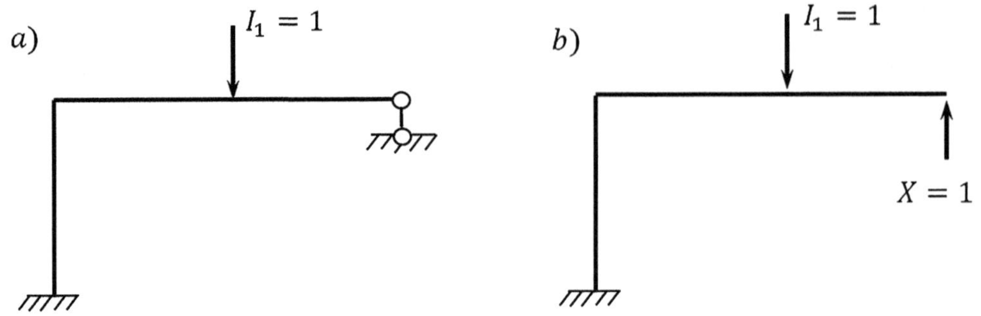

Fig. 9.24 (**a**) Frame structure under the action of inertial force I_1; (**b**) Primary system of the force method

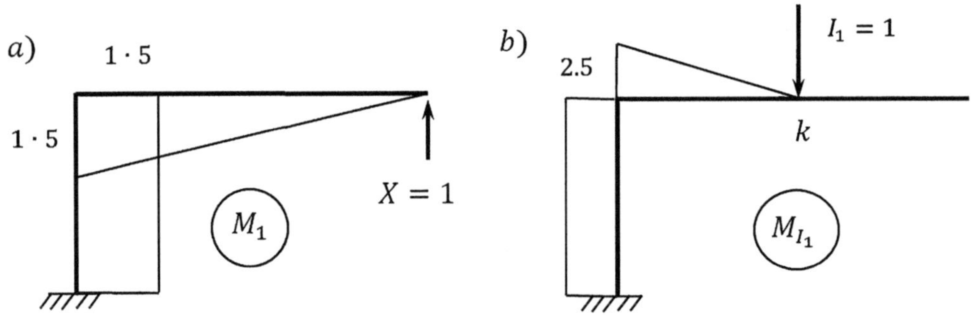

Fig. 9.25 (**a**) Bending moment diagram due to a unit force; (**b**) Bending moment diagram due to a unit inertial force I_1

Plotting a diagram of bending moments from a unit force and from a unit inertial load (Fig. 9.25a, b).

Building a diagram: $\overline{M}_1$

$$\mathrm{EI}\bar{\delta}_{11} = \frac{5 \cdot 5}{2} \cdot \frac{2}{3} 5 \frac{1}{2} + 5 \cdot 3 \cdot 5 = 95.833;$$

$$\mathrm{EI}\Delta_{1I_1} = -\frac{2.5 \cdot 2.5}{2}\left(\frac{2}{3} \cdot 5 + \frac{1}{3} 2.5\right) \cdot \frac{1}{2} - 2.5 \cdot 3 \cdot 5 = -44.012;$$

$$X_1 = -\frac{\Delta_{1I_1}}{\bar{\delta}_{11}} = -\frac{-44.012}{95.833} = 0.459.$$

Where $X = X_1$.

Plotting a diagram of bending moments from a unit force and from a unit inertial load (Fig. 9.26)*:*

$$\overline{M}_1 = M_1 \cdot X_1 + M_{I_1};$$

$$\overline{M}_{1-2} = 5 \cdot 0.459 - 2.5 = -0.204;$$

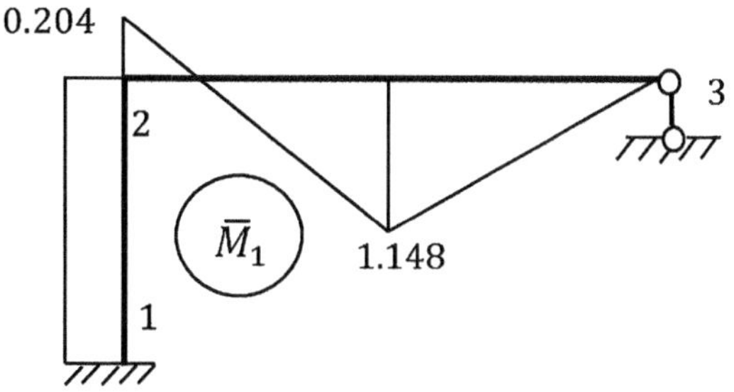

Fig. 9.26 Bending moment diagram due to an inertial force I_1

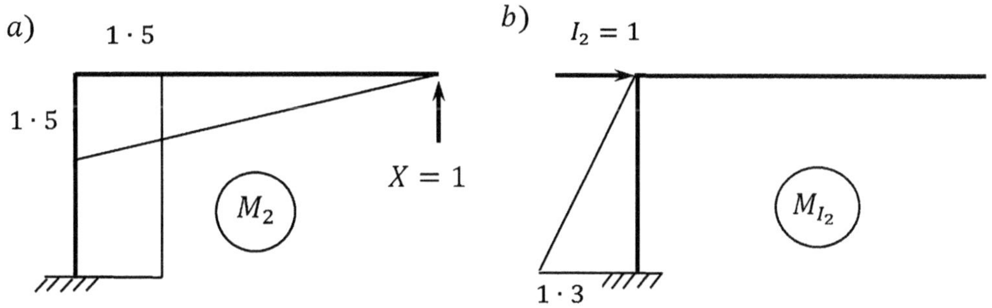

Fig. 9.27 (**a**) Bending moment diagram due to a unit force; (**b**) Bending moment diagram due to a unit inertial force I_2

$$\overline{M}_{2-1} = 5 \cdot 0.459 - 2.5 = -0.204;$$

$$\overline{M}_{2-3} = 5 \cdot 0.459 - 2.5 = -0.204;$$

$$\overline{M}_k = 2.5 \cdot 0.459 = 1.148.$$

Checking the diagram of moment

$$\sum \int \frac{\overline{M}_1 M_1}{\text{EI}} dx = 0,$$

$$\frac{2.5 \cdot 2.5}{2} \cdot \frac{1}{2\text{EI}} \cdot \frac{2}{3} 1.148 - \frac{0.204 \cdot 2.5}{2} \left(\frac{2}{3} 5 + \frac{1}{3} 2.5 \right) \cdot \frac{1}{2\text{EI}} + \frac{1.148 \cdot 2.5}{2} \left(\frac{1}{3} 5 + \frac{2}{3} 2.5 \right) \frac{1}{2\text{EI}}$$
$$- 0.204 \cdot 3 \cdot 5 \cdot \frac{1}{\text{EI}} = 0;$$

Plotting a diagram of bending moments from a unit inertial load I_2 (Fig. 9.27b).

Canonical equation of the chosen method:

Let's designate here that $X = X_2$. Then the canonical equation will take the following form:

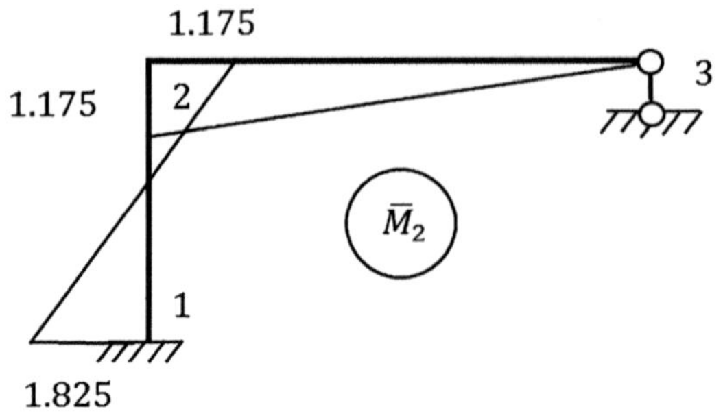

Fig. 9.28 Bending moment diagram due to an inertial force I_2

$$\overline{\delta}_{22}\cdot X_2+\Delta_{2I_2}=0$$

$$\mathrm{EI}\overline{\delta}_{22}=95.833$$

$$\mathrm{EI}\Delta_{2I_2}=-\frac{3\cdot 3}{2}\cdot 5\cdot\frac{1}{\mathrm{EI}}=-22.5$$

$$X_2=-\frac{\Delta_{2I_2}}{\overline{\delta}_{22}}=-\frac{-22.5}{95.833}=0.235.$$

Plotting a diagram of bending moments from a unit force and from a unit inertial load I_2 (Fig. 9.28)*:*

$$\overline{M}_2=M_2\cdot X_2+M_{I_2};$$

$$\overline{M}_{1-2}=5\cdot 0.235-3=-1.825;$$

$$\overline{M}_{2-1}=5\cdot 0.235=1.175;$$

$$\overline{M}_{2-3}=5\cdot 0.235=1.175;$$

$$\overline{M}_{3-2}=0.$$

Checking the diagram of moment

$$\sum\int\frac{\overline{M}_2 M_1}{\mathrm{EI}}dx=0,$$

$$\frac{5\cdot 5}{2}\cdot\frac{2}{3}1.175\frac{1}{2\mathrm{EI}}-\frac{1.825\cdot 3}{2}\cdot 5\cdot\frac{1}{\mathrm{EI}}+\frac{1.175\cdot 3}{2}\cdot 5\cdot\frac{1}{\mathrm{EI}}=0$$

Determining the displacements: δ_{11}, δ_{12}, *and* δ_{22}:

$$\delta_{11}=\sum\int\frac{\overline{M}_1^{\,2}}{\mathrm{EI}}dx=\sum\int\frac{\overline{M}_1M_{I_1}}{\mathrm{EI}}dx=\frac{2.5\cdot2.5}{2}\left(\frac{2}{3}0.204-\frac{1}{3}1.148\right)\frac{1}{2\mathrm{EI}}+$$

$$2.5\cdot3\cdot0.204\cdot\frac{1}{\mathrm{EI}}=\frac{1.145}{\mathrm{EI}}$$

$$\delta_{22}=\sum\int\frac{\overline{M}_2^{\,2}}{\mathrm{EI}}dx=\sum\int\frac{\overline{M}_2M_{I_2}}{\mathrm{EI}}dx=\frac{3\cdot3}{2}\left(\frac{2}{3}1.825-\frac{1}{3}1.175\right)\frac{1}{\mathrm{EI}}=\frac{3.713}{\mathrm{EI}};$$

$$\delta_{12}=\sum\int\frac{\overline{M}_1M_2}{\mathrm{EI}}dx\quad or\quad \delta_{12}=\sum\int\frac{\overline{M}_1M_{I_2}}{\mathrm{EI}}dx=\frac{3\cdot3}{2}\cdot0.204\cdot\frac{1}{\mathrm{EI}}=\frac{0.918}{\mathrm{EI}}.$$

Determining the frequencies ω_1, *and* ω_2*:*

For this case, $m=G/g=40/9.81=4.077$, and from the equation (A''), we obtained:

$$\omega=\frac{\sqrt{2}}{\sqrt{m\left[\delta_{11}+\delta_{22}\pm\sqrt{(\delta_{11}-\delta_{22})^2+4\delta_{12}^2}\right]}}\quad(B'')$$

The displacement values δ_{11}, δ_{22}, and δ_{12} are substituted in (B'')

$$\omega=\frac{1.414\sqrt{\mathrm{EI}}}{\sqrt{4.077\left[1.145+3.713\pm\sqrt{(1.145-3.713)^2+4\cdot0.918^2}\right]}}=$$

$$\frac{1.414\sqrt{\mathrm{EI}}}{\sqrt{4.077(4.858\pm3.157)}}=\frac{1.414\sqrt{\mathrm{EI}}}{\sqrt{19.806\pm12.871)}};$$

Thus, the frequencies of free oscillations are as follows:

$$\omega_1=0.247\sqrt{\mathrm{EI}}\,s^{-1};\quad \omega_2=0.537\sqrt{\mathrm{EI}}\ s^{-1}.\ (C'')$$

Number the frequencies of free oscillations expressed through $\sqrt{\mathrm{EI}}$ from the following condition: $\omega_1<\omega_2$ and then $\omega_1=\omega_{\min.}$ Let us indicate that $\theta=0.8\omega_{\min.}$ The internal forces are determined from the following system of equations:

$$\begin{cases}\left(\delta_{11}-\dfrac{1}{m\cdot\theta^2}\right)I_1+\delta_{12}I_2+\Delta_{1p}=0\\ \delta_{21}I_1+\left(\delta_{22}-\dfrac{1}{m\cdot\theta^2}\right)I_2+\Delta_{2p}=0\end{cases}\quad(D'')$$

Let us write down the obtained values of displacements, mass, and frequencies of oscillations:

$$\delta_{11}=\frac{1.145}{\mathrm{EI}},\qquad \delta_{22}=\frac{3.713}{\mathrm{EI}},\quad \delta_{12}=\delta_{21}=\frac{0.918}{\mathrm{EI}},$$

$$\theta=0.8\omega_{\min}=0.198\sqrt{\mathrm{EI}};\quad \theta^2=\left(0.198\sqrt{\mathrm{EI}}\right)^2=0.040\mathrm{EI};$$

$$\frac{1}{m\cdot\theta^2}=\frac{1}{4.077\cdot 0.040\mathrm{EI}}=\frac{6.135}{\mathrm{EI}};$$

$$\Delta_{1p}=\delta_{11}P_0=\frac{1.145}{\mathrm{EI}}\cdot 15=\frac{17.175}{\mathrm{EI}};$$

$$\Delta_{2p}=\delta_{21}P_0=\frac{0.918}{\mathrm{EI}}\cdot 15=\frac{13.770}{\mathrm{EI}}.$$

Substituting the above values of displacements, frequencies, and mass of the system into (D''), we obtain:

$$\begin{cases}\left(\delta_{11}-\dfrac{1}{m\cdot\theta^2}\right)I_1+\delta_{12}I_2+\Delta_{1p}=0\\ \delta_{21}I_1+\left(\delta_{22}-\dfrac{1}{m\cdot\theta^2}\right)I_2+\Delta_{2p}=0\end{cases}$$

$$\begin{cases}-4.990\,I_1+0.918\,I_2+17.175=0\\ 0.918\,I_1+2.422\,I_2+13.770=0\end{cases}$$

$$I_1=4.825\ \mathrm{kN};\quad I_2=7.514\ \mathrm{kN};$$

Construction of a diagram of moments from the weight of concentrated mass and from the joint action of dynamic load (Fig. 9.29):

$$M=\overline{M}_1(I_1+G+P_0)_1+I_2\overline{M}_2=\overline{M}_1(4.825+40+15)+\overline{M}_2\cdot 7.514=$$

$$\overline{M}_1 59.825+\overline{M}_2\cdot 7.514$$

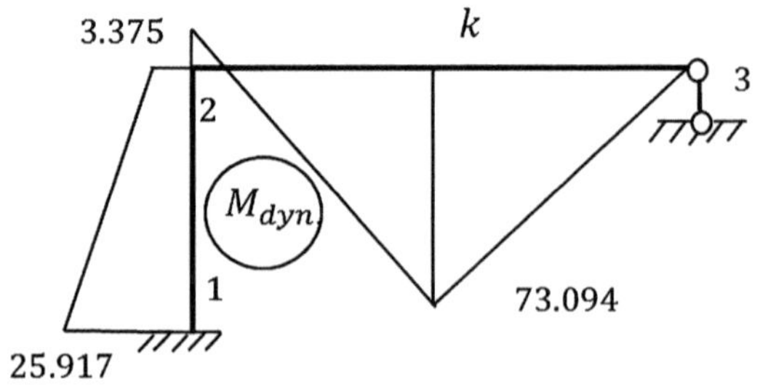

Fig. 9.29 Dynamic diagram of bending moments

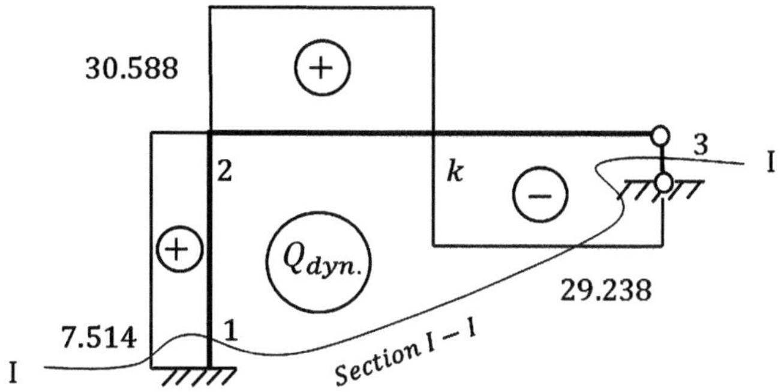

Fig. 9.30 Dynamic diagram of shear forces

$$\overline{M}_{1-2} = -59.825 \cdot 0.204 - 7.514 \cdot 1.825 = -25.917\ \text{kN} \cdot \text{m};$$

$$\overline{M}_{2-1} = -59.825 \cdot 0.204 + 1.175 \cdot 7.514 = -3.375\ \text{N} \cdot \text{m};$$

$$\overline{M}_{2-3} = -59.825 \cdot 0.204 + 1.175 \cdot 7.514 = -3.375\ \text{N} \cdot \text{m};$$

$$\overline{M}_{3-2} = 0;$$

$$\overline{M}_k = 59.825 \cdot 1.148 + \ 7.514 \cdot 0.5875 = 73.094\ \text{N} \cdot \text{m}$$

Construction of a shear force diagram (Fig. 9.30):

$$Q = \frac{dM}{dx} = \tan \alpha;$$

$$Q_{1-2} = Q_{2-1} = \frac{25.917 - 3.375}{3} = 7.514\ \text{kN},$$

$$Q_{2-k} = \frac{3.375 + 73.094}{2.5} = 30.588\ \text{kN},$$

$$Q_{k-3} = \frac{73.094}{2.5} = 29.238\ \text{kN}$$

Building a diagram of normal forces (Fig. 9.32)

Consider the equilibrium of the frame nodes (Fig. 9.31):

$$\sum X = 0; \quad -Q_{1-2} - N_{k-2} + 7.514 = 0; \ N_{k-2} \ = 0$$

$$\sum Y = 0; \quad N_{1-2} - Q_{2-k} = 0; \quad N_{1-2} \ = Q_{2-k} = 30.588\ \text{kN}$$

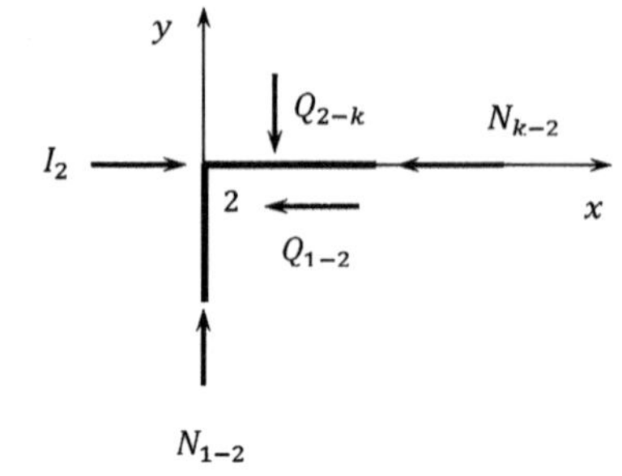

Fig. 9.31 Equilibrium of frame nodes. Node 2

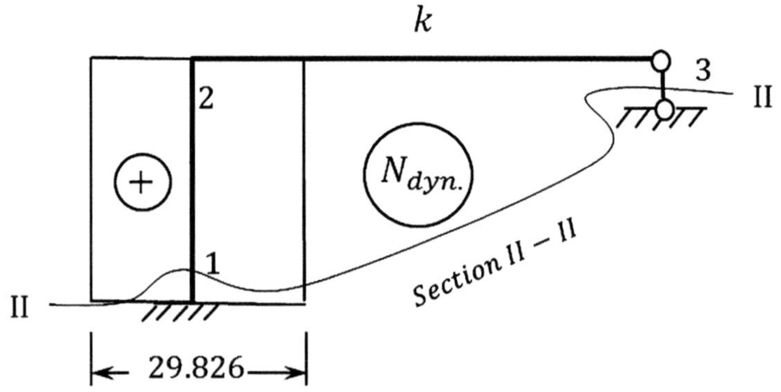

Fig. 9.32 Dynamic diagram of normal forces

Checking the diagrams of shear forces and normal forces based on section methods:

$$\sum Y = 0;\quad Q_{2-k} + N_{1-2} - (G + P_0 + I_1) = 0$$

$$30.588\ \text{kN} + 29.238\ \text{kN} - 40\ \text{kN} - 15\ \text{kN} - 4.825\ \text{kN} = 0$$

References

1. Petrosian, L. G. (1982). The selection of the rational primary systems for frame calculations and designs. Methodical instructions for the course and diploma research projects design. In *Structural mechanics and calculation of structures* (p. 58p). Department of Structural Mechanics, Armenian National University of Architecture & Construction.
2. Levonian L. H. (1988), Hambartsumian V.A. *Calculation of frames on the action of dynamic load. Methodical instructions*. L. G. Petrosian (Ed.) (25p). Department of Structural Mechanics, Armenian National University of Architecture & Construction.

10 Vibrations of Systems with a Finite Number of Degrees of Freedom

Abstract

This chapter is devoted to the study of vibrations of a system with finite number of degrees of freedom. The equation of motion in matrix formulation is discussed. The determination of displacements of rod systems in a matrix form is presented. General provisions, Simpson formula, pliability matrix, quasi-diagonal compliance (pliability) matrix of whole system, canonical equations in matrix form, vector matrix, etc., are discussed. The chapter presents a matrix form of determining the frequencies of free vibrations of a system with concentrated masses. Derivation of resolving equations, matrix form of determining the structure's frequencies on vibration load, calculation of frames on vibration load by the forced method, calculation of frames on vibration load by the displacement method, derivation of resolving equations numbers of problems and solutions are discussed.

10.1 Equations of Motion in a Matrix Formulation

Currently, matrix calculus is widely used in various fields of mathematics, theoretical physics, mechanics, etc. Modern structural mechanics is closely related to linear algebra and operations on matrices. Matrices are a convenient computing apparatus because they allow the expression of the problem being solved in the most convenient form for a computer. The apparatus of matrix algebra unveils great opportunities for transferring all labor-intensive calculations of structural mechanics problems to computers. The above-mentioned especially applies to the calculation of structures for stability and dynamic impacts, since the calculation of these structures, even with a low degree of static indeterminacy, turns into a very complex problem. Thus, matrix calculation allows the engineer to use his/her efforts, knowledge, and time more efficiently, mainly concentrating on the compilation of initial data, keeping a sequential record of matrix operations, as well as to the development of basic prerequisites, assumptions, that are aimed at improving the accuracy of solving problems of structural mechanics.

The above indicates that the activity of a civil engineer in modern conditions is unthinkable without deep and fundamental knowledge of methods of solving problems of structural mechanics. This knowledge should be adapted and aimed at the implementation of computer programs.

Among the numerous publications over the last two decades devoted to the matrix analysis of structures, I would like to point out the following works: W.M.C. McKenzie [1], F.Y. Cheng [2],

L. G. Petrosian, *Dynamics of Structures*, https://doi.org/10.1007/978-3-031-63539-7_10

A. *Kassimali [3],* Paz M., and Leigh W. [4], M. Paradrakakis, and E.J. Sapountzakis [5] Paz M. [6], Praveen Nagarajan [7], J,F. Doyle [8], A.R. Rzhanitsin [9], G.K. Klein, L.P., and Portaev [10], A. Ph. Smirnov, A.V. Alekcandrov [11], while not ignoring the significance and importance of the research of the authors of other publications.

The peculiarity of the presentation in this book is that we are not presenting any proofs. Instead, numerical examples are given that illustrate one or another assumption or position. A concrete numerical implementation is more useful to students and practicing engineers than an abstract proof. An understanding of the dynamic behavior of the main components of structures (rods, beams, arches, columns, etc.), the ability to feel the work of a structure itself, and the ability to solve problems are essential for students and practicing engineers. When presenting numerical methods, as many examples as possible should be given. This way of studying computational mathematics is very promising.

This chapter deals with analysis of beam and frame structures using the matrix method.

Let us consider the compilation of the equations of motion of a discrete system performing free oscillations on the example of a weightless elastic system carrying point masses. In a system that has n degrees of freedom, let us distinguish n linearly independent parameters of displacements: $y_1, y_2, \ldots,$ y_n. The load will be considered along the directions of these displacements. The components of the displacements will be designated accordingly: $Q_1,\ Q_2,\ \ldots\ ,\ Q_n$. Forces of inertia $-m_i\ddot{y}_i$ resist displacements y_i, where m_i is the mass value corresponding to displacement y_i. In the case when the parameter of displacement is the rotation of the element around a certain point, then instead of the mass, the moment of inertia of the mass relative to this point should be taken. Thus, the external forces are the following quantities:

$$F_i = Q_i - m_i\ddot{y}_i \qquad (i = 1, 2, \ \ldots, n) \tag{10.1}$$

These quantities in the expression for external forces (10.1) are the components of the vector,

$$\vec{F} = \vec{Q} - \left(\vec{m}\,\ddot{\vec{y}}\right), \tag{10.2}$$

which is decomposed in the directions: y_i. Here $\ddot{\vec{y}}$ represents a vector of accelerations with components $\ddot{y}_1,\ \ddot{y}_2,\ \ldots, \ddot{y}_n$, and a vector $\vec{m}$—vector of masses with components: $m_1,\ m_2,\ \ldots,\ m_n$. The parentheses denote the product of vectors.

The relationship between external and internal forces is expressed by equilibrium equations:

$$\boldsymbol{A}\vec{N} + \vec{F} = 0, \tag{10.3}$$

where $\boldsymbol{A}$ is a matrix of coefficients of equilibrium equations with n rows and m columns. Where m is the number of internal forces in the system elements; $\vec{N}$ is the vector of internal forces with their respective components: $N_1, N_2,\ \ldots, N_m$. At the beginning of the book, in Chap. 2: “The Fundamentals of the Theory of Calculation of Elastic Systems and Variation Principles of Structural Mechanics”, it was shown that between displacements y_i ($i = 1,\ 2,\ \ldots,\ n$), and deformations λ_j ($j = 1,\ 2,\ \ldots,\ m$) in linearly deformable systems, there is a relation. The equations of compatibility of deformations can be represented in the following matrix form:

$$\boldsymbol{A}^{\boldsymbol{T}}\ \vec{u} + \vec{\lambda} = 0; \quad \vec{\lambda} = -\boldsymbol{A}^{\boldsymbol{T}}\vec{u}, \tag{10.4}$$

where $\boldsymbol{A}^T$ is a transposed matrix $\boldsymbol{A}$ with m rows and n columns; $\vec{\lambda}$ is a deformation vector with corresponding components: λ_j ($j = 1, 2, \ldots, m$), and there is the following relation between deformation vectors $\vec{\lambda}$ and internal forces $\vec{N}$:

$$\vec{N} = \boldsymbol{C}\vec{\lambda}, \tag{10.5}$$

where $\boldsymbol{C}$ is the stiffness or rigidity matrix for internal forces.

Substituting (10.5) and (10.4) into Eq. (10.3), we obtain:

$$\boldsymbol{ACA}^T\vec{u} = \vec{F} \tag{10.6}$$

and based on (10.2) we have:

$$\boldsymbol{ACA}^T\vec{u} + \left(\vec{m}\,\ddot{\vec{y}}\right) = \vec{Q}. \tag{10.7}$$

In expanded form these equations are:

$$\begin{array}{l} \delta_{11}y_1 + \delta_{12}y_2 + \cdots + \delta_{1n}y_n + m_1\ddot{y}_1 = Q_1 \\ \delta_{21}y_1 + \delta_{22}y_2 + \cdots + \delta_{2n}y_n + m_2\ddot{y}_2 = Q_2 \\ \cdots\cdots\cdots\cdots\cdots\cdots\cdots \\ \delta_{n1}y_1 + \delta_{n2}y_2 + \cdots + \delta_{nn}y_n + m_n\ddot{y}_n = Q_n \end{array} \tag{10.8}$$

here $\delta_{ik} = \delta_{ki}$ are coefficients of matrix $\boldsymbol{ACA}^T$.

Equation (10.8) can also be obtained using the displacement method applied to a static system loaded by forces F_i. In this case, it is necessary to consider the state of equilibrium of the primary system which we can obtain by introducing rigid links against all displacements y_i to the given system. The coefficients δ_{ik} represents the reactions within the links r_{ik} introduced in the direction of displacement i from the unit of displacement $y_k = 1$.

10.2 Determination of Displacements of Rod Systems in Matrix Form

10.2.1 General Provisions

Determination of displacements in static and dynamic calculation of rod systems from the action of external loads is an important stage of the calculation. As is known, for plane rod systems with only bending deformations, the displacements are determined with the help of the moment element of Mohr's formula.

$$\Delta = \sum \int \frac{M_k M_p}{\mathrm{EI}} ds \tag{10.9}$$

Let us divide the system into separate sections and consider the i-th section of the element of this system. The integral (10.9) is defined using the Simpson formula:

$$\int_a^b F(s)\,ds = \frac{l}{6}\left(F^a + 4F^c + F^b\right). \tag{10.10}$$

Here F^a, F^b, and F^c are the values of the integrals at the beginning, at the end, and in the middle of the segment. Then the integral (10.9) for the ith section will take the following form [12]:

$$\int_{S_i} \frac{M_k M_p}{\mathrm{EI}} ds = \frac{l_i}{6}\left(\frac{M^a_{ik} M^a_{ip}}{\mathrm{EI}^a_i} + 4\frac{M^c_{ik} M^c_{ip}}{\mathrm{EI}^c_i} + \frac{M^b_{ik} M^b_{ip}}{\mathrm{EI}^b_i}\right),$$

and in the matrix form the integral (10.9) has the following form:

$$\int_{S_i} \frac{M_k M_p}{\mathrm{EI}} ds = \left\| M^a_{ik} \; M^c_{ik} \; M^b_{ik} \right\| \cdot \frac{l_i}{6\mathrm{EI}} \begin{bmatrix} A^a_i & 0 & 0 \\ 0 & 4A^c_i & 0 \\ 0 & 0 & A^b_i \end{bmatrix} \cdot \begin{bmatrix} M^a_{ip} \\ M^c_{ip} \\ M^b_{ip} \end{bmatrix}, \tag{10.11}$$

or

$$\int_{S_i} \frac{M_k M_p}{\mathrm{EI}} ds = \overline{M}'_{ik} \cdot B_i \cdot \vec{M}_{ip}. \tag{10.12}$$

Here

$$B = \frac{l_i}{6\mathrm{EI}} \begin{bmatrix} A^a_i & 0 & 0 \\ 0 & 4A^c_i & 0 \\ 0 & 0 & A^b_i \end{bmatrix}, \tag{10.13}$$

B_i is the matrix of the pliability of the ith part; $A_i = \mathrm{EI}/\mathrm{EI}_i$ is the relative pliability of the sections; $\overline{M}'_{ik}$ is the transposed matrix of the row relative to $\overline{M}_{ik}$, which is composed of the ordinates of the unit diagrams; $\vec{M}_{ip}$ is the vector composed of the ordinates of the bending moments of the load diagram. At $A_i = \mathrm{EI}/\mathrm{EI}_i = 1$, the matrix of pliability will take the following view:

$$B = \frac{l_i}{6\mathrm{EI}} \begin{bmatrix} 1 & 0 & 0 \\ 0 & 4 & 0 \\ 0 & 0 & 1 \end{bmatrix}. \tag{10.14}$$

If the unit and the load diagrams for a given section are linear, then the middle ordinates of these diagrams can be expressed through the edge (extreme) ordinates. The order of the matrix can be omitted using this technique. So, for example, for matrix M we can write:

$$\left[M^a_{ik} \; M^c_{ik} \; M^b_{ik}\right] = \left[M^a_{ik} \;\; M^b_{ik}\right] \cdot \begin{bmatrix} 1 & 0.5 & 0 \\ 0 & 0.5 & 1 \end{bmatrix}, \tag{10.15}$$

for the matrix of column M we get

$$\begin{bmatrix} M^a_{ip} \\ M^c_{ip} \\ M^b_{ip} \end{bmatrix} = \begin{bmatrix} 1 & 0 \\ 0.5 & 0.5 \\ 0 & 1 \end{bmatrix} \cdot \begin{bmatrix} M^a_{ip} \\ M^b_{ip} \end{bmatrix}. \tag{10.16}$$

Let us substitute matrices (10.15) and (10.16) in (10.11):

$$\int_{S_i} \frac{M_k M_p}{\mathrm{EI}} ds = \begin{bmatrix} M_{ik}^a & M_{ik}^b \end{bmatrix} \cdot \begin{bmatrix} 1 & 0.5 & 0 \\ 0 & 0.5 & 1 \end{bmatrix} \cdot \frac{l_i}{6\mathrm{EI}} \begin{bmatrix} A_i^a & 0 & 0 \\ 0 & 4A_i^c & 0 \\ 0 & 0 & A_i^b \end{bmatrix} \cdot \begin{bmatrix} 1 & 0 \\ 0.5 & 0.5 \\ 0 & 1 \end{bmatrix} \cdot \begin{bmatrix} M_{ip}^a \\ M_{ip}^b \end{bmatrix}. \tag{10.17}$$

Then after multiplying the matrices, we obtain:

$$\int_{S_i} \frac{M_k M_p}{\mathrm{EI}} ds = \begin{bmatrix} M_{ik}^a & M_{ik}^b \end{bmatrix} \cdot \begin{bmatrix} 1 & 0.5 & 0 \\ 0 & 0.5 & 1 \end{bmatrix} \cdot \frac{l_i}{6\mathrm{EI}} \begin{bmatrix} (A_i^a + A_i^c) & A_i^c \\ A_i^c & (A_i^c + A_i^b) \end{bmatrix} \cdot \begin{bmatrix} M_{ip}^a \\ M_{ip}^b \end{bmatrix}. \tag{10.18}$$

Here the matrix of pliability B has the form:

$$B_i = \frac{l_i}{6EI} \begin{bmatrix} (A_i^a + A_i^c) & A_i^c \\ A_i^c & (A_i^c + A_i^b) \end{bmatrix} \tag{10.19}$$

If the cross-section of the rod is constant in the considered part, then,

$$B_i = \frac{l_i}{6EI} \begin{bmatrix} 2 & 1 \\ 1 & 2 \end{bmatrix}. \tag{10.20}$$

Thus, we have determined the value of the Mohr's integral for the ith segment. To determine the total displacement Δ_{kp}, it is necessary to sum the obtained values for all sections, that is (i.e.):

$$\Delta_{kp} = \sum \overline{M}_{ik}' \cdot B_i \cdot \vec{M}_{ip} = \begin{bmatrix} \overline{M}_{1k}' & \overline{M}_{2k}' & \ldots & \overline{M}_{mk}' \end{bmatrix} \times \begin{bmatrix} B_1 & & & & \\ & B_2 & & & \\ & & \cdot & & \\ & & & \cdot & \\ & & & & \cdot \\ & & & & & B_m \end{bmatrix} \begin{bmatrix} \vec{M}_{1p} \\ \vec{M}_{2p} \\ \cdot \\ \cdot \\ \cdot \\ \vec{M}_{mp} \end{bmatrix}. \tag{10.21}$$

Or

$$\Delta_{kp} = M_k' \cdot B \cdot \vec{M}_p. \tag{10.22}$$

Here B is the quasi-diagonal compliance (pliability) matrix of the whole system. The compliance matrix is composed of the compliance matrices of the individual parts $(B_1, B_2, \ldots B_m)$:

$$B=\begin{bmatrix} B_1 & & & & \\ & B_2 & & & \\ & & \cdot & & \\ & & & \cdot & \\ & & & & \cdot \\ & & & & & B_m \end{bmatrix}. \tag{10.23}$$

Let us consider the sequence of matrix calculation of frames using the method of forces. The canonical equations of force method in matrix formulation have the following form:

$$\begin{bmatrix} \delta_{11} & \delta_{12}\cdot\ \cdot\ \cdot & \delta_{1n} \\ \delta_{21} & \delta_{21}\cdot\ \cdot\ \cdot & \delta_{21} \\ \cdot\ \cdot\ \cdot & \cdot\ \cdot\ \cdot\ \cdot\ \cdot\ \cdot\ \cdot & \cdot\ \cdot \\ \delta_{n1} & \delta_{n2}.\ \ .\ \ . & \delta_{nn} \end{bmatrix}\begin{bmatrix} X_1 \\ X_2 \\ \cdot \\ X_n \end{bmatrix}+\begin{bmatrix} \Delta_{1p} \\ \Delta_{2p} \\ \cdot \\ \Delta_{np} \end{bmatrix}=0, \tag{10.24}$$

or in brief:

$$D\cdot\vec{X}+\vec{\Delta}_p=0. \tag{10.25}$$

Here D is a square matrix which is composed of the coefficients of the canonical equations; $\vec{X}$ is a vector of unknowns; $\vec{\Delta}_p$ is a column matrix that is composed of free members of the canonical equations.

$$D=\begin{bmatrix} \delta_{11} & \delta_{12}\cdot\ \cdot\ \cdot & \delta_{1n} \\ \delta_{21} & \delta_{21}\cdot\ \cdot\ \cdot & \delta_{21} \\ \cdot\ \cdot\ \cdot & \cdot\ \cdot\ \cdot\ \cdot\ \cdot\ \cdot\ \cdot & \cdot\ \cdot \\ \delta_{n1} & \delta_{n2}.\ \ .\ \ . & \delta_{nn} \end{bmatrix};\quad \vec{X}=\begin{bmatrix} X_1 \\ X_2 \\ \cdot \\ X_n \end{bmatrix};\quad \vec{\Delta}_p=\begin{bmatrix} \Delta_{1p} \\ \Delta_{2p} \\ \cdot \\ \Delta_{np} \end{bmatrix}. \tag{10.26}$$

Here, only bending deformations are considered when determining displacements. If the primary system is divided into separate parts, then:

$$D=\overline{M}'\cdot B\cdot\overline{M}, \tag{10.27}$$

Where

$$\overline{M}=\begin{bmatrix} M_{11} & M_{12}\cdot\ \cdot\ \cdot & M_{1n} \\ M_{21} & M_{21}\cdot\ \cdot\ \cdot & M_{21} \\ \cdot\ \cdot\ \cdot & \cdot\ \cdot\ \cdot\ \cdot\ \cdot\ \cdot\ \cdot & \cdot\ \cdot \\ M_{t1} & M_{t2}.\ \ .\ \ . & M_{tn} \end{bmatrix},$$

t is the number of sections in which the bending moments are determined; $\overline{M}$ is a rectangular matrix, the columns of which are composed of the ordinates of unit diagrams M; B is a quasi-diagonal compliance matrix which is composed of compliance matrices of individual frame elements:

$$B = \begin{bmatrix} B_1 & & & & \\ & B_2 & & & \\ & & \cdot & & \\ & & & \cdot & \\ & & & & \cdot \\ & & & & B_t \end{bmatrix}.$$

$\overline{M}'$ is the transposed matrix $\overline{M}$; $\vec{\Delta}_p$ is a matrix column of free terms:

$$\vec{\Delta}_p = \overline{M}' \cdot B \cdot M_p^0,$$

here M_p^0 is the matrix column of moments on separate parts of the primary system:

$$M_p^0 = \begin{bmatrix} M_{1p}^0 \\ M_{2p}^0 \\ \cdot \\ M_{tp}^0 \end{bmatrix}.$$

We obtain the vector of unknowns by solving Eq. (10.25):

$$\vec{X} = -D^{-1} \cdot \vec{\Delta}_p \quad = \quad -\left(\overline{M}' \cdot B \cdot \overline{M}\right)^{-1} \cdot \left(\overline{M}' \cdot B \cdot M_p^0\right), \tag{10.28}$$

where D^{-1} is the inverse matrix with respect to D.

Bending moments in the given system from a load are determined after determining the vector $\vec{X}$:

$$\vec{M}_p = \overline{M} \cdot \vec{X} + \vec{M}_p^0 \tag{10.29}$$

Verification of the resulting diagram of bending moments in matrix form is carried out according to the following formula:

$$(M_p)(M_s) = M_s' \cdot B \cdot \vec{M}_p, \tag{10.30}$$

here M_s is a matrix column that is composed of columns of total unit diagrams.

10.3 Matrix Form of Determining the Frequencies of Free Vibrations of a System with Concentrated Masses

10.3.1 Derivation of Resultant Equations

Let us derive the resultant equations that determine the frequencies of free vibrations of an elastic system with concentrated masses. As an example, let's choose a single-span beam with hinged supports. The beam carries n concentrated masses. Figure 10.1. When deriving equations, only beam-bending deformations are taken into account.

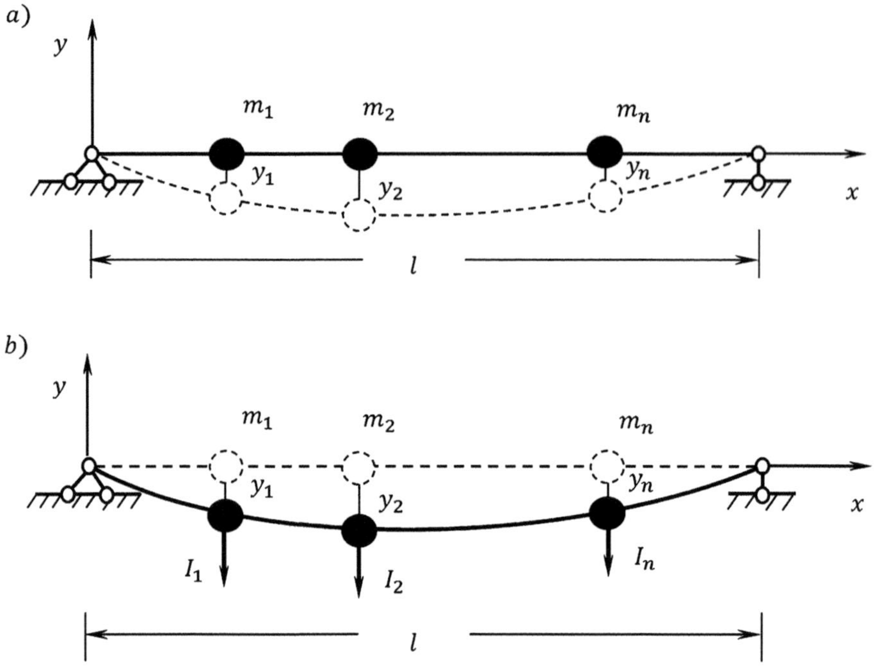

Fig. 10.1 (**a**, **b**) Weightless beam with m_n discrete masses

Let all the concentrated masses of the system m_i be taken out of the state of equilibrium and left to themselves. After the initial deflection, each of the masses m_i will perform free oscillations. The displacements of the concentrated masses at time t will be equal to:

$$\vec{y} = \begin{Vmatrix} y_1 \\ y_2 \\ \cdot \\ \cdot \\ \cdot \\ y_n \end{Vmatrix}. \tag{10.31}$$

In the process of oscillation in the direction of movement of each mass, inertia forces I_i arise. Inertial forces $-m_i\frac{d^2y_i}{dt^2}$ are applied at the points of location of the masses. In this particular case for the concentrated masses:

$$I_1 = -m_1\frac{d^2y_1}{dt^2}; \qquad I_2 = -m_2\frac{d^2y_2}{dt^2}; \qquad I_n = \; -m_n\frac{d^2y_n}{dt^2}.$$

The inertial forces of the system are written in the following matrix form:

$$-m\vec{\ddot{y}} = -\begin{Vmatrix} m_1 & & & & \\ & m_2 & & & \\ & & \cdot & & \\ & & & \cdot & \\ & & & & \cdot \\ & & & & & m_n \end{Vmatrix} \cdot \begin{Vmatrix} \ddot{y}_1 \\ \ddot{y}_2 \\ \cdot \\ \cdot \\ \cdot \\ \ddot{y}_n \end{Vmatrix}. \tag{10.32}$$

The vector of the ordinate of the bending moment diagram from the inertial forces in the given system can be represented in the following form:

$$\vec{M} = L_{mI} m \vec{\ddot{y}}, \tag{10.33}$$

where L_{mI} is the matrix of the influence of bending moments from unit inertia forces. The influence matrix is composed of the ordinates of moments diagrams at the points of concentrated masses 1, 2, . . ., n. Beam deflections are found from the known relationship:

$$\vec{y} = L_{mI}^T B \vec{M}, \tag{10.34}$$

where L_{mI}^T is the matrix transposed with respect to L_{mI}, B is the quasi-diagonal matrix of the system's pliability. The number of sub-matrices b_i of the quasi-diagonal matrix depends on the number of sections into which the given system is divided.

$$B = \begin{Vmatrix} b_1 & & & & \\ & b_2 & & & \\ & & \cdot & & \\ & & & \cdot & \\ & & & & \cdot \\ & & & & & b_n \end{Vmatrix}. \tag{10.35}$$

At constant stiffness (rigidity) along the length of the i-th section, the pliability (compliance) matrix for this section has the following form:

$$b_i = \frac{l_i}{6EI_i} \begin{Vmatrix} 2 & 1 \\ 1 & 2 \end{Vmatrix}, \tag{10.36}$$

where l_i, I_i are, respectively, the length and moment of inertia of the i-th section. The pliability (compliance) matrix with a large number of sections in the system turns out to be very cumbersome and the calculation becomes exceedingly complicated. The order of the matrix should be reduced in order to facilitate calculation. The vector of the ordinate of the bending moment diagram (10.33) is substituted in (10.34), we obtain:

$$\vec{y} = -L_{mI}^T B L_{mI} m \vec{\ddot{y}}. \tag{10.37}$$

A system of homogeneous differential equations can be obtained if we denote the product of four matrices:

$$D = L_{mI}^T B L_{mI} m. \tag{10.38}$$

Then the system of homogeneous differential equations will take the following form:

$$\vec{y} = -D\,\ddot{\vec{y}}. \tag{10.39}$$

The system of Eq. (10.39) has a general solution of the following form:

$$\vec{y} = \vec{A}\,\sin(\omega t + \varphi_0) = \begin{Vmatrix} A_1 \\ A_2 \\ \cdot \\ \cdot \\ \cdot \\ A_n \end{Vmatrix} \sin(\omega t + \varphi_0), \tag{10.40}$$

where A_i is the amplitude of the oscillations of the i-th mass ($i = 1, 2, \ldots, n$); ω is the frequency of free oscillations of the mass; φ_0 is the phase shift of oscillations at the initial moment of time.

Let us substitute (10.40) into Eq. (10.39) and reduce it by $\sin(\omega t + \varphi_0)$, then instead of differential equations we obtain a matrix equation of amplitudes:

$$\vec{A} = \omega^2 D \vec{A}. \tag{10.41}$$

Now, let us denote

$$\lambda = \frac{1}{\omega^2}, \tag{10.42}$$

then (10.41) will take the following form

$$(D - \lambda E)\vec{A} = 0, \tag{10.43}$$

where E is a unit matrix.

Matrix equation (10.43) in the presence of oscillations, has a nontrivial solution only if the determinant is zero:

$$\text{Det} = |D - \lambda E| = \begin{vmatrix} m_1\delta_{11} - \lambda & m_2\delta_{12} & \ldots & m_n\delta_{1n} \\ m_1\delta_{21} & m_2\delta_{22} - \lambda & \ldots & m_n\delta_{2n} \\ \ldots & \ldots & \ldots & \ldots \\ m_1\delta_{n1} & m_2\delta_{n2} & \ldots & m_n\delta_{nn} - \lambda \end{vmatrix} = 0. \tag{10.44}$$

Which corresponds to nonzero values of coefficients a_i. The determinant is called the secular equation. The solution of the secular equation is such values of characteristic numbers λ_i at which the determinant (10.44) turns to zero. Expanding the determinant (10.44) we obtain a polynomial with respect to λ:

$$\lambda^n + k_1\lambda^{n-1} + k_2\lambda^{n-2} + \cdots + k_n = 0 \tag{10.45}$$

The degree of this polynomial is equal to the order of the matrix for which the secular equation is composed. The solution of Eq. (10.45) gives n real values of characteristic numbers $\lambda_1, \lambda_2, \ldots, \lambda_n$. Each of λ_i corresponds to the eigenvector $\vec{A}_i$. The vector $\vec{A}$ and λ_i satisfy the equality:

$$D \cdot \vec{A}_i = \lambda_i \cdot \vec{A}_i.$$

To determine the coordinates of the eigenvector $\vec{A}_i$ with a known value of λ_i, we need to solve a system of joint equations with $n - 1$ unknowns. This system is obtained by removing one of the linearly dependent equations of system (10.43), assuming that $a_{ii} = 1$:

$$\begin{cases} (d_{11} - \lambda_i)a_{i1} + d_{12}a_{i2} + \cdots + d_{1n}a_{in} = -d_{1i} \\ d_{21}a_{i1} + (d_{22} - \lambda_i)a_{i2} + \cdots + d_{2n}a_{in} = -d_{2i} \\ \cdots\cdots\cdots\cdots\cdots\cdots \\ d_{n1}a_{i1} + d_{n2}a_{i2} + \cdots + (d_{nn} - \lambda_i)a_{in} = -d_{ni} \end{cases} \tag{10.46}$$

The solution of the resulting system of joint equations gives the coordinates of the eigenvector i which corresponds to the characteristic number λ_i:

$$\vec{A}_i = \begin{Vmatrix} a_{i1} \\ a_{i2} \\ \vdots \\ a_{ii} = 1 \\ \vdots \\ a_{in} \end{Vmatrix}.$$

Then the eigenvector matrix is obtained by performing the same procedure for all $\lambda_i (i = 1, 2, \ldots, n)$:

$$A = \begin{Vmatrix} 1 & a_{12} & \cdots & a_{1n} \\ a_{21} & 1 & \cdots & a_{2n} \\ \cdots & \cdots & \cdots & \cdots \\ a_{n1} & a_{n2} & \cdots & 1 \end{Vmatrix}. \tag{10.47}$$

The presented matrix has a characteristic feature. This matrix transforms the matrix D to a diagonal form:

$$ADA^{-1} = \begin{Vmatrix} \lambda_1 & & & & & \\ & \lambda_2 & & & & \\ & & \cdot & & & \\ & & & \cdot & & \\ & & & & \cdot & \\ & & & & & \lambda_n \end{Vmatrix}. \tag{10.48}$$

The characteristic numbers of matrix D have special properties: the sum λ_i is equal to the sum of the diagonal elements of the given matrix D (the trace of the matrix SpD), and the product of all characteristic numbers is equal to the determinant of the matrix D. These properties can be used to check the calculated values of the characteristic numbers. Finding the roots of the polynomial (10.45) in closed form is very difficult when $n > 3$. Many different exact methods, as well as approximate methods, have been developed to find the roots of a polynomial, the iteration method being one of them. The iteration method allows the determination of the maximum value of characteristic number λ_{max}. In most cases, it is sufficient to know the lowest frequency of vibration when assessing the strength of a structure. Since the frequency of vibration is the inverse value of the characteristic number, the minimum frequency corresponds to λ_{max}.

The iteration method for determining the largest characteristic number is quite simple and at the same time allows you to calculate λ_{max} with any required accuracy. Simultaneously with the characteristic number, an eigenvector is determined that corresponds to this number and satisfies the following equality:

$$D\vec{y} = \lambda\vec{y}. \tag{10.49}$$

Sequential approximations begin with the fact that an arbitrary vector is specified, for which one of the coordinates, for example, the first one, is taken equal to 1.

$$\vec{y}_0 = \begin{Vmatrix} 1 \\ y_{02} \\ \vdots \\ y_{0n} \end{Vmatrix}.$$

Let's substitute this vector into the left side of Eq. (10.49). As a result, we obtain a new vector. The first element of this vector is set equal to 1 as well.

$$D\begin{Vmatrix} 1 \\ y_{02} \\ \vdots \\ y_{0n} \end{Vmatrix} = \begin{Vmatrix} y_{11} \\ y_{12} \\ \vdots \\ y_{1n} \end{Vmatrix} = \lambda_1^* \begin{Vmatrix} 1 \\ y_{12}^* \\ \vdots \\ y_{1n}^* \end{Vmatrix} = \lambda_1^* \vec{y}_1^{\,*}. \tag{10.50}$$

The resulting new vector $\vec{y}_1^{\,*}$ is substituted into the left side of the equality (10.49). As a result, we get a new value λ_2^* and a new vector $\vec{y}_2$. The process is repeated until, after the specified substitution, the last two vectors are equal. That is, the same value of λ_k^*, in two adjacent approximations will not repeat. This value will correspond to the maximum characteristic number λ_{max}.

Let us apply the iteration method to determine λ_{max} using the example of matrix D:

$$D = \begin{Vmatrix} 4 & 1 & 1 \\ 3 & 1 & 0 \\ 3 & 0 & 1 \end{Vmatrix}$$

The computation is convenient to conduct in a tabular form (Table. 10.1) [10]

Thus, the filling out of the table begins with the assignment of an arbitrary zero vector in the following form:

$$\vec{y}_0 = \begin{Vmatrix} 1 \\ 1 \\ 1 \end{Vmatrix}.$$

Vector $\vec{y}_0$ is written in the first row. Then vector $\vec{y}_0$ is multiplied by the matrix D providing the elements of the new vector $\vec{y}_1$, which are written in the second row. In the next row, we write vector

Table 10.1 Tabular form

Given Matrix	4 3 3	1 1 0	1 0 1	λ^*_{max}
$\vec{y}_0$ $\vec{y}_1$	1 6	1 4	1 4	
$\vec{y}_1^{\,*}$ $\vec{y}_2$	1 5.3333	0.6666 3.6666	0.6666 3.6666	
$\vec{y}_2^{\,*}$ $\vec{y}_3$	1 5.3750	0.6875 3.6875	0.6875 3.6875	5.33333
$\vec{y}_3^{\,*}$ $\vec{y}_4$	1 5.372093	0.6860465 3.6860465	0.6860465 3.6860465	5.375
$\vec{y}_4^{\,*}$ $\vec{y}_5$	1 5.3702164	0.6851082 3.6851082	0.6851082 3.6851082	5.372093
$\vec{y}_5^{\,*}$ $\vec{y}_6$	1 5.3724244	0.6862122 3.6862122	0.6862122 3.6862122	5.3702164
$\vec{y}_6^{\,*}$ $\vec{y}_7$	1 5.3722714	0.6861357 3.6861357	0.6861357 3.6861357	5.3724244
$\vec{y}_7^{\,*}$ $\vec{y}_8$	1 5.3722820	0.6861410 3.6861410	0.6861410 3.6861410	5.3722714
$\vec{y}_8^{\,*}$ $\vec{y}_9$	1 5.3722812	0.6861406 3.6861406	0.6861406 3.6861406	5.3722820
$\vec{y}_9^{\,*}$	1	0.6861406	0.6861406	5.3722812

$\vec{y}_1^{\,*}$ that was obtained from $\vec{y}_1$, by setting the first element to 1. The calculation process is repeated until the elements of vector $\vec{y}_k^{\,*}$ and λ_{max} stop changing. The value by which all elements of vector $\vec{y}_k$ are divided in the process of bringing it to the vector $\vec{y}_k^{\,*}$, is an approximate value of the characteristic number λ_{max} in the k–th approximation.

Table 1 shows that in the ninth approximation all coordinates of the eigenvector repeat the corresponding values of the eighth approximation. Therefore, we can assume that the process of sequential approximations has been completed.

The highest value of the characteristic number $\lambda_{max} = 5.3722812$, the exact value of $\lambda_{max} = 5.3722813$. The determination of the following eigenvalues is based on a sequential decrease in the order of matrix D. For this, a new matrix D_1 is created:

$$D_1 = \vec{y}_1^{\,*} \qquad d_{11}\, d_{12} \cdots d_{1n},$$

where $\vec{y}_1^{\,*}$ is the eigenvector that corresponds to λ_{max}. The characteristic number λ_{max} was obtained in the previous cycle of sequential approximations; $d_{11}\, d_{12} \cdots d_{1n,}$ is the first row of the given matrix D. Then, let us subtract matrix D from the given matrix D_1 and get the following matrix $F = D - D_1$:

$$F = \begin{Vmatrix} d_{11} & d_{12} & \cdots & d_{1n} \\ d_{21} & d_{22} & \cdots & d_{2n} \\ \cdots & \cdots & \cdots & \cdots \\ d_{n1} & d_{n2} & \cdots & d_{nn} \end{Vmatrix} - \begin{Vmatrix} d_{11} & d_{12} & \cdots & d_{1n} \\ d_{21}\vec{y}_{12} & d_{22}\vec{y}_{12} & \cdots & d_{2n}\vec{y}_{12} \\ \cdots & \cdots & \cdots & \cdots \\ d_{n1}\vec{y}_{1n} & d_{n2}\vec{y}_{1n} & \cdots & d_{nn}\vec{y}_{1n} \end{Vmatrix}$$

$$= \begin{Vmatrix} 0 & 0 & \cdots & 0 \\ f_{21} & f_{22} & \cdots & f_{2n} \\ \cdots & \cdots & \cdots & \cdots \\ f_{n1} & f_{n2} & \cdots & f_{nn} \end{Vmatrix}.$$

In matrix F, the first row and the first column are eliminated, and matrix F_1 of order $n - 1$ is obtained. The characteristic numbers from λ_2 to λ_n of matrices F_1 and D are the same. The iterative method described above is used to find $\lambda_2 = \lambda_{2\max}$. Then, the order of matrix F_1 lowered by 1 once again, and matrix k_1 of order $n - 2$ is found. This process continues until all the characteristic numbers of matrix D are found.

The method of determining the frequencies of free oscillations in statically determined systems will be shown in the following examples.

Problem 10.1
Calculate the frequencies of free oscillations of a single-span beam with concentrated masses shown in Fig. 10.2a. The solutions of problems given below belong to Refs. [10, 12].

Solution
The location points of the beam's concentrated masses are numbered: 1, 2, 3, 4, 5 from left to right (Fig. 10.2a). The boundary points of the beam are not numbered since the bending moments in these sections are equal to zero.

Based on the principle of independent action of forces, a unit concentrated force of inertia $I = 1$ will be applied sequentially at the points where the discrete masses of the beam are located. The diagrams of bending moments $\overline{M}_1$, $\overline{M}_2$, $\overline{M}_3$, $\overline{M}_4$, and $\overline{M}_5$ (Fig. 10.2b) from a unit concentrated force are acting at the points of application of the masses.

Let us enter the ordinates of unit moments along the columns into the influence matrix. The influence matrix will take the following form:

$$A = \begin{Vmatrix} 5 & 4 & 3 & 2 & 1 \\ 4 & 8 & 6 & 4 & 2 \\ 3 & 6 & 9 & 6 & 3 \\ 2 & 4 & 6 & 8 & 4 \\ 1 & 2 & 3 & 4 & 5 \end{Vmatrix}$$

Note: *Here, the bending moment diagrams have been shown from the side of the stretched zones of the elements of the beam (Fig. 10.2b).*

Note: *Here, the bending moment diagrams have been shown from the side of the compressed zones of the elements of the beam (Fig. 10.2b*)*

The quasi-diagonal matrix of pliability B in this case will consist of six sub-matrices b_i $(i = 1, 2, 3, \ldots, 6)$:

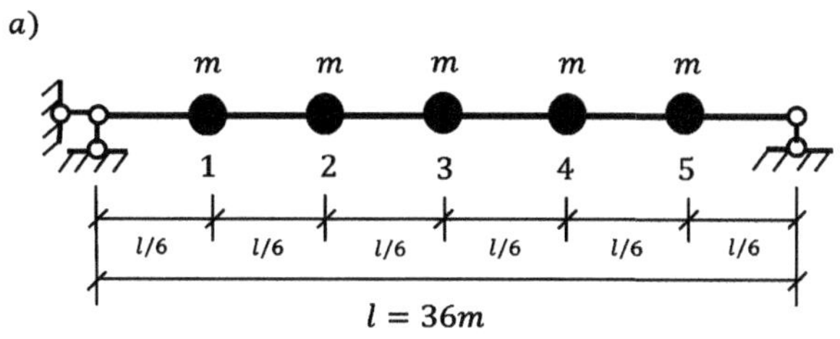

Note: *Here, the bending moment diagrams have been shown from the side of the stretched zones of the elements of the beam (figure 10. 2, b).*

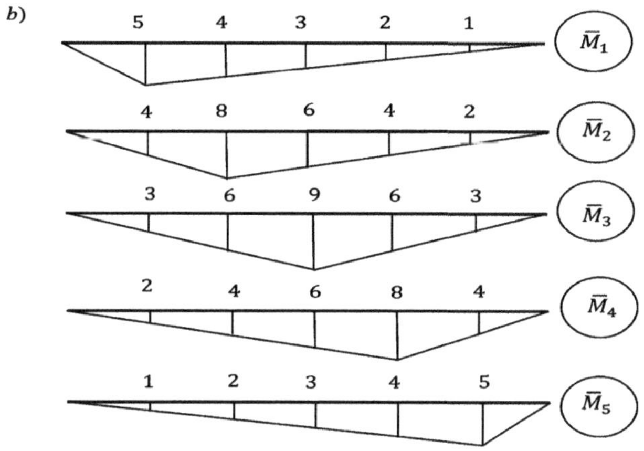

Note: *Here, the bending moment diagrams have been shown from the side of the compressed zones of the elements of the beam (figure 10.2, b*)*

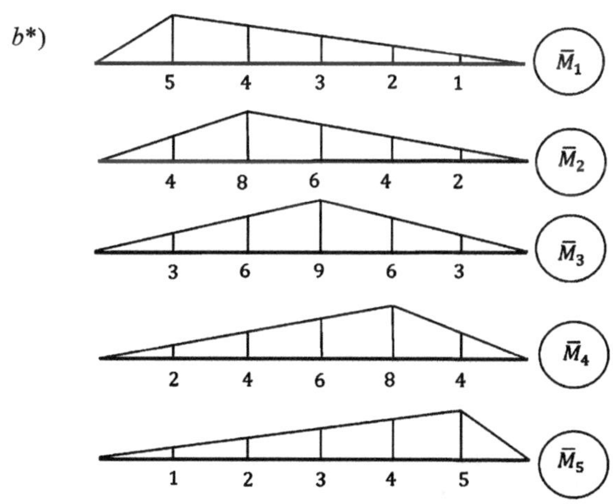

Fig. 10.2 (a, b, b*) Bending moment diagrams due to unit forces, applied at the points of concentrated masses of the beam system

$$B = \begin{Vmatrix} b_1 & & & & & \\ & b_2 & & & & \\ & & b_3 & & & \\ & & & b_4 & & \\ & & & & b_5 & \\ & & & & & b_6 \end{Vmatrix}$$

Since all the sections of the beam division are the same, the pliability matrix of each section will take the form:

$$b_1 = b_2 = b_3 = b_4 = b_5 = b_6 = \frac{l}{36\mathrm{EI}} \begin{Vmatrix} 2 & 1 \\ 1 & 2 \end{Vmatrix} = \frac{1}{\mathrm{EI}} \begin{Vmatrix} 2 & 1 \\ 1 & 2 \end{Vmatrix}.$$

Since all unit diagrams $\overline{M}_i$ are continuous on the boundaries of all sections, and the moments at the boundary points of the beam are equal to zero, we can reduce the order of matrix B. To do this, all sub-matrices b_i are diagonally shifted. At the same time, the first and last rows and the first and last columns of the matrix are discarded. As a result, we obtain matrix B in the following form:

$$B = \frac{1}{\mathrm{EI}} \begin{Vmatrix} 4 & 1 & 0 & 0 & 0 \\ 1 & 4 & 1 & 0 & 0 \\ 0 & 1 & 4 & 1 & 0 \\ 0 & 0 & 1 & 4 & 1 \\ 0 & 0 & 0 & 1 & 4 \end{Vmatrix}.$$

Now let us calculate matrix $D = L_{mI}^T \cdot B \cdot L_{mI} \cdot m$:

$$D = \begin{Vmatrix} 5 & 4 & 3 & 2 & 1 \\ 4 & 8 & 6 & 4 & 2 \\ 3 & 6 & 9 & 6 & 3 \\ 2 & 4 & 6 & 8 & 4 \\ 1 & 2 & 3 & 4 & 5 \end{Vmatrix} \frac{1}{\mathrm{EI}} \begin{Vmatrix} 4 & 1 & 0 & 0 & 0 \\ 1 & 4 & 1 & 0 & 0 \\ 0 & 1 & 4 & 1 & 0 \\ 0 & 0 & 1 & 4 & 1 \\ 0 & 0 & 0 & 1 & 4 \end{Vmatrix} \cdot \begin{Vmatrix} 5 & 4 & 3 & 2 & 1 \\ 4 & 8 & 6 & 4 & 2 \\ 3 & 6 & 9 & 6 & 3 \\ 2 & 4 & 6 & 8 & 4 \\ 1 & 2 & 3 & 4 & 5 \end{Vmatrix}$$

$$\cdot m \begin{Vmatrix} 1 & 0 & 0 & 0 & 0 \\ 0 & 1 & 0 & 0 & 0 \\ 0 & 0 & 1 & 0 & 0 \\ 0 & 0 & 0 & 1 & 0 \\ 0 & 0 & 0 & 0 & 1 \end{Vmatrix} = = \frac{m}{\mathrm{EI}} \begin{Vmatrix} 300 & 456 & 468 & 372 & 204 \\ 456 & 768 & 828 & 736 & 372 \\ 468 & 828 & 972 & 828 & 469 \\ 372 & 736 & 828 & 768 & 456 \\ 204 & 372 & 468 & 456 & 300 \end{Vmatrix}.$$

The secular equation is: $|D - \lambda E| = 0$.

The characteristic numbers of matrix D that were calculated iteratively are equal to the following values:

$$\lambda_1 = 2874.14 \frac{m}{\mathrm{EI}}; \quad \lambda_2 = 180 \frac{m}{\mathrm{EI}}; \quad \lambda_3 = 36 \frac{m}{\mathrm{EI}}; \quad \lambda_4 = 12 \frac{m}{\mathrm{EI}};$$
$$\lambda_5 = 5.86 \frac{m}{\mathrm{EI}}$$

The obtained values of the characteristic numbers $\lambda_1, \lambda_2, \ldots, \lambda_5$ are verified by:

$$S_pD = \frac{m}{\mathrm{EI}}(300 + 768 + 972 + 768 + 300) = 3108\frac{m}{\mathrm{EI}};$$

$$\sum_{i=1}^{s}\lambda_i = \frac{m}{\mathrm{EI}}(2874.14 + 180 + 36 + 12 + 5.86) = 3108\frac{m}{\mathrm{EI}};$$

The frequencies of free oscillations are determined using the formula:

$$\omega_i = \sqrt{\frac{1}{\lambda_i}};$$

$$\omega_1 = 0.018652\sqrt{\frac{\mathrm{EI}}{m}}; \quad \omega_2 = 0.074535\sqrt{\frac{\mathrm{EI}}{m}}; \quad \omega_3 = 0.166666\sqrt{\frac{\mathrm{EI}}{m}};$$

$$\omega_4 = 0.288671\sqrt{\frac{\mathrm{EI}}{m}}; \qquad \omega_5 = 0.413096\sqrt{\frac{\mathrm{EI}}{m}};$$

Problem 10.2

Find the frequencies of free vibrations of the frame shown in Fig. 10.3a. The following geometric characteristics of the system are given: $l_1 = 3$ m; $l_2 = 4$ m; $h = 3$ m. This system has two degrees of freedom. The design scheme of the dynamic calculation is presented in Fig. (10.3b). The diagrams of bending moments from the unit inertial forces are shown in Fig. (10.4a, b).

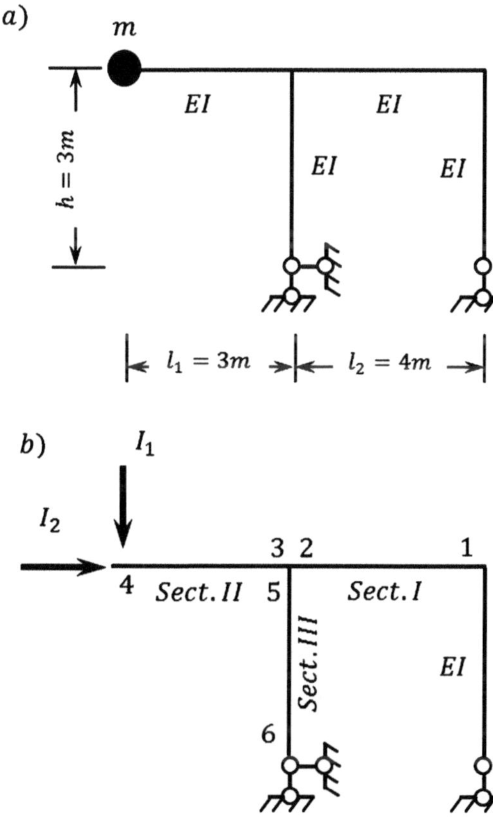

Fig. 10.3 (a) Frame with two degrees of freedom; (b) Design scheme of the frame

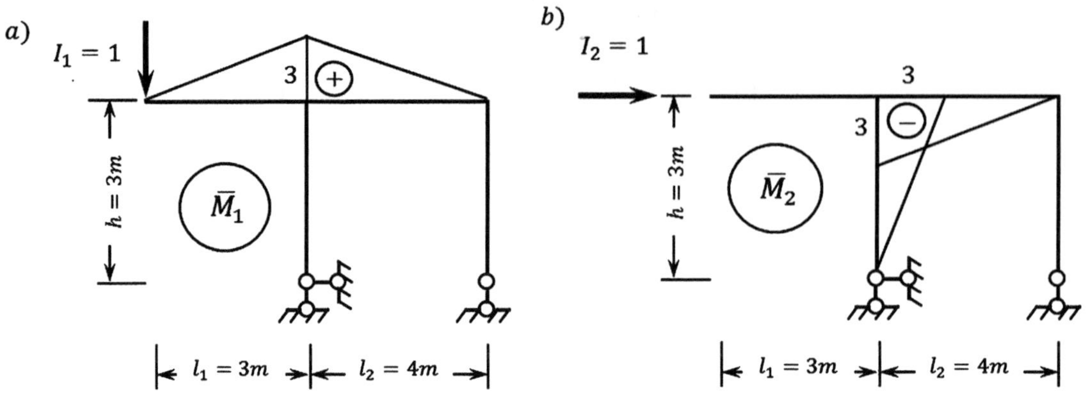

Fig. 10.4 (**a**, **b**) Bending moment diagrams due to unit inertial forces

Note: *Here, the bending moment diagrams have been shown from the side of the stretched zones of the elements of the frame.*

In order to compose the influence matrix, the given frame will be divided into three sections. Since the moments in sections 1, 4, and 6 are equal to zero, the order of matrix L_{mI} can be reduced. The matrix of influence can be written as follows:

$$L_{mI} = \begin{Vmatrix} 3 & - & 3 \\ 3 & & 0 \\ 0 & - & 3 \end{Vmatrix}.$$

The quasi-diagonal matrix B will consist of three matrices: b_1, b_2, b_3. The pliability matrix for each section will be composed separately:

$$b_1 = \frac{4}{6\text{EI}} \begin{Vmatrix} 2 & 1 \\ 2 & 1 \end{Vmatrix} = \frac{1}{6\text{EI}} \begin{Vmatrix} 8 & 4 \\ 4 & 8 \end{Vmatrix};$$

$$b_2 = \frac{3}{6\text{EI}} \begin{Vmatrix} 2 & 1 \\ 2 & 1 \end{Vmatrix} = \frac{1}{6\text{EI}} \begin{Vmatrix} 6 & 3 \\ 3 & 6 \end{Vmatrix};$$

$$b_3 = \frac{3}{6\text{EI}} \begin{Vmatrix} 2 & 1 \\ 2 & 1 \end{Vmatrix} = \frac{1}{6\text{EI}} \begin{Vmatrix} 6 & 3 \\ 3 & 6 \end{Vmatrix}.$$

The order of matrix B eliminating the first, fourth, and sixth rows as well as the first, fourth, and sixth columns will be equal to three:

$$B = \frac{1}{6EI} \begin{Vmatrix} 8 & 0 & 0 \\ 0 & 6 & 0 \\ 0 & 0 & 6 \end{Vmatrix}.$$

In sections 1, 4, and 6, the bending moment is zero. Let us calculate matrix D:

$$D=\begin{Vmatrix} 3 & 3 & 0 \\ -3 & 0 & -3 \end{Vmatrix} \frac{1}{6\mathrm{EI}} \begin{Vmatrix} 8 & 0 & 0 \\ 0 & 6 & 0 \\ 0 & 0 & 6 \end{Vmatrix} \cdot \begin{Vmatrix} 3 & - & 3 \\ 3 & & 0 \\ 0 & - & 3 \end{Vmatrix} \cdot \begin{Vmatrix} m & 0 \\ 0 & m \end{Vmatrix}$$

$$= \frac{m}{\mathrm{EI}} \begin{Vmatrix} 21 & - & 12 \\ -12 & & 21 \end{Vmatrix}.$$

The characteristic numbers of matrix D are equal, respectively:

$$\lambda_1 = 33\frac{m}{\mathrm{EI}}; \qquad \lambda_2 = 9\frac{m}{\mathrm{EI}};$$

The obtained values of the characteristic numbers λ_1,and λ_2 are verified (checked) by:

$$\sum_{i=1}^{2} \lambda_i = \frac{m}{\mathrm{EI}}(33+9) = 42\frac{m}{\mathrm{EI}}; \qquad S_pD = \frac{m}{\mathrm{EI}}(21+21) = 42\frac{m}{\mathrm{EI}};$$

$$\Pi_{i=1}^{2}\lambda_i = (33\cdot 9)\frac{m^2}{(\mathrm{EI})^2} = 297\frac{m^2}{(\mathrm{EI})^2}; \quad DetD = (21\cdot 21 - 12\cdot 12)\frac{m^2}{(\mathrm{EI})^2} = 297\frac{m^2}{(\mathrm{EI})^2}$$

The frequencies of free oscillations are equal:

$$\omega_1 = 0.17408\sqrt{\frac{\mathrm{EI}}{m}}; \qquad \omega_2 = 0.33333\sqrt{\frac{\mathrm{EI}}{m}};$$

To compose the influence matrix L_{mI} from the unit inertial forces in statically indeterminate systems, it is first necessary to calculate a given system by some method of structural mechanics. The matrix of influence in the calculation of a given system by the method of forces is determined as follows:

$$L_{mI} = L_{mI}^{0} - L_{m}^{0}\left(L_{m}^{0\,T}\cdot B\cdot L_{m}^{0}\right)^{-1}\cdot L_{m}^{0\,T}\cdot B\cdot L_{mI}^{0} \tag{10.51}$$

where L_{mI}^{0} is a matrix of the influence of bending moments in the primary system from inertial forces $\overline{I}_i = 1$; L_m^0 is a matrix of the influence of bending moments in the primary system from unit forces X_1, X_2, ..., X_n instead of discarded removed extra links bonds (n is the degree of static indeterminacy of the given system); $L_m^{0\,T}$ is the matrix transposed with respect to L_m^0; B is the pliability matrix of individual sections.

Let's substitute expression (10.51) into (10.38) and obtain the expression for matrix D:

$$D = L_m^{0\,T}\cdot B\left\{L_{mI}^{0} - L_{m}^{0}\left(L_{m}^{0\,T}\cdot B\cdot L_{m}^{0}\right)^{-1}\cdot L_{m}^{0\,T}\cdot B\cdot L_{mI}^{0}\right\}m \tag{10.52}$$

When we calculate a statically indeterminate system by displacement method, matrix D is calculated in the following way:

$$D = L_m^{0\,T}\cdot B\left\{L_{mI} + L_m\left(L_m^T\cdot B\cdot L_m\right)^{-1}\cdot L_m^T\cdot B\cdot L_{mI}^0\right\}m \tag{10.53}$$

where L_m is a matrix of the influence of bending moments in the primary system of the method of displacements from unit displacements of additional links $\overline{Z}_n = 1$. Matrix L_m contains n columns and m rows. The number of columns is equal to the number of unit displacements, that is, equal to the degree of kinematic indeterminacies of a given system. The number of rows equals the number of sections in which the internal forces are calculated; L_m^T is the matrix transposed with respect to L_m; L_{mI}^0 is the matrix of bending moments in any statically determined system from inertial forces $\overline{I}_i = 1$; L_{mI} is the matrix of bending moments in the primary system of displacement method from inertial forces $\overline{I}_i = 1$.

Problem 10.3

Determine the frequencies of free oscillations of the frame shown in Fig. 10.5a. The given frame is statically indeterminate.

Note: *Here, the bending moment diagrams have been shown from the side of the stretched zones of the elements of the frame.*

Solution

The primary system is obtained by replacing the horizontal link on the right support of the frame with horizontal reaction $\overline{X}_1 = 1$. The bending moment diagrams from force $\overline{X}_1 = 1$, Fig. 10.5b, and from inertial forces $\overline{I}_1 = \overline{I}_2 = 1$, (see Fig. 10.4a, b in the primary system are constructed.

Fig. 10.5a Statically indeterminate frame

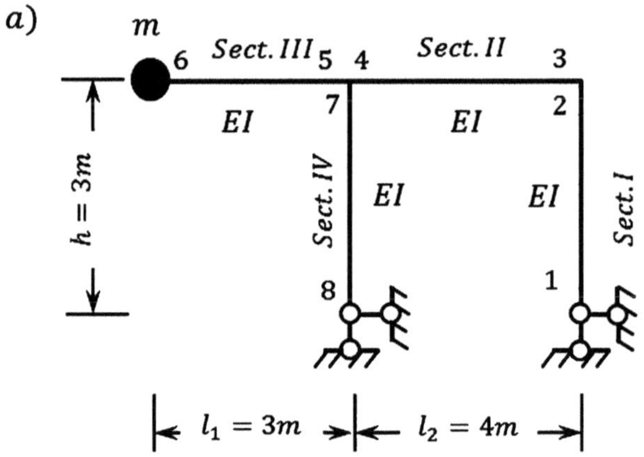

Fig. 10.5b Primary system and bending moment diagram due to $\overline{X}_1 = 1$

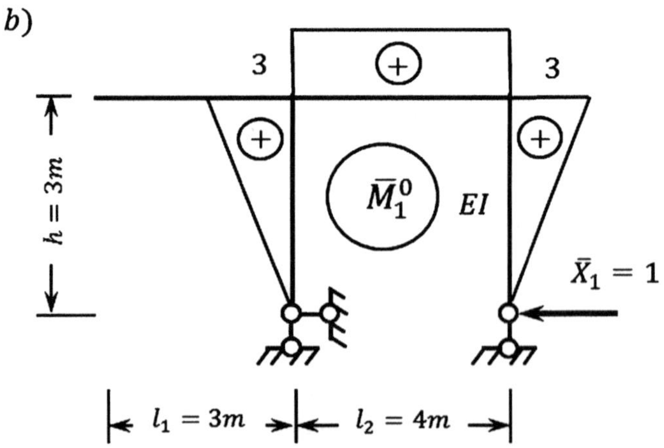

Let us divide the frame into four sections. Since all the bending moment diagrams are continuous on the boundary of sections I and II in cross-sections 2 and 3, the values of the bending moments will be fixed in the influence matrices L_m^0 and L_{mI}^0 . Moreover, since the bending moments in cross-sections 1, 6, and 8 are equal to zero, the matrix L_m^0 will consist of one column by the number of forces $\overline{X}_1$ and four rows by the number of characteristic cross-sections:

$$L_m^0 = \begin{Vmatrix} 3 \\ 3 \\ 0 \\ -3 \end{Vmatrix}.$$

The matrix L_{mI}^0 will consist of two columns by the number of inertial forces $\overline{I}_1,\ \overline{I}_2$ and four rows. The first column corresponds to the diagram of bending moments from $\overline{I}_1$, and the second column from $\overline{I}_2$:

$$L_{mI}^0 = \begin{Vmatrix} 0 & & 0 \\ 3 & - & 3 \\ 3 & & 0 \\ 0 & - & 3 \end{Vmatrix}.$$

The quasi-diagonal matrix B will consist of four sub-matrices b_i. The pliable matrices separately for each section will be:

$$b_1 = b_3 = b_4 = \frac{3}{6\mathrm{EI}} \begin{Vmatrix} 2 & 1 \\ 2 & 1 \end{Vmatrix} = \frac{1}{6\mathrm{EI}} \begin{Vmatrix} 6 & 3 \\ 3 & 6 \end{Vmatrix};$$

$$b_2 = \frac{4}{6\mathrm{EI}} \begin{Vmatrix} 2 & 1 \\ 2 & 1 \end{Vmatrix} = \frac{1}{6\mathrm{EI}} \begin{Vmatrix} 8 & 4 \\ 4 & 8 \end{Vmatrix}.$$

The quasi-diagonal matrix B is composed by taking into account that all moment diagrams are continuous at the boundary of sections I and II. The bending moments in cross-sections 1, 6, and 8 are equal to zero. The quasi-diagonal matrix B will finally take the following form:

$$B = \frac{1}{6\mathrm{EI}} \begin{Vmatrix} 14 & 4 & 0 & 0 \\ 4 & 8 & 0 & 0 \\ 0 & 0 & 3 & 0 \\ 0 & 0 & 0 & 6 \end{Vmatrix}.$$

Now, let's calculate matrix D:

$$D=\begin{Vmatrix} 0 & 3 & 3 & 0 \\ 0 & -3 & 0 & 3 \end{Vmatrix} \frac{1}{6EI}\begin{Vmatrix} 14 & 4 & 0 & 0 \\ 4 & 8 & 0 & 0 \\ 0 & 0 & 3 & 0 \\ 0 & 0 & 0 & 6 \end{Vmatrix}$$

$$\cdot\left\{\begin{Vmatrix} 0 & 0 \\ 0 & -3 \\ 3 & 0 \\ 0 & -3 \end{Vmatrix} - \begin{Vmatrix} 3 \\ 3 \\ 0 \\ -3 \end{Vmatrix} \cdot \left(\begin{Vmatrix} 3 & 3 & 0 & -3 \end{Vmatrix} \cdot \right.\right.$$

$$\left.\frac{1}{6EI}\begin{Vmatrix} 14 & 4 & 0 & 0 \\ 4 & 8 & 0 & 0 \\ 0 & 0 & 3 & 0 \\ 0 & 0 & 0 & 6 \end{Vmatrix} \cdot \begin{Vmatrix} 3 \\ 3 \\ 0 \\ -3 \end{Vmatrix}\right)^{-1} \cdot \begin{Vmatrix} 3 & 3 & 0 & -3 \end{Vmatrix} \frac{1}{6EI}\begin{Vmatrix} 14 & 4 & 0 & 0 \\ 4 & 8 & 0 & 0 \\ 0 & 0 & 3 & 0 \\ 0 & 0 & 0 & 6 \end{Vmatrix}$$

$$\left.\cdot\begin{Vmatrix} 0 & 0 \\ 0 & -3 \\ 3 & 0 \\ 0 & -3 \end{Vmatrix}\right\} \cdot \begin{Vmatrix} m & 0 \\ 0 & m \end{Vmatrix}$$

$$= \begin{Vmatrix} 0 & 3 & 3 & 0 \\ 0 & -3 & 0 & -3 \end{Vmatrix} \frac{1}{6EI}\begin{Vmatrix} 14 & 4 & 0 & 0 \\ 4 & 8 & 0 & 0 \\ 0 & 0 & 3 & 0 \\ 0 & 0 & 0 & 6 \end{Vmatrix} \cdot \begin{Vmatrix} -1 & 1.5 \\ 2 & -1.5 \\ 3 & 0 \\ 1 & 1.5 \end{Vmatrix} \cdot \begin{Vmatrix} m & 0 \\ 0 & m \end{Vmatrix}$$

$$= \frac{m}{EI}\begin{Vmatrix} 15 & -3 \\ -3 & 7.5 \end{Vmatrix}.$$

The characteristic numbers of matrix D are equal to, respectively:

$$\lambda_1 = 16.05\frac{m}{EI}; \qquad \lambda_2 = 6.45\frac{m}{EI};$$

The frequencies of free oscillations are equal to:

$$\omega_1 = 0.2496\sqrt{\frac{EI}{m}}; \qquad \omega_2 = 0.3938\sqrt{\frac{EI}{m}};$$

10.4 Matrix Form of Determining Structures Frequencies on Vibration Loads

10.4.1 Derivation of Resultant Equations

Let the vibrational dynamic load act on an arbitrary elastic system with n concentrated masses. The vibrational dynamic load varies over time according to the following law:

$$\vec{P}(t) = P\sin\theta t = \begin{Vmatrix} P_1 \\ P_2 \\ \vdots \\ P_k \end{Vmatrix} \sin\theta t, \tag{10.54}$$

where P_k is the amplitude value of the harmonic load; θ is the frequency of forced oscillations.

Let us add the terms from the harmonic load (10.54) to the equations of displacement (10.37) resulting in a system of linear nonhomogeneous differential equations:

$$\vec{y} = -c \cdot B \cdot L_{mI} \cdot m \cdot \vec{\ddot{Y}} + \vec{\Delta}_{pI(t)}, \tag{10.55}$$

where $\vec{\Delta}_{pI(t)}$ is the vector of displacements in a given system from an external load $\vec{P}(t)$ in all directions: $I_1, I_2, \ldots, I_n$. Since external loads are applied simultaneously, the matrix $\vec{\Delta}_{pI(t)}$ contains one column. The number of rows of the matrix $\vec{\Delta}_{pI(t)}$ is equal to the number of directions of inertial forces: $I_1, I_2, \ldots, I_n$.

The system of Eq. (10.55) has a particular solution for a steady process of oscillations.

$$\vec{Y}(t) = \vec{A}_i \sin\theta t = \begin{Vmatrix} A_1 \\ A_2 \\ \vdots \\ A_n \end{Vmatrix} \sin\theta t, \tag{10.56}$$

where A_i is the amplitude of forced oscillations of the $i-$ th mass.

The displacement vector under the action of the load which is given by condition (10.54) has the following form:

$$\vec{\Delta}_{pI(t)} = \vec{\Delta}_{pI} \cdot \sin\theta t, \tag{10.57}$$

where $\vec{\Delta}_{pI}$ is the amplitude value of the displacements of the masses m_i from the harmonic load.

Expressions (10.56) and (10.57) are substituted in equation (10.55). Then, the resulting equation is divided by sinθt. The matrix equation will have the following form:

$$L_{mI}^T \cdot B \cdot L_{mI} \cdot m \cdot \theta^2 \cdot \vec{A} + \vec{A} + \vec{\Delta}_{pI} = 0. \tag{10.58}$$

The product $m \cdot \theta^2 \cdot \vec{A}$ is denoted by the vector of the amplitude values of the inertia forces $\vec{I}$:

$$\vec{I} = \begin{Vmatrix} I_1 \\ I_2 \\ \vdots \\ I_n \end{Vmatrix} = m \cdot \theta^2 \cdot \vec{A}\ . \tag{10.59}$$

Expression (10.59) is substituted into Eq. (10.58). As a result, we obtain the matrix equation of the force method:

$$\left(C_\delta - \frac{1}{\theta^2} m^{-1}\right)\vec{I} + \vec{\Delta}_{pI} = 0 \tag{10.60}$$

where $C_\delta = L_m^{0\,T} \cdot B \cdot L_{mI}$ is a matrix of displacements in a given system from all inertial forces $\vec{I}_n = 1$; $\vec{\Delta}_{pI} = L_m^{0\,T} \cdot B \cdot \vec{M}_p$ is a vector of mass displacements from the amplitude values of the harmonic load in a given system; m^{-1} is the inverse matrix of m, which has the following form:

$$m^{-1} = \left\| \begin{matrix} \frac{1}{m_1} & & & & \\ & \frac{1}{m_2} & & & \\ & & \cdot & & \\ & & & \cdot & \\ & & & & \frac{1}{m_n} \end{matrix} \right\|.$$

The vector of bending moments from the amplitude values of vibration load in a statically determined system is defined by the formula:

$$\vec{M}_p = L_{mp}^0 \cdot \vec{P} \tag{10.61}$$

where L_{mp}^0 is a matrix of bending moments from amplitude values of external forces $\vec{P}_1 = \vec{P}_2 = \cdots = \vec{P}_m = 1$.

The vector of bending moments $\vec{M}_p$ in a statically indeterminate system can be determined by one of the classical methods of structural mechanics, for example: the method of forces, the method of displacements, the combined or mixed method. The choice of this or that method is dictated by the simplicity of the calculation, the rational primary system, and the least number of calculations.

The vector $\vec{M}_p$ in the calculation by the force method is defined as follows:

$$\vec{M}_p = \left\{L_{mp}^0 - L_m^0\left(L_m^{0\,T} \cdot B \cdot L_m^0\right)^{-1} L_m^{0\,T} \cdot \mathrm{B} \cdot L_{mp}^0\right\}\vec{P} = \vec{M}_p^0 - L_m^0 \cdot A_\delta^{-1} \cdot \vec{\Delta}_p, \tag{10.62}$$

where L_m^0 is the matrix of the influence of bending moments in a statically determined system from unit forces $\vec{X}_1 = \vec{X}_2 = \cdots = \vec{X}_k = 1$, which are applied separately (here k is the number of unknowns forces method); $\vec{M}_p^0 = L_{mp}^0 \cdot \vec{P}$ is the vector of bending moments in the statically determined system from the amplitude values of the external load; $A_\delta = L_m^{0\,T} \cdot B \cdot L_m^0$ is the matrix of displacements in a statically determined system in the direction of action of all forces $\vec{X}_1 = \vec{X}_2 = \cdots = \vec{X}_k = 1$ from the separate action of all unit forces $\vec{X}_k = 1$; A_δ is a square symmetric matrix that has order k.

The vector of bending moments $\vec{M}_p$ in the displacement method is determined by the formula:

$$\vec{M}_p = \left\{L_{mp} - L_m\left(L_m^T \cdot B \cdot L_m\right)^{-1} L_m^T \cdot \mathrm{B} \cdot L_{mp}^0\right\}\vec{P} = \vec{M}_p^n - L_m \cdot R_r^{-1} \cdot \vec{R}_p, \tag{10.63}$$

where L_{mp} is the matrix of the influence of bending moments in the primary system of the method of displacements from external forces $\vec{P}_m = 1$; L_m is the matrix of the influence of bending moments in the primary system of the method of displacements from unit displacements of additional links

$\vec{Z}_1 = \vec{Z}_2 = \cdots = \vec{Z}_k = 1$, (here k is the number of unknowns of displacement method); L^0_{mp} is the matrix of the influence of bending moments in any statically determinate system from external forces $\vec{P}_m = 1$; $R_r = L^T_m \cdot B \cdot L_m$ is a matrix of reactions in the input of additional links in the primary system from unit displacements of these links (here R_r is a square symmetric matrix of order k). $\vec{M}^R_p$ is the vector of bending moments in the primary system of displacement method from the amplitude values of the vibration load; $\vec{R}_p = L^T_m \cdot B \cdot L_{mp} \cdot \vec{P}$ is the vector of reactive forces in additional links from amplitude values of a given load.

The final values of the bending moments in a given system are found from the solution of the equation:

$$\vec{M} = \vec{M}_p + L_{mI} \cdot \vec{I} \tag{10.64}$$

where $\vec{I}$ is the vector of amplitude values of inertial forces; the vector of amplitude values of the inertial forces is determined by the solution of the matrix equation (10.60), which has the following form:

$$\vec{I} = -\left(C_\delta - \frac{1}{\theta^2} m^{-1}\right) \vec{\Delta}_p. \tag{10.65}$$

10.4.2 Calculation of Frames on Vibration Load by the Forces Method

The method of calculating frames for vibration load will be explained in the following examples:

Problem 10.4

A dynamic diagram of bending moments is required to be built for the frame shown in Fig. 10.6. The frame is under the action of a harmonic force $P(t) = 3 \sin \theta t$ as well as a distributed harmonic load

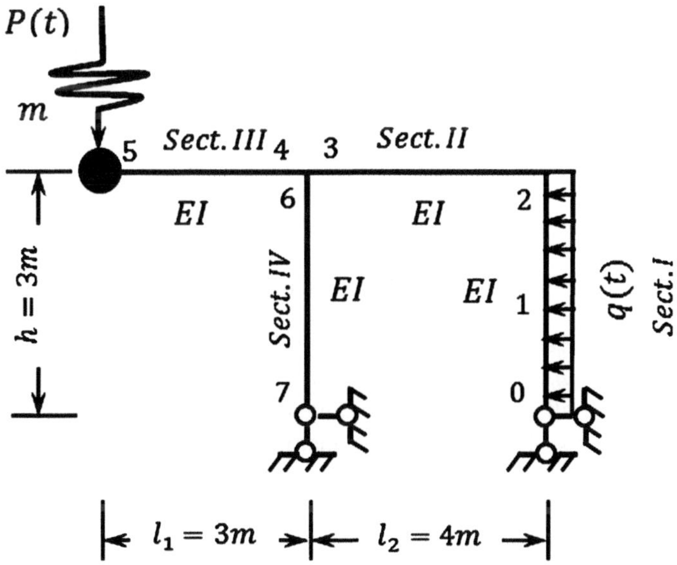

Fig. 10.6 Frame structure under the action of harmonic force $P(t)$ and distributed harmonic load $q(t)$

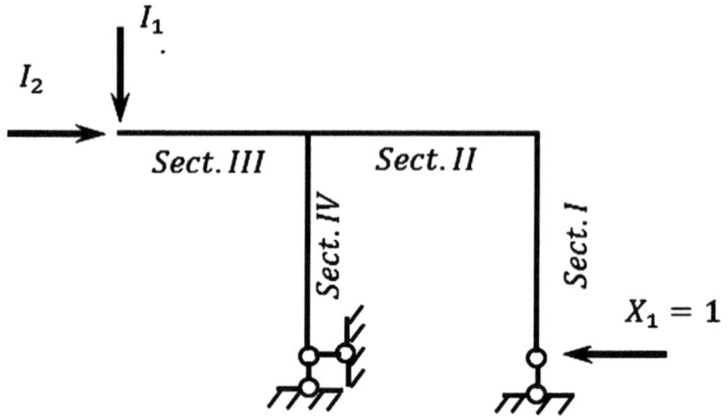

Fig. 10.7 The primary system of method of forces

$q(t) = 3 \sin \theta t$. The frequency of forced oscillations $\theta = 0.8\omega_1$, where ω_1 is the minimum frequency of free vibrations of the frame. In this case the minimal frequency is equal to $\omega_1 = 0.2495\sqrt{\frac{\mathrm{EI}}{m}}$ (see Problem 10.3), then $\theta = 0.1997$.

Solution

The matrix of displacements in the given system C_δ was also calculated in the same example/problem

$$C_\delta = \frac{1}{\mathrm{EI}} \begin{Vmatrix} 15 & -3 \\ -3 & 7.5 \end{Vmatrix}.$$

The vector of bending moments from the amplitude values of external loads $\vec{M}_p$ is determined by the forces method. The primary system of the method of forces is presented in Fig. 10.7.

The diagrams of bending moments from a unit force $X_i = 1$ and the amplitude values of external loads in the primary system of the method of forces are shown in Fig. 10.8c, d.

When compiling the matrix of pliability of individual sections of the frame, let's take into account the following: the load diagram has the form of a square parabola in the first section of Fig. 10.8e, so the pliability matrix for first section b_1 has the third order; all the other diagrams are linear.

$$b_1 = \frac{h}{6\mathrm{EI}} \begin{Vmatrix} 1 & 0 & 0 \\ 0 & 4 & 0 \\ 0 & 0 & 1 \end{Vmatrix} = \frac{3}{6\mathrm{EI}} \begin{Vmatrix} 1 & 0 & 0 \\ 0 & 4 & 0 \\ 0 & 0 & 1 \end{Vmatrix} = \frac{1}{6\mathrm{EI}} \begin{Vmatrix} 3 & 0 & 0 \\ 0 & 12 & 0 \\ 0 & 0 & 3 \end{Vmatrix};$$

$$b_2 = \frac{1}{6\mathrm{EI}} \begin{Vmatrix} 8 & 4 \\ 4 & 8 \end{Vmatrix}; \quad b_3 = \frac{1}{6\mathrm{EI}} \begin{Vmatrix} 6 & 3 \\ 3 & 6 \end{Vmatrix}; \quad b_4 = \frac{1}{6\mathrm{EI}} \begin{Vmatrix} 6 & 3 \\ 6 & 6 \end{Vmatrix}.$$

Note: *Here, the bending moment diagrams have been shown from the side of the stretched zones of the elements of the frame.*

Since all diagrams of bending moments at the boundary of the first and second sections are continuous, then the order of matrices B, can be reduced by shifting the sub-matrices b_1 and b_2 diagonally.

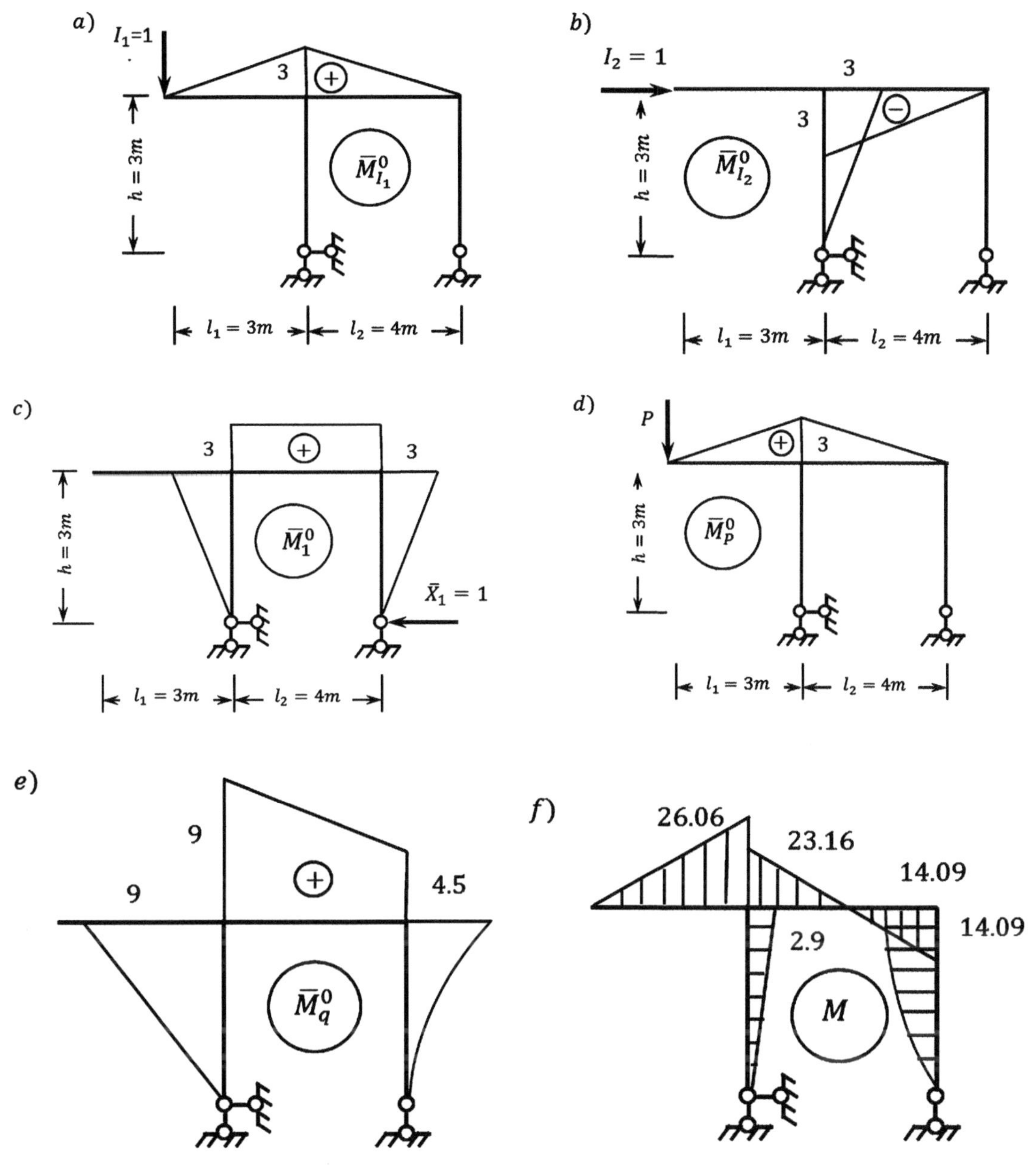

Fig. 10.8 (**a**–**d**) Bending moment diagrams due to a unit force and amplitude values of external loads; (**e**) Load diagram; (**f**) Bending moment diagram due to dynamic loads

$$B = \frac{1}{6\mathrm{EI}}\left\|\begin{matrix} 12 & 0 & 0 & 0 & 0 \\ 0 & 11 & 4 & 0 & 0 \\ 0 & 4 & 8 & 0 & 0 \\ 0 & 0 & 0 & 6 & 0 \\ 0 & 0 & 0 & 0 & 6 \end{matrix}\right\|.$$

The bending moment influence matrix L^0_m will contain five rows:

$$L_m^0 = \begin{Vmatrix} 1.5 \\ 3 \\ 3 \\ 0 \\ -3 \end{Vmatrix}.$$

The matrix of the influence of bending moments from external forces L_{mp}^0 equal to one consists of two columns according to the number of loads:

$$L_{mp}^0 = \begin{Vmatrix} 0 & 1.125 \\ 0 & 4.5 \\ 3 & 9 \\ 3 & 0 \\ 0 & -9 \end{Vmatrix}.$$

The vector of the external load in this case has the following form:

$$\vec{P} = \begin{Vmatrix} P \\ q \end{Vmatrix} = \begin{Vmatrix} 3 \\ 1 \end{Vmatrix}.$$

Let us define the bending moment vector from the amplitude values of external load $\vec{M}_p^0$:

$$\vec{M}_p^0 = L_{mp}^0 \cdot \vec{P} = \begin{Vmatrix} 0 & 1.125 \\ 0 & 4.5 \\ 3 & 9 \\ 3 & 0 \\ 0 & -9 \end{Vmatrix} \cdot \begin{Vmatrix} 3 \\ 1 \end{Vmatrix} = \begin{Vmatrix} 1.125 \\ 4.5 \\ 18 \\ 9 \\ -9 \end{Vmatrix}.$$

Now let's define the product $L_m^{0\,T} \cdot B$:

$$L_m^{0\,T} \cdot B = \| 1.5 \quad 3 \quad 3 \quad 0 \quad -3 \| \frac{1}{6\mathrm{EI}} \begin{Vmatrix} 12 & 0 & 0 & 0 & 0 \\ 0 & 11 & 4 & 0 & 0 \\ 0 & 4 & 8 & 0 & 0 \\ 0 & 0 & 0 & 6 & 0 \\ 0 & 0 & 0 & 0 & 6 \end{Vmatrix} = \frac{1}{6\mathrm{EI}} \| 18 \quad 45 \quad 36 \quad 0 \quad -18 \|.$$

The displacement matrix A_δ will take the following form:

$$A_\delta = L_m^{0\,T} \cdot B \cdot L_m^0 = \| 18 \quad 45 \quad 36 \quad 0 \quad -18 \| \cdot \begin{Vmatrix} 1.5 \\ 3 \\ 3 \\ 0 \\ -3 \end{Vmatrix} \frac{1}{6EI} = \frac{1}{6\mathrm{EI}} \|54\|.$$

The displacement vector $\vec{\Delta}_p$ from the external load will be written:

$$\vec{\Delta}_p = L_m^{0\,T} \cdot B \cdot L_{mp}^0 \cdot \vec{P} = \|\, 18 \quad 45 \quad 36 \quad 0 \quad -18 \,\| \cdot \begin{Vmatrix} 1.125 \\ 4.5 \\ 18 \\ 9 \\ -9 \end{Vmatrix} \cdot \frac{1}{6\mathrm{EI}} = \frac{1}{6\mathrm{EI}} \|172.125\|.$$

The bending moments from the amplitude values of the external load will be calculated:

$$\vec{M}_p = \begin{Vmatrix} 1.125 \\ 4.5 \\ 18 \\ 9 \\ -9 \end{Vmatrix} - \begin{Vmatrix} 1.5 \\ 3 \\ 3 \\ 0 \\ -3 \end{Vmatrix} \cdot \|54\| \cdot \|172.125\| = \begin{Vmatrix} -3.655 \\ -5.06 \\ 8.44 \\ 9 \\ 0.56 \end{Vmatrix}.$$

The matrix of bending moment from the unit inertial forces in the primary system has the form Fig. 10.8a, b:

$$L_{mI}^0 = \begin{Vmatrix} 0 & 0 \\ 0 & 0 \\ 3 & -3 \\ 3 & 0 \\ 0 & 3 \end{Vmatrix}.$$

The matrix of mass displacements from the amplitude values of the given loads $\vec{\Delta}_{pI} = L_{mI}^{0\,T} \cdot B \cdot \vec{M}_p$ is calculated:

$$\vec{\Delta}_{pI} = \begin{Vmatrix} 0 & 0 & 3 & 3 & 0 \\ 0 & 0 & -3 & 0 & 3 \end{Vmatrix} \cdot \frac{1}{6\mathrm{EI}} \begin{Vmatrix} 12 & 0 & 0 & 0 & 0 \\ 0 & 11 & 4 & 0 & 0 \\ 0 & 4 & 8 & 0 & 0 \\ 0 & 0 & 0 & 6 & 0 \\ 0 & 0 & 0 & 0 & 6 \end{Vmatrix} \cdot \begin{Vmatrix} -3.655 \\ -5.06 \\ 8.44 \\ 9 \\ 0.56 \end{Vmatrix} = \frac{1}{\mathrm{EI}} \begin{Vmatrix} 50.64 \\ -21.96 \end{Vmatrix}.$$

The matrix of unknown inertial forces is found from Eq. (10.60):

$$I = \left(\begin{Vmatrix} 15 & -33 \\ -3 & 7.5 \end{Vmatrix} \frac{1}{\mathrm{EI}} - \frac{25.08}{\mathrm{EI}} \begin{Vmatrix} 1 & 0 \\ 0 & 1 \end{Vmatrix} \right)^{-1} \cdot \frac{1}{\mathrm{EI}} \begin{Vmatrix} 50.64 \\ -21.96 \end{Vmatrix} = \begin{Vmatrix} 5.697 \\ -2.23 \end{Vmatrix}.$$

The matrix of the influence of bending moments from single inertial forces in the given system

Let's construct a matrix of the influence of bending moments from unit inertial forces in the given system to determine the dynamic diagram of bending moments.

The diagrams of moments in a given statically indeterminate system from these forces are shown in Fig. 10.9a, b.

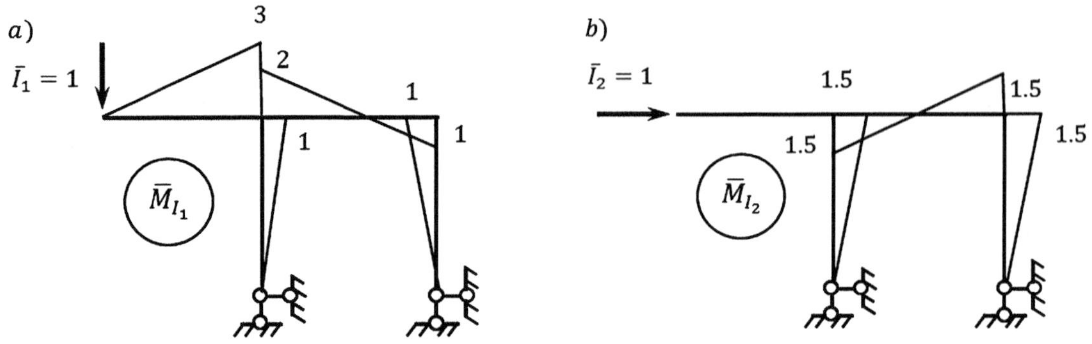

Fig. 10.9 (**a**) Bending moment diagrams due to $\bar{I}_1 = 1$; and (**b**) Bending moment diagrams due to $\bar{I}_2 = 1$

$$L_{mI} = \begin{Vmatrix} -0.5 & 0.75 \\ -1 & 1.5 \\ 2 & -1.5 \\ 3 & 0 \\ 1 & 1.5 \end{Vmatrix}.$$

The vector of bending moments from the action of dynamic loads is equal to:

$$\vec{M}_{\text{dyn.}} = \vec{M}_p + L_{mI} \cdot \vec{I},$$

$$\vec{M}_{\text{dyn}} = \begin{Vmatrix} -3.655 \\ -5.06 \\ 8.44 \\ 9 \\ 0.56 \end{Vmatrix} + \begin{Vmatrix} -0.5 & 0.75 \\ -1 & 1.5 \\ 2 & -1.5 \\ 3 & 0 \\ 1 & 1.5 \end{Vmatrix} \cdot \begin{Vmatrix} 5.697 \\ -2.23 \end{Vmatrix} = \begin{Vmatrix} -8.17 \\ -14.09 \\ 23.16 \\ 26.06 \\ 2.9 \end{Vmatrix}.$$

The diagram of bending moments from dynamic loads is constructed according to the obtained vector $\vec{M}_{\text{dyn.}}$, Fig. 10.8f.

10.4.3 Calculation of Frames on Vibration Load by the Displacement Method

The method of calculating frames for vibration load will be shown in the following examples:

Problem 10.5

A dynamic diagram of bending moments is required to be built for the frame shown in Fig. 10.10. The frame is under the action of a harmonic force $P(t) = 16 \sin \theta t$. The frequency of forced oscillations $\theta = 0.8\omega_1$, where ω_1 is the minimum frequency of free vibrations of the frame. According to the conditions of the problem: $m_1 = m_2 = 2m$; the bending stiffness for the vertical and horizontal elements of the frame is equal to EI.

Fig. 10.10 Statically indeterminate frame under action of a harmonic load

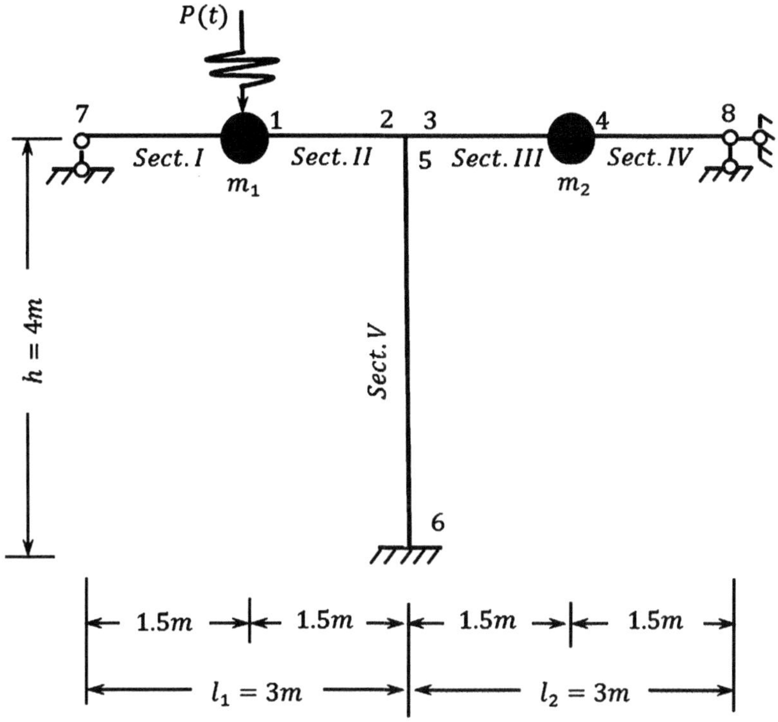

Solution

The given frame is once kinematically indeterminate. The primary system will be chosen by introducing additional links. These links will prevent the angular rotation of the middle node of the frame. The diagram of bending moments in the primary system from inertial forces $\overline{I}_1, \overline{I}_2, \overline{Z}_1$ and from external load P will be constructed. The bending moment diagram in the primary system of the displacement method is shown in Fig. 10.11. The bending moment diagram in the primary system of the force method is shown in Fig. 10.12. The influence matrices will have the following form:

$$L_m = \mathrm{EI}\begin{Vmatrix} 0.5 \\ 1 \\ -1 \\ -0.5 \\ 1 \\ -0.5 \end{Vmatrix}; \quad L_{mI}^{0} = \begin{Vmatrix} 0 & 0 \\ 1.5 & 1.5 \\ 1.5 & -1.5 \\ 0 & 0 \\ 0 & -3 \\ 0 & -3 \end{Vmatrix}.$$

Note: *Here, the bending moment diagrams have been shown from the side of the stretched zones of the elements of the frame.*

The influence matrices from the external load have the following form:

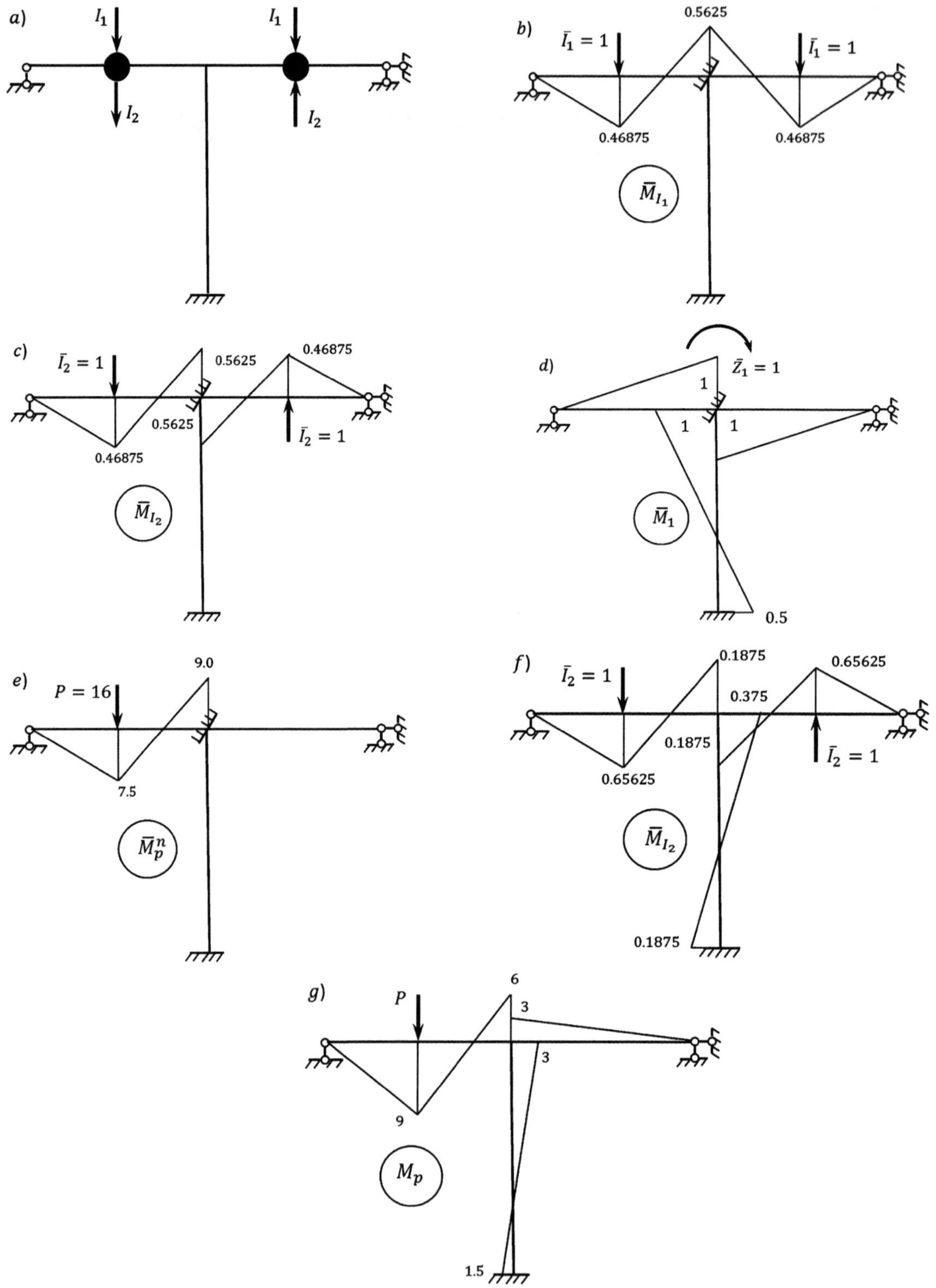

Fig. 10.11 Primary system of displacement method and bending moment diagrams

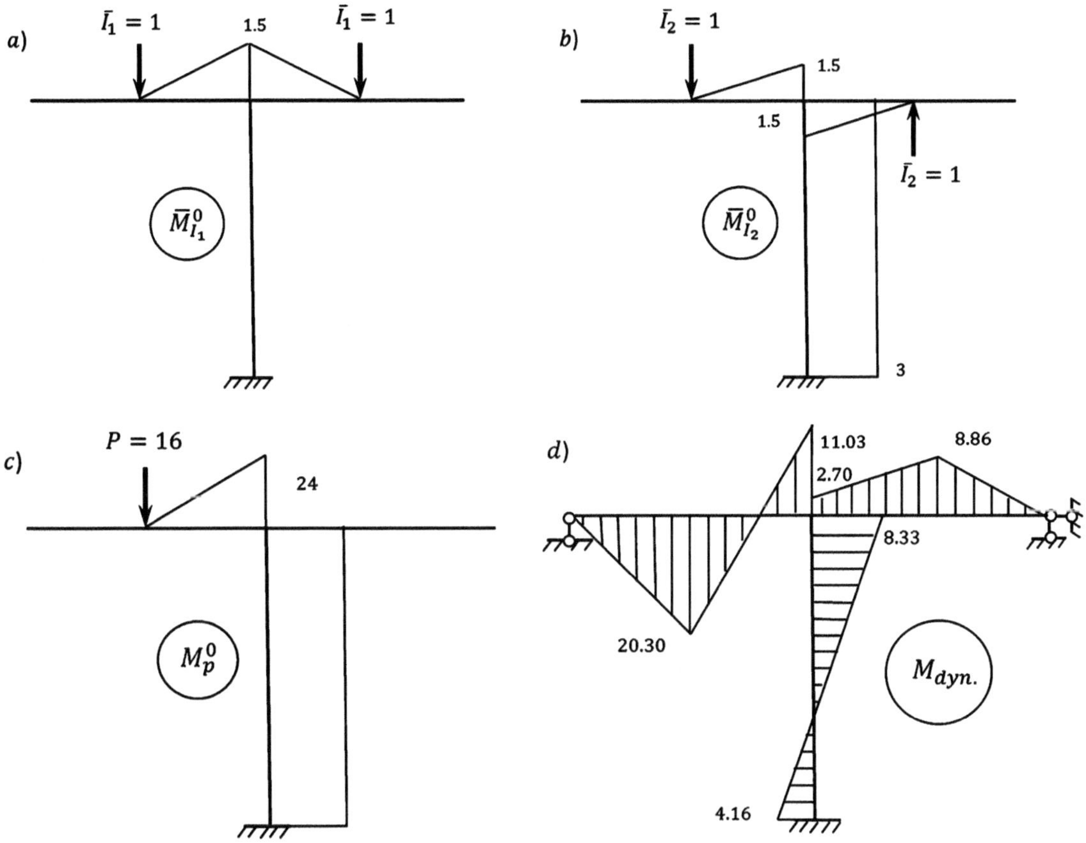

Fig. 10.12 Primary system of force method and bending moment diagrams

$$L_{mp} = \mathrm{EI}\left\| \begin{matrix} -7.5 \\ 9 \\ 0 \\ 0 \\ 0 \\ 0 \end{matrix} \right\|; \qquad L^0_{mp} = \mathrm{EI}\left\| \begin{matrix} 0 \\ 24 \\ 0 \\ 0 \\ -24 \\ -24 \end{matrix} \right\|.$$

The influence matrix of bending moments in the given system from unit inertial forces is equal to:

$$L_{mI} = L^0_{mI} + L_m\left(L^T_m \cdot B \cdot L_m\right)^{-1} L_m \cdot B \cdot L^0_{mI},$$

$$L_{mI} = \left\| \begin{matrix} -0.46875 & -0.65625 \\ 0.5625 & 0.1875 \\ 0.5625 & -0.1875 \\ -0.46875 & 0.65625 \\ 0 & -0.375 \\ 0 & 0.1875 \end{matrix} \right\|.$$

The obtained diagrams of bending moments in the given system are shown in Fig. 10.11.

The matrix of pliability B consists of five sub-matrices b_i:

$$b_1 = b_2 = b_3 = b_4 = \frac{l}{6\text{EI}}\begin{Vmatrix} 2 & 1 \\ 2 & 1 \end{Vmatrix} = \frac{1.5}{6\text{EI}}\begin{Vmatrix} 2 & 1 \\ 1 & 2 \end{Vmatrix} = \frac{1}{4\text{EI}}\begin{Vmatrix} 2 & 1 \\ 1 & 2 \end{Vmatrix};$$

$$b_5 = \frac{l_5}{6\text{EI}}\begin{Vmatrix} 2 & 1 \\ 2 & 1 \end{Vmatrix} = \frac{3}{6\text{EI}}\begin{Vmatrix} 2 & 1 \\ 1 & 2 \end{Vmatrix} = \frac{1}{4\text{EI}}\begin{Vmatrix} 4 & 2 \\ 2 & 4 \end{Vmatrix}.$$

Taking into account the continuity of moments on the borders of sections 1, 2, 3, and 4, and the equality to zero of moments in sections 7 and 8, matrix B will have the following form:

$$B = \frac{1}{4\text{EI}}\begin{Vmatrix} 4 & 1 & 0 & 0 & 0 & 0 \\ 1 & 2 & 0 & 0 & 0 & 0 \\ 0 & 0 & 2 & 1 & 0 & 0 \\ 0 & 0 & 1 & 4 & 0 & 0 \\ 0 & 0 & 0 & 0 & 4 & 2 \\ 0 & 0 & 0 & 0 & 2 & 4 \end{Vmatrix}.$$

Let us define the displacement matrix A_δ:

$$A_\delta = \frac{1}{\text{EI}}\begin{Vmatrix} 0.4921875 & 0 \\ 0 & 0.9140625 \end{Vmatrix};$$

and also define matrix $C = A_\delta \cdot m$

$$C = \frac{1}{\text{EI}}\begin{Vmatrix} 0.4921875 & 0 \\ 0 & 0.9140625 \end{Vmatrix} m \begin{Vmatrix} 1 & 0 \\ 0 & 1 \end{Vmatrix} = \frac{m}{\text{EI}}\begin{Vmatrix} 0.4921875 & 0 \\ 0 & 0.9140625 \end{Vmatrix}.$$

The characteristic numbers of the matrix C are equal to:

$$\lambda_{\max} = 0.9140625\frac{m}{\text{EI}}; \qquad \lambda_{\min} = 0.4921875\frac{m}{\text{EI}}$$

The minimum frequency of natural vibrations is determined by the formula:

$$\omega_1 = 1.046\sqrt{\frac{\text{EI}}{m}}.$$

The vector of bending moments in the given system from the amplitude value of vibration load $\vec{M}_p$ is determined by the following formula:

$$\vec{M}_p = L_{mp} + L_m\left(L_m^T \cdot B \cdot L_m\right)^{-1} L_m^T \cdot B \cdot L_{mp}^0;$$

$$\vec{M}_p = \begin{Vmatrix} -9 \\ 6 \\ 3 \\ 1.5 \\ -3 \\ 1.5 \end{Vmatrix}.$$

The diagram of bending moments from the amplitude value of the load in the given system is shown in Fig. 10.11g.

The vector of bending moments from the dynamic load is equal to:

$$\vec{M}_{\text{dyn.}} = \vec{M}_p - L_{ml}\left(A_\delta - \frac{1}{\theta^2}m\right)^{-1} L^*_{ml} \cdot B \cdot \vec{M}_p$$

$$\vec{M}_{\text{dyn.}} = \begin{Vmatrix} -20.30 \\ 11.03 \\ 2.70 \\ 8.86 \\ -8.33 \\ 4.16 \end{Vmatrix}$$

The dynamic diagram of bending moments is shown in Fig. 10.12d.

References

1. McKenzie, W. M. C. (2006). *Example in structural analysis* (797p). London, New York.
2. Cheng, F. Y. (2001). *Matrix analysis of structural dynamics applications and earthquake engineering* (1016p). CRC Press.
3. Kassimali, A. (2012). *Matrix analysis of structures* (2nd ed., 657p). Global Engineering: Southern Illinois University.
4. Paz, M., & Leigh, W. (2001). *Integrated matrix analysis of structures: Theory and computation* (330p). Springer.
5. Paradrakakis, M., & Sapountzakis, E. J. (2018). *Matrix method for advanced structural analysis* (455p). Institute of Structural Analysis and Research National Technical University of Athens. Butterworth-Heinemann.
6. Paz, M. *Matrix structural analysis & dynamics theory and computation* (485p). Sped Engineering School University of Louisville.
7. Nagarajan, P. (2018). *Matrix method of structural analysis* (361p). CRC Press, Taylor & Francis Group.
8. Doyle, J. F. (1991). Static and dynamic analysis of structures with an emphasis on mechanics and computer matrix methods. In *Solid mechanics and its application* (442p). Springer.
9. Rzhanitsin, A. R. (1982). *Structural mechanics* (398p). Vischaya shkola.
10. Klein, G. K., Portaev, L. P., & Petrakov, A. A. (1980). *Iterative and matrix forms of calculating frames or stability and vibrations* (64p). MESI.
11. Smirnov, A. P., Alekcandrov, A. V., Lashchenikov, B. Y., & Shaposhnikov, N. N. (1982). *Structual mechanics. Dynamics and stability* (411p). Stroyizdat.
12. Petrosian, L. G. (1987). *Calculation of frames of industrial buildings using computers. Methodological guidelines and calculation examples* (45p). Armenian National University of Architecture & Construction. Department: Structural Mechanics.

Vibrations of Systems with an Infinite Number of Degrees of Freedom

11

Abstract

This chapter is devoted to systems with an infinite number of degree of freedom. Free vibrations of beams with uniformly distributed mass, problems and solutions, fundamental functions of the equation of vibration of beams, longitudinal vibrations of rods with a distributed mass, problems and solutions, frequencies of free vibrations of trusses, problems, and solutions are discussed in the present chapter.

11.1 Free Vibrations of Beams with Uniformly Distributed Mass

Numerous publications are devoted to the vibration of continuum dynamic systems, i.e., systems with infinite number of degree of freedom. These include the following works: Clough and Penzien [1], Copra [2], Kabe and Sako [3], Katsikadelis [4], Craig and Kurdila [5], Busby et al. [6], Anderson and Naeim [7], Mario Paz and Young Hoon Kim [8], Clain [9], Zhihui Zhou et al. [10], and many others.

As an example of a system with an infinite number of degrees of freedom, let us consider the vibrations of a beam with a uniformly distributed load. Let us imagine that the distributed load is an equivalent to the mass of the beam with an intensity $q = \overline{m}g$, and with linear stiffness: *EI*. (Fig. 11.1a).

The frequency spectrum of such a system consists of an infinite number of frequencies. To determine these frequencies, let us proceed from the differential equation of the elastic line of the beam:

$$\frac{\partial^2 y}{\partial x^2} = \frac{M}{EI}. \tag{11.1}$$

Let us remember that the curvature of a beam is approximately equal to the second derivative of the beam's deflection.

$$\frac{1}{\rho} = y''. \tag{11.2}$$

It is also known that

L. G. Petrosian, *Dynamics of Structures*, https://doi.org/10.1007/978-3-031-63539-7_11

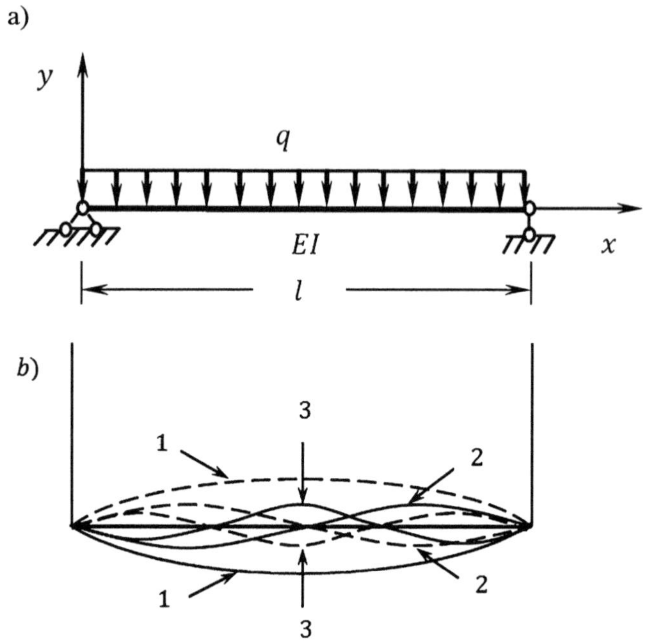

Fig. 11.1 (**a**) Vibration of beam with uniformly distributed load; (**b**) main modes of vibrations

$$\frac{1}{\rho} = \frac{M}{EI}, \tag{11.3}$$

the curvature of the bent axis of the beam is directly proportional to the bending moment and inversely proportional to the stiffness of the beam.

$$\frac{M}{EI} = \pm y'', \tag{11.4}$$

Here M is the bending moment in an arbitrary section of the beam, which can be expressed through the intensity of the inertial load:

$$\frac{\partial^2 M}{\partial x^2} = q; \tag{11.5}$$

Taking into account (11.1), we obtain:

$$EI \frac{\partial^4 y}{\partial x^4} = q \tag{11.6}$$

The notations in partial derivatives must be applied because the displacement y is a function of the two variables coordinate x, and time t.

From expressions (11.1) and (11.5) follows:

$$q = EI \frac{\partial^4 y}{\partial x^4} \tag{11.7}$$

At the same time, the intensity of the inertial load of the beam can be expressed through the linear mass $\overline{m}$ and through the acceleration $\frac{\partial^2 y}{\partial t^2}$ during vibrations.

$$q = -\overline{m}\frac{\partial^2 y}{\partial t^2} \tag{11.8}$$

The differential equation of free oscillations of the beam will be obtained by equating expressions for loads (11.7) and (11.8):

$$\frac{\partial^2 y}{\partial t^2} + \frac{EI}{\overline{m}}\frac{\partial^4 y}{\partial x^4} = 0 \tag{11.9}$$

Thus, obtained equation is homogeneous, linear of the fourth order in partial derivatives.

Let us limit ourselves to finding only such solutions of the differential equation that define standing waves, i.e., a mode of bending independent of time. With this mode of oscillations/vibrations, the solution of the differential Eq. (11.9), based on Fourier method, can be represented as a product of two functions, each of which depends on one variable:

$$y = X(x)T(t). \tag{11.10}$$

To find the functions $X(x)$ and $T(t)$, we differentiate expression (11.10).

$$\begin{aligned} \frac{\partial^2 y}{\partial t^2} &= X(x)\frac{d^2 T(t)}{dt^2}, \\ \frac{\partial^4 y}{\partial x^4} &= T(t)\frac{d^4 X(x)}{dx^4}. \end{aligned} \tag{11.11}$$

On the right side of these expressions are ordinary differentials, each of which depends on only one variable. Let us substitute expression (11.11) into (11.9), as a result we obtain:

$$X(x)\frac{d^2 T(t)}{dt^2} + \frac{EI}{\overline{m}}T(t)\frac{d^4 X(x)}{dx^4} = 0. \tag{11.12}$$

Let us regroup the terms in Eq. (11.12). In the left part, we select the terms that depend only on x, and in the right part of the equation, the terms that depend only on t.

$$\frac{\frac{EI}{\overline{m}}\frac{d^4 X(x)}{dx^4}}{X(x)} = -\frac{\frac{d^2 T(t)}{dt^2}}{T(t)} = \omega^2 \tag{11.13}$$

In Eq. (11.13), the left part depends only on x, and the right part depends only on t, but since the two sides are equivalent, it follows that both parts are independent of both x and t. That is, each of the expressions is a constant number. This constant number is denoted by ω^2. Thus the integration of the partial differential Eq. (11.9) can be reduced to the integration of two ordinary differential equations:

$$\begin{aligned} &\frac{d^2 T(t)}{dt^2} + \omega^2 T(t) = 0 \\ &\frac{d^4 X(x)}{dx^4} - \frac{\overline{m}}{EI}X(x)\omega^2 = 0 \end{aligned} \tag{11.14}$$

For the first differential equation, the solution is known and it has the form:

$$T(t) = A\sin(\omega t + \varphi_0). \tag{11.15}$$

This equation is equivalent to the harmonic oscillations of a system with single degree of freedom. The second differential equation can be written in the following form:

$$EIy^{IV} - m\omega^2 y = 0. \tag{11.16}$$

Let us consider the second equation in expression (11.14). Let us introduce the following notation:

$$\frac{\overline{m}}{EI}\omega^2 = \left(\frac{\psi}{l}\right)^4. \tag{11.17}$$

From here we determine the oscillation frequency ω:

$$\omega = \left(\frac{\psi}{l}\right)^2 \sqrt{\frac{EI}{\overline{m}}} \tag{11.18}$$

The notation (11.17) is substituted in the second equation of the expression (11.14), resulting in the following differential equation:

$$\frac{d^4 X(x)}{dx^4} - \left(\frac{\psi}{l}\right)^4 X(x) = 0 \tag{11.19}$$

The solution of this differential equation is as follows:

$$X(x) = B_1 \sin\left(\frac{\psi}{l}\right)x + B_2 \cos\left(\frac{\psi}{l}\right)x + B_3 \sinh\left(\frac{\psi}{l}\right)x + B_4 \cosh\left(\frac{\psi}{l}\right)x \tag{11.20}$$

Here B_1, B_2, B_3, and B_4, are constants that are defined from the boundary conditions of the problem. The general integral of Eq. (11.9) is represented as an infinite series:

$$y(x,t) = \sum_{n=1}^{\infty} \left(B_{1n} \sin \frac{\psi}{l}x + B_{2n} \cos \frac{\psi}{l}x + B_{3n} \sinh \frac{\psi}{l}x + B_{4n} \cosh \frac{\psi}{l}x\right) \cdot \sin(\omega_n t + \varphi_n) \tag{11.21}$$

As an example, consider a hinged beam on two supports under the following boundary conditions:

$$X(0) = X''(0) = X(l) = X''(l) = 0 \tag{11.22}$$

Taking into account the expressions for hyperbolic functions,

$$\text{sinhx} = \frac{e^x - e^{-x}}{2}; \quad \text{coshx} = \frac{e^x + e^{-x}}{2}$$

we obtain:

1. $B_2 + B_4 = 0$;
2. $-B_2 + B_4 = 0$.
   $2B_4 = 0$, therefore $B_4 = 0$, $B_2 = 0$
3. $B_1 \sin \psi + B_2 \cos \psi + B_3 \sinh \psi + B_4 \cosh \psi = 0$
4. $-B_1 \sin \psi - B_2 \cos \psi + B_3 \sinh \psi + B_4 \cosh \psi = 0$

$$B_1 \sin \psi + B_3 \sinh \psi = 0$$

$$-B_1 \sin \psi + B_3 \sinh \psi = 0$$

A trivial solution to the problem will be obtained when $B_1 = B_3 = 0$, but then the system will be at rest and there will be no oscillations/vibrations. Nonzero values for B_1 and for B_3 will be obtained if the determinant composed of the coefficients turns zero, that is:

$$\begin{vmatrix} \sin\psi & \sinh\psi \\ -\sin\psi & \sinh\psi \end{vmatrix} = 0 \tag{11.23}$$

From here we obtain

$$2\sin\psi\sinh\psi = 0$$

If $\sinh\psi = 0$, then $X = 0$, which cannot be since the system will be at rest and will not oscillate. Therefore, the characteristic equation is: $\sin\psi = 0$. As a result, we have $\psi = \pi, 2\pi, \ldots, n\pi$. Then the expression for finding the frequency of oscillations/vibrations can be written in the following form:

$$\omega_1 = \frac{\pi^2}{l^2}\sqrt{\frac{EI}{\overline{m}}}, \omega_2 = \frac{4\pi^2}{l^2}\sqrt{\frac{EI}{\overline{m}}}, \ldots, \omega_n = \frac{n\pi^2}{l^2}\sqrt{\frac{EI}{\overline{m}}} \tag{11.24}$$

Each frequency has its own mode of vibration. Thus, the lowest frequency (of the fundamental tone) for the hinged supported beam will be:

$$\omega_{\min} = \omega_1 = \frac{\pi^2}{l^2}\sqrt{\frac{EI}{\overline{m}}} = \frac{\pi^2}{l^2}\sqrt{\frac{EIg}{q}}, \tag{11.25}$$

where $q = \overline{m}g$ is the intensity of the uniform load on the beam.

Figure 11.1b shows standing wave modes that correspond to the three lowest frequencies.

Each mode of oscillation vibration corresponds to its frequency. Each frequency value corresponds to its own standing wave shape/mode:

$$X_n(x) = B_n \sin\frac{n\pi}{l}x \tag{11.26}$$

The free vibration frequency of a beam with one concentrated force P in the middle of the span is expressed by the following formula:

$$\omega = \sqrt{\frac{g}{y_{st}}} = \sqrt{\frac{48EIg}{Pl^3}} \tag{11.27}$$

Let us equate expressions (11.25) and (11.27), from where we find the concentrated force $P_{\text{equiv.}}$ which will be equivalent to a uniformly distributed load at the frequency of free oscillations of a hinged supported beam:

$$P_{\text{equiv.}} = \frac{48}{\pi^4}ql = 0.493ql \approx 0.5ql \tag{11.28}$$

Thus, to find the frequency of the fundamental oscillations/vibrations of a beam with a uniformly distributed load, this load can be replaced by half of its resultant load, which is applied in the middle of the beam span.

Now, let us consider the vibrations of a built-in ends beam with a uniformly distributed load (Fig. 11.2).

The boundary conditions for the beam shown in Fig. 11.2 have the following form:

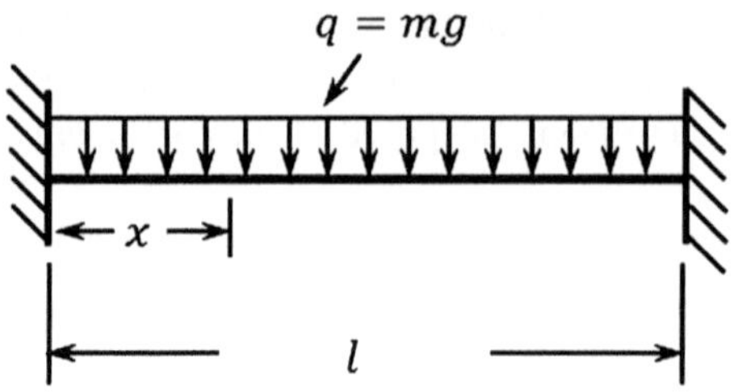

Fig. 11.2 A built-in ends beam with a uniformly distributed load

$$X(0)=X'(0)=X(l)=X'(l)=0 \tag{11.29}$$

From the solution of differential Eq. (11.20), we obtain:

1. $B_2 + B_4 = 0$;
2. $B_1 + B_3 = 0$.
   $B_2 + B_4 = B_1 + B_3$, therefore $B_1 = -B_3$, $B_2 = -B_4$
3. $B_1 \sin\psi + B_2 \cos\psi + B_3 \sinh\psi + B_4 \cosh\psi = 0$
4. $B_1 \cos\psi - B_2 \sin\psi + B_3 \cosh\psi + B_4 \sinh\psi = 0$

The values B_3 and B_4 are substituted into the last two equations:

$$B_1(\sin\psi - \sinh\psi) + B_2(\cos\psi - \cosh\psi) = 0$$
$$B_1(\cos\psi - \cosh\psi) - B_2(\sin\psi + \sinh\psi) = 0$$

The existence of nonzero solutions for B_1 and B_2 is possible only if the determinant of this system of equations is zero, that is:

$$\left(\sin^2\psi - \sinh^2\psi\right) + (\cos\psi - \cosh\psi)^2 = 0,$$
$$\sin^2\psi - \sinh^2\psi + \cos^2\psi - 2\cos\psi\cosh\psi + \cosh^2\psi = 0.$$

Since, $\sin^2\psi + \cos^2\psi = 1$, and $\cosh^2\psi - \sinh^2\psi = 1$ we will obtain:

$$2(1 - \cos\psi\cosh\psi) = 0.$$

Hence, ψ will be determined from the following equation:

$$\cos\psi\cosh\psi = 1.$$

The solutions of this equation are found by choosing the circular and hyperbolic functions. From the tables of circular and hyperbolic functions, we find the first values for ψ:

$$\psi_1 = 4.730, \quad \psi_2 = 7.853, \quad \psi_3 = 10.996, \ \ldots$$

For the first three oscillation frequencies, we obtain the following expressions:

$$\omega_1 = \frac{22.37}{l^2}\sqrt{\frac{EI}{\overline{m}}}, \quad \omega_2 = \frac{61.67}{l^2}\sqrt{\frac{EI}{\overline{m}}}, \quad \omega_3 = \frac{120.91}{l^2}\sqrt{\frac{EI}{\overline{m}}}$$

The frequency of vibrations of built-in ends beam under the action of an equivalent concentrated load $P_{equiv.}$ in the middle of the span is:

Fig. 11.3 Cantilever beam with uniformly distributed load

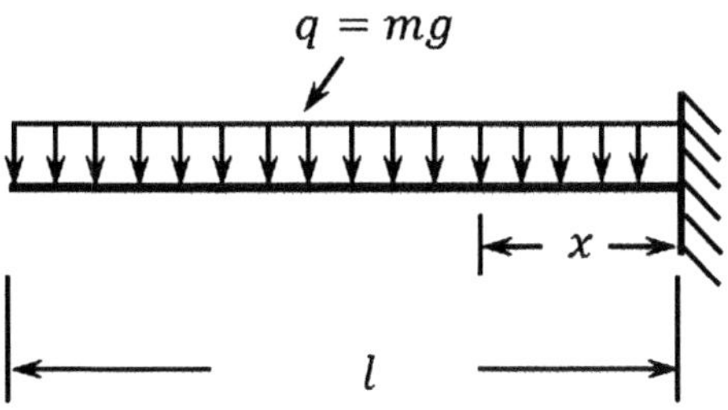

$$\omega = \sqrt{\frac{192EIg}{P_{\text{equiv.}}\, l^3}}$$

Let's equate this expression with the previous one and find the value of the equivalent concentrated load.

$$P_{\text{equiv.}} = 0.385ql \approx 0.4ql.$$

Problem 11.1
It is required to determine the frequencies of free vibrations (oscillations) of a cantilever beam with a uniformly distributed load shown in Fig. 11.3.

Solution
In this case, to determine B_1, B_2, B_3, and B_4, which are included in the solution of the differential Eq. (11.19), we have the following boundary conditions:

$$X(0) = X'(0) = X''(l) = X'''(l) = 0 \tag{11.30}$$

From the solution of differential Eq. (11.19), we obtain:

1. $B_2 + B_4 = 0$;
2. $B_1 + B_3 = 0$.
3. $-B_1 \sin\psi - B_2 \cos\psi + B_3 \sinh\psi + B_4 \cosh\psi = 0$
4. $-B_1 \cos\psi + B_2 \sin\psi + B_3 \cosh\psi + B_4 \sinh\psi = 0$.

Since $B_2 + B_4 = B_1 + B_3$, therefore $B_3 = -B_1$, $B_4 = -B_2$. The values B_3 and B_4 are substituted into the last two equations:

$$B_1(\sin\psi + \sinh\psi) + B_2(\cos\psi + \cosh\psi) = 0$$
$$B_1(\cos\psi + \cosh\psi) + B_2(-\sin\psi + \sinh\psi) = 0$$

The trivial solution $B_1 = B_2 = 0$ does not interest us. Therefore, in order to find nonzero solutions, the determinant composed of coefficients B_1, B_2 must be equal to zero.

$$\begin{vmatrix} \sin\psi + \sinh\psi & \cos\psi + \cosh\psi \\ \cos\psi + \cosh\psi & -\sin\psi + \sinh\psi \end{vmatrix} = 0$$

Let us reveal the determinant, as a result we obtain:

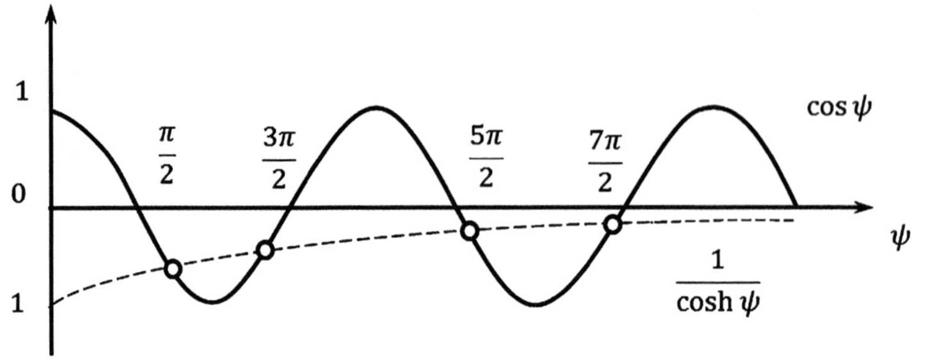

Fig. 11.4 A graph of the solution of transcendental equation

$$-\sin^2\psi + \sinh^2\psi - \cos^2\psi - \cosh^2\psi - 2\cos\psi \cdot \sinh\psi = 0,$$

since

$$\sin^2\psi + \cos^2\psi = \cosh^2\psi - \sinh^2\psi = 1$$

we obtain:

$$\cos\psi \cdot \cosh\psi = 1$$

The solution of the resulting transcendental equation can be represented graphically (Fig. 11.4).

$$\psi_1 = 1.87, \quad \psi_2 = 4.68, \quad \psi_3 = 7.86$$

Let us equate the expression of the frequency of the fundamental tone (frequency of the basic tone) of vibrations to the frequency of vibrations of a beam with one equivalent concentrated mass $m = P_{\text{equiv.}}/q$ or load at the free end, and find the magnitude of the equivalent load:

$$\frac{3.515}{l^2}\sqrt{\frac{EIg}{q}} = \frac{1.73}{l^3}\sqrt{\frac{EIg}{P_{\text{equiv.}}}l}$$

$$P_{\text{equiv.}} = 0.2449ql \approx 0.25ql$$

Figure 11.5 shows the modes of oscillations which correspond to the natural (own) oscillations.

Table 11.1 shows the values of equivalent dynamic loads for single-span beams with different boundary conditions.

Problem 11.2

Find an equation for determining the natural frequencies of the beam shown in Fig. 11.6 using the differential equation for the transverse vibrations of the rod.

Solution

To solve the problem, we use the differential equation for the transverse vibrations of the rod:

$$y = X(x)T(t) = \left[B_1 \sin\left(\frac{\psi}{l}\right)x + B_2\cos\left(\frac{\psi}{l}\right)x + B_3\sinh\left(\frac{\psi}{l}\right)x + B_4\cosh\left(\frac{\psi}{l}\right)x\right] \cdot \cdot (A\cos\omega t + B\sin\omega t) = X \cdot T \tag{11.31}$$

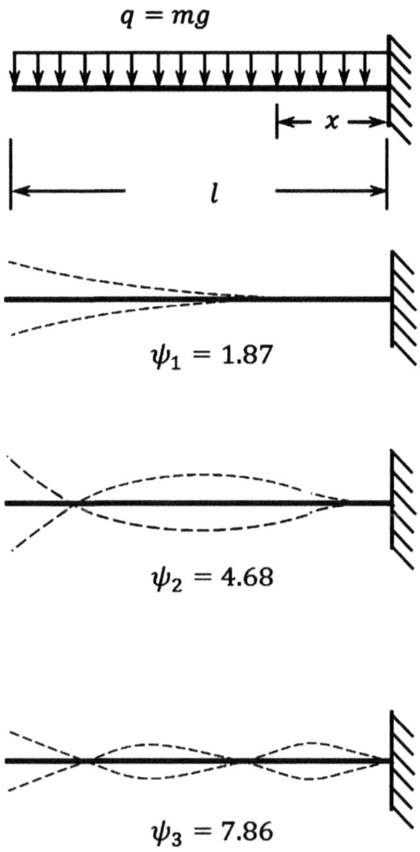

Fig. 11.5 First three modes of oscillations of a cantilever beam

Let us write down the boundary conditions for this problem:

$$\begin{aligned} &\text{At } x = 0: \quad X(0) = X'(0) = 0; \\ &\text{At } x = l: \quad X''(l) = 0; EIX'''(l) = r_{11}X. \end{aligned} \tag{11.32}$$

The boundary conditions (11.32) give four equations for determining arbitrary constants B_i:

1. $B_2 + B_4 = 0$;
2. $B_1 + B_3 = 0$.
3. $-k^3 B_1 \cos\psi + k^2 B_2 \sin\psi + k^3 B_3 \cosh\psi + k^3 B_4 \sinh\psi$

$$= \frac{r_{11}}{EI}(B_1 \sin\psi + B_2 \cos B_2 + B_3 \sinh\psi + B_4 \cosh\psi); \tag{11.33}$$

4. $-B_1 k^2 \sin\psi - B_2 k^2 \cos\psi + B_3 k^2 \sinh\psi + B_4 k^2 \cosh\psi = 0$

Equation (11.33) lead to two equations of the following form:

Table 11.1 The equivalent dynamic load

	The equivalent dynamic load	
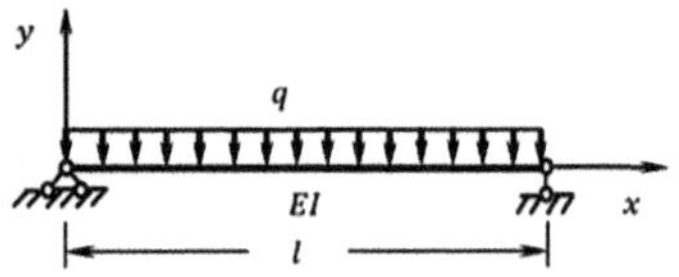	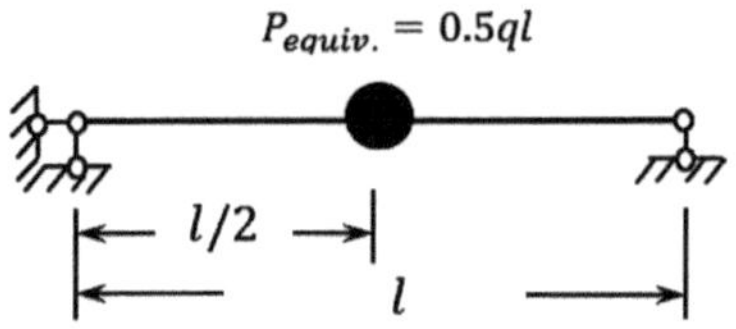	$P_{equiv.} = 0.4939ql \approx 0.5ql$
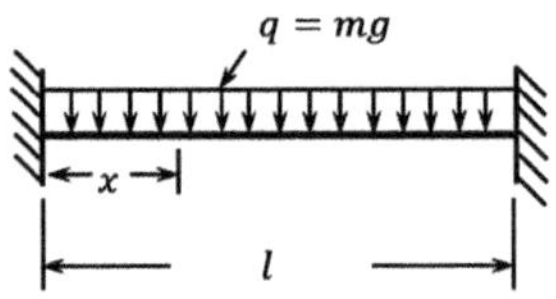	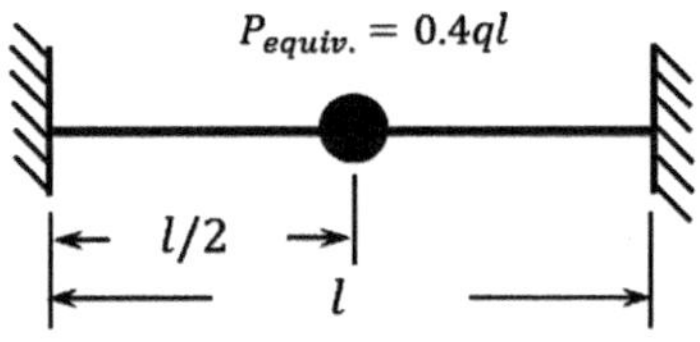	$P_{equiv.} = 0.3859ql \approx 0.4ql$
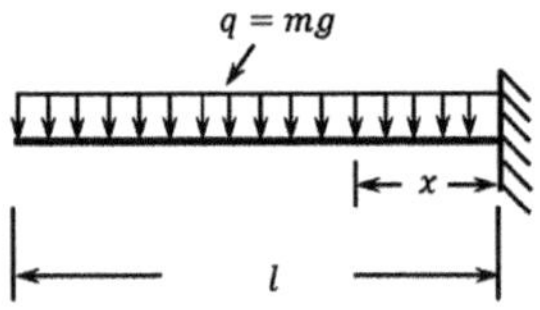	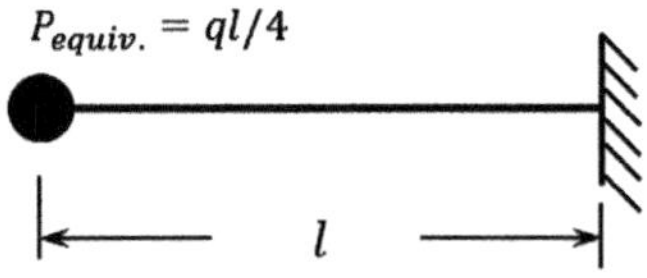	$P_{equiv.} = 0.2449ql \approx 0.25ql$

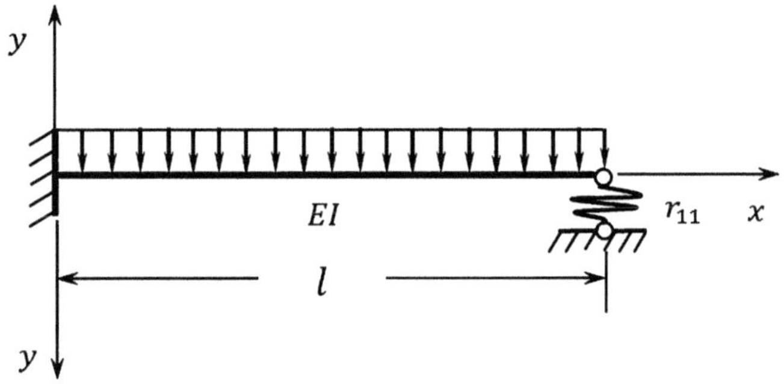

Fig. 11.6 Vibration of beam with uniformly distributed load

$$\begin{aligned}&\left[-k^3(\cos\psi + \cosh\psi) + \frac{r_{11}}{EI}(\sinh\psi - \sin\psi)\right]B_1 + \\ &\left[k^3(\sin\psi - \sinh\psi) + \frac{r_{11}}{EI}(\cosh\psi - \cos\psi)\right]B_2 = 0, \\ &(\sin\psi + \sinh\psi)B_1 + (\cos\psi + \cosh\psi)B_2 = 0.\end{aligned} \tag{11.34}$$

Let us compose the determinant of Eq. (11.34) and equate it to zero. As a result, we obtain an equation for determining the frequency spectrum:

$$\frac{r_{11}}{EI}l^3 = -\psi^3 \frac{1 + \cos\psi\cosh\psi}{\sin\psi\cosh\psi - \cos\psi\sinh\psi}, \tag{11.35}$$

where

$$\psi = kl$$

At $r_{11} = \infty$, we obtain hinged support of the beam boundary condition;
At $r_{11} = 0$, we obtain a free end of the beam boundary condition.

Frame systems are one of the most common structures in the construction industry. A frame is a system with distributed and discrete masses. The exact method of calculation of a system with infinitely many degrees of freedom is extremely complex and cumbersome. Therefore, in practice, various approximation methods are used almost exclusively. Among them is the method of reducing the frame to a system with a finite number of degrees of freedom.

The reduction of the frame to a system with a finite number of degrees of freedom is carried out by replacing the mass distributed along the length of the rod with discrete masses. Thus, the replacement of a system with an infinitely large number of degrees of freedom with an equivalent system with a finite number of degrees of freedom in the dynamics of structures is called the method of mass reduction. Half of the mass of each element of the system (frame) is considered concentrated and applied at the midpoint of its length, while the second half of the mass is transferred by quarters to the nodes of the frame to which this element is adjoined.

Since beams are one of the main components of frames under vibrations, a thorough study of their behavior under various boundary conditions is an important step in the analysis of structures under dynamic loads.

Examples of calculating frames with concentrated masses as systems with one and two degrees of freedom were given earlier in Chaps. 9 and 10.

11.2 Fundamental Functions of the Equation of Vibration of Beams

The functions by which the modes of natural vibrations of an elastic beam with a uniformly distributed mass under various boundary conditions are called fundamental functions. The general solution of these functions is given in solution (11.20). Fundamental functions have two important properties that allow them to be used in solutions of a number of problems not related to the vibration of beams.

The first property of fundamental functions *y(x)* is that their fourth derivative differs from the function itself only by a simple multiplier:

$$y^{IV} = \beta y \tag{11.36}$$

This follows from the differential Eq. (11.16). This equation satisfies fundamental functions. Here we can note that:

$$\beta = m\omega^2/EI$$

The second property is called orthogonality and has the following form

$$\int_0^l y_i y_k dx = 0, \tag{11.37}$$

where y_i, y_k are different fundamental functions that correspond to the same boundary conditions. To prove the orthogonal property, we substitute functions y_i and y_k into Eq. (11.16):

$$\begin{aligned} EIy_i^{IV} - m\omega_i^2 y_i &= 0; \\ EIy_k^{IV} - m\omega_k^2 y_k &= 0. \end{aligned} \tag{11.38}$$

Let us multiply the first equation by y_k, and the second equation by y_i, and. integrate the obtained products:

$$\begin{aligned} EI\int_0^l y_i^{IV} y_k dx &= m\omega_i^2 \int_0^l y_i y_k dx \\ EI\int_0^l y_k^{IV} y_i dx &= m\omega_k^2 \int_0^l y_k y_i dx \end{aligned} \tag{11.39}$$

Equation (11.39) is integrated twice in parts, as a result we obtain:

$$\int_0^l y_i^{IV} y_k dx = \left[y_i''' y_k\right]_0^l - \int_0^l y_i''' y_k' dx = \left[y_i''' y_k - y_i'' y_k'\right]_0^l + \int_0^l y_i'' y_k'' dx;$$

$$\int_0^l y_k^{IV} y_i dx = \left[y_k''' y_i - y_k'' y_i'\right]_0^l + \int_0^l y_k'' y_i'' dx$$

Note that the following expression

$$y_i''' y_k - y_i'' y_k' = [-1/(EI)]\left(Q_i y_k - M_i y_k'\right)$$

is equal to:

$$[-1/(EI)]\left(Q_k y_i - M_k y_i'\right) = y_k''' y_i - y_k'' y_i'.$$

at any end of the beam with any methods of its support based on the reciprocity theorem.

Therefore, the left parts of Eq. (11.39) are equivalent, it follows that the right parts will also be equivalent.

$$m\omega_i^2 \int_0^l y_i y_k dx = m\omega_k^2 \int_0^l k_k y_i dx; \quad m\left(\omega_i^2 - \omega_k^2\right) \int_0^l y_i y_k dx = 0,$$

Since $\omega_i \neq \omega_k$, orthogonality condition (11.37) is satisfied for various fundamental functions. Fundamental functions are usually normalized by choosing proportionality coefficient from the following condition:

$$\int_0^l y_i^2 dx = 1. \tag{11.40}$$

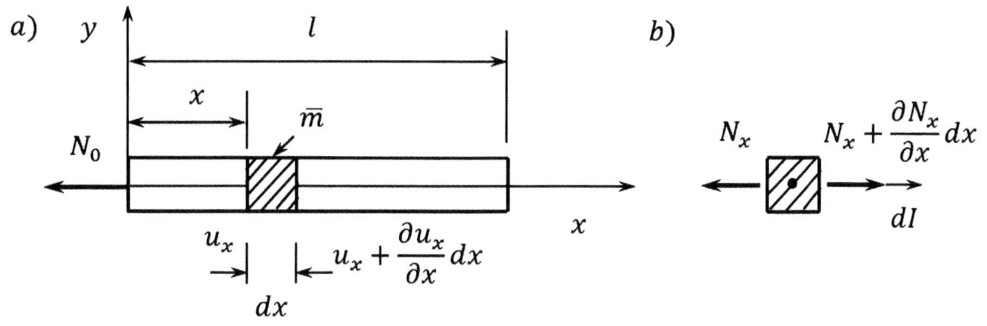

Fig. 11.7 (**a**), and (**b**) Rectilinear rod with a constant cross-section

11.3 Longitudinal Vibrations of Rods with a Distributed Mass

According to the law of equilibrium and the D'Alembert principle, and taking into account all of the assumptions that are made for a rod in static tension and compression, from all of the differential equations of equilibrium one is preserved:

$$\frac{\partial \sigma_x}{\partial x} + I = 0 \tag{11.41}$$

where σ_x and I are functions of two variables x and time t; I is the inertial force per unit volume.

Equation (11.41) can also be obtained based on the equilibrium of the individual elements of the rod. We will cut out a small rod element of length dx at distance x from the origin (Fig. 11.7a). This element is subject to continuously changing longitudinal forces of resistance N_{xt} and longitudinal forces of inertia with the intensity of I Fig. 11.7b.

By the condition $\sum X = 0$, we have:

$$-N_{xt} + \left[N_{xt} + \frac{\partial N_{xt}}{\partial x} dx\right] + I_{xt} A E dx = 0, \tag{11.42}$$

From which we obtain Eq. (11.41) at $N_{xt}/A = \sigma_x$.

Let us consider natural longitudinal vibrations of a rectilinear rod of constant cross-section Fig. 11.7a.

The longitudinal displacement at an arbitrary point at distance x from the origin u_{xt}, denote the relative elongation ε_{xt}, and the longitudinal force N_{xt}. According to Hooke's law for elastic deformations, we have:

$$\varepsilon_{xt} = N_{xt}/AE \ . \tag{11.43}$$

Here, the following assumptions are made:

- Hypothesis of plane section is used.
- During vibrations, lateral deformations are ignored.
- The inertia forces from transverse displacements of the rod are small in comparison to the inertia forces from longitudinal displacements.

Therefore, the general spatial stress state of the rod is reduced to a case of uniaxial stress state. For this task, we only use one of all the Cauchy dependencies.

It is known that: $\varepsilon_{xt} = \partial u_{xt}/\partial x$.

Therefore, according to expression (11.43).

$$N_{xt} = EF\varepsilon_{xt} = AE\frac{\partial u_{xt}}{\partial x}; \qquad \frac{\partial N_{xt}}{\partial x} = AE\frac{\partial^2 u_{xt}}{\partial x^2}. \tag{11.44}$$

The mass of a unit of length of the rod is denoted by $\overline{m}$. Then, the inertia force applied to an infinitely small element of the rod will have the following form:

$$dI = -(\overline{m}dx)\frac{\partial^2 u_{xt}}{\partial t^2}. \tag{11.45}$$

Following D'Alembert principle, the equilibrium condition of the element (Fig. 11.7b) has the following form:

$$dI + \frac{\partial N_{xt}}{\partial x}dx = 0. \tag{11.46}$$

Let us substitute values of N_{xt} and $\partial N_{xt}/\partial x$ from formula (11.44) into equilibrium Eq. (11.45)

$$\frac{\partial^2 u_{xt}}{\partial t^2} = \frac{AE}{\overline{m}}\frac{\partial^2 u_{xt}}{\partial x^2}. \tag{11.47}$$

The following notation is introduced:

$$\frac{AE}{\overline{m}} = v^2 \tag{11.48}$$

where v is the speed of propagation of the elastic wave of deformation at oscillation. The differential equation of free longitudinal vibrations of a rod with a distributed mass will take the final form:

$$\frac{\partial^2 u_{xt}}{\partial t^2} = v^2\frac{\partial^2 u_{xt}}{\partial x^2} \tag{11.49}$$

The general solution of the linear differential Eq. (11.49) is:

$$u_{xt} = \sum_{i=1}^{\infty} u_i = \sum_{i=1}^{\infty} u_{xi} \cdot \sin(\omega_i t + \lambda_i), \tag{11.50}$$

where u_i is the $i-$th mode of longitudinal oscillations. It is obvious that ω_i is the oscillation frequency which corresponds to the $i-$th mode; $u_{xi}-$ is the expression of the $i-$th normal form of longitudinal vibrations. This function depends only on x.

Let us substitute one term from (11.50) into the differential equation of free longitudinal vibrations of a rod with a distributed mass (11.49). In this case, we omit the index i, and as a result obtain:

$$-\omega^2 u_x \sin(\omega t + \lambda) = v^2\frac{d^2 u_x}{dx^2}\sin(\omega t + \lambda) \tag{11.51}$$

or after reduction by $\sin(\omega t + \lambda)$ we will have:

$$v^2\, c = -\omega^2 u_x. \tag{11.52}$$

Let us find the expressions for the normal mode of oscillations using the differential Eq. (11.52). To do this, we divide both parts of the expression (11.52) by v^2 and substitute the expression (11.48) for v^2

$$k^2 = \omega^2/v^2 = \omega^2 \overline{m}/AE, \tag{11.53}$$

we finally obtain the expression for the differential equation of the normal form of vibrations:

$$\frac{d^2 u_x}{dx^2} = -k^2 u_x. \tag{11.54}$$

The initial deflection $u_{x=0} = u_0$ and the initial value of the longitudinal force N_0 at $x = 0$ determines the initial value of the relative elongation during vibration/oscillation.

$$\varepsilon_0 = |\partial u/\partial x|_{x=o} = N_0/AE. \tag{11.55}$$

Let us integrate the differential Eq. (11.54) with the method of initial parameters:

$$u_x = u_{0i}\ \cos k_i x + \varepsilon_{0i} \frac{\sin k_i x}{k_i}, \tag{11.56}$$

Thus, the solution of the differential Eq. (11.49) for the i-th mode of oscillation has the form:

$$u_{xt} = \left(u_{0i}\ \cos k_i x + c \frac{\sin k_i x}{k_i} \right) \sin(\omega_i t + \lambda_i). \tag{11.57}$$

Problem 11.3

Find the frequencies of longitudinal vibrations of a rod of length l with free ends (Fig. 11.7).

Solution

The longitudinal forces at the boundary points of the rod are equal to zero. Therefore, the following boundary conditions have the form:

$$\varepsilon_0 = |\partial u/\partial x|_{x=o} = 0;$$
$$\varepsilon_l = |\partial u/\partial x|_{x=l} = 0.$$

These boundary conditions are valid for any $i-$th normal mode of vibrations/oscillations u_x:

$$u_x = u_{0i}\ \cos k_i x + \varepsilon_{0i} \sin k_i x/k_i$$

Thus, according to the first boundary condition for u_x we have: $u_x = u_{0i}$ cos $k_i x$. The expression for elongation/lengthening is written as follows:

$$\varepsilon_x = \frac{\partial u_x}{\partial x} = -k_i u_{0i} \sin k_i x.$$

When we substitute the expression for elongation/lengthening to the second boundary condition we obtain the following:

$$0 = -k_i u_{0i} \sin k_i l$$

For $k_i \neq 0$ we have:

$$\sin k_i l = 0,$$

where $k_i l = n\pi \quad (n = 1, 2, 3, 4)$.

The oscillation/vibration frequencies are found from expression (11.53):

$$\omega_i = \frac{n\pi}{l}\sqrt{\frac{AE}{m}}$$

11.4 Frequencies of Free Vibrations of Trusses

One of the simplest and most reliable ways for determining the frequency of free oscillations/vibrations of trusses is to replace a truss with a beam of equivalent stiffness and same geometric characteristics. The moment of inertia of the beam, which is equivalent in stiffness of the truss, is usually determined in the midpoint of the span of the beam and truss assuming the condition of equality of deflections is satisfied.

The deflection in the middle of a truss from a uniformly distributed load and discrete loads distributed at the nodes of the truss (Fig. 11.8) is determined by Mohr formula:

$$f_{\text{truss}} = \sum \frac{N_i N_p}{AE} l \tag{11.58}$$

here N_p is the force in an arbitrary truss bar from an external load; N_i is the force in an arbitrary truss rod from the action of a unit force $P = 1$ in the middle of the truss span. The deflection of the beam from the action of the same load in the middle of the span is determined by the formula:

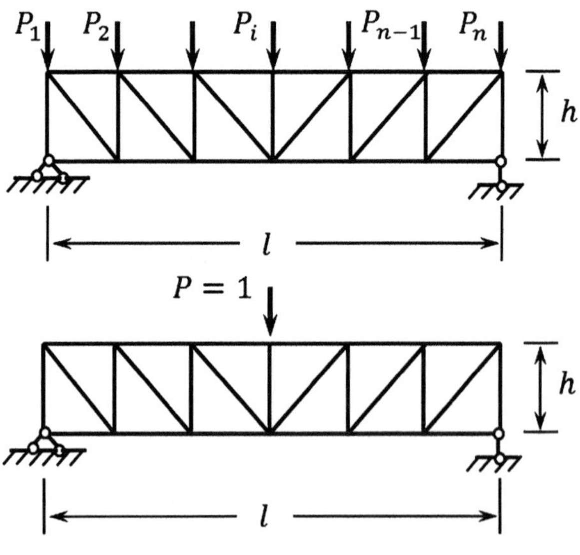

Fig. 11.8 Pratt truss under external load $P_1, \ldots, P_i, \ldots, P_n$, and under unit force $P = 1$.

$$f_{\text{beam}} = \frac{5}{384}\frac{Pl^4}{EI} = f_{\text{truss}} \tag{11.59}$$

from here

$$I = \frac{5}{384}\frac{Pl^4}{Ef_{\text{truss}}} \tag{11.60}$$

from here we determine the corresponding value of the equivalent stiffness of the beam. To do that, we determine the frequency of free oscillations from a uniformly distributed load. For the case when the beam is freely supported at the ends we have:

$$\omega = \frac{\pi^2}{l^2}\sqrt{\frac{EIg}{P}} = \frac{\pi^2}{l^2}\sqrt{\frac{Eg}{P}\frac{5}{384}\frac{Pl^4}{Ef_{\text{truss}}}} = \pi^2\sqrt{\frac{5}{384}}\sqrt{\frac{g}{f_{\text{truss}}}} = 1.13\sqrt{\frac{g}{f_{\text{truss}}}} \tag{11.61}$$

For the case of a cantilever truss we have the following formula:

$$\omega = 1.21\sqrt{\frac{g}{f_{\text{truss}}}} \tag{11.62}$$

The presented method gives good results, especially for beam trusses with parallel chords.

Problem 11.4

Determine the frequency of free vibrations of the beam truss shown in Fig. 11.9 using a practical method of calculation. The load $P = 1000$ kg acts in the truss nodes. The elements of the metal truss ($E = 2.1 \cdot 10^6$ kg/cm^2) have the following values (see Table 11.2).

Solution

Using the exact method of calculation to determine oscillation frequencies usually requires solving the determinants of high orders, which from a practical point of view is often not feasible.

Here we will use formula (11.61):

$$\omega = 1.13\sqrt{\frac{g}{f_{\text{truss}}}}$$

The deflection of the truss is determined by the formula:

$$f_{\text{truss}} = \sum \frac{N_i N_p}{AE} l$$

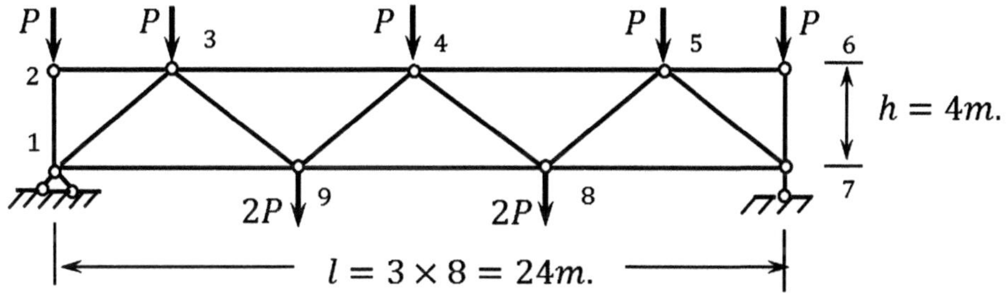

Fig. 11.9 Warren truss

Table 11.2 Geometric characteristics of truss elements and force values in truss rods

# Bar	l (cm)	A (cm^2)	$\frac{l}{A}$	$\frac{N_p}{P}$	N_1	$\frac{N_p N_1}{P} \frac{l}{A}$
2–3	400	80	5	0	0	0
3–4	800	160	5	−6.00	−1.00	30
4–5	800	160	5	−6.00	−2.00	30
5–6	400	80	5	0	0	0
1–9	800	160	5	3.5	0.5	8.8
9–8	800	160	5	6.5	1.5	48.7
8–7	800	160	5	3.5	0.5	8.8
1–2	400	80	5	−1.0	0	0
1–3	556	111.2	5	−4.95	−0.707	17.5
3–9	556	55.6	10	3.54	0.707	25.0
9–4	556	55.6	10	−0.71	−0.707	5.0
4–8	556	55.6	10	−0.71	−0707	5.0
8–5	556	55.6	10	3.54	0.707	25.0
5–7	556	55.6	5	−4.95	−0.707	17.5
6–7	400	80	5	−1.00	0	0
∑ # Bars						**221.3**

Here N_i and N_p are the forces in the truss bars (rods), from the unit load applied in the middle of the span and from the action of the external load, respectively. The forces in all elements of the truss (bottom chords, top chords, and diagonal bracing) from the external load applied in individual nodes of the truss are determined. For convenience, we will take the external load acting in the nodes of the truss as a unit, and then multiply by the actual value of the external forces.

We compile the calculated data in Table 11.2 which shows the calculation results for determining the forces in the bars (rods) of the truss from the action of a unit force applied in the middle of the span as well as from the action of an external load.

Let us determine the deflection in the middle of the truss span based on the data presented in Table 11.2.

$$f_{\text{truss}} = \sum \frac{N_i N_p}{AE} l = 221.3 \frac{P}{E} = \frac{221.3 \cdot 1000}{2.1 \cdot 10^6} = 0.111\,\text{cm}$$

The deflection in the middle of the span of the truss is substituted into the formula for the frequency of free oscillations:

$$\omega = 1.13 \sqrt{\frac{g}{f_{\text{truss}}}} = 1.13 \sqrt{\frac{981}{0.111}} = 106.2\ \text{s}^{-1}$$

This method has standard error of 3–5% compared to the exact method, which is usually an acceptable error margin for most practical problems.

References

1. Clough, R. W., & Penzien, J. (1993). *Dynamics of structures* (2nd ed., 738p). McGraw-Hill College.
2. Copra, A. K. (2016). *Dynamics of structures* (992p). Pearson.
3. Kabe, A. M., & Sako, B. H. (2020, June 17). *Structural dynamics fundamentals and advanced applications, volume I* (1st ed., 970p). Academic Press.

4. Katsikadelis, J. T. (2020, June 23). *Dynamic analysis of structures* (1st ed., 922p). Academic Press.
5. Craig, R. R. J. R., & Kurdila, A. J. (2011). *Fundamentals of structural dynamics* (2nd ed., 752p). John Wiley & Sons, Inc.
6. Busby, H. R., & Staab, G. H. (2017). *Structural dynamics: Concepts and applications* (568p). CRC Press.
7. Anderson, J. C., & Naeim, F. (2012). *Basic structural dynamics* (280p). John Wiley & Sons.
8. Paz, M., & Kim, Y. H. (2018). *Structural dynamics: Theory an computation* (6th ed., 634p). Springer.
9. Clein, G. K. (1982). Fundamentals of the dynamics of rod system. In *The book guide to practical exercises in structural mechanics* (pp. 124–220). Vishaya shkola.
10. Zhou, Z., Wen, Y., et al. (2021). *Fundamentals of structural dynamics* (286p). Elsevier.

Vibration of Beams Under Moving Loads

12

Abstract

This chapter focuses on vibrations of beams under a moving load. Differential equations of motion based on the static method will be presented as problems and solutions for the following two cases:

Case 1. Equations of dynamic equilibrium which differ from equations of static equilibrium in that the forces of inertia in the form of the product of masses or their moments of inertia and accelerations (i.e., the second derivatives of linear or angular displacements over time with respect to time) are taken into account.

Case 2. Differential equations of motion or Lagrange equations of motion based on the energy method will be presented. The energy method is based on the law of conservation of energy, according to which for a conservative system that vibrates without energy loss.

12.1 Equation of Motion of the Beam

The study of vibration of constructions under the influence of moving loads has many great practical values. Transport engineering structures are subjected to moving loads that vary in both time and space. Currently, loads and speeds of traffic continuously grow, and therefore increase the dynamic impacts on constructions. An increase in traffic volume, the weight of traveling loads, and the simultaneous decrease in the mass of structures is the contemporary tendency in bridge construction. Because of this the need for using precise dynamic design schemas, diagrams, and models increases. Significant contributions to the solution of general problems of vibrations of constructions and bridge structures under the influence of mobile loadings were made by C.E, M. Becker, C.E. Inglis, A.N Kryloff, S.P. Timoshenko, S.T. Odman, A. Schallenkamp, H. Steuding, G. Stokes, R. Willis, W. Witting, Clein G.K., V.V. Bolotin, and others. Among the numerous publications of the last two decades, devoted to the vibration of structures from the action of a moving load, the following works should be noted [1–10] in particular.

Here we consider the motion of a concentrated load of a constant value P along a beam with a linear mass $\overline{m}$. The mass of the load will be considered incommensurably small compared to the mass of the beam. The speed of movement v of the load is assumed to be so large that the problem cannot be solved by considering it as static. If we assume that at the initial time $t_0 = 0$ the load enters the beam, then at

L. G. Petrosian, *Dynamics of Structures*, https://doi.org/10.1007/978-3-031-63539-7_12

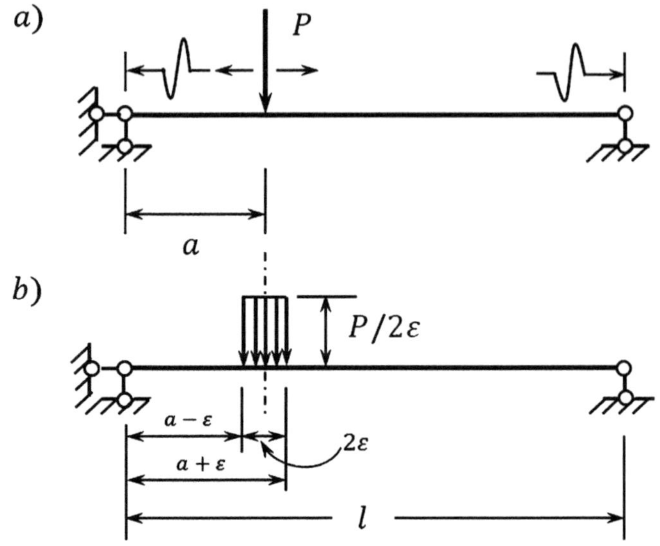

Fig. 12.1 (**a**) Beam under moving load; (**b**) Moving load distributed uniformly over a small segment

time t the load will be at a distance $a = vt$ from the left support. Figure 12.1a g is acceleration of gravity.

Vibration Equation Vibration equation can be obtained from the differential equation of vibration of a beam with a distributed mass (11.9) by adding in the right part $P(x, t)$.

$$EI\frac{\partial^4 y}{\partial x^4} + \overline{m}\frac{\partial^2 y}{\partial t^2} = P(x,t) \tag{12.1}$$

The magnitude of the force in motion remains constant. However, the components of the load along for the main modes of oscillations continuously change their magnitude; therefore, it is assumed: $P(x, t)$.

To find these constituents, let's decompose the force $P(x, t)$ into a trigonometric series by sinuses:

$$P(x,t) = \sum_{n=1}^{\infty} A_n(t)\sin\frac{n\pi x}{l} \tag{12.2}$$

To calculate the coefficients of series $A_n(t)$ it is necessary to represent force P distributed uniformly over a small segment of length 2ε (from $a - \varepsilon$ to $a + \varepsilon$) (Fig. 12.1b). Then the coefficients of the series will be expressed by integrals:

$$\begin{aligned} A_n &= \frac{2}{l}\int_0^l P(x,t)\sin\frac{n\pi x}{l}dx = \frac{2}{l}\int_{a-\varepsilon}^{a+\varepsilon}\frac{P}{2\varepsilon}\sin\frac{n\pi x}{l}dx = \frac{P}{\varepsilon\, l}\int_{a-\varepsilon}^{a+\varepsilon}\sin\frac{n\pi x}{l}dx = \\ &= -\frac{Pl}{\varepsilon l n\pi}\cos\frac{n\pi x}{l}\bigg|_{a-\varepsilon}^{a+\varepsilon} = \frac{2\pi}{\varepsilon n\pi}\sin\frac{n\pi a}{l}\sin\frac{n\pi\varepsilon}{l} \end{aligned} \tag{12.3}$$

Let us reduce ε to a limit equal to zero, and considering, that $a = vt$, we obtain:

$$A_n = \frac{2P}{l}\sin\frac{n\pi vt}{l}.$$

Then the force can be represented by the following expression:

$$P(x,t) = \frac{2P}{l}\sum_{n=1}^{\infty}\sin\frac{n\pi vt}{l}\sin\frac{n\pi x}{l} \tag{12.4}$$

Let us substitute the n-th term of this series into the oscillation eq. (12.1) instead of force P:

$$EI\frac{\partial^4 y_n}{\partial x^4} + \overline{m}\frac{\partial^2 y_n}{\partial t^2} = \frac{2P}{l}\sin\frac{n\pi vt}{l}\sin\frac{n\pi x}{l} \tag{12.5}$$

The solution of this differential equation, corresponding to the n-th main mode of oscillation, can be represented as follows:

$$y_n = T_n(t)\sin\frac{n\pi x}{l} \tag{12.6}$$

Then:

$$\frac{\partial^4 y_n}{\partial x^4} = T_n(t)\frac{n^4\pi^4}{l^4}\sin\frac{n\pi x}{l}$$
$$\frac{\partial^2 y_n}{\partial t^2} = \frac{\partial^2 T_n}{\partial x^2}\sin\frac{n\pi x}{l}.$$

Let us substitute these expressions into the oscillation Eq. (12.1) and reduce by $\sin\frac{n\pi x}{l}$, as a result, we obtain the following differential equation:

$$\frac{n^4\pi^4}{l^4}EIT_n + \overline{m}\frac{d^2T_n}{dt^2} = \frac{2P}{l}\sin\frac{n\pi vt}{l} \tag{12.7}$$

The solution of this equation can be reduced (represented) to the form:

$$T_n = B_n\sin(\omega_n t + \lambda_n) + \frac{2P}{\overline{m}l}\frac{\sin\psi_n t}{\omega_n^2 - \psi_n^2}, \tag{12.8}$$

where

$$\omega_n = \frac{n^2\pi^2}{l^2}\sqrt{\frac{EI}{\overline{m}}}; \tag{12.9}$$

$$\psi_n = \frac{n\pi v}{l}. \tag{12.10}$$

The constants B_n and λ_n can be found from the boundary conditions of the problem: at $t = 0$, and $y_n = 0$, then $B_n \sin \lambda_n = 0$, and since $B_n \neq 0$, then $\lambda_n = 0$; at $t = 0$: and $\frac{\partial y_n}{\partial t} = 0$. Hence

$$B_n = -\frac{2P\psi_n}{\overline{m}l\omega_n}\frac{1}{\omega_n^2 - \psi_n^2}.$$

Therefore:

$$y_n = -\frac{2P\psi_n}{\overline{m}l\omega_n(\omega_n^2 - \psi_n^2)}\sin\omega_n t\sin\frac{n\pi x}{l} + \frac{2P}{\overline{m}l(\omega_n^2 - \psi_n^2)}\sin\psi_n t\sin\frac{n\pi x}{l}. \tag{12.11}$$

The first term of this sum expresses free oscillations of the beam with frequency ω_n, and the second term expresses forced oscillations with frequency ψ_n. The coincidence of frequencies $\omega_n = \psi_n$ leads to resonance. Resonance is possible at the following critical speed:

$$v = \frac{n\pi}{l}\sqrt{\frac{EI}{m}} = \frac{l}{\pi}\omega_n n. \qquad (12.12)$$

However, at $n = 1$ this speed is so high that resonance is practically impossible. The largest value of deflection under the load P standing in the middle of the span turns out to be equal to:

$$y = y_{st}\mu, \qquad (12.13)$$

where y_{st} is the static deflection of the beam under load; μ is the dynamic coefficient of the moving load. The dynamic coefficient is expressed by the following formula:

$$\mu = \frac{1}{1 - \frac{vl}{\pi}\sqrt{\frac{m}{EI}}} = \frac{1}{1 - \frac{v}{\omega}\frac{\pi}{l}} \qquad (12.14)$$

Problem 12.1
It is required to determine the dynamic coefficient of the moving load, which moves along the beam with a span of $l = 20$ m. with a velocity $v = 100$km/hr $= 22.8$m/s. (Fig. 12.1). The frequency of natural oscillations of the beam is $\omega = 25\text{s.}^{-1}$. Also, let us determine the dynamic deflection from the load in an arbitrary section of the beam and the greatest dynamic deflection in the middle of the span.

Solution
The value of the dynamic coefficient is determined by the formula:

$$\mu = \frac{1}{1 - \frac{22.8 \cdot 3.14}{25 \cdot 20}} = 1.17$$

If the mass of the load m is large compared to the mass of the structure, then the obtained value of the dynamic coefficient is not accurate enough. The critical speed, which corresponds to the resonance at $n = 1$, is:

$$v = \frac{l}{\pi}\omega = \frac{20}{3.14}25 = 159\ m/c = 573km/h$$

The deflection under the load

$$y = \frac{R}{EI}\frac{x^2(l-x)^2}{3l}$$

where R is the dynamic pressure under the load. In the expression of pressure

$$R = P - m\frac{\partial^2 y}{\partial t^2} = P\left(1 - \frac{1}{g}\frac{\partial^2 y}{\partial t^2}\right),$$

let us substitute the formula of deflection:

$$y = \frac{P}{EI} \frac{x^2(l-x)^2}{3l}$$

The greatest dynamic deflection in the middle of the span

$$y_{x=l/2}^{max} \frac{Pl^2}{48EI} \mu = \frac{Pl^3}{48EI} \left(1 + \frac{v^2}{g} \frac{Pl}{3EI}\right).$$

where g is the acceleration of gravity; EI is the stiffness of the beam at bending—transverse oscillations of the beam.

If on the contrary, the mass of the beam can be considered incommensurably (disproportionately) small compared to the mass m of the moving load. Then the effect of the moving load on the beam is variable. The dynamic coefficient reaches its maximum value when the load is in the middle of the span of the beam. The dynamic coefficient for the section of a beam located in the middle of the span is determined by the following formula:

$$\mu = 1 + \frac{v^2 lm}{3EI} = 1 + \left(\frac{4v}{\omega l}\right)^2. \tag{12.15}$$

Problem 12.2

The differential equation of forced vibration for a weightless beam with single mass, which is located in the middle of the beam span, needs to be compiled (constructed). The force P(t) acts on the mass m_1 (Fig. 12.2). According to the condition of the problem, the differential equation of forced oscillations should be compiled using the energy principles of solving problems for oscillations/vibrations.

Since, according to the condition of the problem, the differential equation of forced vibrations must be compiled using the energy principles of solving problems for vibrations. Let's write the Lagrange differential equations of the second kind, which describe the oscillations/vibrations of the system:

$$\frac{d}{dt}\left(\frac{\partial K}{\partial \dot{q}_i}\right) - \frac{\partial K}{\partial \dot{q}_i} = Q_i \tag{12.16}$$

Here we recall that the number of these equations is equal to the number of generalized coordinates, i.e., the number of degrees of freedom of the system.

When deriving Eq. (12.16), the generalized forces Q_i were not subjected to restrictions. The generalized forces could be constant, variable in time, and so on. Let us consider the case when the generalized forces have a potential. The work of such forces is a function only of the final displacement

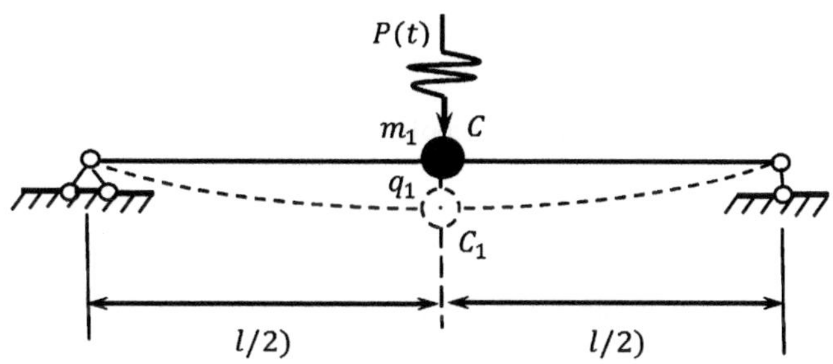

Fig. 12.2 A weightless beam with single discrete mass under the action of force $P(t)$

of their points. The internal resistance forces of a deforming system belong to the forces that have potential. The potential energy of the system is denoted by U. Here, we take into account, that the work on a possible displacement is equal to the decrease in potential energy, and as a result we obtain:

$$Q_1\delta q_1 + Q_2\delta q_2 + \cdots + Q_i\delta q_i + \cdots + Q_n\delta q_n = -\frac{\partial U}{\partial q_1}\delta q_1 - \frac{\partial U}{\partial q_2}\delta q_2 - \cdots$$
$$\cdots \frac{\partial U}{\partial q_i}\delta q_i - \cdots - \frac{\partial U}{\partial q_n}\delta q_n$$

Since infinitesimal displacements $\delta q_1,\ \delta q_2, \ldots, \delta q_n$ are independent, then

$$Q_1 = -\frac{\partial U}{\partial q_1}, Q_2 = -\frac{\partial U}{\partial q_2}, \cdots, Q_i = -\frac{\partial U}{\partial q_i}, \cdots, Q_n = -\frac{\partial U}{\partial q_n}, \tag{12.17}$$

and Eq. (12.16) will take the following form:

$$\frac{d}{dt}\left(\frac{\partial K}{\partial \dot{q}_i}\right) - \frac{\partial K}{\partial \dot{q}_i} + \frac{\partial U}{\partial q_i} = 0 \tag{12.18}$$

In general, when a system is affected by forces that have potential, as well as other external forces, the Eq. (12.16) can be represented as:

$$\frac{d}{dt}\left(\frac{\partial K}{\partial \dot{q}_i}\right) - \frac{\partial K}{\partial \dot{q}_i} + \frac{\partial U}{\partial q_i} = \overline{Q}_i \tag{12.19}$$

Solution

This system has a single degree of freedom (Fig. 12.2). The deflection q_1 can be represented as a generalized coordinate. The expression of kinetic energy can be written in the following form:

$$K = \frac{1}{2}m_1 q_1^2 \tag{12.20}$$

and the expression of potential energy can be represented respectively in the form,

$$U = \frac{1}{2}mR_1 q_1 \tag{12.21}$$

where R_1 is the force equivalent to the total internal resistance forces. This force is equal to the sum of the shear forces at point "C." This force can be expressed through the generalized coordinate q_1 .

The deflection value can be represented as $q_1 = \delta_{11}R_1$, then $R_1 = q_1/\delta_{11}$ Using the expression for R_1, we obtain:

$$U = \frac{q_1^2}{2\delta_{11}} \tag{12.22}$$

here δ_{11} is the deflection at point "C" at $R_1 = 1$;

$$\delta_{11} = \frac{l^3}{48EI}.$$

Next, let us substitute the values (12.20) and (12.22) into (12.19) obtaining the following result,

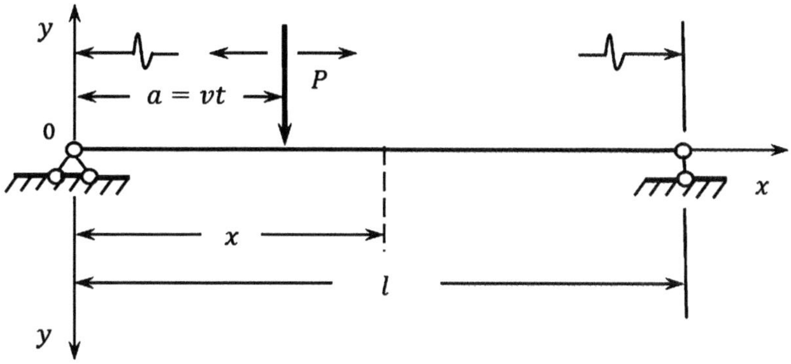

Fig. 12.3 Beam under a moving load

$$m_1\ddot{q}_1+\frac{q_1}{\delta_{11}}=P(t),\qquad or\quad \ddot{q}_1+\omega^2 q_1=\overline{P}(t), \tag{12.23}$$

where $\overline{Q}_i=P(t)$; $\omega=\sqrt{\frac{1}{m_1\delta_{11}}}$ is the natural oscillation/vibration frequency of the system;

$$\overline{P}(t)=\frac{P(t)}{m_1};\qquad \ddot{q}_1=\frac{d^2q_1}{dt^2}.$$

This kind of equation of motion or equation of oscillations (12.23) is usually derived from equilibrium conditions on the basis of the Dalembert principle.

Problem 12.3

Compose a differential equation of forced vibrations of a beam along which a load (cargo) moves at a constant speed *v*. The weight of the beam is incommensurably greater than the weight of the moving load *P* (Fig. 12.3).

Solution

Let us consider function *T*(*t*) in the equation of beam deflections *w*(*x*, *t*) = *T*(*t*)*w*(*x*) as a generalized coordinate q_i. The shape of the bend is taken in the form of beam standard functions $\sin\frac{i\pi x}{l}$ that satisfy the boundary conditions of the problem: at $x=0$, and at $x=l$:

$$w(x,t)=0;\qquad EI\frac{\partial^2 w(x,t)}{\partial t^2}=0. \tag{12.24}$$

Thus, the equation of beam deflections is:

$$w(x,t)=\sum_{i=1}^{i=\infty} q_i\sin\frac{i\pi x}{l}. \tag{12.25}$$

Since the normal functions are orthogonal, here we obtain:

$$U=\frac{1}{2}EI\int_0^l\left[\frac{\partial^2 w(x,t)}{\partial t^2}\right]^2 dx=\frac{EI}{2}\sum_{i=1}^{i=\infty}q_i^2\int_0^l\frac{i^4\pi^4}{l^4}\sin^2\frac{i\pi x}{l}dx=\frac{EI\pi^4}{4l^3}\sum_{i=1}^{\infty}i^4q_i^2 \tag{12.26}$$

and the expression for kinetic energy will take the following form:

$$K=\frac{1}{2}\int_0^l m\left[\frac{\partial w(x,t)}{\partial t}\right]^2 dx=\frac{m}{2}\sum_{i=1}^{n}\dot{q}_i^2\int_0^l \sin^2\frac{i\pi x}{l}dx=\frac{ml}{4}\sum_{i=1}^{\infty}\dot{q}_i^2 \tag{12.27}$$

The generalized force is obtained from the condition that the product of the generalized force on the displacement δq_i gives work.

If an infinitesimal increment δq_i is given to the beam deflection Eq. (12.25) of the generalized coordinate q_i then we obtain:

$$\delta w(x.t)=\delta q_i \sin\frac{i\pi x}{l}, \tag{12.28}$$

and the work of the external force P on this displacement at $x = a$ will be:

$$P\delta w(x.t)=P\delta q_i \sin\frac{i\pi a}{l} \tag{12.29}$$

From here we can find a generalized force:

$$Q_i=P\sin\frac{i\pi a}{l} \tag{12.30}$$

Since $a = vt$, the formula of the generalized force will take the form:

$$Q_i=P\sin\frac{i\pi vt}{l} \tag{12.31}$$

Expressions (12.31), (12.27), and (12.26) will be substituted into (12.19). As a result, the differential equation for q_i will be obtained:

$$\ddot{q}_i+\frac{EI^4\pi^4}{ml^4}q_i=\frac{2P}{ml}\sin\frac{i\pi vt}{l} \tag{12.32}$$

The solution of Eq. (12.23) in the stage of stable oscillations has the following form:

$$q_i=\frac{2P}{ml\omega_i}\int_0^t \sin\frac{i\pi v\tau}{l}\sin\omega_i(t-\tau)d\tau, \tag{12.33}$$

where $\omega_i=\frac{i^2\pi^2}{l^2}\sqrt{\frac{EI}{m}}$ is the expression for finding the frequency of natural oscillations.

References

1. Fryba, L. (1999). *Vibration of solids and structures under moving load* (495p). Institute of Theoretical and Applied Mechanics Academy of Sciences of Czech Repablic. Thomas Telford.
2. Bajer, C. I., & Dyniewicz, B. (2012). *Numerical analysis of vibrations of structures under moving inertial loads* (300p). Springer.
3. Dimitrovová, Z. (2011). Chapter 4: Dynamic analysis of beam structures under moving loads: A review of the modal expansion UNIC, Department of Civil Engineering Universidade Nova de Lisboa, Portugal. In B. H. V. Topping & Y. Tsompanakis (Eds.), *Civil and structural engineering computational technology* (pp. 99–130). Saxe-Coburg Publications. https://doi.org/10.4203/csets.28.4

4. Dimitrovová, Z. (2010). A general procedure for the dynamic analysis of finite and infinite beams on piece-wise homogeneous foundation under moving loads. *Journal of Sound and Vibration, 329*, 2635–2653.
5. Zehsaz, M., Sadeghi, M. H., & Ziaei Asl, A. (2009). Dynamic response of railway under a moving load. *Journal of Applied Sciences, 9*(8), 1474–1481.
6. Choros, J., & Adams, G. G. (1979). A steadily moving load on an elastic beam resting on a tensionless Winkler foundation. *ASME Journal of Applied Mechanics, 46*, 175–180.
7. Kumar, C. P. S., Sujatha, C., & Shankar, K. (2015). Vibration of simply supported beams under a single moving load: A detailed study of cancellation phenomenon. *International Journal of Mechanical Sciences, 99*, 40–47.
8. Museros, P., Moliner, E., & Martínez-Rodrigo, M. D. (2013). Free vibrations of simply-supported beam bridges under moving loads: Maximum resonance, cancellation and resonant vertical acceleration. *Journal of Sound and Vibration, 332*(2), 326–345.
9. Pesterev, A. V., Yang, B., Bergman, L. A., & Tan, C. A. (2003). Revisiting the moving force problem. *Journal of Sound and Vibration, 261*(1), 75–91.
10. He, W. (2018). Vertical dynamics of a single-span beam subjected to moving mass-suspended payload system with variable speeds. *Journal of Sound and Vibration, 418*, 36–54.

References

13 Dynamic Calculation of Frames with Distributed Mass on the Action of Harmonic Load

Abstract

This chapter provides the definition of static and dynamic prerequisites for the calculation of frames and rods with distributed mass as systems with an infinite number of degrees of freedom. The calculation of such frames uses the displacement method whereby the frame is transformed (converted) into a system made up of single-span beams with the introduction of additional links at the nodes with different boundary conditions. A brief overview of the theory of dynamic calculation of frames by the displacement method is provided along with a presentation of calculation sequence. A table of correction coefficients for support reactions and an example of calculating frames using the displacement method for vibration loads are given. The appendix at the end of the book contains tables with numerical values of correction coefficients for the dynamic calculation of frames.

13.1 Determination of Displacements and Reactions from Dynamic Load

In dynamic calculations, frame structures with masses continuously distributed along the length of their rods are considered as systems with an infinite number of degrees of freedom. An accurate dynamic calculation of such a system, within the limits of the selected design scheme and characteristics of the material, should be performed according to the deformed state, taking into account all types of inertial forces and deformations. It should be noted here that the exact calculation in this formulation of the problem is very difficult. Dynamic calculation of frames as a system with an infinite number of degrees of freedom can be simpler if additional initial static and dynamic assumptions (prerequisites) are introduced to simplify the calculation of frames and its constituent elements. Let's write down the initial static and dynamic prerequisites for the calculation of a frame and its constituent elements:

- only transverse vibrations of the frame elements are considered.
- the inertia of rotation of the frame elements, i.e., the frame rods, is not taken into account.
- longitudinal vibrations of the frame rods are not taken into account.
- inertial forces of the rods directed along the axis of the rods in frames with shifting nodes are taken into account in dynamic calculations.

L. G. Petrosian, *Dynamics of Structures*, https://doi.org/10.1007/978-3-031-63539-7_13

- vibrations of all frame elements are assumed to coincide in phase.
- During initial vibration in a system with a harmonic load of constant intensity and frequency only the process of forced steady state is considered.

As is known, the coefficients at unknowns and free terms of canonical equations of the force method are determined by the formula:

$$\Delta_{ik} = \sum \int \frac{\overline{M}_i M_k}{\mathrm{EI}} \, ds \tag{13.1}$$

where $\overline{M}_i$ is the expression of bending moments from the force; M_k is the expression of bending moments from the dynamic load taking into account the forces of inertia.

In this case, difficulties arise. The main difficulties arise precisely in determining the bending moments from the dynamic load taking into account the forces of inertia using formula (13.1). Therefore, the force method is not very effective for the dynamic calculation of frames taking into account the distributed mass of the frame elements. Thus, here we will not dwell on the application of the method of forces to determine the frequencies of vibration, but only note that the resulting equation of frequencies of free vibration is transcendental and complex. Selection is the only way to solve the problem.

Let us first illustrate the dynamic calculation on the example of a rod or a beam with a distributed mass on the action of a harmonic load, under various conditions of fixing the rod or beam at the ends, and then proceed to the calculation of the frames [1].

The differential equation of dynamic equilibrium of a rod of constant cross-section under arbitrary load has the following form:

$$\mathrm{EI}\frac{\partial^4 y}{\partial x^4} + \rho \frac{\partial^2 y}{\partial t^2} = P(x.t), \tag{13.2}$$

here E is the modulus of elasticity; I is the moment of inertia of the cross-section, and ρ is the linear mass of the rod or beam.

Under the action of a harmonic load $P(x.\ t) = q(x) \sin \theta t$, the equation of the elastic line of a rod/beam can be represented in the form: $y(x.\ t) = w(x) \sin \theta t$. Let us substitute these values $P(x.\ t)$ and $y(x.\ t)$ into Eq. (13.2), and divide by $\sin\theta t$ and by EI, we obtain:

$$\mathrm{EI}\frac{d^4 w(x)}{dx^4} - \lambda^4 w(x) = \frac{q(x)}{\mathrm{EI}}, \tag{13.3}$$

where $\lambda^4 = \rho\theta^2/\mathrm{EI}$.

If in Eq. (13.3) the right part is assumed to be equal to zero, then we obtain the equation of the elastic line for free vibrations. The solution of this equation has the following form:

$$w(x) = A \cosh \lambda x + B \sinh \lambda x + C \cos \lambda x + D \sin \lambda x, \tag{13.4}$$

where A, B, C, and D are arbitrary constants of the solution of the differential equation. The arbitrary constants can be determined from the boundary conditions at $x = 0$ and $x = l$.

After determining arbitrary constants, the solution of the differential equation (13.3) expressed through the initial parameters of the rod/beam at the left end: deflection w_0, angle of rotation α_0, bending moment M_0, shear force Q_0 and distributed load q_0, will take the form:

$$w(x) = w_0 A_{\lambda x} + \frac{\alpha_0}{\lambda} B_{\lambda x} - \frac{M_0}{\lambda^2 \mathrm{EI}} C_{\lambda x} - \frac{Q_0}{\lambda^3 \mathrm{EI}} D_{\lambda x} - \frac{q_0}{\lambda^4 \mathrm{EI}} (A_{\lambda x} - 1). \tag{13.5}$$

If $q_0 = 0$ is taken in Eq. (13.5), then we obtain the equation of the elastic line of the rod/beam for free vibrations. The functions $A_{\lambda x}$, $B_{\lambda x}$, $C_{\lambda x}$, and $D_{\lambda x}$ in (13.5) have the following form:

$$\begin{aligned} A_u &= \frac{\cosh u + \cos u}{2}, \qquad B_u = \frac{\sinh u + \sin u}{2}, \\ C_u &= \frac{\cosh u - \cos u}{2}, \qquad D_u = \frac{\sinh u - \sin u}{2}, \qquad u = \lambda x. \end{aligned} \tag{13.6}$$

The angle of rotation $w'(x)$, bending moment $M(x) = -\mathrm{EI}w''(x)$ and shear force $Q(x) = -\mathrm{EI}w'''(x)$ can be found by successive differentiation of expression (13.5). This expression and its derivatives are valid for continuous loading along the entire length of the rod/beam.

$$w'(x) = \lambda w_0 D_{\lambda x} + \alpha_0 A_{\lambda x} - \frac{M_0}{\lambda \mathrm{EI}} B_{\lambda x} - \frac{Q_0}{\lambda^2 \mathrm{EI}} C_{\lambda x} - \frac{q_0}{\lambda^3 \mathrm{EI}} D_{\lambda x}, \tag{13.7}$$

$$M(x) = -\mathrm{EI}w''(x) = -\lambda^2 w_0 \mathrm{EI} C_{\lambda x} - \lambda \alpha_0 \mathrm{EI} D_{\lambda x} + M_0 A_{\lambda x} + \frac{Q_0}{\lambda} B_{\lambda x} + \frac{q_0}{\lambda^2} C_{\lambda x} \tag{13.8}$$

$$Q(x) = -\mathrm{EI}w'''(x) = -\lambda^3 w_0 \mathrm{EI} B_{\lambda x} - \lambda^2 \alpha_0 \mathrm{EI} C_{\lambda x} + \lambda M_0 D_{\lambda x} + Q_0 A_{\lambda x} + \frac{q_0}{\lambda} B_{\lambda x} \tag{13.9}$$

As noted previously, the above expressions are valid under continuous loading along the entire length of the beam. In the case when there are discontinuities in the cross-section $\Delta w(a_i)$, $\Delta w'(a_i)$, $\Delta M(a_i)$, $\Delta Q(a_i)$, and $\Delta q(a_i)$, with coordinate $x = a_i$, then to expression (13.5) and its derivatives (13.7, 13.8, and 13.9) for sections $x > a_i$, provided that the distance from the discontinuity to the section $x - a_i$, should be added:

$$\begin{aligned} \Delta w(x) = {} & \Delta w(a_i) A_{\lambda(x-a_i)} + \frac{\Delta w'(a_i)}{\lambda} B_{\lambda(x-a_i)} - \frac{\Delta M(a_i)}{\lambda^2 \mathrm{EI}} C_{\lambda(x-a_i)} - \frac{\Delta Q(a_i)}{\lambda^3 \mathrm{EI}} D_{\lambda(x-a_i)} - \\ & - \frac{\Delta q(a_i)}{\lambda^4 \mathrm{EI}} \left[A_{\lambda(x-a_i)} - 1 \right]; \end{aligned} \tag{13.10}$$

$$\begin{aligned} \Delta w'(x) = {} & \lambda \Delta w(a_i) D_{\lambda(x-a_i)} + \Delta w'(a_i) A_{\lambda(x-a_i)} - \frac{\Delta M(a_i)}{\lambda \mathrm{EI}} B_{\lambda(x-a_i)} - \frac{\Delta Q(a_i)}{\lambda^2 \mathrm{EI}} C_{\lambda(x-a_i)} - \\ & - \frac{\Delta q(a_i)}{\lambda^3 \mathrm{EI}} D_{\lambda(x-a_i)}; \end{aligned} \tag{13.11}$$

$$\begin{aligned} \Delta M(x) &= -\mathrm{EI}\Delta w''(x) \\ &= -\lambda^2 \mathrm{EI} \Delta w(a_i) C_{\lambda(x-a_i)} - \lambda \mathrm{EI} \Delta w'(a_i) D_{\lambda(x-a_i)} + \Delta M(a_i) A_{\lambda(x-a_i)} \\ &\quad + + \frac{\Delta Q(a_i)}{\lambda} B_{\lambda(x-a_i)} + \frac{\Delta q(a_i)}{\lambda^2} C_{\lambda(x-a_i)}; \end{aligned} \tag{13.12}$$

$$\begin{aligned} \Delta Q(x) &= -\mathrm{EI}\Delta w'''(x) \\ &= -\lambda^3 \mathrm{EI} \Delta w(a_i) B_{\lambda(x-a_i)} - \lambda^2 \mathrm{EI} \Delta w'(a_i) C_{\lambda(x-a_i)} + \lambda \Delta M(a_i) D_{\lambda(x-a_i)} + \\ &\quad + \Delta Q(a_i) A_{\lambda(x-a_i)} + \frac{\Delta q(a_i)}{\lambda} B_{\lambda(x-a_i)}. \end{aligned} \tag{13.13}$$

where $\Delta w(a_i)$ is a discontinuity in the elastic line, that is a discontinuity in the displacement function; $\Delta w'(a_i)$ is a discontinuity in the first derivative, in the angle of rotation; $\Delta M(a_i)$ is a discontinuity in the bending moment function; $\Delta Q(a_i)$ is a discontinuity in the shear force function; $\Delta q(a_i)$ there is a discontinuity in the load function.

The presented expressions (13.5), (13.7), (13.8), and (13.9) and additional expressions (13.10, 13.11, 13.12, and 13.13) allow us to determine the kinematic characteristics of beam vibrations $w(x)$, $w'(x)$ and internal forces $w''(x)$, $w'''(x)$ which are caused by the action of a harmonic load $q_0 \sin \theta t$.

Thus, the problem of determining the kinematic characteristics of beam vibrations and internal forces at any number of discontinuities (under any boundary conditions) is reduced to the problem of determining only two unknown initial parameters, since the other two initial parameters will be zero. Discontinuities $\Delta w(a_i)$ and $\Delta w'(a_i)$ are only possible when the beam consists of two parts along its length, the links between which allow for such discontinuities.

For free vibrations of the beam in (13.5) should be taken $q_0 = 0$. Using the boundary conditions at the left and right ends of the rod/beam, the frequency equations can be obtained. The solution (13.4) can also be used to compose the frequency equation. If there is an intermediate support with stiffness r in the beam span at a distance a_i from the left support, then this is equivalent to the presence of a jump $\Delta Q(a_i)$, equal to $\Delta Q(a_i) = rw(a_i)$. Then to the solution (13.5) and its derivatives for the sections $x \geq a_i$ we should add

$$\Delta w(x) = -\frac{\Delta Q(a_i)}{\lambda^3 EI} D_{\lambda(x-a_i)} \tag{13.14}$$

or the corresponding derivatives.

If there is a concentrated (discrete) mass m in the span at a distance a_i from the left support, then this is equivalent to the presence of a concentrated inertia force, the amplitude value of which is equal to $I = m\omega^2 w(a_i)$. If we assume that $\omega^2 = \lambda^4 \frac{\mathrm{EI}}{\rho}$, then we obtain:

$$I = \frac{m\lambda^4 \mathrm{EI}}{\rho} w(a_i) \tag{13.15}$$

The presence of an inertia force I is equivalent to a jump:

$$\Delta Q(a_i) = \frac{m\lambda^4 \mathrm{EI}}{\rho} w(a_i). \tag{13.16}$$

In this case, for sections at $x \geq a_i$, the following expression

$$\Delta w(x) = -\frac{\Delta Q(a_i)}{\lambda^3 \mathrm{EI}} D_{\lambda(x-a_i)}$$

should be added to the solution of (13.5).

The frequency equation for a beam with various types of support at the ends can be obtained using the boundary conditions at the left and right ends of the beam. The roots of these equations $u_n = \lambda_n l$ make it possible to determine the frequencies of natural vibrations:

$$\omega_n = \frac{u_n^2}{l^2} \sqrt{\frac{\mathrm{EI}}{\rho}} \tag{13.17}$$

The equations for determining the frequencies and roots of these equations u_n are given in Table 13.1.

Table 13.1 Frequency equations and roots of these equations

Types of beam	Frequency equation	Roots of frequency equation
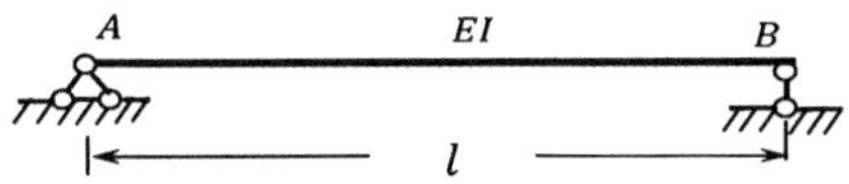	$\sin u = 0$	$u_1 = \pi, \quad u_2 = 2\pi, \ldots, u_n = n\pi$
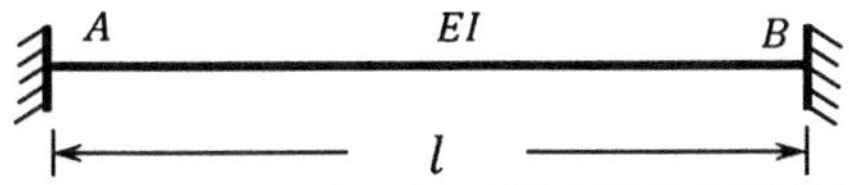	$\cos u \cdot \cosh u = 1$	$u_1 = 4.730;\ u_2 = 7.859;\ u_3 = 10.996$ $u_4 = 14.137;\ u_n = \pi/2(2n + 1)$
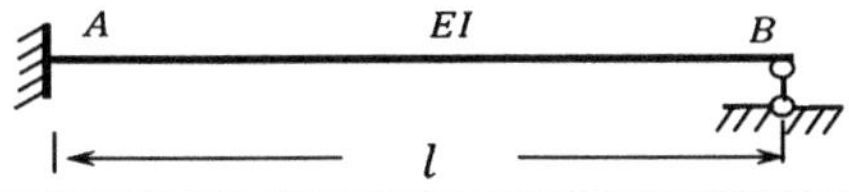	$\tan u = \tanh u$	$u_1 = 3.927;\ u_2 = 7.069;\ u_3 = 10.210;$ $u_4 = 13.352;\ u_n = \pi/4(4n + 1)$
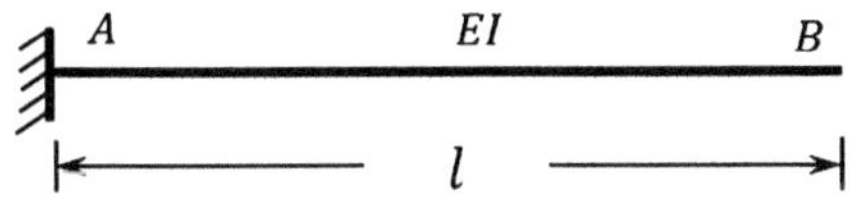	$\cos u \cdot \cosh u = -1$	$u_1 = 1.875;\ u_2 = 4694;\ u_3 = 7.855$ $u_4 = 10.996;\ u_n = \pi/2(2n - 1)$

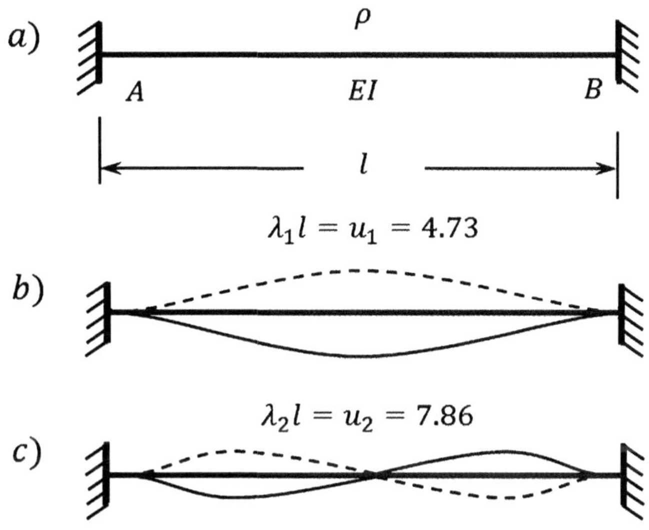

Fig. 13.1 Beam with built-in, fixed supports on both sides

Problem 13.1

Compose a frequency equation for a single-span beam with built-in, fixed supports on both sides and determine the frequency spectrum in Fig. 13.1.

Solution

Let us write down the solution of the elastic line equation of free vibrations (13.4):

$$w(x) = A\cosh \lambda x + B\sinh \lambda x + C\cos \lambda x + D\sin \lambda x,$$

Let's write down the boundary conditions: at $x = 0$, $w(0) = 0$, $w'(0) = 0$; at $x = l$, $w(l) = 0$, $w'(l) = 0$. At $x = 0$, $w(0) = A + C = 0$, whence $C = -A$; $w'(0) = B + D = 0$, whence $D = -B$. At $x = l$

$$w(l) = A\cosh\lambda l + B\sinh\lambda l + C\cos\lambda l + D\sin\lambda l = 0,$$

$$w'(l) = \lambda\,(A\sinh\lambda l + B\cosh\lambda l - C\sin\lambda l + D\cos\lambda l) = 0$$

Let us use the equalities $C = -A$ and $D = -B$, and also denote $\lambda l = u$, then we obtain the following equations:

$$w(l) = A(\sinh u - \cos u) + B(\sinh u - \sin u) = 0,$$

$$\frac{1}{\lambda}w'(l) = A(\sinh u + \sin u) + B(\cosh u - \cos u) = 0,$$

The frequency equation for determining the frequencies of free oscillations is:

$$\Delta = \begin{vmatrix} \cosh u - \cos u & \sinh u - \sin u \\ \sinh u + \sin u & \cosh u - \cos u \end{vmatrix} = 0.$$

Furthermore, let us expand the determinant:

$$(\cosh u - \cos u)^2 - (\sinh u + \sin u)(\sinh u - \sin u) = 0,$$

$$\cosh^2 u - 2\cosh u\cos u + \cos^2 u - \sinh^2 u + \sin^2 u = 0.$$

Since $\cosh^2 u - \sinh^2 u = 1$, and $\sin^2 u + \cos^2 u = 1$, we obtain $2 - 2\cosh u\cos u = 0$, or $\cosh u\cos u = 1$, whence:

$$\cos u = \frac{1}{\cosh u} = \sinh u.$$

The solution of the obtained equation can be represented graphically (Fig. 13.2):

$$u_1 = 4.73, \qquad u_2 = 7.86, \qquad u_n = \frac{3}{2}\pi + (n-1)\pi.$$

The frequencies of natural oscillations are determined by formula (13.17)

$$\omega_n = \frac{u_n^2}{l^2}\sqrt{\frac{EI}{\rho}}.$$

In Fig. 13.2, the modes of oscillations at u_1 and u_2 are given.

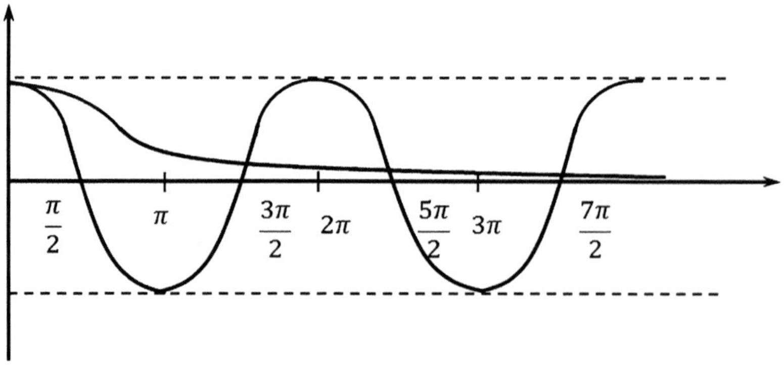

Fig. 13.2 Modes of oscillations at u_1 and u_2 are given

Below please see an illustration of the dynamic calculation of frames with a distributed mass using the displacement method.

As is known, the primary system for dynamic calculation of frames by the displacement method is obtained by adding additional links at the nodes [2]. These links should prevent angular and linear displacements of nodes. Unknowns are dynamic displacements of nodes $Z_1(t) = \overline{Z}_1 \sin \theta t,\ Z_2(t) = \overline{Z}_2 \sin \theta t$, where $\overline{Z}_1$ and $\overline{Z}_2$ are amplitudes of displacements of nodes.

The condition of absence of full dynamic reactions of the introduced links is expressed by the system of canonical equations of the method of displacements:

$$\begin{aligned} r_{11}\overline{Z}_1 + r_{12}\overline{Z}_2 + R_{1p} &= 0 \\ r_{21}\overline{Z}_1 + r_{22}\overline{Z}_2 + R_{2p} &= 0 \end{aligned} \tag{13.18}$$

The unknowns in this system of equations are the amplitudes of vibration displacements $\overline{Z}_1,\ \overline{Z}_2$.

The amplitude values of reactions of the introduced links from unit vibration displacements are coefficients of the unknowns. The amplitude reactions of the links from the vibration load are free terms of the canonical equations of the displacement method. The coefficients of the unknowns and free terms differ from the corresponding values used in static calculations in that when determining them, the inertia forces of the uniformly distributed masses of the frame rods are also taken into account. This is done by using correction functions to formulas for static reactions. The amplitudes of dynamic reactions for rods with uniformly distributed mass along their length are expressed by the formulas given in Table B.1 [3], Appendix B. The influence of inertia forces of distributed masses is captured by adding correction functions ψ (u) to the formulas for static reactions. In Table B.1, Appendix B, the moment diagrams correspond to vibrations with frequency $\theta < \omega_1$, namely, to the value $u < u_1$. The numerical values of these functions depend on the argument u, which is represented by the following formula:

$$u = l\sqrt[4]{\frac{\rho\theta^2}{\mathrm{EI}}} = \sqrt[4]{\frac{l^3\rho\theta^2}{i}} \tag{13.19}$$

where l is the length of the rod; ρ is the linear mass of the rod; EI is the rigidity of the rod; i is the linear stiffness of the rod; θ is the frequency of forced oscillations, which is equal to the frequency of the acting loads.

The free terms in the canonical equations that characterize free vibrations are zero. Equation (13.18) will take the form:

$$\begin{aligned} r_{11}\overline{Z}_1 + r_{12}\overline{Z}_2 &= 0; \\ r_{21}\overline{Z}_1 + r_{22}\overline{Z}_2 &= 0. \end{aligned} \tag{13.20}$$

The system of Eq. (13.20) has a trivial solution $\overline{Z}_1 = \overline{Z}_2 = 0$. The trivial solution corresponds to the state of rest. Solutions that are different from zero for $\overline{Z}_1$ and $\overline{Z}_2$ can be obtained under the condition when the determinant D of the system of Eq. (13.20) is equal to zero:

$$D = \begin{vmatrix} r_{11} & r_{12} \\ r_{21} & r_{22} \end{vmatrix} = 0; \quad r_{11}r_{22} - r_{12}^2 = 0. \tag{13.21}$$

Thus, the roots of the frequency equation (13.21) correspond only to nontrivial solutions of the homogeneous equations of free oscillations (13.20).

Equation (13.21) is transcendental, and this equation cannot be solved in general form. The solution of Eq. (13.21) is found approximately by selection. The argument u is found from the solution of

Eq. (13.21). The values of the frequencies of natural oscillations can be found after determining the values of the argument u by the following formula:

$$\omega_k = \left(\frac{u_k}{l}\right)^2 \sqrt{\frac{\mathrm{EI}}{\rho}}. \tag{13.22}$$

As is known, the roots of Eq. (13.21) correspond only to nontrivial solutions of homogeneous equations of free oscillations (13.20).

The final dynamic bending moment diagram is constructed using the following formula:

$$M = M_1\overline{Z}_1 + M_2\overline{Z}_2 + \cdots + M_n\overline{Z}_n + M_p^0, \tag{13.23}$$

where

$\overline{M}_1,\ \overline{M}_2,\ \ldots,\ \overline{M}_n$ are diagrams of bending moments in the primary system from the corresponding unit displacements; M_p^0 is the diagram of bending moments in the primary system from the amplitude values of loads.

For symmetric frames with symmetrically arranged distributed masses, the unknown displacements are grouped together. Symmetrical and inversely symmetrical vibration modes are determined separately, independently of each other.

Below is the sequence of dynamic calculation of a frame with a distributed mass under the action of a vibration load by the displacement method.

1. The primary system for calculating the displacement method is selected. The frequency equation is composed. The ratio of linear stiffness and the argument u for various frame elements is found [1].
2. The unit moment diagrams are constructed. The correction functions $\psi_i(u)$ and their numerical values: correction coefficients, correction factors ψ_i are taken from the Table B.1, Amplitudes of Dynamic Reactions, Appendix B.
3. The coefficients r_{ii}, r_{ik}, at the unknowns of the canonical equations are determined. The longitudinal force of inertia of the rod mass should be included when determining the reactions. These reactions are caused by a unit of linear displacement of the frame nodes. The value of the longitudinal inertia force of the mass of the rod is equal to: $I^* = -\rho l\omega^2$, here l is the length of the rod, the length of the frame elements.
4. The frequency equation is solved, the argument u is found, and then the frequency of forced vibrations is determined.
5. The canonical equation of forced oscillations is compiled.
6. Argument u for the given frequencies, and then the coefficients r_{ii}, r_{ik}, for unknown are found.
7. A diagram of bending moments for the primary system due to the load is constructed M_p^0. The ordinate values are taken according to the formulas given in the table in Appendix B. The numerical values of the correction coefficients ψ_{ip} and ψ_{iq} are determined from the Table B.2 in Appendix B.
8. The free terms of the canonical equations are determined based on the diagram M_p^0.
9. The canonical equations are solved.
10. The final dynamic bending moment diagram M_{pq} is constructed according to formula (13.23).

Problem 13.2

Perform a dynamic calculation of a statically indeterminate frame with a distributed mass shown in Fig. 13.3a under the action of a given harmonic load.

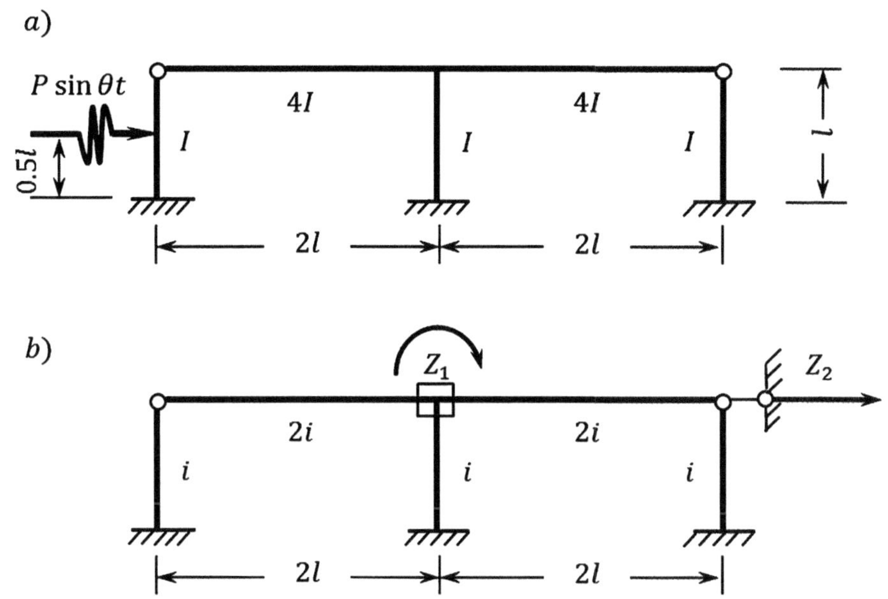

Fig. 13.3 (**a**) Frame with a distributed masses under the action of a harmonic load; (**b**) Primary system of the displacement method

Solution

The primary system of the displacement method is shown in Fig. 13.3b. This problem was also considered in Refs. [2, 3]. Here we will use Eqs. (13.20) and (13.21):

$$r_{11}\overline{Z}_1 + r_{12}\overline{Z}_2 = 0$$

$$r_{21}\overline{Z}_1 + r_{22}\overline{Z}_2 = 0$$

$$D = \begin{vmatrix} r_{11} & r_{12} \\ r_{21} & r_{22} \end{vmatrix} = 0; \quad r_{11}r_{22} - r_{12}^2 = 0$$

The linear stiffness of the frame columns (struts) and the frame beams are equal according to the problem condition:

$$i_c = \frac{\text{EI}}{l} = i; \qquad i_b = \frac{4\text{EI}}{l} = 2i$$

The argument u, or frequency coefficient of the frame struts:

$$u_c = l\sqrt[4]{\frac{\rho\omega^2}{\text{EI}}} = u.$$

The argument u, or frequency coefficient of the frame of the horizontal elements, beams:

$$u_b = 2l\sqrt[4]{\frac{4\rho\omega^2}{4\text{EI}}} = 2u.$$

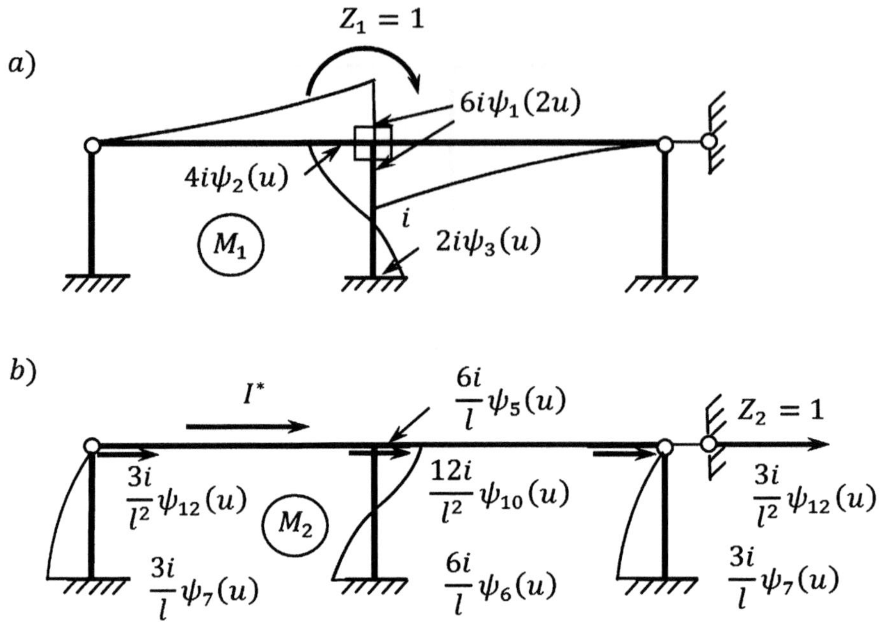

Fig. 13.4 (**a**) Unit diagram due to the rotation of the middle frame node; (**b**) Unit diagram due to the horizontal displacement of the beam

The unit diagram from the rotation of the middle frame node Z_1 and from the horizontal displacement of the beam Z_2, respectively, M_1 and M_2 are shown in Fig. 13.4a, b.

The correction coefficients ψ_i for diagrams M_1, and M_2 are taken according to Table B.1. Appendix B. The numerical values of these coefficients are taken from the Tables B.2 in Appendix B.

The reactions r_{ii} and r_{ik} of the canonical equations are determined from the diagram M_1:

$$r_{11} = 2 \cdot 6i\psi_1(2u) + 4i\psi_2(u) = [12\psi_1(2u) + 4\psi_2(u)]i;$$

$$r_{12} = -\frac{6i}{l}\psi_5(u).$$

The force of inertia of the beam mass at unit displacement $\overline{Z}_2 = 1$, of the horizontal frame element is formed. The force of inertia of the beam mass is directed along the beam axis in the positive direction $\overline{Z}_2$.

The longitudinal force of inertia I^* in the considered problem is equal to:

$$I^* = (4\rho L)\omega^2 \cdot 1 = (4\rho \cdot 2 \cdot 2l)\omega^2 \cdot 1 = 16\rho l\omega^2;$$

Since

$$L = 2 \cdot 2l = 4l.$$

Because $\omega = \frac{u^2}{l^2}\sqrt{\frac{EI}{\rho}}$, and $\omega^2 = \frac{u^4 EI}{l^4 \rho}$, then $I^* = 16\rho l \frac{u^4 EI}{l^4 \rho} = 16\frac{iu^4}{l^2}$.

Let's cut out the upper chord of the frame for clarity in determining r_{22}. Figure 13.5

To determine the unit reaction from diagram M_2, Fig. 13.4b, we obtain the following expression:

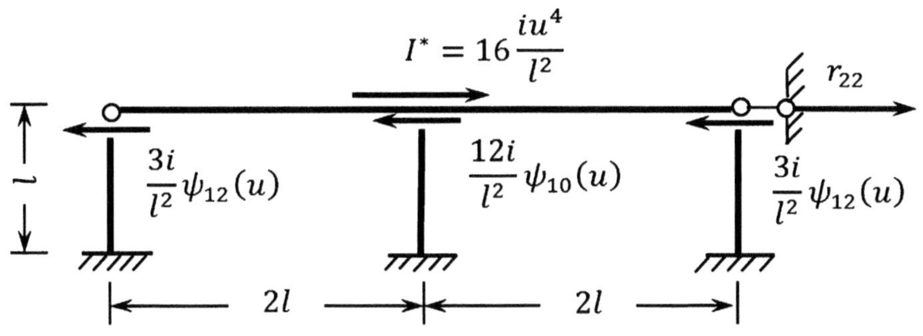

Fig. 13.5 Using the section method for upper chord of frame

Table 13.2 Results of calculations of the frequency equations & the function $D = f(u)$

u	$\psi_1(2u)$	$\psi_2(u)$	$r_{11}\frac{l}{i}$	$\psi_5(u)$	$r_{12}\frac{l}{i}$	$\psi_{10}(u)$	$\psi_{12}(u)$	$r_{22}\frac{l^2}{i}$	$D\frac{l^2}{i^2}$
0	1.0000	1.0000	16.00	1.0000	−6.00	1.0000	1.0000	18.00	252.0000
0.5	0.9936	0.9998	15.9224	0.9994	−5.9964	0.9981	0.9951	16.9478	233.8928
0.8	0.9573	0.9990	15.4836	0.9964	−5.9784	0.9873	0.9678	11.1008	136.1391
0.9	0.9306	0.9984	15.1608	0.9943	−59558	0.9797	0.9484	6.9492	69.8839
1.0	0.8919	0.9976	14.6922	0.9913	−5.9478	0.9690	0.9212	1.1552	−18.4027
1.4	0.4886	0.9908	9.8264	0.9663	−5.7978	0.8806	0.6955	−46.7254	−497.7569
1.6	−0.1325	0.9842	2.3468	0.9422	5.6532	0.7958	0.4772	−92.4448	−248.9081
1.7	−0.8150	0.9798	−58608	0.9262	−5.5572	0.7393	0.3309	−122.7766	688.6866

$$r_{22} = 2 \cdot \frac{3i}{l^2}\psi_{12}(u) + \frac{12i}{l^2}\psi_{10}(u) - \frac{16iu^4}{l^2} = \frac{i}{l^2}\left[6\psi_{12}(u) + 12\psi_{10}(u) - 16u^4\right].$$

Let's substitute the values of unit reactions $r_{11},\ r_{22},\ r_{12}$ into the frequency equation. As a result, we obtain:

$$D = r_{11}r_{22} - r_{12}^2 = i[12\psi_1(2u) + 4\psi_2(u)] \cdot \frac{i}{l^2}\left[6\psi_{12}(u) + 12\psi_{10}(u) - 16u^4\right] - \left[-6\psi_5(u)\right]^2 \frac{i^2}{l^2} = 0.$$

The functions $\psi_i(u)$ and $\psi_i(2u)$ are transcendental. In general, it is difficult to obtain a solution to the frequency equation. Therefore, we will look for an approximate solution. To do this, let's draw a graph $D = f(u)$. The points where the curve crosses $D = f(u)$ the abscissa axis will correspond to the solution of the frequency equation. The results of calculating the function $D = f(u)$ are given in Table 13.2.

According to these calculations, let's draw a graph $D = f(u)$ (Fig. 13.6). In this graph, the curves intersect the x-axis in the interval: $u_1 = 0.9 \div 1.0$, and $u_2 = 1.6 \div 1.7$. By linear interpolation, we find the frequency coefficients $u_1 = 0.9792$ and $u_2 = 1.6265$, which corresponds to the first ω_1 and second ω_2 frequencies of natural oscillations:

$$\omega_1 = \frac{0.9588}{l^2}\sqrt{\frac{EI}{\rho}}; \quad \omega_2 = \frac{2.6455}{l^2}\sqrt{\frac{EI}{\rho}}$$

Now, let's determine the coefficients $r_{ii},\ r_{ik}$ for forced vibration. The calculation will be performed for frequencies that correspond to frequency coefficients $u_1 = 0.9$ and $u_2 = 1.0$. The frequencies of forced vibrations will be:

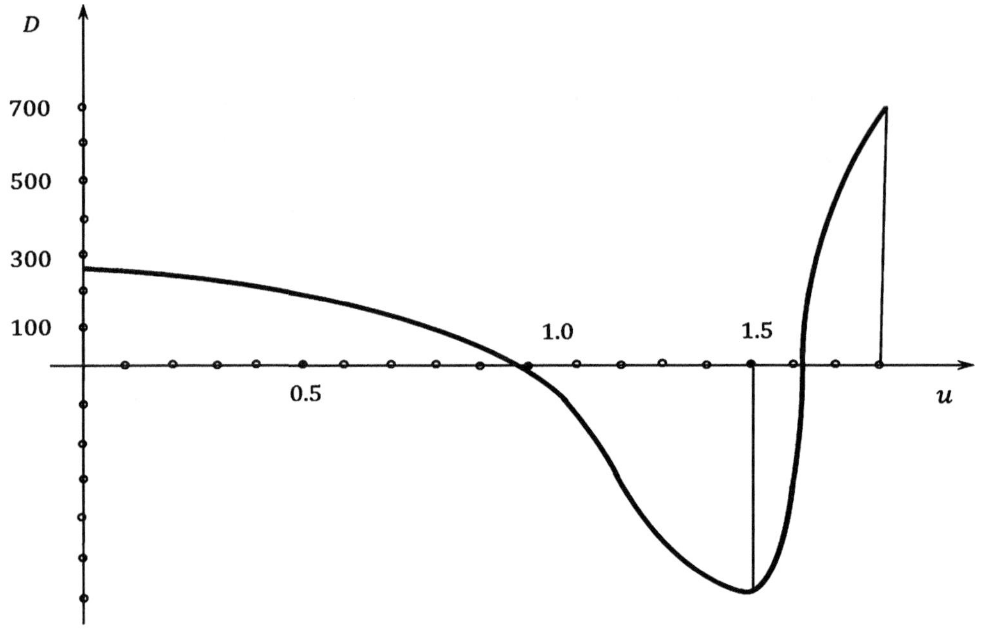

Fig. 13.6 A graph of the solution of transcendental equation

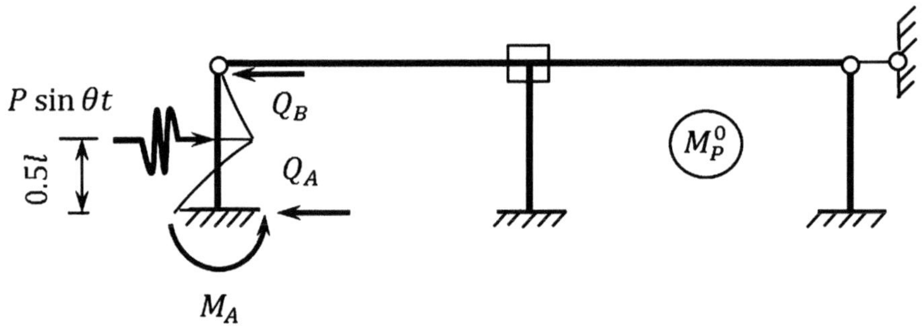

Fig. 13.7 Bending moment diagram due to the given load

$$\theta_1 = \frac{0.81}{l^2}\sqrt{\frac{\text{EI}}{\rho}} < \omega_1; \quad \theta_2 = \frac{1.0}{l^2}\sqrt{\frac{\text{EI}}{\rho}} > \omega_1.$$

Reactions that correspond to these frequencies:

At $u = 0.9$ $r_{11} = 15.1608i$; $r_{12} = -5.9558\frac{i}{l}$; $r_{22} = 6.9492\frac{i}{l^2}$;
At $u = 1.0$ $r_{11} = 14.6932i$; $r_{12} = -5.9478\frac{i}{l}$; $r_{22} = 1.1552\frac{i}{l^2}$.

Now, let's build (construct) a diagram of bending moments M_p^0 from a given vibration load (Fig. 13.7).

The bending moment diagram M_p^0 from given load is constructed according to the data in Table B.1 of Appendix B.

$$M_A = \frac{3}{16} Pl\psi_{3p}(u); \quad Q_A = \frac{11}{16} P\psi_{4p}(u); \quad Q_B = \frac{5}{16} P\psi_{5p}(u).$$

The following coefficients $\psi_{3p}(u)$, $\psi_{4p}(u)$, and $\psi_{5p}(u)$ according to the table of Appendix B are equal to:

At $u = 0.9$ $\psi_{3p}(0.9) = 1.0029$; $\psi_{4p}(0.9) = 1.0030$; $\psi_{5p}(0.9) = 1.0044$;
At $u = 1.0$ $\psi_{3p}(1.0) = 1.0043$; $\psi_{4p}(1.0) = 1.0046$; $\psi_{5p}(1.0) = 1.0067$;

Let's move on to determining the free terms of the canonical equations and solving them:

At $u = 0.9$ $R_{1p} = 0$; $R_{2p} = -Q_B = -\frac{5}{16}P \cdot 1.0044 = -0.3139P$;
At $u = 1.0$ $R_{1p} = 0$; $R_{2p} = -Q_B = -\frac{5}{16}P \cdot 1.0067 = -0.3146P$;

Solution of canonical equations at $u = 0.9$

$$15.1608i\overline{Z}_1 - 5.9558\frac{i}{l}\overline{Z}_2 = 0;$$

$$-5.9558\frac{i}{l}\overline{Z}_1 + 6.9492\frac{i}{l^2}\overline{Z}_2 - 0.3139P = 0;$$

$$\overline{Z}_1 = 0.0268\frac{Pl}{i}; \qquad \overline{Z}_2 = 0.0681\frac{Pl^2}{i}.$$

Solution of canonical equations at $u = 1.0$

$$14.6922\,\overline{Z}_1 - 5.9478\frac{i}{l}\overline{Z}_2 = 0;$$

$$-5.9478\frac{i}{l}\overline{Z}_1 + 1.1552\frac{i}{l^2}\overline{Z}_2 - 0.3146P = 0;$$

$$\overline{Z}_1 = -0.1017\frac{Pl}{i}; \qquad \overline{Z}_2 = -0.2512\frac{Pl^2}{i}.$$

Since $u = 0.9$ corresponds to the frequency $\theta_1 < \omega_1$, and $u = 1.0$ corresponds to the frequency $\theta_2 > \omega_1$, the displacements of $\overline{Z}_1$ and $\overline{Z}_2$ during the transition through the resonance have flipped the sign. Since the frequency of forced vibration θ is closer to the frequency of natural oscillations ω_1, the absolute numerical values of displacements $\overline{Z}_1$ and $\overline{Z}_2$ at $u = 1.0$ are significantly more $\overline{Z}_1$ and $\overline{Z}_2$ than at $u = 0.9$.

The diagram of dynamic moments is constructed by formula (13.23):

$$M_{\text{dyn}} = M_1\overline{Z}_1 + M_2\overline{Z}_2 + M_p^0$$

1. At a frequency θ_1 that corresponds to $u = 0.9$.
   Let's build (construct) the following diagrams: M_p^0, $M_1\overline{Z}_1$, $M_2\overline{Z}_2$ and M_{dyn}, Fig. 13.8a–d, and check the equilibrium of moments in the frame node:

$$\sum M = Pl(0.2993 - 2 \cdot 0.1496) = 0.0001Pl$$

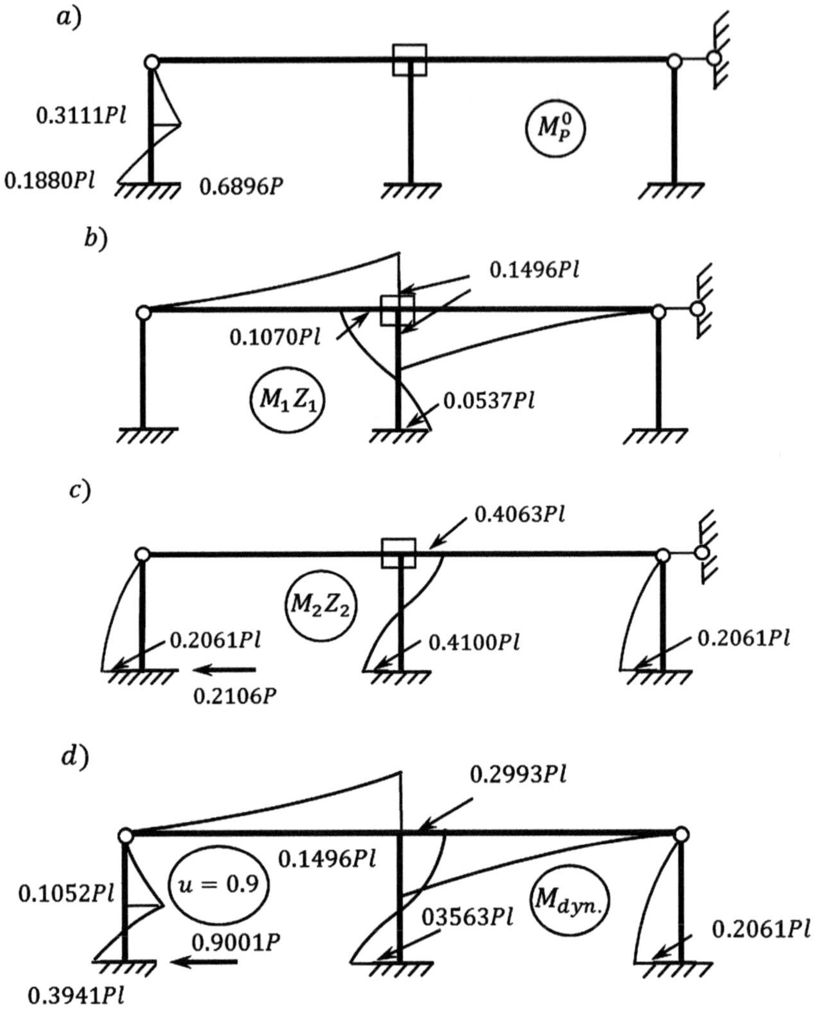

Fig. 13.8 Bending moment diagrams: M_p^0, $M_1\overline{Z}_1, M_2\overline{Z}_2$, and M_{dyn}

The error is:

$$\Delta = \frac{0.0001}{0.2993} 100\% = 0.033\%$$

2. At a frequency θ_1 that corresponds to $u = 1.0$.
   Let's build the following diagrams: M_p^0, $M_1\overline{Z}_1$, $M_2\overline{Z}_2$ and M_{dyn}, Fig. 13.9a–d, and check the equilibrium of moments in the frame node:

$$\sum M = Pl(2 \cdot 0.5442 - 1.0883) = 0.0001Pl$$

The error is:

$$\Delta = \frac{0.0001}{1.0883} 100\% = 0.009\%$$

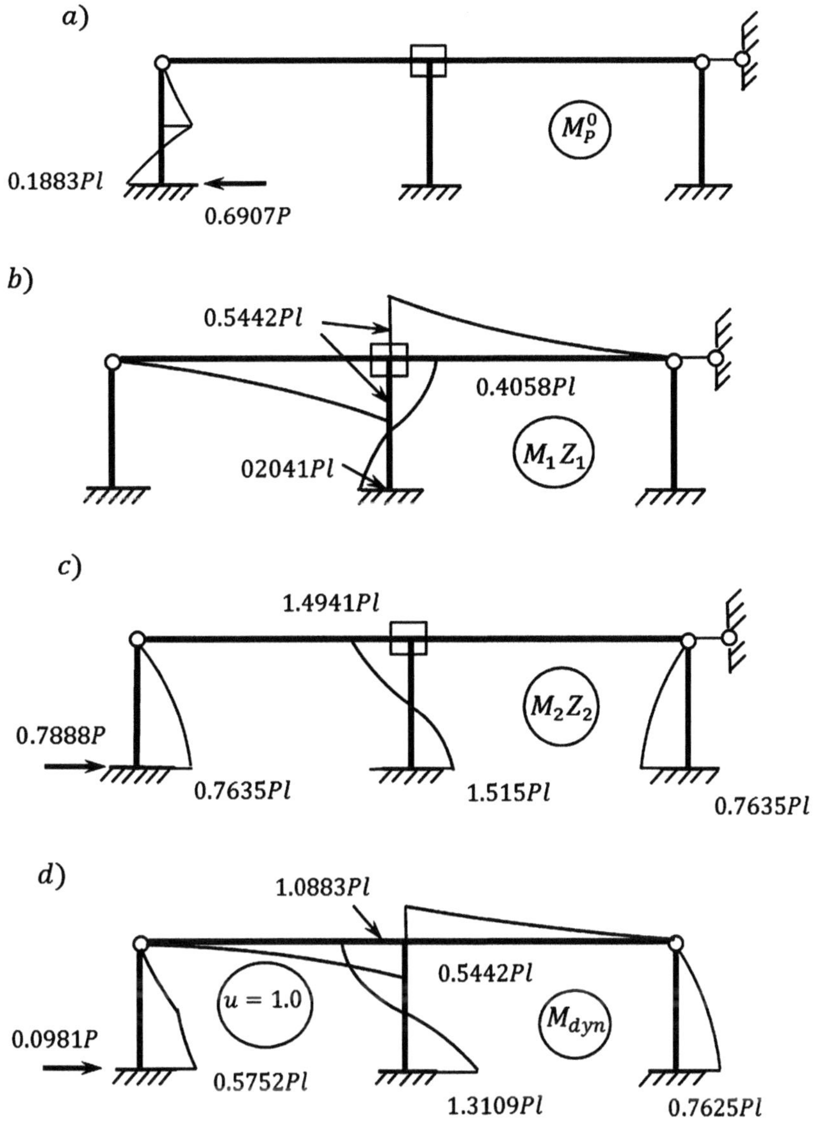

Fig. 13.9 Bending moment diagrams: M_P^0, $M_1\overline{Z}_1$, $M_2\overline{Z}_2$, and M_{dyn}

References

1. Petrosian, L. G. (1982). The selection of the rational primary systems for frame calculations and designs. Methodical instructions for the course and diploma research projects design. In *Structural mechanics and calculation of structures* (p. 58p).
2. Petrosian, L. G. (1989). Calculation of frames of industrial buildings using computer. In *Manual guidelines & instructions. Dynamics of structures: Structural mechanics and calculation of structures* (47p).
3. Koshelev, L. I., & Kosheleva, A. A. (1989). *Dynamic calculation of frames using a computer. Training Manual* (100p). Department of Structural Mechanics, MESI.

Methods of Dynamic Design of Structural Elements

14

Abstract

This chapter presents methods of dynamic design of structural elements. This chapter shows the main methods used in the dynamic analysis of rods, plates, and shells. The application of this method is described in the examples of solving problems of free vibration of rectangular and circular membranes. The range of problems that can be solved accurately, in a closed form, is limited and, therefore, there are often variational methods used in practice. The variational methods are discussed. The use of integral equations, in particular the reduction of problems with oscillation of one-dimensional structures to the integral equations of Volterra of the second kind and Fredholm of the second kind are demonstrated. The methods used for numerical solutions of problems of dynamics are described. These are the finite element method (FEM) and finite difference method (FDM). In either the FEM or FDM, the whole domain of the partial differential equations (PDEs) requires discretization. The FEM is presented as an example of a plane problem of the theory of elasticity. A diagram of solving a problem of nonstationary oscillations of a cantilever plate is shown. The oscillations of a compressed rod/bar are solved using the finite difference method (FDM). This chapter describes both a general approach to the construction of various types of boundary equations and a spectral method of boundary elements (SMBE) method, which allows to obtain effective solutions of various boundary value problems of structural mechanics. The application of SMBE in this book is further illustrated by solving several problems on the structural analysis of the design of construction on the static and dynamic impacts. The generalized functions in structural mechanics, generalized method of integral transformation (GMIT), delta-transform method (DTM), and compensating loads method (CLM) are also presented in this chapter.

14.1 The Method of Separation Variables

This chapter presents the main methods used in the dynamic analysis of robs, plates, and shells. Historically, one of the first methods of solving equations of mathematical physics is the method of separation of variables. The application of this method will be described in the examples of solving problems of free oscillations/vibrations of rectangular and circular membranes.

Let us briefly present the substance of the Fourier method or a method of separation of variables. The *separation of variables* is the process in which solutions are found for separable differential equations. The method of separation of variables relies upon the assumption that a function of the form

L. G. Petrosian, *Dynamics of Structures*, https://doi.org/10.1007/978-3-031-63539-7_14

$u(x, t) = X(x)T(t)$ will be a solution to linear homogeneous partial differential equations (PDEs) in x and t. Product of two function is the solution of the problem, and provided the boundary conditions are also linear and homogeneous, this will also satisfy the boundary conditions. This technique works because if the product of functions of independent variables is a constant, each function must separately be a constant as well. Success requires choice of an appropriate coordinate system and may not be attainable at all depending on the equation. Consider the problem of oscillations if the string is fixed at the ends. The problem is reduced to the solution of the equations of hyperbolic type:

$$\frac{\partial^2}{\partial} \frac{u}{t^2}(x,t) = a^2 \frac{\partial^2}{\partial} \frac{u}{x^2} \quad (x,t) \tag{14.1}$$

Let u (x, t) denote the vertical displacement of the string from the x axis at position x and time t. In the derivation of Eq. (14.1), it was assumed that the extension of individual sections of the string during the oscillation does not occur and, therefore, according to Hooke's law, the tension $T_0 = |\boldsymbol{T}|$ does not depend on time or on x . The axis x coincides with the direction of a string in equilibrium position. The string is a thin string that does not resist bending and is not associated with a change in its length.

The string has length l : $(0 < x < l)$. The function $u(x, t)$ satisfies the initial conditions. The conditions that specify the initial state of the system are:

$$\begin{aligned} u(x,0) &= u_0(x) \\ \frac{\partial u}{\partial t} u(x,0) &= u_1(x) \end{aligned} \tag{14.2}$$

The conditions at the boundary of the problem are:

$$u(0,t) = 0 \qquad u(l,t) = 0 \tag{14.3}$$

The idea of separation of variables is simple. The idea is to assume that the solution to the PDE (14.1) has a special form of solution, namely, $u(x, t) = X(x)T(t)$. We attempt to convert the given PDE into several ordinary differential equations. When a problem is posed, such as our problem for $u(x, t)$, one can look for a product solution in the form of $u(x, t) = X(x)T(t)$. The solution can be done by inserting $X(x)T(t)$ into the PDE for the variable u, and then separating the variables so that each side of the equation depends on only one variable. Once the equation has been broken up into separate equations of one variable, the problem can be solved like an ordinary differential equation. We will first seek particular solutions of Eq. (14.1) that are not equal to zero and satisfy the boundary conditions (14.3), in the form of:

$$u(x,t) = X(x)T(t). \tag{14.4}$$

Substituting Eq. (14.4) in (14.1), we come to equations:

$$T''(t) + a^2 \lambda T(t) = 0; \tag{14.5}$$

$$X''(x) + \lambda X(x) = 0, \tag{14.6}$$

where λ is constant. The value of λ is not specified in the equation, finding the values of λ for which there exists a nontrivial solution of Eq. (14.1) satisfying the boundary conditions. For obtaining the nontrivial solutions of Eq. (1.4), it is necessary to find nontrivial solutions satisfying the conditions:

$$X(0)=0, \qquad X(l)=0. \tag{14.7}$$

This brings us to the problem of Sturm–Liouville or eigenvalue problem. In mathematics, a certain class of PDEs are subject to extra constraints, known as boundary values, on the solution.

The eigenvalues of this problem are the numbers:

$$\lambda_\kappa=\left(\frac{\pi\kappa}{l}\right)^2 \qquad (k=1,2,\ldots).$$

These eigenvalues correspond to the normalized unique eigenfunctions:

$$X_k(x)=\sqrt{\frac{2}{l}}\ \sin\frac{\pi\kappa x}{l}.$$

When $\lambda=\lambda_\kappa$, Eq. (14.5) has the general solution

$$T_k(t)=a_k\ \cos\frac{\kappa\pi at}{l}+b_k\sin\frac{k\pi at}{l},$$

Therefore the function

$$u_k(x,t)=X_k(x)T_k(t)=\left(a_k\ \cos\frac{\kappa\pi at}{l}+b_k\sin\frac{k\pi at}{l}\right)\sin\frac{k\pi x}{l},$$

satisfies Eq. (14.1) and the boundary conditions (14.3) for any a_k and b_k. The solution of Eq. (14.1), satisfying the conditions (14.2) and (14.3), is found in the form of a series

$$u(x,t)=X_k(x)T_k(t)=\sum_{k=1}^{\infty}\left(a_k\ \cos\frac{\kappa\pi at}{l}+b_k\sin\frac{k\pi at}{l}\right)\sin\frac{k\pi x}{l} \tag{14.8}$$

If this series converges uniformly and can be twice differentiated term by term, the sum of the series will satisfy Eq. (14.1) and the boundary conditions (14.3). Determining constants a_k and b_k, such that the sum of the series (14.8) satisfies the initial condition (14.2), the following equations are obtained:

$$u_0(x)=\sum_{k=1}^{\infty}a_k\sin\frac{k\pi x}{l}, \tag{14.9}$$

$$u_1(x)=\sum_{k=1}^{\infty}\ b_k\frac{k\pi a}{l}\sin\frac{k\pi x}{l}; \tag{14.10}$$

Formulas (14.9) and (14.10) give a decomposition of functions $u_0(x)$ and $u_1(x)$ in the Fourier sine series in the interval (0, l) The coefficients of these decompositions are calculated from the known formulas:

$$a_k=\frac{2}{l}\int_0^l u_0(x)\sin\frac{k\pi x}{l}dx, \qquad b_k=\frac{2}{k\pi a}\int_0^l u_1(x)\sin\frac{k\pi x}{l}\,dx,$$

So, while the sum (14.8) can be very complicated, each term, is quite simple. For each fixed t, and for each fixed x, the term $\sin\frac{k\pi x}{l}\left(a_k\cos\frac{\kappa\pi at}{l}+b_k\sin\frac{k\pi at}{l}\right)$ is just a constant times $\sin\frac{k\pi x}{l}$, and constant time cos $\frac{k\pi at}{l}$ plus a constant time sin $\frac{k\pi at}{l}$ for each fixed x. As x runs from 0 to l, the argument of sin $\frac{k\pi x}{l}$ runs from 0 to $k\pi$, which is k half-periods of sin. If the string has density ρ and tension T, then $a = \sqrt{\frac{T}{\rho}}$.

The method of separation of variables is one of the most widespread methods of the solution of tasks on oscillations of rods, plates, shells, membranes, and other constructive elements.

14.1.1 Free Oscillations of a Rectangular Membrane

We will illustrate this method with an example of a problem of free oscillations of a rectangular membrane. In the Cartesian coordinates, oscillations/vibrations of a membrane are described by the wave equation of the type:

$$\frac{\partial^2 w}{\partial x^2} + \frac{\partial^2 w}{\partial y^2} - \frac{1}{\alpha^2}\frac{\partial^2 w}{\partial t^2} = 0 \ ; \tag{14.11}$$

$$\alpha^2 = \frac{T}{\rho}; \qquad 0 \le x \le a; \qquad 0 \le y \le b,$$

where $W(x, y, t)$ is displacement of the membrane; T is tension of the membrane; ρ is density per unit area; a and b are dimensions of the membrane in the plan.

Boundary and initial conditions are as follows:

$$\begin{gathered} W(0, y, t) = W(a, y, t) = W(x, 0, t) = W(x, b, t) = 0 \\ W(x, y, 0) = \varphi(x, y,) \end{gathered} \tag{14.12}$$

$$\frac{dw}{dt}(x, y, 0) = \psi(x, y,)$$

The particular solutions of equation of (14.11) we are looking for are in the form of:

$$W(x, y, t) = T(t) \times X(x) \times Y(y). \tag{14.13}$$

Substituting (14.13) into (14.11) and dividing on $T(t) \times X(x) \times Y(y)$, we will obtain:

$$\frac{X''}{X} + \frac{Y''}{Y} - \frac{1}{\alpha^2}\frac{T''}{T} = 0. \tag{14.14}$$

Designating (denoting):

$$\frac{X''}{X} = -k_1^2; \quad \frac{Y''}{Y} = -k_2^2; \quad \frac{T''}{\alpha^2 T} = -\omega^2, \tag{14.15}$$

we obtain the following differential equations:

$$\begin{aligned} T'' + (\alpha\omega)^2 T &= 0 \\ X'' + k_1^2 X &= 0 \\ Y'' + k_2^2 Y &= 0 \end{aligned} \tag{14.16}$$

With a view to (14.15) from (14.14), we will obtain the interrelation

$$\omega^2 = k_1^2 + k_2^2 \tag{14.17}$$

The solutions of Eq. (14.16) will be

$$\begin{aligned} T(t) &= A\cos\alpha\omega t + B\sin\alpha\omega t \\ X(x) &= C\cos k_1 x + D\sin k_1 x \\ Y(y) &= E\cos k_2 y + F\sin k_2 y \end{aligned} \tag{14.18}$$

Using boundary conditions, we find:

$$C = E = 0$$

$$\sin k_1 a = 0 \quad \sin k_2 b = 0$$

$$k_{1m} = \frac{m\pi}{a}; \quad k_{2n} = \frac{n\pi}{b};$$
$$(m = 1, 2, \ldots) \quad (n = 1, 2, \ldots).$$

From Eq. (14.17) follows that:

$$\omega_{mn}^2 = k_{1m}^2 + k_{2n}^2 = \pi^2\left(\frac{m^2}{a^2} + \frac{n^2}{b^2}\right) \tag{14.19}$$

The general solution of the task bearing in the mind (14.13) is represented in the form of:

$$W(x, y, t) = \sum_{m=1}^{\infty}\sum_{n=1}^{\infty}(A_{mn}\cos\alpha\omega_{mn}t + B_{mn}\sin\alpha\omega_{mn}\ t)\sin\frac{m\pi}{a}x\sin\frac{n\pi}{b}y \tag{14.20}$$

The coefficients A_{mn}; B_{mn} are determined from the initial conditions (14.12), using the orthogonality of the eigenfunctions of $X_{mn}(x)$ and $Y_{mn}(y)$ in the intervals $0 \le x \le a$; $0 \le y \le b$.

Thus, the use of the method of separation of variables made it possible to present the solution of the problem in the form of a series in which each term is the product of three functions depending on x, y, and t.

14.1.2 Free Oscillations of a Circular Membrane

We will consider the free oscillations of a circular membrane. The differential equation of oscillations of a circular membrane in polar coordinates has the form:

$$\frac{\partial^2 w}{\partial r^2} + \frac{1}{r}\frac{\partial w}{\partial r} + \frac{1}{r^2}\frac{\partial^2 w}{\partial \theta^2} - \frac{1}{\alpha^2}\frac{\partial^2 w}{\partial t^2} = 0, \tag{14.21}$$

where r, θ are polar coordinates.

Using the method of separation of variables, we search for the solution in the form of:

$$W(r, \theta, t) = T(t)\ V(\ r)\Phi(\ \theta) \tag{14.22}$$

Substituting Eq. (14.22) into Eq. (14.21) and dividing by $T(t)\ V(\ r)\Phi(\ \theta)$, we will obtain:

$$\frac{V'' + \frac{1}{r}V'}{V} + \frac{1}{r^2}\frac{\Phi''}{\Phi} - \frac{1}{\alpha^2}\frac{T''}{T} = 0. \tag{14.23}$$

Denoting:

$$\frac{\Phi''}{\Phi} = -n^2; \quad \frac{T''}{\alpha^2 T} = -\omega^2.$$

And then we can write:

$$\Phi'' + n^2\Phi = 0$$

$$T'' + \alpha^2\omega^2 T = 0$$

$$V'' + \frac{1}{r}V' + \left(\frac{\omega^2}{\alpha^2} - \frac{n^2}{r^2}\right)V = 0 \tag{14.24}$$

Solutions of these equations are:

$$\begin{aligned} \Phi(\theta) &= A\cos n\theta + B\sin n\theta \ ; \\ T(t) &= C\sin \alpha\omega t + D\cos \omega\alpha t \ ; \\ V(r) &= E\mathrm{I}_n\left(\frac{\omega r}{\alpha}\right) + FY_n\left(\frac{\omega r}{\alpha}\right), \end{aligned} \tag{14.25}$$

where $\mathrm{I}_n\left(\frac{\omega r}{\alpha}\right)$; $Y_n\left(\frac{\omega r}{\alpha}\right)$ are, respectively, Bessel functions of nth order of the first and second kind. If a round membrane is considered that of the condition of periodicity on θ, it follows that coefficient n is an integer ($n = 0, 1, 2, \ldots$).

If we have a membrane having a sector shape in plan, the value of the coefficient is defined from the boundary conditions given on the edges $\theta = 0, \ \theta_1$. So, for example, if it is given that:

$$W(r, 0, t) = W(r, \theta_1, t) = 0 \tag{14.26}$$

using (14.22), (14.25), and (14.26) we find that

$$n = \frac{k\pi}{\theta_1}; \qquad (k = 0, 1, 2, 3\ldots) \tag{14.27}$$

Since n in this case is not a whole number, then the solution of third Eq. (14.24) is represented as:

$$V(r) = EI_n\left(\frac{\omega r}{\alpha}\right) + FI_{-n}\left(\frac{\omega r}{\alpha}\right) \tag{14.28}$$

Using the obtained particular solutions (14.25) as in the case of a rectangular membrane, it is possible to compose the general solution that can satisfy the boundary and initial conditions.

14.2 The Variational Methods

Let us apply the Hamilton's principle to the elastic oscillation body, according to which:

$$\delta\Gamma = \delta\int_{t_0}^{t_1}(\mathrm{U} - \mathrm{K})dt = 0, \tag{14.29}$$

where

U, K are potential and kinetic energy of the system. According to this principle, in the time interval between $t = t_0$ and $t = t_1$, the system displacements are such that a variation of δ integral is zero. In terms of static loads from condition (14.29), the following condition of the minimum of potential energy has the form:

$$\delta \mathrm{U} = 0 \quad \text{or} \quad \mathrm{U} = \min \tag{14.30}$$

The variation method provides a general prescription for improving any wave function with a parameter by minimizing that function with respect to the parameter.

The Hamilton Functional has the form for a bar transverse oscillations:

$$\Gamma = \frac{1}{2}\int_{t_0}^{t_1}\int_0^l \left[EI\left(\frac{\partial^2 y}{dx^2}\right)^2 - \rho(x)\left(\frac{\partial y}{\partial t}\right)^2\right] dxdt, \tag{14.31}$$

where E, ρ are the module of elasticity and the linear density of the rod/bar, respectively; I is inertia moment of transverse section of the rod/bar; l is length of the rod/bar; and y is transverse displacement of the rod /bar.

Consider the problem of free oscillations, taking:

$$y(x,t) = Y(y)\sin \omega t, \tag{14.32}$$

substituting in Eq. (14.31) and integrating within one period, we obtain:

$$\Gamma = \frac{\pi}{2\omega}\int_0^l \left[EI(Y'')^2 - \rho(x)\omega^2 Y^2\right] dx. \tag{14.33}$$

One of the commonly used variational methods is the Ritz method. The application of this method uses Hamilton's functional of the form (14.33). The variational problem in applying the Ritz method is reduced to search the extremum of function of many variables. The scheme for the solution of a problem of this method consists of the following: solution of the problem is sought in the form of:

$$y(x) = \sum_{i=1}^{n} a_i \varphi_i(x), \tag{14.34}$$

where $\varphi_i(x)$ are fundamental functions, satisfying the boundary conditions of the problem; and a_i are parameters, whose values are determined in the process of solving the problem. After substituting (14.34) in (14.33), the Hamilton functional becomes a function of the variables: $a_i (i = 1, 2, \ldots, n)$. The function's extremum conditions $L(a_i)$:

$$\frac{dL}{da_i} = 0, \quad i = 1, 2, \ldots n \tag{14.35}$$

give n equations, from which can be defined the values a_i .

In the case of transverse vibrations/oscillations of a beam, substituting (14.34) in (14.33), the functional (14.33) gives the following expression:

$$\Gamma(a_i) = \sum_{i=1}^{n}\sum_{k=1}^{n} W_{ik}a_i a_k - \omega^2 \sum_{i=1}^{n}\sum_{k=1}^{n} T_{ik}a_i a_k,$$

where

$$W_{ik} = \int_0^l EI\varphi_i'' \varphi_k'' dx,$$
$$T_{ik} = \int_0^l \rho\varphi_i\varphi_k dx. \tag{14.36}$$

Condition (14.35) take the form:

$$\sum_{k=1}^{n} \left[W_{ik} - \omega^2 T_{ik}\right] a_k = 0; \quad i = 1, 2, \ldots n \tag{14.37}$$

The homogeneous system of Eqs. (14.37) has a nonzero solution if the determinant comprised of the coefficients of unknowns is equal to zero. That brings to the equation of frequencies:

$$\left\| W_{ik} - \omega^2 T_{ik} \right\| = 0, \qquad i, k = 1, 2, \ldots\ n; \tag{14.38}$$

The special case of a method of Ritz is the method of Rayleigh by which the frequency of the main tone can be determined. Assuming, that $i = 1$, $k = 1$, from (14.38) we will obtain:

$$\omega_1^2 = \frac{W_{11}}{T_{11}} = \frac{\int_0^l EI\varphi_1''^2 dx}{\int_0^l \rho\varphi_1^2 dx}. \tag{14.39}$$

Expression (14.39) presents Rayleigh formula for transverse vibrations of a beam. In the case of longitudinal oscillations of a rod/bar, the frequency equation has the same form (14.34), except that the value of W_{ik} will be determined by the formula:

$$W_{ik} = \int_0^l EF\varphi_i'\varphi_k' dx \tag{14.40}$$

At torsional or shift oscillations of the bar in expression (14.40), modulus of rigidity at twisting and shifting GI_r, and kFG, should be used instead of rigidity of longitudinal deformation EF. Here G—modules of shear; I_r—polar moment of inertia; F—cross-sectional area; k—coefficient, considering uneven distribution of shearing stresses over the cross-section of the rod/bar [1, 2].

One of the most often applied variational methods is the method of Galerkin; let us express the essence of this method. Assume that differential equation of one-dimensional oscillations is given in the form of:

$$L\,(Y) = 0 \tag{14.41}$$

The solution to the problem we search for is in the form of Eq. (14.34). The essence of the method is the requirement that the function $L\,(Y)$ was orthogonal with all functions $\varphi_i(x)$; $i = 1, 2, \ldots, n$, that the condition was satisfied:

$$\int_0^l L\,(Y)\varphi_i(x) dx = 0, \qquad i = 1, 2, \ldots .n \tag{14.42}$$

If $Y(x)$ would be the exact solution of the original Eq. (14.41), then the conditions (14.42) would be satisfied identically. At approximate problem solving we make decision as (14.34) and substituting (14.42), we will obtain a system of n equations regarding functions a_i. Defining a_i we will obtain the desired displacement of the system. In the case of transverse oscillations of the rod/bar, the system (14.42) will have an appearance:

$$\sum_{k=1}^{n}\left[V_{ik}-\omega^2 T_{ik}\right]a_k=0, \qquad i=1,2,\ \ldots n; \tag{14.43}$$

where T_{ik} are defined by expression (14.36)

$$V_{ik}=\int_0^l \left(EI\varphi_i''\right)\varphi_k dx, \tag{14.44}$$

The respecting equation of frequencies has the form:

$$\left\|V_{ik}-\omega^2 T_{ik}\right\|=0 \qquad i=1,2,\ \ldots n; \tag{14.45}$$

If we compare the Ritz and Galerkin Methods, it should be noted that the range of the application of Galerkin method is wider since its application is not connected to the existence of Hamilton's type functional that is applied with the use of Ritz's method.

14.3 Integral Equations and Integral Transforms Methods

One of the most effective ways of solving boundary value problems is methods based on the use of integral equations. The problem of oscillations of rods, bars, plates, and shells can be reduced to integral equations of Volterra or Fredholm. We will show the methodology of reducing an ordinary differential equation to the integral equation of Volterra of the second kind [3].

Let us consider a differential equation of n order:

$$y^n(x)+a_1(x)y^{(n-1)}(x)+\cdots+a_n(x)y(x)=f(x) \tag{14.46}$$

where: $y^n(x)=\frac{d^n y}{dx^n}$; $a_1(x),\ \ldots a_n(x),\ \ f(x)$- given functions.

The equation of the type (14.46) reduces the consideration of different tasks, in particular, the problem about established vibrations/oscillations of the beam of variable section. This equation has the following form:

$$EI(x)y^{IV}(x)+2EI^{I}(x)y^{III}(x)+EI^{II}(x)y^{II}(x)-mF(x)\omega^2 y(x)=q(x), \tag{14.47}$$

where E is module of elasticity; $I(x)$ is moment of inertia, variable along the length; $F(x)$ is cross-sectional area; m is mass of uniform length; ω is circular frequency; and $q(x)$ is distributed dynamic load.

Dividing all members of (14.47) on $EI(x)$ we find that Eq. (14.47) is a special case of (14.46). In this case:

$$n=IV; \qquad a_1(x)=2\frac{I^{I}(x)}{I(x)}; \qquad a_2(x)=\frac{I^{II}(x)}{I(x)};$$
$$a_3(x)=0; \quad a_4(x)=-m\omega^2\frac{F(x)}{EI(x)}; \quad f(x)=\frac{q(x)}{EI(x)}.$$

Assuming that the initial conditions of Eqs. (14.46) are given as:

$$y(b) = y_0; \qquad y^I(b) = y_0{}^I; \qquad y^{n-1}(b) = y_0{}^{n-1}$$

Denote:

$$y^n(x) = \psi(x) \tag{14.48}$$

Integrating (14.48), we find

$$y^{(n-1)}(x) = \int_b^x \psi(\xi)\, d\xi + y^{(n-1)}(b) \tag{14.49}$$

Integrating (14.49), we obtain:

$$y^{(n-2)}(x) = \int_b^x \int_b^x \psi(\xi) d\xi^2 + y^{(n-1)}(b)(x-b) + y^{(n-2)}(b); \tag{14.50}$$

$$y^{(n-m)}(x) = \int_b^x \int_b^x \cdots \int_b^x \psi(\xi)\, d\xi^m + \sum_{k=1}^{m} y^{(n-k)}(b) \frac{(x-b)^{m-k}}{(m-k)!}; \tag{14.51}$$

$$y(x) = \int_b^x \int_b^x \cdots \int_b^x \psi(\xi) d\xi^n + \sum_{k=1}^{n} y^{(n-k)}(b) \frac{(x-b)^{n-k}}{(n-k)!}. \tag{14.52}$$

Substituting expressions (14.49–14.52) for $y(x)$ into Eq. (14.46), and using the formula of multiple integrals to the sample

$$\int_b^x \int_b^x \cdots \int_b^x \psi(\xi) d\xi^m = \int_b^x \frac{(x-\xi)^{m-1}}{(m-1)!} \int_b^x \psi(\xi) d\xi, \tag{14.53}$$

we will obtain:

$$\psi(x) + \int_b^x \sum_{m=1}^{n} a_m(x) \frac{(x-\xi)^{m-1}}{(m-1)!} \psi(\xi) d\xi = F(x). \tag{14.54}$$

where

$$F(x) = f(x) - \sum_{m=1}^{n} \sum_{k=1}^{m} y^{(n-k)}(b) a_m(x) \frac{(x-b)^{(m-k)}}{(m-k)!}.$$

Designating

$$\sum_{m=1}^{n} a_m(x) \frac{(x-\xi)^{m-1}}{(m-1)!} = K(x, \xi).$$

Equation (14.54) reduces to the integral equation of Volterra of the second kind.

$$\psi(x) + \int_b^x K(x, \xi) \psi(\xi) d\xi = F(x) \tag{14.55}$$

The solution of Eq. (14.55) can be found as an infinite series, using known methods of solving the integral equations of Volterra of the second kind.

Let us consider oscillation of the beam with a constant cross-section. The equation of oscillations has the form (14.47), where the following is: $I^{I} = I^{II} = 0$. The motion of this beam can be described by a Fredholm integral equation of the second kind in the following form:

$$y(x) = \int_0^l \left[mF\omega^2 y(\xi) + q(\xi)\right] G(x,\xi)d\xi, \tag{14.56}$$

where $G(x, \xi)$ is Green's function or influence function, representing displacement in the cross-section x from static unit single concentrated load applied transversely at section ξ. Eq. (14.56) can be obtained from the following considerations. To determine the displacement in cross-section x, Green's function is multiplied by the values of the inertia forces and external loads of those acting on the elementary length and summarized—integrated over the entire length of the beam. The Green's function $G(x, \xi)$ for a beam simply supported at each end has the form:

$$G(x,\xi) = \begin{cases} -\dfrac{(l-\xi)x}{6lEI}\left(x^2 - 2\xi l + \xi^2\right), & 0 \le x \le \xi \le l \\ -\dfrac{(l-x)\xi}{6lEI}\left(\xi^2 - 2xl + x^2\right); & 0 \le \xi \le x \le l \end{cases} \tag{14.57}$$

From expression (14.57), it follows that $G(x, \xi) = G(\xi, x)$, which is the consequence of the reciprocity theorem of the displacement (Betty Theorem). For a cantilever beam fixed in the cross-section $x = 0$, the Green's function has the form:

$$G(x,\xi) = \begin{cases} -\dfrac{x^2(3\xi - x)}{6EI}; & 0 \le x \le \xi \le l \\ -\dfrac{\xi^2(3x - \xi)}{6EI} & 0 \le \xi \le x \le l \end{cases} \tag{14.58}$$

The equations of type (14.56) can be composed for plates and shells. In these cases, the motion will be described by the two-dimensional equations of Fredholm of the second kind.

In applied mathematics, the use of integral decomposition for the solution of various differential and integral equations is called the method of integral transform. Fourier's integral is the simplest and the most widely used integral decomposition

$$f(x) = \frac{1}{2\pi}\int_{-\infty}^{\infty} e^{-ix\xi}\int_{-\infty}^{\infty} e^{i\xi\lambda} f(\lambda)d\lambda d\xi. \tag{14.59}$$

If function $f(x)$ is integrated in the interval $(-\infty, \infty)$ and any finite interval has finite variation, then the right part of formula (14.59) is equal to $f(x)$ at each point of continuity of this function, and is equal to $1/2[f(x + 0) + f(x - 0)]$ at each point of discontinuity. We will designate

$$\frac{1}{2\pi}\int_{-\infty}^{\infty} e^{i\xi x} f(x)dx = F(\xi). \tag{14.60}$$

We can write the integral expression (14.59) of function $f(x)$ in the following form:

$$f(x) = \frac{1}{2\pi}\int_{-\infty}^{\infty} e^{-i\xi x} F(\xi)d\xi. \tag{14.61}$$

The function $F(\xi)$ is called integral Fourier transform or transform of function $f(x)$, and expression (14.61) by the inversion formula for the integral Fourier transform [4]. Formulas (14.60) and (14.61) call dual or considering that $f(x)$ and $F(\xi)$ are Fourier's transform of each other and also are formulas of inversion. If $f(x)$ is an even function, then formulas (14.60) and (14.61) will take the form:

$$F(\xi)=\sqrt{\frac{2}{\pi}}\int_0^{\infty}\cos\xi x f(x)dx;$$
$$f(x)=\sqrt{\frac{2}{\pi}}\int_0^{\infty}\cos\xi x F(\xi)d\xi. \tag{14.62}$$

Formula (14.62) define a Fourier cosine-transform. Another transformation is Fourier sine-transform obtained from (14.60) and (14.61) with the odd $f(x)$.

$$F(\xi)=\sqrt{\frac{2}{\pi}}\int_0^{\infty}f(x)\sin\xi x dx\ ;$$
$$f(x)=\sqrt{\frac{2}{\pi}}\int_0^{\infty}F(\xi)\sin\xi x d\xi\ . \tag{14.63}$$

14.3.1 Application of the Method of Integral Transform on the Example of Transfers Vibrations of an Infinite Rod/Bar

Let us examine the application of a method of integral transform to the solution of PDEs based on the example problem about the lateral oscillations of infinite bar. The equation of oscillations of a rod/bar can be written as:

$$EI\frac{\partial^4 w}{\partial x^4}+m\frac{\partial^2 w}{\partial t^2}=q(x,t), \tag{14.64}$$

where EI is stiffness of the rod / bar; m is mass per unit length; q is external load; w is deflection.

Boundary and initial conditions are:

$$w(\pm\infty,t)=\frac{\partial w}{\partial x}(\pm\infty,t)=\ \ .\ .\ .\ =0;$$
$$w(x,0)=\varphi(x),\quad \frac{\partial w}{\partial t}(x,0)=\psi(x). \tag{14.65}$$

Let us multiply the left and right parts of Eq. (14.64) to $(2\pi)^{-1/2}e^{i\xi x}$ and will integrate over x from $-\infty$ to $+\infty$, we will obtain

$$\frac{EI}{\sqrt{2\pi}}\int_{-\infty}^{\infty}\frac{\partial^4 w}{\partial x^4}e^{i\xi x}dx+\frac{m}{\sqrt{2\pi}}\int_{-\infty}^{\infty}\frac{\partial^2 w}{\partial x^2}e^{i\xi x}dx=\frac{1}{\sqrt{2\pi}}\int_{-\infty}^{\infty}q(x,t)e^{i\xi x}dx. \tag{14.66}$$

We will designate the transforms

$$W(\xi,t)=(2\pi)^{-1/2}\int_{-\infty}^{\infty}w(x,t)e^{i\xi x}dx;$$
$$Q(\xi,t)=(2\pi)^{-1/2}\int_{-\infty}^{\infty}q(x,t)e^{i\xi x}dx.$$

Expression (14.66) integrates by (in) parts, and in the result, we will find

$$\frac{EIe^{i\xi x}}{\sqrt{2\pi}}\left[\frac{\partial^3 w}{\partial x^3} - i\xi\frac{\partial^2 w}{\partial x^2} - \xi^2\frac{\partial w}{\partial x} + i\xi^3 w\right]_{-\infty}^{\infty} + EI\xi^4 W(\xi,t) + m\frac{d^2 W}{dt^2} = Q(\xi,t). \tag{14.67}$$

Using boundary conditions (14.65), we have

$$\frac{\partial^2 W}{\partial t^2} + \frac{EI\xi^4}{m}W = \frac{Q}{m}. \tag{14.68}$$

Thus, with an aid of the Fourier transformation, the equation in the partial derivatives (14.64) is reduced to the ordinary differential equation. After determination from Eq. (14.68) of transform $W(\xi, t)$, we will obtain the solution to the problem according to the inversion formula (14.61)

$$w(x,t) = \frac{1}{\sqrt{2\pi}}\int_{-\infty}^{\infty} W(\xi,t)e^{-i\xi x}dx. \tag{14.69}$$

Note that for solving Eq. (14.68) it is necessary to enter Fourier's transforms for the functions, entering in the initial conditions:

$$W(\xi,0) = \frac{1}{\sqrt{2\pi}}\int_{-\infty}^{\infty}\varphi(x)e^{i\xi x}dx = \Phi(\xi);$$
$$\frac{dW}{dt}(\xi,0) = \frac{1}{\sqrt{2\pi}}\int_{-\infty}^{\infty}\psi(x)e^{i\xi x}dx = \Psi(\xi).$$

This is the basic scheme for solving differential equations by the method of integral transformations. The multiplication of the left and right parts of the equation in a certain function (kernel of integral transform) and further integration of domain by changing the variable, which is excluded from the equation as a result of the application of transformation, is characteristic of this scheme. Sometimes under the integral transformations the classical method of Fourier based on finding solutions of problems in a form of the integral of Fourier's type, with unknown density, determined as a result of the substitution of integral into the initial differential equation is described. As we see, both methods are equivalent; however, the first one, which is the common method of integral transformations, is more convenient, especially for solutions of nonhomogeneous equations with inhomogeneous boundary conditions.

Thus, the method of integral transformations is based on the decomposition of the solution into an integral by a certain set of functions, chosen based on a form of differential equation, variable range within domain, and boundary conditions. In addition to the above-described Fourier transforms, applied mathematics uses other integral transformation; in particular the transforms of Laplace, Mellin, Hankel, Melera–Foka, Meyer, Kontorovich–Lebedev, and Weber. All these transforms are introduced on basis of the corresponding integral decompositions. Since the transforms for them are obtained as a result of the integration in the infinite intervals, they are called transforms with infinite limits. On the other hand, it is possible to introduce the integral transforms based on the decomposition in series. If the functions $f(x)$ of a certain class are presented in the form of series of orthogonal system $\{u_n(x)\}$, i.e.,

$$f(x) = \sum_n F_n u_n(x) \qquad (a \le x \le b), \tag{14.70}$$

where the coefficients of decomposition F_n are determined from formula

$$F_n = \int_a^b f(x)u_n(x)dx, \tag{14.71}$$

Of course such interpretation does not add anything to the possibilities of common method of series or method of separation of variables; however, the procedure itself of determination of transforms, i.e., the coefficients of the series, is typical of the conventional method of integral transforms and is more convenient. These advantages are especially evident in the solutions of problems with the inhomogeneous boundary conditions. The specified transformations are called transformations with final limits Sneddon [5], or final transformations, since in (14.71) integration extends to the finite interval. Division into the transformations with finite and infinite limits conditionally and does not reflect the principle of the matter. The domain of change of the variable in terms of which transformation is achieved is not essential here and consequently neither is the interval of integration, but the nature of inversion formulas is determined by the form of the assumed as basis method of decomposition discrete or continuous is of importance. Exceptionally continuous integral decompositions were used in the classical method of integral transformation. However, the tendency to expand the circle solved by this method's problems led to the start in the integral transformations of discrete and discrete-continuous decompositions (in series). The general theory of integral transformations, which covers from unified point of view an entire diversity of the questions, connected with the substantiation, by application, and by the development of method, is not yet completed. The most general approach in this sense is associated with the use of a spectral theory of linear operators. If necessary, this approach is especially effective in the building of new transformations and corresponding integral decompositions.

14.4 The Finite Element Method

The finite element method (FEM) is a method of analysis in which a structure is discretized into elements connected at nodes; the shape of the element displacement field is assumed, partial or complete compatibility is maintained among the element interfaces, and nodal displacements are determined by using energy variational principles or equilibrium methods.

During application of the method of finite elements, a continuous medium is broken into subdomains called finite elements. In the case of a two-dimensional problem, the finite elements take the form of a triangle or rectangle; in a three-dimensional case—the type is that of a tetrahedron or rectangular prism. Depending on the geometry of the problem, finite elements can have a more complicated (complex geometry) form. The breakdown of the medium into finite elements substitutes the system with an infinite number of degrees of freedom with a system with a finite number of degrees of freedom. The displacements of the nodes of the final elements are taken as the unknown.

The function, which determines displacements inside the finite element, expressed through the displacements of nodal points is selected. Knowing the nodes of displacements we can determine both the strains and stresses within the limits of final element and on its boundaries. To determine the nodes of displacements, either the condition of the equilibrium of the nodes of finite elements or condition of the minimum of the functional of the corresponding variational problem must be adapted.

It should be noted that historically one of the first prototypes of the application of a method of finite elements is the method of deformations (displacements) of the calculation of plane frame structures to static load, and subsequently to stability and oscillations/vibrations. The finite elements in this case are beams and columns of the frame structure. The angular and linear displacements of the frame nodes are unknowns. In this particular case, the function describing deformations of a finite element is defined precisely (within the theory of rods/bars), and therefore this method gives the exact solution of the problem.

Let us proceed to the derivation of the equations of finite elements method using the example of the plane problem of the theory of elasticity. Displacement at any point of the finite element |*f*| is a function of coordinates *x*, *y*, and is determined as a vector:

$$|f| = \begin{vmatrix} U(x,y) \\ V(x,y) \end{vmatrix} \tag{14.72}$$

U(*x*, *y*), *V*(*x*, *y*) are horizontal and vertical displacements of the point with coordinates *x*, *y*. The displacement at any point inside the triangular finite element is defined as:

$$|f| = |N||\delta|^e = |N_{i,}N_{j,}N_{m,}| \begin{vmatrix} \delta_i \\ \delta_j \\ \delta_m \end{vmatrix}, \tag{14.73}$$

where

$|\delta|^e$ is matrix of the displacements of nodal points at any point inside the triangular finite element with the index "e,"

δ_i is displacement of node "*i*" of finite element,

$$\delta_i = \begin{vmatrix} U_i \\ V_i \end{vmatrix};$$

U_i, V_i are projections of the displacement of node "*i*" in the directions of *x* and *y*; and

|*N*| is matrix, which characterizes the distribution of displacements in the finite element.

In order to obtain the concrete type of this matrix, we assume a certain distribution of displacements within the finite element. The linear polynomials are the simplest representation of displacements [6].

$$\begin{aligned} U &= \alpha_1 + \alpha_2 x + \alpha_3 y \\ V &= \alpha_4 + \alpha_5 x + \alpha_6 y \end{aligned} \tag{14.74}$$

If we substitute into the first equation of the coordinates of the points *i*, *j*, and *m*, then we will obtain:

$$\begin{aligned} U_i &= \alpha_1 + \alpha_2 x_i + \alpha_3 y_i \\ U_j &= \alpha_1 + \alpha_2 x_j + \alpha_3 y_j \\ U_m &= \alpha_1 + \alpha_2 x_m + \alpha_3 y_m \end{aligned} \tag{14.75}$$

Determining α_1, α_2, and α_3 from this system and substituting in the first Eq. (14.74), we will obtain:

$$U = \frac{1}{2\Delta}\left\{(a_i + b_i x + c_i y)U_i + \left(a_j + b_j x + c_j y\right)U_j + (a_m + b_m x + c_m y)U_m\right\} \tag{14.76}$$

where

$$\begin{aligned} a_i &= x_j y_m - x_m y_j \\ b_i &= y_j - y_m, \\ c_i &= x_m - x_j \end{aligned} \tag{14.77}$$

$$2\Delta = \begin{vmatrix} 1 & X_i & Y_i \\ 1 & X_j & Y_j \\ 1 & X_m & Y_m \end{vmatrix} \tag{14.78}$$

Similarly, it is possible to obtain the dependence of the vertical displacement V using the displacements of the nodes of the finite element.

$$V = \frac{1}{2\Delta}\left\{(a_i + b_i x + c_i y)V_i + \left(\alpha_j + b_j x + c_j y\right)V_j + (\alpha_m + b_m x + c_m y)V_m\right\} \tag{14.79}$$

Relationships (14.76) and (14.79) can be represented in the matrix form:

$$|f| = \begin{vmatrix} U \\ V \end{vmatrix} = |N||\delta|^e = \left[\mathrm{I}N_i^I, \mathrm{I}N_j^I, \mathrm{I}N_m^I\right]|\delta|^e$$
$$N_i^I = \frac{a_i + b_i x + c_i y}{2\Delta} \tag{14.80}$$

The values N_j^I, N_m^I are obtained by replacement of index i with j and m.

I is the unit matrix of the dimensionality of 2 x 2.

Using the known displacements (14.80), it is possible to determine the values of the deformation in dependence on the displacement of nodes. This dependence can be presented in the matrix form:

$$|\varepsilon| = |B|\ |\delta|^e = \left|B_i\ B_j\ B_m\right|\ |\delta|^e. \tag{14.81}$$

For definition of specific type of the matrix |*B*|, we use known relationships [6]:

$$|\varepsilon| = \begin{vmatrix} \varepsilon_x \\ \varepsilon_y \\ \gamma_{xy} \end{vmatrix} = \begin{vmatrix} \dfrac{\partial U}{\partial x} \\ \dfrac{\partial V}{\partial y} \\ \left(\dfrac{\partial U}{\partial y} + \dfrac{\partial V}{\partial x}\right) \end{vmatrix} \tag{14.82}$$

Substituting (14.80) to (14.82), we will obtain:

$$|\varepsilon| = \frac{1}{2\Delta}\begin{vmatrix} b_i & 0 & b_j & 0 & b_m & 0 \\ 0 & c_i & 0 & c_j & 0 & c_m \\ c_i & b_i & c_j & b_j & c_m & b_m \end{vmatrix} |\delta|^e = \frac{1}{2\Delta}\left|B_i\ B_j\ B_m\right|\ |\delta|^e = |B|\ |\delta|^e, \tag{14.83}$$

$$b_i = \frac{\partial N_i^I}{\partial x}; \qquad c_i = \frac{\partial N_i^I}{\partial y}$$

b_j, c_j, b_m, and c_m are obtained by the replacement of index i with the index j, and m.

Under the plane strain conditions, the relationship between the strains and stresses is the following [6]:

$$\varepsilon_x = \frac{1}{E}\left(\sigma_x - \nu\sigma_y - \nu\sigma_z\right), \qquad \varepsilon_y = \frac{1}{E}\left(-\nu\sigma_x + \sigma_y - \nu\sigma_z\right),$$
$$\gamma_{xy} = \frac{2(1+\nu)}{E}\sigma_{xy}, \qquad \varepsilon_z = \frac{1}{E}\left(-\nu\sigma_x - \nu\sigma_y + \sigma_z\right), \tag{14.84}$$

Let us show the relationship between the stresses and strains in a matrix form. Excluding σ_z, we will determine the dependence of stresses from the deformations:

$$|\sigma| = |D|\ |\varepsilon|$$
$$\begin{vmatrix} \sigma_x \\ \sigma_y \\ \sigma_z \end{vmatrix} = |D| \begin{vmatrix} \varepsilon_x \\ \varepsilon_y \\ \gamma_{xy} \end{vmatrix} \tag{14.85}$$

where:

$$|D| = \frac{E(1-\nu)}{(1+\nu)(1-2\nu)} \begin{vmatrix} 1 & \frac{\nu}{1-\nu} & 0 \\ \frac{\nu}{1-\nu} & 1 & 0 \\ 0 & 0 & \frac{1-2\nu}{2(1-\nu)} \end{vmatrix}$$

The simplest and most efficient way to compose the equations of the method of finite elements is to use the virtual work principle. According to the virtual work principal, with the arbitrary nodal displacement, the sum of the external and internal work accomplished by different forces and stresses on this displacement is equal to zero. Suppose that $d|\delta|^e$ is a virtual displacement of node, then (14.73) and (14.81) can be presented in the form:

$$d|f| = |N|d\ |\delta|^e, \quad d|\varepsilon| = |B|d\ |\delta|^e. \tag{14.86}$$

The work of nodal forces is equal to the product of each force component on the corresponding displacement, that is:

$$(d\ |\delta|^e)^T\ |F|^e \tag{14.87}$$

Here through $(d\ |\delta|^e)^T$—the corresponding transposed matrix is designated.

The work of internal forces is equal to the work of stresses and distributed forces $|P|$ per unit volume:

$$d|\varepsilon|^T|\sigma| - d|f|^T|P| \tag{14.88}$$

Having (14.81) and (14.73) and using a formula of the transposition of the matrix product

$$(|A|\ |B|)^T = |B|^T\ |A|^T \tag{14.89}$$

we can present (14.88) in the form of:

$$(d\ |\delta|^e)^T\ (|B|^T|\sigma| - |N|^T|P|) \tag{14.90}$$

Equating the work of external forces (14.87) to the work of internal forces, which is obtained by integrating (14.90) by the volume, we obtain:

$$(d\ |\delta|^e)^T\ |F|^e = d\ |\delta|^e)^T\left(\int |B|^T|\sigma|dV - \int |N|^T|P|dV\right) \tag{14.91}$$

Since this equation is correct at any arbitrary displacement, we can equate the expressions before $d\ |\delta|^e$ of the left and right parts of Eq. (14.91). Then substituting (14.81) and (14.85), we obtain:

$$|F|^e = \left(\int |B|^T |D| |B| dV \right) |\delta|^e - \int |N|^T |P| dV \tag{14.92}$$

Eq. (14.92) is the condition of equality of the external and internal forces acting on the finite element with the number "*e*". Equation (14.92) can be represented as:

$$|F|^e = |K|^e\ |\delta|^e + |F|_p^e \tag{14.93}$$

where

$|K|^e$ is stiffness matrix. The types of matrix $|K|^e$ and $|F|_p^e$ are the same, comparing (14.93) with (14.92). In the expression of the works of internal forces, sometimes the influence of initial deformations and stresses is considered as well. In that case (14.93) is written in the form of:

$$|F|^e = |K|^e\ |\delta|^e + |F|_p^e + |F|_{\varepsilon_0}^e + |F|_{\sigma_0}^e \tag{14.94}$$

where:

$$\begin{aligned} |K|^e &= \int |B|^T |D| |B| dV, \\ |F|_p^e &= -\int |N|^T |P| dV, \\ |F|_{\varepsilon_0}^e &= -\int |B|^T |D| |\varepsilon_0| dV, \\ |F|_{\sigma_0}^e &= \int |B|^T |\sigma_0| dV. \end{aligned} \tag{14.95}$$

$|\varepsilon_0|$, $|\sigma_0|$ are matrixes, which characterize initial strains and initial stresses [6].

Thus, the solution of the static problem of the theory of elasticity by the method of finite elements is reduced to the composition of the equations of equilibrium for each finite element of the following form:

$$|F|^e = 0; \qquad e = 1, 2, \ldots . \tag{14.96}$$

After the determination of the nodal displacements $|\delta|^e$, $e = 1, 2, \ldots$, displacements and stresses can be determined by the formulas given above.

For obtaining the equations of dynamic equilibrium in expression (14.92) instead of $|P|$ accept the following:

$$\overline{|P|} - \rho \frac{\partial^2}{\partial t^2} |f| - \mu \frac{\partial}{\partial t} |f| \tag{14.97}$$

where ρ is mass of unit volume, μ is coefficient accounting for damping according to the hypothesis of a linear damping (Voigt model). Substituting (14.97) in (14.92) and taking into account (14.96) we obtain:

$$|K|^e\ |\delta|^e + |C|^e \frac{\partial}{\partial t} |\delta|^e + |M|^e \frac{\partial^2}{\partial t^2} |\delta|^e + \overline{|F|}^e = 0 \tag{14.98}$$

where

$$|C|^e = \mu \int |N|^T |N| dV; \quad |M|^e = \rho \int |N|^T |N| dV; \quad \overline{|F|}^e = -\int |N|^T \overline{|P|} dV.$$

$|C|^e$ and $|M|^e$ – the matrixes of damping and masses of the element, respectively.
According to Eq. (14.80), we have:

$$|N| = \left[\mathrm{I}N_i^I, \mathrm{I}N_j^I, \mathrm{I}N_m^I \right] \tag{14.99}$$

where:

$$\mathrm{I} = \begin{vmatrix} 1 & 0 \\ 0 & 1 \end{vmatrix}$$

The matrix $|M|^e$ can be represented in the form:

$$|M|^e = \rho t\, \mathrm{I} \iint |N|^T |N| dxdy \tag{14.100}$$

or

$$|M|^e = \frac{\rho t \Delta}{3} \begin{vmatrix} c & d & d \\ d & c & d \\ d & d & c \end{vmatrix},$$

where c, d are sub matrixes expressed in the following forms:

$$c = \begin{vmatrix} 1/2 & 0 \\ 0 & 1/2 \end{vmatrix}; \quad d = \begin{vmatrix} 1/4 & 0 \\ 0 & 1/4 \end{vmatrix},$$

t is thickness of a plate, Δ is area of a triangle.
The method of finite elements is successfully applied in nonstationary problems as well.

14.4.1 Nonstationary Vibrations of Cantilever Plate

Let us consider the nonstationary vibrations of cantilever plate, which is either a plane stress or plane strain condition. If the low of vibration/accelerations of the ground $y_0^{II}(t)$ is known, then vibration of the system can be represented as:

$$|M| \frac{\partial^2}{\partial t^2} |\delta| + |C| \frac{\partial}{\partial t} |\delta| + |K||\delta| = -|M| y_0^{II}(t). \tag{14.101}$$

Next the problem of free oscillations of the system is considered, that is, the system of the following type is solved:

$$\left(|K| - \omega^2 |M| \right) |\delta_0| = 0 \tag{14.102}$$

and the natural frequencies of the system ω_i, $i = 1, 2 \ldots$, and their modes of oscillation $|\delta_0|_i$, $i = 1, 2 \ldots$ are determined.

Any nonstationary motion of the system can be sought in the form:

$$|\delta| = \left[|\delta_0|_1, \ |\delta_0|_2, \ \ldots \ |\delta_0|_n \right] \ |q(t)| = |\Delta_0| \ |q|, \tag{14.103}$$

wherein $|\Delta_0|$ includes the normalized eigenfunctions, and $|q|$—the generalized coordinates to be determined. Substituting (14.103) in (14.101) and multiplying all members of the equation by $|\Delta_0|^T$, we will obtain:

$$\begin{gathered}|\Delta_0|^T|M|\,|\Delta_0|\frac{\partial^2}{\partial t^2}\,|q|+|\Delta_0|^T|C|\,|\Delta_0|\,\frac{\partial}{\partial t}|q|+\\+|\Delta_0|^T|K|\,|\Delta_0||q|=-|\Delta_0|^T|M|y_0^{II}\end{gathered} \tag{14.104}$$

For eigenfunctions $|\delta_0|_i$, the property of orthogonality takes place [6, 7]:

$$|\delta_0|_i^T|M||\delta_0|_j=\begin{cases}0 & i\neq j\\ 1 & i=j\end{cases} \tag{14.105}$$

From (14.102) follows that

$$|K||\delta_0|_i=\omega_i^2\;|M||\delta_0|_i$$

Hence formula (14.105) can be presented also in the form:

$$|\delta_0|_i^T|K||\delta_0|_j=\begin{cases}0 & i\neq j\\ \omega_i^2 & i=j\end{cases} \tag{14.106}$$

Since the matrix $|C|$ is proportional to $|M|$, we have orthogonality of eigenfunctions of the following form:

$$|\delta_0|_i^T|C||\delta_0|_j=\begin{cases}0 & i\neq j\\ 2n_i\omega_i^2 & i=j\end{cases} \tag{14.107}$$

Using conditions (14.105), (14.106), and (14.107) leads the system of ordinary differential equations to the following form:

$$\frac{\partial^2}{\partial t^2}q_i+2n_i\omega_i\frac{d}{dt}q_i+\omega_i^2\;q_i=-|\delta_0|_i^T|M|y_0^{II} \tag{14.108}$$

The solution is:

$$q_i=-|\delta_0|_i^T|M|\int\limits_0^t y_0^{II}(\xi)\;e^{-n_i\omega_i(t-\xi)}\sin\omega_i(t-\xi)d\xi \tag{14.109}$$

Then the displacement of any nodal point of the structure can be determined using (14.103).

14.5 The Finite Difference Method

The finite difference method (FDM) is a method of analysis in which the governing differential equation is satisfied at discrete points of a structure. The finite difference approximations for derivatives are one of simplest and oldest methods of solving differential equations. The essence of the finite difference method consists in the replacement of the derivatives in the differential equation with finite differences, using differential quotients, as a result of which these equations are reduced to the system of linear algebraic equations.

14.5.1 Application of the Finite Difference Method on the Example of Transverse Vibrations of a Beam

We will consider application of the finite difference method on the example of transverse oscillation of a compressed beam of a constant cross-section. The differential equation of free oscillation of such beam is given by:

$$\frac{\partial^4 y}{\partial x^4} + \lambda \frac{d^2 y}{dx^2} - \alpha^2 y = 0 \tag{14.110}$$

$$\lambda = \frac{N}{EI}; \quad \alpha^2 = \frac{\omega^2 \rho}{EI}$$

where y is displacement, N is normal compressive force.

We will divide the beam into n equal parts.

$$l = n \cdot \Delta x; \qquad \Delta x = \frac{l}{n} = l\Delta\xi \ ; \qquad \Delta\xi = \frac{1}{n}.$$

Derivatives for point k will replace with the ratio of differences:

$$\begin{aligned} \frac{\partial^4 y}{\partial \xi^4} \to \frac{\Delta^4 y}{\Delta \xi^4} \Bigg|_k &= n^4 \left(y_{k-2} - 4y_{k-1} + 6y_k - 4y_{k+1} + y_{k+2}\right) \\ \frac{\partial^2 y}{\partial \xi^2} \to \frac{\Delta^2 y}{\Delta \xi^2} \Bigg|_k &= n^2 \left(y_{k-1} - 2y_k + y_{k+1}\right); \quad \Delta\xi = \frac{x}{l} \end{aligned} \tag{14.111}$$

Substituting (14.111) in (14.110) gives:

$$\begin{aligned} & y_{k-2} - y_{k-1}\left(4 - \frac{\lambda}{n^2}\right) + y_k\left(6 - \frac{2\lambda}{n^2} - \frac{\alpha^2}{n^4}\right) - \\ & - y_{k+1}\left(4 - \frac{\lambda}{n^2}\right) + y_{k+2} = 0; \quad k = 1, 2, \ldots, n-1 \end{aligned} \tag{14.112}$$

For the hinge-supported beam, boundary conditions take the form:

$$y(0) = Y(l) = 0; \qquad \frac{\partial^2 y}{\partial \xi^2}(0) = \frac{\partial^2 y}{\partial \xi^2}(l) = 0. \tag{14.113}$$

Dividing the beam into n parts, the point $n = 0$ corresponds to the left support, and the point—n corresponds to the right support. The boundary conditions (14.113), considering (14.111), will take the form:

$$y_0 = 0 \ ; \quad y_n = 0; \quad y_{-1} - 2y_0 + y_1 = 0; \quad y_{n-1} - 2y_n + y_{n+1} = 0; \tag{14.114}$$

Assuming that $n = 5$, then $n = 1, 2, 3, 4$. Now, let us compose Eqs. (14.112), considering (14.114):

$$\left(5-\frac{2\lambda}{n^2}-\frac{\alpha^2}{n^4}\right)y_1-\left(4-\frac{\lambda}{n^2}\right)y_2+y_3=0$$
$$-\left(4-\frac{\lambda}{n^2}\right)y_1+\left(6-\frac{2\lambda}{n^2}-\frac{\alpha^2}{n^4}\right)y_2-\left(4-\frac{\lambda}{n^2}\right)y_3+y_4=0$$
$$y_1-\left(4-\frac{\lambda}{n^2}\right)y_2+\left(6-\frac{2\lambda}{n^2}-\frac{\alpha^2}{n^4}\right)y_3-\left(4-\frac{\lambda}{n^2}\right)y_4=0 \tag{14.115}$$
$$y_2-\left(4-\frac{\lambda}{n^2}\right)y_3+\left(5-\frac{2\lambda}{n^2}-\frac{\alpha^2}{n^4}\right)y_4=0$$

Equating the determinant of the coefficients y_1, y_2, y_3, and y_4 to zero, we obtain the equation of the fourth degree with respect to α^2. The solution of the equation gives: four values of the frequencies of free oscillations of a beam. In this case accurate solution of Eq. (14.112) can be obtained in the form of:

$$y_k = A\sin\beta k, \qquad k = 1, 2, \ldots n-1. \tag{14.116}$$

This satisfies the boundary conditions at the left end of the beam, i.e., the following conditions:

$$y_0 = 0, \qquad y_1 = -y_{-1};$$

Now, let us satisfy the boundary conditions at the right end of the beam:

$$y_n = 0, \qquad y_{n+1} = -y_{n-1}.$$

We will obtain:

$$\sin\beta n = 0; \qquad \beta = \frac{\pi i}{n}; \qquad i = 1, 2, \ldots, n$$

and therefore:

$$y_k = A\ \sin\frac{\pi i}{n}k \tag{14.117}$$

Substituting (14.117) in (14.112), we will obtain:

$$\omega_i = \frac{n^2}{l^2}\sqrt{\frac{EI}{\rho}}\cdot\sqrt{4\left(1+\cos\frac{\pi i}{n}\right)^2-\frac{2\lambda}{n^2}\left(1+\cos\frac{\pi i}{n}\right)} \tag{14.118}$$
$$i = 1, 2, 3\ldots \qquad n = 1, 2\ldots$$

Using this expression, we can determine the frequency of the ith tone (mode) of the beam oscillations, which is divided into n parts. When $n \to \infty$, the oscillation frequency ω_i tends to an exact number. Alternatively, with $\lambda = 0$, the smaller is n the less accurate is the value of the oscillation frequency ω_i. With $\lambda = 0$, expression (14.118) for the beam divided into five parts $n = 5$ provides a 5% error in determination of the frequency of fundamental tone ω_1.

14.6 The Generalized Functions in Structural Mechanics

In the various problems of mechanics and physics, it is necessary to use functions whose analytical description is not possible within the framework of classical applied analysis. The simplest examples of such functions are concentrated force, moment, point source, etc. In particular, the concentrated force is defined as the final load. The load intensity is assumed to be zero everywhere except at the point where it takes on an infinite value. The properties of functions of this kind are not known. In the framework of classical mathematical analysis, it is impossible to choose such a function. Therefore, the concentrated force is replaced by a distributed force over a finite interval. Then, in solving the problem, a limit transition is carried out. In modern engineering practice, this method is considered time-consuming and inefficient.

These difficulties disappeared with the construction of the theory of generalized functions. The theory of generalized functions allowed us to solve some fundamental problems. For example, various operations with divergent integrals and series, differentiation of discontinuous functions, and changing the order of integration and summation have received a strict (rigorous) mathematical foundation. Generalized functions play an important role when using (in the use of) the method of integral transformations. Since various operations with divergent integrals and series should be understood only in the sense of generalized functions in the method of integral transformations. The main results of the theory of generalized functions, which have the greatest application in problems of structural engineering and structural mechanics, are briefly given below.

Here, we will limit (restrict) ourselves to a minimum of mathematical definitions, and present the basic properties of generalized functions, as well as some useful values of integrals and series expansions in applications. This section provides basic information from the theory of generalized functions (Tseitlin [8], Tseitlin and Petrosian [9], and Petrosian [10]). The reader can find more detailed information about generalized functions and actions on them in the following works [11, 12].

Generalized functions do not have values at certain (individual) points and are characterized integrally, unlike classical ones. Generalized functions are understood as linear continuous functionals f acting over a certain space of basic functions. Each regular function corresponds to a functional in the space of basic functions. As the basic (main) functions, we take real-valued functions $\varphi(x)$ that have continuous derivatives of all orders and turn to zero outside of some limited domain. A generalized function, that is, a functional f assigns (sets) a certain number (f, φ), to each basic function $\varphi(x)$. In this case, the following conditions must be met:

(a) for any numbers α_1, α_2 and any basic functions φ_1, φ_2 there is equality $(f, \alpha_1 \varphi_1 + \alpha_2 \varphi_2) = \alpha_1(f,$ f, α_1 φ_1 + α_2 φ_2) = α_1(f, φ_1) + α_2(f, φ_2) *(functional linearity* f *);*
(b) if the sequence of basic functions $\varphi_n(x)$ converges to zero, then $\lim_{n\to\infty}(f, \varphi_n) = 0$ (Continuity of functional).

For example, a delta function is a functional $\delta(x - x_0)$, acting according to the formula:

$$(\delta(x - x_0),\ \varphi(x)) = \varphi(x_0) \tag{14.119}$$

and in particular, for $x_0 = 0$

$$(\delta(x), \varphi(x)) = \varphi(0). \tag{14.120}$$

The jump function or the Heaviside step function $\theta(x - x_0)$ is another example of a generalized function

$$(\theta(x - x_0), \varphi(x)) = \int_{x_0}^{\infty} \varphi(x)dx \tag{14.121}$$

If the function $f(x)$ is locally integrable, then it is obvious that the integral

$$(f,\varphi)=\int_{-\infty}^{\infty} f(x)\varphi(x)dx \tag{14.122}$$

defines some generalized function. Such generalized functions are called regular. Thus, the jump function above is an example of a regular generalized function. The Heaviside function corresponds to the regular function θ, equal to zero at $x < x_0$ and one at $x > x_0$.

Generalized functions that are not represented in integral form with a locally integrable function $f(x)$, are called singular functions. An example of a singular generalized function is a delta function for which the following representation is valid.

$$(\delta(x-x_0),\ \varphi(x))=\int_{-\infty}^{\infty} \delta(x-x_0)\varphi(x)dx=\varphi(x_0). \tag{14.123}$$

Generalized functions have certain local properties. For example, we can assume that the generalized function, $f(x)$, is equal to zero in the neighborhood (vicinity) of a point x_1, if for any basic function $\varphi(x)$ other than zero only in the neighborhood (vicinity) of x_1, $(f, \varphi) = 0$.

The basic function is called any real finite function $\varphi(x)$, defined over $-\infty < x < \infty$, and having derivatives of any order, i.e., turning to zero outside of some finite domain, or interval.

Any function $f(x)$, defined over $-\infty < x < \infty$, and taking real values and locally summable, is called a regular function.

It is obvious that the delta function $\delta(x - x_0)$ is equal to zero in the neighborhood (vicinity) of any point except the point x_0. This point supports a real-valued function $\varphi(x)$ and is not mapped to zero, and called essential. The set of all essential points of a generalized function is called its support. A single point is the support of delta function. The delta function is a linear function that maps every function to its value at zero [12].

For example, for the jump function or Heaviside step function $\theta(x - x_0)$, all points $x > x_0$ are essential/support. Hence the semiaxis $x > x_0$ is its support.

For generalized functions, various operations can be defined:

addition and subtraction

$$(f \pm g,\varphi)=(f,\varphi)\pm(g,\varphi); \tag{14.124}$$

differentiation

$$\left(f^{(q)},\varphi\right)=(-1)^q\left(f,\varphi^{(q)}\right); \tag{14.125}$$

integration

$$(f_1,\varphi')=-(f,\varphi), \qquad \text{where} \qquad f_1'=f. \tag{14.126}$$

In this case, the primitive generalized function f_1 is determined with accuracy to constant term. For an unambiguous definition of a function f_1, it is necessary to select some basic function ψ, such that

$$\int_{-\infty}^{\infty} \psi(x)dx \neq 0 \tag{14.127}$$

and input $(f_1, \psi) = C$.

In particular, we have

$$\theta'(x - x_0) = \delta(x - x_0), \quad \text{and} \tag{14.128}$$

$$\int_{-\infty}^{\infty} \delta^{(q)}(x - x_0)\varphi(x)dx = (-1)^q \varphi^{(q)}(x_0). \tag{14.129}$$

By definition, generalized functions are infinitely differentiable. Let us show the transformations in the domain of independent variables of generalized functions by the example of variable replacement:

1. $(f(x - h), \varphi) = (f, \varphi(x + h),)$—Shift by the value of h;
2. $(f(-x), \varphi(x)) = (f(x), \varphi(-x))$—Reflection from the beginning of the coordinates;
3. $(f(x/\alpha), \varphi(x)) = \alpha(f(x), \varphi(\alpha x))$—Transformation of similitude (similarity).

From the operation of reflection from the beginning of coordinates (2.), the evenness properties of the delta functions important for the subsequent representation can be obtained. Setting $\delta^{(q)}(x_0 - x) = \delta^{(q)}(-x - x_0)$ and using property 2, we obtain:

$$\delta^{(q)}(x_0 - x) = (-1)^q \delta^{(q)}(x - x_0), \tag{14.130}$$

i.e., the delta function and its even derivatives are even functions of the complete argument, and the odd derivatives are odd functions of the complete argument.

If a function $r(x)$ has only simple zeros x_h and is infinitely differentiable, then:

$$\begin{gathered} \delta[r(x)] = \sum_h \frac{\delta(x - x_h)}{|r'(x_h)|} \\ \delta^{(q)}[r(x)] = \sum_h \frac{1}{|r'(x_h)|}\left(\frac{1}{r'(x_h)}\frac{d}{dx}\right)^q \delta(x - x_h) \end{gathered} \tag{14.131}$$

A sequence of generalized functions $\{f_m\}$ is called convergent to a generalized function f if $\lim_{m \to \infty}(f_m, \varphi) = (f, \varphi)$.

If such a sequence of generalized functions converges to a delta function, then it is called a delta-image sequence. For example, when $t \to 0$

$$\begin{gathered} \frac{1}{\pi}\frac{t}{x^2 + t^2} \to \delta(x) \\ \frac{1}{2\sqrt{\pi t}} exp\left(\frac{x^2}{4t}\right) \to \delta(x) \\ \frac{1}{\pi}\frac{\sin x/t}{x} \to \delta(x) \end{gathered} \tag{14.132}$$

If the delta- image sequence of functions is symmetric with respect to the point x_0, then it is assumed that:

$$\int_{x_0}^{b} \varphi(x)\delta(x - x_0)dx = \int_{a}^{x_0} \varphi(x)\delta(x - x_0)dx = \frac{1}{2}\varphi(x_0) \tag{14.133}$$

Generalized functions can be expanded into series according to different systems of functions. Here are some useful expansions/decompositions in applications:

$$\sum_{n=-\infty}^{\infty} e^{inx} = 2\pi \sum_{n=-\infty}^{\infty} \delta(x - 2\pi n);$$
$$\sum_{n=-\infty}^{\infty} \cos nx = -\frac{1}{2} + \pi \sum_{n=-\infty}^{\infty} \delta(x - 2\pi n);$$
$$\sum_{n=1}^{\infty} \sin nx = \frac{1}{2} \cot \frac{x}{2}; \tag{14.134}$$
$$\sum_{n=1}^{\infty} n^q \cos\left(nx + \frac{q\pi}{2}\right) = \pi \sum_{n=-\infty}^{\infty} \delta^{(q)}(x - 2\pi n);$$
$$\sum_{n=1}^{\infty} n^{q-1} \cos\left(nx + \frac{q\pi}{2}\right) = -\left\{ \ln\left|2 \sin \frac{x}{2}\right| \right\}^q;$$

When solving problems by the method of integral transformations, it is often necessary to deal with various divergent integrals. In the method of integral transformations, the use of generalized functions allows us to give meaning to various divergent integrals. An algorithm that allows assignment of a certain numerical value to divergent integrals is called regularization. Let the function $f(x)$ have a non-integrable singularity at the point x_0, then the integral

$$\int_{-\infty}^{\infty} f(x)\varphi(x)dx \tag{14.135}$$

diverges if the function $\varphi(x)$ is not equal to zero in the neighborhood/vicinity of the point x_0. Any functional f that acts on basic functions $\varphi(x)$ equal to zero in a neighborhood of a point x_0, by formula (14.135) is called a regularization of the divergent integral (14.135), or a regularization of function $f(x)$. The same function can have different regularizations. The regularization is called canonical and the corresponding functional (f, φ) is denoted by $c.\ r.\ f(x)$, if it satisfies the following conditions:

$$c.r.[\alpha_1 f_1 + \alpha_2 f_2] = \alpha_1 c.r.f_1 + \alpha_2 c.r.f_2;$$
$$c.r.\left[\frac{d}{dx} f\right] = \frac{d}{dx}[c.r.f];$$
$$c.r.[r(x)] = r(x) c.r.f,$$

where $r(x)$ is an infinitely differentiated function.

In particular, the integral

$$\int_0^{\infty} x^{\lambda}\varphi(x)dx = \left(x_+^{\lambda}, \varphi\right)$$

is regularized as follows:

at $Re\lambda > -n-1,\ \lambda \neq -1, -2, \ldots, -n$

$$
\begin{aligned}
(x_+^\lambda, \ \varphi) = & \int_0^1 x^\lambda \left[\varphi(x) - \varphi(0) - x\varphi'(0) - \cdots - \frac{x^{n-1}}{(n-1)!} \varphi^{(n-1)}(0) \right] dx \\
& + \int_1^\infty x^\lambda \varphi(x) dx + \sum_{k=1}^{n} \frac{\varphi^{(n-1)}(0)}{(k-1)!(\lambda + k)}
\end{aligned}
\tag{14.136}
$$

at $-n - 1 < Re\lambda < -n$

$$
(x_+^\lambda, \varphi) = \int_0^\infty x^\lambda \left[\varphi(x) - \varphi(0) - x\varphi'(0) - \cdots - \frac{x^{n-1}}{(n-1)!} \varphi^{(n-1)}(0) \right] dx \tag{14.137}
$$

For the integral

$$
\int_{-\infty}^{\infty} x^\lambda \varphi(x) dx = (x^\lambda, \varphi)
$$

we have: for $\lambda = -2m$ (m is an integer)

$$
\begin{aligned}
(x^{-2m}, \ \varphi) = & \int_0^\infty x^{-2m} \left\{ \varphi(x) - \varphi(-x) - 2\left[\varphi(0) + \frac{x^2}{2!}\varphi''(0) + \cdots \right.\right. \\
& \left.\left. \ldots + \frac{x^{2m-2}}{(2m-2)!} \varphi^{(2m-2)}(0) \right] \right\} dx,
\end{aligned}
\tag{14.138}
$$

for $\lambda = -2m - 1$

$$
\begin{aligned}
(x^{-2m-1}, \varphi) = & \int_0^\infty x^{-2m-1} \left\{ \varphi(x) - \varphi(-x) - 2\left[x\varphi'(0) + \frac{x^3}{3!}\varphi'''(0) + \cdots \right.\right. \\
& \left.\left. \ldots + \frac{x^{2m-1}}{(2m-1)!} \varphi^{(2m-1)}(0) \right] \right\} dx
\end{aligned}
\tag{14.139}
$$

In particular:

$$
\begin{aligned}
(x^{-1}, \varphi) &= \int_0^\infty \frac{\varphi(x) - \varphi(-x)}{x} dx; \\
(x^{-2}, \varphi) &= \int_0^\infty \frac{\varphi(x) + \varphi(-x) - 2\varphi(0)}{x^2} dx; \\
(x^{-3}, \ \varphi) &= \int_0^\infty \frac{\varphi(x) - \varphi(-x) - 2x\varphi'(0)}{x^3} dx
\end{aligned}
\tag{14.140}
$$

In the case of finite interval at $Re\lambda > -n - 1, \quad \lambda \neq -1, -2, \ldots$ the following regularization takes place:

$$\int_0^b x^\lambda \varphi(x)dx = \int_0^b x^\lambda \left[\varphi(x) - \varphi(0) - \cdots - \frac{x^{n-1}}{(n-1)!}\varphi^{(n-1)}(0)\right]dx$$
$$+\varphi(0)\frac{b^{\lambda+1}}{\lambda+1} + \cdots + \varphi^{(n-1)}(0)\frac{b^{\lambda+n}}{n!(\lambda+n)} \tag{14.141}$$

For regularization of integrals of type

$$\int_b^\infty x^\lambda \varphi(x)dx$$

is assumed $y = x^{-1}$, which leads at, $\lambda \neq -1, 0, 1, \ldots$ to integral

$$\int_b^\infty x^\lambda \varphi(x)dx$$
$$= \int_b^{1/b} y^{-\lambda-2}\varphi\left(\frac{1}{y}\right)dy \tag{14.142}$$

The regularization of this integral is carried out by formula (14.141). Table 14.1 [8] shows some generalized functions and their Fourier transformations. Then the real and imaginary parts are separated in the Fourier transforms for the functions $\theta(x)$ and x_+^1 . The following integrals that are frequently encountered in engineering applications have been obtained:

$$\int_0^\infty \cos \xi x dx = \pi\delta(\xi), \tag{14.143}$$

$$\int_0^\infty \sin \xi x dx = \xi^{-1}, \tag{14.144}$$

Table 14.1 Some generalized functions and their Fourier transforms

Generalized function	Corresponding regular function	Fourier transform $\boldsymbol{F(\xi)}$
Integrable basic function $f(x)$	$f(x)$	$F(\xi) = \int_{-\infty}^{\infty} f(x)e^{-i\xi x}dx$
$\delta(x)$	–	1
$\theta(x)$	$1 \ at \ x > 0; \ 0 \ at \ x < 0$	$i\xi^{-1} + \pi\delta(\xi)$
x_+^n (n – integer)	$x^n \ at \ x > 0; \ 0 \ at \ x < 0$	$i^{n+1} n!\, \xi^{-n-1} + (-i)^n \pi\delta^{(n)}(\xi)$
x_-^n	$0 \ at \ x > 0; \ \lvert x^n \rvert \ at \ x < 0$	$(-i)^{n+1} n!\, \xi^{-n-1} + i^n \pi\delta^{(n)}(\xi)$
x^m	x^m	$2(-i)^m \pi\delta^{(m)}(\xi)$
x^{-m}	x^{-m}	$i^m \frac{\pi}{(m-1)!} \xi^{m-1} \mathrm{sign}\xi$
$\delta^{2m}(x)$	–	$(-1)^m \xi^{2m}$
$\delta^{(2m+1)}(x)$	–	$(-1)^{m+1} i\xi^{2m+1}$

$$\int_{0}^{\infty} x\cos\xi x dx = -\xi^{-2}, \tag{14.145}$$

$$\int_{0}^{\infty} x\sin\xi x dx = -\pi\delta'(\xi), \tag{14.146}$$

These results can be appropriately extended to the case of generalized functions of several variables. Let $r = \sqrt{x_1^2 + x_2^2 + \cdots + x_n^2}$, then the formula for the n—dimensional delta function has the form

$$\delta(r) = \frac{1}{\omega_n}\frac{\delta(|r|)}{|r|^{n-1}}, \tag{14.147}$$

where ω_n is the surface area of n—dimensional unit sphere. In particular, for $n = 2$ the formula for two-dimensional delta function has the form:

$$\delta(r) = \frac{\delta(|r|)}{2\pi|r|} \tag{14.148}$$

This expression gives: the value of two-dimensional delta functions used in solving problems in polar coordinates. The generalized functions allow extending the well-known Cauchy formula to the case of an arbitrary one $\lambda Re\lambda > 0$.

$$D^{-n}g(x) = \int_{0}^{x}\int_{0}^{\xi_{n-1}} \cdots \int_{0}^{\xi_2}\int_{0}^{\xi_1} g(\xi)d\xi d\xi_1 \ldots d\xi_{n-1}$$
$$= \frac{1}{(n-1)!}\int_{0}^{x} g(\xi)(x-\xi)^{n-1}d\xi, \quad (x > 0),$$

The Cauchy formula reduces the calculation of n multiple of the antiderivative to the computation of a single integral. The Cauchy formula can be written as using the above dependencies

$$D^{-n}g(x) = g_{+}(x) * \frac{x_{+}^{n-1}}{\Gamma(n)},$$

where $\Gamma(n)$ is Gamma function; the sign + means that the given function at $x < 0$ is taken to be zero; and the sign * indicates the convolution.

Assuming any exponent of λ instead of n, and defining the antiderivative of the λ order of the function $g(x)$ as a convolution.

$$D^{-\lambda}g(x) = g_{+}(x) * \frac{x_{+}^{\lambda-1}}{\Gamma(\lambda)},$$

This formula is valid for $Re\lambda > 0$. For the rest of the values of λ, the function $x_{+}^{\lambda-1}/\Gamma(\lambda)$ must be considered as a generalized function. In particular, with $\lambda = -k (k = 0, 1, 2, \ldots)$ we have

$$D^k g(x) = \frac{d^k}{dx^k} g(x)$$

The derivative of order λ of the generalized function $g(x)$ is defined as

$$D^\lambda g(x) = \frac{x_+^{-\lambda-1}}{\Gamma(-\lambda)}. \tag{14.149}$$

For the operators $I^\lambda = D^{-\lambda}$, acting by the formula

$$I^\lambda g(x) = \frac{1}{\Gamma(\lambda)} \int_0^x g(\xi)(x-\xi)^{\lambda-1} d\xi \ ,$$

the following relations are true: $I^\lambda I^\beta g = D^{-\alpha} D^{-\beta} g = I^{\alpha+\beta} g = D^{-\alpha-\beta} g$. These operators are called Riemann–Liouville fractional integrals of order λ [11].

The fractional Weyl integral of order λ is also considered

$$K^\lambda g(x) = \frac{1}{\Gamma(\lambda)} \int_x^\infty g(\xi)(\xi - x)^{\lambda-1} d\xi,$$

along (together) with the Riemann–Liouville integrals. In the space of generalized functions, the operators I^λ and K^λ are conjugate (adjoint), i.e., the following formulas are valid for two functions $\varphi(x)$ and $\psi(x)$:

$$\int_x^\infty \psi(x) I^\lambda \varphi(x) dx = \int_0^\infty \varphi(x) K^\lambda \psi(x) dx.$$

If the Mellin transform operator [5] is defined by M

$$M_s[g] = \int_0^\infty g(x) x^{s-1} dx,$$

then:

$$M_z\left[I^\lambda g\right] = \frac{\Gamma(1-\lambda-z)}{\Gamma(1-z)} M_{z+\lambda}[g];$$

$$M_z\left[K^\lambda g\right] = \frac{\Gamma(z)}{\Gamma(1+z)} M_{z+\lambda}[g].$$

Fractional integration operators find applications in solving various problems in mechanics and mathematical physics.

In structural mechanics, the theory of the generalized functions does not yet apply as thoroughly as in other sections of mechanics and physics. In this book and also in Ref. [13], it is shown that the theory of generalized functions allows us to obtain broad generalizations of the classical methods of structural mechanics, simplify solutions of many problems, and identify the relationships among the methods.

Known simplifications can be obtained for the solutions of boundary and initial-boundary problems using an apparatus of the theory of the generalized functions.

The most frequently used functions are Dirac and Heaviside. Dirac function or delta function provides the analytical description to the concentrated parameters (forces, moments, concentrated masses, etc.) and is determined by the following its own basic properties

$$=f(x) \quad (a<c<b)$$

$$\int_a^b f(x)\delta(x-c)dx=$$

$$=0 \quad (c\langle a, \ c\rangle b)$$

The Heaviside step function $\theta(x)$ is equal to zero when $x<0$, and 1 if $x \geq 0$. Both these functions are connected together by the following differential dependence

$$\frac{d\theta}{dx}=\delta(x).$$

Clearly, the step function can describe a load and other physical functions different from zero in a certain interval. For example, if a certain function $f(x)$ is other than zero in an interval $(a, \ b)$, then it can be written on the entire x axis as:

$$\bar{f}(x)=[\theta(x-a)-\theta(x-b)]f(x)$$

Delta function can be used to describe a unit single/concentrated force, $\delta^{I}(x)$—will designate a unit moment, and $\delta^{II}(x)$—angular deformation, and $\delta^{III}(x)$—jump of the angle of rotation, determined with the accuracy to EI.

The use of generalized functions allows us to formalize the transition from concentrated impacts, determining the right part of the differential equations, to the boundary or initial conditions and, vice versa, the transition from inhomogeneous boundary and initial conditions to the homogeneous with the transfer of heterogeneity to the right part of the equation.

It should be noted that the theory of the generalized functions in its modern form is still insufficiently well incorporated into structural mechanics, although attempts to create a similar apparatus for calculating beams, plates, and other constructions were undertaken a long time ago (Gersevanov [14]). In their time, the functional interrupters–breakers (the limits of specially selected functions: step functions) were important achievements; however, after the creation of sufficiently strict theory of the generalized functions, the necessity to use them disappeared. The generalized functions unify analytical operations above the continuous and discontinuous functions and give mathematical sense to such purely technical concept as concentrated force and moment. It is evident that a concentrated load at the point x_0 can be written as

$$q(x)=q_0\delta^{(n)}(x-x_0) \tag{14.150}$$

where q_0 is intensity of load, with the dimensions of unit of force per unit length; $n=0$ in case of force, and $n=1$ in the case of moment. In structural mechanics, uploading by the moments of higher orders is also considered. For example, in [15, 16], and [8, 17] concentrated deformations load was used, which is equivalent to the dual and triple differentiation of the solutions, obtained by the action of concentrated force. It is easy to see that in this case in formula (14.150) it is possible to place $n=2$ when loading with the angular deformations and $n=3$ when loading with linear deformations (displacements).

Let us keep in mind that during the integration of generalized functions the dimensionality of results changes per unit of length, for example, when the concentrated force

$$\int_{-\infty}^{\infty} q(x)dx = q_0 \int_{-\infty}^{\infty} \delta(x)dx = q_0,$$

where q_0, unlike (14.150), already has dimension of force.

In the integration of derivatives of delta function the dimensionality of results changes, again, because of the differentiation of functions in accordance with the formula

$$\int_{-\infty}^{\infty} \delta^{(q)}(x - x_0)\varphi(x)dx = (-1)^q \varphi^{(q)}(x_0) \qquad (14.151)$$

In the Table 14.2, loads and corresponding to them generalized functions are given, based on an example of a beam [8]. The intensity of a load for the delta function and its derivatives is determined as a result of the integration of the equation of the equilibrium of the beam

$$EIy^{IV}(x) = q_0\delta^{(n)}(x - a) \qquad (n = 0, 1\ 2, 3) \qquad (14.152)$$

n + 1 times. Equation (14.152) may conveniently be written in the so-called quasi-derivatives (see formula (14.160) in the next Sect. 14.7)

$$y^{[IV]}(x) = q_0\delta^{[n]}(x - a) \qquad (n = 0, 1, 2, 3), \qquad (14.153)$$

where,

$$\begin{gathered} y^{[0]}(x) = y(x), \quad y^{[I]}(x) = D_x y(x) = y'(x); \\ y^{[II]}(x) = EIy''(x); \; y^{[III]}(x) = -EIy^{[''']}(x); \; y^{[IV]}(x) = EIy'''(x). \end{gathered} \qquad (14.154)$$

The solution to Eq. (14.153) satisfies certain boundary conditions and is expressed through the generalized Green's functions corresponding to boundary-value problems. We will designate the solution of Eq. (14.153) through $G_n(x, a)$ when $q_0 = 1$. The generalized Green's function $G_n(x, a)$ represents a single displacement from considered above force (force, moment) and kinematic (the jump-step function of the angle of rotation the jump-step function of the displacement) impacts. Quasi-derivatives $G_n^{[m]}(x, a)$ are a unit of angles of rotation and unit of reactions of the beam (bending moment and shear force). Table 14.3 shows the unit of displacements and unit of reactions, expressed through the generalized Green's function. The top index (superscript) designates the parameter, which is described by Green's function, and the bottom index denotes parameter that undergoes a jump due to an applied external impact. As an example $G_3^{[II]}(x, a) = G_y^{[M]}(x, a)$ is a bending moment in the section with the current coordinate x, due to unit jump of displacement applied in the section $x = a$. The following formula is given in Tseitlin [8] that describes the generalized Green's functions:

$$G_n^{[M]}(x, a) = D_a^{[n]} G_0^{[m]}(x, a). \qquad (14.155)$$

14.7 The Generalized Method of Integral Transformation

The rapidly developed apparatus of boundary integral equations, used for solving different boundary value problems and design constructions, is a highly effective way of obtaining numerical results, competing with such widespread methods as finite element and finite difference methods. The wide

Table 14.2 Loads and, corresponding to them, generalized functions

Loading diagram	Generalized function	Loading intensity (*M*—unit length)
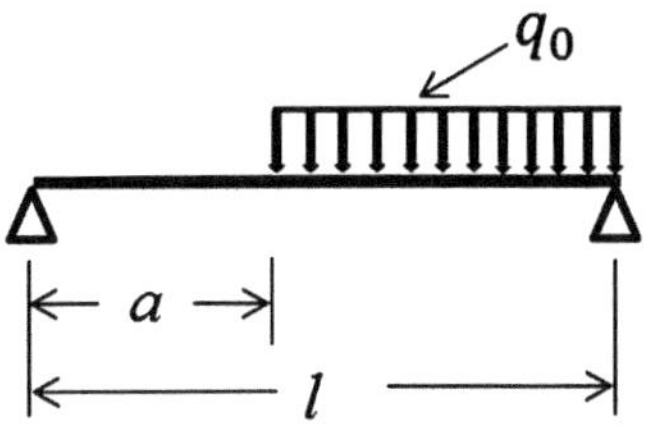	$q = q_0\theta(x-a)$	q_0
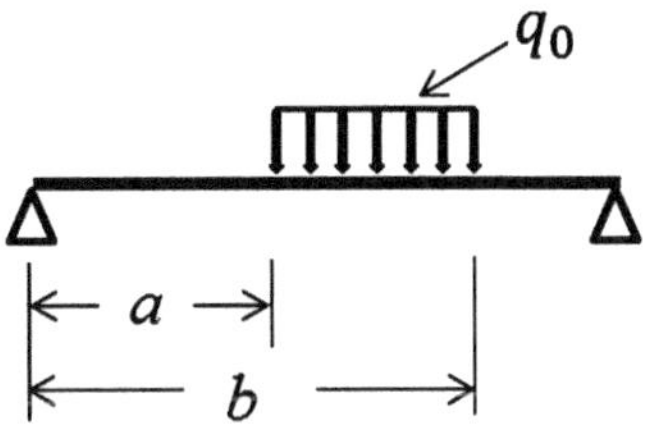	$q = q_0[\theta(x-a) - \theta(x-b)]$	q_0
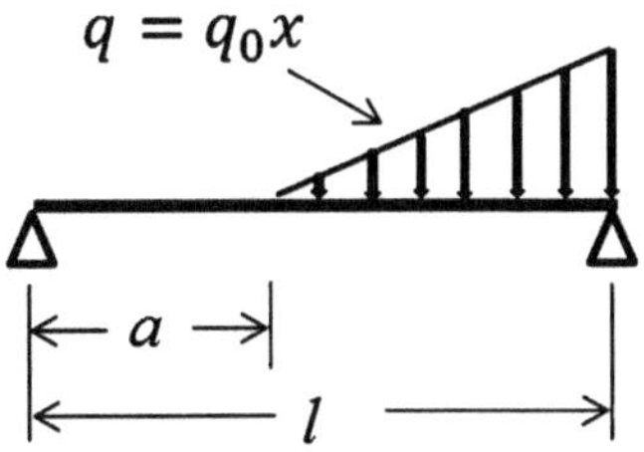	$q = q_0(x-a)_+$	q_0
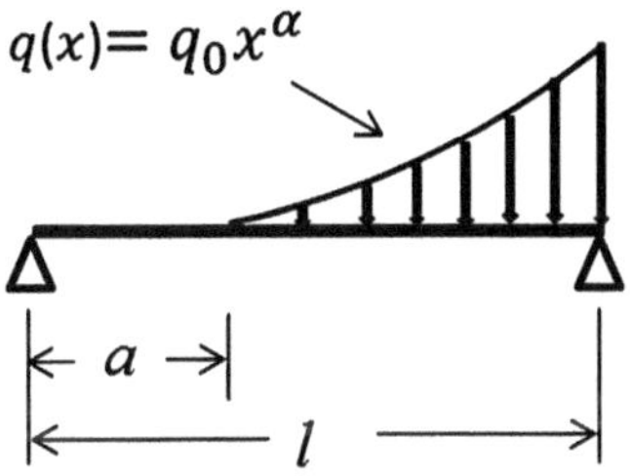	$q = q_0(x-a)^{\alpha}{}_{+}$	q_0
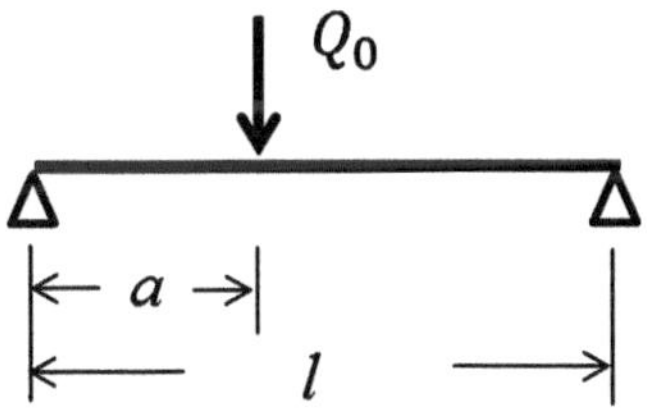	$q = q_0\delta(x-a)$	$Q_{0/M}$
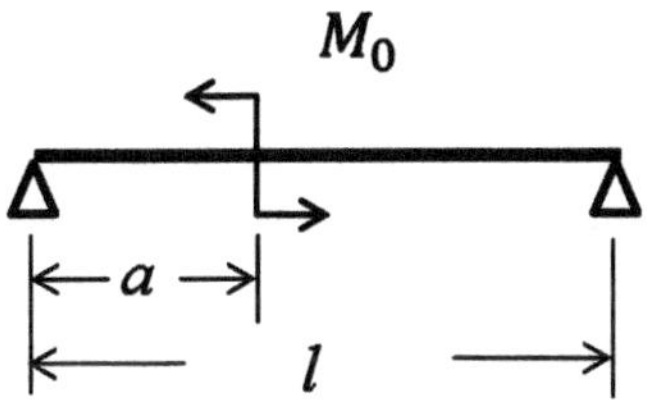	$q = q_0\delta^{[']}(x-a)$	M_{0/M^2}
	$q = q_0\delta^{['']}(x-a)$	$\varphi_0 EI_{/M^3}$

Loading diagram	Generalized function	Loading intensity (*M*—unit length)
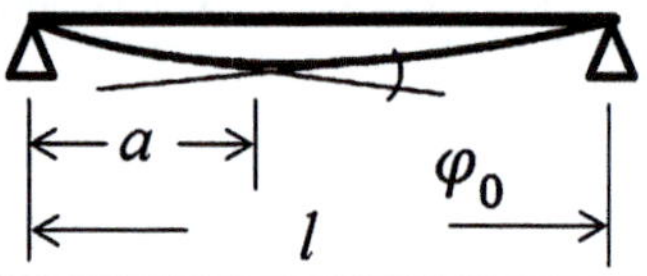		
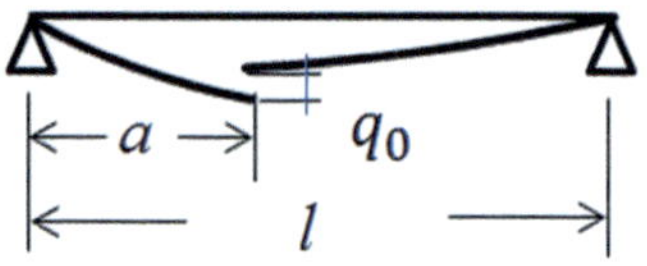	$q = q_0\delta^{[''']}(x-a)$	$y_0 EI_{/M^4}$

Table 14.3 Unit displacements and unit reactions expressed through generalized green functions

	External unit impact			
Unit factor	Force δ	Moment $\delta^{[']}$	Rotation angle $\delta^{['']}$	Displacement $\delta^{[''']}$
Displacement *y*	$G_0 = G_Q^y$	$G_1 = G_M^y$	$G_2 = G_\varphi^y$	$G_3 = G_y^y$
Angle of Rotation φ	$G_0^{[I]} = G_Q^\varphi$	$G_1^{[I]} = G_M^\varphi$	$G_2^{[I]} = G_\varphi^\varphi$	$G_3^{[I]} = G_y^\varphi$
Bending Moment *M*	$G_0^{[II]} = G_Q^M$	$G_1^{[II]} = G_M^M$	$G_2^{[II]} = G_\varphi^M$	$G_3^{[II]} = G_y^M$
Shear Force *Q*	$G_0^{[III]} = G_Q^Q$	$G_1^{[III]} = G_M^Q$	$G_2^{[III]} = G_\varphi^Q$	$G_3^{[III]} = G_y^Q$

spectrum of questions, which can be united under the name of boundary equations (methods of potentials, integral equations, compensating loads, extended domain, delta transform, etc.), should include the classical methods of structural mechanics, also based on composition of boundary equations.

The classical integral transforms of Fourier, Hankel, Laplace, Mellin, and others can be considered as decomposition of eigenfunctions of one-dimensional self-conjugate and non-self-conjugate differential operators. For a given boundary-value problem the appropriate transformation is dictated by the differential operator, the fundamental domain, and the boundary conditions prescribed. Based on the foregoing a theory of all known, integral transformations can be built.

The development of methods of integral transformation has to follow two pathways. The first pathway is followed through on expansion of the range of kernels, used for solving boundary-value problems of mathematical physics, and mechanics. The second pathway is through as along the line of the application of the method to the boundary-value problem, which is different from generating problems under conditions of domain definition, boundary conditions, presence of internal conditions, and so on. Examples of such use of integral transformations in the extended domain can be found in [18]. The integral transformations in the extended domain and similar methods were applied to areas with internal cuts [19, 20], more complex internal conditions [21, 22], and in problems with mixed boundary conditions [23, 24]. Popov for the first time combined and generalized these approaches, extending them to the multi-coherent domains and heterogeneous media [25, 26]. Another trend in the development of integral transformations is application of non-orthogonal kernels Tseitlin, Petrosian [3, 27], and [5].

Let us first consider integral transformations as decomposition of eigenfunctions of linear differential operators. We present some general knowledge from the theory of these operators below [28]. The linear differential operator, known as the operator $L = L(y)$, is originated by the differential expression:

$$l(y) = p_0(x)y^{(n)} + p_1(x)y^{(n-1)} + \cdots + p_n(x)y \tag{14.156}$$

and edge (boundary) conditions

$$U_s(y) = 0, \qquad s = 1, 2, \ldots, m \leq 2n, \tag{14.157}$$

where function $p(x)$—coefficients of differential expression; n—order of the differential equation; $U_s(y)$– linearly independent forms relative to the values of functions y, and its first $n - 1$ derivatives at the boundary points a, b of the interval $[a, b]$, where differential expression (14.156) operates. The differential expression adjoints (conjugates) to $l(y)$ and is named:

$$l^*(y) = (-1)^n (p_0 y)^{(n)} + (-1)^{(n-1)} (p_1 y)^{(n-1)} + \cdots + p_n y \tag{14.158}$$

Differential expression is called self-adjoint if $l(y) = l^*(y)$. Any self-adjoint differential expression with real coefficients can be reduced to the following form:

$$l(y) = \left(p_0 y^{(n)}\right)^n + \left(p_1 y^{(n-1)}\right)^{(n-1)} + \cdots + p_n y \tag{14.159}$$

The differential expression can be conveniently represented by so-called quasi-derivatives $l(y) = y^{[2n]}$, which are defined by the formulas:

$$\begin{aligned} &y^{[0]} = y, \quad y^{[k]} = \frac{d^k y}{dx^k}, \quad y^{[n]} = p_0 \frac{d^n y}{dx^n} \quad (k = 1, 2, \ldots, n-1); \\ &y^{[n+k]} = p_k \frac{d^{n-k} y}{dx^{n-k}} - \frac{d}{dx}\left(y^{[n+k-1]}\right) \qquad (k = 1, 2, \ldots, n) \end{aligned} \tag{14.160}$$

The boundary conditions expressed through quasi-derivatives according to the formulas of the type $y^{[k]}(x) = 0$, is called canonical.

If y and z are two arbitrary functions, having continuous derivatives up to the n– th order inclusive on (a, b) interval, for such arbitrary functions y, and z then the Lagrange's identity takes place

$$\int_a^b l(y)\bar{z}dx - \int_a^b y\, l^*(\bar{z})dx = \left[\sum_{s=1}^{2n} U_s(y) V_{2n-s+1}(\bar{z})\right]_a^b = [y, \bar{z}]_a^b, \tag{14.161}$$

where V—linearly independent forms relative to the values of functions z, and its first $n - 1$ derivatives; $\bar{z}$—designation of the conjugate (adjoint) value.

Boundary (edging) conditions of $V_{2n-s+1}(\bar{z}) = 0$ are called conjugate (adjoint) to the conditions (14.157), and if they are equivalent to the boundary conditions (14.157), then they are called self-adjoint.

A differential operator generated by self-conjugate (self-adjoint) differential expression and self-conjugate boundary (edging) conditions is also called self-conjugate (self-adjoint).

For a self-adjoint operator, Lagrange's identity takes the following form:

$$\int_a^b l(y)zdx - \int_a^b yl(z)dx = [y, z]_a^b. \tag{14.162}$$

A differential operator is called regular if the interval $[a, b]$ is finite and the function $p_0^{-1}(x)$ is summable on it, otherwise the operator is called singular.

The nontrivial solution of the equation is called the eigenfunction of operator L:

$$Ly(x, \lambda) = \lambda y(x, \lambda) \tag{14.163}$$

Parameter values of λ with which eigenfunctions are in the domain of definition of operator L are called eigenvalues. One eigenvalue may correspond to one or more (up to n) of the eigenfunctions; their number is called the multiplicity of the eigenvalue. The eigenvalues of the adjoint operators are the adjoint (conjugate) numbers. If the eigenfunctions $y(x, \lambda)$, and $z(x, \mu)$ of adjoint operators L and L^* are orthogonal, i.e., $\lambda \neq \overline{\mu}$

$$\int_a^b y\,(x, \lambda)\overline{z}(x, \overline{\mu})dx = 0 \tag{14.164}$$

If we assume with $\lambda = \overline{\mu}$,

$$\int_a^b y\,(x, \lambda)\overline{z}(x, \overline{\mu})dx \ = 1$$

(condition of standardizing /normalization the eigenfunctions), then

$$\int_a^b y\,(x, \lambda)\overline{z}(x, \overline{\mu})dx = \delta_{\lambda\overline{\mu}} \tag{14.165}$$

where

$\delta_{\lambda\overline{\mu}}$—Kronecker's symbol: $\delta_{\lambda\overline{\mu}} = 1(\lambda - \overline{\mu})$; $\quad \delta_{\lambda\overline{\mu}} = 0 \quad (\lambda \neq \overline{\mu}\,)$.

In case of the self-adjoint operator:

$$\int_a^b y\,(x, \lambda)y(x, \mu)dx = \delta_{\lambda\mu}, \tag{14.166}$$

where y—normalized eigenfunctions.

The differential expression (14.156) considered on the interval (a, b), can originate various operators.

The operator L with a maximum domain of definition in $L_2(a, b)$, i.e., a square summable on (a, b), is an operator which operates according to expression (14.156) to all functions $y(x)$ from $L_2(a, b)$, having absolutely continuous quasi-derivatives up to $2n - 1$ th order inclusive, where $y^{[2n]}$ belongs to $L_2(a, b)$.This space of functions is denoted D in [28].

Another operator L_0 is defined on the set of all functions $y(x) \in D$, which satisfy conditions:

$$y^{[k]}\big|_{x=a} = y^{[k]}\big|_{x=b} = 0 \qquad (k = 0,\ 1,\ 2,\ \ldots,\ 2n - 1)$$

in the regular case, and $[y, z]_a^b = 0$ in the singular case for all $z \in D$. In the case of one singular end, the domain of definition of operator L_0 consists only of those functions $y(x) \in D$, which satisfy the conditions:

$$y^{[k]}\big|_{x=a} = 0 \qquad (k = 0,\ 1\ ,2,\ \ldots, 2n - 1)$$
$$[y, z]_a^b = 0 \qquad (z \in D).$$

Operator L_0 allows various self-adjoint extensions. The domain of definition of D' of any self-adjoint operator L_u consists of those functions $y(x) \in D$, which satisfy the following conditions:

if $y, \quad z \in D'$, then $[y, z]_a^b = 0$;
if $z \in D$ and for all $y(x) \in D'$, $[y, z]_a^b = 0$, then $z \in D'$.

Note that $L_0 = L^*$, i.e., the operators L_0 and L are adjoint (conjugate).

The pair of numbers which corresponds to the number of linearly independent solutions of the equations

$$A^* u = \lambda u, \quad A^* u = \bar{\lambda} u \quad [Im\, \lambda > 0, \quad u \in L_2 \quad (a, b)], \tag{14.167}$$

are called index of the defect operator of A. The index of defect of operator L_0 is defined by a pair of identical integers (m, m), moreover since $0 \leq m \leq 2n$, it is designated by one number m. If the index of operator L_0 is equal $2n$, then the spectra of all its self-adjoint extensions are discrete. In a general case, the spectrum can be discrete-continuous, and the continuous part of all self-adjoint extensions is identical (same). If one end of the interval (a, b) is regular, then $n \leq m \leq 2n$. In particular, if functions $(p_0^{-1})', p_1, p_2, \ldots p_n$ are summarized in $[0, \infty)$, and $\lim_{x \to +\infty} p_0(x) > 0$; then the index of defect (deficiency) of operator L_0 is (n, n). The same index has an operator L_0, when $(p_0^{-1})', p_1, p_2, \ldots, p_{n-1}, p(x)$ are summarized on interval $[0, \infty)$, $\lim_{x \to +\infty} p_0(x) > 0$; $p_n(x) = p(-x) + q(x)$, where $q(x)$—measured and substantially limited in interval $[0, \infty)$, function, for example, when

$$l(y) = (-1)^n \frac{d^{2n} y}{dx^{2n}} + q(x) y.$$

If the index of defect (deficiency) of the operator is n, then

$$[y, z]_b = 0 \tag{14.168}$$

for all $y, \quad z \in D$.

Any material λ, with which the number of belonging to $L_2(a, b)$ linearly independent solutions of equation

$$l(y) = \lambda y$$

is lower than the deficiency number of operator L_0, and belongs to the discrete or continuous part of the spectrum of any self-adjoint extension of operator L_0.If in this case one of the ends is regular, then λ belongs to the continuous part of the spectrum.

Any self-adjoint operator of L_u can be determined with the aid of a certain operator function P_λ as follows: if y belongs to the domain of definition of operator L_u, then

$$L_u y = \int_{-\infty}^{\infty} \lambda d[P_\lambda y]. \tag{14.169}$$

Operator function P_λ is called the spectral function of the operator L_u. Spectral function is related to resolvent of the self-adjoint operator by

$$R_\lambda = \int_{-\infty}^{\infty} \frac{dP_\mu}{\mu - \lambda}, \tag{14.170}$$

where λ is a regular point of the operator.

We note, that the resolvent of any self-adjoint operator L_u is an integral operator; the rezolvent's kernel $G(x, \xi, \mu)$ builds as Green's function of the operator $L_u - \mu E$, i.e., as the solution of the equation

$$l(y) - \mu y = \delta(x - \xi). \tag{14.171}$$

In the case of one regular end (a), the kernel of a rezolvent satisfies the conditions:

$$\int_a^b |G(x, \xi, \mu)|^2 d\xi < \infty; \quad \int_a^b |G(x, \xi, \mu)|^2 dx < \infty.$$

Any function $f(x)$ from $L_2(a, b)$ can be decomposed according to the eigenfunctions of the self-adjoint differential operator. Let L_u—be the self-adjoint operator, generated by differential equation $l(y)$ in the interval (a, b), and $\{u_k(x, \lambda)\}_{k=1}^{2n}$ – solutions of equation $l(y) = \lambda y$, which satisfy the following conditions:

$$u_j^{[\nu-1]}\big|_{x=a^*} = \delta_{j\nu}, \quad (a < a^* < b), \tag{14.172}$$

then occur the following inversion formulas:

$$F_j(\lambda) = \int_a^b f(x) u_j(x, \lambda) dx; \tag{14.173}$$

$$f(x) = \int_{-\infty}^{\infty} \sum_{j,k=1}^{2n} F_j(\lambda) u_k(x, \lambda) d\tau_{jk}(\lambda). \tag{14.174}$$

Here $\|\tau_{jk}(\lambda)\|$—matrix distribution function.

The decomposition of the kernel of the resolvent of operator L_u takes the following form:

$$G(x, \xi, \mu) = \int_{-\infty}^{\infty} \sum_{j,k=1}^{2n} \frac{u_k(x, \lambda) u_j(\xi, \lambda) d\tau_{jk}(\lambda)}{\lambda - \mu}. \tag{14.175}$$

In the case when one end of the interval (for example, a) is regular, the order of the matrix distribution function can be lowered. In particular, if the index of defect of operator L_0 is (n, n), then summing in formulas (14.174) and (14.175) the given above is produced only up to n.

If functions $\left(p_0^{-1}\right)', p_1, p_2, \ldots, p_n$ are summarized on interval $[0, \infty)$, and $\lim_{x \to \infty} p_0(x) > 0$, then for the operator L_u, acting on the semi-axis $[0, \infty)$ and does not have discrete spectrum, the invasion formulas take the form:

$$\begin{aligned} F(\lambda) &= \int_0^{\infty} f(x) u(x, \lambda) dx; \\ f(x) &= \int_0^{\infty} F(\lambda) u(x, \lambda) u(x, \lambda) d\lambda, \end{aligned} \tag{14.176}$$

where $u(x, \lambda)$ is the solution of the equation $l(u) = \lambda u$, which satisfies corresponding self-adjoint boundary conditions at the end of $x = 0$.

The spectral function of distribution can be easily determined through the kernel of the resolvent of operator. Directing x to $a - 0$ and ξ to $a + 0$, in the formula of decomposition of a kernel of the operator L_u (14.175) we obtain:

$$G(a-0,a+0,\mu)=\int_{-\infty}^{\infty}\sum_{j,k=1}^{2n}\frac{d\tau_{jk}(\lambda)}{\lambda-\mu}. \tag{14.177}$$

We will use further the inversion formula of Stieltjes, according to which the integral equation

$$\varphi(\mu)=\int_{-\infty}^{\infty}\frac{d\omega(t)}{t-\mu}\qquad(\mu=\alpha+i\beta), \tag{14.178}$$

has a solution

$$\omega(a)-\omega(b)=\frac{\sin\beta}{2\pi i}\lim_{\beta\to 0}\int_a^b[\varphi(\mu)-\varphi(\overline{\mu})]\,d\alpha. \tag{14.179}$$

The last formula (14.179) can be transformed and brought to a different form. For example, it is possible to write down

$$\omega(a+\delta)-\omega(\delta)=-\frac{\sin\beta}{\pi}\lim_{\beta\to 0}\int_{\delta}^{a+\delta}Im\varphi(\alpha+i\beta)d\alpha. \tag{14.180}$$

At the points of the continuity of function ω, the inversion formula takes the form:

$$\omega'(a)=-\frac{\sin\beta}{\pi}\lim_{\beta\to 0}Im\varphi(\alpha+i\beta). \tag{14.181}$$

Designating $G_{q-1,\nu-1}(a-0,\ a+0,\ \mu)=M_{\nu q}(\mu)$, where indexes designate quasi-derivatives of x and ξ, respectively, from the formula of spectral functions of distribution (14.177) and from the formula of transformed solution of integral Eq. (14.179), it is possible to obtain [377].

$$\tau_{jk}(\lambda)=\frac{1}{\pi}\lim_{\delta\to+0}\lim_{\varepsilon\to+0}\int_{\delta}^{\lambda+\delta}ImM_{jk}(\lambda+i\varepsilon)d\lambda. \tag{14.182}$$

Matrix M_{jk} is called the characteristic matrix of operator L_u .

Thus, integral transformation can be built on the basis of eigenfunctions decomposition of regular and singular differential equations.

In principle, any expansion can be interpreted as the integral transformation, only if the analytical connection or numerical procedure is known, which makes it possible to establish a correspondence between transforms (coefficients of decomposition) and original (initial) functions.

For example, the system of functions $u(\xi,\ x)$ is the solution to the equation

$$Lu-\lambda\sigma' u=0, \tag{14.183}$$

where L—some linear self-adjoint operator; σ'—weighting function is considered in the general case as the generalized function; λ—spectral parameter.

Let the eigenfunctions $u(\xi,\ x)$ satisfy the conditions of orthogolality and completeness in the form:

$$\begin{aligned}\sigma'(x)\int_{-\infty}^{\infty}u(\lambda,x)u(\lambda,x_1)d\tau(\lambda)=\delta(x-x_1);\\ \tau'(\lambda)\int_{-\infty}^{\infty}u(\lambda,x)u(\lambda_1,x)d\sigma(x)=\delta(\lambda-\lambda_1),\end{aligned} \tag{14.184}$$

where $\tau(\lambda)$—spectral function of distribution.

Then there are integral transformations with symmetrical inversion formulas in the form of Stieltjes integrals:

$$f(x)=\int_{-\infty}^{\infty}F(\lambda)u(\lambda,x)d\tau(\lambda);$$
$$F(\lambda)=\int_{-\infty}^{\infty}f(x)u(\lambda,x)d\sigma(x). \tag{14.185}$$

If the spectral function $\tau(\lambda)$ and $\sigma'(x)$ function are continuous, then Stieltjes's formulas (14.185) define standard integral transformation with infinite limits.

With the discrete spectrum of the decomposition, when the spectral function $\tau(\lambda)$ is a function of the jumps, and means spectral density $\tau'(\lambda)$ is a sequence of delta functions, the inversion formulas define finite integral transformation, convenient for solving boundary value problems with discrete spectrum. But, if in this case the weighting function is a function of the jumps, then transformation becomes discrete, and it can be applied to the solution of discrete boundary-value problems of those described by equations in finite differences.

Below we will consider further generalizations of the method of integral transformations in particular, we will use Lagrange's identity within the classical scheme of integral transformation by extending and narrowing of the domain and adding kernels which are eigenfunctions of close operators and non-orthogonal kernels Petrosian [10], Tseitlin and Petrosian [27]. Let us present these results for the purpose of the substantiation (validation) of the methods of building the boundary equations.

First, we will consider integral transformations with the orthogonal kernels which are eigenfunctions of some one-dimensional linear differential operator L and satisfy the equation:

$$lu(x,\lambda)=\lambda u(x,\lambda) \qquad (x_1<x< \ x_2) \tag{14.186}$$

It is assumed that the self-adjoint operator L originated (created) by differential operation of order $2n$

$$lu(x)=\sum_{m=0}^{n}\left(p_m u^{(n-m)}(x)\right)^{(n-m)} \tag{14.187}$$

with n times differentiable in the domain $D=(x_1,\ x_2)$ by coefficients $p_m(x)$ and by the decomposed boundary conditions:

$$A_j u(x_1)=0;\quad A_j u(x_2)=0 \quad (j=1,2,\ldots,s;\quad n\le s\le 2n) \tag{14.188}$$

$A_j u$ are understood to be linearly independent actual forms relative to the values of function $u(x)$ and its derivatives up to $2n-1$ order, inclusive. The boundary conditions assigned at each end of the interval equal to s, while different forms of A_j at both ends of the interval are limited from n to $2n$. When $s<2n$, let us supplement the set of boundary conditions with some linear independent forms to the complete system of $2n$ forms. We will write down the dual formulas decomposition of eigenfunctions of operator L with some spectral function $\tau(\lambda)$:

$$\int_{x_1}^{x_2}f(x)u(x,\lambda)dx=(f,u)=F(\lambda)$$
$$\int_{-\infty}^{\infty}F(\lambda)u(x,\lambda)d\tau(\lambda)=f(x) \tag{14.189}$$

which determines u—transformation of function $f(x)$ and its inversion. Here $\tau(\lambda)-$ nondecreasing function, continuous with continuous spectrum and stepped (with countable number of jumps) at the discrete spectrum of decomposition.

By a jump j of a function $g(x)$ at a point x_0 we mean the difference between the right-hand and left-hand limits of $g(x)$at x_0; that is $j = g(x_0 + 0) - g(x_0 - 0)$..

Formulas (14.189) cover integral transforms with the infinite limits, usually having a continuous spectrum, and with the finite limits, when spectrum is discrete.

In the classical scheme of integral transformation, formula (14.189) is applied for solving the boundary-value problems, generated by the differential operation (14.187) on $(x_{1,}x_2)$ with the homogeneous or inhomogeneous boundary conditions. Let us extend this method to the more general case of the boundary-value problems, generated operators close to L. Assume, that it is assigned the boundary-value problem

$$l_* w(x) = q(x) \qquad (a < x < b), \tag{14.190}$$

with close operation l_*, and boundary conditions:

$$\begin{gathered} B_J w(a+0) = g_{ja}; \qquad B_J w(b-0) = g_{jb}; \\ (j = 1, 2, \ldots, s; n \le s \le 2n) \end{gathered} \tag{14.191}$$

For the sake of simplicity, we will consider that the order of operations l and l_* is identical. However, a case, when operation l_* has order multiple with respect to, could be analogously considered. In accordance with the standard schema, let us multiply the left and right sides of Eq. (14.190) by $u(x, \lambda)$, and integrate over the interval $(x_{1,}x_2)$. If the $(x_{1,}x_2)$ is wider than (a, b), then $w(x)$ and $q(x)$ must be prolonged somehow out of the given interval (a, b). We have:

$$(u, l_* \, w) = (u, q) \tag{14.192}$$

The method of integral transformations further follows partial integration. It is more convenient, however, to use Lagrange's formula directly, on each interval of continuity (α, β), instead of the integration by parts

$$(u, l_* \, w) = (l_* u, w) + \left[\sum_{j=1}^{2n} B_j w(x) B_{2n-j+1} u(x, \lambda) \right]_{\alpha}^{\beta},$$

assuming that the function $w(x)$ is prolonged out of (a, b) in such a way that it satisfies the boundary conditions (14.188), when $x = x_1$ and $x = x_2$, we obtain:

$$(l_* u, w) = (u, q) + \Gamma(a, \lambda) + \Gamma(b, \lambda) \tag{14.193}$$

where

$$\begin{gathered} \Gamma(a, \lambda) = \sum_{j=1}^{2n} \Delta B_j w(x) B_{2n-j+1} u(x, \lambda) \Bigg|_{x=a}, \\ \Delta B_j w(x) = B_j w(x+0) - B_j w(x-0). \end{gathered}$$

If $l_* = l, \quad B_j = A_j$ and (a, b) coincides (x_1, x_2) that of (14.193) taking into account (14.189) follows standard result for the transformation of the required function

$$W(\lambda) = \lambda^{-1}(u, q) \tag{14.194}$$

From where we will obtain according to the inversion formula

$$w(x) = \int_{-\infty}^{\infty} \lambda^{-1}(u, q)u(x, \lambda)d\tau(\lambda). \tag{14.195}$$

Let us consider separately three cases of the possible generalizations of the method of the integral transforms, when these conditions are not satisfied:

- transformation in the extended $(x_1, x_2) \supset (a, b)$ and
- the narrowed $(x_1, x_2) \subset (a, \ b)$ domain;
- transformation with close kernel $l_* \neq l$;
- transformation with close boundary conditions $A_j \neq B_j$

1. In the case of the extended domain formula (14.193) with $l_* = l$ gives:

$$\lambda W(\lambda) = (u, q) + \Gamma(a, \lambda) + \Gamma(b, \lambda)$$

from where we will obtain according to the inversion formula (14.189)

$$\begin{aligned} w(x) = w_q(x) + \sum_{j=1}^{2n} \Delta B_j w(a) R_{2n-j+1}(x, a) \\ + \sum_{j=1}^{2n} \Delta B_j w(b) R_{2n-j+1}(x, b) \end{aligned} \tag{14.196}$$

where

$$w_q(x) = \int_{-\infty}^{\infty} \lambda^{-1} u(x, \lambda)(u, q) d\tau(\lambda)$$

There is a solution to the problem about the action of the given load in the extended domain (area);

$$R_{2n-j+1}(x, a) = \int_{-\infty}^{\infty} \lambda^{-1} u(x, \lambda) B_{2n-j+1} u(x_1, \lambda)\Big|_{x_1 = a} d\tau(\lambda).$$

Using boundary conditions (14.191), we will obtain the system of the boundary algebraic equations

$$\begin{aligned} & B_j w_q(a+0) + \\ & + \sum_{j=1}^{2n} \left[\Delta B_j w(a) B_j R_{2n-j+1}(x, a)\big|_{x=a} + \Delta B_j w(b) B_j R_{2n-j+1}(x, b)\big|_{x=a} \right] = g_{ja} \\ & B_j w_q(b-0) + \\ & + \sum_{j=1}^{2n} \left[\Delta B_j w(a) B_j R_{2n-j+1}(x, a)\big|_{x=b} + \Delta B_j w(b) B_j R_{2n-j+1}(x, b)\big|_{x=b} \right] = g_{jb} \end{aligned} \tag{14.197}$$

After solving this system relative to $\Delta B_j w(a)$, $\Delta B_j w(b)$ according to the formula (14.196) we will obtain a final solution of the problem. During the narrowing of the domain, which can be connected with the presence of internal conditions or localization of the solutions close to the zone of indignation, integration can be implemented on invariable intervals [24] or on a few intervals covering the standard for this transforms domain.

In multidimensional problems with a complex configuration, it is possible to use transformation in the narrowed domain, which is obtained as a result of the partition of the given domain of boundary value problem into several standard sub-domains. A similar method in series was examined, in particular in [29].

2. Transformation with a close kernel is convenient when the solution of a boundary value problem with close differential operation and the similar boundary conditions as in the given problem are known. Implementing transformation on a given interval (a, b), according to (14.193) we will obtain:

$$(l_* \, u, w) = (u, q) - G(\lambda), \tag{14.198}$$

where $G(\lambda)$—the known function, obtained from the bilinear Lagrange form.

We will present differential expression l_* in the form of $l_* = l + \tilde{l}$, where $\tilde{l}$ is an additional operator. Then from (14.198) according to the conversion formula (14.189), we obtain the integral equation of the second kind relative to $w(x)$

$$w(x) = \int_a^b K(x, x_1) w(x_1) dx_1 + v(x), \tag{14.199}$$

where

$$K(x, x_1) = -\int_{-\infty}^{\infty} \lambda^{-1} \, u(x, \lambda) \, \tilde{l}u(x_1, \lambda) d\tau(\lambda);$$

$$v(x) = \int_{-\infty}^{\infty} \lambda^{-1} [\, (u, q) - G(\lambda)] u(x, \lambda) d\tau(\lambda).$$

If (a, b) is finite interval and spectrum of boundary-value problem is discrete, then:

$$\begin{aligned} K(x, x_1) &= -\sum_{i=1}^{\infty} \lambda_i^{-1} u(x, \lambda_i) \tilde{l}u(x_1, \lambda_i) \tau_i; \\ v(x) &= \sum_{i=1}^{\infty} \lambda_i^{-1} [(u, q) - G(\lambda_i)] u(x, \lambda_i) \tau_i, \end{aligned} \tag{14.200}$$

where τ_i—normalizing numbers of discrete decomposition.

Being limited to finite number terms of series m in (14.200), the solution of Eq. (14.199) can be obtained according to the formula

$$w(x) = \sum_{i=1}^{m} C_i \lambda_i^{-1} u(x, \lambda_i) + v(x), \tag{14.201}$$

where C_i—solution of the system of the algebraic equations:

$$C_i = \sum_{j=1}^{m} K_{ij}\, C_j + f_i;$$
$$K_{ij} = \lambda_i^{-1} \tau_j \int_a^b u(x, \lambda_i) \tilde{l}\, u(x, \lambda_i) dx;$$
$$f_i = \tau_i \int_a^b v(x) \tilde{l}\, u(x, \lambda_i) dx$$

From Eq. (14.198) with the discrete spectrum, it is also possible to obtain a system of algebraic equations, a similar system to the method of Bubnov–Galerkin

$$\sum_{i=1}^{m} (u_i, w) \left(\tilde{l}\, u_j, u_i \right) \tau_i = (u_j, q) - G(\lambda_j) \qquad (j = 1, 2, \ldots, m), \tag{14.202}$$

if we substitute in (14.198) the decomposition

$$w(x) = \sum_{i=1}^{\infty} (u_i, w) u_i \tau_i, \tag{14.203}$$

being limited in it by first m terms of series. After solving this system relative to $W_i = (u_i, w)$ according to formula (14.203) it is possible to obtain the required solution of the problem.

3. In many cases the difficulties of solving the boundary-value problem are connected to the inconvenient boundary conditions. If a solution of a similar problem with any other boundary conditions is known, then the use of transformation with its own kernel of this problem reduces to the boundary equations, whose solution is the simpler, the closer the boundary conditions are to the original and close problems.

Let $l_* = l;\quad x_1 = a - 0;\quad x_2 = b + 0;\quad A_j \neq B_j$. Then, assuming for all considering j

$$B_j w(x_1) = B_j w(x_2) = 0 \tag{14.204}$$

we will obtain

$$\begin{aligned} w(x) = w_q(x) &+ \sum_{j=1}^{2n} B_j w(a+0) R_{2n-j+1}(x, a) \\ &- \sum_{j=1}^{2n} B_j w(b-0) R_{2n-j+1}(x, b) \end{aligned} \tag{14.205}$$

Using boundary conditions (14.191), it is possible to exclude $2n$ boundary expressions on the right side of (14.205). The number remained unknowns $B_j w(a + 0);\quad B_j w(b - 0)$, can be reduced, if $u(x, \lambda)$ satisfies some of the boundary conditions (14.188).

For determining the unknowns, boundary equations are used obtained as a result of substitution (14.205) into the boundary conditions (14.191). The number of equations will be equal to the number of incongruent differential A_j and B_j.

These generalizations of the method of the integral transforms also known as generalized integral transformation [25, 26], are especially effective in solving multi-dimensional problems. When choosing the kernel that satisfies the differential equation of a problem, the application of the generalized integral transforms reduces the problem to the solution of boundary equations—integral with the continuous spectrum and algebraic with the discrete. If the kernel of transformation satisfies only boundary conditions, then we come to the integral or algebraic equations, obtained in the entire domain of definition of the operator of considered problem. The given generalizations of the method of integral transforms do not go beyond the ideas, connected with the use of the boundary-value problems of orthogonal or biorthogonal decompositions with the discrete and continuous spectrum. The generalizations substantially different in their basis can be obtained by means of the input into the procedure of the integral transformation of non-orthogonal kernels. As such kernels, we will use Green's functions and fundamental solutions of close problems, as well as delta function. Let us first consider a boundary-value problem (14.190), (14.191) in the finite interval (a, b).

We will accept as the kernel of transformation, the Green's function $G(x_*, x)$ of the closed boundary-value problem, described by the differential Eq. (14.190) and boundary conditions (14.191). We will multiply the left and right sides of Eq. (14.190) by $G(x_*, x)$ and will integrate over the extended domain. (x_1, x_2) As a result, we find:

$$\begin{aligned} w(x_*) = w_q(x_*) + \sum_{j=1}^{2n} \Big[\Delta B_j w(a) B_{2n-j+1} G(x_*, x)\Big|_{x=a} \\ + \Delta B_j w(b) B_{2n-j+1} G(x_*, x)\big|_{x=b}\Big], \end{aligned} \tag{14.206}$$

where

$$w_q(x_*) = \int_{x_1}^{x_2} q(x) G(x_*, x) dx. \tag{14.207}$$

From (14.186) we will obtain the system of boundary equations:

$$\begin{aligned} B_j w_q(a+0) + B_j[\Gamma_1(x_*, a) + \Gamma_1(x_*, b)]_{x_*=a+0} = g_{ja}, \\ B_j w_q(b-0) + B_j[\Gamma_1(x_*, a) + \Gamma_1(x_*, b)]_{x_*=b-0} = g_{jb}, \\ \Gamma_1(x_*, x) = \sum_{j=1}^{2n} \Delta B_j w(a) B_{2n-j+1} G(x_*, x). \end{aligned} \tag{14.208}$$

In (14.206) and (14.208), there are $4n$ unknown values of $\Delta B_j w(a)$; $\Delta B_j w(b)$. From them $2n$ unknowns can be arbitrarily assigned a value, for example, equal to zero. The selection of these values is determined by the substitution of boundary conditions at points $x = a - 0$ and $x = b + 0$ with the extension of domain and the design of the Green function.

Design of the Green function by itself in many cases represents a complex task, therefore, it is desirable to select such close operator, for which Green function either can be obtained in closed form or constructed with an aid of a sufficiently simple algorithm. Determining $G(x, x_1)$ in the infinite interval is the simplest solution of boundary-value problems. In this case, as you can see, the method considered is the analog to the classical method of potential. The fundamental solution of Eq. (14.190) can serve as another non-orthogonal kernel. Due to the different form of boundary conditions, which can satisfy the fundamental solution, in formula (14.208) can remain these or other forms of boundary conditions. In particular, taking a fundamental solution, which satisfies the system of $2n$ homogeneous boundary conditions at one end of the interval (a, b), as a kernel of transformation we will obtain an analog of the method of initial values (initial parameters), and in the multidimensional case—a method of initial functions. Finally, as the non-orthogonal kernel of integral transformation, delta function can be accepted (Tseitlin [30]). The method of delta-transform is examined in detail in the following section, where we will dwell on the various solutions, and then we will trace their connection to the classical methods of structural mechanics.

Before proceeding let us pause to establish the following fundamental definitions and concepts pertaining to the extended domain, boundary-value and fundamental solutions, and the choice of a primary system. Thus, for a solution of a problem, we can choose various intervals of integration $[A, B]$, boundary conditions on $[A, B]$, fundamental solutions, and any arbitrary constant to be determined.

Extended Domain An interval of integration $[A, B]$, can coincide with (a, b), or to be infinite. In multidimensional problems, there can be a convenient choice of a finite interval of the integration including (a, b). The unlimited extended domain is characteristic for the method of compensating loads, generalized solutions, and methods of the type of potential.

Boundary conditions for $[A, B]$ can be taken arbitrarily, in particular so that the building of fundamental solutions would not cause difficulties. The choices of an interval of integration, boundary conditions, and discontinuities of the boundaries on (a, b) determine a problem, similar to the examined, mechanical interpretation of which can be connected to the widely used in structural mechanics concept of a primary system. In the case of unlimited domain $[A, B]$, the usual conditions of limitedness are established, or conditions of the type of radiation.

Fundamental Solutions As a fundamental solution, it is more convenient to accept Green's function for the extended domain. In an interval of integration that coincides with $[a, b]$, the fundamental solution can be built so that it becomes zero when $x < a$, while the generalization of the method of initial parameters for the arbitrary systems, described by one-dimensional differential operators is obtained.

Primary System The choice of a primary system predetermines the scheme of the solutions and the type of the resolving system of equations. If jumps (discontinuities) remain on the borders of (a, b) and at internal points $x = x_i$ corresponding to the introduction of additional force factors, then the resolving system of equations will be an analog of the canonical equations of the method of force, and the fundamental solutions—analog of unit displacements. While preserving discontinuities, corresponding to the natural conditions, we come to the generalization of the method of displacement. The fundamental solutions in this case will correspond to the unit reactions. With other combinations of jumps (discontinuities, breaks) we will obtain an analogue of the mixed method.

14.8 The Method of Delta-Transform

The method of delta- transform arose as the apparatus, justifying the application of the generalized functions for constructing boundary equations of different types. On the basis of this method and the method of the generalized integral transforms, it is possible to give a sufficiently general scheme of construction and classification of all existing methods and types of boundary equations. It should be noted that a number of subtle, specific questions, solutions of which this scheme could convert into a strict theory of boundary equations, still requires in-depth analysis and relevant proofs. Nevertheless, the method of delta-transform makes it possible to connect virtually all methods of constructing boundary integral equations into a united system. Moreover, the method of delta-transform gives a clear mechanical interpretation of the various methods, and highlights the close connections between them, in particular, between the methods of potentials and *boundary integral equations method* (BIEM).

Below you will find the results of delta and Green transformation schemas [9, 17, 30]. First, we will discuss the usage of delta transformation for solving general type of self-adjoint (self-conjugate) boundary value problems. Secondly, we will show the connection between this method and Green transformation method [17, 25].

First, we will consider an application of delta-transform to the solution of a self-adjoint (conjugate) boundary-value problem of general view, and then we will point to a connection of this method to the integral transformation with kernel in a form of the Green's function of a considered or close boundary-value problem.

One-dimensional boundary-value problem for the operator L generated by the linear self-adjoint differential operation

$$lw(x) = \sum_{m=0}^{n} \left(p_m(x) w^{(n-m)}(x) \right)^{(n-m)}, \qquad (a < x < b), \tag{14.209}$$

and by the boundary conditions

$$A_j w(x)\big|_{x=a+0} = 0; \quad A_j w(x)\big|_{x=b-0} = 0, \qquad (j = 1, 2, \ldots, s), \tag{14.210}$$

which we for certainty will also consider self-adjoint is described by the equation

$$lw(x) = q(x) \qquad (a < x < b), \tag{14.211}$$

and boundary conditions (14.210).

Let us apply to Eq. (14.211) of delta-transform in the maximally extended length, i.e., integral transform with the kernel in a form of a delta function on an infinite interval. For this purpose, we will multiply both sides of Eq. (14.211) on $\delta(\xi - x)$ and will integrate along entire axis $(-\infty < x < \infty)$, continuing $w(x)$ and $q(x)$ out of (a, b). In accordance with (14.162) we obtain with $u = \delta(\xi - x)$:

$$\begin{aligned} \int_{-\infty}^{\infty} w(x) l\delta(\xi - x) dx = q(\xi) &+ \left[\sum_{j=1}^{2n} \Delta B_j w(x) B_{2n-j+1} \delta(\xi - x) \right]_{x=a} \\ &+ \left[\sum_{j=1}^{2n} \Delta B_j w(x) B_{2n-j+1} \delta(\xi - x) \right]_{x=b} \end{aligned} \tag{14.212}$$

The integral on the left side of the formula (14.212) on the entire real axis, except for points $x = a$, $x = b$, according to the basic property of delta function (Gelfaund and Shilov [12].) equals to $lw(\xi)$. Therefore, we will rewrite (14.212) in the following more simplified form:

$$lw(\xi) = q(\xi) + \sum_{j=1}^{2n} \Delta B_j w(x) B_{2n-j+1} \delta(\xi - x) \Bigg|_{x=a,b.} \qquad (|\xi| < \infty) \tag{14.213}$$

The analysis of Eq. (14.213) presents a certain interest. First, along with differential expression, this equation contains the operators of boundary conditions of the delta functions. Here we equalize the boundary forms B_j with differential forms of boundary conditions (14.210), i.e., assume $B_j = A_j$. Secondly, it is set in the extended interval. Thirdly, Eq. (14.213) contains $4n$ unknown step (jump) functions $\Delta B_j w(a)$, and $\Delta B_j w(b)$, for determining where there are only $2n$ boundary conditions (14.210). Finally, the right-hand side of Eq. (14.213) includes differential operators of the delta functions representing the generalized forces and kinematic actions. Let us examine this question in more detail. It is known that during the design of beams the force and moment can be described mathematically as a delta function and its first derivative. Similarly, higher moments can be introduced also. So, the second derivative of the delta function, that is the first derivative of the moment, describes the kinematic actions, corresponding to the unit step of the angle of rotation in the section $x = \xi$ and the third derivative of the delta function corresponding to the unit step of displacement (with the accuracy to the multiplier of EI).

The central issue in the method of delta-transform is the choice of unknown jump functions $\Delta B_j w(a)$, and $\Delta B_j w(b)$. Of the $4n$ step functions $2n$ can be set to zero in order to define the remaining $2n$ step functions from $2n$ boundary conditions. For this purpose, the following mechanical considerations can be used in classical methods of structural mechanics when choosing the primary system. Assuming any jump function $\Delta B_j w$ equals to zero, we fix one or extend the considered system through boundary points. For example, if we accept the first $n - 1$ continuous derivatives of the function $w(x)$, corresponding to geometrical factors (shape), then we continue construction through the boundary sections as continuous. In this case, unknowns become the jump functions of the higher derivatives, which correspond to the forces factors. Hence, in this case, the primary system is chosen as an infinite continuous construction, described by Eq. (14.211), and loaded in the boundary sections by external forces and moments. An alternative approach is to accept continuous forces factors in the boundary sections, assuming the jump functions of the corresponding boundary forms equal to zero. Such an approach will correspond to the continuation of construction through the boundary section, where only kinematic impact is set, i.e., in boundary sections jump functions of displacements and angles of rotations are assigned.

These two methods of extension of domain and choosing the primary system are similar to classical methods of forces and displacements in structural mechanics. If we keep different jumps $\Delta B_j w(x)$ in the sections $x = a$, $x = b$, then we will obtain the analog of the mixed method of structural mechanics. It should be noted that not all jumps can be to assumed to be zero, and consequently not all of them can be kept in the solution of a problem. In fact, the choice of force and kinematic factors, determined by the jumps of the differential forms, must provide linear independence of the solution of the homogeneous equation, forming Green function and its derivatives. Therefore, the jumps of adjoint differential forms cannot be saved or omitted if, for example, we assume $\Delta B_j w(x) = 0$, in one of the boundary sections, then for this section, it is already impossible to accept $\Delta B_{2j - j + 1} w(x)$. Besides the two indicated variants of the choice of the unknown jumps, determining the force and kinematic impacts, it is possible to indicate the following possible loads of boundary sections. One of such loads is connected to the preservation of all jumps, i.e., all possible types of force and kinematic impacts; the other, contrary to this, is connected to the complete absence of any compensating impacts in the boundary points. In the latter case, in order to obtain nontrivial solution of a problem, it is necessary to place the compensating force and kinematic impacts in the extended domain outside (beyond) of the boundaries of the interval (a, b). This case is equivalent to the modification of the method of compensating loads proposed by the Korenev [19].

With the retention of all jumps of differential forms, additional conditions, setting the relationship between the unknown jumps, so that the number of jumps $4n$ would correspond to the number of boundary conditions $2n$ and the number of additional conditions to $2n$, must be set.

Thus, for systematization of the equations of the method of delta-transform, we should examine three basic cases when the compensating actions are located at the boundary points.

I. $2n$ jumps of geometric factors described by differential forms to $n - 1$th order inclusive are set to equal to zero. The external force impacts are unknown. Solving Eq. (14.213) and satisfying the boundary conditions (14.210), we come to the system of $2n$ boundary equations relative to the jumps of the force factors. This scheme of solution corresponds to the common form of the force method, method of compensating loads, and indirect method of boundary elements.

II. $2n$ jumps of force factors described by differential forms, which contain the derivatives of order n and above are set to equal to zero. The external kinematic impacts are unknown. After solving Eq. (14.213) and satisfying boundary conditions (14.210) we come to the system of $2n$ boundary equations relative to the jumps of the kinematic parameters in the boundary sections. In this case, the higher derivatives of the delta function in the right part of Eq. (14.213) remain, which reduces to the irregular equations and, in the case of using the transformations leads to divergent decomposition. The obtained equations can be regularized through the isolation (separation) of delta function and its derivatives, making the equation easily solvable. In its mechanical content, this method is the analog of the method of displacements.

III. All jumps of differential forms remain in the solution. In this case, they must be set, as mentioned above, to $2n$ additional conditions, connecting these jumps. In particular, choosing any n of the boundary forms, corresponding to any self-adjoint boundary conditions in the section $x = a$ and n of the possible other forms, which also correspond to some self-adjoint boundary conditions in the section $x = b$, has us assume them to be equal to the values of the corresponding forms on the internal boundaries of interval (a, b). This gives $2n$ additional conditions:

$$\Delta B_j w(a) = B_j w(a+0), \quad \Delta B_j w(b) = B_j w(b-0).$$

Then, respectively,

$$B_j w(a-0) = 0, \quad B_j w(b+0) = 0$$

Since we can set $q(x) = 0$ with $x < a,\ x > b$ and boundary conditions are homogeneous, then the solution of Eq. (14.209) in the closed interval $[a, b]$ must be trivial; that is, identically equal to zero. Therefore, the remaining boundary forms $B_j w(a + 0),\ B_j w(b - 0)$ are also equal to zero, and therefore the jumps, which we for distinction will designate through B_i, are equal

$$\Delta B_i w(a) = B_i w(a+0), \quad \Delta B_i w(b) = B_i w(b-0).$$

Thus, all jumps are equal to the values of the corresponding differential forms on the internal boundaries (borders) of the interval (a, b). Therefore, instead of (14.213) case we can write down the following:

$$lw(\xi) = q(\xi) + \sum_{j=1}^{2n} B_j w(x) B_{2n-j+1} \delta(\xi - x) \Big|_{x=a,b.} \tag{14.214}$$

Furthermore, it is possible to satisfy $2n$ conditions in Eq. (14.210) since $(B_j = A_j)$, and the last equation will contain only $2n$ unknowns, making it possible to solve any correctly stated problem. This method of composition of boundary equations is equivalent to the direct method of boundary elements (DBEM).

For solving Eq. (14.213) we can recommend two methods—Green's function and the method of integral transformation. In the first case, it is assumed that Green's function of boundary value problem (14.209) in the extended domain is known. Further, we will consider generalized Green's functions, which are solutions to the following equations.

$$l_\xi \qquad G_k(\xi, x) = B_k^\xi \delta(\xi - x). \tag{14.215}$$

Here, the index ξ means, that the differential operations are taken on variable ξ. These solutions can be easily built, if the operators l_ξ^{-1} and B_k^ξ are permutable. Applying operation l_ξ^{-1} to the left and right sides of Eq. (14.215), we obtain

$$G_k(\xi, x) = B_k^\xi l_\xi^{-1} \delta(\xi - x).$$

Since Green's function satisfied the equation:

$$l_\xi G_k(\xi, x) = \delta(\xi - x),$$

and, consequently,

$$l_\xi^{-1} \delta(\xi - x) = G(\xi, x),$$

then,

$$G_k(\xi, x) = B_k^\xi G(\xi, x). \tag{14.216}$$

Taking into account (14.215), the solution of Eq. (14.213) can be written down in the form of:

$$w(\xi) = w_q(\xi) + \sum_{j=1}^{2n} \Delta B_j w(x) G_{2n-j+1}(\xi, x) \bigg|_{x=a,b}, \tag{14.217}$$

where $w_q(\xi)$—is the solution of $lw(\xi) = q(\xi)$, satisfying some boundary conditions at the ends of the extended domain.

A similar equation can be obtained by direct application of the Green's function transformation to Eq. (14.209). The Green's function transformation, as mentioned above, will be called integral transformation with the kernel in a form of Green's function of the considered or close boundary-value problem.

It is not difficult to see that assuming $u = G(\xi, x)$, in Eq. (14.193) we will obtain (14.217). The advantage of the delta-transform, in comparison with the transformation of Green's function, is the clear mechanical interpretation of compensating impacts, and the use of other methods, such as the integral transformation for solving Eq. (14.213), but not reducing it to Eq. (14.217).

It should be noted that the type of representation in (14.217), analogous to Somigliana's formula in the theory of elasticity, plays an important role in the boundary elements and boundary equations methods. In particular, direct methods of boundary conditions are built on the basis of the use of Green's function for an infinite domain and integral identity of Betty within the limits of the given domain (a, b), which reduces to the type of formula in (14.217), where we have values of these forms at the boundary points instead of jumps $\Delta B_j w(x)$, As we see, unlike the DBEM, the application of the method of delta-transform or Green's function transformation provides more opportunities for constructing boundary equations. Instead of one scheme of direct MBE, the delta-transform and Green's function transformation methods lead to a wide variety of boundary equations and give a unified scheme for composing them.

14.8.1 General Approaches to Constructing Boundary Equations

The application of the methods of boundary equations and boundary elements to the solution of boundary-value problems is the most effective when considering two-dimensional and three-dimensional problems. Nevertheless, in the simplest one-dimensional problems the application of these methods is often quite convenient. Therefore, first we will consider the reduction of one-dimensional boundary-value problems to boundary equations of different types, which will establish a link between the different methods of the type of boundary elements and to a certain extent, classify these methods. On the other hand, the following examples, along with an illustrative function, can represent an independent interest. Let us examine a linear self-adjoint differential expression of order $2n$:

$$lw(x) = \sum_{m=0}^{n} \left(p_m(x) w^{(n-m)}(x)\right)^{(n-m)}, \quad (a < x < b) \tag{14.218}$$

with $n - m$ times differentiable on the interval (a, b) coefficients $p_m(x)$, where$p_0(x) \neq 0$.

We will join expression (14.218) with the decomposed boundary conditions in a form of the linear independent substantial forms, relative to the values of the function $w(x)$ and its derivatives up to $2n -$ 1 orders, inclusive in the boundary points $x = a + 0$ and $x = b - 0$:

$$A_j w(x)\big|_{x=a+0} = 0; \quad A_j w(x)\big|_{x=b-0} = 0, \quad (j = 1, 2, \ldots, s) \tag{14.219}$$

The boundary conditions assigned at each end of the interval (a, b) equal to s, while different forms of A_j at both ends of the interval are limited from n to $2n$.

Thus, the total of both ends of the interval can meet $s$various forms A_j. If the total number of various modes (forms) $s < 2n$, then we will complement the system of boundary conditions with some linearly independent forms up to the complete system of $2n$ modes (forms).

On each section of continuity (α, β) the functions $w(x)$, $v(x)$ and their derivatives up to $2n$ order inclusive for differential operation (14.218) are valid per the Lagrange-Green formula (see Naimark [28])

$$\int_{\alpha}^{\beta} v(x) lw(x) dx = \int_{\alpha}^{\beta} w(x) lv(x) dx + [v, w]_{\alpha}^{\beta}, \tag{14.220}$$

where the Lagrange's brackets $[\,v, w]$ have the form:

$$[v, w]_{\alpha}^{\beta} = \sum_{j=1}^{2n} B_j v(x) B_{2n-j+1}^{*} w(x) \Bigg|_{x=\alpha}^{x=\beta}. \tag{14.221}$$

Here B_j—the arbitrary system of the linearly independent boundary modes (forms); B_{2n-j+1}^{*}—adjoint (conjugate) system.

A one-dimensional boundary-value problem for an operator L, generated by the linear self-adjoint differential operation (14.218) and by the boundary conditions (14.219), which we certainly will also consider as a self-adjoint, is described by the equation:

$$lw(x) = q(x) \qquad (a < x < b), \tag{14.222}$$

and boundary conditions (14.219).

The general solution of Eq. (14.222) is equal to the sum of any particular solution of the nonhomogeneous Eq. (14.222) and the general solution of the homogeneous equation $lw(x) = 0$, satisfying the boundary conditions (14.219). Let us choose as the particular solution of the inhomogeneous equation any solution of Eq. (14.222) in any domain Ω, including (a, b).

At the maximum such domain (henceforth an extended domain) is the entire real axis $(-\infty, \infty)$; at the maximum is a section $[a, b]$, which is obtained from (a, b) by inclusion of the boundary points $x = a - 0$ and $x = b + 0$. We will designate this particular solution through $w_q(x)$. Then the general solution of Eq. (14.222) can be represented in the form:

$$w(x) = w_q(x) + \sum_{j=1}^{2n} C_j \varphi_j(x), \tag{14.223}$$

where $\varphi_j(x)$ are linearly independent solutions of the homogeneous equation

$$l\varphi(x) = 0 \tag{14.224}$$

C_j—arbitrary constants.

Determining arbitrary constants from the boundary conditions in (14.219) and completing building the solutions for problem presents a system of equations, which can be named boundary, since these equations are built for the boundary points. This is, of course, an elementary way to find solutions. The methods of boundary equations use essentially the same procedure, but they operate in this case with the solutions of problems only in the extended domain. Any system of $2n$ linearly independent solutions of Eq. (14.224) is called a fundamental system, and the fundamental solution of Eq. (14.222) is called any solution of equation

$$lE(x, \xi) = \delta(x - \xi) \tag{14.225}$$

If it satisfies some homogeneous boundary conditions, then it is the function of Green $G(x, \xi)$ of the corresponding boundary-value problem. Green's function for extended unlimited domain represents the greatest interests, since in this case it can easily be built, and usually has a simple structure. In this case, $G(x, \xi)$ along with Eq. (14.225) satisfies conditions of limitedness at infinity. Green's function is the convenient tool for analyzing boundary-value problems and is the basis for most methods of the type of boundary equations. In calculations of structures, Green's function is usually the displacement of the system at any section x due to unit concentrated force at point ξ. Let us consider the equation:

$$lG_k(x, \xi) = \frac{d^k}{d\xi^k} \delta(x - \xi), \tag{14.226}$$

then $G_k(x, \xi)$ will represent the displacement of the system under the action of unit moments and kinematic perturbation of the type of unit displacement and unit angels of rotation (with the accuracy to functional coefficients).

The Green's function of self-adjoint operators, which we examine here, they are symmetrical, i.e.:

$$G(x, \xi) = G(\xi, x)$$

As it is known from [9, 17, 30, 31], the Green's function can be constructed from the homogeneous solution,

$$G(x,\xi)=\begin{cases}\sum_{i,j=1}^{2n}\alpha_{ij}\varphi_i(x)\varphi_j(\xi) & (x\leq\xi),\\ \sum_{i,j=1}^{2n}\alpha_{ji}\varphi_j(x)\varphi_i(\xi) & (\xi\leq x),\end{cases}\tag{14.227}$$

where α_{ij}—some constant coefficients, and $\varphi_j(\xi)$—linearly independent solutions of Eq. (14.224).

If Green's function of the problem (14.222), (14.219) is known, then the solution can be obtained by the formula

$$w(x)=\int_a^b G(x,\xi)q(\xi)d\xi.\tag{14.228}$$

Thus, the solution of nonhomogeneous Eq. (14.222) can be obtained, if any particular solution and fundamental system of the solutions of the homogeneous equation is known. For obtaining a particular solution, corresponding to the right side $q(x)$, it is convenient to use Green's function of the operator l in any extended domain. In particular, the Green's function for differential operators in an unlimited domain is known. Therefore, using the Green's function for the extended domain of Ω we can obtain a particular solution of Eq. (14.222) in the form of:

$$w_q(x)=\int_\Omega G(x,\xi)q(\xi)d\xi.\tag{14.229}$$

The idea behind the method of boundary equations allows to obtain the general solution (14.223) with the same Green's function, used in (14.219) without determining the fundamental system. Alternatively, in addition to the Green's function, the solution in (14.222) in the extended domain can be used. The simplest mechanical representation of the stress-strain condition of the construction, which is undergone the action of static or dynamic loads can be used. The foregoing indicates the opportunity of a dual determination of fastening and boundary conditions of a corresponding boundary-value problem, by counter-proposing some value of energy conjugate (adjoint) factor to each given geometrical or force factor, i.e.: force–displacement, moment–angle of rotation, and vice versa. For example, if a beam is given a boundary condition corresponding to a built-in at one end (cantilever beam), then this condition can be considered as equal to zero displacement and angle of rotation in the end section or as action in this section such as force and moment, which together with the given load become zero displacement and angle of rotation. In the latter case, the problem is solved relative to the boundary factors, force, and moment for the beam, having the arbitrary conditions of fastening of end section.

This fact has quite obvious mathematical basis. Let us examine as an illustration a beam with the built-in ends, which lies on the elastic Winkler foundation. The equation of the equilibrium of beam can be written as:

$$EIy^{IV}(x)+ky(x)=q(x),\tag{14.230}$$

where EI, l—rigidity and span of beam; k—coefficient of bed; $y(x)$—deflection; $q(x)$—external load. The boundary conditions will be set on the internal borders of the interval $(0, l)$.

$$y(+0)=y'(+0)=0,\qquad y(l-0)=y'(l-0)=0.\tag{14.231}$$

The general solution of Eq. (14.230) can be obtained by adding the general solution of the corresponding homogeneous equation $y_0(x)$, and the particular solution of nonhomogeneous equation $y_n(x)$. We will accept as a particular solution Eq. (14.230) for any domain Ω , including (0, l) with any boundary conditions on the border of Ω. The domain Ω , or the extended domain, can represent the entire real axis, or any limited domain up to [0, l], with the attached points of external border of this interval $x = -0$ and $x = l + 0$. Assume that in the extended domain, are assigned the boundary conditions, corresponding to the free ends of the beam are assigned, for example:

$$y_n^{II}(-0) = y_n^{III}(-0) = y_n^{II}(l+0) = y_n^{III}(l+0) = 0. \tag{14.232}$$

Then,

$$y_n(+0) \neq 0, \qquad y_n^{I}(+0) \neq 0, \quad \text{etc.}$$

Therefore, on the boundary of the given domain (0, l) the following conditions derived from (14.231) must be satisfied:

$$\begin{aligned} y_0(+0) &= -y_n(+0); & y_0^{I}(+0) &= -y_n^{I}(+0); \\ y_0(l-0) &= -y_n(l-0); & y_0^{I}(l-0) &= -y_n^{I}(l-0). \end{aligned} \tag{14.233}$$

The general solution of the homogeneous equation can be obtained as the solution of the following equation:

$$EIy_0^{IV}(x) + ky_0(x) = 0, \tag{14.234}$$

with the nonhomogeneous boundary conditions:

$$\begin{aligned} y_0^{II}(-0) &= \frac{M_0}{EI}; & y_0^{III}(-0) &= \frac{P_0}{EI} \\ y_0^{II}(l+0) &= \frac{M_l}{EI}; & y_0^{III}(l+0) &= \frac{P_l}{EI}, \end{aligned} \tag{14.235}$$

and then in Eq. (14.233), the unknowns will be the boundary factors P_0, P_l, M_0, M_l on which the solution $y_0(x)$ will linearly depend. Thus, (14.233) is the system of "boundary equations", since these equations are assigned at the boundary points of considering boundary-value problem.

It should be noted that the general solution for a homogeneous equation can be obtained by solving a nonhomogeneous equation with the homogeneous boundary conditions in (14.232) and the right side being other than zero outside of the interval (0, l) only, in particular by force and moment, applied at the points $x = -0$, and $x = l + 0$. Thus, the general solution of the problem will consist of two solutions of nonhomogeneous equation with boundary conditions (14.232), as well as given and some additional loads. Thus, the boundary equations contain solutions of unassigned problem (14.230), (14.231), and some other problem, (14.230) in the extended domain. Usually, an extended domain and boundary conditions on its boundary are chosen in such a way that the corresponding solution could be obtained in the simplest way. In this case, instead of solving the homogeneous equation with boundary conditions (14.235), the Green's function can be used to obtain the solution of the given problem in a reliable and quite convenient form.

Therefore, the extension of the domain when considering a problem and obtaining solutions of nonhomogeneous equations in the extended domain and corresponding given load and load which is applied at the boundary points, allows to build boundary equations without determining the fundamental system. This circumstance is especially important when solving multidimensional problems.

A question about transition from the homogeneous differential equation with nonhomogeneous boundary conditions (14.235) to the nonhomogeneous differential equation with homogeneous boundary conditions (14.232) requires a special consideration [17, 32]. It is known that concentrated impact such as force, moment, jump in the angle of rotation, and jump in the displacement can be represented in the form of generalized functions $\delta(x)$, $\delta^{I}(x)$, $\delta^{II}(x)/EI$, and $\delta^{III}(x)/EI$, respectively, where $\delta(x)$—delta function. Therefore, for such an equation as in (14.234), this transition does not cause difficulties. In the more general case, to pass from heterogeneous (or inhomogeneous) boundary conditions to the nonhomogeneous equation, it is possible to transfer heterogeneity from boundary conditions to the right part of equation [32].

14.9 The Spectral Method of Boundary Elements

As Sect. 14.7 shows, the application of generalized integral transformations (transforms) to solve boundary-value problems when eigenfunctions of operators close to the operator of the problem, are used as kernels is reduced to boundary equations, we have one less dimensionality than the initial (original) equation of equilibrium or motion. In Sects. 14.7 and 14.8 various generalizations of integral transformations associated with the use of classical transforms in non-canonical domains and with non-canonical boundary conditions, as well as transformations with close kernel and non-orthogonal kernels were considered. Below, the method of building of integral and algebraic equations bases of the generalized integral transformations will be presented in detail. Following Popov [25, 26], we will examine orthogonal and biorthogonal decompositions on eigenfunctions of close operators. Some extensions of this method, being the spectral analogue of BEM and method of delta transformation are given. So, we will call this method the spectral method of boundary elements (SMBE) [33]. The essence of the SMBE for one- dimensional problems briefly presented in paragraphs 14.7, and 14.8 consists of building of boundary equations, not with the aid of Green's function, but on the basis of the spectral decomposition of the differential operators considered in the expanded area (domain). In particular for this purpose, the transformations of Fourier and Hankel, along with widely used structural mechanics decomposition on beam functions, may be used. The spectral method is especially convenient when constructing a Green's function is complicated or can be carried out only with the aid of an integrated or discrete decomposition. In general form the boundary equations of SMBE for the one-dimensional problem take form (14.197) and contain twice as many unknown jumps as the available boundary conditions for their determination. Depending on the choice of these jumps, as in the case of a delta-transformation method, we will obtain the spectral analogs of direct and indirect methods of boundary elements. In particular, if we assume that jumps of all differential forms, which contain derivatives up to the nth order, are equal to zero, then this approach will be equivalent to either an indirect method of the BEM or the method of potentials with lower multiplicity, corresponding to force impacts (for the equations of the second order simple layer). But if we assume that jumps of the differential forms, which contain the derivatives of order higher than $n - 1$, are equal to zero, then we will obtain an analog of the method of potential with the use of layers of the highest multiplicity (for the equations of the second order dual layer). With the retention of all jumps it is necessary to use an additional condition, for example, equating the displacements out of the interval (a, b) to zero, ensuring for example, displacements out of the interval (a, b), are equal to zero which provide the sole solution of the problem. In this case, SMBE is equivalent to the direct method of boundary elements or to method of boundary integral equations. For the purpose of clarity, let us illustrate a scheme of a spectral method to solve the simplest model problem, in particular we will consider the problem described in Sect. 14.8.1, with bent free beam on elastic Winkler foundation. The boundary conditions can be written as:

$$y''(+0) = y'''(+0) = 0, \qquad y''(l-0) = y'''(l-0) = 0. \tag{14.236}$$

In addition, we will assume the displacement of the beam, together with the first three derivatives as equal to zero at infinity. To solve equation in (14.234) with the boundary conditions (14.236) we will use Fourier transformation. For this purpose, we will prolong $y(x)$ and $q(x)$ to the entire axis, and will proceed in the left and right parts of the equation to the Fourier transforms. After multiplication (14.234) by $\frac{1}{\sqrt{2\pi}}e^{i\xi x}$ and integration, we will obtain:

$$\left(EI\xi^4 + k\right)Y(\xi) = Q(\xi) - \frac{EI}{\sqrt{2\pi}}\left[\Delta y'''(x) - i\xi\Delta y''(x) - \xi^2\Delta y'(x) + i\xi^3\Delta y(x)\right]_0^l,$$

where

$$Y(\xi) = \frac{1}{\sqrt{2\pi}}\int_{-\infty}^{\infty} y(x)e^{i\xi x}dx; \qquad Q(\xi) = \frac{1}{\sqrt{2\pi}}\int_{-\infty}^{\infty} q(x)e^{i\xi x}dx.$$

Similarly to the method of delta-transformation, and green-transformation, here we have twice as many unknown jumps as boundary conditions (eight unknown jumps with four boundary conditions (14.236)). Therefore, there are variants of continuation of function $y(x)$ to the entire axis. We will consider the following two cases of choice of the jumps:

1. Let us assume that displacements and angles of rotations in cross-sections $x = 0$ and $x = l$ of an infinite beam in its expanded domain are continuous. Loading is achieved by force factors: force and moment. In this case $\Delta y(0) = \Delta y'(0) = \Delta y(l) = \Delta y'(l) = 0$ and

$$Y(\xi) = \frac{Q(\xi) - (Q_0 - i\xi M_0 - Q_l + i\xi M_l)/\sqrt{2\pi}}{EI\xi^4 + k}, \tag{14.237}$$

where M_0, M_l, Q_0, and Q_l are jumps of the moment and shear force in the central section and in the section $x = l$. Using the inversion formula, we have:

$$y(x) = y_q(x) - \frac{Q_0 - Q_l}{\pi}\int_0^{\infty} \frac{\cos\xi x d\xi}{EI\xi^4 + k} - \frac{(M_0 - M_l)}{\pi}\int_0^{\infty} \frac{\xi\sin\xi x d\xi}{EI\xi^4 + k}, \tag{14.238}$$

where

$$y_q(x) = \frac{1}{\sqrt{2\pi}}\int_{-\infty}^{+\infty} \frac{Q(\xi)e^{-i\xi x}d\xi}{EI\xi^4 + k}$$

is displacement of infinite beam due to load $q(x)$.

We will obtain boundary equations, satisfying the conditions (14.236)

$$
\begin{aligned}
&y_q''(0)+\lim_{x\to+0}\left[\frac{(Q_0-Q_l)}{\pi}\int_0^\infty\frac{\xi^2\cos\xi xd\xi}{EI\xi^4+k}+\frac{(M_0-M_l)}{\pi}\int_0^\infty\frac{\xi^3\sin\xi xd\xi}{EI\xi^4+k}\right]=0,\\
&y_q''(l)+\lim_{x\to l-0}\left[\frac{(Q_0-Q_l)}{\pi}\int_0^\infty\frac{\xi^2\cos\xi xd\xi}{EI\xi^4+k}+\frac{(M_0-M_l)}{\pi}\int_0^\infty\frac{\xi^3\sin\xi xd\xi}{EI\xi^4+k}\right]=0,\\
&y_q'''(0)+\lim_{x\to+0}\left[-\frac{(Q_0-Q_l)}{\pi}\int_0^\infty\frac{\xi^3\sin\xi xd\xi}{EI\xi^4+k}+\frac{(M_0-M_l)}{\pi}\int_0^\infty\frac{\xi^4\cos\xi xd\xi}{EI\xi^4+k}\right]=0\\
&y_q'''(l)+\lim_{x\to l-0}\left[-\frac{(Q_0-Q_l)}{\pi}\int_0^\infty\frac{\xi^3\sin\xi xd\xi}{EI\xi^4+k}+\frac{(M_0-M_l)}{\pi}\int_0^\infty\frac{\xi^4\cos\xi xd\xi}{EI\xi^4+k}\right]=0.
\end{aligned}
\tag{14.239}
$$

For calculation of the integrals in the boundary Eq. (14.239), let us note that with $q(x) = \delta(x)$, and $M_0 = Q_0 = 0$ solution (14.238) gives the expression of Green's function about the infinite beam problem.

$$
G_0(x)=\frac{1}{\pi}\int_0^\infty\frac{\cos\xi xd\xi}{EI\xi^4+k}
\tag{14.240}
$$

We will obtain the derivatives of Green's function $G_n(x)=\frac{d^n}{dx^n}G(x)$, by formal differentiation:

$$
\begin{aligned}
G_1(x)&=-\frac{1}{\pi}\int_0^\infty\frac{\xi\sin\xi xd\xi}{EI\xi^4+k},\\
G_2(x)&=-\frac{1}{\pi}\int_0^\infty\frac{\xi^2\cos\xi xd\xi}{EI\xi^4+k},\\
G_3(x)&=\frac{1}{\pi}\int_0^\infty\frac{\xi^3\sin\xi xd\xi}{EI\xi^4+k},\\
G_4(x)&=\frac{1}{\pi}\int_0^\infty\frac{\xi^4\cos\xi xd\xi}{EI\xi^4+k}.
\end{aligned}
\tag{14.241}
$$

Integrals, representing functions $G_3(x)$, and $G_4(x)$, disperse and do not exist in a classical sense. Therefore, they have to be understood as the generalized functions, which consist in a general case of regular and irregular components. The last integral transforms as follows:

$$
\begin{aligned}
\frac{1}{\pi}\int_0^\infty\frac{\xi^4\cos\xi xd\xi}{EI\xi^4+k}&=\frac{1}{\pi EI}\int_0^\infty\frac{\left(EI\xi^4+k-k\right)\cos\xi xd\xi}{EI\xi^4+k}\\
&=\frac{1}{\pi EI}\int_0^\infty\cos\xi xd\xi-\frac{k}{\pi EI}\int_0^\infty\frac{\cos\xi xd\xi}{EI\xi^4+k}
\end{aligned}
$$

It is known from the theory of generalized functions [12], that

$$
\int_0^\infty\cos\xi xd\xi=\pi\delta(x)
\tag{14.242}
$$

$$
\int_0^\infty\frac{\sin\xi x}{\xi}d\xi=\frac{\pi}{2}\operatorname{sign}x
\tag{14.243}
$$

Multiplying and then dividing the integrand expression in (14.242) and (14.243) by $EI\xi^4 + k$, we will find

$$G_3(x) = -\frac{2\lambda^4}{k}\,\text{sign}x e^{-\lambda|x|}\cos\lambda x,$$
$$G_4(x) = \frac{1}{EI}[\delta(x) - kG_0(x)] = \frac{1}{EI}\left[\delta(x) - \frac{\lambda}{2}e^{-\lambda|x|}(\cos\lambda x + \sin\lambda x)\right]. \tag{14.244}$$

Further satisfaction of the boundary conditions taking into account only regular components of $G_4(x)$, leads to boundary equations, which are similar to the equations of the method of delta-transformation considered in Sect. 14.8, and also to equations of the indirect method of boundary elements. It should be noted that the issues of improvement of convergence of integrals of the type (14.241), and similar integrals, which are encountered in the close problems for the beams, lying on linearly deformable base, were considered by many authors, in particular Korenev [15], Popov [25], Travush [21, 22], and Tseitlin and Petrosian, Atadjanov [34], Tsudik [35], etc.

2. Let us assume that the force factors $\Delta y''(0) = \Delta y'''(0) = 0$ in sections $x = 0$, and $x = l$ are continuous, and the kinematic impacts $\Delta y(0) \neq 0$, $\Delta y'(0) \neq 0$ are applied. In this case

$$Y(\xi) = \frac{Q(\xi) + \left[i\xi(y_0 - y_l) - \xi^2(\varphi_0 - \varphi_l)\right]/\sqrt{2\pi}}{EI\xi^4 + k} \tag{14.245}$$

where y_0, y_l, φ_0, and φ_l are jumps of displacement and angles of rotation in sections $x = 0$, $x = l$. By the inversion of (14.245), and taking into account (14.241), we will have:

$$y(x) = y_q(x) - \frac{(y_0 - y_l)}{\sqrt{2\pi}}G_3(x) + \frac{(\varphi_0 - \varphi_l)}{\sqrt{2\pi}}G_2(x) \tag{14.246}$$

The further submission (14.246) into (14.236) using the limiting transitions gives

$$\begin{aligned}
&y_q''(0) + \lim_{x\to+0}\left[-\frac{y_0 - y_l}{\sqrt{2\pi}}G_3''(x) + \frac{\varphi_0 - \varphi_l}{\sqrt{2\pi}}G_2''(x)\right] = 0,\\
&y_q''(l) + \lim_{x\to l-0}\left[-\frac{y_0 - y_l}{\sqrt{2\pi}}G_3''(x) + \frac{\varphi_0 - \varphi_l}{\sqrt{2\pi}}G_2''(x)\right] = 0,\\
&y_q'''(0) + \lim_{x\to+0}\left[-\frac{y_0 - y_l}{\sqrt{2\pi}}G_3'''(x) + \frac{\varphi_0 - \varphi_l}{\sqrt{2\pi}}G_2'''(x)\right] = 0,\\
&y_q'''(l) + \lim_{x\to l-0}\left[-\frac{y_0 - y_l}{\sqrt{2\pi}}G_3'''(x) + \frac{\varphi_0 - \varphi_l}{\sqrt{2\pi}}G_2'''(x)\right] = 0,
\end{aligned} \tag{14.247}$$

Integrals in expression (14.247) can be transformed the same as integral $G_4(x)$ in (14.241). In particular:

$$\begin{aligned}
G_3''(x) &= \frac{1}{\pi}\int_0^\infty \frac{\xi^5 \sin\xi x d\xi}{EI\xi^4 + k} = \frac{1}{\pi EI}\int_0^\infty \frac{\xi(EI\xi^4 + k - k)\sin\xi x d\xi}{EI\xi^4 + k}\\
&= \frac{1}{\pi EI}\frac{d}{dx}\int_0^\infty \cos\xi x d\xi +\\
&\quad + \frac{k}{\pi EI}\frac{d}{dx}\int_0^\infty \frac{\cos\xi x d\xi}{EI\xi^4 + k} = -\frac{\delta'(x) - kG'(x)}{EI}.
\end{aligned}$$

Lowering singular components of delta function and its derivatives we will obtain boundary equations.

14.10 The Compensate Loads Method

The method of the compensating loads [15, 19] arose as a generalization of methods of structural mechanics associated with the consideration of the infinite primary system, as applied to the analysis of beams on elastic Winkler foundation. The method of compensating loads proposed by B.G. Korenev provides precise satisfaction of boundary conditions on the entire surface of the construction with finite dimensions. So, we have Fredholm equations of the first kind, and we also have Fredholm equations of the second kind. However, it is impossible to obtain exact solutions even for canonical bodies because of mathematical difficulties. Nevertheless, the compensation loads method (CLM) is widely known as an approximate method. As a rule, researchers build the system of integral equations, and then they apply to it an approximate method of solution and reduce the problem to the system of linear algebraic equations. In the method of compensating loads the boundary impacts, which are the unknown functions of spatial coordinates (compensating loads), to be determined in the process of solving the problem, are assigned at the outline of domain in the form of forces and moments including the moments of the highest orders corresponding to jumps of kinematic factors—displacements and angles of rotation. As an indirect method of boundary elements, the method of compensating loads demands expansion of the domain to some canonical, in particular, not limited application on its boundary of some compensating loads. The solution of the problem is built as the sum of the basic and compensating solutions in the extended domain. Satisfaction of boundary conditions leads to the boundary equations. The method of the compensating loads was developed in two directions. The first direction is connected with the use of the compensating loads only on the border of the domain, occupied by construction. In this case, the extraction of special features on the border makes it possible to reduce the problem to regular and singular equations of the second kind recorded for the boundary of the construction. The second direction [36, 37] is associated with a full realization of the opportunities presented by CLM, in particular with the arbitrary choice of the compensating loads and their locations relative to the boundary of the construction, and it is, in a certain sense, a generalization of the method. Then the boundary-value problem, generated by elliptical differential operator can be reduced to the integral equations of the first kind with the continuous kernels relative to the compensating loads, located out of the considered construction. B.G Korenev considered a wide range of questions relating to the method of compensating loads: solution one- and two-dimensional problems, the connection of the method of the compensating loads with the method of the initial parameters; the interpretation of the compensating loads of the high orders - which are the discrete analogs (analogs) of multiple layers; the calculation of the constructions by the method of the compensating loads taking into account plastic deformations; and the solution of dynamic problems—calculation of plates of arbitrary shape on an elastic foundation. etc.

An important result of B.G.Korenev's proposal is the removal of compensating loads beyond the limits of the considered domain. In this case, this deteriorates the structure of boundary equations, i.e., on the way of practical application of several essential difficulties, associated with an incorrectness of the resolving equations are encountered, and Tikhonov and Arsenin [38]. As a consequence, the need for the building of the special regularizing algorithms for the numerical application of the method appears; however, they possess properties of regularity that are very necessary for simplification of the calculations. Further works of mathematical justification and numerical realization of this modification of the method of the compensating loads belong to Ventsel [37].

Let us use the scheme of the method of the compensating loads for the solution of a three-dimensional problem of the theory of elasticity, for the purposes of its comparison to BEM and the method of potential. We will consider a little more general linear boundary–value problem for the equation

$$Lu(p)=\varphi(p); \qquad p=(x_1,x_2)\in D \tag{14.248}$$

after joining to its boundary conditions

$$\alpha u(p)+\beta lu(p)=f(p) \qquad (p\in\Gamma) \tag{14.249}$$

where α, β—constants. With $\beta=0,\ \alpha=1$ from (14.249) we obtain the boundary conditions of the first basic problem, and with $\alpha=0,\ \beta=1$– the second.

Here L—elliptical operator on the order of $2n$. Any system of $2n$ linearly independent solutions of the equation

$$Lu(p)=0 \tag{14.250}$$

is a fundamental system, and the fundamental solution of Eq. (14.248) is any solution of the equation:

$$LE(p,q)=\delta(p-q) \tag{14.251}$$

Here $l=(l_1,l_2,\ldots,l_n)$—differential operator of boundary conditions; $\varphi=(\varphi_1,\varphi_2,\ldots.\varphi_n)$ are known on the border Γ vector functions.

We will introduce the expanded domain Ω, which includes D ($D\cup\Gamma\subseteq\Omega\subseteq R$) into space R. The minimum expansion of the domain D is the accession to it of the border, and the maximum—the accession of the space R. In accordance with CLM, consider a wider domain Ω with boundary Γ'

$$d(\Gamma,\Gamma')=\inf|p-q|=d_\varepsilon>0, \qquad (p\in\Gamma, q\in\Gamma') \tag{14.252}$$

The approximate solution of the problem (14.248) and (14.249) is found in the form of

$$u_q(p)=\int_D E(p,q)\varphi(q)ds_q+\int_{\Gamma'}\sum_{i=0}^{n-1}\frac{\partial^i}{\partial n_q{}^i}E(p,q)X_{i+1}(q)dl_q, \tag{14.253}$$

where n_q—normal to Γ'at point q; $X(q)=(X_1(q), X_2(q),\ldots, X_n(q))$—unknown vector function of compensating loads, determined at Γ'. The function $u_q(p)$ satisfies Eq. (14.248). Substituting it into boundary conditions (14.249), we obtain the resolving system of the integral equations of the first kind with the continuous kernels due to (14.252) relatively to compensating loads $X(q)$

$$\begin{aligned}&\sum_{i=1}^{n-1}\int_{\Gamma'}\left\{\left[\frac{\partial^i}{\partial n_q{}^i}E(p,q)\right]\right\}X_{i+1}(q)dl_q=\\&=f(p)-\int_D[lE(p,q)]\,\varphi(q)ds_q\end{aligned} \tag{14.254}$$

or in the operator's form

$$E\cdot X=K \tag{14.255}$$

The numerical solution of resolving integral Eqs. (14.254) is unstable and requires the application of the regularizing algorithms [38]. It should be noted that many publications in resent decade considered the problems solved by the method of compensating loads with the presence of all specific stages and procedures for indirect boundary elements method (BEM), including dividing (partition) of the boundary and satisfaction of boundary conditions in selected nodal points, considerably moved the scientific works on (BEM), ahead.

References

1. Collar, A. R. (1953). The effect of shear flexibility and rotatory inertia of the bending vibrations of beams. *Quarterly Journal of Mechanics and Applied Mathematics, 6*, 186–222. London.
2. Timoshenko, S. P. (1967). *Oscillations in Engineering* (444p). Nauka.
3. Mikhlin, S. G. (1967). *The Application of Integral Equations to the Some Problems of Mechanics, Mathematical Physics and Technology* (304p). Gostexizdat.
4. Fox, C. (1961). The G and H functions as symmetrical Fourier kernels. *Transactions of the American Mathematical Society, 98*, 395–429.
5. Sneddon, I. (1955). *Fourier Transform* (667p).
6. Zienkiewicz, O. C. (1975). *Finite Element Method in Techniques (in Russian)* (541p). Mir.
7. Zienkiewicz, O. C., & Taylor, R. L. (1991). The finite element method. In *Soil and fluid mechanics, dynamics and non-linearity* (Vol. 2). McGraw-Hill Book Co.
8. Tseitlin, A. I. (1984). *The applied methods of solving the boundary-value problems of structural mechanics* (334p). Stroyizdat.
9. Tseitlin, A. I., & Petrosian, L. G. (1987). *The methods of boundary elements in structural mechanics* (200p). Luis.
10. Petrosian, L. G. (1984). *The generalized method of finite integral transformations in the calculation of Rod systems* (pp. 32–38). YPTU. Series of Technical Sciences.
11. Brychkov, Y. A., & Prudnilov, A. P. (1977). *Integral transformations of generalized functions* (286p). Science.
12. Gelfand, I. M., & Shilov, G. E. (1959). *The generalized functions and operations over them* (470p). Fizmatgiz.
13. Petrosian, L. G., & Ambartsumian, V. A. (2020). Static and dynamic analysis of engineering structures. In *Incorporating the boundary element method* (527p). Wiley.
14. Gersevanov, N. M. (1948). *Collection of scientific articles* (373p). Stroyvoenmorizdat.
15. Korenev, B. G. (1962). *The calculation of beams and plates on an elastic foundation* (335p). Goststroyizdat.
16. Korenev, B. G., & Chernigovskaya, E. I. (1964). *The calculation of plates on an elastic foundation* (232p). Goststroyizdat.
17. Petrosian, L. G. (1990). The methods of boundary equations in the problems of calculating the structures on an elastic foundations. The dissertation on competition of a scientific degree of Doctor of Technical Sciences (Post Doctorate Research). In *The thesis of Doctor of Technical Sciences* (310p). CSRICS.
18. Lurie, A. I. (1950). *Operational calculus and its application to the problems of mechanics* (432p). Gostexisdat.
19. Korenev, B. G. (1980). The method of compensating loads in the application to the problems of equilibrium, vibration, and stability of plates and membranes. *Applied Mathematics and Mechanics, 4*(5–6).
20. Serebryaniy, R. V. (1962). *The calculation of thin hinged joined plates on an elastic foundation* (64p). Gostroyizdat.
21. Travush, V. I. (1975). The rectangular uninsulated plate lying on linear-deformed base. In *Structural mechanics and calculation of construction* (Vol. 3, pp. 17–22).
22. Travush, V. I. (1982). The method of generalized solutions in the problems of bending of plates on the linear-deformed base. In *Structural mechanics and calculation of construction* (Vol. 1, pp. 24–29).
23. Goldstein, R. V., Ryskov, I. N., & Salganik, R. L. (1969). The central cross-section crack in an elastic medium. *Izvestia AH CCCP, Ser MTT, 4*, 97–105.
24. Slepian, L. I. (1982). *Non-stationary elastic waves* (373p). Sudostroenie.
25. Popov, G. J. (1982). *Concentration of elastic stresses near the dies, sections, and thin inclusions* (342p). Nauka.
26. Popov, G. J. (1982). *Contact problems for linear-deformed bases* (168p). Bisha shkola.
27. Tseitlin, A. I., & Petrosian, L. G. (1984). About some generalizations of the method of integral transformations and their connection with the method of boundary equations. In *Structural Mechanics and Calculation of Construction* (Vol. 3, pp. 18–23).
28. Naimark, M. A. (1969). *Linear differential operators* (526p). Nauka.
29. Sheynin, I. S., Vilin, G. V., & Kalitseva, I. P. (1976). The method of solution of the plane problems about the oscillations of constructions on the border with the region of liquid of complex form. In B. E. Vedeneeva (Ed.), *Dynamic of Energy – Construction. Theses* (Vol. 5, pp. 3–17). VNIIG.
30. Tseitlin, A. I. (1980). The method of delta-transformation. In *Engineering problems of structural mechanics* (pp. 139–147). Stroyizdat.
31. Petrosian, L. G. (2022). Analysis of structures on elastic foundation. In *Incorporating the spectral method of boundary elements* (556p). CRC Press/Taylor & Frances Group.
32. Petrosian, L. G. (1986). About reduction of boundary-value and initial problems to the << Standard>> form with the aid of delta-transform method. Izvestia AS arm SSR. Structural Mechanics. BINITI, #3965- B86, 12p.
33. Petrosian, L. G. (1986). The spectral method of boundary elements. In *Structural mechanics and calculation of construction* (Vol. 4, pp. 45–50).

34. Tseitlin, A. I., Petrosian, L. G., & Atadzhanov, D. R. (1987). About the methods of calculation, based on the application of boundary equations. In *Study in the dynamics of construction* (pp. 4–18). TsNIISK named Kucherenko.
35. Tsudik, E. (2012). *Structures on elastic foundations* (600p). J. Ross publishing.
36. Chander, S., Donaldson, B. K., & Negm, H. M. (1979). Improved extended field method numerical results. *Journal of Sound and Vibration, 66*(1), 39–51.
37. Ventsel, E. S. (1984). Construction of integral equations of thin plates with complex shape using the method of compensating loads. *Zbornik Trudov Moscovskogo Instituta Ingenerov selskokoz proisvodstva, Vol. II.*
38. Tikhonov, A. N., & Arsenin, V. J. (1986). *The methods of solving incorrect problems* (287p). Science.

15 Plates Vibration

Abstract

This chapter is devoted to plates' vibrations covering the following topics:

Technical theory of bending, small vibrations of elastic plates; application of the refined theory of bending in plate vibrations; anisotropic plates; flexible plates; methods of solving equations of vibrations of plates; free vibrations of rectangular plates; forced vibrations of plates; vibration of plates on elastic foundation; forced vibrations of a rectangular plate on an elastic foundation, problems and solutions are provided. As an example of the application of the Spectral Method of Boundary Elements (SMBE) to dynamic problems, the oscillations of a plate of an arbitrary form, which lies on the elastic Winklerian base, under the action of a load $q(x, y, t)$ is presented. This chapter presents solutions to plate vibration problems which were developed by Korenev and Tseitlin (Plates vibrations, Stroyisdat, pp. 213–249, 1982), Petrosian (About one elastic foundation model. In: Structural mechanics and calculation, Vol. 5, pp. 7–11, n.d.), and Petrosian (Analysis of structures on elastic foundation. Incorporating the spectral method of boundary elements, p. 556, 2022).

15.1 General Provisions

A cylindrical or prismatic body whose height (thickness) is small compared to other dimensions is called a plate. The plate's form depends on the configuration of its base, i.e., rectangular, circular, trapezoidal, triangular, and so on. Plates made of a material with the same mechanical properties in all directions are called isotropic. Plates made of a material with different mechanical properties in different directions are called anisotropic. This type can also include plates with frequently located ribs with oriented reinforcement, also called structurally anisotropic plates.

Depending on the ratio of thickness to the smallest dimension (size) of the base, three main types of plates and slabs are usually considered: thick plates and slabs; thin plates or rigid plates; flexible plates. The plate is considered to be thick if the ratio of thickness to the smallest dimension (size) of the base exceeds 1/5. The stressed state of thick plates is described by the general equations of the theory of elasticity. Thin plates or rigid plates have a smaller ratio than 1/5 and their deflections are small compared to the thickness. The stress-strain state of such plates is described by the technical theory of bending.

L. G. Petrosian, *Dynamics of Structures*, https://doi.org/10.1007/978-3-031-63539-7_15

A flexible plate is characterized by a small ratio of its thickness to its area and a significantly large deflections relative to its thickness. In this case, longitudinal forces appear that have a significant influence on the stress state of the plate.

The above classification of plates is conditional. The main value for determining the nature of the deformation of a plate is the ratio of thickness to the smallest wavelength of deformation. Therefore, in the study of free vibrations of plates, the determination of low frequencies and vibration modes is usually based on the application of the technical theory of bending. Refined theories or general equations of elasticity theory are used to determine the high frequencies and vibration modes of the plates. The application of one or another theory of plate bending is based on the type of plate under consideration and may also depend on the nature of the loading of the plate as well as the parameters of the stress-strain state. The stress-strain state of the plate or slab is determined through a calculation.

15.2 Technical Theory of Bending: Small Vibrations of Elastic Plates

The technical theory of plate bending is based on the following assumptions:

(a) When the plate is deformed, the normal to the median plane remain straight, i.e., the hypothesis of straight normal is accepted.
(b) Normal stresses on the areas which are parallel to the median surface of the plate are absent.
(c) It is also assumed that the deformation of the plate during bending remains small, elastic, and obeying Hooke's law.

The equations of motion of an elementary prism, which is selected near the point of a median plane with coordinates x, y (Fig. 15.1), have the following form [1]:

$$\begin{aligned} &\frac{\partial Q_x}{\partial x} + \frac{\partial Q_y}{\partial y} = m\frac{\partial^2 w}{\partial t^2} - q; \\ &\frac{\partial M_x}{\partial x} - \frac{\partial M_{yx}}{\partial y} = Q_x; \\ &\frac{\partial M_y}{\partial x} + \frac{\partial M_{xy}}{\partial x} = Q_y; \end{aligned} \tag{15.1}$$

where w is the deflection (displacement) of the plate; q is the external load; m is the mass of the plate per unit area; M_x, M_y are bending moments; M_{xy}, M_{yx} are torque moments; Q_x, Q_y are shear forces.

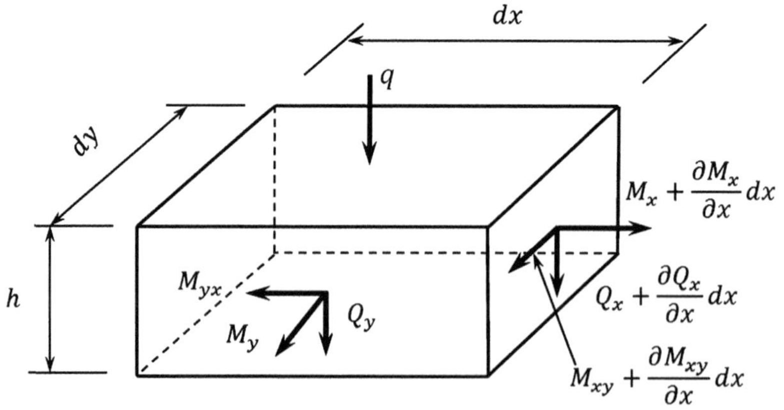

Fig. 15.1 Infinitesimal element of plate in Cartesian coordinates

The internal forces in the plate using Hooke's law can be expressed in terms of its deflection according to the following formulas:

$$\begin{aligned} M_x &= -D\left(\frac{\partial^2 w}{\partial x^2} + \nu\frac{\partial^2 w}{\partial y^2}\right); \\ M_y &= -D\left(\frac{\partial^2 w}{\partial y^2} + \nu\frac{\partial^2 w}{\partial x^2}\right); \\ M_{xy} &= -M_{yx} = -D(1-\nu)\frac{\partial^2 w}{\partial x \partial y}; \\ Q_x &= -D\frac{\partial}{\partial x}\Delta w, Q_y = -D\frac{\partial}{\partial y}\Delta w. \end{aligned} \tag{15.2}$$

Here $D = \frac{Eh^3}{12(1-\nu^2)}$ is the cylindrical stiffness/rigidity of the plate; E is the modulus of elasticity of the plate material; ν is the Poisson's ratio; h is the plate height; and $\Delta = \frac{\partial^2}{\partial x^2} + \frac{\partial^2}{\partial y^2}$ is the Laplace operator in rectangular coordinates.

Let us exclude the shear forces in Eqs. (15.1), we obtain:

$$\frac{\partial^2 M_x}{\partial x^2} + 2\frac{\partial^2 M_{xy}}{\partial x \partial y} + \frac{\partial^2 M_y}{\partial y^2} = m\frac{\partial^2 w}{\partial t^2} - q. \tag{15.3}$$

The expression (15.2) is substituted into expression (14.31). In cases when the plate has variable stiffness, we can obtain the following expression:

$$\begin{aligned} &D\Delta\Delta w + 2\frac{\partial D}{\partial x}\cdot\frac{\partial}{\partial x}\Delta w + 2\frac{\partial D}{\partial y}\cdot\frac{\partial}{\partial y}\Delta w + \Delta D\cdot\Delta w - \\ &-(1-\nu)\left(\frac{\partial^2 D}{\partial x^2}\cdot\frac{\partial^2 w}{\partial y^2} - 2\frac{\partial^2 D}{\partial x \partial y}\cdot\frac{\partial^2 w}{\partial x \partial y} + \frac{\partial^2 D}{\partial y^2}\cdot\frac{\partial^2 w}{\partial x^2}\right) = m\frac{\partial^2 w}{\partial t^2} + q. \end{aligned} \tag{15.4}$$

If the plate has a constant stiffness/rigidity, then the equation of vibrations/oscillations can be presented in the following form:

$$D\Delta\Delta w + m\frac{\partial^2 w}{\partial t^2} = q(x, y, t). \tag{15.5}$$

If longitudinal forces act in the plane of a homogeneous plate, then additional terms $N_x\frac{\partial^2 w}{\partial x^2}$ and $N_y\frac{\partial^2 w}{\partial y^2}$ will appear in the expression (15.5):

$$D\Delta\Delta w + m\frac{\partial^2 w}{\partial t^2} \pm N_x\frac{\partial^2 w}{\partial x^2} \pm N_y\frac{\partial^2 w}{\partial y^2} = q(x, y, t), \tag{15.6}$$

where N_x and N_y are the intensity of the longitudinal forces that are parallel to the x and y axes, respectively. Here the plus sign is taken for compressive forces, and the minus sign for tensile forces.

When using polar coordinates (Fig. 15.2), internal forces are expressed through plate displacements by the following formulas:

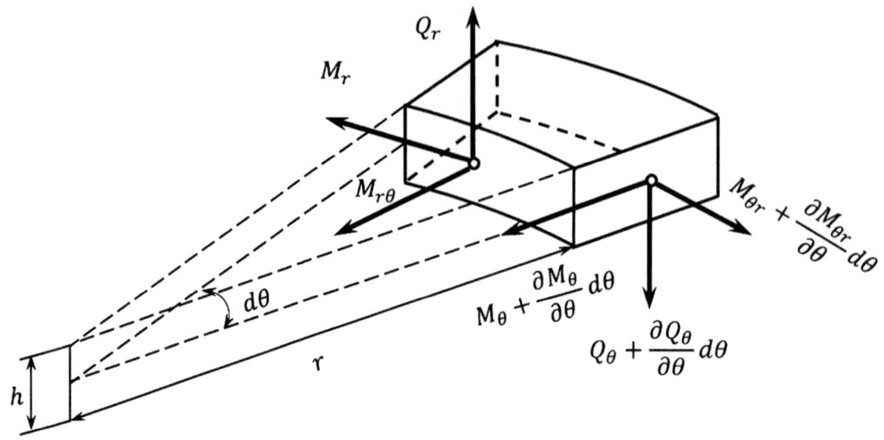

Fig. 15.2 Elementary prism of plate in polar coordinates

$$\begin{aligned} M_r &= -D\left[\frac{\partial^2 w}{\partial r^2} + \nu\left(\frac{1}{r}\cdot\frac{\partial w}{\partial r} + \frac{1}{r^2}\cdot\frac{\partial^2 w}{\partial \theta^2}\right)\right]; \\ M_\theta &= -D\left(\frac{1}{r}\cdot\frac{\partial w}{\partial r} + \frac{1}{r^2}\cdot\frac{\partial^2 w}{\partial \theta^2} + \nu\frac{\partial^2 w}{\partial r^2}\right); \\ M_{r\theta} &= -D(1-\nu)\frac{\partial}{\partial r}\left(\frac{1}{r}\cdot\frac{\partial w}{\partial \theta}\right); \\ Q_r &= -D\frac{\partial}{\partial r}\Delta_r w;\, Q_\theta = -D\frac{1}{r}\cdot\frac{\partial}{\partial \theta}\Delta_r w, \end{aligned} \tag{15.7}$$

where $\Delta_r = \frac{\partial^2}{\partial r^2} + \frac{1}{r}\cdot\frac{\partial}{\partial r} + \frac{1}{r^2}\cdot\frac{\partial^2}{\partial \theta^2}$ is the Laplace operator in polar coordinates.

The equation of vibrations/oscillations of a plate of constant cross section in polar coordinates has the form:

$$D\Delta_r\Delta_r w + m\frac{\partial^2 w}{\partial t^2} = q(r,\theta,t). \tag{15.8}$$

With (At) axisymmetric vibrations/oscillations of a plate, i.e., when the external load or the initial perturbation (disturbance) that caused the oscillations does not depend on the polar angle. In this case, the Eq. (15.8) is simplified and takes the following form:

$$D\left(\frac{\partial^2}{\partial r^2} + \frac{1}{r}\cdot\frac{\partial}{\partial r}\right)^2 \cdot w + m\frac{\partial^2 w}{\partial t^2} = q(r,t). \tag{15.9}$$

Moments and share forces are determined by the formulas:

$$\begin{aligned} M_r &= -D\left(\frac{\partial^2 w}{\partial r^2} + \frac{\nu}{r}\cdot\frac{\partial w}{\partial r}\right); M_\theta = -D\left(\nu\frac{\partial^2 w}{\partial r^2} + \frac{1}{r}\cdot\frac{\partial w}{\partial r}\right); \\ Q_r &= -D\frac{\partial}{\partial r}\left(\frac{\partial^2 w}{\partial r^2} + \frac{1}{r}\cdot\frac{\partial w}{\partial r}\right); Q_\theta = M_{r\theta} = 0. \end{aligned} \tag{15.10}$$

Under the action of radial forces that are applied to the contour in the plane of the plate, the oscillation equation takes the form:

$$D\Delta_r\Delta_r w \pm N\Delta_r w + m\frac{\partial^2 w}{\partial t^2} = q(r, \theta, t) \tag{15.11}$$

where N is intensity of longitudinal force on the plate contour; plus sign is taken for compressive forces, and—minus sign is taken for tensile forces.

The solution of plate vibrations/oscillations equation must satisfy the boundary conditions. The boundary conditions of the problem depend on how the edges of the plate are fixed. If restrictions are imposed on displacements and rotation angles at the edges of the plate, then the boundary conditions are called geometric. If restrictions are imposed on moments and share/transverse forces at the edges of the plate, then the boundary conditions are called dynamic.

If the rectangular plate is considered in Cartesian coordinates, then the boundary conditions on the edges $x = a$, or $y = b$ are set as follows:

(1) Hinge-supported edge (deflection/displacement, and bending moment are equal to zero)

$$\begin{aligned} w = 0; \quad & \frac{\partial^2 w}{\partial x^2} + \frac{\partial^2 w}{\partial y^2} = 0 \quad at \quad x = a; \\ w = 0; \quad & \frac{\partial^2 w}{\partial y^2} + \nu\frac{\partial^2 w}{\partial x^2} = 0 \quad at \quad y = b. \end{aligned} \tag{15.12}$$

From the condition $w = 0$ for $x = a$ it follows that: $\left.\frac{\partial^2 w}{\partial y^2}\right|_{x=a} = 0$. Therefore the boundary conditions Eq. (15.12) can be simplified:

$$\begin{aligned} w = \frac{\partial^2 w}{\partial x^2} = 0 \quad at \quad x = a; \\ w = \frac{\partial^2 w}{\partial y^2} = 0 \quad at \quad y = b. \end{aligned} \tag{15.13}$$

(2) Built-in end, cantilever edge (deflection/displacement, and rotation angle are equal to zero):

$$\begin{aligned} w = \frac{\partial w}{\partial x} = 0 \quad at \quad x = a; \\ w = \frac{\partial w}{\partial y} = 0 \quad at \quad y = b. \end{aligned} \tag{15.14}$$

(3) Free edge (bending moment and shear force are equal to zero):

$$Q_x + \frac{\partial M_{xy}}{\partial y} \; or \quad Q_x + \frac{\partial M_{yx}}{\partial x}$$

$$\begin{aligned} \frac{\partial^2 w}{\partial x^2} + \nu\frac{\partial^2 w}{\partial y^2} = 0; \; \frac{\partial^3 w}{\partial x^3} + (2-\nu)\frac{\partial^3 w}{\partial x \partial y^2} = 0 \; at \; x = a; \\ \frac{\partial^2 w}{\partial y^2} + \nu\frac{\partial^2 w}{\partial x^2} = 0; \; \frac{\partial^3 w}{\partial y^3} + (2-\nu)\frac{\partial^3 w}{\partial y \partial x^2} = 0 \; at \; y = b. \end{aligned} \tag{15.15}$$

(4) Displaceable/shifted but non-rotating edge (sliding build-in) (the angle of rotation and the reduced shear force are equal to zero):

$$\frac{\partial w}{\partial x} = \frac{\partial^3 w}{\partial x^3} = 0 \quad \text{at} \quad x = a;$$
$$\frac{\partial w}{\partial y} = \frac{\partial^3 w}{\partial y^3} = 0 \quad \text{at} \quad y = b. \tag{15.16}$$

The listed boundary conditions cover the main cases of fixing the edges of the plate. More complex cases of boundary conditions are also possible. For example, when linear combinations of force and kinematic parameters are given.

The presented boundary conditions are called homogeneous. If kinematic or force parameters are set on the contour of the plate, then in the right parts of the corresponding equalities the given functions are inserted. Such boundary conditions are called inhomogeneous.

The boundary conditions at the edge of $r = a$ for circular/round and ring/annular plates, which are considered in the polar coordinate system, have the following form:

(a) Hinge-supported edge

$$w(a, \theta, t) = 0; \left[\frac{\partial^2 w}{\partial r^2} + \frac{\nu}{r} \cdot \frac{\partial w}{\partial r}\right]_{r=a} = 0 \tag{15.17}$$

(b) Build-in end, cantilever edge

$$w(a, \theta, t) = \frac{\partial w}{\partial r}(a, \theta, t) = 0 \tag{15.18}$$

(c) Free edge

$$\left[\frac{\partial^2 w}{\partial r^2} + \nu\left(\frac{1}{r} \cdot \frac{\partial w}{\partial r} + \frac{1}{r^2} \cdot \frac{\partial^2 w}{\partial \theta^2}\right)\right]_{r=a} = 0;$$
$$\left[\frac{\partial^3 w}{\partial r^3} + \frac{1}{r} \cdot \frac{\partial^2 w}{\partial r^2} - \frac{1}{r^2} \cdot \frac{\partial w}{\partial r} - \frac{(2-\nu)}{r^3} \cdot \frac{\partial^2 w}{\partial \theta^2} + \frac{\nu}{r^2} \cdot \frac{\partial^3 w}{\partial r \partial \theta^2}\right]_{r=a} = 0 \tag{15.19}$$

At axisymmetric vibrations\oscillations in the boundary conditions "c" the polar angle derivatives with respect to the polar angle disappear. For solid plates that do not have a hole at the origin, it is necessary to set conditions at point $r = 0$.

If there is no concentrated load at this point, the deflection, the angle of rotation, the bending moment, and the shear force are limited and continuous. If a concentrated force $P(t)$ acts at the point $r = 0$, then it should be assumed

$$2\pi r Q|_{r=0} = -P(t),$$

and therefore:

$$2\pi D \lim_{r\to 0}\left[r\left(\frac{\partial^3 w}{\partial r^3}+\frac{1}{r}\cdot\frac{\partial^2 w}{\partial r^2}-\frac{1}{r^2}\cdot\frac{\partial w}{\partial r}\right)\right]=P(t). \tag{15.20}$$

For a plate fixed in the center, the condition must be satisfied:

$$w(0,\theta,t)=0. \tag{15.21}$$

The potential energy of deformations of the plate at vibrations/oscillations is calculated by the formulas:

Potential energy of deformations in rectangular coordinates

$$U=\frac{1}{2}\iint_A D\left\{(\Delta w)^2-2(1-\nu)\left[\frac{\partial^2 w}{\partial x^2}\cdot\frac{\partial^2 w}{\partial y^2}-\left(\frac{\partial^2 w}{\partial x\partial y}\right)^2\right]\right\}dA \tag{15.22}$$

Potential energy of deformations in polar coordinates

$$\begin{aligned}U=\frac{1}{2}\iint_A D\Bigg\{&\left(\frac{\partial^2 w}{\partial r^2}+\frac{1}{r}\cdot\frac{\partial w}{\partial r}+\frac{1}{r^2}\cdot\frac{\partial^2 w}{\partial\theta^2}\right)^2\\&-2(1-\nu)\left[\frac{\partial^2 w}{\partial r^2}\left(\frac{1}{r}\cdot\frac{\partial w}{\partial r}+\frac{1}{r^2}\cdot\frac{\partial^2 w}{\partial\theta^2}\right)\right.\\&\left.-\frac{\partial}{\partial r}\left(\frac{1}{r}\cdot\frac{\partial w}{\partial\theta^2}\right)^2\right]\Bigg\}dA.\end{aligned} \tag{15.23}$$

The kinetic energy of the plate at oscillations can be represented as follows:

$$K=\frac{1}{2}\iint_A m\left(\frac{\partial w}{\partial t}\right)dA, \tag{15.24}$$

where A is the area of the plate; $dA = dxdy$ is the area of the elementary plate in rectangular coordinates; $dA = rdrd\theta$ is an area of the elementary plate in polar coordinates.

15.3 Application of the Refined Theory of Bending in Plate Vibrations/Oscillations

The technical theory of bending in plate vibration applications is based on the Kirchhoff-Love hypothesis. The Kirchhoff-Love theory of plates is a two-dimensional mathematical model that is used to determine the stresses and deformations in thin plates subjected to forces and moments. This technical theory gives reasonably accurate results for cases where the ratio of the plate thickness to the smallest wave length of deformation does not exceed 1/10. When determining high frequencies and vibration modes, the plate should be considered as a three-dimensional elastic body. In this case, the vibrations/oscillations are described by the equations of elasticity theory. The solution of the three-dimensional dynamic problem of the theory of elasticity faces great difficulties. Therefore, various authors have made attempts to construct equations of an intermediate character\nature, i.e., between the equations of technical theory and the exact equations of elasticity theory. These intermediate modified equations, which have been proposed by various researchers, allow taking into account shift\shear, rotation inertia, and so on [1–3].

The refined equations of plate vibrations/oscillations in rectangular coordinates x, y , and z can be represented in the following general form [1]:

$$\begin{aligned}
\Delta(\varphi+w) &= -\frac{2(1+\nu)}{\beta Eh}\left(q-\rho h\frac{\partial^2 w}{\partial t^2}\right);\\
\Delta\varphi &= -\frac{\nu(1+\nu)}{Eh}\cdot\frac{\alpha}{\beta}\left(q-\frac{\rho h}{6}\cdot\frac{\partial^2 w}{\partial t^2}\right)+\\
&+\frac{6\beta(1-\nu)}{h^2}(\varphi+w)+\frac{(1-\nu^2)}{E}\rho\frac{\partial^2\varphi}{\partial t^2};\\
\Delta\psi &= \frac{12\beta}{h^2}\psi+\frac{2(1+\nu)}{E}\cdot\frac{\partial^2\psi}{\partial t^2},
\end{aligned} \tag{15.25}$$

where w is the plate deflection; φ, ψ are functions defined by the following relations:

$$\begin{aligned}
\frac{\partial\varphi}{\partial x}+\frac{\partial\psi}{\partial y} &= -\frac{12}{h^3}\int_{-h/2}^{h/2} uzdz\\
\frac{\partial\varphi}{\partial y}-\frac{\partial\psi}{\partial x} &= -\frac{12}{h^3}\int_{-\frac{h}{2}}^{\frac{h}{2}} vzdz.
\end{aligned} \tag{15.26}$$

The displacement components u, v in the direction of the x and y axes are related to the functions φ, and ψ by formulas:

$$\begin{aligned}
u &= -[z+f_1(z)]\left(\frac{\partial\varphi}{\partial x}+\frac{\partial\psi}{\partial y}\right)+f_1(z)\frac{\partial w}{\partial x}\\
v &= -[z+f_1(z)]\left(\frac{\partial\varphi}{\partial y}-\frac{\partial\psi}{\partial x}\right)+f_1(z)\frac{\partial w}{\partial y}.
\end{aligned} \tag{15.27}$$

The above formulas take into account the average rotation and curvature of the element normal. The stress components are expressed through bending moment, torque moment, and shear forces according to the formulas:

$$\begin{aligned}
\sigma_x &= -\frac{12}{h^3}\left\{M_x z+f_1\left[M_x-D\left(\frac{\partial^2 w}{\partial x^2}+\nu\frac{\partial^2 w}{\partial y^2}\right)+\frac{\alpha\nu h^3\rho}{12\beta(1-\nu)}\cdot\frac{\partial^2 w}{\partial t^2}\right]\right\};\\
\sigma_y &= -\frac{12}{h^3}\left\{M_y z+f_1\left[M_y-D\left(\frac{\partial^2 w}{\partial y^2}+\nu\frac{\partial^2 w}{\partial x^2}\right)+\frac{\alpha\nu h^3\rho}{12\beta(1-\nu)}\cdot\frac{\partial^2 w}{\partial t^2}\right]\right\};\\
\tau_{xy} &= -\frac{12}{h^3}\left\{M_{xy} z+f_1\left[M_{xy}-D(1-\nu)\frac{\partial^2 w}{\partial x\partial y}\right]\right\};\\
\tau_{xz} &= -\frac{(1+\nu)GF'(z)}{EA(h/2)}Q_x;\\
\tau_{xz} &= -\frac{(1+\nu)GF'(z)}{EA(h/2)}Q_y.
\end{aligned} \tag{15.28}$$

Here

$$\begin{aligned}
M_x &= D\left[\frac{\partial^2\varphi}{\partial x^2}+\frac{\partial^2\psi}{\partial x\partial y}+\nu\left(\frac{\partial^2\varphi}{\partial y^2}+\frac{\partial^2\psi}{\partial x\partial y}\right)\right]- \\
&\quad -\frac{\alpha\nu h^2}{12\beta(1-\nu^2)}\left[q-(1-\beta)\rho h\frac{\partial^2 w}{\partial t^2}\right]; \\
M_y &= D\left[\frac{\partial^2\varphi}{\partial y^2}-\frac{\partial^2\psi}{\partial x\partial y}+\nu\left(\frac{\partial^2\varphi}{\partial x^2}+\frac{\partial^2\psi}{\partial x\partial y}\right)\right]- \\
&\quad -\frac{\alpha\nu h^2}{12\beta(1-\nu^2)}\left[q-(1-\beta)\rho h\frac{\partial^2 w}{\partial t^2}\right]; \\
M_{xy} &= D\frac{(1-\nu)}{2}\left(\frac{\partial^2\psi}{\partial y^2}+2\frac{\partial^2\varphi}{\partial x\partial y}-\frac{\partial^2\psi}{\partial x^2}\right); \\
Q_x &= \frac{\beta Eh}{2(1+\nu)}\left(\frac{\partial\varphi}{\partial x}+\frac{\partial\psi}{\partial y}-\frac{\partial w}{\partial x}\right) \\
Q_y &= \frac{\beta Eh}{2(1+\nu)}\left(\frac{\partial\varphi}{\partial y}-\frac{\partial\psi}{\partial x}-\frac{\partial w}{\partial y}\right).
\end{aligned} \tag{15.29}$$

The functions $f_1(z)$, $F(z)$ have the following relationship,

$$f_1(z)=\frac{h^3F(h/2)}{12\int_{-h/2}^{h/2} zF(z)dz}-z. \tag{15.30}$$

But may have a different form. It is usually assumed that

$$F'(z)=1-\frac{4z^2}{h^2}.$$

In this case, $\beta = 5/6$, and α is taken either equal to zero if the stress is ignored σ_z, or equal to one. The boundary conditions have the form (as an example at the edge of $x = const.$):

(a) Hinge supported edge

$$w=0; M_x=0; \frac{\partial\varphi}{\partial y}-\frac{\partial\psi}{\partial x}=0$$

(b) Build-in end, cantilever edge

$$w=0; \frac{\partial\varphi}{\partial x}+\frac{\partial\psi}{\partial y}=0; \frac{\partial\varphi}{\partial y}-\frac{\partial\psi}{\partial x}=0$$

(c) Free edge

Table 15.1 The values of dimensionless frequencies of natural oscillations

Type of equations	Number of half-waves of the oscillation form							
	$m=1$ $n=1$	$m=1$ $n=2$	$m=1$ $n=3$	$m=2$ $n=2$	$m=2$ $n=3$	$m=3$ $n=3$	$m=7$ $n=7$	$m=10$ $n=10$
Equations of technical theory of bending	0.0642	0.160	0.321	0.257	0.417	0.577	3.14	6.40
Refined equations of technical theory of bending: $\alpha=0;\ \beta=2/3$	0.0639	0.159	0.316	0.254	0.409	0.564	2.80	5.21
$\alpha=0;\ \beta=5/6$	0.0640	0.159	0.317	0.254	0.411	0.566	2.84	5.36
$\alpha=1;\ \beta=5/6$	0.0640	0.159	0.317	0.254	0.411	0.566	2.86	5.38
Three-dimensional equations of the theory of elasticity	0.0640	0.159	0.317	0.254	0.411	0.5666	2.86	5.39

$$M_x=0; M_{xy}=0; Q_x=0.$$

Table 15.1 compares the dimensionless natural oscillations/vibrations frequencies $\frac{ap}{\pi}\sqrt{\frac{2\rho(1-\nu^2)}{E}}$ calculated from the equations of the technical bending theory, the refined bending theory equations, and the three-dimensional equations of the theory of elasticity [4]. As an example, a square plate hinged with sides $a=40h$ with $\nu=0.3$ was considered.

15.4 Anisotropic Plates

The relationship between strains/deformations and stresses in the material of an anisotropic plate is described by the generalized Hooke's law:

$$\begin{aligned}\varepsilon_x &= a_{11}\sigma_x + a_{12}\sigma_y + a_{16}\tau_{xy};\\ \varepsilon_y &= a_{12}\sigma_x + a_{22}\sigma_y + a_{26}\tau_{xy};\\ \gamma_{xy} &= a_{16}\sigma_x + a_{26}\sigma_y + a_{66}\tau_{xy}.\end{aligned} \tag{15.31}$$

The bending moments, torques (torsional moments) and shear forces of the plate in a rectangular coordinate system are determined by the following formulas:

$$\begin{aligned}M_x &= -\left(D_{11}\frac{\partial^2 w}{\partial x^2} + 2D_{16}\frac{\partial^2 w}{\partial x\partial y} + D_{12}\frac{\partial^2 w}{\partial y^2}\right);\\ M_y &= -\left(D_{12}\frac{\partial^2 w}{\partial x^2} + 2D_{26}\frac{\partial^2 w}{\partial x\partial y} + D_{22}\frac{\partial^2 w}{\partial y^2}\right);\\ M_{xy} &= -\left(D_{16}\frac{\partial^2 w}{\partial x^2} + 2D_{66}\frac{\partial^2 w}{\partial x\partial y} + D_{26}\frac{\partial^2 w}{\partial y^2}\right);\end{aligned} \tag{15.32}$$

$$Q_x = -\left[D_{11}\frac{\partial^3 w}{\partial x^3} + 3D_{16}\frac{\partial^3 w}{\partial x^2\partial y} + (D_{12}+2D_{66})\frac{\partial^3 w}{\partial x\partial y^2} + D_{26}\frac{\partial^3 w}{\partial y^3}\right];$$

$$Q_y = -\left[D_{16}\frac{\partial^3 w}{\partial x^3} + (D_{12}+2D_{66})\frac{\partial^3 w}{\partial x^2\partial y} + 3D_{26}\frac{\partial^3 w}{\partial x\partial y^2} + D_{22}\frac{\partial^3 w}{\partial y^3}\right],$$

where,

$$\begin{aligned} D_{11} &= \frac{h^3}{12\Delta_0}\left(a_{22}a_{66}-a_{26}^2\right); D_{22}=\frac{h^3}{12\Delta_0}\left(a_{11}a_{66}-a_{16}^2\right);\\ D_{12} &= \frac{h^3}{12\Delta_0}\left(a_{16}a_{26}-a_{12}a_{66}\right);\quad D_{16}=\frac{h^3}{12\Delta_0}\left(a_{12}a_{26}-a_{22}a_{16}\right);\\ D_{26} &= \frac{h^3}{12\Delta_0}\left(a_{12}a_{16}-a_{16}a_{26}\right); D_{66}=\frac{h^3}{12\Delta_0}\left(a_{11}a_{22}-a_{12}^2\right); \end{aligned} \tag{15.33}$$

Δ_0 is a matrix of coefficients standing on the right side of the expression (15.31). D_{11} and D_{22} are bending stiffness around the x and y axes; D_{66} is the torsional stiffness/regidity. The ratio of stiffnesses

$$\nu_1=\frac{D_{11}}{D_{22}};\nu_2=\frac{D_{12}}{D_{11}} \tag{15.34}$$

is called reduced Poisson's coefficients.

The equation of vibration/oscillations of an anisotropic plate has the following form:

$$\begin{aligned} &D_{11}\frac{\partial^4 w}{\partial x^4}+4D_{16}\frac{\partial^4 w}{\partial x^3\partial y}+2(D_{12}+2D_{66})\frac{\partial^4 w}{\partial x^2\partial y^2}+4D_{26}\frac{\partial^4 w}{\partial x\partial y^3}+\\ &+D_{22}\frac{\partial^4 w}{\partial y^4}+m\frac{\partial^2 w}{\partial t^2}=q(x,y,t). \end{aligned} \tag{15.35}$$

The expression of the potential energy of deformation of the plate has the following form:

$$\begin{aligned} U=\frac{1}{2}\iint\limits_A &\left[D_{11}\left(\frac{\partial^2 w}{\partial x^2}\right)^2+2D_{12}\frac{\partial^2 w}{\partial x^2}\cdot\frac{\partial^2 w}{\partial y^2}+D_{22}\left(\frac{\partial^2 w}{\partial y^2}\right)^2+\right.\\ &\left.+4D_{66}\left(\frac{\partial^2 w}{\partial x\partial y}\right)^2+4\left(D_{16}\frac{\partial^2 w}{\partial x^2}+D_{26}\frac{\partial^2 w}{\partial y^2}\right)\frac{\partial^2 w}{\partial x\partial y}\right]dA. \end{aligned} \tag{15.36}$$

In particular, the equation of vibration/oscillations for an orthotropic plate for which the properties are the same in two mutually perpendicular directions parallel to the x and y axes has the following form:

$$D_1\frac{\partial^4 w}{\partial x^4}+2D_3\frac{\partial^4 w}{\partial x^2\partial y^2}+D_2\frac{\partial^4 w}{\partial y^4}+m\frac{\partial^2 w}{\partial t^2}=q(x,y,t), \tag{15.37}$$

where,

$$\begin{aligned} D_1&=\frac{E_1h^3}{12(1-\nu_1\nu_2)}; & D_2&=\frac{E_2h^3}{12(1-\nu_1\nu_2)};\\ D_3&=D_1\nu_2+2D_k; & D_k&=\frac{Gh^3}{12}. \end{aligned} \tag{15.38}$$

The bending moments, torsional moments, and shear forces of an orthotropic plate are determined by the formulas:

$$\begin{aligned} M_x &= -D_1\left(\frac{\partial^2 w}{\partial x^2} + \nu_2 \frac{\partial^2 w}{\partial y^2}\right); \\ M_y &= -D_2\left(\frac{\partial^2 w}{\partial y^2} + \nu_1 \frac{\partial^2 w}{\partial x^2}\right); \\ M_{xy} &= -2D_k \frac{\partial^2 w}{\partial x \partial y} \\ Q_x &= -\frac{\partial}{\partial x}\left(D_1 \frac{\partial^2 w}{\partial x^2} + D_3 \frac{\partial^2 w}{\partial y^2}\right); \\ Q_y &= -\frac{\partial}{\partial y}\left(D_3 \frac{\partial^2 w}{\partial x^2} + D_2 \frac{\partial^2 w}{\partial y^2}\right). \end{aligned} \tag{15.39}$$

The potential energy of deformations of the orthotropic plate has the form:

$$\begin{aligned} U = \frac{1}{2}\iint_A \Bigg[D_1\left(\frac{\partial^2 w}{\partial x^2}\right)^2 + 2D_1\nu_2 \frac{\partial^2 w}{\partial x^2} \cdot \frac{\partial^2 w}{\partial y^2} + \\ +D_2\left(\frac{\partial^2 w}{\partial y^2}\right)^2 + 4D_k\left(\frac{\partial^2 w}{\partial x \partial y}\right)^2\Bigg] dA. \end{aligned} \tag{15.40}$$

In some cases, the stiffness coefficients D can be determined by approximate formulas. These formulas give sufficiently accurate results for practical calculations [5].

1. Reinforced concrete slabs that are reinforced in two directions parallel to the x and y axes. In this case we assume that there are no cracks in the reinforced concrete slab:

$$\begin{aligned} D_1 &= \frac{E_c}{1-\nu_c^2}\left[I_c^{(x)} + \left(\frac{E_r}{E_c} - 1\right) I_r^{(x)}\right]; \\ D_2 &= \frac{E_c}{1-\nu_c^2}\left[I_c^{(y)} + \left(\frac{E_r}{E_c} - 1\right) I_r^{(y)}\right]; \\ D_3 &= \sqrt{D_1 D_2}, \end{aligned} \tag{15.41}$$

here E_c, E_r is the modulus of elasticity of concrete and reinforcement; ν_c is the Poisson's ratio for concrete; $I_c^{(x)}$ is the moment of inertia of the concrete section $x =$ constant; $I_r^{(x)}$ is the moment of inertia of the reinforcement relative to the neutral axis in section $x =$ constant; $I_c^{(y)}$ is the moment of inertia of the concrete section $y =$ constant; $I_r^{(y)}$ is the moment of inertia of the reinforcement relative to the neutral axis in section $y =$ constant;

The expression $y_1 = y(D_1/D_2)^{1/4}$ is substituted into the equation of oscillations (15.37). With the values of stiffness (15.41) this equation is reduced to the solution of the equation of vibrations of an isotropic plate (15.5) with cylindrical stiffness $D = D_1$.

15.5 Flexible Plates

A plate is considered flexible when its deflections from the transverse load exceed 1/4–1/5 of the plate thickness. For flexible plates the technical theory of bending gives distorted results. The theory of flexible plates is based on taking into account the deformation of the plate in its plane, along with shear/ transverse bending. The equations of oscillations/vibrations of a flexible plate in a rectangular coordinate system have the form:

$$D\Delta\Delta w - h\left(\frac{\partial^2\Phi}{\partial y^2}\cdot\frac{\partial^2 w}{\partial x^2}+\frac{\partial^2\Phi}{\partial x^2}\cdot\frac{\partial^2 w}{\partial y^2}-2\frac{\partial^2\Phi}{\partial x\partial y}\cdot\frac{\partial^2 w}{\partial x\partial y}\right)+$$
$$+m\frac{\partial^2 w}{\partial t^2}=q(x.y.t); \tag{15.42}$$
$$\Delta\Delta\Phi=\mathrm{E}\left[\left(\frac{\partial^2 w}{\partial x\partial y}\right)^2-\frac{\partial^2 w}{\partial x^2}\cdot\frac{\partial^2 w}{\partial y^2}\right],$$

where Φ is a stress function. The stress function is related to the stresses in the middle surface of the plate by the following relations:

$$\sigma_x=\frac{\partial^2\Phi}{\partial y^2};\sigma_y=\frac{\partial^2\Phi}{\partial x^2};\tau_{xy}=\frac{\partial^2\Phi}{\partial x\partial y}\quad . \tag{15.43}$$

These equations are called nonlinear Kármán equations. The boundary conditions for Eqs. (15.42) have restrictions that are imposed on the displacement function w and its derivatives. The boundary conditions for Eqs. (15.42) also contain parameters that are expressed through the stress function. The following conditions are added to the boundary conditions (15.12)–(15.16) with the appropriate fixing methods:

Boundary condition: the edge of the plate is not fixed from displacements along the normal to it. The normal stress is equal to zero.

$$\begin{aligned}&\frac{\partial^2\Phi}{\partial y^2}=0;\quad \text{at}\quad x=a;\\&\frac{\partial^2\Phi}{\partial x^2}=0;\quad \text{at}\quad y=b.\end{aligned} \tag{15.44}$$

Boundary condition: the edge of the plate from longitudinal displacements along the edge is not fixed. The tangential stress is equal to zero.

$$\frac{\partial^2\Phi}{\partial x\partial y}=0;\text{at}\quad x=a;\text{and at}\quad y=b. \tag{15.45}$$

Boundary condition: the parallel edges of the plate are fixed in such a way that their mutual displacement along the common normal is equal to zero. In this case, the boundary condition has an integral form:

$$\begin{aligned}&\int_0^a\left[\frac{\partial^2\Phi}{\partial y^2}-\nu\frac{\partial^2\Phi}{\partial x^2}-\frac{E}{2}\left(\frac{\partial w}{\partial x}\right)^2\right]dx=0\\&\text{or}\\&\int_0^b\left[\frac{\partial^2\Phi}{\partial x^2}-\nu\frac{\partial^2\Phi}{\partial y^2}-\frac{E}{2}\left(\frac{\partial w}{\partial y}\right)^2\right]dy=0.\end{aligned} \tag{15.46}$$

In the case of a polar coordinate system, the basic differential equations for the oscillations of a flexible plate take the following form:

$$\begin{aligned} D\Delta_r\Delta_r w - hL(w,\Phi) + m\frac{\partial^2 w}{\partial t^2} &= q(r,\theta,t); \\ \Delta_r\Delta_r\Phi + \frac{E}{2}L(w,\Phi) &= 0. \end{aligned} \tag{15.47}$$

Here, the notation is introduced through $L(w, \Phi)$:

$$\begin{aligned} L(w,\Phi) = {} & \frac{\partial^2 w}{\partial r^2}\left(\frac{1}{r}\cdot\frac{\partial\Phi}{\partial r}+\frac{1}{r^2}\cdot\frac{\partial^2\Phi}{\partial\theta^2}\right)+\left(\frac{1}{r}\cdot\frac{\partial w}{\partial r}+\frac{1}{r^2}\cdot\frac{\partial^2 w}{\partial\theta^2}\right)\frac{\partial^2\Phi}{\partial r^2} - \\ & -2\frac{\partial}{\partial r}\left(\frac{1}{r}\cdot\frac{\partial\Phi}{\partial\theta}\right)\frac{\partial}{\partial r}\left(\frac{1}{r}\cdot\frac{\partial w}{\partial\theta}\right). \end{aligned} \tag{15.48}$$

For axisymmetric vibrations/oscillations:

$$L(w,\Phi) = \frac{1}{r}\left(\frac{\partial\Phi}{\partial r}\cdot\frac{\partial^2 w}{\partial r^2}+\frac{\partial w}{\partial r}\cdot\frac{\partial^2\Phi}{\partial r^2}\right). \tag{15.49}$$

The boundary conditions for circular and ring plates are written in the same way as it was shown for rectangular plates. If the edge of circular plate is fixed from radial displacements, then $u = 0$. If the edge of circular plate is not fixed from radial displacements, i.e., radial displacements are not zero, then $\sigma_r = 0$. In addition, over the entire area of the plate, σ_r should be limited. This condition requires that:

$$\lim_{r\to 0}\frac{\partial\Phi}{\partial r} = 0.$$

15.5.1 Methods of Solving Equations of Vibrations/Oscillation of Plates

Differential equations of plate vibrations/oscillations allow exact solutions only in some cases. Therefore, well-developed approximate calculation methods are often used to solve problems of this type. Among these methods, variation and finite-difference methods of calculation are most widely used.

In studying the vibrations of plates, the equation of modes of free vibrations is of great importance:

$$D\Delta\Delta W - mp^2 W = 0 \tag{15.50}$$

as well as the equation of modes of forced harmonic vibrations:

$$D\Delta\Delta W - m\omega^2 W = q. \tag{15.51}$$

We will show the application of various approximate methods of calculation on the example of two equations mentioned above (15.50), (15.51). The solution of Eqs. (15.50) and (15.51) by exact methods is given in Sects. 15.6 and 15.7.

Variation methods, as it is known, refer to the approximate methods of structural analysis. Variation methods of calculation which are used in the theory of plates depend on determining the coefficients c_{mn} of decomposition/expansion of the solution in series on some system of functions φ_{mn}. These functions are called basis or coordinate functions:

$$W = \sum_{m,n=1}^{k} c_{mn}\varphi_{mn} \tag{15.52}$$

By the Ritz method, minimizing form Eq. (15.52) is substituted into the expression $K_{max} - U_{max} = 0$ and the coefficients c_{mn} are found from the condition of the minimum of this expression.

$$\frac{\partial}{\partial c_{mn}}(K_{max} - U_{max}) = 0, (m, n = 1, 2, \ldots, k) \tag{15.53}$$

The product of two functions is usually taken as a minimizing form. One of these functions depends only on one coordinate, and the second function depends only on the other coordinate. The basal functions, which are included in the minimizing form, must form a complete system of linearly independent functions and satisfy the boundary conditions of the problem.

In the Ritz method, at the very least, the satisfaction of geometrical boundary conditions is required. It is most convenient to take polynomials or beam functions as basal/basic functions when solving problems in Cartesian coordinates.

(a) Application of the Ritz method in studying the modes of free vibrations of plates Eq. (15.50). For Eq. (15.50), the system of homogeneous algebraic Eqs. (15.53) has a non-trivial solution if the determinant composed of its coefficients is equal to zero. This condition leads to a frequency equation. The solution of this equation gives approximate values of natural frequencies. When solving Eq. (15.51) a system of inhomogeneous algebraic equations with respect to the decomposition/expansion coefficients c_{mn} is obtained. The Ritz method for natural frequencies gives an upper approximation.
(b) According to the Bubnov-Galerkin method, form Eq. (15.52) is substituted directly into the oscillation equation, for example, in Eq. (15.50) the following equation is obtained as a result:

$$\sum_{m,n=1}^{k} c_{mn}\left(D\Delta\Delta - mp^2\right)\varphi_{mn} = 0, . \tag{15.54}$$

If the function on the left side Eq. (15.54) is continuous, then the condition of equality to zero can be considered as the requirement that this function be orthogonal to any of the linearly independent functions that form a complete orthogonal system. Therefore, to determine the k^2 coefficients of c_{mn}, we can use the orthogonal condition of the left part Eq. (15.54) to the first k^2 functions of φ_{mn}. The Eqs. (15.54) is multiplied by φ_{ij} and integrated over the entire area of the plate, we get k^2 equations:

$$\iint_A \varphi_{ij}(x, y) \sum_{m,\ n=1}^{k} c_{mn}\left(D\Delta\Delta - mp^2\right)\varphi_{mn} dA = 0. \tag{15.55}$$

The condition of existence of nontrivial solutions of the system of Eqs. (15.55) leads to the frequency equation. When using the Galerkin method in the problems of plate vibrations/oscillations, the basic functions must satisfy all boundary conditions of the problem.

Finite difference methods are based on replacing derivatives in differential Eqs. (15.50) and (15.51) with finite differences with varying degrees of accuracy. In this case, the area occupied by the plate is divided by some grid. The differential and other operations are expressed through the values of the functions at the nodes of the grid. The result is a system of algebraic equations. For the case of free oscillations/vibrations, we will have a homogeneous system of algebraic equations. For the case of forced oscillations, we obtain an inhomogeneous system of algebraic equations. Recently, finite-difference equations have been greatly developed due to the fact that they are especially convenient when using computers.

15.6 Free Vibrations/Oscillations of Rectangular Plates

In the absence of an external dynamic load, a plate can perform free vibrations/oscillations, which are caused by initial perturbations/disturbance. The initial perturbations/disturbance are given in the form of initial conditions:

$$w(x,y,0) = \varphi(x,y); \frac{\partial w}{\partial t}(x,y,0) = \psi(x,y). \tag{15.56}$$

Assuming in the equation of oscillations (15.5) $q = 0$, and $w = W(x, y) \sin(pt + \varepsilon)$, we obtain the differential equation of the modes of free (natural) vibrations/oscillations of the plate:

$$\Delta\Delta W - \frac{mp^2}{D} W = 0 , \tag{15.57}$$

Where p is the circular frequency of natural vibrations/oscillations.

Using the method of separation of variables, the solution of the Eq. (15.57) is found (sought) as:

$$W(x,y) = X(x)Y(y)$$

The variables are separated if

$$X'' = -\alpha_m^2 X; X'' = -\alpha_m^2 X''; \tag{15.58}$$

or

$$Y'' = -\beta_n^2 Y; Y^{IV} = -\beta_n^2 Y''; \tag{15.59}$$

where α_m and β_n are some constants. The first condition, in particular, corresponds to:

$$X(x) = A \sin \alpha_m x + B \cos \alpha_m x. \tag{15.60}$$

Function $W = Y(y)(A \sin \alpha_m x + B \cos \alpha_m x)$ can only satisfy the conditions of hinged support of the edges $x = 0,\ x = a$ ($B = 0$) or a sliding built-in (sliding cantilever) at these edges ($A = 0$), in this case $\alpha_m = m\pi/a,\ m = 1, 2, 3\ldots$.

Similar expressions are also obtained when condition (15.59) is satisfied.

If $X(x)$ is determined by the formula (15.60), then the following equation is obtained for $Y(y)$:

$$Y^{IV} - 2\alpha_m^2 Y'' - \left(r^4 - \alpha_m^4\right) Y = 0; r^4 = \frac{mp^2}{D}. \tag{15.61}$$

The solution to this equation can be represented as:

$$Y(y) = Y(0)\overline{A}(y) + Y'(0)\overline{B}(y) + Y''(0)\overline{C}(y) + Y'''(0)\overline{D}(y) \tag{15.62}$$

where,

$$
\begin{aligned}
\overline{A}(y) &= \frac{1}{2r^2}\left(\delta_1^2 \cosh \delta_2 y + \delta_2^2 \cos \delta_1 y\right);\\
\overline{B}(y) &= \frac{1}{2r^2}\left(\frac{\delta_1^2}{\delta_2} \sinh \delta_2 y + \frac{\delta_1^2}{\delta_1} \sin \delta_1 y\right);\\
\overline{C}(y) &= \frac{1}{2r^2}(\cosh \delta_2 y - \cos \delta_1 y);\\
\overline{D}(y) &= \frac{1}{2r^2}\left(\frac{\sinh \delta_2 y}{\delta_2} - \frac{\sin \delta_1 y}{\delta_1}\right);\\
\delta_1 &= r^2 - \alpha_m^2; \delta_2 = r^2 + \alpha_m^2.
\end{aligned}
\tag{15.63}
$$

Using the solution (15.62) and the boundary conditions on the sides of the plate $y = 0$, $y = b$, we obtain the frequency equation. For example, if these sides are built-in, then $Y(0) = Y'(0) = 0$; $Y(b) = Y'(b) = 0$, and hence:

$$
\overline{C}^2(b) - \overline{D}(b)\overline{C}'(b) = 0. \tag{15.64}
$$

The equation of frequencies (15.64) can be reduced to the following two equations:

$$
\delta_2 \tanh \frac{\delta_2 b}{2} + \delta_1 \tan \frac{\delta_1 b}{2} = 0 \tag{15.65}
$$

and

$$
\delta_2 \coth \frac{\delta_2 b}{2} - \delta_1 \cot \frac{\delta_1 b}{2} = 0. \tag{15.66}
$$

The presented equations correspond to symmetric (15.65) and oblique-symmetric skew-symmetric (15.66) modes of natural vibrations/oscillations. Each value of α_m corresponds to an infinite sequence of the roots of Eqs. (15.65) and (15.66). Thus, it is possible to determine consecutive values of r_{mn}, as well as the frequency of its natural vibrations/oscillations by the formula:

$$
p_{mn} = r_{mn}^2 \sqrt{\frac{D}{m}}. \tag{15.67}
$$

The modes of natural vibrations/oscillations of the plate will be:

$$
W_{mn}(x, y) = X_m(x)\left[\overline{C}_n(y) - \frac{\overline{D}_n(b)}{\overline{C}_n(b)}\overline{D}_n(y)\right] \tag{15.68}
$$

Obviously, they are determined up to a constant factor (cofactor, multiplier, multiplicand).

The modes of natural oscillations (that is, the eigenfunctions) of the plate have the property of orthogonality, that is:

$$
\iint_A W_{mn}(x, y) W_{ij}(x, y) dxdy = 0, \tag{15.69}
$$

if $m \neq i$ or $n \neq j$. Normalized modes of natural vibrations/oscillations are defined, while assuming that:

$$
\iint_A \mathrm{W}_{mn}^2(x, y) dxdy = 1. \tag{15.70}
$$

The general solution of Eq. (15.57) is obtained as a result of summing all partial solutions of the type (15.68).

$$w(x, y, t) = \sum_{m,n=1}^{\infty} A_{mn} W_{mn}(x, y) \sin(p_{mn} t + \varepsilon_{mn}) \tag{15.71}$$

The remaining unknown constants A_{mn} and ε_{mn} are determined from the initial conditions. To do this, the functions $\varphi(x, y)$ and $\psi(x, y)$ should be represented as series of eigenfunctions:

$$\begin{aligned} \varphi(x, y) &= \sum_{m,n=1}^{\infty} a_{mn} W_{mn}(x, y); \\ \psi(x, y) &= \sum_{m,n=1}^{\infty} b_{mn} W_{mn}(x, y). \end{aligned} \tag{15.72}$$

The coefficients of the series (15.72) can be determined by the formulas:

$$\begin{aligned} a_{mn} &= \iint_A \varphi(x, y) \mathrm{W}_{mn}(x, y) dxdy \\ b_{mn} &= \iint_A \psi(x, y) \mathrm{W}_{mn}(x, y) dxdy \end{aligned} \tag{15.73}$$

In this case the eigenfunctions are assumed to be normalized.
Then

$$\varepsilon_{mn} = p_{mn} \tan^{-1} \frac{a_{mn}}{b_{mn}}; A_{mn} = \frac{a_{mn}}{\sin \varepsilon_{mn}} = \frac{b_{mn}}{p_{mn} \cos \varepsilon_{mn}} . \tag{15.74}$$

For a plate that is supported along the entire contour,

$$X(x) = \sin \frac{m\pi x}{a}, \quad Y(y) = \sin \frac{n\pi y}{b} ,$$

the frequency equation has the form:

$$\sin \delta_1 b = 0. \tag{15.75}$$

From here we have:

$$r_{mn}^2 = \left(\frac{m\pi}{a}\right)^2 + \left(\frac{n\pi}{b}\right)^2. \tag{15.76}$$

Thus, if two opposite edges of a rectangular plate are hinged or have a sliding built-in (sliding cantilever), then it is possible to obtain an exact solution of the problem in the form of series in terms of eigenfunctions or modes of natural vibrations/oscillations (eigenmodes). For any other boundary condition only an approximate solution can be obtained. However, even in these cases, the modes of oscillations can be taken as: $\mathrm{W}_{mn} = A_{mn} X_m(x) Y_n(y)$, where $X_m(x)$ and $Y_n(y)$ are beam functions that satisfy the boundary conditions of built in (fixing) of the edges of the plate. The natural vibrations/oscillation frequencies and coefficients A_{mn} are determined using approximate methods (Sect. 15.5.1).

When the plate vibrates/oscillates along one of the natural modes, lines are formed on its surface that connect the points at which the deflection of the plate is equal to zero. These lines are called nodal lines. The nodal lines can also include the edge of the plate fixed from displacement.

When the edges of a rectangular plate are fixed in an arbitrary manner, the frequencies of natural vibrations/oscillations can be determined by the formula:

$$p_{mn} = \frac{k_{mn}^2}{a^2}\sqrt{\frac{D}{m}}, \tag{15.77}$$

where

$$k_{mn}^2 = \pi^2\left\{A_m^4 + \frac{a^4}{b^4}A_n^4 + 2\frac{a^2}{b^2}\left[\nu B_m B_n + (1-\nu)C_m C_n\right]\right\}^{1/2}, \tag{15.78}$$

Here, the expression for k_{mn}^2 gives the first approximation by the Ritz method. If two opposite sides of the plate are hinge supported, this value is accurate. Here m, n is the number of nodal lines, including fixed edges that are parallel to the x and y axes, respectively, a, b are the lengths of the sides of the plate.

If longitudinal forces act in the middle plane of the plate, then:

$$k_{mn}^2 = c\left(\overline{k}_{mn}^2 \pm N_1 C_m \pm N_2 C_n \frac{a^2}{b^2}\right)^{1/2}, \tag{15.79}$$

were

$$N_x = \frac{q_x a^2}{D\pi^2};\, N_y = \frac{q_y b^2}{D\pi^2}$$

q_x, q_y is the intensity of longitudinal forces in the direction of the x and y axes; $\overline{k}_{mn}^2$ is determined by the formula (15.78) without the coefficient π^2.

Frequencies of natural vibrations/oscillations of orthotropic rectangular plates can be determined by the formula (15.77), in which it should be put:

$$k_{mn}^2 = \pi^2\left\{D_1 A_m^4 + D_2\frac{a^4}{b^4}A_n^4 + 2\frac{a^2}{b^2}\left[\nu_2 D_1 B_m B_n + 2D_k C_m C_n\right]\right\}^{\frac{1}{2}},$$

here

$$D_1 = \frac{E_1 h^3}{12(1-\nu_1\nu_2)};\quad D_2 = \frac{E_2 h^3}{12(1-\nu_1\nu_2)};\quad D_k = \frac{Gh^3}{12};\quad D_3 = D_2\nu_1 + 2D_k;$$

E_1, E_2 are modulus of elasticity in the axial directions of the plate; G is the shear modulus. The values of coefficients A, B, C, D, and E are given in Table 15.2 [6].

The values of the natural oscillation frequencies of rectangular built-in (cantilever) and free plates can be determined using Table 15.3. For square plates with such fixing conditions at $m - n = \pm 2, 4, 6, \ldots$ vibrational forms of the type $m/n \pm n/m$ are possible, in which the nodal lines are not parallel to the edges of the plate.

The approximate values of the coefficients k_{mn} for a square plate, as well as the shapes of the nodal lines, are given in Table 15.4 [1].

Table 15.2 The values of the coefficients $A,\ B,\ C,\ D,\ E.$ [6]

Conditions for fixing edges	A				B			C		
	$m=1$		$m=2$	$m\geq 3$	$m=1$	$m=2$	$m\geq 3$	$m=1$	$m=2$	$m\geq 3$
Supported edges	0	1	$m-1$	0	1	A^2	0	1	A^2	
Supported edges built-in	0	1.25	$m-0.75$	0	$A^2-\frac{A}{\pi}$	$A^2-\frac{A}{\pi}$	0	B	B	
Supported free edges	0	1.25	$m-0.75$	0	$A^2-\frac{A}{\pi}$	$A^2-\frac{A}{\pi}$	$\frac{3}{\pi}$	$A^2+\frac{3A}{\pi}$	$A^2+\frac{3A}{\pi}$	
Built-in edges	0	1.506	$m-0.5$	0	1.248	$A^2-\frac{2A}{\pi}$	0	1.248	B	
Built-in free edges	0.597	1.494	$m-0.5$	0.087	1.347	$A^2-\frac{2A}{\pi}$	0.471	3.284	$A^2+\frac{2A}{\pi}$	
Free edges	0	1.506	$m-0.5$	0	1.248	$A^2-\frac{2A}{\pi}$	$\frac{12}{\pi^2}$	5.017	$A^2+\frac{6A}{\pi}$	
Conditions for fixing edges	D					E				
	$m=1$	$m=2$	$m=3$		$m=4$	$m=1$	$m=2$	$m=3$	$m=4$	
Supported edges—built-in	−8.139	−25.7	−52.13		−89.17	0	0	0	0	
Supported edges—free edges	0.1274	0.0706	0.049		0.0374	0.9991	0.9998	1	1	
Built-in edges	−0798	30.91	58.78		162	0	0	0	0	
Built-in—free edges	6.802	−44.06	121.2		6.647	8.711	1.928	2.003	2	

15.7 Forced Vibrations/Oscillations of Plates

There are several methods for solving problems of forced vibrations of plates. These methods are associated with two principle ways of solving the problem. The first way is based on the exclusion of time from the equation of vibrations/oscillations and obtaining a quasi-static equation in which only spatial coordinates appear. The solution of this equation is achieved by the usual exact or approximate methods that are typical for static problems. The integral Laplace transform is used to exclude time. The classical operational calculus, or the integral Fourier transform, as well as the expansion (decomposition) of the load in a Fourier series in time, is closely related to the Laplace transform. The second way is related to the exclusion of spatial coordinates from the equation of vibrations/oscillations of the plate and obtaining an ordinary differential equation with respect to time or a system of such equations. To exclude spatial coordinates, one can use expansions (decompositions) in a series of eigenfunctions (modes of natural vibrations), integral transformations, variational, finite difference, and other methods.

Let us consider the equation of forced vibrations of the plate:

$$DLw + m\frac{\partial^2 w}{\partial t^2} = q, \tag{15.80}$$

where L is a biharmonic operator (double Laplace operator) in rectangular or polar coordinates. For simplicity, we assume that the initial (essential) conditions are zero. The application of Laplace to Eq. (15.80) gives [7]:

$$DL\overline{w} + mp^2\overline{w} = \overline{q}, \tag{15.81}$$

Table 15.3 Values of the coefficients k_{mn}^2 for the built-in and free rectangular plates [1]

$\frac{a}{b}$	m	n 1	2	3	4	5	6
Plate with built-in edges							
1	1	36	73.41	131.9	210.5	309	428
1.5		27.01	65.5	126	206	303	422
2		24.58	64.1	124	204	302	421
1	2	73.41	108.2	165	242.7	340.6	458.3
1.5		41.72	79.81	138.6	218	316.1	436
2		31.83	71.08	130.4	210	308.1	427
1	3	131.9	165	220	296.4	393.4	509.9
1.5		66.53	103	161.2	241	339	457
2		44.78	83.2	142.4	221	320.1	439
1	4	210.5	242.7	296.4	371.4	467.3	593.8
1.5		100.8	136.1	193.2	271.2	369.3	488
2		63.34	100.8	159.5	238.4	337. 1	456
1	5	309	340.6	393.4	467.3	562.2	676
1.5		144.2	178	234.7	312	409	529
2		87.26	124.2	181.8	261	358	478
1	6	428	458.3	509.9	583.8	676	792.5
1.5		195	230	285.4	361.9	456	576.6
2		117	151.9	209.6	287.5	382	504.3
Plate with free edges ($\nu = 1/6$)							
1	1	14.92	37.28	75.95	134.1	214.1	309.1
1.5		9.905	30.36	69.56	127.7	205.1	302.1
1		7.374	26.52	65.17	123.3	200.7	298.9
1	2	37.28	67.59	110.6	170	248.1	345.7
1.5		22.25	46.65	86.03	145.2	222.1	320.4
2		17.61	37.59	75.05	132.9	210	307.3
1	3	75.95	110.6	159.3	222.7	302.8	396.8
1.5		40.34	68.39	111.5	160.5	250	347.8
2		27.03	51.7	91.96	149.6	226.4	324.7
1	4	134.1	170	222.7	290.4	374	474.6
1.5		66.31	97.82	143.5	204.8	283.7	382.6
2		42.25	70.01	11.58	171	248.9	346
1	5	214.1	248.1	302.8	374	461	565.5
1.5		100.9	133.4	182.2	245.9	326.6	425.6
2		61.63	91.78	135.8	196.6	274.6	372.9
1	6	292.4	345.7	399.2	474.6	562.6	671
1.5		144.5	177.6	226.2	294.3	374.8	476.9
2		85.56	117.3	162.7	223.5	303.2	403

where

$$\overline{w} = \int_0^\infty e^{-pt} w dt; \overline{q} = \int_0^\infty e^{-pt} q dt. \tag{15.82}$$

The Eq. (15.81) does not include time, it coincides with the equation of bending of the plate, which lies on an elastic Winkler base with a bed coefficient $k = mp^2$. The solution of quasi-static Eq. (15.81)

Table 15.4 The values of the coefficient k_{mn}^2

Plate with built-in edges		Plate with free edges	
Nodal lines	k^2	Nodal lines	k^2
	35.99	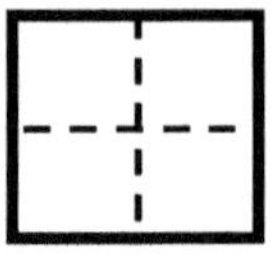	13.47
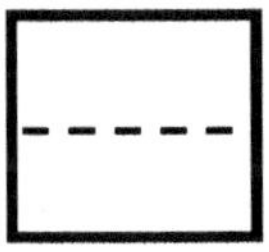	73.41		19.6
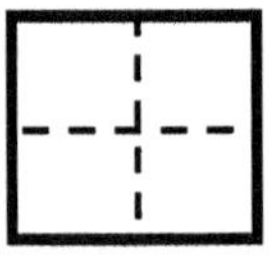	108.27		24.27
	131.64	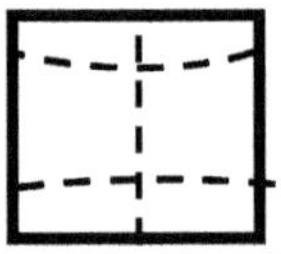	34.8
	132.25	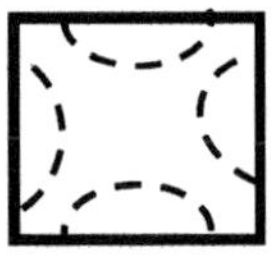	61.09
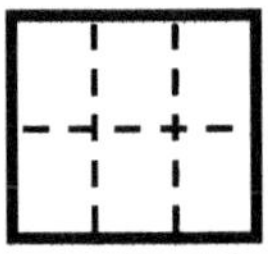	165.15	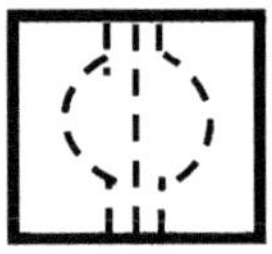	63.69

can be obtained by the exact or approximate methods. The inversion of the transform (15.82) is carried out according to the following formula:

$$w = \frac{1}{2\pi i} \int_{\gamma - i\infty}^{\gamma + i\infty} e^{pt} \overline{w} dp, \tag{15.83}$$

where γ is the chosen real number that should ensure convergence of the integral (15.83).

Time from Eq. (15.80) for the case when the load is harmonic or polyharmonic is excluded quite simply. For example, let

$$q = \sum_{j=1}^{k} q_j e^{i\omega_j t}$$

Assuming that

$$w = \sum_{j=1}^{k} W_j e^{i\omega_j t}$$

we obtain the equation:

$$DLW_j - m\omega_j^2 W_j = q_j, \ (j = 1, 2, \ \ldots, k). \tag{15.84}$$

If the load is periodic, it can be decomposed into a Fourier series and limited to the first few terms of the series. For each term, we will have an equation like (15.84). The general solution is found by (through) the superposition of individual solutions.

To exclude the spatial coordinate from Eq. (15.80), the solution of the equation and the load can be represented as a series of some system of functions. For example, for rectangular coordinates:

$$\begin{aligned} w(x, y, t) &= \sum_{m,n} c_{mn}(t) w_{mn}(x, y); \\ q(x, y, t) &= \sum_{m,n} b_{mn}(t) W_{mn}(x, y). \end{aligned} \tag{15.85}$$

If the solution of the homogeneous Eq. (15.80) is known, i.e., the modes of natural vibrations/oscillations are known, then it is most convenient to take these eigenfunctions as functions of $W_{mn}(x, y)$. This results in an ordinary differential equation for determining the coefficients c:

As a result of this, to determine the coefficients c_{mn}, an ordinary differential equation is obtained:

$$\begin{aligned} \frac{\partial^2}{\partial t^2} c_{mn} + p_{mn}^2 c_{mn} &= \frac{b_{mn}}{m}; \\ b_{mn} &= \iint_S q W_{mn} dx dy, \end{aligned} \tag{15.86}$$

where p_{mn} is the frequency of natural vibrations of the plate.

The solution of Eq. (15.86) under zero initial (essential) conditions has the form:

$$c_{mn}(t) = \frac{1}{m p_{mn}} \int_0^t b_{mn}(\tau) \sin p_{mn}(t - \tau) d\tau. \tag{15.87}$$

In the case of harmonic oscillations $\left(b_{mn}(t) = b_{mn}^0 \cos \omega_0 t\right)$, the solution of the Eq. (15.86) corresponding to the forced steady-state vibrations/oscillations has the following form:

$$c_{mn}(t) = \frac{b_{mn}^0 \cos \omega_0 t}{m\left(p_{mn}^2 - \omega_0^2\right)} \ . \tag{15.88}$$

It follows from Eq. (15.88) that at resonance, that is, when one of the natural frequencies p_{mn} is equal to the frequency of the disturbing load ω_0, the plate deflection and internal forces turn to infinity. Taking into account the damping of vibrations/oscillations makes it possible to obtain finite deflections and forces. If we use the complex theory and introduce an internal inelastic resistance, then the equation of vibrations/oscillations of the plate can be written as:

$$D^*Lw^* + m\frac{\partial^2 w^*}{\partial t^2} = qe^{i\omega t}, \tag{15.89}$$

where $D^* = D(u + iv)$; w^* is a complex displacement ($\text{Re}w^* = w$); u, v are parameters that characterize the internal inelastic resistance. Decomposition (expansion) into the modes of natural vibrations/oscillations gives:

$$w = w' \cos \omega t + w'' \sin \omega t, \tag{15.90}$$

where

$$c = \sum_{m,n} \frac{b_{mn} W_{mn}\left(1 - \frac{\omega^2}{p_{mn}^2}\right)}{p_{mn}^2\left[\left(1 - \frac{\omega^2}{p_{mn}^2}\right)^2 + \gamma^2\right]}; \tag{15.91}$$

$$w'' = \gamma \sum_{m,n} \frac{b_{mn} W_{mn}}{p_{mn}^2\left[\left(1 - \frac{\omega^2}{p_{mn}^2}\right)^2 + \gamma^2\right]}, \gamma = \frac{\delta}{\pi};$$

δ is the logarithmic decrement of vibrations/oscillations.

The deflection amplitude is determined by the following formula:

$$w_0 = \sqrt{[w']^2 + [w'']^2}. \tag{15.92}$$

The amplitudes of bending moments, torques moments, and shear forces can be determined by the formula (15.92) if the eigenfunctions W_{mn} in the expressions for w' and w'' are replaced by the corresponding expressions according to the formulas that relate the deflections of the plate to the internal forces in it.

As already mentioned, the modes of natural vibrations/oscillations of a rectangular plate can be obtained by an exact method, i.e., when two opposite edges of the plate, for example $x = 0$, $x = a$, are hinged supported or fixed in a sliding built-in. In this case $X_m(x) = \begin{matrix} \sin \\ \cos \end{matrix} \frac{m\pi x}{a}$, and $Y_n(y)$ is the solution of the Eq. (15.62) which satisfies one or the other of the conditions at the edges: $y = 0$, $y = b$. At the same time, in the expansions/decompositions (15.83), summation over all the roots of the frequency equation is performed.

In the case of circular and ring plates, the modes of natural vibrations/oscillations have the form:

$$W_{mn}(\rho, \theta) = \begin{matrix} \cos \\ \sin \end{matrix} n\theta[A_{mn} I_n(k_{mn}\rho) + B_{mn} I_n(k_{mn}\rho) + C_{mn} Y_n(k_{mn}\rho) + D_{mn} K_n(k_{mn}\rho)]. \tag{15.93}$$

For plates of finite dimensions, the orthonormal form of natural oscillations $W_{mn}(x, y)$ or $W_{mn}(\rho, \theta)$ can be considered as the kernel of the integral transformation. The transform of this integral transformation for the function w has the following form:

$$c_{mn}(t) = \iint_A w W_{mn} dA \tag{15.94}$$

and the inversion formula looks like:

$$w = \sum_{m,n} c_{mn} W_{mn}. \tag{15.95}$$

When using the finite integral transformation, the procedure for obtaining the solution differs from the eigenfunction expansion only in that the transformant, i.e., the coefficient in the expansion (15.95), is found directly from the vibration/oscillation equation. For this purpose, the left and right parts of the vibration/oscillation equation are multiplied by W_{mn} and the integration over the entire plate area is performed:

$$D\iint_A W_{mn} L w dA + m\frac{\partial^2}{\partial t^2}\iint_A W_{mn} w dA = \iint_A q W_{mn} dA. \tag{15.96}$$

Since W_{mn} is the eigenfunctions of the operator L , then

$$\iint_A W_{mn} L w dA = \iint_A w L W_{mn} dA = k^4 \iint_A W_{mn} w dA = k^4 c_{mn},$$

and hence Eq. (15.96) transforms to Eq. (15.86). Under inhomogeneous boundary conditions, additional functions appear in the last expression. In this case, finite transformations are especially convenient. For unlimited plates, the decomposition (15.95) (the inverse formula) has an integrative form:

$$w = \iint_A c(\xi, \eta) W(\xi, \eta) dA \tag{15.97}$$

the integration is carried out over the entire spectrum of the operator L. In particular, for an unlimited plate without fixtures, a Fourier transform can be applied to each of the coordinates in the direction in which the plate is unbounded. In the polar coordinate system for such plates, the Hankel transformation is applied. For semi-infinite plates, the Fourier transform is applied along the coordinate in the direction of which the plate is not limited (unbounded). In this case, the transformations introduced in [8] can be applied along the direction of the second coordinate. For circular and ring plates, it is possible to use the final transformation of the Fourier by the polar angle and the transformation which corresponds to decomposition by the modes of natural vibrations/oscillations (15.96).

In the case of arbitrary support of the edges of the plate, any suitable functions are taken as W_{mn}, and the coefficients of $c_{mn}(t)$ are determined by approximate methods described in paragraph 5. For example, by multiplying the left and right parts of (15.80) by w_{ij}, and then integrating over the entire plate area and using (15.87), we come to a system of ordinary differential equations of the Bubnov-Galerkin method with respect to functions c_{mn}. In this case, the functions W_{mn}, must satisfy all boundary conditions. Other approximate methods can be applied to solve Eq. (15.80), such as: Kantorovich-Vlasov method, mesh method, collocation method, etc.

In the practical calculation of rectangular plates with arbitrary fixing of edges to forced vibrations/oscillations when using the Ritz method, Bubnova Galerkin method,etc., can be taken: $W_{mn}(x, y) = X_m(x)Y_n(y)$, where X_m, Y_n are the beam functions. Here we can limit the first approximation. As a result, the differential Eq. (15.86) is obtained for plate deflection decomposition coefficients. In particular, for harmonic excitation the displacement amplitudes can be determined by the formula (15.92).

In many cases, the solution of problems on forced vibrations/oscillations of circular and ring plates can be obtained in a closed form; in this case, the application of the method of compensating loads [9] appears to be quite convenient. In particular, the solution of the axisymmetric problem of forced vibrations of a circular plate, which is loaded in the center by a concentrated force $P\cos\omega t$ (without taking into account internal inelastic resistance), can be represented as:

$$w(\xi,t)=W(\xi)\cos wt, \tag{15.98}$$

where

$$W(\xi)=w_0(\xi)+A_0J_0(\xi)+B_0I_0(\xi);$$

$$w_0(\xi)=-\frac{P}{8D\lambda_1^2}\left[Y_0(\xi)+\frac{2}{\pi}K_0(\xi)\right];\lambda_1=\sqrt[4]{\frac{m\omega^2}{D}};\xi=\lambda_1 r.$$

The function w_0 is called the basic solution. This function has in the particular case under consideration at $\xi=0$ a singularity of the concentrated force type, i.e., its decomposition (expansion) in the vicinity of the point $\xi=0$ has a term of the form $\xi^2\ln\xi$. The constants A_0 and B_0 are determined from the boundary conditions on the plate contour. If the plate contour $r=r_0=\beta/\lambda_1$ is built-in, then at $\xi=\beta$ we have $w(\beta)=0$;

$$\frac{dw}{d\xi}(\xi)=0,$$

hence follows:

$$A_0=\frac{P}{8D\lambda_1^2}\cdot\frac{I_1(\beta)Y_0(\beta)+I_0(\beta)Y_1(\beta)+\frac{2}{\pi\beta}}{J_0(\beta)I_1(\beta)+J_1(\beta)I_0(\beta)} \tag{15.99}$$

$$B_0=\frac{P}{4\pi D\lambda_1^2}\cdot\frac{I_1(\beta)K_0(\beta)-I_0(\beta)K_1(\beta)+\frac{1}{\beta}}{J_0(\beta)I_1(\beta)+J_1(\beta)I_0(\beta)} \tag{15.100}$$

If a circular plate is loaded uniformly $q\cos\omega t$ over a circle of radius $r=\alpha/\lambda_1$, then the function $w(\xi,t)$ can also be represented in the form (15.98). The function $w_0(\xi)$, in this case, should at $\xi=a$ give a discontinuity in the shear force by value q and has the following form:

at $\xi\leq\alpha$

$$w_0(\xi)=-\frac{\pi q\alpha}{4D\lambda_1^3}\left[J_0(\xi)Y_0(\alpha)+\frac{2}{\pi}I_0(\xi)K_0(\alpha)\right]; \tag{15.101}$$

at $\xi\geq\alpha$

$$w_0(\xi)=-\frac{\pi q\alpha}{4D\lambda_1^3}\left[Y_0(\xi)J_0(\alpha)+\frac{2}{\pi}K_0(\xi)I_0(\alpha)\right]. \tag{15.102}$$

In this case, with a built-in edge of the plate, we will have:

$$A_0=\frac{\pi q\alpha}{4D\lambda_1^3}\cdot\frac{J_0(\alpha)[I_1(\beta)Y_0(\beta)+I_0(\beta)Y_1(\beta)]+\frac{2}{\pi\beta}I_0(\alpha)}{I_0(\beta)I_1(\beta)+I_1(\beta)I_0(\beta)}; \tag{15.103}$$

$$B_0 = \frac{q\alpha}{2D\lambda_1^3} \cdot \frac{I_0(\alpha)[I_1(\beta)K_0(\beta) - J_0(\beta)K_1(\beta)] + \frac{1}{\beta}J_0(\alpha)}{I_0(\beta)J_1(\beta) + I_1(\beta)I_0(\beta)}. \tag{15.104}$$

If the load is uniformly distributed over the ring with reduced radii α_1, α_2, where $\alpha_1 < \alpha_2$, then the function w_0 has the form:

at $\xi \le \alpha_1$

$$\begin{aligned} w_0(\xi) = -\frac{\pi q_1}{4D\lambda_1^4}\Big\{&[\alpha_2 Y_1(\alpha_2) - \alpha_1 Y_1(\alpha_1)]J_0(\xi) - \frac{2}{\pi}[\alpha_2 K_1(\alpha_2) - \\ &- \alpha_1 K_1(\alpha_1)]I_0(\xi)\Big\}; \end{aligned} \tag{15.105}$$

at $\xi \ge \alpha_2$

$$\begin{aligned} w_0(\xi) = -\frac{\pi q_1}{4D\lambda_1^4}\Big\{&[\alpha_2 J_1(\alpha_2) - \alpha_1 J_1(\alpha_1)]Y_0(\xi) + \frac{2}{\pi}[\alpha_2 I_1(\alpha_2) - \\ &- \alpha_1 I_1(\alpha_1)]K_0(\xi)\Big\}; \end{aligned} \tag{15.106}$$

at $\alpha_2 \ge \xi \ge \alpha_1$

$$\begin{aligned} w_0(\xi) = -\frac{\pi q_1}{4D\lambda_1^4}\Big[&\frac{4}{\pi} - \alpha_1 J_1(\alpha_1)Y_0(\xi) - \frac{2}{\pi}\alpha_1 I_1(\alpha_1)K_0(\xi) + \\ &+ \alpha_2 Y_1(\alpha_2)J_0(\xi) - \frac{2}{\pi}\alpha_2 K_1(\alpha_2)I_0(\xi)\Big], \end{aligned} \tag{15.107}$$

and the numbers A_0 and B_0 are determined from the boundary conditions.

For ring plates, the $W(\xi)$ function is:

$$W(\xi) = w_0(\xi) + A_0 J_0(\xi) + B_0 I_0(\xi) + A_0^* Y_0(\xi) + B_0^* K_0(\xi). \tag{15.108}$$

The determination of the constants A_0, B_0, A_0^*, B_0^* becomes time-consuming. To solve the problem of forced vibrations of circular plates, it is convenient to apply the method of initial parameters. For this purpose, it is necessary to construct a system of functions similar to the Krylov functions in the problem of beam vibrations. These functions and a detailed presentation of the calculation scheme are considered in [9, 10].

In the case of non-axisymmetric vibrations/oscillations of circular plate, which are caused by concentrated force $P \cos \omega t$, solution by the method of compensating loads can be presented in the form:

$$w(\xi, \theta) = w_0(\xi, \theta) + w_k(\xi, \theta), \tag{15.109}$$

here the concentrated force is applied at the point with coordinates: r_1, $\theta_1 (r_1 = \alpha_1/\lambda_1)$, where

$$w_0(\xi,\theta) = -\frac{P}{8D\lambda_1^2}\left[Y_0(z) + \frac{2}{\pi}K_0(z)\right];$$
$$z = \sqrt{\alpha_1^2 + \xi^2 - 2\alpha_1\xi\cos(\theta_1 - \theta)}\ .$$

ξ, θ are the coordinates of the point at which the deflection is determined;

$$w_k(\xi,\theta) = \sum_{n=0}^{\infty\prime} [A_n J_n(\xi) + B_n I_n(\xi)] \cos n(\theta_1 - \theta).$$

Here the prime symbol means that at $n = 0$ the corresponding term of the series is multiplied by 1/2. The coefficients A_n and B_n are determined from the conditions on the contour of the plate. To actually obtain the solution, it is necessary to use the formulas for the addition of Bessel functions. If the plate is built-in along the contour, then

$$A_n = \frac{P}{4D\lambda_1^2} \cdot \frac{J_n(\alpha)\left[I_n(\beta)Y_n'(\beta) - I_n'(\beta)Y_n(\beta)\right] - \frac{2}{\pi\beta}I_n(\alpha)}{I_n(\beta)J_n'(\beta) - J_n(\beta)I_n'(\beta)} \tag{15.110}$$

$$B_n = \frac{P}{4D\lambda_1^2} \cdot \frac{\frac{2}{\pi}I_n(\alpha)\left[J_n'(\beta)K_n(\beta) - J_n(\beta)\ K_n'(\beta)\right] - \frac{2}{\pi\beta}J_n(\alpha)}{I_n(\beta)J_n'(\beta) - I_n(\beta)I_n'(\beta)}. \tag{15.111}$$

The presented solution makes it possible to consider the problem of free and forced vibrations of a circular plate with concentrated additional masses or point supports. This solution can also be used to obtain a solution to the problem of forced vibrations/oscillations of a girder less slab (beamless overlap).

The vibration of plates on elastic foundation and forced vibrations of a rectangular plate on an elastic foundation will be discussed in the next chapter.

15.8 Vibration of the Plates on Elastic Foundation

The method of integral transformation is especially effective for solutions of multidimensional and dynamic problems; therefore, it is reasonable to expect that Spectral Method of Boundary Element (SMBE) can also be successfully applied to dynamic calculation of structures [11].

As an example of the application of the SMBE to dynamic problems, let us consider the oscillations of a plate of an arbitrary form which lies on the elastic Winklerian base and which is under the action of a load $q(x, y, t)$. The equation of oscillations we will write in the form of:

$$\mathrm{D}\nabla^2\nabla^2 w(x,y,t) + m\ddot{w}(x,y,t) + kw(x,y,t) = q(x,y,t) \tag{15.112}$$

where the designations are conventional. We will accept the boundary conditions on a contour of a plate as those corresponding to those at the fixed end (built in) the edge

$$\begin{aligned} &w(x,y,t) = 0,\\ &(x, y \in \Gamma^+\ ),\\ &\frac{\partial w}{\partial n}(x,y,t) = \beta_n(x,y,t) = 0, \end{aligned} \tag{15.113}$$

where $n-$ normal to the contour of the plate. This choice of boundary conditions does not harm the generality of the consideration, since other boundary conditions can absolutely similarly be taken into account. We assume that the initial conditions are zero. We will consider, furthermore, that the contour of the plate $\Gamma-$ is the boundary of the domain D^+, which is formed by a smooth curve with no angular points. The presence of angular points does not introduce fundamental difficulties into the solution of the problem, but only increases the number of the unknowns in algebraic equations due to the angular forces [12]. Therefore, for simplicity, we will not take them into account. Furthermore, the rounding of plate angles is possible, eliminating the angular forces.

Let us designate through n_x, and n_y the directing cosines of n normal with the axes x, and y; S_x, S_y- the directing cosines of tangential vector S (Figs. 15.3 and 15.4).

The expressions for normal M_n moment and for tangential M_s moment will take the form [13]:

$$\begin{aligned} M_n &= n_x M_1 + n_y M_2, \\ M_S &= -n_y M_1 + n_x M_2 \end{aligned} \tag{15.114}$$

where

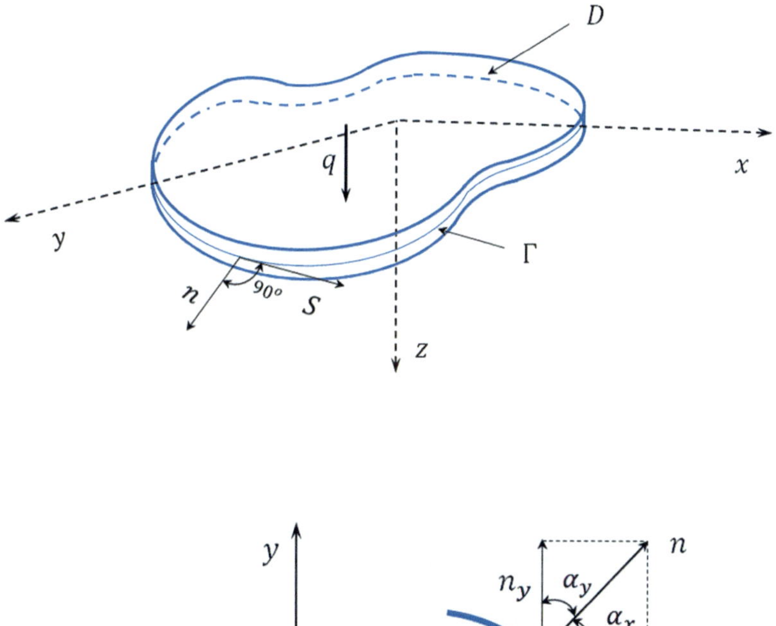

Fig. 15.3 Scheme of a plate

Fig. 15.4 Element of the boundary of a plate

$$M_1 = n_x M_x + n_y M_{xy} \ ,$$
$$M_2 = n_x M_{xy} + n_y M_y \ ,$$

M_x, M_y– bending moments; M_{xy}– torque moment.
Let us also introduce the designations for the angle of rotations:

$$\beta_x = -\frac{dw}{dx}; \beta_y = -\frac{dw}{dy}; \beta_n = -\frac{dw}{dn}; \beta_S = -\frac{dw}{dS} \tag{15.115}$$

and the given shear force

$$H = n_x Q_x + n_y Q_y + \frac{dM_S}{dS}. \tag{15.116}$$

Here Q_x, Q_y– shear forces. The moments and forces are determined using the standard formulas of the technical theory of plates

$$M_x = -D\left(\frac{\partial^2 w}{\partial x^2} + \nu\frac{\partial^2 w}{\partial y^2}\right), M_x = -D\left(\frac{\partial^2 w}{\partial y^2} + \nu\frac{\partial^2 w}{\partial x^2}\right),$$
$$M_{xy} = -D(1-\nu)\frac{\partial^2 w}{\partial x dy}$$
$$Q_x = -D\frac{\partial}{\partial x}\nabla^2 w, Q_y = -D\frac{\partial}{\partial y}\nabla^2 w,$$

where ν—Poisson's coefficient (ratio). At the indicated designations the generalized formula of Green –Betty is fair [13, 14].

$$D\int_\Omega \nabla^2\nabla^2 ww^* d\Omega = D\int_\Omega \nabla^2\nabla^2 w^* w d\Omega + \int_\Gamma (M_n\beta_n^* + M_S\beta_S^* + \\ + Hw^* - M_n^*\beta_n - M_S^*\beta_S - H^* w)d\Gamma. \tag{15.117}$$

Let us apply a dual Fourier transformation of the spatial coordinates x, y to the left and right sides of the Eq. (15.112)

$$\frac{D}{2\pi}\iint_{-\infty}^{\infty} e^{i(\xi x+\eta y)}\nabla^2\nabla^2 w(x,y,t)dxdy + m\ddot{W}(\xi,\eta,t) + kW(\xi,\eta,t) = Q(\xi,\eta,t),$$

where $W(\xi, \eta, t)$ and $Q(\xi, \eta, t)$– Fourier transforms of functions W and q. Since on the plate border Γ, the function $w(x, y, t)$ and its derivatives can have gaps, apply the formula of Green-Betty (15.117) in the domain D^+ and D^- separately. We will obtain as a result of the addition of both results

$$\left[D(\xi^2+\eta^2)^2 + k\right]W(\xi,\eta,t) + m\ddot{W}(\xi,\eta,t) = Q(\xi,\eta,t) + \\ \int_\Gamma (M_n^*\Delta\beta_n + H^*\Delta w - \Delta M_n\beta_n^* - \Delta H w^*)d\Gamma. \tag{15.118}$$

Here we must accept

$$w^* = \frac{1}{2\pi}e^{i(\xi x+\eta y)}$$

and respectively:

$$\begin{aligned}
&\beta_n^* = -\beta_x^* n_x = -\frac{in_x\xi}{2\pi} e^{i(\xi x+\eta y)};\\
&\beta_S^* = -\beta_x^* S_x = -\frac{iS_x\xi}{2\pi} e^{i(\xi x+\eta y)};\\
&M_n^* = D\left[n_x^2(\xi^2+\nu\eta^2)+2n_xn_y(1-\nu)\xi\eta+n_y^2(\eta^2+\nu\xi^2)\right]e^{i(\xi x+\eta y)};\\
&H^* = D\left[i\xi n_x(\xi^2+\eta^2)+i\eta n_y(\xi^2+\eta^2)+S_x(\xi^2+\nu\eta^2)\right].
\end{aligned} \tag{15.119}$$

Here, it is also taken into account that

$$\frac{dM_S}{dS} = M_x S_x$$

The solution of the Eq. (15.118) with zero initial conditions takes the form:

$$W(\xi,\eta,t) = \frac{1}{mp(\xi,\eta)}\int_0^t F(\xi,\eta,\tau)\sin p(\xi,\eta)(t-\tau)d\tau, \tag{15.120}$$

where

$$\begin{aligned}
p(\xi,\eta) &= \sqrt{\left[D(\xi^2+\eta^2)+k\right]m^{-1}},\\
F(\xi,\eta,t) &= Q(\xi,\eta,t) - \int_\Gamma (M_n^*\Delta\beta_n + H^*\Delta w - \Delta M_n\beta^* - \Delta H w^*)d\Gamma.
\end{aligned}$$

Using the formula of conversion for the dual Fourier transform, we obtain

$$w(x,y,t) = \frac{1}{2\pi m}\iint_{-\infty}^{\infty} p^{-1}\int_0^t e^{-i(\xi x+\eta y)}F(\xi,\eta,\tau)\sin p(\xi,\eta)(t-\tau)d\tau d\xi d\eta.$$

Let us satisfy the boundary conditions of problem, having preliminary selected any jumps on Γ. For example, we will consider that the displacements and angle of rotations are continuous on Γ. Then $F(\xi, \eta, t)$ will take the form:

$$F(\xi,\eta,t) = Q(\xi,\eta,t) + F_1(\xi,\eta,t), \tag{15.121}$$

where

$$F_1(\xi,\eta,t) = \int_\Gamma (\Delta M_n\beta_n^* + \Delta H w^*)d\Gamma.$$

For determination of two unknown functions $\Delta M_n(\xi, \eta, t)$ and $\Delta H(\xi, \eta, t)$ we have two equations, obtained as a result of the satisfaction of the boundary conditions

$$\begin{aligned}
&\frac{1}{2\pi m}\iint_{-\infty}^{\infty}\int_0^t p^{-1}e^{-i(\xi x+\eta y)}F_1(\xi,\eta,\tau)\sin p(t-\tau)d\tau d\xi d\eta = -w_q(x,y,t),\\
&-\frac{n_x i}{2\pi m}\iint_{-\infty}^{\infty}\int_0^t \xi e^{-i(\xi x+\eta y)}p^{-1}F_1(\xi,\eta,\tau)\sin p(t-\tau)d\tau d\xi d\eta =\\
&= -n_x\frac{\partial}{\partial x}w_q(x,y,t),\ (x,y\in\Gamma).
\end{aligned} \tag{15.122}$$

Here

$$w_q(x, y, t) = \frac{1}{2\pi m} \iint_{-\infty}^{\infty} \int_0^t e^{-i(\xi x + \eta y)} Q(\xi, \eta, \tau) \sin p(t - \tau) d\tau d\xi d\eta.$$

Equation (15.122) must be satisfied with each t. By dividing the interval (0, t) into sections, in the limits of which we can disregard the changes in the integrand function or use any interpolation, it is possible to obtain the step-by-step procedure for solving the problem by the method of sequential solutions of the quasi-stationary system of boundary Eqs. (15.122). The solution of this system may be preferable and less time-consuming, than, the use of transform of Laplace or Fourier, over time for example. Despite the apparent cumbersomeness of the Eqs. (15.122), which, after the substitution of the expression $F(\xi, \eta, t)$ must be converted by replacement of the order of integration, there are opportunities of significant simplifications based on the calculations of some integrals on ξ or on η in a closed form. In particular, it is possible to use the results of the work [15], where such integrals are calculated. Since the above equations are obtained for dynamic calculations of a plate, a particular case of the static equilibrium of a plate on the Winklerian (Winkler) foundation presents a great deal of interest. In this case, in Eq. (15.118), it is possible to assume that $m = 0$ and having the continuity of moments and shear forces on the contour of plate Γ, write down the following:

$$W(\xi, \eta) = \frac{F_1(\xi, \eta) + Q(\xi, \eta)}{D(\xi^2 + \eta^2) + k}. \tag{15.123}$$

Hence,

$$w(x, y) = \frac{1}{2\pi} \iint_{-\infty}^{\infty} e^{-i(\xi x + \eta y)} \frac{F_1(\xi, \eta) + Q(\xi, \eta)}{D(\xi^2 + \eta^2) + k} d\xi d\eta. \tag{15.124}$$

As a result of the satisfaction of the boundary conditions we obtain:

$$\begin{aligned} &\frac{1}{2\pi} \iint_{-\infty}^{\infty} \frac{F_1(\xi, \eta) e^{-i(\xi x + \eta y)}}{D(\xi^2 + \eta^2) + k} d\xi d\eta = w_q(x, y), \\ &\frac{i n_x}{2\pi} \iint_{-\infty}^{\infty} \frac{F_1(\xi, \eta) e^{-i(\xi x + \eta y)}}{D(\xi^2 + \eta^2) + k} d\xi d\eta = -n_x \frac{\partial}{\partial x} w_q(x, y), \\ &(x, y \in \Gamma), \end{aligned} \tag{15.125}$$

where

$$w_q(x, y) = \frac{1}{2\pi} \iint_{-\infty}^{\infty} e^{-i(\xi x + \eta y)} \frac{Q(\xi, \eta)}{D(\xi^2 + \eta^2) + k} d\xi d\eta.$$

It should be noted that calculating double integrals in Eq. (15.125) can be a very difficult task since it requires the separation of the regular part of the integrals. For this procedure, the allocation (separation) of delta functions and its derivatives, represented in the form of Fourier integrals is recommended.

15.9 The Forced Vibrations of a Rectangular Plate on an Elastic Foundation

The spectral method of boundary elements (SMBE), closely related to the theory of integral transformations, can be a highly effective means for the solution of boundary-value problems with complex boundary conditions. In the present paragraph, the SMBE is used for the solution of the problem of harmonic oscillations of a rectangular isotropic plate lying on an elastic foundation of a Winklerian type. This problem is encountered when inertia of base, which plays a significant role in the general picture of dynamic deformation of plates, can be accounted for within the framework of a discrete model for dynamic calculations of pavements of industrial buildings, power (thrust) plates, slabs, roads, and airports. As a special case, from the obtained solution follows a solution of a static problem of the bend of a rectangular plate on an elastic foundation. Similar static problems for plates with free edges were examined in the works of E. A. Palatnikov [16], V.I. Travush, and V.K. Sangadzhiev [17]. However, in contrast to the above works, the following solution gives a sufficiently simple algorithm, and the unknowns concentrated at the angular points that can be obtained with the use of a SMBM are automatically included. The problem in this case is a specific example of the application of the SMBE to the solution of boundary-value problems since the area occupied by the plate is a rectangle, it therefore does not reveal all possibilities of this method in comparison with the domains of complex configuration. The solution to the problem can be obtained by applying generalized, finite integral transformations. Let us write down the equation of harmonic oscillations of a rectangular plate, which lies on a discrete inertia base under the action of a load $q(x, y)e^{i\omega t}$ in the form of [18]:

$$D\nabla^2\nabla^2 w(x,y,t) + m\frac{\partial^2 w(x,y,t)}{\partial t^2} + \\ + \int_{-\infty}^{t} R(t-\tau)w(x,y,t)d\tau = q(x,y)e^{i\omega t}, \tag{15.126}$$

where $\nabla^2 = \frac{\partial^2}{\partial x^2} + \frac{\partial^2}{\partial y^2}$ – is the Laplace operator in Cartesian coordinates; k– the coefficient of the bed of the base; $w(x, y, t)$– the displacement of the plate; $R(t-\tau)$– the kernel of the discrete model, which describes the elastic, inertial, and dissipative properties of the base; $q(x, y)$– the intensity of harmonic load; ω– circular frequency of the load; D, m– cylindrical rigidity and mass per unit area of the plate. Assuming that $w(x, y, t) = w(x, y)e^{i\omega t}$, and excluding time, we will obtain:

$$D\nabla^2\nabla^2 w(x,y) + k_0(\omega)w(x,y) = q(x,y), \tag{15.127}$$

where

$$k_0(\omega) = k(\omega) - m\omega^2; \quad k(\omega) = \int_0^{\infty} R(z)e^{-i\omega z}dz.$$

For simplicity we will consider $k(\omega)$ in the future as a real function, disregarding the dissipation of energy in the base. Let us introduce designations $|k_0(\omega)|/D = l^{-4}$ and switch over to dimensionless coordinates:

$$x = x_* l, y = y_* l, a = a_* l, b = b_* l.$$

Then the Eq. (5.3.2) takes the form

$$\left[\nabla^2\nabla^2 + sign k_0(\omega)\right]w(x,y) = q_1(x,y), \tag{15.128}$$

where

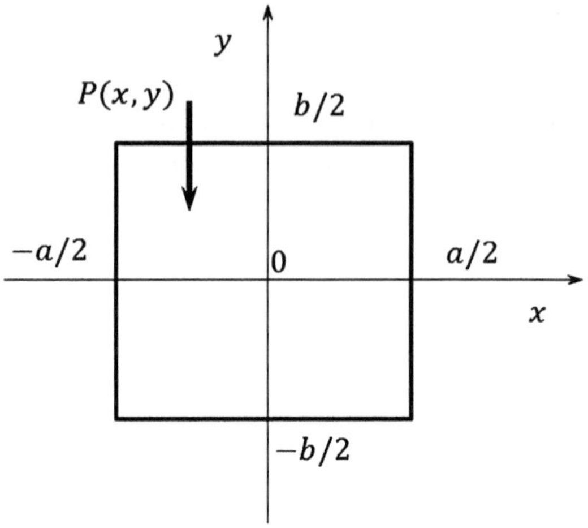

Fig. 15.5 The design scheme of a plate on an elastic foundation

$$q_1(x,y) = D^{-1} l^4 q(x_*, y_*).$$

In this case, indices with x_* and y_* for simplicity are omitted. The SMBE can be used for calculating the plate with arbitrary boundary conditions. In all cases, together with distributed boundary densities, the method makes it possible to obtain angular singularity. Subsequently, let us limit the examination to the most important case of plates with free edges, which have the greatest applications in civil engineering [19, 20]. The boundary conditions for the free lying plate with the selected system of coordinates (Fig. 15.5) take the form of

$$L_i w(x,y)\big|_{x=\pm a/2} = 0, \qquad L_{i+2} w(x,y)\big|_{y=\pm b/2} = 0, \qquad (i = 1, 2) \tag{15.129}$$

where

$$L_1 = \frac{\partial^2}{\partial x^2} + \nu \frac{\partial^2}{\partial y^2}, \; L_2 = \frac{\partial^3}{\partial x^3} + (2-\nu)\frac{\partial^3}{\partial x \partial y^2},$$

L_3 and L_4 are obtained by the replacement of x on y in L_i and vice versa.

If an arbitrary load acts on a plate, then it is possible to decompose the load into symmetric and skew-symmetric components (Fig. 15.6) and design the plate to the action of each component separately. The general solution is obtained by superposition of the particular solutions. Cosine transformation should be used for the solution of problems with symmetric loading, sine transformation for problems with an inversely symmetric loading, and cosine and sine transformations for mixed cases. To arrive at the resolving system of boundary equations, we must consider both the boundary conditions (15.129) and the condition of equilibrium:

$$\int_{-\frac{a}{2}}^{\frac{a}{2}}\int_{-\frac{b}{2}}^{\frac{b}{2}} k_0(\omega) w(x,y)\,dx\,dy = \int_{-\frac{a}{2}}^{\frac{a}{2}}\int_{-\frac{b}{2}}^{\frac{b}{2}} q(x,y)\,dx\,dy. \tag{15.130}$$

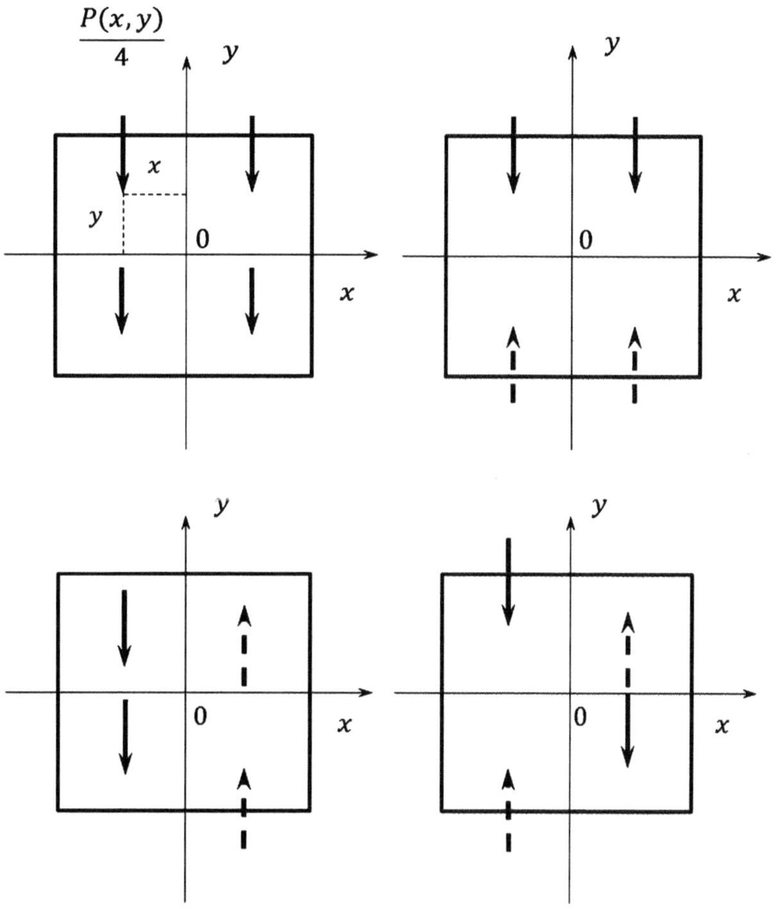

Fig. 15.6 The decomposition of the load on (into) symmetric and skew-symmetric components

Let us examine a case of symmetric loading relative to both axes *Ox* and *Oy* (Fig. 15.6), and we will limit ourselves to considering the domain occupying a quarter of the plate accordingly.

For the solution of the Eq. (15.128) we will apply the two-dimensional finite cosine transformation. Let us first multiply both parts of the equation with $\cos\frac{2m\pi x}{a}\cos\frac{2n\pi y}{b}$ and integrate the limits (0, *a*/2), (0, *b*/2) we have:

$$\int_0^{a/2}\int_0^{b/2}\left(\nabla^2\nabla^2 + signk_0\right)w(x,y)\cos\alpha_m x\cos\beta_n y dxdy = \\ = \int_0^{a/2}\int_0^{b/2} q_1(x,y)\cos\alpha_m x\,\cos\beta_n y dxdy. \tag{15.131}$$

An integration in parts of the left side of the Eq. (15.131) results in:

$$\left[\left(\alpha_m^2+\beta_n^2\right)^2+signk_0\right]F_c[w(x,y)]=q(m,n)-\sum_{i=1}^{3}Q_i, \tag{15.132}$$

where

$$\alpha_m=\frac{2m\pi}{a},\beta_n=\frac{2n\pi}{b},$$

Q_i– "compensating" actions (impact, effects) containing unknown densities in the form of some differential expressions from $w(x, y)$ and distributed along the contour of the plate, including angular points take the form:

$$\begin{aligned}
Q_1(m,n,a) &= (-1)^m W_x'''\left(\frac{a}{2},n\right)-(-1)^m\alpha_m^2 W_x'\left(\frac{a}{2},n\right),\\
Q_2(m,n,b) &= (-1)^n W_y'''\left(\frac{b}{2},m\right)-(-1)^n\beta_m^2 W_y'\left(\frac{b}{2},n\right),\\
Q_3(m,n,a,b) &= 2(-1)^{m+n}W_{xy}''\left(\frac{a}{2},\frac{b}{2}\right)+2(-1)^m\beta_m^2 W_x'\left(\frac{a}{2},n\right)+\\
&\quad +2(-1)^n\alpha_m^2 W_y'\left(\frac{b}{2},m\right),\ (m,n=0,1,2,\ldots,\infty)
\end{aligned} \tag{15.133}$$

where

$$\begin{aligned}
W_x^{(j)}\left(\frac{a}{2},n\right) &= \int_0^{b/2} W_x^{(j)}\left(\frac{a}{2},y\right)\cos\beta_n y dy,\\
W_y^{(j)}\left(\frac{b}{2},m\right) &= \int_0^{a/2} W_y^{(j)}\left(\frac{b}{2},x\right)\cos\alpha_m x dx.
\end{aligned} \tag{15.134}$$

Using the boundary conditions:

$$L_2 w(x,y)\big|_{x=\frac{a}{2}}=0,\quad L_4 w(x,y)\big|_{y=\frac{b}{2}}=0,$$

According to Eq. (15.129) we have

$$\begin{aligned}
w_x'''\left(\frac{a}{2},y\right) &= (\nu-2)\frac{\partial^3 w\left(\frac{a}{2},y\right)}{\partial x\partial y^2},\\
w_y'''\left(\frac{b}{2},x\right) &= (\nu-2)\frac{\partial^3 w\left(\frac{b}{2},x\right)}{\partial x^2\partial y}.
\end{aligned} \tag{15.135}$$

We multiply both parts of the first Eq. (15.135) by $\cos\beta_n y$, and both parts of the second equation by $\cos\alpha_m x$, and integrate respectively with the limits $(0, b/2)$ and $(0, a/2)$. After integrating by parts we obtain

$$\int_0^{b/2} w_x'''\left(\frac{a}{2},y\right)\cos\beta_n y dy = (-1)^n(\nu-2)\frac{\partial^2 w\left(\frac{a}{2},\frac{b}{2}\right)}{\partial x\partial y} -$$
$$-(\nu-2)\beta_n^2\int_0^{b/2} w_x'\left(\frac{a}{2},y\right)\cos\beta_n y dy;$$
$$\int_0^{a/2} w_y'''\left(\frac{b}{2},x\right)\cos\alpha_m x dx = (-1)^m(\nu-2)\frac{\partial^2 w\left(\frac{a}{2},\frac{b}{2}\right)}{\partial x\partial y} -$$
$$-(\nu-2)\alpha_m^2\int_0^{\frac{b}{2}} w_y'\left(\frac{b}{2},x\right). \tag{15.136}$$

Let us designate according to Eq. (15.134)

$$A(n) = \int_0^{b/2} w_x'\left(\frac{a}{2},y\right)\cos\beta_n y dy;$$
$$B(m) = \int_0^{a/2} w_y'\left(\frac{b}{2},x\right)\cos\alpha_m x dx; \qquad c = \frac{\partial^2 w\left(\frac{a}{2},\frac{b}{2}\right)}{\partial x\partial y}. \tag{15.137}$$

Taking into account designations (15.137) and equality (15.136) in Eq. (15.133), we obtain:

$$\begin{aligned}
&Q_1(m,n,a,b) = (-1)^{m+n}(\nu-2)c - (-1)^m\left[(\nu-2)\beta_n^2+\alpha_m^2\right]A(n),\\
&Q_2(m,n,a,b) = (-1)^{m+n}(\nu-2)c - (-1)^n\left[(\nu-2)\alpha_m^2+\beta_n^2\right]B(m)\\
&Q_3(m,n,a,b) = (-1)^{m+n}\cdot 2c + 2(-1)^m\beta_n^2 A(n) + 2(-1)^n\alpha_m^2 B(m),\\
&m,n = 0,1,2, \ \ldots \ ,\infty.
\end{aligned} \tag{15.138}$$

As can be easily seen, Eq. (15.138) contains three unknown values, which are determined from the boundary conditions and the conditions of equilibrium. Let us substitute Eq. (15.138) into Eq. (15.132) and after inversion, (conversion, transformation) we will obtain the solution of the equation of motion in the form:

$$\begin{aligned}
w(x,y) = &\frac{16}{ab}\sum_{m=0}^{\infty'}\sum_{n=0}^{\infty'}\frac{1}{\left(\alpha_m^2+\beta_n^2\right)^2 + sign\ k_0}\left[q(m,n)+\right.\\
&+ (-1)^m(\nu-4)\left(\alpha_m^2+\beta_n^2\right)A(n) + (-1)^n(\nu-4)\left(\alpha_m^2+\beta_n^2\right)B(m)\\
&\left.+(-1)^{m+n}(2-2\nu)c)\right]\cos\alpha_m x\cos\beta_n y,
\end{aligned} \tag{15.139}$$

where the prime in the sum symbol means that an additional multiplier 1/2 is entered (introduced) for the zeroth member. Inserting Eq. (15.139) into boundary conditions (15.129) with operators L_1 and L_3, and into the conditions of equilibrium we will obtain the following system of algebraic equations:

$$\sum_{m=0}^{\infty'}\frac{(\alpha_m^2+\nu\beta_n^2)(-1)^m}{\left(\alpha_m^2+\beta_n^2\right)^2+sign\ k_0}\{q(m,n)+$$
$$+(\nu-4)\left(\alpha_m^2+\beta_n^2\right)[(-1)^m A(n)+(-1)^n B(m)]\}=0,$$
$$\sum_{n=0}^{\infty'}\frac{(\beta_n^2+\nu\alpha_m^2)(-1)^n}{\left(\alpha_m^2+\beta_n^2\right)^2+sign\ k_0}\{q(m,n)+$$
$$+(\nu-4)\left(\alpha_m^2+\beta_n^2\right)[(-1)^m A(n)+(-1)^n B(m)]\}=0$$
$$q(0,0)+(2-2\nu)c=\int_0^{\frac{a}{2}}\int_0^{\frac{b}{2}} q_1(x,y)dxdy, \tag{15.140}$$
$$q(m,n)=\int_0^{\frac{a}{2}}\int_0^{\frac{b}{2}} q_1(x,y)\cos\alpha_m x\cos\beta_n y dxdy,$$
$$q_1(x,y)=D^{-1}l^4 q(x,y),\quad \alpha_m=\frac{2m\pi}{a},\quad \beta_n=\frac{2n\pi}{b}\ .$$
$$m,\ n=0,\ 1,\ 2,\ \ldots\ ,\infty\ .$$

From the third equation of system (15.140) we will obtain $c = 0$. Let us designate

$$\left(\alpha_m^2+\beta_n^2\right)^2+sign\ k_0=A$$
$$a_1(n)=\sum_{m=0}^{\infty'}\frac{\left(\alpha_m^2+\beta_n^2\right)\left(\alpha_m^2+\nu\beta_n^2\right)(\nu-4)}{A},$$
$$b_1(m,n)=\frac{\left(\alpha_m^2+\nu\beta_n^2\right)\left(\alpha_m^2+\beta_n^2\right)(\nu-4)(-1)^{m+n}}{A},$$
$$c_1(n)=-\sum_{m=0}^{\infty'}\frac{\left(\alpha_m^2+\nu\beta_n^2\right)(-1)^m q(m,n)}{A}, \tag{15.141}$$
$$a_2(m,n)=\frac{\left(\beta_n^2+\nu\alpha_m^2\right)\left(\alpha_m^2+\beta_n^2\right)(\nu-4)(-1)^{m+n}}{A},$$
$$b_2(m)=\sum_{n=0}^{\infty'}\frac{\left(\beta_n^2+\nu\alpha_m^2\right)\left(\alpha_m^2+\beta_n^2\right)(\nu-4)}{A},$$
$$c_2(m)=-\sum_{n=0}^{\infty'}\frac{\left(\beta_n^2+\nu\alpha_m^2\right)(-1)^n q(m,n)}{A}.$$

Taking into account the designations (15.141), the system (15.140) takes the following form:

$$a_1(n)A(n)+\sum_{m=0}^{\infty'} b_1(m,n)B(m)=c_1(n),$$
$$\sum_{n=0}^{\infty'} a_2(m,n)A(n)+b_2(m)B(m)=c_1(m), \tag{15.142}$$
$$m,\ n=0,\ 1,\ \ldots\ ,\ \infty.$$

The coefficients with the unknowns $A(n)$ and $B(m)$ in the system (15.142) are represented by the divergent series. Therefore it is necessary to regularize these series after isolating (extracting) generalized functions from them. Let us examine an expression for $a_1(n)$

$$a_1(n) = \left[\sum_{m=0}^{\infty'} \frac{(\alpha_m^2 + \beta_n^2)(\alpha_m^2 + \nu\beta_n^2)(\nu - 4)\cos\frac{2m\pi x}{a}(-1)^m}{(\alpha_m^2 + \beta_n^2)^2 + sign\ k_0} \right]_{x=a/2}$$

and let us isolate from its regular and singular parts. For this we will use the formula, known from the theory of generalized functions [21].

$$\sum_{m=1}^{\infty} \cos m\alpha = -\frac{1}{2} + \pi \sum_{m=-\infty}^{\infty} \delta(\alpha - 2\pi m).$$

Let us assume in this formula that $\alpha = z - \pi$, then we will obtain

$$\sum_{m=1}^{\infty} (-1)^m \cos mz = -\frac{1}{2} + \pi \sum_{m=-\infty}^{\infty} \delta[z - \pi(2m + 1)].$$

Let us switch over further to a new variable $z = \frac{2\pi x}{a}$ and take into account known property of delta function [21].

$$\delta[r(x)] = \sum_h \frac{\delta(x - x_h)}{|r'(x_h)|},$$

where x_h- is a simple zero of function $r(x)$. With $r(x) = \frac{2\pi x}{a} - \pi(2m + 1)$, $x_h = \frac{a}{2}(2m + 1)$, and consequently, we have

$$\delta\left[\frac{2\pi x}{a} - \pi(2m + 1)\right] = \frac{a\delta\left[x - \frac{a}{2}(m + 1)\right]}{2\pi}$$

Hence

$$\sum_{m=1}^{\infty} (-1)^m \cos\frac{2m\pi x}{a} = -\frac{1}{2} + \frac{a}{2} \sum_{m=-\infty}^{\infty} \delta\left[x - (2m + 1)\frac{a}{2}\right], \tag{15.143}$$

where $a = 2\pi$, $x \in (-\pi, \pi)$.

We will obtain

$$a_1(n) = \left[(\nu - 4)\sum_{m=1}^{\infty} (-1)^m \cos\frac{2m\pi x}{a} + \frac{\nu - 4}{2} + \right.$$
$$\left. + \sum_{m=0}^{\infty'} \frac{\left[(\nu - 1)(\beta_n^4 + \alpha_m^2\beta_n^2) - 1\right](\nu - 4)\cos\frac{2m\pi x}{a}(-1)^m}{(\alpha_m^2 + \beta_n^2)^2 + \text{sign}\ k_0} \right]_{x=\frac{a}{2}} =$$
$$= \frac{\pi a}{2}(\nu - 4) \lim_{\substack{x \to \frac{a}{2} \\ x < a/2}} \sum_{m=-\infty}^{\infty} \delta\left[x - (2m + 1)\frac{a}{2}\right] +$$
$$+ \sum_{m=0}^{\infty'} \frac{\left[(\nu - 1)(\beta_n^4 + \alpha_m^2\beta_n^2) - \text{sign}\ k_0\right](\nu - 4)}{(\alpha_m^2 + \beta_n^2)^2 + \text{sign}\ k_0}.$$

As mentioned above, a strict formulation of boundary conditions requires the determination of the boundary values of the solution of the boundary-value problem on the inner contour of the domain. The value of delta function in the last expression inside the domain and in the limit on its boundary can be considered equal to zero (although formally, the delta function at point is not defined). In view of this fact, the singular component can be neglected in the expression for determining the coefficient $a_1(n)$. After discarding the singular part, the expression for $a_1(n)$, it takes the form

$$a_1(n) = (\nu - 4)\sum_{m=0}^{\infty}{}' \frac{(\nu-1)\left(\beta_n^4 + \alpha_m^2\beta_n^2\right) - signk_0}{A},$$

With the aid of similar operations the expression for $b_2(m)$ takes the form

$$b_2(m) = (\nu - 4)\sum_{n=0}^{\infty'} \frac{(\nu-1)\left(\alpha_m^4 + \alpha_m^2\beta_n^2\right) - \mathrm{sign}k_0}{\left(\alpha_m^2 + \beta_n^2\right)^2 + 1}.$$

Thus, the series for all coefficients of systems converge more rapidly than

$$\sum_{m=1}^{\infty} \frac{1}{\alpha_m^2} or \sum_{n=1}^{\infty} \frac{1}{\beta_n^2}.$$

Their calculation is more expedient to conduct with an aid of computer software. The choice of the upper limit of the sum depends on the speed of the decrease of the coefficients $B(m)$ included in the system, and therefore it is necessary to conduct a study of their asymptotic behavior with $m \to \infty$, which is also more conveniently carried out with the aid of computer software. After assigning the upper limit of summation and truncation of the system of Eqs. (15.142) to s of order, brings this system to the form:

$$\sum_{n=0}^{s'} a_2(m,n)\frac{c_1(n) - \sum_{m=0}^{t'} b_1(m,n)B(m)}{a_1(n)} + b_2(m)B(m) = c_2(m),$$

or

$$\sum_{m=0}^{t'} b_1(m,n)\frac{c_2(n) - \sum_{n=0}^{s'} a_2(m,n)A(n)}{b_2(m)} + a_1(n)A(n) = c_1(n).$$

For square plate ($B(m) = A(m)$) in particular we have:

$$a_1(n)A(n) + \sum_{m=0}^{t'} b_1(m,n)A(m) = c_1(n), \quad m,\ n = 0,\ 1,\ 2,\ \ldots,\ t. \tag{15.144}$$

Where $a_1(n)$ and $b_1(m, n)$ are determined from formula (15.142), by replacing β_n to α_n. After the solution of system of Eqs. (15.142), the function $w(x, y)$ is determined by the formula (15.139). Further the calculation of internal forces M_x, Q_x, M_y, Q_y, and H_{xy} at any point of the plate is achieved according to the known formulas of the theory of plates:

$$\begin{aligned}
M_x(x,y) &= -\frac{D}{l^2}\left(\frac{\partial^2}{\partial x^2}+\nu\frac{\partial^2}{\partial y^2}\right)w(x,y),\\
M_y(x,y) &= -\frac{D}{l^2}\left(\frac{\partial^2}{\partial y^2}+\nu\frac{\partial^2}{\partial x^2}\right)w(x,y),\\
Q_x(x,y) &= -\frac{D}{l^3}\frac{\partial}{\partial x}\nabla^2 w(x,y),\\
Q_y(x,y) &= -\frac{D}{l^3}\frac{\partial}{\partial y}\nabla^2 w(x,y),\\
H_{xy}(x,y) &= -\frac{D}{l^2}(1-\nu)\frac{\partial^2 w}{\partial x\partial y}\quad .
\end{aligned} \tag{15.145}$$

The substitution of (15.139) into formula (15.145) leads to a divergent series, similar to (15.140), the regularization of which can be carried out according to the formula (15.143). For example, for the square plate the bending moment is described by the expression

$$\begin{aligned}
M(x,y) = \frac{16}{a^2}\sum_{m=0}^{\infty'}\sum_{n=0}^{\infty'}\{(\alpha_m^2+\nu\alpha_n^2)q(m.n)+(\nu-4)(\nu-1)\left(\alpha_m^2+\alpha_n^2\right)\times\\
\times\left[(-1)^m\alpha_n^2A(n)+(-1)^n\alpha_m^2A(m)\right]-signk_0-(\nu-4)[(-1)^m A(n)+\\
+(-1)^n A(m)]\}\frac{\cos\alpha_m x\cos\alpha_n y}{\left(\alpha_m^2+\alpha_n^2\right)^2+signk_0}.
\end{aligned} \tag{15.146}$$

As an example, let us examine a square plate with a given size 2 × 2, with the following types of load:

(1) Force concentrated in the center of the plate $P = 1$;
(2) Uniform load $q_1 = 100$, distributed along the area of the plate of size 0.1 × 0.1 in the center of the plate

The diagrams of static deflections and bending moments in the sections throughout the diagonal $x = y$ and along the axis $x = 0$ are obtained by taking into account the 25 terms of the series, and are shown in Figs. 15.7 and 15.8. Moreover, with the last version of the load of the value of deflection $w = 0.272$ and bending moment $M = 0.2987$ in the center of plate differ from those obtained by alternative method in the work of [17] by values less than for 1%.

The analysis of the dependence of the deflections of the plate on the frequency of the load was conducted for the last version of the load. An increase in the frequency of load leads to a decrease of the given rigidity $k_0(\omega)$, and consequently to an increase in the absolute value of the reduced length l . At the same time, the sign $k_0(\omega)$ changes in the transition into the "transresonant" regime ($\omega > k(\omega)/m$).

Figure 15.9 shows the dependence of the deflection of the center of a plate on the parameter $\xi = lsignk_0$, which shows that this transition is accompanied by a change in the phase of oscillations. The calculations show that the use of the spectral method of boundary element expansions with discrete spectrums leads to very effective algorithms of other methods connected with the eigenfunction expansions.

The specific advantage of SMBE in such problems is the generality of the procedure used and the opportunity of direct use of generalized functions. In the problems for domains with complex boundary, SMBE becomes particularly effective. This work is the first publication that provides solutions to problems for non-isolated beams and plates using SMBE.

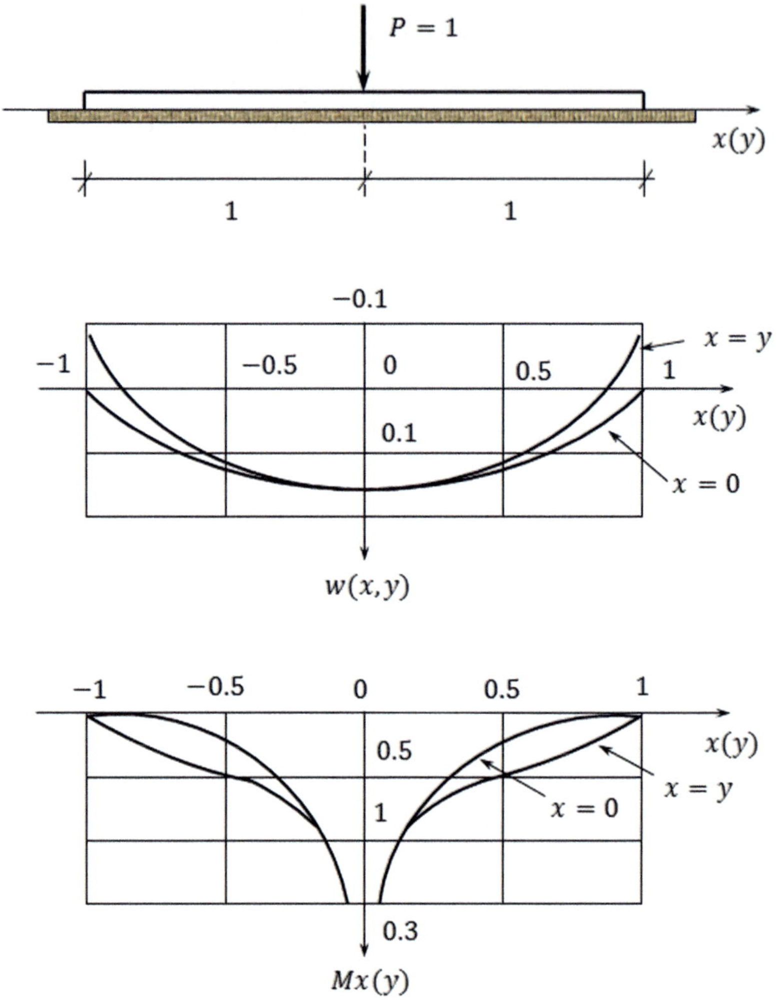

Fig. 15.7 The diagram of deflections and bending moments

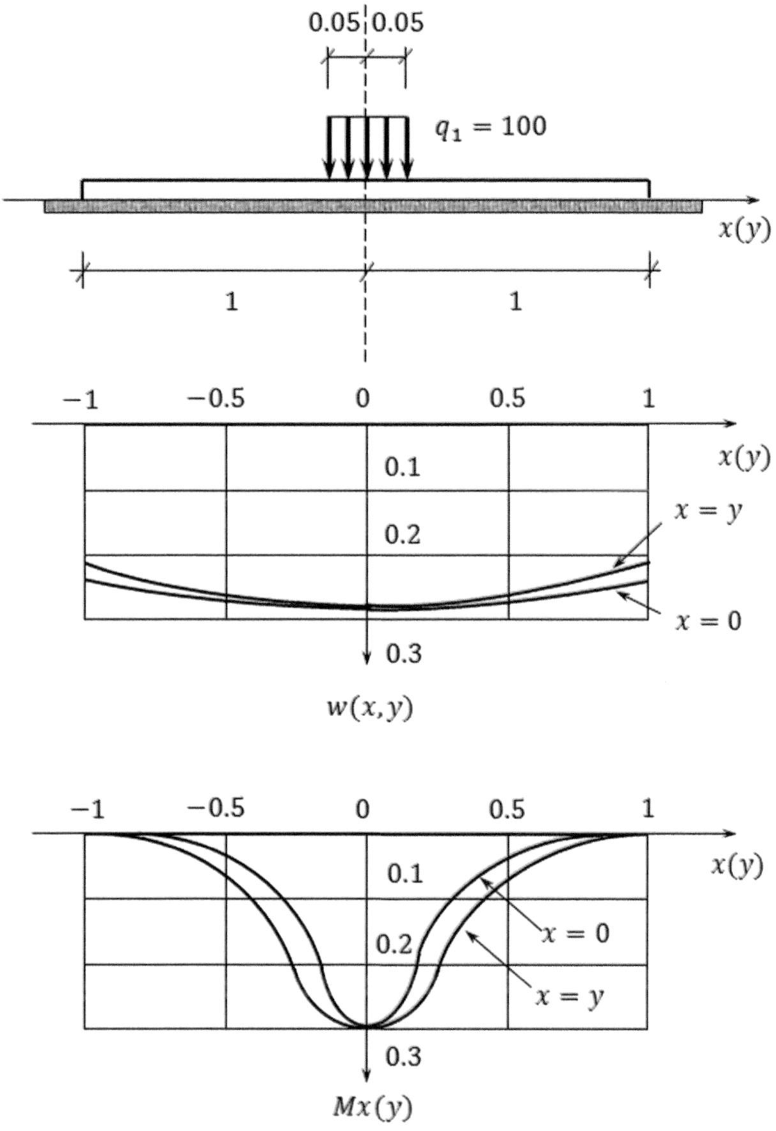

Fig. 15.8 The diagram of deflections and bending moments

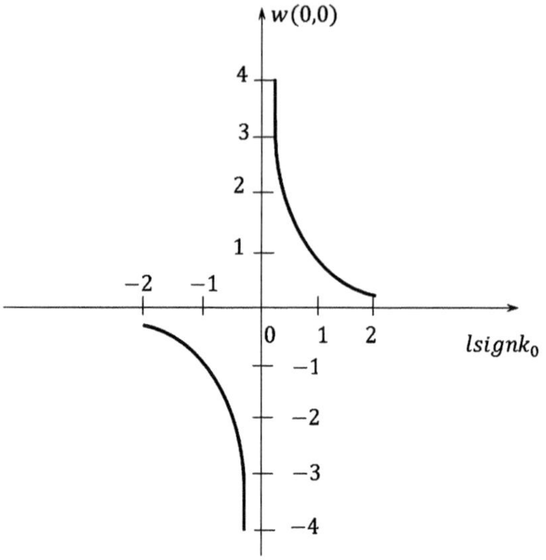

Fig. 15.9 The dependence of the deflection of the center of the plate on frequences of load

References

1. Korenev, B. G., & Tseitlin, A. I. (1982). *Plates vibrations* (pp. 213–249). Reference Book. Dynamics of Structures. Stroyisdat.
2. Kaul, R. K., & Cadambe, V. (1956). The natural frequencies of thin skew plates. *Aeron Quart, 74*, 337.
3. Mindlin, R. D. (1951). Influence of rotatory inertia and shear n flexural motion of isotropic elastic plates. *Journal of Applied Mechanics, 18*, 1.
4. Moskalenko, V. H. (1961). Application of refined theories of plate bending in the eigenvalue problem. *Engineering Magazine, 1*, 3.
5. Timoshenko, S. P., & Woinovsky-Krieger, S. (1989). *Theory of plates and shells* (p. 611). McGraw-Hill Book Company.
6. Bolotin, V. V., & Moskalenko, V. N. (1968). Plates vibrations. Reference book. In *Strength, stability and oscillations* (Vol. 3). Mashinostroenie.
7. Lurie, A. I. (1970). *Operational calculus and its application to the problems of mechanics* (p. 432). Gostexizdat.
8. Tseitlin, A. I. (1965). *Integral transformations related to the biharmonic problem of half-plane and half-space and their application to the problems of the theory of elasticity*. Izvestia AN USSR, OTH, Mechanics, #1.
9. Korenev, B. G. (1960). *Some tasks of the theory of elasticity and thermal conductivity solving in Bessel functions* (p. 458). Fizmatgiz.
10. Watson, G. H. (1979). *The theory of Bessel functions* (p. 797). Part III.Moscow: Izdatelstvo Inostranoi Literaturi.
11. Petrosian, L. G. (1986). The spectral method of boundary elements. In *Structural mechanics and calculation of construction* (Vol. 4, pp. 45–50). SMBE.
12. Banerjee, P. K., & Butterfield, R. (1984). *Boundary element methods in applied science* (p. 494). Mir.
13. Brebbia, C. A., Tells, Z., & Wrobel, L. (1987). *Boundary element methods* (p. 524). Mir.
14. Brebbia, C. A., & Walker, C. (1982). *The application of boundary elements methods in the engineering* (p. 248). Mir.
15. Travush, V. I. (1969). Bending uninsulated plates lying on linear-deformed bases. In *Theory of structures* (Vol. 17, pp. 73–84). Stroyizdat.
16. Palatnikov, E. A. (1964). *Rectangular plate on an elastic foundation* (p. 236). Stroyizdat.
17. Travush, V. I., & Sangadzhiev, V. K. (1984). Bending of rectangular plates with a free contour (outline) on an elastic base. In *Structural mechanics and calculation of construction* (Vol. 6, pp. 37–40).
18. Petrosian, L. G. (1988). *Static and dynamic design problems for structures on an elastic foundation* (p. 69). Luis, BBK38.112.

19. Tseitlin, A. I., Petrosian, L. G., & Basilaya, V. M. (1985). The dynamic calculation of plates on an elastic foundation by the method of boundary equations. In *Theses of reports of All-Union Conference on the dynamic bases, foundations, and underground structures*.
20. Petrosian, L. G. (n.d.). About one elastic foundation model. In *Structural mechanics and calculation* (Vol. 5, pp. 7–11).
21. Gelfand, I. M., & Shilov, G. E. (1959). *The generalized functions and operations over them* (p. 470). Fizmatgiz.
22. Petrosian, L. G. (2022). Analysis of structures on elastic foundation. *Incorporating the spectral method of boundary elements* (p. 556). CRC Press Taylor & Francis Group, UK, USA.

Vibrations of Combined Arch Systems: Rigid Beam-Flexible Arch 16

Abstract

This chapter describes free vibrations of combined systems such as "rigid beam-flexible arch," considering the dynamic thrust when the axes of the arch are outlined by a square parabola and a circular arch. Problems and solutions are discussed. A methodology and results of a dynamic testing model of rigid beam-flexible arch and an experimentally theoretical method for calculating combined systems taking into account joint dynamic work of the main span to adjacent bridge trestles using factor "increasing" frequency are presented. A study and analysis devoted to the oscillations of a combined system accounting for its extent at a given harmonic motion base is given. A method of seismic analysis of combined systems with regard to their length using real seismograms of powerful earthquakes and general equations of motion for a system with a finite number of degrees of freedom in a spatial deformation structure are also discussed.

16.1 Free Vibrations of Combine Arch System

The problem of dynamics of combined systems, which are subject to the impact of moving loads, is a real problem of structural mechanics. Free oscillations of combined systems were investigated by many scientists: one of the first was the work of M. Becker. Free and forced oscillations of combined systems are also investigated by M.E. Phillips, C.E. Inglis, S.T. Odman, G. Stokes, R. Willis, G. Stokes, C.E. Inglis, W. Witting, and others. Free and forced oscillations of combined bridge systems are also studied by A. Ph. Smirnov [1], S.I. Konashenko [2], V.A. Smirnov [3], S.A. Ilyasevich [4], G.Y. Kunnos [5], I.A. Kolesnik [6], and others.

We attempted to determine the free (natural) oscillations of combined systems accounting for deformations, not only of the lowest frequencies, but also of a range of frequencies for the combined system, such as a flexible arch and rigid beam, that contain the spectrum of frequencies [7, 8]. In these systems, our interest lies in discerning the cases in which the lowest frequency is arch—reverse-symmetrical, and the cases in which the lowest frequency is beam—symmetrical. The theoretical results we obtained were tested on models and on real bridge structures.

The basic preconditions (prerequisites) for the derivation of a differential equation of oscillations/vibrations of this system are the same as for the static calculation: the axis of the arch is outlined along a square parabola, the posts (pillars) of the overhead arch structure are not compressible, the stiffness of the beam is constant along the entire length, and the mass of the span is distributed along the length of

L. G. Petrosian, *Dynamics of Structures*, https://doi.org/10.1007/978-3-031-63539-7_16

the beam. Thus, the system with an infinite number of degrees of freedom is obtained. The position of the system is determined by the elastic line. The elastic line under dynamic loads is a function of two variables: the coordinates of the cross-section x and time t. The horizontal forces of inertia are negligible. Based on the above assumptions and based on the differential equation of bending, we write the following expression:

$$E_b I_b \frac{d^2 y_0}{dx^2} = -M \tag{16.1}$$

Let us differentiate the differential bending equation twice, and also substitute the values M and Q. As a result, we obtain the equilibrium equation for the static action of forces:

$$E_b I_b \frac{d^4 y_0}{dx^4} = q + H\frac{d^2 y}{dx^2} \pm H\frac{d^2 y_0}{dx^2} \tag{16.2}$$

here:

y_0 is a static deflection
y is an arch axis ordinate in a section x
H is a static thrust
E_b is the modulus of elasticity of the beam
I_b is the cross-sectional moment of inertia of the beam.

During oscillations/vibrations, Eq. (16.2) is transformed: inertial forces appear, as well as a dynamic addition to the thrust $\Delta H = \Delta H(t)$ from inertia forces and dynamic deflection $\eta = \eta(x, t)$ occur (arise). Let us subtract from the transformed equation of oscillations the Eq. (16.2), which characterizes the system under the static action of forces. Then neglecting the term

$$\Delta H = \frac{\partial^2}{\partial x^2}(y_0 + \eta) \tag{16.3}$$

we obtain the differential equation [9]:

$$\frac{\partial^4 \eta}{dx^4} + \frac{H}{E_b I_b} \cdot \frac{\partial^2 \eta}{dx^2} - \frac{\Delta H}{E_b I_b} \cdot \frac{\partial^2 y}{dx^2} + \frac{m}{E_b I_b} \cdot \frac{\partial^2 \eta}{dt^2} = 0. \tag{16.4}$$

Let us examine the free vibrations/oscillations of this combined system shown on Fig. 16.1. This system is described by the integro-differential equation of motion (16.4). This equation is obtained

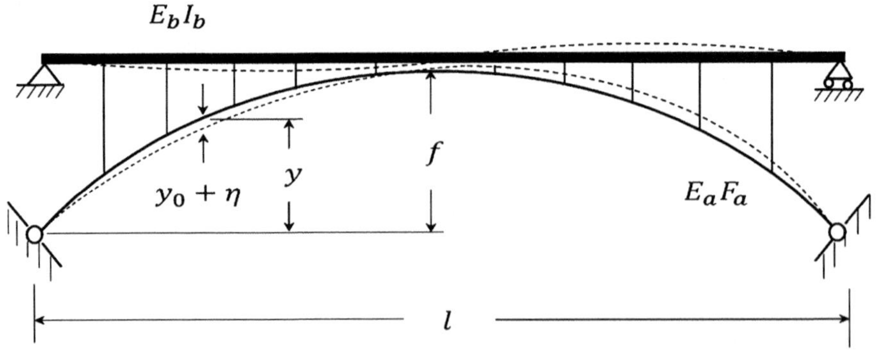

Fig. 16.1 The scheme of a combined system of a rigid-beam type and flexible arch with a ride on the top

with the following assumptions: the axis of the arch is outlined on the square parabola, the posts are not stretchable or compressible, the rigidity of beam is constant on the entire span, and the mass of bridge span is distributed along the length of the rigid beam. The dynamic part of the thrust, which is obtained with the examination of the strained state of the element of arch, takes the following form:

$$\Delta H = \frac{\frac{\partial^2 y}{dx^2}}{\frac{l}{E_a F_a}(1+8n^2)} \int_0^l \eta dx, \tag{16.5}$$

where $n = f/l$, and the arch axis is outlined on the square parabola:

$$y = \frac{4f}{l^2} x(l - x)$$

E_a is the modulus of elasticity of the arch; F_a is the cross-sectional area of the arch;

For solving integro-differential Eq. (16.4), we assume:

$$\eta = \eta_x A \sin(pt + \nu), \quad \frac{\partial^2 y}{dx^2} = -\frac{q}{H}. \tag{16.6}$$

In Eqs. (16.5) and (16.6) $\Delta H(t)$ is expressed in the form of:

$$\Delta H(t) = -\left(\frac{qE_a F_a}{Hl(1+8n^2)} A \int_0^l \eta_x dx\right) \sin(pt+\nu) = \Delta HA \sin(pt+\nu). \tag{16.7}$$

Now let us examine the equation for η_x. From Eqs. (16.4), (16.5), and (16.7) we obtain:

$$E_b I_b \frac{\partial^4 \eta_x}{dx^4} + H \frac{\partial^2 \eta_x}{dx^2} - \frac{q^2 E_a F_a}{H^2 l(1+8n^2)} \int_0^l \eta_x dx - mp^2 \eta_x = 0. \tag{16.8}$$

The form of the standing wave is approximated by Fourier series:

$$\eta_x = \sum_{j=1}^{\infty} C_j \frac{\sqrt{2l}}{j\pi} \sin j \frac{\pi x}{l} \tag{16.9}$$

Then multiplying (16.8) by $\sin i \frac{\pi x}{l}$ and integrating from 0 to l, we obtain:

$$\begin{aligned} &E_b I_b C_i \frac{\sqrt{2l}}{j\pi} \left(\frac{i\pi}{l}\right)^4 \int_0^l \sin^2 \frac{i\pi x}{l} dx - HC_i \left(\frac{i\pi}{l}\right)^2 \frac{\sqrt{2l}}{i\pi} \int_0^l \sin^2 \frac{i\pi x}{l} dx - \\ &- \frac{q^2 E_a F_a}{H^2 l(1+8n^2)} \sum_{j=1}^{\infty} C_j \frac{\sqrt{2l}}{j\pi} \left(\int_0^l \sin \frac{j\pi x}{l} dx\right) \times \int_0^l \sin \frac{i\pi x}{l} dx - \\ &- mp^2 C_i \frac{\sqrt{2l}}{i\pi} \int_0^l \sin^2 \frac{i\pi x}{l} dx = 0. \end{aligned} \tag{16.10}$$

Bearing in mind the values of the following integrals

$$\int_0^l \sin^2\frac{i\pi x}{l}dx = \frac{1}{2}l, \text{ and } \int_0^l \sin\frac{i\pi x}{l}dx = -\frac{l[(-1)^i - 1]}{i\pi},$$

the Eq. (16.10) takes the form:

$$E_bI_bC_i\left(\frac{i\pi}{l}\right)^4\frac{l}{2} - HC_i\left(\frac{i\pi}{l}\right)^2\frac{l}{2} - \frac{q^2E_aF_a}{H^2l(1+8n^2)}\times$$
$$\times\sum_{j=1}^{\infty}C_j\frac{l^2[(-1)^i-1]}{j\pi}\frac{[(-1)^i-1]}{i\pi} - mp^2C_i\frac{l}{2} = 0$$

or after simple conversions (transformations)

$$E_bI_bC_i\left(\frac{i\pi}{l}\right)^4 - HC_i\left(\frac{i\pi}{l}\right)^2 - \frac{2q^2E_aF_a}{H^2(1+8n^2)}\frac{[(-1)^i-1]}{i\pi}\times$$
$$\times\sum_{j=1}^{\infty}C_j\frac{[(-1)^i-1]}{j\pi} - mp^2C_i = 0. \tag{16.11}$$

We will introduce the following designations for solving Eq. (16.11):

$$A = E_bI_b\left(\frac{\pi}{l}\right)^4; \tag{16.12}$$

$$B = H\left(\frac{\pi}{l}\right)^2;\quad C = \frac{2q^2E_aF_a}{H^2(1+8n^2)}; \tag{16.13}$$

$$a_i = \frac{1-(-1)^i}{i\pi}. \tag{16.14}$$

By increasing the degree of strengthening of the beam by the arch, parameter C is increased, and a simple beam corresponds to the value of $C = 0$. Then Eq. (16.11) is rewritten as (where i are odd numbers)

$$Ai^4C_i - C_iBi^2 - Ca_i\sum_{j=1}^{\infty}C_ja_j - mp^2C_i = 0.$$

Let us designate:

$$x = Ca_i\sum_{j=1}^{\infty}C_ja_j,$$

then

$$C_iAi^4 - C_iBi^2 - Ca_ix - mp^2C_i = 0$$

$$C_i = \frac{Ca_ix}{Ai^4 - Bi^2 - mp^2}.$$

Then we have (all i are odd numbers, otherwise $a_i = 0$)

$$x=\sum_{j=1}^{\infty} C_{2j-1}a_{2j-1}=C_x\sum_{j=1}^{\infty}\frac{a^2{}_{2j-1}}{A(2j-1)^4-B(2j-1)^2-mp^2}$$

or $x=0$, then $C_{2j-1}A(2j-1)^4-C_{2j-1}B(2j-1)^2=mp^2C_{2j-1}$, i.e., C_{2j-1} differs from zero for the value of two of j. If the value of j is one, then $x\neq 0$, therefore it is necessary to determine whether expression $A(2j-1)^4-B(2j-1)^2$ can have an identical positive value for two different natural j. It depends on the values of A and B, but it is impossible since according to Vieta's formula it can be written:

$$\frac{mp^2}{A}=-(2j_1-1)^2\cdot(2j_2-1)^2<0,$$

where j_1 and j_2 are sought values of j, and means $x\neq 0$, and

$$1=C\sum_{j=1}^{\infty}\frac{a^2{}_{2j-1}}{A(2j-1)^4-B(2j-1)^2-mp^2}. \tag{16.15}$$

Let us calculate sum

$$\sum_{j=1}^{\infty}\frac{a^2{}_{2j-1}}{A(2j-1)^4-B(2j-1)^2-mp^2}. \tag{16.16}$$

According to Eq. (16.14) this sum takes the form:

$$\frac{4}{\pi^2}\sum_{j=1}^{\infty}\frac{1}{(2j-1)^2\left[A(2j-1)^4-B(2j-1)^2-mp^2\right]}. \tag{16.17}$$

We will designate $y=(2j-1)^2$, then we will obtain

$$\frac{1}{y^2\left(Ay^2-By-mp^2\right)}=\frac{\gamma_1}{y-y_1}+\frac{\gamma_2}{y-y_2}+\frac{\gamma_3}{y},$$

where γ_1, γ_2, and γ_3— are some coefficients, and y_1 and y_2— are the roots of the equation

$$Ay^2-By-mp^2=0. \tag{16.18}$$

After reduction to the common denominator, let us write

$$1=\gamma_1(y-y_2)y+\gamma_2(y-y_1)y+\gamma_3(y-y_1)(y-y_2).$$

Taking into account that $(y-y_1)(y-y_2)=y^2-\frac{B}{A}y-\frac{mp^2}{A}$ and equating coefficients, with the identical degrees of y on the left and right sides, we will obtain:

$$y^2 \quad \gamma_1+\gamma_2+\gamma_3=0;$$

$$y \quad -\gamma_1 y_1-\gamma_2 y_2-\gamma_3\frac{B}{A}=0; \qquad \begin{cases}\gamma_1+\gamma_2=-\dfrac{A}{mp^2};\\ y_1\gamma_1-y_2\gamma_2=\dfrac{B}{mp^2}\end{cases}$$

$$1 \quad -\gamma_3\frac{mp^2}{A}=1; \qquad \gamma_2(y_1-y_2)=\frac{Ay_1-B}{mp^2}$$

$$\gamma_3=-\frac{A}{mp^2};\ \gamma_2=\frac{Ay_1-B}{(y_1-y_2)mp^2}.$$

Finally:

$$\gamma_3=-\frac{A}{mp^2},\ \gamma_2=\frac{Ay_1-B}{(y_1-y_2)mp^2},\ \gamma_1=\frac{Ay_2-B}{(y_2-y_1)mp^2}. \tag{16.19}$$

Now sum (16.17) takes the form:

$$\frac{4}{\pi^2}\left(\gamma_1\sum_{j=1}^{\infty}\frac{1}{(2j-1)^2-y_1}+\gamma_2\sum_{j=1}^{\infty}\frac{1}{(2j-1)^2-y_2}+\gamma_3\sum_{j=1}^{\infty}\frac{1}{(2j-1)^2}\right). \tag{16.20}$$

For arbitrary z let us calculate sum

$$\sum_{j=1}^{\infty}\frac{1}{(2j-1)^2-z^2}=\Phi(z). \tag{16.21}$$

For the following cases:

(a) $\Phi(z)$ has simple poles at the points $\pm(2j-1)$, $j=1,\ldots$, the function $\tan\frac{\pi}{2}z$ at the same points has poles;

(b) $\Phi(z)=\frac{1}{2z}\sum\limits_{j=1}^{\infty}\left(\frac{1}{z-(2j-1)}+\frac{1}{z+(2j-1)}\right).$

From here it is visible that the function $\Phi(z)2z$ is periodic with period 2, and at the points of the form of $2j-1$, $j=0,\ \pm 1,\ldots$ has deductions, equal to one. Function $\tan\frac{\pi}{2}z$ at point $z=2j-1$ has a deduction $\lim\limits_{z\to 2j-1}\left(\tan\frac{\pi}{2}z\right)\cdot(z-2j-1)$, equal to $\lim\limits_{z\to 2j-1}-\frac{1}{\frac{\pi}{2}\frac{1}{\sin^2\frac{\pi}{2}z}}=-\frac{2}{\pi}$, moreover function $\tan\frac{\pi}{2}z$ is periodic with the period 2.

From the three statements above, it follows that the function $2z\Phi(z)+\frac{\pi}{2}\tan\frac{\pi}{2}z$ does not have poles (they are reduced). This function is periodic with period 2 and the strip $0<R_c z<2$ is limited, therefore it is limited in the entire complex plane. According to theorem of Liuvilya whereby a bounded function, analytical in the entire complex plane, is a constant. Hence, we obtain:

$$2z\Phi(z)+\frac{\pi}{2}\tan\frac{\pi}{2}z=const$$

and since at the point of $z=0$ the left side of equation becomes zero, the constant $=0$, i.e., $\Phi(z)=\frac{\pi}{4z}\tan\frac{\pi}{2}z$, now (16.20) takes the form:

$$\frac{1}{\pi}\left(\gamma_1\frac{\tan\frac{\pi}{2}\sqrt{y_1}}{\sqrt{y_1}}+\gamma_2\frac{\tan\frac{\pi}{2}\sqrt{y_2}}{\sqrt{y_2}}+\gamma_3\frac{\pi}{2}\right)$$

and Eq. (16.15) is rewritten in the following form:

$$-\frac{\pi}{C}=\gamma_1\frac{\tan\frac{\pi}{2}\sqrt{y_1}}{\sqrt{y_1}}+\gamma_2\frac{\tan\frac{\pi}{2}\sqrt{y_2}}{\sqrt{y_2}}+\gamma_3\frac{\pi}{2}. \tag{16.22}$$

This transcendental equation relative to p_i. We convert Eq. (16.22) for more convenient use.

$$-\frac{\pi}{C}\left(\cos\frac{\pi}{2}\sqrt{y_1}\right)=\gamma_1\frac{\sin\frac{\pi}{2}\sqrt{y_1}}{\sqrt{y_1}}+\cos\frac{\pi}{2}\sqrt{y_1}\left[\gamma_2\frac{\tan\frac{\pi}{2}\sqrt{|y_2|}}{\sqrt{|y_2|}}+\gamma_3\frac{\pi}{2}\right]. \tag{16.23}$$

Substituting into (16.23) the explicit expressions for y_1, and y_2, the roots of Eq. (16.18), the coefficients A, B, and C from (16.12) and (16.13), and also γ_1, γ_2, and γ_3 from (16.19) will obtain the exact values for p_i.

As an example, let us determine the free vertical oscillations of a combined steel bridge system: rigid beam—flexible arch with a rigid beam riding on top. The bridge being investigated overlaps the canyon river Arpa in the city of Jermuk, Armenia. The basic design parameters of the bridge are given below:

$$l=120\ m;\ \ f=18\ m;\ \ E_a=E_b=210\ GPA;\ \ H=4025,3\ kN$$
$$F_a=0.0527\ m^2;\ \ F_b=0.13373\ m^2;\ \ I_b=0.04356\ m^4$$

The calculation of the Eq. (16.23) was produced using an algorithmic language “FORTRAN-IV.” Table 16.1 gives the values of the first seven frequencies for considering this example, calculated according to proposed formula (16.23) and according to the formula proposed by S.I. Konashenko [2]. The table shows that including the calculations of deformations somewhat reduces the natural frequencies of the vertical oscillations [9].

The modes of free oscillations were represented by expression (16.9) in the form of trigonometric series

$$\eta=\sum_{j=1}^{\infty}C_j\frac{\sqrt{2l}}{j\pi}\sin\frac{j\pi x}{l}=\sum_{j=1}^{\infty}C_j\varphi_j(x). \tag{16.24}$$

Here $j-$ integers. Of course, the more inclusive the sum (16.24), the greater the accuracy of η amplitides. Coefficients C_j depend on parameters ,B, C, and the circular frequency p_i .

Table 16.1 The values of the first seven frequencies of vibrations/oscillations

Number of frequency #	The Frequency of vertical oscillations p_i (s^{-1})	
	Proposed formula (16.23)	Formula S.I. Konashenko [2]
1	3.850	4.102
2	7.504	–
3	10.504	9.968
4	25.409	25.930
5	49.592	50.351
6	82.834	83.190
7	123.899	124.233

We will consider the modes of free oscillations, limited by two members of the sum of (16.24). Then for the first mode of the oscillations, we have:

$$\eta_1 = C_1\varphi_1(x) + C_3\varphi_3(x) = C_1\left[\varphi_1(x) + \frac{C_3}{C_1}\varphi_3(x)\right] \tag{16.25}$$

For different values of frequencies p_i we will have different values of C_{1i}/C_{3i} . Obviously, without the initial conditions of motion, it is not possible to find C_1 and C_3; only their ratio C_1/C_3 or C_3/C_1 the so-called characteristic of the main modes of oscillations, can be determined. For the given combined system, the characteristic of the first mode of oscillations was determined using Eq. (16.11), which has the following value:

$$\left(\frac{C_3}{C_1}\right) = -3.910.$$

Consequently Eq. (16.25) takes the following form:

$$\eta_1 = C_1\frac{\sqrt{2l}}{\pi}\left[\sin\frac{\pi x}{l} - 1.303\sin\frac{3\pi x}{l}\right]. \tag{16.26}$$

The crossing of the curve of oscillations with the axis of abscissa is determined from the conditions of equality zero of η_1, which leads to the equation:

$$x = \frac{l}{2\pi}\sqrt{3\left(1 + \frac{C_1}{C_3}\right)}.$$

For the considered combined system $x = 0.237l$. The first symmetrical mode of natural vertical oscillations is shown graphically in Fig. 16.2.

The calculations showed that with the decrease in values of B/A and C/A, the modes of oscillations of the first and second symmetrical tones approached beamed modes. Natural mode of oscillation of bridges depends on the relative rigidity of its main components: arches and beams. The mode of oscillation will be identical to whichever element's rigidity is meaningfully higher, i.e., if the rigidity of the arch largely exceeds that of the beam, the mode of natural oscillations of the bridge will be identical to that of the arch mode of oscillations and the reverse is true.

Let us examine free oscillations of a combined system, when the arch is outlined on a circular arc and is very gently sloping, i.e., the ratio of the arrow of the lifting arch to the flight (the ratio of the boom lift to the arch span) is $f/l = 0.15$. The basic preconditions for differential equation of the oscillations for this system are the same as for the static design: posts are not compressed, rigidity (stiffness) of the beam E_bI_b is constant throughout the entire length; the mass of bridge span is evenly distributed along the length of the beam; the mass of the arch (parts of a bridge trusses: posts, joints, upper and lower chords, diagonals, etc.) and the mass of posts (pillars) and top braces (pillars and upper joints) are related to the beam stiffness; and horizontal inertial forces are negligible.

Since the arch is gently sloping, the difference in the forces in the posts (pillars) is small and can therefore be disregarded. Thus, we obtain a system with an infinitely large number of degrees of freedom. Its position is determined by an elastic curve, which, with dynamic loads, is a function of two variables—coordinate x and time t. We start with integro-differential equation of oscillations (16.4) for a system with a rigid beam and flexible arch with the ride on top. We will write down the equation of the axis of the arch $\sqrt{R^2 - (x - l/2)^2} + (R - f)$. There is no doubt that with oscillations of arched systems, we must take into account the dynamicity of thrust. Here we will show how to determine the

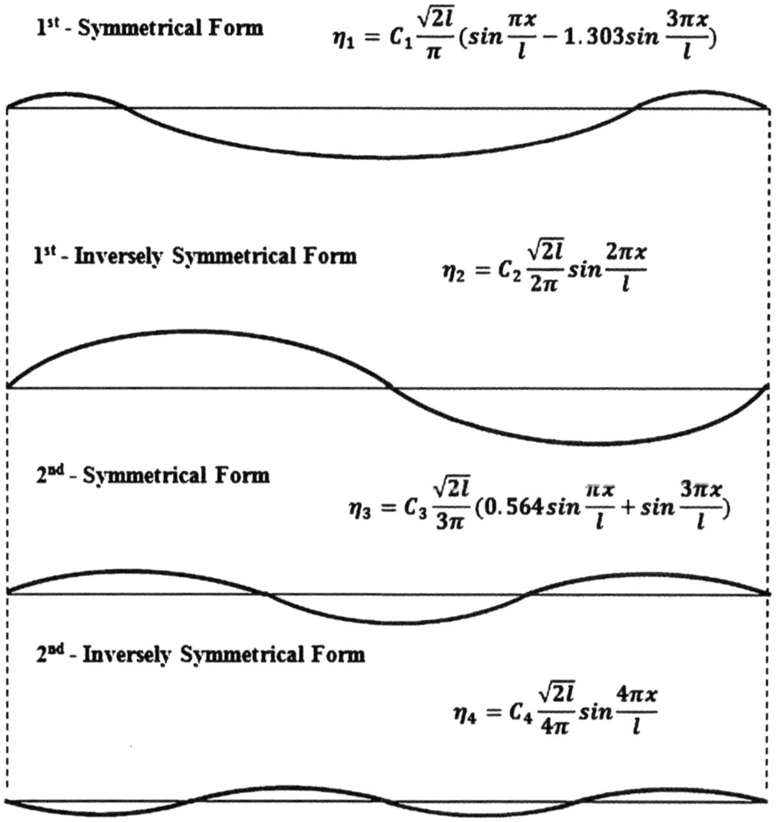

Fig. 16.2 Modes of natural vertical oscillations

dynamic part of the thrust—the dynamic "additive." The dynamic part of thrust is obtained from the examination of the strained state of an element of the arch

$$d\xi = \frac{ds}{dx}\Delta ds - \frac{dy}{dx}\Delta d\eta. \tag{16.27}$$

Since

$$\Delta ds = -\frac{\Delta H}{E_a F_a \cos\varphi} ds, \tag{16.28}$$

we will obtain

$$d\xi_a = -\frac{\Delta H}{E_a F_a}\frac{dx}{\cos^3\varphi} - \frac{dy}{dx}d\eta,$$

$\int_0^l d\xi_a -$ is the divergence (represents a discrepancy) of the thrust sections and it is equal to zero.

$$\xi_a = \int_0^l d\xi_a = 0. \tag{16.29}$$

Substituting (16.28) in (16.29) we will obtain:

$$\frac{\Delta H}{E_a F_a} \int_0^l \frac{dx}{\cos^3\varphi} + \int_0^l \frac{dy}{dx} d\eta = 0. \tag{16.30}$$

The value of the first integral:

$$\frac{\Delta H}{E_a F_a} \int_0^l \frac{dx}{\cos^3\varphi} = \frac{2R\Delta H}{E_a F_a} \cdot \frac{l}{\sqrt{4R^2 - l^2}}.$$

The value of the second integral:

$$\int_0^l \frac{dy}{dx} d\eta = \frac{dy}{dx}\eta \Big|_0^l - \int_0^l \frac{d^2y}{dx^2}\eta dx = -\int_0^l \frac{d^2y}{dx^2}\eta dx.$$

Substituting the values of these integrals into (16.30), we will obtain:

$$\frac{2R\Delta H}{E_a F_a} \cdot \frac{l}{\sqrt{4R^2 - l^2}} - \int_0^l \frac{d^2y}{dx^2}\eta dx = 0.$$

The dynamic part of the thrust:

$$\Delta H = \frac{1}{\frac{2R}{E_a F_a} \cdot \frac{l}{\sqrt{4R^2 - l^2}}} \int_0^l \frac{d^2y}{dx^2}\eta dx. \tag{16.31}$$

To solve Eq. (16.4), similar to the beginning of Sect. 16.1, we assume:

$$\begin{aligned} \eta(x,t) &= \eta_x A \sin(pt + \nu); \\ \Delta H(t) &= \Delta H A \sin(pt + \nu), \end{aligned} \tag{16.32}$$

where p is the natural frequency of oscillations; ν is the initial phase of oscillations. We approximate the form of standing wave of oscillations (vibrations) by Fourier series,

$$\eta_x = \sum_1^\infty b_n \sin\frac{n\pi x}{l}, \tag{16.33}$$

where b_n—coefficient, can be expressed as follows per recommendations of S.A. Ilyasevich [6]:

$$b_n = c_n \frac{\sqrt{2l}}{n\pi}$$

here c_n—the coefficient, depending on load. Substituting (16.32) into (16.4) we will obtain:

$$E_b I_b \frac{d^4\eta_x}{dx^4} + H\frac{d^2\eta_x}{dx^2} - \Delta H\frac{d^2y}{dx^2} - mp^2\eta_x = 0. \tag{16.34}$$

Next, substituting m (16.31) and (16.33) into 16.1.34), we will obtain:

$$E_bI_b\sum_{n=1}^{\infty}b_n\frac{n^4\pi^4}{l^4}\sin\frac{n\pi x}{l}-H\sum_{n=1}^{\infty}b_n\frac{n^2\pi^2}{l^2}\sin\frac{n\pi x}{l}+ \\ +\frac{y_0\int_0^l y''\left(\sum_{n=1}^{\infty}b_n\sin\frac{n\pi x}{l}\right)dx}{\frac{l2R}{E_aF_a\sqrt{4R^2-l^2}}}-mp^2\sum_{n=1}^{\infty}b_n\sin\frac{n\pi x}{l}=0. \quad (16.35)$$

We introduce the following designations:

$$B=\frac{2R}{E_aF_a\sqrt{4R^2-l^2}}$$

$$\int_0^l y''\left(\sum_{n=1}^{\infty}b_n\sin\frac{n\pi x}{l}\right)dx=\sum_{n=1}^{\infty}b_n\int_0^l y''\sin\frac{n\pi x}{l}=\sum_{n=1}^{\infty}b_na_n,$$

where

$$a_n=\int_0^l y''\sin\frac{n\pi x}{l}dx.$$

Then Eq. (16.35) takes the following form:

$$E_bI_b\sum_{n=1}^{\infty}b_n\frac{n^4\pi^4}{l^4}\sin\frac{n\pi x}{l}-H\sum_{n=1}^{\infty}b_n\frac{n^2\pi^2}{l^2}\sin\frac{n\pi x}{l}+ \\ +\frac{y''\sum_{n=1}^{\infty}b_na_n}{lB}-mp^2\sum_{n=1}^{\infty}b_n\sin\frac{n\pi x}{l}=0. \quad (16.36)$$

Multiplying both parts of the Eq. (16.36) by $\sin\frac{n\pi x}{l}$, integrating from 0 to l, and taking into account, that:

$$\int_0^l \sin\frac{n\pi x}{l}\sin\frac{i\pi x}{l}dx=\begin{cases}0 \text{ when } i\neq n;\\ \frac{l}{2} \text{ when } i=n,\end{cases}$$

we will obtain:

$$E_bI_bb_i\frac{i^4\pi^4}{l^4}\cdot\frac{l}{2}-Hb_i\frac{i^2\pi^2}{l^2}\cdot\frac{l}{2}+\frac{a_i\sum_{n=1}^{\infty}a_nb_n}{lB}-mp^2b_i\frac{l}{2}=0. \quad (16.37)$$

Instead of $\sum_{n=1}^{\infty}a_nb_n$, let us limit ourselves by the first four members of series, and taking into account that the even coefficients of $a_{2n}=0$, we arrive at a system of homogeneous linear equations relative to b_1, b_2, b_3, and b_4. From the first and third equations, we obtain:

$$p_{1,3}=\frac{\pi^2}{l^2}\sqrt{\frac{E_bI_bg}{q}\left[41-5\overline{H}_1+(a_1{}^2-a_3{}^2)S_1\pm40\sqrt{\left[1+0.1\overline{H}_1+\frac{(a_3{}^2-a_1{}^2)}{40}\right]^2+\frac{a_1{}^2a_3{}^2S_1{}^2}{400}}\right]} \quad (16.38)$$

here

$$\overline{H}_1 = \frac{l^2}{\pi^2 E_b I_b} H; \qquad S_1 = \frac{E_a F_a}{E_b I_b} \cdot \frac{l^2 \sqrt{4R^2 - l^2}}{2R};$$
$$a_1 = \frac{R^2 \int_0^l \sin \frac{\pi x}{l} dx}{\left[R^2 - (x - l/2)^2\right]^{3/2}}; \quad a_3 = R^2 \int_0^l \frac{\sin \frac{3\pi x}{l} dx}{\left[R^2 - (x - l/2)^2\right]^{3/2}}. \tag{16.39}$$

From the second and fourth equations we will determine:

$$p_2 = \frac{4\pi^2}{l^2} \sqrt{\frac{E_b I_b g}{q}} \left(1 - \overline{H}_2\right);$$
$$p_4 = \frac{16\pi^2}{l^2} \sqrt{\frac{E_b I_b g}{q}} \left(1 - \overline{H}_4\right), \tag{16.40}$$

where

$$\overline{H}_2 = \frac{l^2}{4\pi^2 E_b I_b} H, \ \overline{H}_4 = \frac{l^2}{16\pi^2 E_b I_b} H.$$

In order to compare the results of the calculation carried out according to formulas (16.23), (16.38), and (16.40), let us determine the natural (free) frequencies for a combined system with geometric and mass parameters, which coincide with the construction of the Jermuk bridge. The list of necessary design data is given at the beginning of this paragraph. Using this data, we can determine the values for a_1, and a_3 from formula (16.39):

$$a_1 = 109^2 \cdot 0.0000650 = 0.772265, \ \ {a_1}^2 = 0.569$$
$$a_3 = 109^2 \cdot 0.0000300 = 0.356430, \ \ {a_3}^2 = 0.127$$

and values for S_1 and $\overline{H}_1$

$$S_1 = 108.746, \ \overline{H}_1 = 0.4665.$$

Further, using (16.38) and (16.40), we can determine the frequencies of free (natural) oscillations of the given combined system:

$$p_1 = 4.5215 \ s^{-1}; p_2 = 9.016 \ s^{-1}; p_3 = 16.08 \ \ s^{-1}; p_2 = 29.255 \ s^{-1}$$

With the calculation of p_2 and p_4 the value $\overline{H}_2 = 0.118$, and the value $\overline{H}_4 = 0.00024$.

16.2 Results of Dynamic Testing Model of Arch Bridge System

The study of the oscillations of constructions under the influence of moving loads has a great practical value. Currently, loads and speeds of traffic continuously grow, and therefore increase the dynamic impacts on constructions. An increase in traffic volume, the weight of traveling load, and the simultaneous decrease in the mass of bridges is the contemporary tendency in bridge construction. Because of this, using precise dynamic design schemas, diagrams, and models becomes essential.

Traffic excitation causes nonstationary oscillations of bridge spans. In general, the oscillations of bridge spans represent a transient process, i.e., the combination of forced and accompanying free oscillations. The degree of manifestation of the latter depends, in essence, on the inelastic resistance of the material of the construction. A large number of studies are devoted to the oscillations/vibrations of bridge spans. Significant contributions to the solution of general problems of oscillation/vibrations of constructions and bridge structures under the influence of mobile loadings were made by C.E. M. Becker, C.E. Inglis, A.N. Kryloff, S.P. Timoshenko, S.T. Odman, A. Schallenkamp, H. Steuding, G. Stokes, R. Willis, W. Witting, V.V. Bolotin, A.A. Alexandrov, N.G. Bondar, I.N. Goldenblat, I. A Kolesnik, S. I. Konashenko, B.G. Korenev, A.B. Morgaevsky, G.B. Muravsky, I.G. Panovko, A.A. Petropavlovsky, V.A. Smirnoff, A.P. Philipov, I.I. Kazey, and others. The number of experimental studies is considerably lower when compared to the number of theoretical works on the dynamics of steel and reinforced concrete beam and arch bridges with movable mobile loads. The first works devoted to this question were experimental (M. Becker) or experimentally theoretical (R. Willis, and G. Stokes). After earlier theoretical and experimental studies analyzed the dynamic behavior of brides due to moving loads, as performed by C.E. Inglis [10] in 1934, various newly developed and revised moving loads and bridge models were applied to the moving load analysis. D.S. Carder (1937) conducted a field investigation and observed vibrations of the San Francisco-Oakland Bay Bridge and the Golden Gate Bridge. These studies were conducted to determine the probability of damage caused by resonance during seismic excitation. An early summary of dynamic testing of highway bridges in the USA performed between 1948 and 1965 was presented by R.F. Varney and C.F. Galambos (1966). T. Iwasaki, J Penzien, and R. Clough (1972) summarized tests performed in Japan between 1958 and 1969 to determine the dynamic properties of bridge structures. Another summary of field and laboratory model tests on bridge systems was presented by H.S. Ganga Rao. In 1984, Wang carried out research on the impact of freight trains on concrete railway bridges. C.R. Farrar, W.E. Baker, T.M. Bell, et al. (1994) report ambient vibration test on the I-40 Bridge over the Rio Grande prior to a damage detection study. Traffic excitation on the bridge of interest was first used as the ambient vibration source. After traffic had been removed from the bridge, traffic on an adjacent bridge was used as the excitation source. O.S. Salawu and C. Williams (1995) provided review of full-scale dynamic testing of bridges, where methods of excitation were examined and reasons for performing dynamic tests were summarized. The analysis of experimental studies performed in the former USSR, were conducted by N.G. Bondar, I.I. Kazey, Y. G. Kozmin, and B.F. Lesokhin [11]. The extensive use of testing in the evaluation of bridges has resulted in the American Society of Civil Engineers (ASCE) Committee on Bridge Safety publishing a guide for field-testing of bridges in 1980. This guide includes an extensive reference list of papers summarizing previous bridge tests. Also discussed are static and dynamic load application methods, instrumentation, data acquisition, and methods for measuring material characteristics. The presented standards addressed both static and dynamic testing of bridges. Topics that are summarized include categories of load tests, test preparation, test procedures, evaluation of load steps, and evaluation of load tests and reporting of results. G.P. Tilley discusses different methods of measuring damping in bridges. Field test methods discussed include excitation methods such as driving a test vehicle over the bridge, step-relaxation, single-pulse loading (eccentric mass shakers, people jumping in unison, and pulling on ropes), and ambient excitation. C.R. Farrar, T.A. Duffey, and P. J. Cornwell [12], along with S.W. Doebling, presented a report in 1999 which summarized methods of excitation that have been used in past dynamic testing of bridge structures.

The attributes and difficulties associated with the various excitation methods have been discussed in a very general manner. Although there does not appear to be consensus that no one particular method is better than the other, for large bridges, the ambient excitation methods are the only practical method of exciting the structure. The report contains a more thorough list of references pertaining to the excitation methods and dynamic testing of bridges in general. Despite the large number of publications on vibrations (oscillations) of rod/bar structures (systems), bridge structures under the influence of moving loads, this issue still requires further development.

In our opinion, the most expedient methods of analysis of the dynamic effect of mobile inertial loads are the methods of C.E. Inglis [10], V.V. Bolotin [13, 14], A.P Filipov [15], S.S. Kokhmanyuk, O.A. Goroshko and the method of A. Schallenkamp and I.I.Kazey [16]. It is necessary to improve upon the existing methods of dynamic analysis and to establish the limits of their applicability for solving specific problems put forward by the practice of using numerical methods. The study of the dynamic processes, revealed by traveling loads, can be conducted in two directions:

(1) A study of the behavior of a bridge structure under a given load, and
(2) A behavioral inquiry into an integrated system of a bridge and vehicle, i.e., the study of an interaction of a bridge span and a traveling load, and development of parameters needed to model the dynamics of a bridge-vehicle interaction. Despite the large number of publications on oscillations of rod/bar structures and combined bridge structures such as rigid beam—flexible arch under the influence of moving loads, this problem still requires further development.

Research on the dynamic analysis of a vehicle-bridge coupled vibration system is an important issue in civil engineering. In general, three methods are used to address this problem: the theoretical derivation, dynamic loading test, and numerical simulation. In recent years, with the development of powerful computer software, numerical simulation studies are widely used.

Analytical studies of the dynamic interaction of a span-vehicle coupled model for complicated bridges, such as combined systems "rigid beam-flexible arch" with various types of running vehicles, present enormous difficulties and, in most cases, do not provide sufficiently reliable results for practice. That is why the experimental theoretical method of study is the most appropriate method of research. The basis of this method is the assumption of a design schema (mathematical model), whose parameters are determined by comparing analytical results to experimental data.

A unique long-span (large-span) steel bridge in the form of a combined system "rigid beam (stiffening girder)-flexible arch" with the travel on top and the overlapping canyon of the river Arpa in the city of Jermuk, Armenia, was put in operation in 1973 (Fig. 16.3).

The complex configuration of the arch bridge span and the adjacent overpasses, with significant static indeterminacy, leads to a considerable difficulty in obtaining a precise dynamic calculation of this structure. Under such circumstances, using models, necessary to calculate dynamic characteristics, and parameters, considering the joint work of a bridge span structure and trestle approaches, is preferable and expedient. From structural, design, and methodological considerations, the model of Jermuk's bridge was made geometrically similar to the original, on scale 1:20 (Fig. 16.4).

It is known that the observance of geometric similarity provides the similarity of the original and the model in the qualitative deformations, modes of natural vibration/oscillations, and distribution of the natural oscillatory periods in structures made of homogeneous materials (A.G. Nazarov [17], V.A. Zakarian [18]). In this case, homogeneity of the material of the original structure is disrupted only by roadway pavement of the bridge, possessing a lower modulus of elasticity. Therefore, the plate of the roadway together with the pavement in the model is realized in the form of a plate of rectangular cross-section, width, and thickness, selected from the conditions equivalent to the original structure with respect to weight and lateral stiffness. This condition, together with the geometric similarity of the entire bridge structure, provides the required mechanical similarity of the original and model.

The geometric characteristics of the model are given in Table 16.2.

The axis of the arch is outlined on the square parabola: $y = \frac{4f}{l^2}x(l-x)$. Static tests were conducted for the purpose of development of the character of vertical and lateral deformations of a combined

Fig. 16.3 Overall view of the Jermuk bridge

Fig. 16.4 View of the model of Jermuk bridge

system of "rigid beam-flexible arch," and also to determine the influence of separate elements of a structure on the stiffness in vertical and transverse directions.

A load in the vertical direction is applied to the central cross section of the arch and, with the aid of a side guy-rope on a pulley, a transverse horizontal force was created. The vertical and horizontal displacements of the points were determined by gauges. In order to exclude the possibility of random errors and to obtain reliable data, each test was conducted not less than ten times. For the final results of the dynamic characteristics of the bridge model, the averaged values obtained by careful processing of

Table 16.2 The geometric characteristics of the bridge model

#	Geometrical characteristics of the model of the bridge	
1	The design span of arches	$l = 6\ m$
2	The distance between axes of arches	$B = 0.42\ m$
3	The arrow of the rise (lifting) of the arch	$f = 0.93\ m$
4	The slope of the arch	$f/l = 0.15$
5	The length of the panel of above-arch structure	$K = 0.4\ m$
6	The cross-sectional area of the beam	$F_b = 5.14 \cdot 10^{-4}\ m^2$
7	The cross-sectional area of the arch	$F_a = 1 \cdot 10^{-4}\ m^2$
8	The moment of inertia of beam	$I_b = 42.73 \cdot 10^{-8}\ m^4$
9	The moment of inertia of arch	$I_a = 1.15 \cdot 10^{-8}\ m^4$
10	The modulus of elasticity of beam and arch	$E = 210\ GPa$
11	The ratio of the moment of inertia beam to the arch	$I_b/I_a = 37.16$

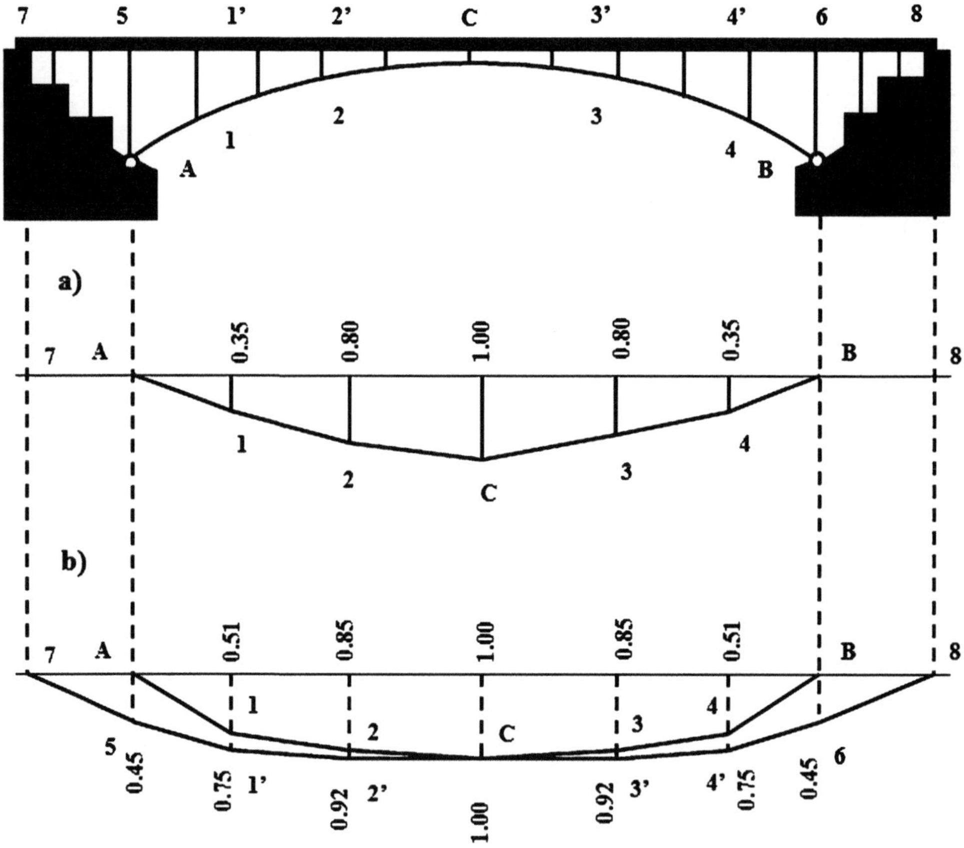

Fig. 16.5 Graphs of static vertical (**a**) and horizontal transverse (**b**) displacements of the points of a model due to forces applied at the crown section of the arch—in the vertical (**a**) and—in the transverse (**b**) the directions (ordinates are represented in arbitrary units)

a series of oscillograms were accepted [19]. The graphs of the displacements of the characteristic points of the system are given in Fig. 16.5.

To evaluate the influence of rigidity of separate elements of the bridge structure, a model testing was conducted without orthotropic deck, an arch, the struts (cross-bars), and a beam structure. Tests on the

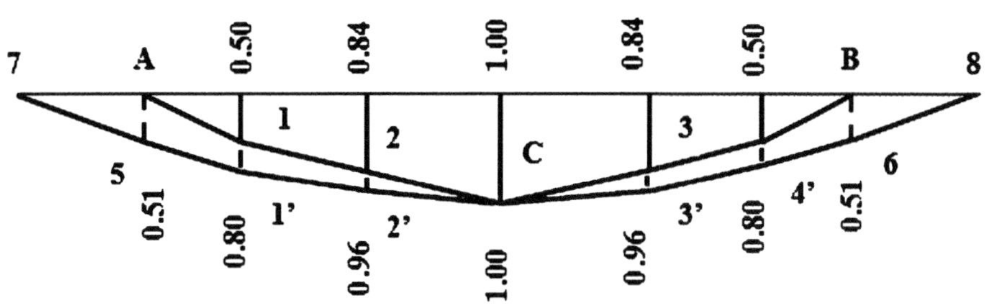

Fig. 16.6 Mode of natural lateral (transverse) oscillations of model

Table 16.3 Periods of free oscillations

#	Studied structure	Periods of free oscillations	
		Vertical, (s)	Lateral, (s)
1	Bridge span	0.0434	0.07
2	Bridge span without orthotropic plate	0.039	0.065
3	Free arch	0.0375	0.0563

model showed that the orthotropic deck substantially influences both the vertical and especially the lateral stiffness of the bridge span structure.

The stiffness of the model decreased approximately fourfold in the transverse direction, and by 15% at removal of the struts (cross-bars). The lateral stiffness of separate flexible arches is approximately 25 times less than the stiffness of bridge span structures. With the dynamic tests, free horizontal lateral oscillations of the model were excited by pulling out followed by a sudden release of the central section of the arches. The oscillations were also caused by the symmetrical dropping of loads in one direction for the excitement of symmetrical oscillations (vibrations) and in the opposite directions for the excitement of antisymmetric oscillations. During the model tests the same scheme of measurements of dynamic characteristics was used, as with the full-scale field bridge testing L.G. Petrosian [19] (see Figs. 16.10 and 16.11). The task of the full-scale field bridge tests was to verify the dynamic characteristics of an operable bridge of the type of "rigid beam-flexible arch" with a travel on top, as well as the comparison with the results, obtained with the theoretical analysis of the given bridge structure. The obtained oscillograms showed that the characteristic points of the model executed synchronous oscillations and the amplitude ratios were constant for various points. The mode shapes of oscillations of the fundamental tone are represented in Fig. 16.6, and Table 16.3 gives the values of their periods.

It is evident from Table 16.3, that the periods of free oscillations of the arch are somewhat lower than for the bridge span. This is explained by the dominating influence on the periods and the increase in the inert mass due to the part of the bridge structure, located above the arch. Experiments confirm that the first frequency of spatial horizontal oscillations of free flexible arches is approximately 1.5 times lower than the frequency of the vertical oscillations. In order to find the influence of trestle parts on the dynamic characteristics of the main bridge span, testing was carried out in four stages (Table 16.4).

Tabulated data show that the frequency of free vertical oscillations (vibrations) diminishes with the decrease of the stiffness of trestle approaches, and its value with the complete removal of the overpasses (see Table 16.4, position 4) coincides with the theoretical calculations [9] with adequate accuracy. In the absence of trestle approaches to the main bridge span structure, the angular frequency

Table 16.4 Angular frequency and period of free oscillations

#	Schemes of bridge model type rigid beam flexible arch	Vertical vibration		Horizontal vibration	
		$P\ (s^{-1})$	$T(s)$	$P\ (s^{-1})$	$T\ (s)$
1	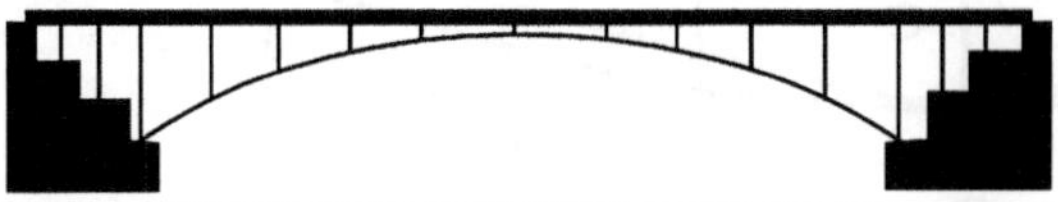	144.69	0.0434	89.71	0.0700
2		132.60	0.0474	87.50	0.0720
3		116.80	0.0538	85.50	0.0734
4		100.48	0.0625	92.50	0.0680

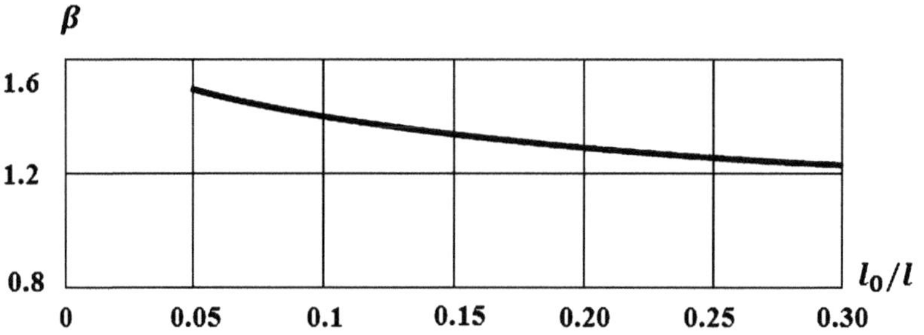

Fig. 16.7 The dependence of coefficient β on l_0/l

of free vertical oscillation of the model is equal to: $P = 100.48s^{-1}$, but the frequency, calculated for the model according to the formula (16.23) gives a close result $P = 102.5\ s^{-1}$.

In a case when the flexible arch with a rigid beam has trestle approaches, the elastic fixed ends seemingly appear in the sections of the contiguity of the overpasses to the main bridge span. To use formula (16.23) the influence of trestle parts on the dynamic characteristics of the entire bridge structure is assumed not to increase the frequency coefficient β, which makes it possible to approximately determine the frequency of free vertical oscillations for such systems. Let us replace the multi-span overpass with an adjustable single-span, creating an elastic fixed-end of the same intensity as the multi-span. Table 16.4 depicts the computed values for the given adjustable spans for two and three equal spans of trestle approaches. For the various ratios of l_0/l, where l– is the main span, the fundamental frequencies of oscillations/vibrations are determined.

The analysis of the experiments showed that for the ratio $l_0/l = 1/6 \div 1/8$, which is more frequently encountered in practice, and with $EI_b/EI_a \geq 30$, the frequency of fundamental tone depends little on these relations. A curve for determining the coefficient of an increase of frequency β (Fig. 16.7) is built from the results of the experiments on the models.

As an example let us determine the fundamental frequency of oscillations of a bridge, which has trestle approaches from both ends of the main span. The angular frequency of the free vertical oscillations, calculated according to the formula (16.23), is equal: $P = 7.50\ s^{-1}$. For three overpasses

Table 16.5 Adjustable overpass single-span

	Adjustable overpass single-span l_0, (m)	
Trestle bays $l_{trestle}$, (m)	Two bays overpass	Three bays overpass
4	3.504	3.468
6	5.256	5.202
8	7.008	6.936
10	8.760	8.670
12	10.512	10.404
14	12.264	12.138
16	14.016	13.872
18	15.768	15.606

Fig. 16.8 Bridge model under dynamic testing

with equal bays $l_{trestle} = 8m$, let us determine the value of the given adjustable span $l_0 = 6.936m$, according to Table 16.5. We will determine the coefficient $\beta = 1.44$ according to graph in Fig. 16.7. Then for the entire bridge structure, the angular frequency of free vertical oscillations will be $P = 10.8\ s^{-1}$, with the corresponding period of the oscillations $T = 0.58\ s$. The period of free vertical oscillations with the full-scale field tests was $T = 0.57\ s$.

The precise calculations with the application of computer programs show that the approximate determination of dynamic characteristics from the proposed method gives a certain overstating of the natural frequencies up to 10%. In Figs. 16.8 and 16.9, separate fragments of the tested model of the bridge are shown.

Summarizing the results of the dynamic testing of the model of a combined system "rigid beam-flexible arch," we come to the following conclusions (L.G. Petrosian [19]):

Fig. 16.9 Part of a bridge model

Fig. 16.10 The motion of the test load along the bridge with various speeds

(1) The dependence of frequencies of natural vertical oscillations of a combined system of the type "rigid beam-flexible arch" with travel on top on an elastic, geometric, and mass parameters is established taking into account the deformation of the entire system. The generated formula (16.23) makes it possible to determine the frequency spectrum of free vertical oscillations for the combined systems. Analysis of the Eq. (16.23) shows, that the enforcement of the beam structure by an arch substantially influences the first frequencies of the symmetrical mode shape of oscillations and free vibrations. The frequencies, which correspond to the antisymmetric mode shape of oscillations do not depend on the degree of the enforcement (strengthening) beam by an arch and the rigid beam works as a simple beam.

Fig. 16.11 The moment of the impact in the characteristic sections of the main arched span of bridge

(a) Taking into account the deformations of a combined system results in lower frequencies of free vertical oscillations vs. when deformation is ignored. Moreover, the greatest deviation is obtained with the first frequency to the corresponding anti-symmetric mode of oscillations. For large spans of 100 m and more the divergence is 7–12%.
(b) We developed a method that makes it possible to determine the fundamental tone of frequencies of vertical oscillations of combined systems with at least 90% accuracy. The method uses the coefficient of "increased" frequency and takes into account joint dynamic work of the main bridge span with the adjacent trestle approaches.
(c) Of the two factors influencing the change periods of oscillations: (i) an increase in rigidity due to the part located above the arch structure, which decreases the oscillations period: (ii) and an increase in the inert mass (mass of inertia) due to the part located above the arch structure which increases the period of oscillations, the latter proved to be the dominant factor.

16.3 Vibrations of Combined Arch System Taking into Account Its Extent at a Given Harmonic Motion Base

According to the current methodology of calculations of buildings and constructions to seismic excitation, all points of base oscillate (vibrate) synchronously with the same shift and acceleration. However, the assumption that the length in the direction of the propagation of excitation is commensurate with the length of seismic waves can lead to significant errors with the determination of the reactions/response of these systems. The long-span arched structures of bridges present complex design systems, which consist both of rectilinear "frame" elements and curvilinear "arched" elements. The seismic calculation of such constructions is associated with complex and labor-intensive computational operations. Therefore, in certain cases it is expedient to replace the curvilinear elements with rectilinear elements, thereby simplify the design scheme of a structure.

Guarantees of reliability of residential, public, and industrial objects and facilities under seismic action are an important national, economic, and public safety imperative. When designing both the objects of mass construction and special construction social, technical, safety, and economic factors must be considered. This in particular relates to the questions of the design of construction undertaken in seismic regions. Our increasing ability to quantify real inertial forces, which appear in extreme natural events such as earthquakes, allows for a more economic and more reliable design. This relates especially to bridges and tunnels, because their damages due to earthquakes cause tragic and material losses, leading to severe consequences, including loss of life and extended traffic disruption. The latter aggravates the disaster for the victims in the affected area. Restoration requires complex and labor-intensive work and becomes one of the first recovery priorities. Therefore, in many countries, the design standards for bridges and other road constructions provide heightened requirements for seismic resistance. The task of the seismic resistance for road engineering structures should be considered as an independent problem, which has a number of special features, namely, the presence of long-span massive elements, possessing significant inertia features, and capable of accomplishing horizontal and spatial vibrations. It is also necessary to take into account the differences in the dynamic features of bridge spans and supports, considerable extension in the plan, and the essential role of the lateral seismic pressure of ground on the abutment of a bridge. Thus, all of the features of road engineering structures mentioned above require the development of specific methods of design and calculation, and they happen to be the component parts of the general theory of seismic resistance (seismic stability). The calculation of seismic resistance of bridge structures as in all engineering constructions, is produced according to the spectral method adapted by the current seismic building code. Unique and especially important construction calculations can be produced according to real accelerograms (ground acceleration, peak ground acceleration), seismograms, or ground velocity (peak ground velocity), which correspond to the seismicity of the region. The accuracy of such calculations, in many respects, depends on the right selection of a dynamic design scheme, which can be represented in the form of the systems with continual or discrete parameters. Schemas in the form of the system with a finite number of degrees of freedom are more convenient with the use of contemporary methods of computer technology, and therefore, are used more often in practical calculations. Only the maximum seismic forces for separate forms of vibrations are calculated according to the spectral curve β. The estimated maximum value of seismic forces is identified as middle-quadratic from the stresses, which appear with various forms of vibrations. In this case, all stresses except the maximum are introduced with the coefficient 0.7. Usually, the first three or five modes of oscillations (vibrations) are considered in calculations. When applying the methods of calculation using the accelerograms of strong earthquakes, the displacements of the system for a discrete scheme (diagram) are determined from the following formula [20]:

$$y_k(t) = -\sum_{i=1}^{n} \frac{\eta_{ik} T_i}{2\pi} \int_0^t \ddot{Y}_0(\tau) e^{-\frac{\gamma\pi}{T_i}(t-\tau)} \sin \frac{2\pi}{T_i}(t-\tau), \tag{16.41}$$

where $k = 1, 2, \ldots, n$.

For a system with distributed, continuous parameters, the displacements are determined from the following formula:

$$y(x,t) = -\sum_{i=1}^{n} \frac{T_i}{2\pi} \eta_{i(x)} \int_0^t \ddot{Y}(\tau) e^{-\frac{\gamma\pi}{T_i}(t-\tau)} \sin \frac{2\pi}{T_i}(t-\tau) d\tau. \tag{16.42}$$

In the given formulas T_i—period of the ith mode of free oscillations (vibrations) of system; $\gamma = \frac{\delta}{\pi}$—coefficient of inelastic resistance of the system; δ—the logarithmic decrement of oscillations (vibrations); and $\ddot{Y}(t)$—the given accelerogram of an earthquake.

In calculations of the bridges exposed to seismic excitations according to the real accelerograms, the value of the coefficient of inelastic resistance starts within the limit (0.025–0.08) [20]. In this case, the lowest value of the coefficient of inelastic resistance corresponds to steel bridge structures, and the maximum value corresponds to reinforced concrete bridge structures. With the study of seismic vibrations of girder (beam, frame) bridges separate transverse and longitudinal vibrations are considered. In preliminary calculations, we can calculate vibrations on the first main (basic) mode only. In order to determine seismic forces both in the longitudinal and in transverse directions it is recommended to consider a temporary load rigidly fastened with the construction and not to ignore the possible phenomenon of slippage. The mode coefficient of η on the level of bridge spans of girder (beam, frame) bridges varies within the limits (1.1–1.3) [20]. The calculations of seismic vibrations/oscillations of bridges in the longitudinal direction is quite difficult because it is necessary to take into account the presence of friction in the mobile supporting parts of the bridge spans, asynchronous oscillations, ground soil under the individual bridge supports, and so forth. The practical design methodology must be based on the use of experimental values of natural frequencies and thus indirectly on account of the work of nodal (joint) links and the friction of bearing supports with vibrations. The horizontal and spatial vibrations of arch bridges with the ride on the top as complex oscillatory systems are of difficult theoretical analysis. The dynamic interaction of bridge spans with supports is one of the difficult factors to consider [21, 22]. The practical design recommendations, which are concerned with the calculations of such bridges to the seismic resistance, have an approximate character and will subsequently be refined in the future in proportion to the accumulation by sufficient theoretical and experimental information. With seismic design of long-span bridges in the longitudinal direction, it is necessary to take into account the asynchronous oscillations/vibrations of the outermost supports. However, the existing design standards do not provide practical recommendations to incorporate this factor, since they are based on the assumption that all points of the base oscillate/vibrate synchronously with one and the same displacement and the same acceleration. This assumption leads to significant errors in the determination of the reactions of these systems. Accounting for asynchronous oscillations of supports or, in other words taking into account the extent of the structure is especially necessary in cases when the length of a seismic wave in the direction of propagation excitement is commensurate with the length of the construction itself.

The influence of the extent of construction of the values of seismic loads is investigated for the first time by G. Hausner [23]. This question is considered by I.L. Korchinsky [24], and V.A. Grossman [25]. Assuming that the law of motion of the base is represented in the form of the sum of the damped sinusoids, it is found that the effect of the extent is greater with rigid than with flexible systems. The influence of the extent with harmonic oscillations of a base with different design diagrams of construction is investigated in the works of V.K. Egupova and T.A. Komandrina [26], S.G. Napetvaridze [27, 28], A.P. Sinitsin [29, 30], and E.E. Khachian, et al. [31]. The statistic aspects of this problem are examined by M.Ph. Barshtein [32]. The translational-rotational oscillations (vibrations), of constructions extended in the plan are investigated in the work of N.A. Amiraslanov and M.Ph., Barshtein [33]. In that work, cases of match and mismatch between the centers of mass and stiffness of structures are examined. The displacements and accelerations of the base were represented as stationary random processes. The authors obtain concrete results about the influence of the accounting of structure extent on the magnitude of the seismic forces for a one-storied frame. So, for example, for a building of $72m \times 144m$ size, with a wave of propagation velocity of $v = 600\ m/s$, accounting for structure extent, results in a decrease of inertial force in one direction by 25.5%, and in the other by 8%.

The reactions of extent structures with the use of instrumental records of earthquakes are investigated in the work of G.L. Nikiporets [34]. The design model of the extended structure is selected in the form of an absolutely rigid beam on elastic supports. Taking simplified records of the equations of motion, the author built the spectra of translational and rotational (torsional) acceleration of constructions with different parameters, which characterize the dimensions of construction in the plan, velocity of propagation of seismic waves, and the nature of distribution of the stiffness of construction. The spectra of reactions are built for six accelerograms of strong earthquakes. The obtained spectra of reactions show the significant influence of the accounting for the extent of construction, especially for comparatively rigid constructions. The research work has important practical value in view of the use of records of real accelerogram. However, the work was limited to only the simplified form of initial equations of motion, while the examination of oscillations of structures with multi-degree-of-freedom was also not possible.

Another study by the same author G.L. Nikiporets, proposes approximation formulas for determining seismic load that acts on extent structures. It is noted, that seismic load can be determined utilizing spectra of reactions, built for construction of non-extent structures in the plan. The extent is considered with the aid of two coefficients, depending on the characteristics of construction and ground.

In the research work of I.L. Korchinsky [24] seismic oscillations on extent structures of a one-story single-span frame with an absolutely rigid cross-bar (collar beam) are examined. The design scheme is given as a system with one degree of freedom. Oscillations of systems in cases when the base oscillates according to the law of damped sinusoid is also investigated. The basic conclusion of the study is reduced to the fact that accounting for extent leads to the decrease of seismic forces in the constructions, the natural period of oscillations is more than 0.15 s. For rigid structures, especially for girder bridges, the account of the length of the structure can lead to an increase in the seismic load. It is noted in [24] that the given analysis and conclusions based on this do not purport to the quantitative assessment of the phenomenon. Nevertheless, the main conclusion of the work is that the extent system should not be designed to be rigid. In another work [35] it is proposed to take into account the extent of construction by calculation of the average value of acceleration throughout the entire length of the building. It accounts for when the direction of the acceleration of the oscillations of the ground is perpendicular to the direction of the motion of seismic wave. In this case, in addition to progressive inertial forces, rotational forces will also act on the structure. Simple formulas for enumerating these loads are proposed.

The research work of A.B. Puhovsky and V.A. Gordin [36] examines the seismic resistance of an arch. The design schema is considered as a system with a finite number of degrees of freedom. The frequencies and modes of free oscillations of an arch are determined by the solution of the corresponding equations with the help of computer software. The internal forces of the structure from the seismic loads are determined by the current norms and standards: "Recommended Provisions for Seismic Regulations for New Buildings and Structures" [37], considering three and six modes of oscillations. It is shown that the internal forces in the structure can be determined with adequate accuracy taking into account three modes of oscillations. A method of determining design forces taking into account the highest modes of oscillations is proposed. The seismic oscillations of the extent girder, arched, and suspension bridges are studied to a considerably smaller degree.

Vertical oscillations of a suspension bridge with horizontal harmonic oscillations of the base have been examined by K. A. Kubo [38]. In this work, it is shown that under the action of a running harmonic wave on the supports of long-span suspension bridges the essential influence renders commensurably the span of structure with the wavelength.

Nonstationary random oscillations of suspension bridges with action on the supports of a running seismic wave are examined in the work of A.A. Petrov, and S.B. Basilevsky [39]. Horizontal seismic acceleration that propagates with a constant velocity at an arbitrary angle to the longitudinal axis of construction is considered. The impact is given in the form of the amplitude of a nonstationary random process. The equations of forced oscillations of the beam of rigid bridge are given. The decomposition in terms of the natural oscillations modes is used for the solution of the problem. The expressions are obtained for horizontal seismic force and torque moment, which are close to the expressions given in the design standards. These formulas contain the coefficients, accounting for commensurability of the span of bridge suspension with the length of seismic wave and small damping of the oscillations of construction and nonstationary of seismic excitation. The authors give the approximate expressions for determining the coefficient that take into account the extent of the construction.

In the works [9, 40], the seismic loads acting on the arch bridge across the river Arpa, (Armenia) with different schemas for the motion of supports are determined. The motion of supports is considered as the statistical independent random processes with one and the same energy spectrum. The seismic loads are determined on the basis of the current design standards. The distribution of the seismic forces with different modes of oscillations is given. The authors carried out a survey of research work on the accounting of extent of a structure with the determination of seismic loads which showed that, despite the numerous works conducted by scientists, this problem is still far from a form that is necessary for practical design and construction. In particular, this is relevant to the work taking into account the extent of seismic oscillations on girder bridges but is even more relevant to the arched long-span bridges. Meanwhile, promising direction of research studies should include the refinement of buildings of three-dimensional, spatial equations of motion of construction, taking into account the final velocity of propagation of seismic waves and the actual distribution of mass and stiffness of the system. The substantiated analysis of this phenomenon assumes knowledge of the necessary seismological information, for instance, the velocity of propagation of seismic waves on the surface of ground, multicomponent seismograms and accelerograms of earthquakes, as well as of specific dynamic characteristics of the construction itself: periods and the modes of free oscillations and the value of decrement of oscillations. The aforesaid relates to long-span bridge constructions, for which the account of extent has fundamental importance and can lead to a substantial change in condition (state) of stress.

Based on this, we have set up the precise equations of motion for the extent of multi-span and multi-story frames and long-span arched bridges, representing systems with multiple degrees of freedom. The dynamic characteristics of a long-span arched steel bridge are determined experimentally for the purpose of comparison with the analytical calculations, and for checking the authenticity of the design schemas given. Thus, in the light of the above, we conducted studies of seismic vibrations the following structures taking into account their extent:

- Long-span arched bridges
- Multi-span frame bridges
- Single-span, one-story frames

When designing objects of massive large-scale construction, and special construction, we must consider sociological, technical, environmental, and economic factors. This relates to the questions of design of construction in seismic zones. The determination of the real inertial forces which appear with earthquakes will contribute to more economical, safer, and more reliable design. Long-span combined bridge systems are unique constructions, and, therefore, seismic loads for them should be determined not only for the acting normative standards and provisions but also for real seismograms or accelerograms of recorded strong earthquakes. Furthermore, in the construction of a considerable range, whose length is commensurate with the impacts of the length of seismic wave, the displacements of the individual supports of construction in the general case will be generally shifted on a phase. These structures can be subject to very different motions along their length due to spatial variability of the input seismic motion. Therefore, during calculations of bridge constructions, we should consider the asynchronicity of the oscillations (vibrations) of individual supports.

The influence of the extent of construction length on the values of seismic loads was investigated initially by G. Hausner [23]. Observations conducted by Hausner during earthquakes have shown that seismic ground motions can vary significantly over a distance of the same order of magnitude as the dimensions of these extended structures. The works of: I. L. Korchinsky, A.A. Grill, A.P. Sinitsina, J.E. Bogdanoff, S.D. Werner, M. Shinozuka, M. Hoshiya, W.J. Hall, N.M. Newmark, M.J. Priestley,

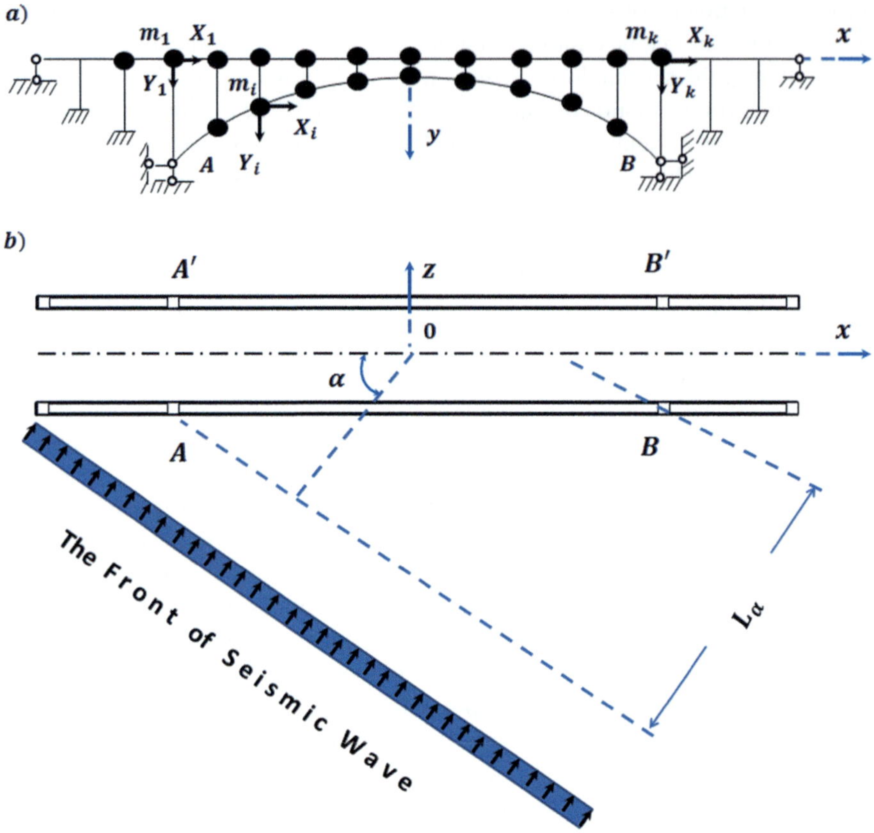

Fig. 16.12 The dynamic design model and the scheme of seismic excitation

R. Richards, D.G. Elms, A.D. Kiureghian, R.S. Harichandran, E.E. Khachian, A. Kareem, L.C. Lee, L.H. Wong, O. Ramadan, E.H. Vanmarcke, J.-N. Yang, J. Wang, J.E. Luco, H.L. Wong, C.E. Taylor, A. Zerva, and others were dedicated to the influence of extent with harmonic oscillations of ground with different design diagrams of construction. According to S.D. Werner, et al. [41] research analyses, the differences in the input ground motions applied to bridge foundations can have significant effects on the bridge response. Also, the nature of structural response to these traveling waves is strongly dependent on the direction of incidence as well as on the excitation frequency of the seismic waves. A study of these authors' works suggests that the use of dynamic characteristics of a structure along with the study of real records of spatial variation in seismic ground motions to understand the response of extended structures, such as beam (girder) and arched bridges, presents a valuable methodology.

Let us propose a method for calculating seismic resistance of combined bridge systems by using real records of ground motion of strong earthquakes. In Fig. 16.12, the dynamic design model and the scheme of seismic excitation are shown. The dynamic design model of long-span combined bridge system of the type rigid beam—flexible arch is taken as a system with a finite number of degrees of freedom.

In this case, each concentrated mass possesses three degrees of freedom, and the position of *i*th mass in the space is characterized by the displacements of the X_i, Y_i in the plane of the bridge and displacement Z_i in the direction perpendicular to the plane of the bridge.

It is assumed that the seismograms of the oscillations of support A in three mutually perpendicular directions $X_A(t)$, $Y_A(t)$, and $Z_A(t)$ are known and so are the value of the velocity of propagation of waves of excitation on the surface of the ground V. The seismograms of the oscillations (vibrations) of support B will be:

$$X_B(t) = X_A\left(t - \frac{L_\alpha}{V}\right); \quad Y_B(t) = Y_A\left(t - \frac{L_\alpha}{V}\right); \quad Z_B(t) = Z_A\left(t - \frac{L_\alpha}{V}\right).$$

It is assumed that the front of seismic excitation is the plane propagating with velocity V, under angle α to the longitudinal axis of the bridge.

The equations of motions of this oscillatory system with asynchronous motions of supports in the general three-dimensional case of oscillations (vibrations) can be represented in the form of:

$$\begin{aligned}
X_j(t) &= -\sum_{i=1}^{k} m_i X_i'' \delta_{ji}^{(xx)} - \sum_{i=1}^{k} m_i Y_i'' \delta_{ji}^{(xy)} - \sum_{i=1}^{k} m_i Z_i'' \delta_{ji}^{(xz)} + \delta_{jA}^{(xx)} X_A + \\
&\quad + \delta_{jB}^{(xx)} X_B + \delta_{jA}^{(xy)} Y_A + \delta_{jB}^{(xy)} Y_B + \delta_{jA}^{(xz)} Z_A + \delta_{jB}^{(xz)} Z_B \\
Y_j(t) &= -\sum_{i=1}^{k} m_i X_i'' \delta_{ji}^{(yx)} - \sum_{i=1}^{k} m_i Y_i'' \delta_{ji}^{(yy)} - \sum_{i=1}^{k} m_i Z_i'' \delta_{ji}^{(yz)} + \delta_{jA}^{(yx)} X_A + \\
&\quad + \delta_{jB}^{(yx)} X_B + \delta_{jA}^{(yy)} Y_A + \delta_{jB}^{(yy)} Y_B + \delta_{jA}^{(yz)} Z_A + \delta_{jB}^{(yz)} Z_B \\
Z_j(t) &= -\sum_{i=1}^{k} m_i X_i'' \delta_{ji}^{(zx)} - \sum_{i=1}^{k} m_i Y_i'' \delta_{ji}^{(zy)} - \sum_{i=1}^{k} m_i Z_i'' \delta_{ji}^{(zz)} + \delta_{jA}^{(zx)} X_A + \\
&\quad + \delta_{jB}^{(zx)} X_B + \delta_{jA}^{(zy)} Y_A + \delta_{jB}^{(zy)} Y_B + \delta_{jA}^{(zz)} Z_A + \delta_{jB}^{(zz)} Z_B
\end{aligned} \tag{16.43}$$

where

m_i—mass concentrated in the i th section; k—a quantity (number) of concentrated masses;

$X_j, Y_j, Z_j,\ X_j'', Y_j'', Z_j''$—respectively displacements and the acceleration of j th mass in three mutually perpendicular directions;

$\delta_{ji}^{(xx)}, \delta_{ji}^{(xy)}, \delta_{ji}^{(xz)}$—displacements of jth mass in the direction of the X unit forces applied at the point i, in the directions X, Y, and Z respectively;

$\delta_{jA}^{(xx)}, \delta_{jA}^{(xy)}, \delta_{jA}^{(xz)}$—displacements of jth mass in the direction of the X due to unit displacement of the support A in the directions X, Y, and Z respectively;

$\delta_{jB}^{(xx)}, \delta_{jB}^{(xy)}, \delta_{jB}^{(xz)}$—displacements of jth mass in the direction of the X due to unit displacement of the support B in the directions X, Y, and Z respectively.

The determination of seismic loads is reduced to the integration of the system of differential Eq. (16.43) and determining the maximum values:

$$m_i X_i''(t), m_i Y_i''(t), \text{and } m_i Z_i''(t).$$

With the use of the known determined law of ground motion, the seismic loads, and the displacements of the system can be determined with the aid of the analytical solution of the system of linear differential equations. When using the real accelerograms, assigned in tabular form, we use the numerical solutions of equations by the methods of Runge-Kutta or Adams. Let us examine a concrete example of determining seismic loads taking into account the extent of construction, operating in a large-span (long-span, wide-span) steel bridge which overlaps the canyon of a river in Armenia. We will determine the reactions of construction, taking into account its extent with assigned horizontal harmonic motion of the base. The influence of asynchronous oscillation of the outer (extreme) supports will be most essential when the front of a seismic wave is perpendicular to the longitudinal axis of bridge x. The dynamic design scheme of construction is taken in the form of a system with three concentrated masses as those applied in the characteristic sections of the combined system of the type of "rigid beam – flexible arch." Since each mass has two degrees of freedom, the entire system will have six degrees of freedom. The equation of motion of this oscillatory system with asynchronous oscillations of supports can be represented in the following form:

$$\begin{aligned} X_j(t) = & -\sum_{i=1}^{3} m_i X_i'' \delta_{ji}^{(xx)} - \sum_{i=1}^{3} m_i Y_i'' \delta_{ji}^{(xy)} + \delta_{jA}^{(xx)} X_A + \\ & + \delta_{jB}^{(xx)} X_B + \frac{\alpha}{\omega_1} (X_j - X_A)' = 0; \\ Y_j(t) = & -\sum_{i=1}^{3} m_i X_i'' \delta_{ji}^{(yx)} - \sum_{i=1}^{3} m_i Y_i'' \delta_{ji}^{(yy)} + \delta_{jA}^{(yx)} X_A + \\ & + \delta_{jB}^{(yx)} X_B + \frac{\alpha}{\omega_1} (Y_j - Y_A)' = 0; \\ & (j = 1, 2, \ \ldots \ 3; \ Y_A = 0). \end{aligned} \tag{16.44}$$

We have accepted upon consideration of internal friction, that with the asynchronous oscillations damping is the same as with the synchronous. This assumption can be considered acceptable, having the low value of damping coefficient. In equations of motion (16.43), damping is taken into account on the equivalent hypothesis of Voight [31] by adding the following members (terms) to the left side of the equations:

$$\frac{\alpha}{\omega_1} (\Delta X_j)'; \ \frac{\alpha}{\omega_1} (\Delta Y_j)'; \ \frac{\alpha}{\omega_1} (\Delta Z_j)'.$$

Here ΔX_j, ΔY_j, ΔZ_j— represent the corresponding deformations of the jth cross-section in the directions x, y, and z. α— the damping coefficient, ω_1— fundamental frequency of free vibrations. The deformation of jth cross-section ΔX_j can be presented in the following form:

$$\Delta X_j = X_j - \delta_{jA}^{(xx)} X_A - \delta_{jB}^{(xx)} X_B - \delta_{jA}^{(xy)} - \delta_{jB}^{(xy)} - \delta_{jA}^{(xz)} - \delta_{jB}^{(xz)}. \tag{16.45}$$

The expression for ΔY_j, and ΔZ_j have a similar appearance. If the oscillations of supports occur synchronously ($X_A = X_B$), and only in the plane of xoy, then from (16.44), follows that $\Delta X_j = X_j - X_A$.

The equations of motion are built with the assigned horizontal motion of the base —$X_A(t)$, $X_B(t)$ only. Determining the seismic reactions of construction requires a preliminary determination of the constant coefficients, entering the system of Eq. (16.43): unit displacements δ_{ji}, the values of the concentrated masses m_i, fundamental tone of frequency ω_1. It is also necessary to have the characteristic of the internal friction of the material of the bridge, the coefficient of damping α, and the characteristics of the influence of seismic action (effect)—seismogram of the ground motion, and velocity of propagation of waves of excitation of the ground surface. The displacements in the multiple statically indeterminate system are determined by the application of Moore's formula taking into account the influence of the bending moment and normal force. We will determine each of these parameters separately. The rods /bars system is a rigid beam—"stiffening girder" being supported by a flexible arch. The calculation of this system is connected with the essential difficulties. In the cases examined by us, rigidity of posts and arches is small in comparison with the rigidity of a beam, therefore, the problem can be simplified considerably by following the recommendation of A. Ph. Smirnov [1], accepting the design scheme shown in Fig. 16.13. The beam of stiffness of the given system works as a seven-spans continuous beam (one support is usually considered as an immovable

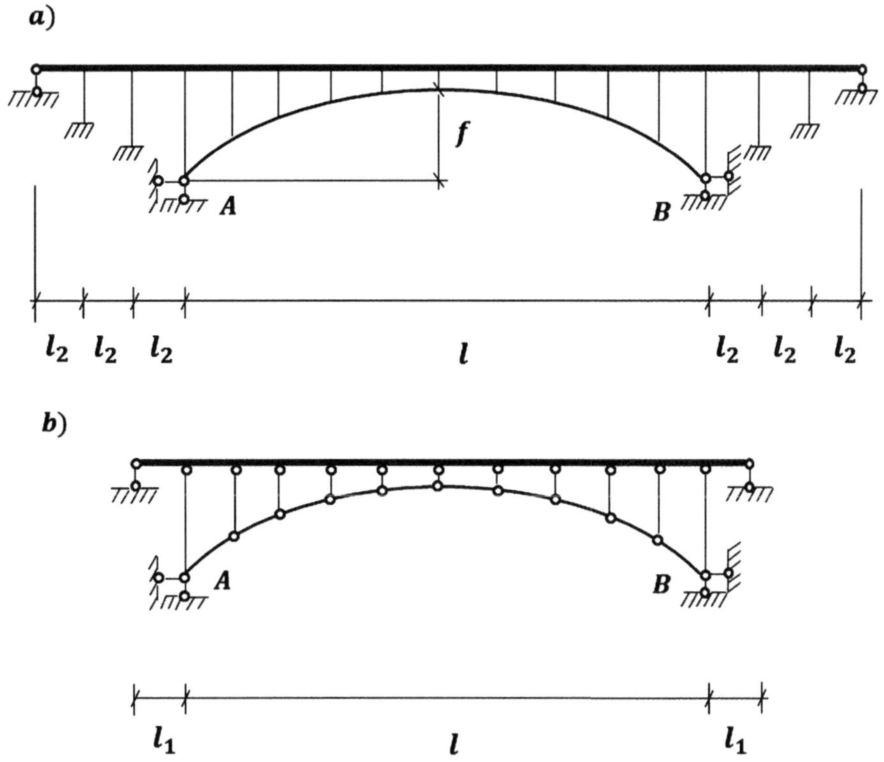

Fig. 16.13 Structural and design schemes of the bridge: (**a**) structural schema; (**b**) design scheme

hinge while the other supports are hinges on rollers), resting on supports of the trestle (overpass). For simplification in the calculations, we accept the three-span scheme in which the first and last spans of the overpass adjacent to the main arch span create the built-in support of the same stiffness as seven-spans. Here the rotation on the support of a three-spans continuous beam with spans l_2 due to a unit moment must be equal to the rotation angle on the support of a single-span beam with the span l_1 with identical rigidity. Equating the angles of rotations of the single-span and three-span continuous beams, we obtain the value of the reduced length l_1, $l_1 = 0.87l_2 = 0.7m$. Represented in this form, the system becomes three times statically indeterminate. We will obtain a simple solution of the problem if as redundant unknowns we select the thrust of one of the supports and as the group of unknowns, the concentrated moments at the sections $A - A'$, and $B - B'$ (Fig. 16.13).

Unit displacements, entering the canonical equations for determination of "redundant" unknowns, are determined by the application of Moore's formula, taking into account the influence of the bending moment and normal force. For determining the displacement δ_{ik} of a statically indeterminate system, first the redundant unknowns due to force $P = 1$ applied in the direction k are determined. Then the values M_i and N_i due to force $P = 1$, applied in the direction i are determined. Thus M_i and N_i can be calculated for any geometrically invariable (unchangeable, immovable), statically indeterminate system and in particular, for this primary system. The following values of the geometrical dimensions and characteristics of a bridge structure were taken, then calculating:

$$l = 120m;\ f = 18m;\ l_2 = 8m;\ I_{Arch} = 0.0012745m^4;\ I_{Beam} = 0.04356m^4$$
$$F_{Arch} = 0.05276m^2;\ F_{Beam} = 0.13373m^2.$$

The results of static analysis and calculations of a bridge, including the values of unit displacements, are given in the research works of L.G. Petrosian [7–9]. The equations of motion (16.44) can be solved both analytically and through using numerical integration with the help of computer software. The decomposition in the natural modes of oscillation is most frequently used in analytical solutions. The knowledge of the fundamental natural frequency is also necessary for numerical integration. Thus, it is necessary to determine natural frequencies and modes of oscillations. The equations of free oscillations of a system are obtained, if we consider that in (16.44) $X_A = X_B = 0$; $\alpha = 0$, and that: $X_j = \overline{X}_j \sin \omega t$; $Y_j = \overline{Y}_j \sin \omega t$. The natural frequencies ω are determined from the frequency equation:

$$\left| A - \frac{1}{\omega^2} E \right| = 0, \tag{16.46}$$

where

$A = \|m_i \delta_{ij}\|$—square matrix of unit displacements;
E—unit matrix.

The coefficients of the modes of oscillations $\overline{X}_j, \overline{Y}_j$, are determined by the solution of the system of the equations:

$$\begin{gathered} -\overline{X}_j + \sum_{i=1}^{k} m_i \omega^2 X_j \delta_{ji}^{(xx)} + \sum_{i=1}^{k} m_i \omega^2 Y_j \delta_{ji}^{(xy)} = 0; \\ -\overline{Y}_j + \sum_{i=1}^{k} m_i \omega^2 X_j \delta_{ji}^{(yx)} + \sum_{i=1}^{k} m_i \omega^2 Y_j \delta_{ji}^{(yy)} = 0, \\ (j = 1, 2, \ldots k) \end{gathered} \tag{16.47}$$

in which one of the amplitudes of oscillations is taken as the given one, for example, equal to "1" (one), and the system is solved with $(2n - 1)$ equations. For determining ω_i, X_i, Y_i, it is necessary to have values of the concentrated masses m_i. These values, calculated according to the working drawings of the bridge and taken with the corresponding reduction coefficients, proved to be equal to:

$$m_1 = m_3 = m_4 = m_6 = 15.1\ \frac{kNs^2}{m};\ m_2 = m_5 = 35.3 \frac{kNs^2}{m}.$$

In this case, most significant is the mass of the rigid beam, which is taken with the reduction coefficient, equal to 0.5 A.P. Philipov [15]. The values of the first six periods of oscillations for the given bridge, obtained by the solution of frequency Eq. (16.46), are:

$$\begin{array}{lll} T_1 = 0.650s.; & T_2 = 0.608s.; & T_3 = 0.261s.; \\ T_4 = 0.242s.; & T_5 = 0.104s.; & T_6 = 0.092s. \end{array}$$

The values of the coefficients of the modes of oscillations of the bridge are given in the Table 16.6.

The results of dynamic analysis, including the geometrical construction of the first six modes of free oscillations which correspond to the symmetrical and inversely symmetrical modes of vertical oscillations, are given in the research works of L.G. Petrosian [9, 40]. The determination of the reactions of the extensive bridge under seismic influences was realized by the numerical solution of

Table 16.6 The values of the coefficients of the modes of free oscillations of the bridge

Modes # Modes Coefficients	1	2	3	4	5	6
X_1	0.892	0.082	−0.074	0.568	−0.038	0.408
X_2	1.000	0	0	1	0	1
X_3	0.901	−0.083	0.075	0.567	0.038	0.409
Y_1	0.129	−0.824	−1.430	0.081	−1.261	0.062
Y_2	0	−1.000	1	0	1	0
Y_3	−0.130	−0.824	1.430	−0.082	1.259	−0.062

differential equations of motion (16.47). For solving the system of Eq. (16.47), with arbitrary law of ground motion in accordance with a seismogram of the $X_A(t)$, $X_B(t)$, (the program of calculation in the algorithmic language FORTRAN –IV was developed prior to using this program), the system of the equations is reduced to its normal form. For this purpose, it is necessary to first solve the Eq. (16.47) relative to the second derivatives. For simplicity in this case, designating $Y_j(t) = X_{j+3}(t),\ j = 1, 2, 3,$ we convert the Eq. (16.47) to the form:

$$X'_{k+6} = X_{k+6};\ k = 1, 2, \ldots, 6;$$
$$X'_{k+6} = -\frac{1}{m_k}\left\{\sum_{j=1}^{6} C_{kj}\left[X_j + \frac{\alpha}{\omega_1}(X_j - X_A)'\right] - d_{kA}X_A - d_{kB}X_B\right\} \tag{16.48}$$
$$k = 1, 2, \ldots, 6.$$

Where the following designations are accepted:

$$C_{kj} = (-1)^{j+k}\frac{M_{jk}}{\Delta};$$

$\Delta = \|\delta_{kj}\|_1^6$ – matrix of unit displacements;
M_{jk}– minor of element δ_{kj} of determinant $|\Delta|$;

$$d_{kA} = \frac{1}{\Delta}\sum_{i=1}^{6}\delta_{iA}(-1)^{i+k}\cdot M_{ik};$$
$$d_{kB} = \frac{1}{\Delta}\sum_{i=1}^{6}\delta_{iB}(-1)^{i+k}\cdot M_{ik},$$
$$k = 1, 2, \ldots, 6.$$

The program provides for the determination of horizontal and vertical accelerations of concentrated masses, i.e., the values of X'_{k+6}; $k = 1, 2, \ldots, 6$, and also the relative displacements of masses—($X_k - \delta_{kA}X_k - \delta_{kB}X_{kB}$); $k = 1, 2, \ldots, 6$.

The Eq. (16.48) were solved with harmonic oscillations of the ground with parameters corresponding to the real seismic influence. The predominant periods of oscillations of the ground during strong earthquakes most frequently vary with the limits: $0.1s. \div 0.4s.$ [31]. The laws of the motion of bridge supports were taken in the form of:

$$X_A = A\sin\omega_0 t;\ \ X_B = A\sin\omega_0\left(t - \frac{l}{v_0}\right)$$

I—option

$$\omega_0 = 31.41s^{-1};\ \ T_0 = \frac{2\pi}{\omega_0} = 0.2s;\ \ A = 0.031m,$$

II—option

$$\omega_0 = 15.70s^{-1};\ \ T_0 = \frac{2\pi}{\omega_0} = 0.4s;\ \ A = 0.031m.$$

Three types of ground were examined with the values of velocities of propagation of the surface waves, equaling to: 152.5*m*/*s*.; 305*m*/*s*.; 610*m*/*s*. [42] respectively. The taken amplitude value of acceleration of the seismic influence approximately corresponds to the maximum amplitude values of the seismograms of real strong earthquakes. The values of velocities 305*m*/*s*., and 610*m*/*s*., correspond to the velocities of surface waves in sandy loams, loams, and clay soil. The ground soil conditions in the canyon of River Arpa are approximately similar. For comparison, the softest soil and filled ground for which $v = 152.5m/s$. were also considered. The decrement of the oscillations of steel bridges, obtained by us according to the results of field full-scale tests, was set equal to $\delta = 0.023$. The coefficient α, in the equations of motion, is equal to: $\alpha = \delta/\pi$. The results of dynamic analysis, including geometrical construction of the dependences of accelerations of concentrated masses on time, are given in the research work of L.G. Petrosian [7, 9]. The dependences of accelerations of concentrated masses on time are obtained ($T = 0.2$ *s*., $T = 0.4$ *s*.). The values of maximum accelerations with the first option of the external seismic excitation ($T = 0.2$ *s*.) are given in Table 16.7.

The values of maximum accelerations with the second option of external seismic excitation ($T = 0.4$ *s*.) are given in Table 16.8.

The given results show that the calculation of extent leads to a decrease of the horizontal inertial forces. With the first option ($T = 0.2s$.), the maximum horizontal accelerations, calculated taking into account the extent of the bridge construction, significantly decrease. In ground with the values of velocities of $v = 152, 5m/s$, $v = 305m/s$, and $v = 610m/s$, accelerations of X_2'' decrease by $\frac{12.00}{11.05} = 1.09$; $\frac{12.00}{9.94} = 1.21$, and $\frac{12.00}{8.97} = 1.34$ times, respectively. In the second option ($T = 0.4s$.), the influence of extent on the degree of the decrease of inertial forces is slightly more pronounced. This is possible to explain by the fact that in the second case fundamental frequency is close to the frequency of the oscillation of the base.

From the given data, another characteristic: regularity—increases in vertical accelerations of the bridge with an increase at asynchronous oscillations of the supports. However, the absolute value of

Table 16.7 Values of the maximum, horizontal and vertical accelerations, acting on the long-span steel bridge ($T = 0.2s$.)

l/v (s) (m/s^2) Accelerations	0	0.1967	0.3934	0.7868
X_1''	0.46	0.42	0.35	0.31
X_2''	12.00	11.05	9.94	8.97
X_3''	0.47	0.41	0.35	0.32
Y_1''	0.17	0.20	0.23	0.25
Y_2''	1.19	1.33	1.42	1.59
Y_3''	0.18	0.21	0.22	0.28

Table 16.8 Values of the maximum, horizontal and vertical accelerations, acting on the long-span steel bridge ($T = 0.4s$.)

l/v (s) (m/s^2) Accelerations	0	0.1967	0.3934	0.7868
X''_1	1.02	0.93	0.71	0.62
X''_2	18.10	16.07	11.98	8.04
X''_3	1.04	0.91	0.72	0.61
Y''_1	0.42	0.53	0.65	0.84
Y''_2	0.98	1.15	1.67	2.21
Y''_3	0.41	0.53	0.62	0.82

the amplitude of vertical acceleration, resulting from horizontal oscillations is significantly small compared to the amplitudes of horizontal accelerations.

Thus, the analysis showed that the inclusion of asynchronous oscillations of supports, when considering bridge spans and ground conditions, result in a reduction of maximum inertial loads, which increases the resistance of the structure to seismic excitation. Summarizing the results of the oscillations of a combined bridge structure rigid beam-flexible arch, taking into account its extent under seismic excitation, we come to the following conclusions:

(1) The design procedure of seismic calculations of long-span bridge structures taking into account their extent with the application of real seismograms of strong earthquakes is proposed.
(2) General equations of motion for the system with finite number of degrees of freedom at spatial deformation of constructions are built.
(3) Analysis of seismic reaction of bridges showed that taking into account the extent, in a ground of medium hardness, can lead to a decrease of inertia loads on average of $1.2 \div 1.5$ times.
(4) Vertical inertia loads, which appear with asynchronous horizontal oscillations of supports, are small and can be disregarded.

References

1. Smirnov, A. P. (1974). *Construction stability and oscillations* (p. 571). Transzheldorizdat.
2. Konashenko, S. I. (1953). *The free and forced oscillations of flexible arch with a stiffening beam* (Vol. 23, pp. 173–197). Dnepropetrovsk Institute of Railroads. Transzheldorizdat.
3. Smirnov, V. A. (1975). *Suspension bridges of large spans* (p. 369). Vishaya Shkola.
4. Kunnos, V. I. (1954). Free oscillations of combined systems. In *Problems of dynamics and dynamics strength* (Vol. 2, pp. 21–58). Academy of Science of Latvian SSR.
5. Kolesnik, I. A. (1977). *The oscillations of combined systems under the action of moving loads* (p. 149). Visha shkola.
6. Ilyasevich, S. A. (1964). *About the oscillations of steel bridges* (p. 135). Voenizdat.
7. Petrosian, L. G. (1978). *About the vibration of combine system* (Vol. XII,5, pp. 39–45). YPTU. Series of Technical Sciences.
8. Petrosian, L. G., & Rubanovich, S. G. (1979). The free oscillations of systems "Flexible Arch-Rigid Beam". *Izv. AS Arm. SSP, Ser. Tech. Sci. XXXII, 4*, 22–28.
9. Petrosian, L. G. (1980). *Research of some aspects of the stress-strain state of long-span structures—Structure mechanics—01.02.03 code* (p. 204p). Dissertation thesis. Technical Sciences (Ph. D).
10. Inglis, C.E. (1934). *Minutes of proceedings of the Institute of Civil Engineers*, 234, Paper 4870.
11. Bondar, N. G., Kazey, I. I., Lesokhin, B. F., & Kozmin, Y. G. (1965). *The dynamics of railroad bridges* (p. 410). Transportizdat.
12. Farrar, C. R., Duffey, P. J., & Cornwell, S. W. (1999). *Dynamic testing of bridge structures a review*. Los Alamos National Laboratory Report in Preparation.
13. Bolotin, V. V. (1965). *Statistical methods in the structural mechanics* (p. 279p). Stroyizdat.

14. Bolotin, V. V. (1976). *The dynamic stability of elastic systems* (p. 600). Gostekhteoretizdat (State Publishing House of Technical and Theoretical Literature).
15. Philipov, A. P. (1971). *The methods of calculation of construction to oscillations* (p. 226). Stoyisdat.
16. Kazey, I. I. (1960). *Dynamic calculation of spans of the railroad bridges* (p. 466). Transzheldorizdat.
17. Nazarov, A. G. (1965). *About the mechanical similarity of the solid bodies being deformed* (p. 214). Academy of Science Arm SSR.
18. Zakaryan, V. A. (1971). Experimental study of the dynamic characteristics of high-rise buildings. *Izv. AH Arm. SSR, Ser. Tech. Nauk. XXIV, 3*, 33–44.
19. Petrosian, L. G. (1979). The results of dynamic model tests of Dzhermuk's bridge. *Izv. AS Arm. CCP, Ser. Tech. Sci. XXXV, 3*, 28–34.
20. Kartsivadze, G. N. (1974). *The seismic stability Bar structures* (p. 264). Transport.
21. Goldenblat, I. I., Kartsivadze, G. N., Napetvaridze, S. G., & Nikolaenko, N. A. (1971). *The design of seismic resistance hydro-technical, transportation and some special structures* (p. 277p). Stroyizdat.
22. Tseitlin, A. I., Neustroev, E. A., & Petrosian, L. G. (1982). Seismic interaction of structures with the ground. In *Theses of reports of All–Union Conference and 3rd scientific conference of the Far-Eastern Section of MSSS* (pp. 95–97).
23. Hausner, G. W. (1957). Interaction of building and ground during an earthquake. *Bulletin of the Seismological Society of America, 47*, 179–186.
24. Korchinsky, I. L. (1962). The influence of the length of the building on the value of the seismic load. In *The seismic resistance of industrial buildings and civil constructions* (pp. 42–69). Gosstroyizdat.
25. Grossman, V. A. (1967). Oscillations of extended systems. In *The seismic stability of buildings and structures* (pp. 26–45). Stroyizdat.
26. Egupov, V. K., & Komandrina, T. A. (1969). *The calculation of buildings to seismic action* (p. 208). Budivelnik.
27. Napetvaridze, S. G. (1959). *Seismic resistance of hydraulic structures*. Gosstroyizdat.
28. Napetvaridze, S. G. (1963). The influence of the length of a structure on the magnitude of the seismic force. *AS Gr. SSR IX*, 213–217.
29. Sinitzin, A. P. (1961). *The influence of a running seismic wave on massive structures* (Vol. 17, pp. 87–99). The Institute of GeoPhysics.
30. Sinitsin, A. P. (1967). *The method of finite elements in the dynamic of structures*. Stroyizdat.
31. Khachiyan, E. E., Ambartsumyan, V. A., & Petrosian, L. G. (1977). *The determination of the reactions of the extended buildings and structures taking into account the initial phase of the passage of seismic wave*. Izvestia. AN Arm. SSR.
32. Barshtein, M. P. (1968). The oscillations of the extended constructions under seismic action. In *Structural mechanics and calculation of construction* (Vol. 6, pp. 30–36).
33. Amiraslanov, N. A., & Barshtein, M. P. (1970). The oscillations of the structures extended in the plane with the earthquake. In *Structural mechanics and calculation of construction* (Vol. 6, pp. 35–40).
34. Nikiporets, G. L. (1977). The spectra of the reactions of the extensive in the plan constructions under seismic actions. *Seismic Resistance, 2*, 5–10. Moscow.
35. Korchinsky, I. L., & Grill, A. A. (1969). The determination of seismic loads of the long-span cable-stayed bridges. In *Design of steel structures* (Vol. VII, pp. 20–42) Moscow.
36. Pukhovsky, A. B., & Gordin, V. A. (1977). *Seismic resisted arch structure covering spans of 62m., with pre- stressed steel panels* (Vol. 3, pp. 3–10). TsINIS/Earthquake Resistant Structures.
37. Khachiyan, E. E. (1985). *Recommendations—Guidelines for determining dynamic characteristics and seismic loads for buildings and structures using accelerograms of earthquakes* (p. 107p). ArmNIISA.
38. Kubo, K. (1960). The seismicity of suspension bridges forced to vibrate longitudinally. In *Proc. SWCEE* (Vol. 7). Kyoto.
39. Petrov, A.A., & Duzinkevich, M.S. (n.d.). *The determination of seismic loads on the steel arched bridge through the Arpa River (Armenia).*
40. Petrosian, L. G. (1984). *The influence of the length of bridge structures on the formation of seismic loads* (pp. 66–69). The Industry of Armenia.
41. Werner, S. D., Lee, L. C., Wong, L. H., & Trifunac, M. D. (1979). Factors involved in the seismic design of bridge abutments. In *Proceeding of a workshop on the earthquake resistance of highway bridges*. Applied Technology Council.
42. Skavutso, R., Bailey, J., & Raftopulos, D. (1971). Horizontal interaction of construction with seismic waves. *Transactions of the American Society of Mechanical Engineers Series E, Journal of applied mechanics, 1*, 123–130.

Vibration of Extended Frame Structures 17

Abstract

This chapter presents the determination of the reactions of the frame, taking into account the initial phase of passage of the seismic wave. This chapter discusses the vibrations of a multistory, multi-span frame structure of considerable extent in the plan. The dynamic design model is accepted as a system with finite number of degrees of freedom. The equations of seismic oscillations of an elastic frame system taking into account extent can be obtained on the basis of the equations of the displacement method with the static horizontal shifting of supports and taking into account the inertial forces in accordance with the Dalember's principle. In the equations of motions, it is assumed that the energy absorption of the first floor at asynchronous oscillations supports/columns is the same as with the synchronous oscillations. Equation of motion, Laplace-Carson transformation, and Equation of motion of a single-span frame, taking into account its length with arbitrary movement of the base, are discussed.

17.1 The Determination of the Reactions of the Extended Buildings and Frame Structures Taking into Account the Initial Phase of Passing (Propagation) of the Seismic Wave

According to the current methodology of calculating the impact of seismic excitation in buildings and other construction, all points of base oscillate synchronously with the same shift and acceleration. However, for buildings and structures, with a length in the direction of propagation of excitation commensurate to the length of seismic waves, this methodology can lead to significant errors when determining the reactions of these systems. Long-span arched constructions of bridges present a complex design systems, which consist both of the rectilinear frame elements and of curvilinear arched elements. Seismic calculation of such constructions is associated with complex and labor-intensive computational operations. Therefore, in certain cases, it is expedient to replace curvilinear elements with rectilinear elements and, thus, simplify the design diagram (scheme) of a structure [1].

Let us examine the vibrations of a multistory, multi-span frame structure of considerable extent in the plan Fig. 17.1. In this particular case, it is possible to obtain the equations of motion of trestle frames from the derivation of the equation of motion. The methodology of derivation of the equations of motion can be extended to frame-arched structures, in which rectilinear elements are substituted for curvilinear elements. The derived equations of motion can be used in the analysis of a stressed state of

L. G. Petrosian, *Dynamics of Structures*, https://doi.org/10.1007/978-3-031-63539-7_17

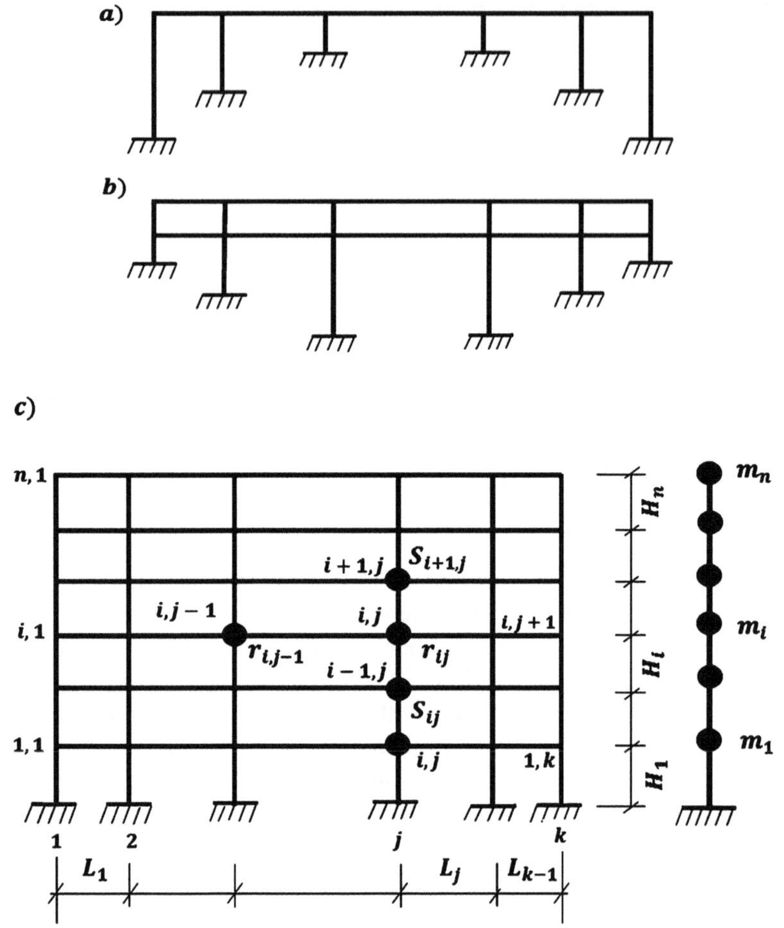

Fig. 17.1 Schemes of multi-span trestle frames (**a**) and (**b**); Dynamic design scheme of the multistory, multi-span frame structure (**c**)

an over-arch pass-through span structure of a frame-arch system of a bridge. In particular, it is possible to consider an over-arch pass-through span structure of one- or two-tier frame Fig. 17.1a, b as a special case of a multistoried, multi-span frame Fig. 17.1c.

The dynamic design model is accepted as a system with finite number of degrees of freedom. The equations of seismic oscillations of an elastic frame system taking into account extent can be obtained on the basis of the equations of the displacement method with the static horizontal shifting of supports and taking into account the inertial forces in accordance with the Dalember's principle.

Equation of Motion The equations obtained in this way for a multistory frame, where each support has a certain displacement, can be represented in the following common form:

$$
\begin{aligned}
&(1)\quad 2S_{ij}\varphi_{i-1}+2r_{i,j-1}\varphi_{i,j-1}+4\left(S_{ij}+r_{i,j-1}+r_{ij}+S_{i+1,j}\right)\varphi_{ij}+\\
&\quad +2r_{ij}\varphi_{i,j+1}+2S_{i+1,j}\varphi_{i+1,j}-6\frac{S_{ij}}{H_i}\left(y_i-y_{i-1}\right)-6\frac{S_{i+1,j}}{H_{i+1}}\left(y_{i+1}-y_i\right)=0;\\
&\quad i=2,3,\ldots,n;\ j=1,2,\ldots,k;\ r_{i0}=r_{ik}=S_{n+1,j}=y_{n+1}=0;\\
&(2)\quad 2r_{1,j-1}+4\left(S_{1j}+r_{1,j-1}+r_{1j}+S_{2j}\right)\varphi_{1j}+2r_{1j}\varphi_{1,j+1}+2S_{2j}\varphi_{2j}-\\
&\quad -6\frac{S_{1j}}{H_1}\left(y_1-y_{0j}\right)-6\frac{S_{2j}}{H_2}\left(y_2-y_1\right)=0;\\
&\quad j=1,2,\ldots,k;\ r_{1,0}=r_{1,k}=0\\
&(3)\quad 6\sum_{j=1}^{k}\frac{S_{ij}}{H_1}\left(\varphi_{ij}+\varphi_{i-1,j}\right)-12\sum_{j=1}^{k}\frac{S_{i,j}}{H_i^2}\left(y_i-y_{i-1}\right)+\sum_{\xi=i}^{n}\left(-m_\xi y_\xi''\right)=0\\
&\quad i=2,3,\ldots,n;\\
&(4)\quad 6\sum_{j=1}^{k}\frac{S_{1j}}{H_i}\varphi_{1j}-12\sum_{j=1}^{k}\frac{S_{1j}}{H_1^2}\left(y_1-y_{0j}\right)+\sum_{\xi=1}^{n}\left(-m_\xi y_\xi''\right)=0
\end{aligned}
\tag{17.1}
$$

where

$S_{ij}=\frac{EI_{ij}^{(s)}}{H_i}$—linear stiffness of jth column ith floor;
$r_{ij}=\frac{EI_{ij}^{(r)}}{L_i}$— linear stiffness of jth beam ith floor;
φ_{ij}— angle of rotation of the node with the coordinates i and j;
y_i— displacement of the ith floor;
E— modulus of elasticity of material;
$I_{ij}^{(s)}$, $I_{ij}^{(r)}$ are the moments of inertia, of jth column or beam ith floor respectively;
H_i— height of ith floor;
L_j—length of jth span;
m_i—mass concentrated at the level of the ith floor;
y_j''—acceleration of the mass of the ith floor;
n—numbers of floors;
k—numbers of floor columns;
y_{0j}—displacement of jth support.

The total number of Eq. (17.1) is equal to $kn + n$. Since the system has n degrees of freedom, then the number of unknowns, which determine the deformed state of the system, must be equal to n. If we solve the part of the system (17.1), representing the conditions of equality of moments in the nodes, with respect to $\varphi_{ij}(i = 1, 2, \ldots, n;\ \ j = 1, 2, \ldots, k)$ and substitute the dependences obtained $\varphi_{ij} = f_{ij}y(-y_i)$, $(i = 1, 2, \ldots, n;\ \ j = 1, 2, \ldots, k)$ into the rest of the equations, then we obtain n equations of motion.

Equation of Motion The equations of motion of a multi-span, one-story (single-tier) frame are a special case of the general Eq. (17.1), and it is possible to present them in the following form:

$$
\begin{aligned}
&4(S_1+r_1)\varphi_1+2r_1\varphi_2-\frac{6S_1}{H}(y-y_{01})=0;\\
&2r_1\varphi_1+4(r_1+r_2+S_2)\varphi_2+-\frac{6S_2}{H}(y-y_{02})=0;\\
&2r_{j-1}\varphi_{j-1}+4(r_{j-1}+S_j)\varphi_j+-\frac{6S_j}{H}(y-y_{0j})=0;\\
&-my''+\sum_{j=1}^{k}\frac{6S_j}{H}\varphi_j-\sum_{j=1}^{k}\frac{12S_j}{H^2}(y-y_{0j})=0,
\end{aligned}
\tag{17.2}
$$

with

$$
t\geq t_{k-1}=\sum_{q=1}^{k-1}\frac{L_q}{v_0}.
$$

The equations of motion of the frame (17.1) are built for a case when each support has a certain displacement y_{0j}. However, a more detailed analysis of the oscillations of the system should also consider the initial moment of passage of seismic waves [2]. With the propagation of excitation, for example, from left to right, the first support will move first, followed by the second support, and after a certain time, the remainder of the supports will be involved in the movement. Here it is assumed that the seismic excitation propagates with a constant velocity v_0, dependent on the soil/ground conditions. Hence, it follows that the displacements of individual supports of a structure have the following dependence:

$$
y_{0j}(t)=y_{01}\left(t-\sum_{q=1}^{j}\frac{L_q}{v_0}\right).
\tag{17.3}
$$

During motion of only the first support, the behavior of the system is described by the system of Eq. (17.1), in which the remaining displacements of supports are taken as equal to zero, i.e., $y_{0j}=0$; $j>1$. The solution of this system is correct with $0<t<L_1/v_0$. With $L_1/v_0<t<(L_1+L_2)/v_0$ when the second support moves, the behavior of the system is described by the system of Eq. (17.1), in which $y_{0j}=0$; for $j>2$. Finally, when all supports are involved in the motion, i.e., with.

$$
t>\sum_{q=1}^{k-1}\frac{L_q}{v_0},
$$

the equations of motions will take the form (17.1), where all displacements of supports y_{0j} are already different from zero. Thus, at the initial moment of the passage of the seismic wave from the first support to the last support, the motion of the system will be described by k different systems of equations. With $0<t<L_1/v_0$ the system of equations of motion must be solved with the initial conditions set to zero. With $L_1/v_0<t<(L_1+L_2)/v_0$ the initial parameters are those values of deformations and velocities which came out during the solution of the first system of equations at point $t=L_1/v_0$. Similarly, the remaining systems of equations must be solved. The solution of a system of equations of motion with a larger number of supports increases the volume of computational operations. However, as will be shown below, with the determination of the reactions of multi-span frames satisfactory results can be obtained, if with the initial passage of a seismic wave we take into account only outer supports, i.e., we accept that the motion of the system is described only by two different systems of equations with.

$$0 < t < \sum_{q=1}^{k-1} \frac{L_q}{v_0}, \quad and \quad t > \sum_{q=1}^{k-1} \frac{L_q}{v_0}.$$

In the case when the stiffness of a superstructure is several times greater than the stiffness of supports (columns, pears), it is possible to assume that the superstructure is not deformed. The equations of motion for this system will take a comparatively simple form. Assuming in (17.1) $\varphi_{ij} = 0$, we obtain:

$$\begin{gathered} \sum_{\xi=i}^{n} m_\xi y''_\xi + 12 \sum_{j=1}^{k} \frac{S_{ij}}{H_i^2}(y_i - y_{i-1}) = 0 \\ (i = 2, 3, \ldots, n) \\ \sum_{\xi=1}^{n} m_\xi y''_\xi + 12 \sum_{j=1}^{k} \frac{S_{1j}}{H_1^2}(y_1 - y_{0j}) = 0. \end{gathered} \tag{17.4}$$

By simple algebraic conversions, we have:

$$\begin{gathered} m_i y''_i - a_{i+1}(y_{i+1} - y_i) + a_i(y_i - y_{i+1}) = 0 \\ (i = 2, 3, \ldots, n) \\ m_1 y''_1 - a_2(y_2 - y_1) + \sum_{j=1}^{k} a_{1j}(y_1 - y_{0j}) = 0 \end{gathered} \tag{17.5}$$

where,

$$a_i = \sum_{j=1}^{k} \frac{S_{ij}}{H_i^2}, \qquad (i = 1, 2, \ldots, n).$$

It is sometimes more convenient in the Eq. (17.5) to substitute the variables. If we take into account the dumping per the equivalent hypothesis of Voigt and designate $\overline{y}_i = y_i - y_{01}$, then it is possible to obtain:

$$\begin{aligned} & m_i(\overline{y}_i + y_{01})'' - a_{i+1}(\overline{y}_{i+1} - \overline{y}_i) - \frac{a_{i+1}}{\omega}\alpha_{i+1}(\overline{y}_{i+1} - \overline{y}_i)' + \\ & + a_i(\overline{y}_i - \overline{y}_{i-1}) + \frac{a_i}{\omega}\alpha_i(\overline{y}_i - \overline{y}_{i-1})' = 0; \\ & m_1(\overline{y}_1 + y_{01})'' - a_2(\overline{y}_2 - \overline{y}_1) - \frac{a_2}{\omega}\alpha_2(\overline{y}_2 - \overline{y}_1)' + \\ & + \sum_{j=1}^{k} a_{1j}(\overline{y}_1 + y_{01} - y_{0j}) + a_1\frac{\alpha_1}{\omega}\overline{y}'_1 = 0, \end{aligned} \tag{17.6}$$

where,

α_k—coefficient of damping of k th floor (story, tier); ω—the fundamental frequency (the frequency of the main tone) of free oscillations.

In the Eq. (17.6), it is assumed that the energy absorption of the first floor at asynchronous oscillations supports/columns is the same as with the synchronous oscillations. This assumption leads to an unessential change in damping of the entire system. However, a significant simplification of the problem with the numerical solution of the equations of motion, when the seismogram or accelerogram of earthquake is given, is obtained. Thus, the equation of motion solves both the absolute displacements y_i and relative displacements $\bar{y}_i$. In the first case, we use an accelerogram, and in the second case, a seismogram and accelerogram of earthquake. From (17.6) with $y_{01} = y_{oj} = y_0$ the equations of motion with the synchronous oscillations of supports are obtained [3].

In the solution of the Eq. (17.6), we search for in the form of:

$$y_i = \sum_{r=1}^{\infty} C_{ir} q_r(t),$$

where.

$q_r(t)$—generalized coordinate; C_{ir}—coefficient, which characterizes the amplitude of free oscillations, of *i*th mass with *r*th mode of oscillations. Substituting the value of y_i into (17.6), and using the orthogonality of the modes of oscillations C_{ir}, we obtain a differential equation relative to generalized coordinate $q_r(t)$. Solving this differential equation, we first find $q_r(t)$, and then $y_i(t)$. The seismic force S_i is found using the expression:

$$S_i = m_i (\bar{y}_i + y_{01})''.$$

For simplicity, the countdown of time starts for each interval equal to zero. Finally, we will obtain that in the interval:

$$\sum_{q=1}^{j} \frac{L_q}{v_0} < t_j < \sum_{q=1}^{j+1} \frac{L_q}{v_0},$$

where *j*—is the number of oscillating supports; and the seismic force of *i*th floor/tiers S_i is determined by the expression:

$$\begin{aligned} S_i = & -m_i \sum_{r=1}^{n} e^{-\frac{\alpha_r}{2}\omega_r t} \left(\omega_r C_{ir} \frac{|C_{ir}|^*}{|C_{ir}|} \sin \omega_r t + \omega_r^2 \frac{C_{ir} |C_{ir}|^{**}}{|C_{ir}|} \cos \omega_r t \right) + \\ & + m_i \sum_{r=1}^{n} \eta_{ir} \omega_r \int_0^t y_0''(\xi) e^{-\frac{\alpha_r}{2}\omega_r (t-\xi)} \cdot \sin \omega_r (t-\xi) d\xi + \\ & + m_i \sum_{r=1}^{n} \bar{\eta}_{ir} \omega_r \sum_{q=1}^{j} a_{iq} \int_0^t \left[y_{0q}(\xi) - y_{01}(\xi) \right] \cdot e^{-\frac{\alpha_r}{2}\omega_r (t-\xi)} \cdot \sin \omega_r (t-\xi) d\xi, \end{aligned} \tag{17.7}$$

where C_{ir}—normalized values of displacements of *i*th floor/tiers with the free oscillations of rth mode; η_{ir}—coefficient of modes of oscillations taken by building codes:

$$\eta_{ir} = \frac{C_{ir} \sum_{i=1}^{n} m_i C_{ir}}{\sum_{i=1}^{n} m_i C_{ir}^2}; \quad \bar{\eta}_{ir} = C_{ir} \frac{C_{ir}}{\sum_{i=1}^{n} m_i C_{ir}^2}$$

$|C_{ir}|$—determinant of the corresponding coefficients (i, r = 1, 2, ..., n); and $|C_{ir}|^*$, $|C_{ir}|^{**}$—determinants in which the *r*th column is replaced with the numbers $y_1(0), \ldots y_n(0)$; $y_1'(0), \ldots y_n'(0)$ respectively. With the synchronous oscillations of supports from (17.7), we obtained the well-known formula for determining seismic forces. As was noted above, the equations of motion of the extended construction can be set up by two methods. In one case, the seismogram enters into the equations of motion, while in the other, both the seismogram and accelerogram of the earthquake enter into the equations of motion. During strong earthquakes, the most frequently obtained records of oscillations of ground instrumentally are in the form of accelerograms. In order to use them with a numerical solution of the Eq. (17.7), it is necessary to integrate them twice to obtain the corresponding seismogram, which is not always possible to successfully obtain because of the large number of errors introduced by the recording mechanism. Therefore, the question arises: is it possible to set up the equations of motion in such a way as to use only an accelerogram to obtain the solution? Below, the opportunity of the accomplishment of the task is analyzed based on the simplest example of a system with a single degree of freedom. The equations of motion, obtained from (17.2) and (17.4) in this case will be:

$$y'' + \omega^2 y = f(t);$$
$$f(t) = \begin{cases} \frac{1}{2}\omega^2 y_0(t) & 0 < t < t_1 = \frac{L}{v_0}; \\ \frac{1}{2}\omega^2 y_0(t) + \frac{\omega^2}{2} y_0(t - t_1)\sigma_0(t - t_1) & t > t_1, \end{cases} \tag{17.8}$$

where $\sigma_0(t - t_1)$—unit retarded (late) function [217]. The solution of the Eq. (17.8) with $t > t_1$ takes the form:

$$y(t) = \frac{\omega}{2} \int_0^l y_0(\xi) \sin \omega(t - \xi) d\xi + \frac{\omega}{2} \int_{t_1}^l y_0(\xi - t_1) \sin \omega(t - \xi) d\xi. \tag{17.9}$$

In order to use the accelerogram of real earthquakes, we will differentiate the Eq. (17.8) twice, and obtain:

$$y^{IV} + \omega^2 y'' = f''(t);$$
$$f''(t) = \begin{cases} \frac{1}{2}\omega^2 y_0''(t); \\ \frac{1}{2}\omega^2 y_0''(t) + \frac{\omega^2}{2} y_0''(t - t_1)\sigma_0(t - t_1) \end{cases} \tag{17.10}$$

The obtained differential equation of the fourth degree is generally not equivalent to the initial Eq. (17.8). For its solution, it is necessary to have four initial conditions. Two "extra" initial conditions should be selected so that the solutions of the Eqs. (17.8) and (17.10) match.

Laplace-Carson Transformation Let us solve the Eq. (17.10) using the Laplace-Carson transformation. The equation relative to images with $t > t_1$ takes the form:

$$Y(p) = \frac{\omega^2 y_0''(p)}{2p^2(p^2 + \omega^2)} + \frac{\omega^2 y_0''}{2p^2(p^2 + \omega^2)} e^{-pt_1} + \frac{p^2 y''(0) + p y'''(0)}{p^2(p^2 + \omega^2)}. \tag{17.11}$$

Passing to the initial functions, we will obtain:

$$\begin{aligned} y(t) = {} & \frac{1}{2} \int_0^t y_0''(\xi)(t - \xi) d\xi + \frac{1}{2} \int_{t_1}^t y_0''(\xi - t_1)(t - \xi) d\xi - \\ & - \frac{1}{2\omega} \int_0^t y''(\xi) \sin \omega(t - \xi) d\xi - \frac{1}{2\omega} \int_{t_1}^t y''(\xi - t_1) \sin \omega(t - \xi) d\xi + \\ & + \frac{y''(0)}{\omega^2}(1 - \cos \omega t) + \frac{y'''(0)}{\omega^2}\left(t - \frac{\sin \omega t}{\omega}\right). \end{aligned} \tag{17.12}$$

As can be seen from (17.10), the obtained solution of the Eq. (17.10) depends thus far on the yet unknown values of the second and third derivative and on the displacement at initial point, i.e., of $y_0''(t)$ and $y_0'''(t)$. Two other initial conditions with the derivation of the formula (17.12) were assumed equal to zero, i.e., $y(0) = y'(0) = 0$. Integrating by parts twice the integrals in (17.12) and comparing the obtained expression $y(t)$ with (17.9), we can see that both solutions are the same, if the condition is satisfied:

$$\begin{aligned} &\frac{y_0(0)}{2}\{[1-\cos\omega t]+[1-\cos\omega(t-t_1]\}=\frac{y''(0)}{\omega^2}(1-\cos\omega t);\\ &\frac{y_0'(0)}{2}\left\{\left[1-\frac{\sin\omega t}{\omega}\right]+\left[(t-t_1)-\frac{\sin\omega(t-t_1)}{\omega}\right]\right\}=\frac{y'''(0)}{\omega^2}\left(t-\frac{\sin\omega t}{\omega}\right). \end{aligned} \tag{17.13}$$

Substituting the obtained from (17.13) value of $y''(0)$ and $y'''(0)$ into (17.12) we finally obtain the expression $y(t)$ with $t > t_1$:

$$\begin{aligned} y(t)=&\frac{1}{2}\int_0^t y_0''(\xi)(t-\xi)d\xi-\frac{1}{2\omega}\int_0^t y_0''(\xi)\sin\omega(t-\xi)d\xi+\\ &+\frac{1}{2}\int_{t_1}^t y_0''(\xi-t_1)(t-\xi)d\xi-\frac{1}{2\omega}\int_{t_1}^t y_0''(\xi-t_1)\sin\omega(t-\xi)d\xi+\\ &+\frac{y_0(0)}{2}\{[1-\cos\omega t]+[1-\cos\omega(t-t_1)]\}+\\ &+\frac{y_0'(0)}{2}\left\{\left[t-\frac{\sin\omega t}{\omega}\right]+\left[(t-t_1)-\frac{\sin\omega(t-t_1)}{\omega}\right]\right\}. \end{aligned} \tag{17.14}$$

With $0 < t < t_1$, $y(t)$ it is determined by the expression:

$$\begin{aligned} y(t)=&\frac{1}{2}\int_0^t y_0''(\xi)(t-\xi)d\xi-\frac{1}{2\omega}\int_0^t y_0''(\xi)\sin\omega(t-\xi)d\xi+\\ &+\frac{y_0(0)}{2}(1-\cos\omega t)+\frac{y_0'(0)}{2}\left(t-\frac{\sin\omega t}{\omega}\right). \end{aligned} \tag{17.15}$$

The seismic force will be equal to $my''(t)$. As can be seen from the expressions (17.14) and (17.15), for determining the reactions of a system with a given accelerogram, it is necessary to know only the initial values, corresponding seismograms, and peak ground velocity (PGV), which can be determined by the approximation of the initial section of an accelerogram by an analytic function and by numerical integration in that section only.

17.2 Equations of Motion of a Single-Span Frame, Taking into Account Its Length with Arbitrary Movement of the Base

When determining seismic forces acting on structures of significant length, of bridges, trestles, single-span industrial buildings and structures, overpasses, and so forth, it is necessary to take into account the influence of extent on the values of the reactions of these systems. Thorough analysis of the influence of extent must be executed by integration of the precise equations of the motion of these systems

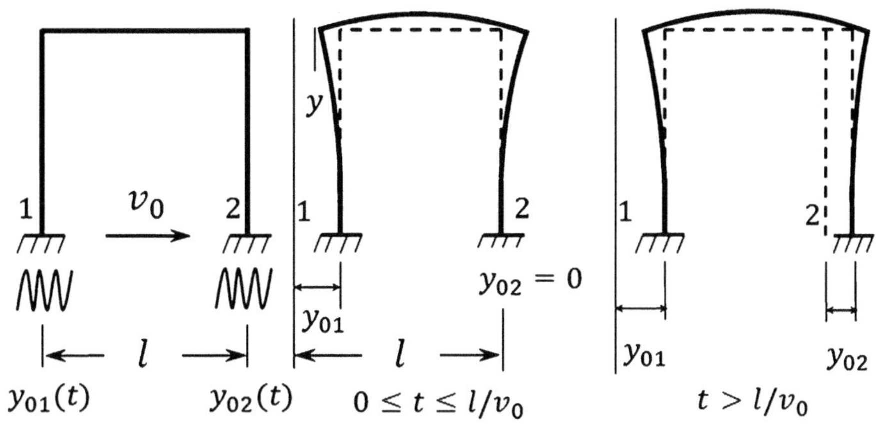

Fig. 17.2 Scheme of motion an extended one-story, singe-span frame

during the motion of the base according to the law close to the actual seismic excitation. In this paragraph, the derivation of the equations of motion of one-story, one-deck frame structures, taking into account their extent with the arbitrary ground motion, is given (Fig. 17.2). For the derivation of these equations, the method of deformation (displacement method) is used, which in this case leads to significant simplifications. With the general calculations of construction during seismic excitation as the design dynamic model, a cantilever bar (rod) with the concentrated masses is taken. Let us examine horizontal oscillations of a single-span frame, where a system with a single degree of freedom is taken as the dynamic design scheme. It is assumed that the law of motion of base and the values of the velocity of propagation of seismic excitation over the surface of ground is given. Let us build the equations of motion of frame in the case when the seismogram of ground $y_0(t)$ and the velocity of the motion of wave v are given. It is more convenient to obtain the equation of motion of a frame using the appropriate static equations of equilibrium with the motion of supports with the addition of inertial terms. With the displacement of the left support on, the equation of motion will take the following form:

$$\begin{gathered}4(S_1+r_1)\varphi_1+2r_1\varphi_2-\frac{6S_1}{H}(y-y_{01})=0;\\ 2r_1\varphi_1+4(r_1+S_2)\varphi_2-\frac{6S_2}{H}y=0;\\ -my''+\frac{6S_1}{H}\varphi_1+\frac{6S_2}{H}\varphi_2-\frac{12S_1}{H^2}(y-y_{01})-\frac{12S_2}{H^2}y=0,\end{gathered} \tag{17.16}$$

where

S_1, S_2—linear stiffness of column/support of frame;
r_1—linear stiffness of beam;
φ_1, φ_2—angles of rotation of the frame nodes;
m—mass concentrated at the level of the first floor;
y, y''—displacement and acceleration of the concentrated mass;
y_{01}—assigned displacement of right column (support);
H—height of the frame;
L—length of span frame;

In Fig. 17.2 scheme of motion on extended one-story, single-span frame is shown.

The Eq. (17.16) is valid with $0 < t < t_1 = l/v$, where l—is the distance between the supports. The Eq. (17.16) is more convenient to modify and to reduce to one differential equation relative to y. For this of the first two equations of the system (17.16) we determine φ_1 and φ_2, and substituting into the third equation we find:

$$my'' + y(a_1 + a_2) - y_{01}a_2 = 0, \tag{17.17}$$

where

$$\begin{aligned} a_1 &= \frac{\left(12S_1S_2^2 + 12S_2^2r_1 + 66S_1S_2r_1 + 36S_2r_1^2\right)}{H^2\left(4S_1r_1 + 4S_1S_2 + 3r_1^2 + 4r_1S_2\right)}; \\ a_2 &= \frac{\left(12S_1^2r_1 + 12S_1^2S_2 + 66S_1S_2r_1 + 36S_1r_1^2\right)}{H^2\left(4S_1r_1 + 4S_1S_2 + 3r_1^2 + 4r_1S_2\right)}. \end{aligned} \tag{17.18}$$

During the motion of both supports on y_{01} and on y_{02}, respectively, the equations of motion will take the form:

$$\begin{aligned} &4(S_1 + r_1)\varphi_1 + 2r_1\varphi_2 - \frac{6S_1}{H}(y - y_{01}) = 0; \\ &2r_1\varphi_1 + 4(r_1 + S_2)\varphi_2 - \frac{6S_2}{H}(y - y_{02}) = 0\ ; \\ -my'' + &\frac{6S_1}{H}\varphi_1 + \frac{6S_2}{H}\varphi_2 + \frac{12S_1}{H^2}(y - y_{01}) - \frac{12S_2}{H^2}(y - y_{02}) = 0, \end{aligned} \tag{17.19}$$

when at $t > t_1$. Excluding φ_1, and φ_2 from (17.18) we will obtain:

$$my'' + (a_1 + a_2)y - a_1y_{01} - a_2y_{02} = 0. \tag{17.20}$$

From (17.18), it is possible to obtain the values a_1 and a_2 with an absolutely rigid beam, see Fig. 17.3. After dividing numerator and denominator of the expression (17.18) by r^2 and then assuming that $r^2 = \infty$ we will obtain:

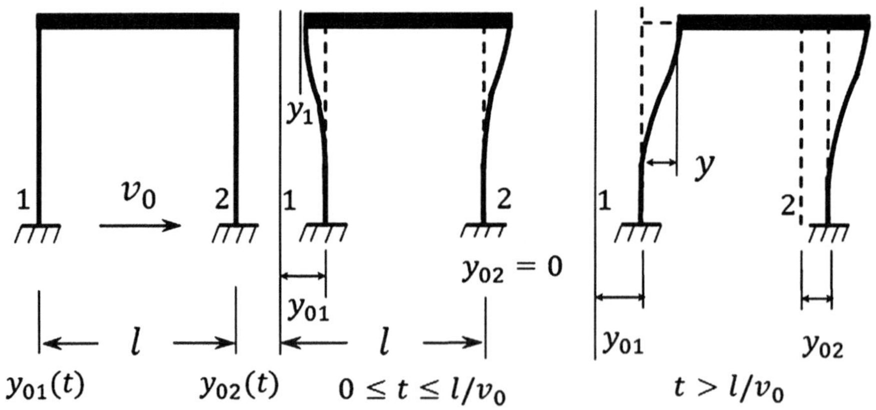

Fig. 17.3 Scheme of motion: an extended one-story, single-span frame with absolute rigid beam

$$a_1 = \frac{36S_2}{3H^2} = \frac{12S_2}{H^2}; \quad a_2 = \frac{36S_1}{3H^2} = \frac{12S_1}{H^2}, \tag{17.21}$$

where a_1, and a_2—stiffness of the first and second support/column, respectively. The Eqs. (17.17) and (17.19) describe the motion of the frame of a significant extent in the plan with the given seismogram of ground $y_0(t)$. If the motion of the left support occurs according to the law $y_{01}(t)$, then the motion of the right support can be taken as equal:

$$y_{02}(t) = y_{01}\left(t - \frac{l}{v}\right). \tag{17.22}$$

Here v represents the velocity of propagation (motion) of seismic wave over the surface of the ground. If there is a real measured value of velocity, then it is possible to use these data, but in general the value v can be accepted as equal to the velocity of propagation of transverse waves—the waves of shift (shifting).

In some cases, it is more convenient to replace variables in the Eq. (17.20). Designating $\bar{y} = y - y_{01}$, and taking into account the damping in Voigt's hypothesis (into somewhat modified type [2, 5], we will obtain:

$$m(\bar{y}_1 + y_{01})'' + a_1\bar{y} + a_2(\bar{y} + y_{01} - y_{02}) + a_1\frac{\alpha_1}{\omega}\bar{y}' = 0, \tag{17.23}$$

where

α_1—damping coefficient;
ω—the fundamental frequency of free oscillations.

In the Eq. (17.23), it is accepted that the energy absorption with the asynchronous oscillations of columns (supports) is the same as with the synchronous oscillations. This assumption leads to an insignificant change in damping of the entire system, and at the same time, the problem is simplified with the numerical solution of equation of motion, where assigned seismogram or accelerogram of earthquakes are given. Thus, the equations of motion can be solved relative to both the absolute displacements y_j, and displacements $\bar{y}_j$. In the first case, the seismogram is used, and in the second case, both the seismogram and the accelerogram of earthquake are used. The equation of motion with the synchronous oscillations of supports is obtained from (17.23) when $y_{01} = y_{0j} - y_0$.

For obtaining the spectrum of the seismic reactions, taking into account the extent of a structure, let us examine a single-span, one-story frame—a system with a single degree of freedom. The equation of motion, which is calculated for the relative displacement $\bar{y}$, for a frame having columns of equal stiffness, will be:

$$\bar{y}'' + \left(\frac{2\pi}{T}\right)^2\bar{y} + \frac{2\pi}{T}\alpha\bar{y}' = y_0''(t) - \frac{1}{2}\left(\frac{2\pi}{T}\right)^2\left[y_0\left(t - \frac{L}{v_0}\right) - y_0(t)\right]. \tag{17.24}$$

This equation is valid with $t > L/v_0$ and is found at zero initial conditions. The Eq. (17.24) was integrated with the assigned law of the motion of the base:

$$y_0''(t) = 100 \sin 10\pi t. \tag{17.25}$$

The period and amplitude of the oscillation (vibration) of the base are respectively equal:

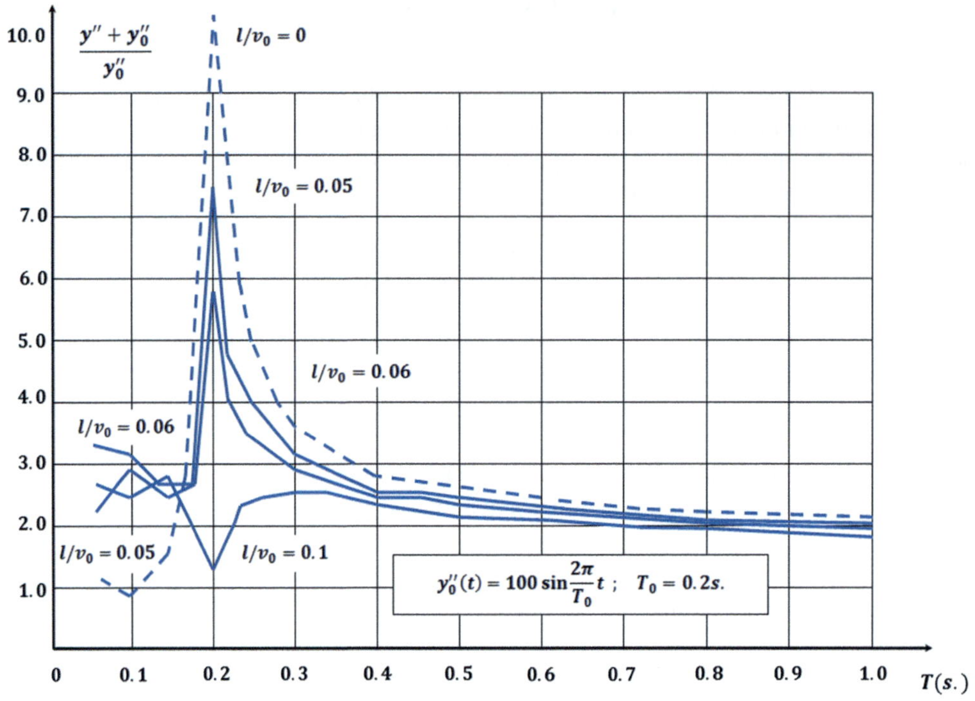

Fig. 17.4 Spectra of the accelerations of the extent structures with harmonic motion of base

$$T = 0.2\ s.; \qquad y''_{max} = 1\ m/s^2.$$

The parameters of oscillations of the base in the first approximation simulate the real accelerograms of strong earthquakes [3, 5]. The damping coefficient is accepted as equal $\alpha = 0.08$, which corresponds to reinforced concrete structures. The constructed spectrum of accelerations shows that accounting for the extent leads to a substantial change in the nature of the resonance curve, Fig. 17.4. For the periods $T < 0.2\ s.$, accounting for the extent increases the dynamic coefficient up to two and more times. For the periods $0.2\ s. < T < 0.4\ s.$, the dynamic coefficient decreases $1.5 \div 1.7$ times and with $T > 0.4\ s.$, the influence of the extent is not substantial. For some values L/v_0, the dynamic coefficient can be equal to one (1); Fig. 17.5. Hence it follows that with the harmonic oscillations of the ground for the given soil conditions, it is possible to choose a length of structure so that the phase shifting of supports would be equal:

$$\frac{L}{v_0} = \frac{T_0}{2} + kT_0, \qquad (k = 0, 1, 2, \ldots) \tag{17.26}$$

where T_0 is the predominant period of oscillation of the ground.

In this case, the construction will not undergo a dynamic impact, since the dynamic coefficient is approaching 1.

Using the formula (17.26) as a first approximation, it is possible to obtain the optimum value of the spans of construction for different types of ground/soil with the predominant periods of base oscillation. Table 17.1 provides optimal values of the extent spans of trestle frames or one-story, singe-span frames.

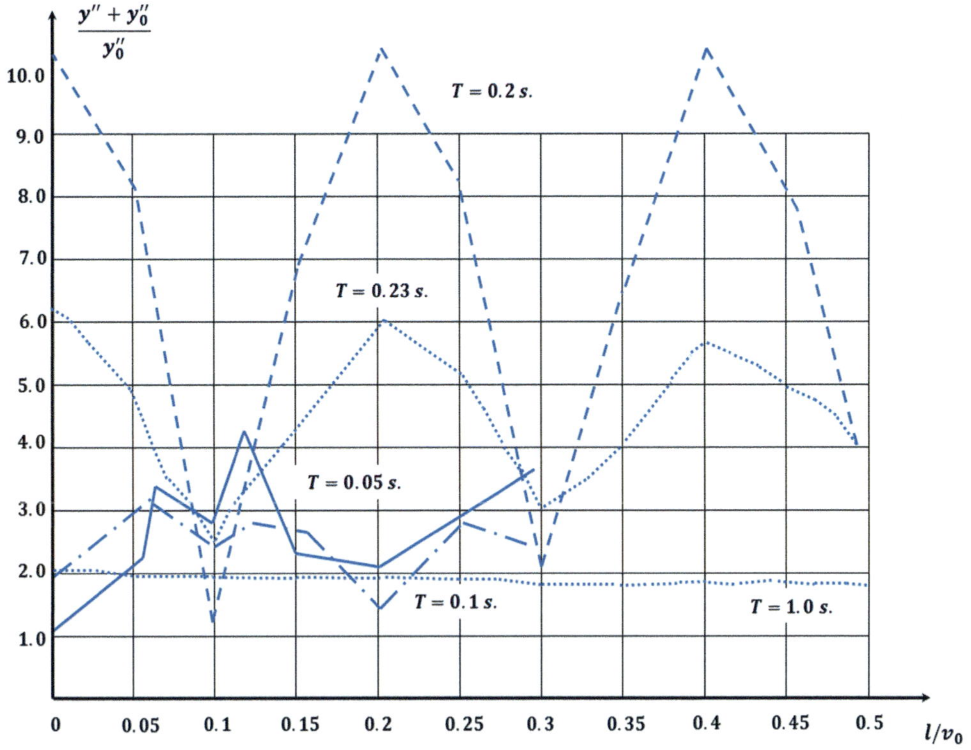

Fig. 17.5 Dependence of dynamic coefficient on the extent structures and ground conditions

Table 17.1 The optimum values of the span length

Period of ground oscillation T_0 (*s*.)		The Velocity of Propagation of the Seismic Waves v_0			
		100 (m/s)	200 (m/s)	500 (m/s)	1000 (m/s)
$T_0 = 0.1$	$k = 0$	5	10	25	50
	$k = 1$	15	30	75	150
	$k = 2$	25	50	125	250
$T_0 = 0.2$	$k = 0$	10	20	50	100
	$k = 1$	30	60	150	300
	$k = 2$	50	100	250	500
$T_0 = 0.3$	$k = 0$	15	30	75	150
	$k = 1$	45	90	225	450
	$k = 2$	75	150	375	750
$T_0 = 0.4$	$k = 0$	20	40	100	200
	$k = 1$	60	120	300	600
	$k = 2$	100	200	500	1000

From Table 17.1, it follows that in ground, with the velocity of propagation of the seismic waves of $v_0 = 200$ *m/s* with the predominant period of $T_0 = 0.2$ *s*., the optimum structure, is one whose length equals to 20 m, 60 m, and 100 m. The influence of the extent on the values of the reactions of multi-span one-story, one-deck frames was investigated. Also examined were one-, two-, three-, four-, and five-span frames, with the same length and the same ground/soil conditions. It shows that the number of intermediate supports/columns has little influence on the magnitude of the seismic forces and that

Table 17.2 Magnitude of seismic forces and number of intermediate supports

Period	The values $(y_0'' + \bar{y}'')_{max} (m/s^2)$ for single-tier, one-story frame with the numbers of supports (columns)				
(*s*.)	1	2	3	4	5
0.2	2.05	2.49	2.05	1.99	2.27
0.4	2.32	2.33	2.32	2.31	2.31
0.6	2.26	2.26	2.26	2.26	2.26
1.0	2.14	2.14	2.14	2.14	2.14

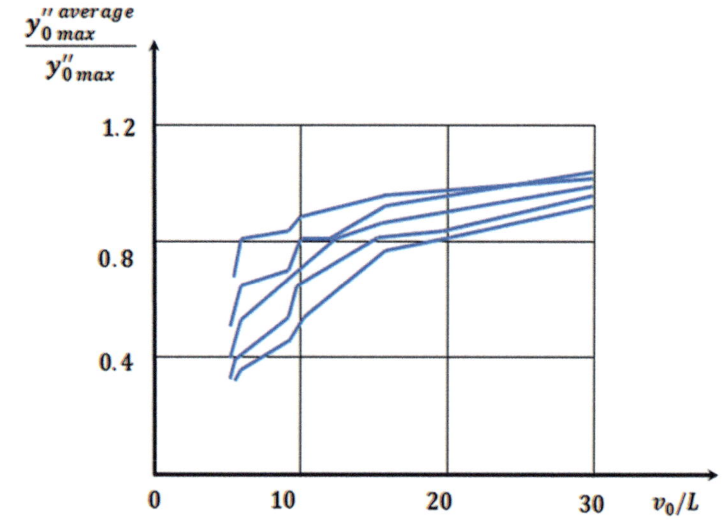

Fig. 17.6 Averaged values of the accelerations of the real accelerogram of five strong earthquakes

the essential factor is the overall length of the structure. The results of the calculations are given in Table 17.2.

In some cases, the problem of the calculation of the extent of a construction is possible to solve by averaging the acceleration of the ground and its extension to the entire base of construction, i.e., as the initial acceleration of the ground to the following expression is accepted [6, 7]:

$$y_0''^{(average)}(t) = \frac{v_0}{L} \int_0^{\frac{L}{v_0}} y_0''(t - \xi) d\xi. \tag{17.27}$$

Using the accelerograms of some earthquakes, we determined the values $y_0^{(average)}$(t) by numerical integration using the expression (17.27).

The averaged values of the accelerations of the real accelerogram of five strong earthquakes are shown in Fig. 17.6.

According to the obtained results in Fig. 17.7, the dependencies of the relations of the averaged maximum and actual maximum values of the accelerations value $y''^{(average)}_{0,max} / y''^{(average)}_{0,max}$ from the length of construction $-L$ with different values of the velocities of propagation of seismic wave v are built. Structures with lengths up to 100 *m* are considered, and grounds/soils for which $v_0 = 100 \div 1000$ *m*/*s*. Let us recall that the minimum speed corresponds to the softest soil (soft, broken shale) and bulk soils, and the maximum speed corresponds to the sedimentary rock and medium hard rock grounds. These

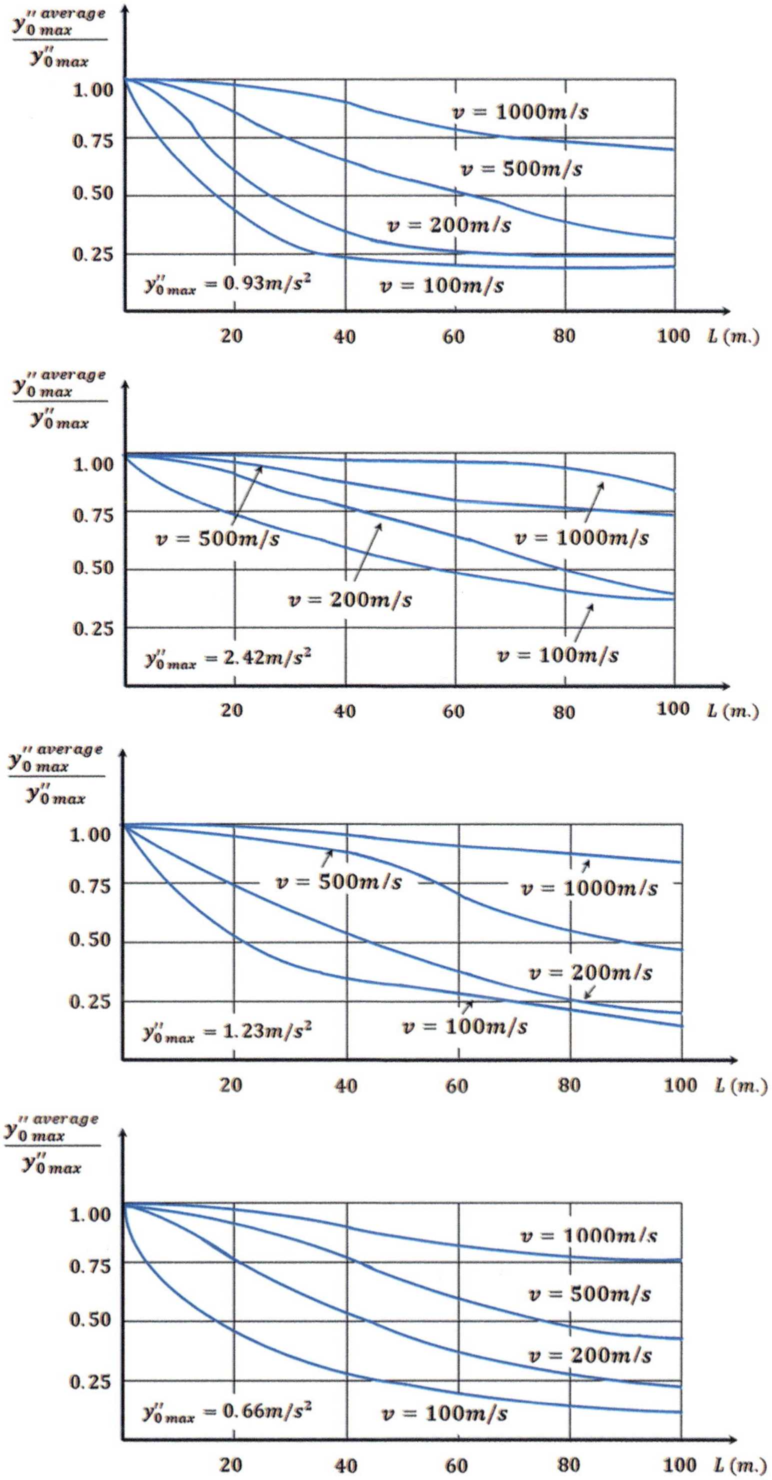

Fig. 17.7 Dependencies of relations of the averaged and maximum accelerations on the length of structure

dependencies are built for four accelerograms of actual earthquakes with maximum accelerations of 0.93 m/s^2, 1.23 m/s^2, 2.42 m/s^2, and 0.66 m/s^2. The characteristics of these accelerograms are given in [2, 3, 5]. The given results show that the influence of the extent is substantial for such construction and ground, for which $v_0/L < 10$.

Since the frequency characteristics of accelerograms change insignificantly with the averaging, the reactions of structures undergo a similar phenomenon.

Summarizing the results of the oscillations of multiple spans of frame bridges, extended buildings, and structures, taking into account their extent under seismic excitation, we come to the following conclusions:

(1) We derived the equations of motion of extent multi-tiered, multistoried structures with asynchronous oscillations of supports. From the general equations as a special case the equations of oscillations of trestle frames are obtained.
(2) The possible applications of the records of acceleration (accelerograms) of real earthquakes are analyzed for investigating the reactions of the extent system.
(3) Analytical values of the reactions of multi-mass systems, taking into account the final velocity of propagation of seismic waves, are obtained. The spectra of accelerations of systems in harmonic influence, having parameters corresponding to real seismic excitation are constructed.
(4) For the very stiff systems $T < 0.2$ *s*., the accounting of extent can increase the value of dynamic coefficient. For the periods $0.2\ s. < T < 0.4\ s.$, the dynamic coefficient decreases $1.5 \div 1.7$ times and for $T > 0.4$ *s*., the influence of the extent is not substantial.
(5) In the first approximation, we obtained the optimal ratio of length of the construction to the value of velocity of seismic wave propagation on the surface of the ground (Table 17.1).
(6) We investigated the possibility of taking into account the influence of the extent of a structure on its reaction by using the averaged acceleration of the ground. The given results show that the influence of the extent is substantial for such construction and ground, when $v_0/L < 10$.

References

1. Petrosian, L. G. (1982). The selection of the rational primary systems for frame calculations and designs. Methodical instructions for the course and diploma research projects design. In *Structural mechanics and calculation of structures* (58p).
2. Khachiyan, E.E., Ambartsumian, V.A., and Petrosian, L.G. (1977). The determination of the reactions of the extended buildings and structures taking into account the initial phase of the passage of seismic wave. *:* Izvestia. AN Arm. SSR 30 (4): 37–47.
3. Khachiyan, E. E. (1973). *Seismic influence on tower buildings and structures* (328p). Hayastan.
4. Lurie, A. I. (1950). *Operational calculus and its application to the problems of mechanics* (432p). Gostexisdat.
5. Petrosian, L. G. (1980). *Research into some aspects of the stress-strain state of long-span structures—Structure mechanics—01.02.03 code* (204p). Dissertation thesis. Technical Sciences (Ph. D).
6. Korchinsky, I. L., & Grill, A. A. (1969). The determination of seismic loads of the long-span cable-stayed bridges. In *Design of steel structures* (Vol. VII, pp. 20–42).
7. Petrov, A.A. & Duzinkevich, M.S. *The determination of seismic loads on the steel arched bridge through the Arpa River* (Armenia).

18 The Special Features of Solution of Dynamic Problems

Abstract

In this chapter describes the features of solving dynamic methods of boundary equations. It is known that most building materials have damping properties, characterized by the frequency's independent loss. However, there is a certain dependency of damping parameters (coefficients loss, decrement) on the frequency of the deformation for soils in the experiments. Frequency-dependent losses are typical for a number of modern materials, such as certain plastics. It is therefore a very important task to describe the dynamic behavior of soil and structural materials possessing both frequency-independent and frequency-dependent internal friction.

This chapter presents a new way to do the calculation of Hilbert transform through the Fourier transformation of corresponding functions, which simplifies the problem due to extensive Fourier transform tables. The presented method of calculating the Hilbert transform allows to comparatively and easily analyze the aprioristic models of the dynamic systems based on the creation of complex rigidity or complex pliability of system. This method makes it possible to analytically determine one of the components of the complex rigidity in accordance with the experimentally defined other component. A few examples of the calculation of the Hilbert transform of delta-functions, Heaviside function, and a power function are examined. The above-described method of calculating the Hilbert transform is applied to the analysis of dynamic models bases of internal friction, described by the hypothesis of Voigt, Sorokin, and Schlippe-Boc. This chapter is devoted to the construction of Green's function for an unlimited plate, lying on Winklerian type elasto-viscous base with an arbitrary dependence of parameters of the complex coefficient of bed on the frequency. In this chapter, the problem of construction of the Green's function for single-dimensional dynamic systems with damping, described by different model frequency-independent and frequency-dependent internal friction, is examined and solved.

18.1 Accounting for Internal Friction in Problems of Structures on an Elastic Foundation

The solution of dynamic problems using the BEM requires building of the Green's function in their spectral (frequency) or temporary representation (view). Accounting for real properties of materials of structures or bases, in particular the internal friction and dependences of elastic characteristics on the frequency perturbations is an extremely important question to be analyzed. Accounting for internal

L. G. Petrosian, *Dynamics of Structures*, https://doi.org/10.1007/978-3-031-63539-7_18

friction not only provides a more adequate description of the dynamic processes in the most important and dangerous resonance modes (regimes) of oscillations but also considerably facilitates the solution of dynamic problems since it eliminates the need for calculating improper integrals with the special features on the real axis during the construction of the corresponding Green's functions. It also eliminates the need to satisfy the additional conditions of emission or their analogs, which ensure the dispersion of energy at infinity. In the theory of structures on an elastic foundation, the dynamic problems, as a rule, invoke taking into account the internal friction in the material of a base. The damping properties in structural materials are expressed substantially less than in ground, and, therefore, they can be neglected. With bases of coherent type, the elastic isotropic half-space is usually used. The damping of oscillations due to internal friction is rarely considered because dissipation of energy in infinity ensures damping of free oscillations and the limitedness of amplitudes at steady processes [1]. With the use of the discrete bases such as Winkler-type of elastic support, the internal friction in the base in the majority of research is accounted for with the aid of the Voight model or the frequency-independent model of E. S. Sorokin. For example, the equation of stationary (steady-state) oscillations of a two-dimensional structure on an elastic foundation is written in the form of:

$$Lw - m\omega^2 w + k(\omega)w = q \tag{18.1}$$

where L—operator of the elastic forces of structure; m —is the mass per unit area; $w(x, y, t)$—vertical displacement; $q(x, y)$—the function of the amplitudes of external action; ω—circular frequency; $k(\omega)$—the complex coefficient of bed. In the case of the Voight's model:

$$k(\omega) = k_0 + ic, \tag{18.2}$$

where k_0—real common coefficient of bed, for a frequency independent model of E.S. Sorokin:

$$k(\omega) = k_0(a + ib), \tag{18.3}$$

where a, and b are parameters of the internal friction, introduced by Sorokin [2].

In recent years, numerous studies of oscillations of dissipative system were devoted mostly to the construction of different models of internal friction and their use in the dynamic design of structures. In the scientific literature and applications, the preference is given to the linear models, used in practical calculations and in scientific research, as they do not cause computational difficulties and allow to correctly estimate the influence of internal friction on the oscillating process. The hypothesis of Voight, elementary frequency-independent models (hypothesis E.S. Sorokin, frequency-independent, elastic-viscous resistance) and hereditary type models, most prevalently described by integral operators in the linear models, have gained the widest application. For construction materials, the most essential experimentally established fact influencing the selection of models of internal friction is the independence of frequency parameters from damping—coefficient of losses γ and logarithmic decrement δ in the common, for real actions, range of frequency of perturbances (impacts, influences). Therefore, in practical calculations the models of frequency-independent internal friction must be used. With the use of this widespread model, such as the Voight's hypothesis, which frequently occurs when solving dynamic problems of the theory of elasticity, the obtained results can be considerably distorted in comparison with the real results as a result of the frequency dependence of dissipative forces and their increasing suppression of the high-frequency components of motion.

When building linear models of frequency-independent internal friction it becomes necessary to observe the fundamental principle of mechanics—the causality principle, requiring, a reaction of a physically realizable system to a certain action occur only after such action is applied to the system. That is, the present output depends only on the past and present inputs, not on the future inputs. The transfer function of a system completely defines its input-output relationship in a steady state. The impulse transient or impulse response function $y_\delta(t)$ of a causal system, being a Fourier transform of the transfer function $\Phi(\omega)$, is other than zero only on the positive semi-axis of time

$$y_\delta(t) = \frac{1}{2\pi} \int_{-\infty}^{\infty} \Phi(\omega) e^{i\omega t} d\omega = 0, \quad (t < 0). \tag{18.4}$$

These constraints are associated with the interdependence of real and imaginary parts of the transfer function. Meanwhile, in the dynamic calculations of construction and machine-building designs, a priori representations of structural behavior with harmonic loads are frequently used, assigning different forms of complex rigidity or pliability. In particular, this approach is used in the well-known hypothesis of internal friction of Bock-Schlippe and E.S. Sorokin, where the complex stiffness (rigidity) is introduced in various modes into the equations of harmonic oscillations of a system without analyzing its conformity (compliance with the principle of causality). From the theory of dynamic systems, it follows that between the real and imaginary parts of the transfer function of a causal system

$$\Phi(\omega) = P(\omega) + iQ(\omega), \tag{18.5}$$

must exist the following dependencies [3]

$$\begin{aligned} P(\omega) &= -\frac{1}{\pi} \int_{-\infty}^{\infty} \frac{Q(\alpha) d\alpha}{\alpha - \omega}, \\ Q(\omega) &= \frac{1}{\pi} \int_{-\infty}^{\infty} \frac{P(\alpha) d\alpha}{\alpha - \omega}, \end{aligned} \tag{18.6}$$

where the integrals are understood to be the principal (main) values. The formulas (18.6) define by the Hilbert transformation (transform) the calculation of which presents certain difficulties. Therefore, a direct verification of the single-valued (unambiguous) relationship between the frequency characteristics $P(\omega)$ and $Q(\omega)$ is difficult for real systems. Below, we present a new way of calculating the transformation of Hilbert using Fourier transformation of the functions $P(\omega)$ and $Q(\omega)$ [4], which simplifies the task as there are extensive tables for Fourier transforms. The Hilbert transformation follows directly from the real modes of the Fourier transform and, therefore, the original $s(\tau)$ and its Hilbert transformation

$$S(t) = \frac{1}{\pi} \int_{-\infty}^{\infty} \frac{s(\tau) d\tau}{\tau - t} \tag{18.7}$$

are bound by the following formulas:

$$F_s(\alpha) = iF_S(\alpha) \mathrm{sign}\alpha, \quad F_S(\alpha) = -iF_s(\alpha) \mathrm{sign}\alpha, \tag{18.8}$$

where

$$F_s(\alpha) = \frac{1}{\sqrt{2\pi}} \int_{-\infty}^{\infty} s(\tau) e^{i\alpha\tau} d\tau,$$
$$F_S(\alpha) = \frac{1}{\sqrt{2\pi}} \int_{-\infty}^{\infty} S(\tau) e^{i\alpha\tau} d\tau, \tag{18.9}$$

are the Fourier transformations of the functions $s(\tau)$ and $S(t)$, respectively. The formula (18.9) makes it possible to calculate the integrals (18.6) in such cases when one of the transformations (18.9) is known. Actually, let the integral $F_S(\alpha)$ be known, then according to the formula (18.8), we can calculate $F_s(\alpha)$, where according to the conversion (inversion) formula

$$s(\tau) = \frac{1}{\sqrt{2\pi}} \int_{-\infty}^{\infty} F_s(\alpha) e^{-i\alpha\tau} d\alpha, \tag{18.10}$$

we can determine the function $s(\tau)$. In many instance calculations of the integral (18.10) can be simpler than calculations of the transformations (18.6). Let us examine several examples of calculating the Hilbert's transformations.

(a) Let $S(t) = \delta(t)$, where $\delta(t)$—Dirac-delta function. Then

$$F_S(\alpha) = \frac{1}{\sqrt{2\pi}}, \quad F_s(\alpha) = i\frac{\text{sign}\alpha}{\sqrt{2\pi}}.$$

According to (18.10), we obtain

$$s(\tau) = \frac{i}{\sqrt{2\pi}} \int_{-\infty}^{\infty} \text{sign}\alpha e^{-i\alpha\tau} d\alpha = \frac{1}{2} \int_{0}^{\infty} \sin \alpha\tau d\alpha = \frac{1}{\pi\tau}.$$

Here we used the formula, known from the theory of the generalized functions [5]

$$\int_{0}^{\infty} e^{i\xi x} d\xi = ix^{-1} + \pi\delta(x),$$

where it follows:

$$\int_{0}^{\infty} \sin \xi x d\xi = \frac{1}{x}, \qquad \int_{0}^{\infty} \cos \xi x d\xi = \pi\delta(x).$$

Substituting the obtained expression for $s(\tau)$ in (18.7), it is possible to obtain an integral representation of the Dirac-delta-function in the form of the convolution of two functions $1/\pi t$

$$\delta(t) = \frac{1}{\pi^2} \int_{-\infty}^{\infty} \frac{d\tau}{\tau(t-\tau)}.$$

It is also not difficult to notice that the formula obtained for $s(\tau)$ directly follows from the basic property of the delta-function. It follows from (18.7) that

$$s(\tau) = -\frac{1}{\pi} \int_{-\infty}^{\infty} \frac{S(t)dt}{t-\tau},$$

and with $S(t) = \delta(t)$ immediately we will obtain

$$s(\tau) = \frac{1}{\pi\tau}.$$

(b) If the function $S(t)$ has the mode (form) of the unit function of Heaviside: $S(t) = H(t)$, where

$$H(t) = \begin{cases} 1 & (t \geq 0) \\ 0 & (t < 0) \end{cases},$$

then

$$F_S(\alpha) = \frac{1}{\sqrt{2\pi}} \int_0^{\infty} e^{i\alpha t} dt = \sqrt{\frac{\pi}{2}}\delta(\alpha) + i\frac{1}{\alpha\sqrt{2\pi}}. \tag{18.11}$$

Hence

$$\begin{aligned} s(\tau) &= \frac{1}{\sqrt{2\pi}} \int_{-\infty}^{\infty} \left[\sqrt{\frac{\pi}{2}}\,\delta(\alpha) + i\frac{1}{\alpha\sqrt{2\pi}}\right] \mathrm{sign}\alpha e^{-i\tau\alpha} d\alpha = \\ &- \frac{1}{\pi} \int_0^{\infty} \frac{\cos\tau\alpha}{\alpha} d\alpha \end{aligned} \tag{18.12}$$

The obtained improper integral diverges and, from the standpoint of classical analysis, does not exist. In order to give this integral meaning, consider the Fourier transformation of the function: $y = |x|^{-2m-1}$. According to [5] page 227, in the designations given above

$$F_y(\alpha) = F_{|x|^{-2m-1}}(\alpha) = \frac{\alpha^{2m}}{\sqrt{2\pi}}\left[C_0^{(2m+1)} - C_1^{(2m+1)}\ln|\alpha|\right], \tag{18.13}$$

where

$$\begin{aligned} C_0^{(2m+1)} &= \frac{2\cdot(-1)^m}{(2m)!}\left[1 + \frac{1}{2} + \ldots + \frac{1}{2m} + \Gamma'(1)\right], \\ C_1^{(2m+1)} &= \frac{2\cdot(-1)^m}{(2m)!}, \end{aligned} \tag{18.14}$$

$\Gamma'(1)$—derivative of the gamma function, the value of which can be determined through the ψ—function, which is the logarithmic derivative of the gamma function

$$\psi = \frac{\Gamma'(z)}{\Gamma(z)}.$$

Thus $\Gamma'(1) = \psi(1)\Gamma(1)$, and therefore [6], $\Gamma'(1) = -\boldsymbol{C}$, where $\boldsymbol{C} = 0.5772\ldots$—Euler's constant. At the same time from (18.13) it follows:

$$F_{|x|^{-2m-1}}(\alpha) = \sqrt{\frac{2}{\pi}} \int_0^\infty \frac{\cos \alpha x d\alpha}{x^{2m+1}}, \qquad (18.15)$$

where when $m = 0$ we obtain the integral (18.12). Hence,

$$s(\tau) = -\frac{1}{2\pi^2}[1 - \boldsymbol{C} - \ln|\alpha|]. \qquad (18.16)$$

Now, let us examine the case $S(t) = t^n$ ($n = 1, 2, 3, \ldots$). We have:

$$F_S(\alpha) = \sqrt{\frac{2}{\pi}} \int_0^\infty t^n \cos \alpha t \; dt. \qquad (18.17)$$

With even n and

$$F_S(\alpha) = i\sqrt{\frac{2}{\pi}} \int_0^\infty t^n \sin \alpha t dt, \qquad (18.18)$$

with odd n

Calculating the integrals (18.17) and (18.18), when n is even, we find [7, 8]:

$$F_S(\alpha) = \sqrt{\frac{2}{\pi}} \frac{\Gamma(\mathrm{n}+1)}{\alpha^{n+1}} \cos \frac{(\mathrm{n}+1)\pi}{2},$$

and with odd n

$$F_S(\alpha) = i\sqrt{\frac{2}{\pi}} \frac{\Gamma(\mathrm{n}+1)}{\alpha^{n+1}} \sin \frac{(\mathrm{n}+1)\pi}{2}.$$

Consequently, and respectively

$$F_s(\alpha) = i\sqrt{\frac{2}{\pi}} \, signa \frac{\Gamma(\mathrm{n}+1)}{\alpha^{n+1}} \cos \frac{(\mathrm{n}+1)\pi}{2},$$

$$F_s(\alpha) = -\sqrt{\frac{2}{\pi}} \, signa \frac{\Gamma(\mathrm{n}+1)}{\alpha^{n+1}} \sin \frac{(\mathrm{n}+1)\pi}{2}.$$

Further when n is even we obtain

$$s(\tau)=\frac{2i\Gamma(\mathrm{n}+1)\cos\frac{(\mathrm{n}+1)\pi}{2}}{\pi}\int_0^\infty\frac{\cos\alpha\tau d\alpha}{\alpha^{n+1}}. \tag{18.19}$$

The integral in (18.19) can be calculated according to the formula (18.13). So, taking into account (18.14) and designating $n = 2m$, we find:

$$\int_0^\infty\frac{\cos\alpha\tau d\alpha}{\alpha^{2m+1}}=\frac{(-1)^m\tau^{2m}}{(2m)!}\left[1+\frac{1}{2}+\ldots+\frac{1}{2m}-C+\ln|\tau|\right].$$

With odd n it is necessary to use the following formula ([5], page 218) with introduction of an additional multiplier through a different designation of the Fourier transform:

$$F_{x^{-2m}}(\alpha)=\frac{(-1)^m}{\sqrt{2\pi}}|\alpha|^{2m-1}\frac{\pi}{(2m-1)!}.$$

Since,

$$F_{x^{-2m}}(\alpha)=\sqrt{\frac{2}{\pi}}\int_0^\infty\frac{\cos\alpha x dx}{x^{2m}},$$

then

$$\int_0^\infty\frac{\cos\alpha x dx}{x^{2m}}=\frac{(-1)^m\pi}{2(2m-1)!}|\alpha|^{2m-1}.$$

Let us further use the aforementioned method to calculate the Hilbert's transformation to analyze dynamic models of a base, the internal friction in which is described by hypothesis of Voight et al. [2].

Let us first examine an elasto-viscous Winkler foundation with complex rigidity (stiffness)

$$K(\omega)=k+ic\omega.$$

The real and imaginary parts of the transfer function of the base take the form:

$$\begin{aligned}\mathrm{P}(\omega)&=ReK^{-1}(\omega)=\frac{k}{k^2+c^2\omega^2},\\ \mathrm{Q}(\omega)&=ImK^{-1}(\omega)=-\frac{c\omega}{k^2+c^2\omega^2}.\end{aligned} \tag{18.20}$$

Let us verify, whether the dependencies in (18.6) comply with the characteristics in (18.20). We will calculate the Fourier transform of $\mathrm{P}(\omega)$ function:

$$F_P(\alpha)=\frac{k}{\sqrt{2\pi}}\int_{-\infty}^{\infty}\frac{e^{i\omega\alpha}d\omega}{k^2+c^2\cdot\omega^2}=\sqrt{\frac{\pi}{2}}C^{-1}e^{-|\alpha|kC^{-1}}.$$

Hence,

$$F_Q(\alpha) = -i\text{sign } \alpha \sqrt{\frac{\pi}{2}} C^{-1} e^{-|\alpha| k C^{-1}},$$

$$Q(\omega) = -\frac{i}{2c} \int_{-\infty}^{\infty} \text{sign } \alpha e^{-|\alpha| k C^{-1} - i\alpha\omega} d\alpha = -\frac{c\omega}{k^2 + \omega^2 c^2}.$$

I.e., we come to the function $Q(\omega)$, given by the formula (18.20). Thus, the Voight's coefficient of the bed satisfies the conditions of the physically realizable dynamic model. The inertia-less base with the internal friction of E.S. Sorokin is described by the differential equation of harmonic oscillation

$$k(u + iv)y = qe^{i\omega t}. \tag{18.21}$$

Where the correctly determined parameters of complex rigidity (stiffness) take the form [9, 10]

$$u = \frac{1}{\sqrt{1 + \gamma^2}}, \quad v = \frac{\gamma}{\sqrt{1 + \gamma^2}}.$$

The parameters u, and v do not depend on the frequency of disturbance (perturbance, excitation). Here $\gamma-$ coefficient of losses. Since $u^2 + v^2 = 1$, then for this model

$$P(\omega) = uk^{-1}, \qquad Q(\omega) = -vk^{-1}. \tag{18.22}$$

It should be noted that the complex rigidity (stiffness) of the corresponding hypothesis of Bock-Schlippe, has a similar form according to which the equation of harmonic oscillations of the base takes the form (mode)

$$\frac{k \cdot \gamma}{\omega} \cdot \dot{y} + ky = qe^{i\omega t}.$$

The complex rigidity (stiffness) in this case

$$K(\omega) = k(1 + i\gamma),$$

coincides with the complex rigidity for the hypothesis E.S. Sorokin with commonly used in dynamic calculations values of parameters $u = 1$, and $v = \gamma$. Let us verify that the dependencies in (18.6) comply with the characteristics in (18.20). We will determine Fourier the transform of $P(\omega)$ function

$$F_P(\alpha) = \frac{u}{k\sqrt{2\pi}} \int_{-\infty}^{\infty} e^{i\omega\alpha} d\omega = \frac{u\sqrt{2\pi}}{k} \delta(\alpha), \tag{18.23}$$

hence

$$F_Q(\alpha) = -iuk^{-1}\sqrt{2\pi} \text{ sign } \alpha\delta(\alpha),$$

and

$$Q(\omega) = -\frac{iu}{k} \int_{-\infty}^{\infty} \text{sign } \alpha e^{-i\omega\alpha} \delta(\alpha) d\alpha = 0.$$

So, as we can see, for these models, the dependencies in (18.6) are not executed, and consequently, these models are not physically realizable. Thus, the presented method of calculating the Hilbert transformation allows us to comparatively easily analyze the aprioristic models of dynamic systems based on creation of complex rigidity or complex pliability of systems. Additionally, this method makes it possible to analytically determine one of the components of complex rigidity according to the experimentally defined other component.

18.2 Construction of Green's Function for Bases Having Frequency-Dependent Internal Friction

With respect to the constructions made from traditional materials like steel, concrete, reinforced concrete, wood, brick, and so on, there is already sufficiently convincing work by many authors that proves the need for describing the internal friction using frequency independent type of models. On the other hand, for ground conditions, the experiments have shown that there are more adequate models with significant dependence of parameters of complex rigidity (stiffness) on frequency.

The works of many authors sufficiently and convincingly proved the need to describe the internal friction by frequency-independent type of models for constructions built from traditional materials like steel, concrete, reinforced concrete, wood, and brick and so on. And for ground conditions, the experiments show that more adequate models are used with the significant dependence of parameters of complex rigidity (stiffness) on the frequency.

The questions of dynamic calculations of structures and bases in non-stationary systems for materials with a sufficiently arbitrary dependency of complex rigidity on the frequency have only been touched upon in the literature.

One of the main methods of research of linear dynamic dissipative systems is construction and analysis of their characteristics in a frequency domain. For example, in the case of transfer functions, their parameters in many instances can be determined experimentally with sufficiently high accuracy, with subsequent transition into the time domain. A similar approach is used during construction of certain models of elastic-viscous media and of internal friction with the introduction of complex modules or complex rigidities based on aprioristic mechanical representations (views). For researching such dynamic systems in the time domain, it is necessary to move to their transient characteristics (features) using the Fourier transformation. This procedure is not simple in all cases. Furthermore, the procedure assumes that the transfer function possesses certain properties, ensuring observance of the principle of causality. Addressing the elastic basis foundations, specifically in the case when its material possesses elasto-viscous characteristics dependent on the frequency of the cyclic deformation, in order to construct Green's function, it is necessary to solve quasi-static task. Please remember that construction of Green's function is the base of BEM. First, we have to solve the problem of steady oscillations and then substitute the complex values instead of the modulus of elasticity E or coefficient of bed k accordingly

$$E(\omega) = E'(\omega) + iE''(\omega) \quad \text{or} \quad k(\omega) = k'(\omega) + ik''(\omega). \tag{18.24}$$

Where $k'(\omega)$ is the elastic component of the complex rigidity and $k''(\omega)$ is the dissipative component of the complex rigidity (stiffness).

According to the theory of dynamical systems, the parameters of complex rigidity must have certain properties. These properties must ensure the physical realizability (feasibility) of the system. For $\omega > 0$ the functions $k'(\omega)$ and $k''(\omega)$ must be positive. Moreover, the elastic component of the complex rigidity $k'(\omega)$ must be an even function, and the dissipative component of the complex rigidity $k''(\omega)$ must be an odd function of frequency ω.

Therefore, the functions $k'(\omega)$ and $k''(\omega)$ can be represented as expansion in a series in even and odd degrees of ω, respectively:

$$k'(\omega) = \sum_n \alpha_{2n}\omega^{2n},$$
$$k''(\omega) = \sum_n \beta_{2n+1}\omega^{2n+1}. \tag{18.25}$$

The loss coefficient has the following view:

$$\gamma(\omega) = \frac{k''(\omega)}{k'(\omega)} = \frac{\sum_n \beta_{2n+1}\omega^{2n+1}}{\sum_n \alpha_{2n}\omega^{2n}}. \tag{18.26}$$

The complex rigidity of the system can be represented as:

$$k(\omega) = |\omega|^{\alpha} k_{*}\omega, \quad (0 < \alpha < 1, \ k_{*}(0) > 0) \tag{18.27}$$

if this system has unlimited creep.

The equation of motion of a system with single degree of freedom with steady harmonic oscillations can be represented as follows:

$$m\ddot{y} + k(\omega)y = Pe^{i\omega t}, \tag{18.28}$$

here m is the mass of the system, y is the displacement of the system; P is the amplitude of the external load.

The solution of the equation of motion for steady oscillations and for $P = 1$ has the form:

$$y_{\omega}(t) = \Phi(\omega)e^{i\omega t}, \tag{18.29}$$

where

$$\Phi(\omega) = \left[-m\omega^2 + k(\omega)\right] \tag{18.30}$$

is the system transfer function.

Let us examine, for example, the construction (building) of Green's function for a plate lying on a Winklerian type elasto-viscous base with the arbitrary dependence of the parameters of the complex coefficient of bed on the frequency [11, 12]. We write the equation for steady (established) oscillations of a plate in complex amplitudes, determining Green's function for a stationary problem

$$D\nabla^2\nabla^2 w(x, y) + \left[k(\omega) - m\omega^2\right]w(x, y) = \delta(x - x_1)\delta(y - y_1), \tag{18.31}$$

where, as well as above

$$k(\omega) = \int_0^{\infty} R(z)e^{-i\omega z}dz$$

$R(z)$—the kernel of the relaxation of the Winklerian model of the base. We will designate

$$
\begin{aligned}
k_0'(\omega) &= \int_0^{\infty} \cos \omega z R(z) dz, \\
k''(\omega) &= \int_0^{\infty} \sin \omega z R(z) dz
\end{aligned}
\tag{18.32}
$$

and in (18.31) will obtain the Fourier transform. Considering that the displacement of the plate together with all derivatives disappears at infinity, which is ensured by the presence of damping, we will obtain:

$$\left[D\left(\xi^2+\eta^2\right)^2+k(\omega)-m\omega^2\right]\overline{W}(\xi,\eta)=(2\pi)^{-1}. \tag{18.33}$$

The transfer function of a system in the transforms space is determined by the formula (let us designate it by index "e")

$$W_e(\xi,\eta,\omega)=\frac{1}{2\pi}\left[D\left(\xi^2+\eta^2\right)^2+k(\omega)-m\omega^2\right]^{-1}. \tag{18.34}$$

The transfer function, determined by the formula (18.34), coincides with the transfer function of a system with single degree of freedom, which has complex rigidity $2\pi k(\omega) = 2\pi[D(\xi^2 + \eta^2)^2 + k(\omega)]$ and mass $2\pi m$. It is sufficient to take Fourier transform from the corresponding transfer function to obtain the impulse transient function in the space of transforms,

$$\overline{W}_\delta(\xi,\eta,t)=\frac{1}{2\pi}\int_0^{\infty} \overline{W}_e(\xi,\eta,\omega)e^{i\omega t}d\omega \tag{18.35}$$

Then using the double inversion of Fourier, from (18.35) we find the impulse transient function in the space of the originals, i.e., the reaction of the plate to the action of a unit instantaneous impulse.

It should be noted, however, that practical calculations of the Fourier transform using (18.35) in a closed mode is not feasible and the use of numerical integration complicates an already complex calculations of the double Fourier integrals for obtaining the originals. Therefore, calculations of the impulse transient function (18.35) in a closed mode is important from the practical point of view.

Thus, knowing the parameters of the complex coefficient of bed, we can obtain the impulse transient function of a plate, and then using the Duhamel's formula, solve for any dynamic load $q(x, y, t)$. However, according to the theory of dynamic systems, the parameters of the complex coefficient of bed must possess the properties described above, ensuring physical realization of the system. In other words, it is necessary to observe the principle of causality. Functions $k_0'(\omega)$ and $k_0''(\omega)$ must be positive with $\omega > 0$, moreover $k_0'(\omega)$ must be an even and $k_0''(\omega)$ odd function of frequency ω. The coefficient of losses:

$$\gamma(\omega)=\frac{k_0''(\omega)}{k_0'(\omega)}. \tag{18.36}$$

The following representation for complex rigidity can be written for systems having unlimited creep

$$k_0(\omega)=|\omega|^{\alpha}k_*(\omega), \quad \left(0<\alpha<1, k_*(0)\neq 0, k_*'(0)>0\right). \tag{18.37}$$

In order to observe (compliance with) the one-to-one (single-validness) relationship between its real and imaginary parts, the transfer function of a causal system should not have poles in the lower half-plane or on the real axis of a complex plane. In addition, because the actual constructions possess limited creep, they belong to the minimum-phase systems, i.e., systems realizing a minimum shift of a phase at any frequency at the same amplitude. Thus, the transfer function, must have not only poles but also zeroes on the entire lower half-plane, including the real axis. In general, for transfer functions we will assume the existence of at least one pole at the beginning of the coordinates (with $\omega = 0$), corresponding to the systems with unlimited creep under static load. For studying free oscillations of a system, it is sufficient to build an impulse transient function (18.35).

$$\overline{W}_{\delta}(\xi,\eta,t) = \frac{1}{4\pi^2} \int\limits_{-\infty}^{\infty} \frac{e^{i\omega t}}{-m\omega^2 + k(\omega)} d\omega \tag{18.38}$$

This can also be represented in the form of the following Mellin's integral by means of the substitution $p = i\omega$.

$$\overline{W}_{\delta}(\xi,\eta,t) = \frac{1}{4\pi^2 i} \int\limits_{-i\infty}^{i\infty} \frac{e^{pt}}{mp^2 + \overline{k}_0'(p) + \overline{k}_o''(p)} dp. \tag{18.39}$$

Here,

$$\overline{k}_0'(p) = k_0'(-ip), \qquad \overline{k}_o''(p) = ik_0''(-ip). \tag{18.40}$$

According to experimental data, for all construction materials $k_0'(\omega)$—smooth function with the small changes in a range of frequencies, which represents an interest in the dynamic calculations, and $k''(\omega) \ll k'(\omega)$. Therefore, the denominator of the integrant expression (18.39) has a degree of freedom not more than the second, and consequently, not more than two roots, different from zero. Furthermore, as was indicated above, it is possible to have a root of $p = 0$. For the analysis of the poles of the integrant in (18.39) let us write down:

$$p = \pm i\overline{\lambda}(p)\sqrt{1 + \overline{\gamma}(p)}, \tag{18.41}$$

where

$$\lambda(p) = \sqrt{\frac{\overline{k}_0'(p)}{m}}, \quad \overline{\gamma}(p) = \frac{\overline{k}_o''(p)}{\overline{k}_0'(p)} = i\gamma(\omega). \tag{18.42}$$

Taking into account that $\overline{\gamma}(p) \ll 1$, it is possible to assume

$$\sqrt{1 + \overline{\gamma}(p)} = 1 + \frac{\overline{\gamma}(p)}{2}.$$

Then

$$p = \pm i\overline{\lambda}(p)\left[1 + \frac{\overline{\gamma}(p)}{2}\right], \tag{18.43}$$

The assumptions adopted above, as applied to the weak damping and the presence of two zero roots in a frequency equation, make it possible to obtain the approximate dependencies from the expression (18.43)

$$p_1 = i\overline{\lambda}(p_0^+)\left[1 + \frac{\overline{\gamma}(p_0^+)}{2}\right],$$
$$p_2 = -i\overline{\lambda}(p_0^-)\left[1 + \frac{\overline{\gamma}(p_0^-)}{2}\right], \tag{18.44}$$

where $i|p_0| = |\omega_0|$—the natural frequency of conservative system with the rigidity (stiffness) $k_0'(\omega)$; ω_0^+, ω_0^-—the solution of equation

$$-m\omega_0^2 + k_0'(\omega_0) = 0 \tag{18.45}$$

or

$$mp^2 + \overline{k}_0'(p_0) = 0. \tag{18.46}$$

The roots of p_0^+ and p_0^- are purely imaginary and also satisfy the equations

$$p_0^+ = i\overline{\lambda}(p_0^+), \quad p_0^- = -i\overline{\lambda}(p_0^-). \tag{18.47}$$

Taking into account that $\overline{\lambda}(p)$—is an even function, and $\overline{\gamma}(p)$—is an odd function, we obtain:

$$p_1 = i\overline{\lambda}(p_0^+)\left[1 + \frac{\overline{\gamma}(p_0^+)}{2}\right],$$
$$p_2 = -i\overline{\lambda}(p_0^-)\left[1 - \frac{\overline{\gamma}(p_0^+)}{2}\right]. \tag{18.48}$$

Turning to $|\omega_0|$, we will obtain two complex-conjugate roots

$$p_1 = -\frac{|\omega_0|\gamma(|\omega_0|)}{2} + i|\omega_0|,$$
$$p_2 = -\frac{|\omega_0|\gamma(|\omega_0|)}{2} - i|\omega_0|. \tag{18.49}$$

Thus, the integrand expression in (18.39) is a multivalued function which has the branch point of $p = 0$, and two poles that lie on the left half-plane. The integral (18.39) can be examined on the first sheet of Riemann (Riemannian) surface with the section (a cut) along the negative semi-axis $Rep < 0$ as

$$\int_{-i\infty}^{i\infty} \overline{\Phi}(p)e^{pt}dp = \lim_{\rho\to 0}\left(\int_{-i\infty}^{-i\rho} + \int_{C_\rho} + \int_{i\rho}^{i\infty}\right)\Phi(p)e^{pt}dp. \tag{18.50}$$

Here $\overline{\Phi}(p) = \left[mp^2 + \overline{k}_0'(p) + \overline{k}_0''(p)\right]^{-1}$; C_ρ—the semicircle of the radius ρ, which envelepes (rounds) to the right of the branching point $p = 0$. Let us select the contour (loop) as shown in Fig. 18.1.

We will consider the integral in the circle C_ρ. The existent assessment:

$$\left|\int_{C_\rho} \frac{e^{pt}dp}{mp^2 + \overline{k}_0'(p) + \overline{k}_0''(p)}\right| = \left|\int_{C_\rho} \frac{p^{-\alpha}e^{pt}dp}{S(p)}\right| \le 2\pi\rho^{1-\alpha} max\left|\frac{e^{pt}}{S(\rho)}\right|, \tag{18.51}$$

where,

Fig. 18.1 Contour of integration

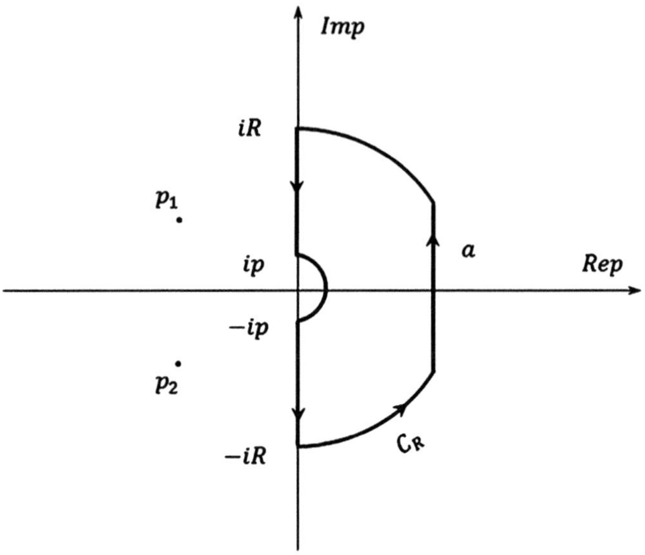

$$S(\rho) = m\left|\rho^{2-\alpha}\right| + \overline{k}_*(\rho). \tag{18.52}$$

The right part of the last inequality with $\rho \to 0$ approaches zero, therefore, the left part also approaches zero. On the other hand, when $R \to \infty$ the integrals along the arcs C_R approach zero due to Jordan's lemma. Since inside the contour, represented in Fig. 18.1, the integrand function does not have singular points, the following is true:

$$\overline{W}_\delta(\xi, \eta, t) = \frac{1}{4\pi^2 i} \int\limits_{a-i\infty}^{a+i\infty} \frac{e^{pt}}{mp^2 + \overline{k}_0'(p) + i\overline{k}_0''(p)} dp. \qquad (a > 0). \tag{18.53}$$

For calculations of the integral (18.53), following the works [13, 14] let us compose another contour (Fig. 18.2). We will present the integral (18.53) in the form:

$$\begin{aligned} &\lim_{\substack{R\to\infty \\ \rho\to 0}} \frac{1}{4\pi^2 i} \int\limits_{a-iR}^{a+iR} \overline{\Phi}(p)e^{pt}dp = \sum_{j=1}^{2} \mathrm{res}\left[e^{p_j t}\overline{\Phi}\ (p_j)\right] + \\ &+ \lim_{\substack{R\to\infty \\ \rho\to 0}} \frac{1}{4\pi^2 i} \left[\left(\int\limits_{l_2} + \int\limits_{l_3} + \int\limits_{C_\rho} + \int\limits_{C_R}\right) \overline{\Phi}(p)e^{pt}dp\right]. \end{aligned} \tag{18.54}$$

As has already been indicated above, integrals in small and large circles tend to approach zero with $\rho \to 0$ and $R \to \infty$, respectively. Hence it follows that

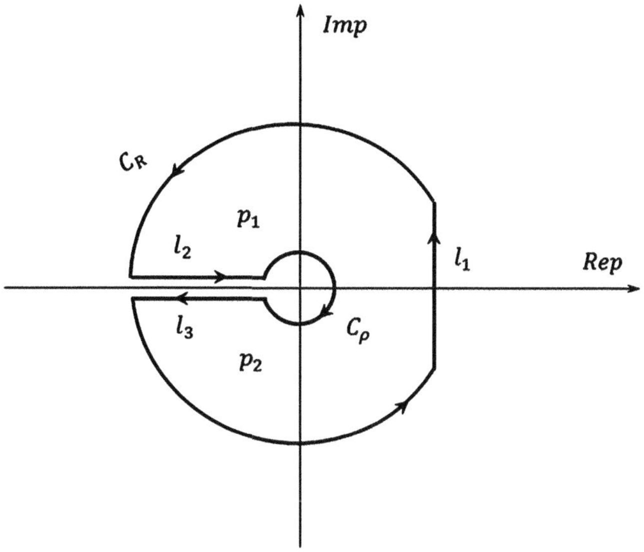

Fig. 18.2 Contour of Integration

$$\overline{W}_e(\xi,\eta,t)=\frac{1}{4\pi^2 i}\int\limits_0^\infty\left[\overline{\Phi}\left(re^{i\pi}\right)-\overline{\Phi}\left(re^{-i\pi}\right)\right]e^{-rt}dr+ \\ +\sum_{j=1}^{2}res\left[e^{p_j t}\overline{\Phi}\left(p_j\right)\right]. \tag{18.55}$$

Taking into account the conjugality of real functions from conjugated arguments, we finally write

$$\overline{W}_\delta(\xi,\eta,t)=\Omega^{-1}\exp[-0,5|\omega_0|\,\gamma\,(|\omega_0|)t]\cos(|\omega_0|t-\varepsilon)+A(t), \tag{18.56}$$

where,

$$\Omega=\frac{\pi}{2}\left\{\left[-m|\omega_0|\,\gamma+Re\frac{d\overline{k}_0'(p_1)}{dp}+Re\frac{d\overline{k}_0''(p_1)}{dp}\right]^2+ \\ +\left[2m|\omega_0|+Im\frac{d\overline{k}_0'(p_1)}{dp}+Im\frac{d\overline{k}_0''(p_1)}{dp}\right]^2\right\}^{\frac{1}{2}}, \tag{18.57}$$

$$\varepsilon=\tan^{-1}\left[\frac{2m|\omega_0|+Im\frac{d\overline{k}_0'(p_1)}{dp}+Im\frac{d\overline{k}_0''(p_1)}{dp}}{-m|\omega_0|\,\gamma+Re\frac{d\overline{k}_0'(p_1)}{dp}+Re\frac{d\overline{k}_0''(p_1)}{dp}}\right]. \tag{18.58}$$

Let us examine in more detail the function $A(t)$. Since

$$\overline{k}_0\left(re^{-i\varphi}\right)=\overline{k}_0^*\left(re^{i\varphi}\right), \tag{18.59}$$

where *—indicates a complex conjugate value, and assuming

$$\overline{k}_0\left(re^{i\pi}\right)=r^{\alpha}[a(r)+ib(r)],$$

where $a(r)$—nondecreasing function and $a(0)\neq 0$, we obtain

$$A(t)=\frac{1}{2\pi^2}\int_0^{\infty}\frac{r^{-\alpha}b(r)e^{-rt}dr}{\left[mr^{2-\alpha}+a(r)\right]^2+b^2(r)}. \tag{18.60}$$

According to (18.60) $A(t)$ is a monotonically decreasing function of time, with the initial value

$$A(0)=\frac{1}{2\pi^2}\int_0^{\infty}\frac{r^{-\alpha}b(r)dr}{\left[mr^{2-\alpha}+a(r)\right]^2+b^2(r)},$$

for which the following approximation can be obtained:

$$A(0)\approx\frac{b_{max}}{2\pi^2(2-\alpha)a^2(0)}\left[\frac{a(0)}{m}\right]^z\frac{\Gamma(z)\Gamma(2-z)}{\Gamma(2)},\quad z=\frac{1-\alpha}{2-\alpha}.$$

Taking into account the properties of the gamma-functions, we find that.

$$A(0)\approx\frac{b_{max}}{2\pi(2-\alpha)^2a^2(0)}\left[\frac{a(0)}{m}\right]^z\sec\frac{\pi\alpha}{4(2-\alpha)},$$

or with $\alpha\ll 1$

$$A(0)\approx\frac{b_{max}}{8\pi a^{3/2}(0)m^{1/2}}.$$

Thus, the system performs free oscillations with the frequency $|\omega_0|$ and logarithmic decrement

$$\delta(|\omega_0|)=\pi\gamma(|\omega_0|), \tag{18.61}$$

relative to the position of the quasi-static equilibrium, described by the function $A(t)$, and the corresponding small deformations of aftereffect. Thus, a closed solution is obtained for the transform of impulse transient function $\overline{W}_\delta(\xi,\eta,t)$. The impulse transient response function of the plate will be determined by the formula

$$w_\delta(x,y,t)=\frac{1}{2\pi}\iint_{-\infty}^{\infty}\overline{W}_\delta(\xi,\eta,t)e^{-i(\xi x+\eta y)}d\xi d\eta. \tag{18.62}$$

Further, it is possible to obtain a solution of a problem for an arbitrary load $q(x,\ y,\ t)$ by convolution. In addition, we studied the behavior of a dynamic system, the material, whose characteristics have an arbitrary dependence on the frequency of the cycles of deformation.

18.3 The Green's Functions of Systems with the Frequency-Independent Internal Friction

Let us examine the Green's functions for one-dimensional dynamic systems with damping, described by different models of frequency-independent and frequency-dependent internal friction. The Green functions in the case of harmonic steady oscillations of a dynamic problem based on the elasto-viscous analogy can be expressed through a Green's function of the corresponding static problem with the replacement of the elastic characteristics with complex modules. For nonstationary problems, the construction of Green's function can be realized (carried out) on the basis of applying integral transformations in the spatial coordinate and reducing the equations of motion to ordinary differential equations with respect to time, similar to the equation of motion of dissipative system with a single degree of freedom.

To solve a system with an arbitrary linear damping, primary attention will be given to the frequency-independent internal frictions, as they are the most relevant to the real properties of traditional building structures. As it relates to the present book describing problems for calculating structures on an elastic foundation, these models are of interest to us for describing the dissipative properties of structures themselves, while the description of the dissipative properties in general, as was noted above, is given more adequately with frequency-dependent models. In dynamics and seismic resistance of structures, a well-known method is used to describe the internal friction. This method is based on the introduction into the differential equations of motion, divided in the generalized coordinates, of dissipative terms (members) with such coefficients which ensure the frequency independence of the decrement of oscillations for each mode of oscillations. In the works [9, 15, 16], the linear models of internal friction are built, one of which described by the integral operator with Abel's kernel, ensures strict frequency independence of decrements of oscillations and coefficients of losses. The other is based on the introduction of fractional operators of elastic and inertial forces and leads exactly to the same results as the above-mentioned artificial technique. The last model, named the model of the elasto-viscous frequency-independent resistance, possesses within a certain conditionality (the coefficients with the dissipative terms depend on the inertia and elastic properties of the system), sufficient simplicity and convenience for solving practical problems and it allows to take into account the frequency-independent energy losses strictly at the level of initial equations of motion. First, let us examine a stationary one-dimensional wave problem based on the example of the longitudinal oscillations of an isotropic uniform unlimited bar, in the central section of which acts a single harmonic load. Alternatively, we will consider a problem based on the example of the lateral oscillations of a string. The equation of oscillations we write down in the form.

$$-(ic)^n \int_{-\infty}^{t} R(t-\tau)\frac{\partial^n u}{\partial x^n}(x,\tau)d\tau - c^2\frac{\partial^2 u}{\partial x^2}(x,t) + \frac{\partial^2 u}{\partial t^2}(x,t) = =\rho^{-1}\delta(x)e^{i\omega t}. \tag{18.63}$$

With boundary conditions:

$$\lim_{|x|\to\infty} u(x,t) = 0. \tag{18.64}$$

Here c—the wave propagation velocity, equal at the lateral oscillations of the string to $c=\sqrt{Sm^{-1}}$ and at the longitudinal oscillations of a bar $c=\sqrt{E_0/\rho}$, where S, m—the tension of the string and its linear mass, E_0, ρ– modulus of the elasticity of the bar and its density; $R(t)$– relaxation kernel. The Eq. (18.63) also describes the propagation of plane waves in the isotropic elastic medium; in this case $c=\sqrt{\mu\rho^{-1}}$ for the transverse waves, and $c=\sqrt{(\lambda+2\mu)\rho^{-1}}$ for the waves of expansion, where λ, μ—Lamé parameters.

The following basic models of a medium are examined: (a) general (common) linear model; (b) the elastic-viscous model of Voight; and (c) the model of the frequency-independent elasto-viscous resistance [9, 17]. For these models respectively we have:

(a) $n=2$
(b) $n=2;\quad R(t-\tau)=\varepsilon\delta'(t-\tau)E_0^{-1}$
(c) $n=1;\quad R(t-\tau)=\gamma\delta'(t-\tau)$

where γ—coefficient of losses; ε—damping coefficient (factor) according to Voight. Let us proceed in the left and right sides of the Eq. (18.63) to the Fourier transforms for the coordinate x and we will be taking into account (18.64).

$$-(ic)^n(-i\xi)^n\int\limits_{-\infty}^{t}R(t-\tau)U(\xi,\tau)d\tau+c^2\xi^2U(\xi,t)+\frac{\partial^2U}{\partial t^2}(\xi,t)==\frac{e^{i\omega t}}{\rho\sqrt{2\pi}}.\tag{18.65}$$

Here $U(\xi,\tau)-$ the transform of displacement.

$$U(\xi,\tau)=\frac{1}{\sqrt{2\pi}}\int\limits_{-\infty}^{\infty}u(x,t)e^{-i\xi x}dx.$$

We will search for the solution of the Eq. (18.65) in the form of [18]

$$U(\xi,\tau)=A_0(\xi)e^{i[\omega t+\varphi(\xi)]}.\tag{18.66}$$

Substituting (18.66) into (18.65) and replacing the variable $t-\tau=z$ in the integrant, we obtain

$$A_0(\xi)\left[-\omega^2+c^2\xi^2+(-ic)^n(-i\xi)^n(R_c-iR_s)\right]=\frac{e^{-i\varphi(\xi)}}{\rho\sqrt{2\pi}},$$

where

$$R_c(\omega)=\int\limits_0^{\infty}R(z)\cos\omega z dz,$$

$$R_s(\omega)=\int\limits_0^{\infty}R(z)\sin\omega z dz.$$

From here

$$\begin{cases}A_0(\xi)=\dfrac{1}{\sqrt{2\pi}\,\rho c^2\xi^2}\left[\left(1+R_c-\dfrac{\omega^2}{c^2\xi^2}\right)^2+R_s^2\right]^{-1/2}\\ \varphi(\xi)=\tan^{-1}\dfrac{R_s}{\left(1+R_c-\dfrac{\omega^2}{c^2\xi^2}\right)}\end{cases},\quad(n=2)\tag{18.67}$$

and

$$\begin{cases}A_0(\xi)=\dfrac{1}{\sqrt{2\pi}\,\rho}\left[\left(c\xi R_c+\omega^2-c^2\xi^2\right)^2+c^2\xi^2R_s^2\right]^{-1/2}\\ \varphi(\xi)=\tan^{-1}\dfrac{c\xi R_s}{\left(c\xi R_c+\omega^2-c^2\xi^2\right)}.\end{cases}\quad(n=1)\tag{18.68}$$

For the models mentioned above we have:

$$\text{b)}\quad R_c = 0, \quad R_s = -\frac{\varepsilon\omega}{E_0};$$
$$\text{c)}\quad R_c = 0, \quad R_s = -\gamma\omega.$$

Therefore,

$$A_0(\xi) = \frac{1}{\sqrt{2\pi}\,\rho c^2 \xi^2}\left[\left(1 - \frac{\omega^2}{c^2\xi^2}\right)^2 + \frac{\varepsilon^2\omega^2}{E_0^2}\right]^{-1/2},$$
$$\varphi(\xi) = \tan^{-1}\frac{\varepsilon\omega}{E_0\left(\frac{\omega^2}{c^2\xi^2} - 1\right)},$$

and

$$A_0(\xi) = \frac{1}{\sqrt{2\pi}\,\rho}\left[\left(\omega^2 - c^2\xi^2\right)^2 + \gamma^2 c^2 \xi^2 \omega^2\right]^{-1/2},$$
$$\varphi(\xi) = \tan^{-1}\frac{\gamma\omega c\xi}{\left(c^2\xi^2 - \omega^2\right)}.$$

Thus

$$u(x,t) = \frac{1}{\sqrt{2\pi}}\int_{-\infty}^{\infty} A_0(\xi) e^{i[\omega t + \varphi(\xi) - \xi x]} d\xi. \tag{18.69}$$

The amplitude of the displacement

$$\overline{U}(x,\omega) = \sqrt{\frac{2}{\pi}}\left(B_c^2 + B_s^2\right),$$
$$B_c = \int_0^{\infty} A_0(\xi)\cos\xi x \cos\varphi(\xi) d\xi,$$
$$B_s = \int_0^{\infty} A_0(\xi)\sin\xi x \sin\varphi(\xi) d\xi.$$

When solving nonstationary boundary problems, it is necessary to determine the nonstationary Green's function—solution of the problem of free oscillations of a system under the impact of an instantaneous impulse. In this case, the equation of motion (18.65) transformed by Fourier takes the form:

$$(-i\xi)^n \int_{-\infty}^{t} R(t-\tau)U(\xi,\tau)d\tau + c^2\xi^2 U(\xi,t) + \frac{\partial^2 U}{\partial t^2}(\xi,t) = \frac{\delta(t)}{\rho\sqrt{2\pi}}. \tag{18.70}$$

To solve (18.70), we can use the standard methods of the theory of viscoelasticity, in particular the Laplace transform. However, we can use the results for the arbitrary linear model, obtained in this paragraph, taking into account that the solutions of the Eqs. (18.70) and (18.65) are connected by Fourier transformations the same way as the transfer function is connected with the impulse transient function (formula 18.38)). Let us write down the transfer function of the system, described by the Eq. (18.65) with the fixed value of the parameter ξ.

$$\Phi(\omega,\xi)=\frac{1}{\rho\sqrt{2\pi}\left[c^2\xi^2-\omega^2+(-c\xi)^n(R_c-iR_s)\right]}. \tag{18.71}$$

By analogy with (18.38) (with an accuracy to a multiplier $\frac{1}{\sqrt{2\pi}}$) we can accept, that:

$$m=\rho$$
$$k'(\omega)=\rho c^2\xi^2+(-1)^n\rho c^n\xi^n R_c^n(\omega),\quad k''(\omega)=(-1)^{n+1}\rho c^n\xi^n R_s(\omega).$$

Now the solution of the Eq. (18.70), considered as an impulse transient function of the same system, can be written as (18.56). The general solution of the problem is obtained by the inverse Fourier transform:

$$\begin{aligned} u(x,t)=\frac{1}{\sqrt{2\pi}}\int_{-\infty}^{\infty} e^{i\xi x}\left\{\Omega^{-1}(\xi)\,exp\left[-\frac{|\omega_0|\gamma|\omega_0|t}{2}\right]\times\right.\\ \left.\times\cos[|\omega_0|t-\varepsilon(\xi)]+A(\xi,t)\right\}d\xi, \end{aligned} \tag{18.72}$$

where $\Omega(\xi)$, $\varepsilon(\xi)$, and $A(\xi, t)$ are determined by the formulas (18.57), (18.58) and (18.60) respectively, with the fixed value of the parameter ξ.

For basic models (Voight's model and the elasto-viscous frequency-independent resistance), the construction of the nonstationary function of Green can be easily realized from the examination of the equations of motion in the differential form. In particular, the differential equation of longitudinal oscillation of a bar, taking into account the frequency-independent elasto-viscous resistance, can be written in the form:

$$\begin{gathered} E_0Lu(x,t)+\gamma\sqrt{E_0\rho}\,L^{1/2}\,\dot{u}(x,t)+\rho\ddot{u}(x,t)==\delta(x)\delta(t),\\ (|x|<\infty,t>0). \end{gathered} \tag{18.73}$$

where $L=-\frac{\partial^2}{\partial x^2}$ – is the operator of elastic forces. We will search for the solution of the Eq. (18.73), which approaches zero at infinity, together with the first derivative of the spatial coordinate. Let us accept the initial conditions as zero. To interpret the operator $L^{1/2}$, we will consider that the domain of definition of operators is the entire real (material) axis. Therefore,

$$L^{1/2}=\pm i\frac{\partial}{\partial x^2},$$

so that the Eq. (18.73) takes the form

$$-E_0u''\pm i\gamma\sqrt{E_0\rho}\;\;\dot{u}'+\rho\ddot{u}==\delta(x)\delta(t), \tag{18.74}$$

where the prime indicates the differentiation with respect to x, and the sign before the dissipative term is selected as the opposite to the sign of the spectral parameter. To solve the Eq. (18.74) we will use Fourier transformation. Moving (18.74) to Fourier's transforms, we obtain:

$$E_0\xi^2U+\gamma|\xi|\sqrt{E_0\rho}\;\dot{U}+\rho\ddot{U}=\frac{\delta(t)}{\sqrt{2\pi}}, \tag{18.75}$$

where

$$U(\xi,t) = \frac{1}{\sqrt{2\pi}} \int_{-\infty}^{\infty} u(x,t)e^{i\xi x}dx. \tag{18.76}$$

The solution of the Eq. (18.76) at zero initial conditions takes the form:

$$U(\xi,t) = \frac{\sin p_1 \xi t}{\sqrt{2\pi}\,\rho p_1 \xi}\, exp\left(-\frac{\gamma c|\xi|t}{2}\right), \tag{18.77}$$

where

$$p_1 = c\sqrt{1-\gamma^2/4}. \tag{18.78}$$

Converting (18.77) according to Fourier, we will find

$$u(x,t) = \frac{1}{\pi\rho p_1} \int_0^{\infty} \frac{\cos \xi x \sin p_1 \xi t \; exp\left(\frac{\gamma c|\xi|t}{2}\right)}{\xi}\, d\xi. \tag{18.79}$$

The integral (18.79) is calculated [6]

$$u(x,t) = \frac{1}{2\pi\rho p_1}\left\{ \tan^{-1}\frac{\gamma c p_1 t^2}{x^2 - p_1^2 t^2 + \frac{\gamma^2 c^2 t^2}{4}} + \begin{matrix} \pi\left(x^2 - p_1^2t^2 + \frac{\gamma^2c^2t^2}{4} < 0\right) \\ + \\ 0\left(x^2 - p_1^2t^2 + \frac{\gamma^2c^2t^2}{4} \geq 0\right) \end{matrix} \right\}. \tag{18.80}$$

By assuming that $\nu_1 = \sqrt{1-\frac{\gamma^2}{2}}$, $\nu_2 = \sqrt{\gamma\sqrt{1-\frac{\gamma^2}{2}}}$, the solution (18.80) leads to the form

$$u(x,t) = \frac{1}{2\pi\rho p_1}\left\{ \tan^{-1}\frac{c^2t^2\nu_2^2}{x^2 - c^2t^2\nu_1^2} + \pi[1 - H(|x| - ct\nu_1)]\right\}, \tag{18.81}$$

where $H(|x| - ct\nu_1)$– the Heaviside unit step function. The first derivative of the displacement, which determines the stress with the longitudinal oscillations of a bar or the angle of rotation of the section of a string will be:

$$\frac{\partial u}{\partial x}(x,t) = \frac{1}{2\pi\rho p_1}\left[\frac{\left(x^2 - c^2t^2\nu_1^2\right)^2}{\left(x^2 - c^2t^2\nu_1^2\right) + c^4t^4\nu_2^2} - \pi\delta(|x| - ct\nu_1)\right]. \tag{18.82}$$

In the conclusion of this section, let us examine problems of construction of Green's function for a nonstationary problem with transverse oscillations of a bar and plate, taking into account the frequency-independent elasto-viscous resistance. The differential equation of oscillations of a bar can be written in the following form:

$$EIy^{IV} - \gamma\sqrt{EIm}\,\dot{y}^{II} + m\ddot{y} = \delta(x)\delta(t), \qquad (|x| < \infty, \quad t > 0). \tag{18.83}$$

Assuming that with $|x| \to \infty$ the displacement of the bar $y(x, t)$ and its derivatives over the coordinate approach zero, and by using the Fourier transform on x in (18.83) we obtain

$$\ddot{Y}(\xi, t) + \gamma p\xi^2\dot{Y}(\xi, t) + p^2\xi^4 Y(\xi, t) = \frac{\delta(t)}{m\sqrt{2\pi}}, \tag{18.84}$$

where

$$Y(\xi, t) = \frac{1}{\sqrt{2\pi}}\int_{-\infty}^{\infty} y(x, t)e^{i\xi x}dx,$$

$$p = \sqrt{EIm^{-1}}.$$

The solution of the Eq. (18.84) with zero initial conditions takes the form (with lower-order accuracy [second-order])

$$Y(\xi, t) = \frac{1}{\sqrt{2\pi}} \cdot \frac{\sin p\xi^2 t}{mp\xi^2}\exp\left(-\frac{\gamma p\xi^2 t}{2}\right). \tag{18.85}$$

Converting (18.85), we find the displacement under the impact of the instant impulse

$$y(x, t) = \frac{1}{\pi mp}\int_0^{\infty} \frac{\cos\xi x \sin p\xi^2 t\, \exp\left(-\frac{\gamma pt\xi^2}{2}\right)}{\xi^2}d\xi. \tag{18.86}$$

As a result of (18.86) for Green's function we have an expression

$$G(x - x_1, t - t_1) = \frac{1}{mp\pi}\int_0^{\infty} \frac{\cos\xi(x - x_1)\sin p\xi^2(t - t_1)\exp\left[-\frac{\gamma p\xi(t - t_1)}{2}\right]}{\xi^2}d\xi. \tag{18.87}$$

In the initial section of $x = 0$, the displacement will be

$$y(0, t) = \frac{1}{\pi mp}\int_0^{\infty} \frac{\sin p\xi^2 t\, \exp\left(-\frac{\gamma pt\xi^2}{2}\right)}{\xi^2}d\xi. \tag{18.88}$$

Let us switch over to the new variable of integration $z = p\xi^2 t$, we obtain

$$y(0, t) = \frac{\sqrt{t}}{2\pi m\sqrt{p}}\int_0^{\infty} \frac{e^{-\frac{\gamma z}{2}}\sin z}{z\sqrt{z}}dz. \tag{18.89}$$

The last integral is calculated [6]

$$y(0,t) = \frac{\left(1+\frac{\gamma^2}{4}\right)\sqrt{t}}{m\sqrt{\pi p}} \sin\left(\frac{1}{2}\tan^{-1}\frac{2}{\gamma}\right). \tag{18.90}$$

For the real constructions $\gamma \ll 0.1$, therefore by disregarding γ^2 in comparison with the unit ("1") and by performing a simple conversion, we will obtain

$$y(0,t) = \frac{\sqrt{(1-\gamma/2)t}}{m\sqrt{2mp}}. \tag{18.91}$$

As we can see, the influence of internal friction in the case of free oscillations of the unlimited beam somewhat decreases the displacements, although their total increase, related to the absence of supporting links, occurs just as well without taking into account damping and is proportional $\sqrt{t}$.

The Green's function of a nonstationary problem with lateral oscillations of an elastic plate with frequency-independent elastoviscous resistance is determined from the equation

$$\begin{gathered} D\nabla^2\nabla^2 w(x,y,t) + \gamma\sqrt{Dm}\,\nabla^2\dot{w}(x,y,t) + m\ddot{w}(x,y,t) = \delta(x)\delta(y)\delta(t), \\ (|x|,|y|<\infty, \quad t>0). \end{gathered} \tag{18.92}$$

Assuming as we did above, that at infinity the displacement of the plate and its derivatives disappear and performing a dual Fourier transformation in (18.92), we obtain an equation similar to (18.84)

$$\ddot{W}(\xi,\eta,t) + \gamma p\left(\xi^2+\eta^2\right)\dot{W}(\xi,\eta,t) + p^2\left(\xi^2+\eta^2\right)^2 W(\xi,\eta,t) = \frac{\delta(t)}{2\pi m}, \tag{18.93}$$

where

$$\begin{gathered} W(\xi,\eta,t) = \frac{1}{2\pi}\iint\limits_{-\infty}^{\infty} w(x,y,t)e^{i(\xi x+\eta t)}, \\ p = \sqrt{Dm^{-1}}. \end{gathered} \tag{18.94}$$

With the accuracy to the small of the second order

$$W(\xi,\eta,t) = \frac{1}{2\pi}\cdot\frac{\sin p\left(\xi^2+\eta^2\right)t}{mp\left(\xi^2+\eta^2\right)}\,exp\left(-\frac{\gamma p\left(\xi^2+\eta^2\right)t}{2}\right). \tag{18.95}$$

Converting (18.95), we obtain

$$w(x,y,t) = \frac{1}{mp\pi^2}\iint\limits_{0}^{\infty} \cos\xi x\cos\eta y\frac{\sin\left(\xi^2+\eta^2\right)t}{\left(\xi^2+\eta^2\right)}e^{\left(-\frac{\gamma p\left(\xi^2+\eta^2\right)t}{2}\right)}d\xi d\eta. \tag{18.96}$$

Green's function, which is the function of difference arguments $x - x_1$, $y - y_1$, $t - t_1$, can be represented in the form

$$
\begin{aligned}
G(x-x_1, \quad y-y_1, \quad t-t_1) = \\
= \frac{1}{mp\pi^2} \iint_0^\infty \cos\xi(x-x_1)\cos\eta(y-y_1)\frac{\sin(\xi^2+\eta^2)t}{(\xi^2+\eta^2)} \times \\
\times\, exp\left(-\frac{\gamma p(\xi^2+\eta^2)t}{2}\right)d\xi d\eta.
\end{aligned} \tag{18.97}
$$

Further calculation of the Green's function can be performed by the decomposition of cosines in exponential series. Thus, in the case of the dynamic calculation of structures on the elastic foundation, the results of the present paragraph make it possible to take into account the dependency of the parameters of stiffness (rigidity) and damping on frequency both for the structures and for the base. These dependencies are reflected in the construction of the corresponding functions of Green, separately for the structures and bases, or for the general Green function of the entire system: "structure-base/foundation."

Summarizing the special features of the solutions of dynamic problems by the boundary equations method, we come to the following conclusions:

A new method of the analysis of physical realization of dynamic systems, based on calculations of the conversions (transformations, transforms) of Hilbert of real and imaginary frequency characteristics through their Fourier transformations is proposed. The method is illustrated based on the example of the elasto-viscous base, which possesses the internal friction, described by the models of Voigt and E.S. Sorokin.

The solution to a dynamic problem for a base with frequency-dependent elasto-viscous characteristics is obtained for the first time. Based on the example of construction of Green's function for a plate which lies on Winklerian type elasto-viscous base, the opportunity of obtaining a close solution of the equation of oscillations for the generalized coordinate in the time domain is shown if experimentally determined parameters of complex rigidity of the base are known.

We obtained closed expressions for Green's functions for a number of one-dimensional and two-dimensional dynamic problems, taking into account the frequent-independent internal friction.

18.4 The Construction of the Green's Function of the Dynamic Stationary Problem for the Elasto-Viscous Half-Plane

Construction of the Green's function is the most important stage when solving boundary-value problems using the methods of boundary integral equations. Boundary equations are built on the basis of the Green's Function. Usually, Green's functions are used for unlimited domains since they take the simplest form and can be obtained in a closed form. However, during the solution of many applied problems, the use of Green's function for an unlimited domain can lead to superfluous operations. In particular, for problems of dynamics of structure and the theory of seismic resistance, it is necessary to examine limited spaces, which are a part of or an addition to a half-space (half-plane). Hence, the Green's function is more effective in a case of a semi-infinite domain, which frees us from the need for solving additional integral equations on the boundary of the half-space (half-plane). In the present paragraph, we offer a solution to the problem of steady oscillations of an isotropic homogeneous elasto-viscous half-plane. These oscillations are caused by the action of an arbitrarily oriented concentrated force, changing in time according to the harmonic law [19].

A similar problem for the elastic half-space and elastic half-plane was examined in the works of Case and Hazeltine [20], Ilichev and Schechter [21], and Petrosian [22]. It is assumed that the material

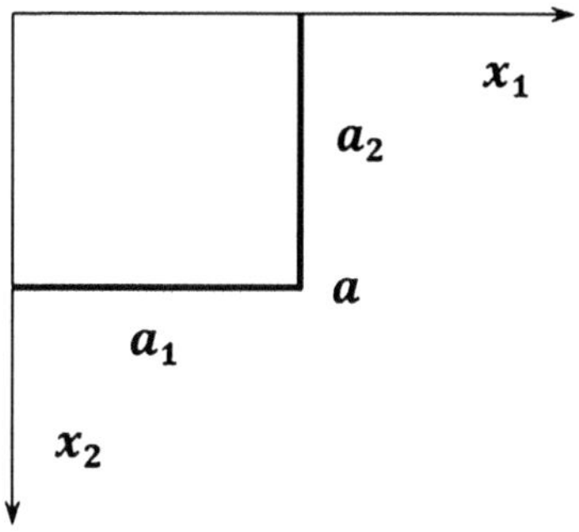

Fig. 18.3 Half-plane in Cartesian coordinates

of the medium possesses internal friction, which is described by the model of the frequency-independent and relative damping by Tseitlin [9]. In this case the complex module takes the form of:

$$E(\omega) = E_1|\omega|^{\alpha}\left(\cos\frac{\pi\alpha}{2} + i\text{sign}\omega\sin\frac{\pi\alpha}{2}\right), \tag{18.98}$$

and Poisson's ratio σ is taken as real and does not depend on the frequencies of harmonic oscillations ω. Here $E_1 = E(1)$; $\alpha = 2\pi^{-1}\tan^{-1}\gamma$; $\gamma-$ loss factors (coefficient of loses). This model provides the ideal frequency independence of energy loss on the entire frequency axis, making it possible to investigate any dynamic processes; however, this excludes the consideration of a static load as a result of an unlimited creep with $\omega = 0$. Let us examine a half-plane $x_2 > 0$ in, Fig. 18.3, in the Cartesian coordinates $\boldsymbol{x}(x_1, x_2)$. Assume that at a certain point of half-plane $\boldsymbol{a}(a_1, a_2)$ a unit concentrated force, changing in time according to the harmonic low is applied.

The solution of the problem of steady oscillations of a half-plane under the action of this load, after the exclusion of time, is described by the matrix function of Green $\boldsymbol{g} = [g_{ij}]$, where $g_{ij}(i, j = 1, 2)$—displacement along the ith coordinate from the action of the unit force, directed along jth coordinate. The Green's function satisfies the equations

$$\boldsymbol{\nabla}^{*}\boldsymbol{g}(\boldsymbol{x};\boldsymbol{a}) + \rho\omega^2\boldsymbol{g}(\boldsymbol{x};\boldsymbol{a}) + \boldsymbol{\delta}(\boldsymbol{x};\boldsymbol{a}) = 0, \tag{18.99}$$

where $\boldsymbol{\nabla}^{*} = \left[\nabla^{*}_{ij}\right]$ is the matrix differential operator

$$\nabla^{*}_{ij} = \mu\delta_{ij}\nabla + (\lambda+\mu)\frac{\partial^2}{\partial x_i \partial x_j}; \qquad (i = 1, 2),$$

$$\nabla = \frac{\partial^2}{\partial x_1^2} + \frac{\partial^2}{\partial x_2^2};$$

δ_{ij}—Kronecker's delta; $\boldsymbol{\delta}(\boldsymbol{x};\boldsymbol{a})$—two-dimensional Dirac delta function, concentrated at point $\boldsymbol{a}$

$$\boldsymbol{\delta}(\boldsymbol{x};\boldsymbol{a}) = \begin{bmatrix} \delta(x_1 - a_1)\delta(x_2 - a_2) & 0 \\ 0 & \delta(x_1 - a_1)\delta(x_2 - a_2) \end{bmatrix};$$

ρ—material density of half-plane; λ—Lame's constants

$$\lambda = \frac{E\sigma}{(1+\sigma)(1-2\sigma)}, \qquad \mu = \frac{E}{2(1+\sigma)}.$$

As a result of (18.98), we assume:

$$\lambda = \lambda^* (E' + iE''), \quad \mu = \mu^* (E' + iE'')$$

where

$$\lambda^* = \frac{\sigma}{(1+\sigma)(1-2\sigma)}, \qquad \mu^* = \frac{1}{2(1+\sigma)},$$
$$E'(\omega) = E_1 |\omega|^\alpha \cos\frac{\pi\alpha}{2}, \quad E''(\omega) = E_1 |\omega|^\alpha sign\omega \sin\frac{\pi\alpha}{2}.$$

Boundary of the half-plane is free of stresses and has the following boundary conditions:

$$\boldsymbol{T_n g}(\boldsymbol{x}_1 + 0, a) = 0 \qquad (n_1 = 0, \;\; n_2 = 1) \tag{18.100}$$

where
$\boldsymbol{T_n} = [T_{nij}]$—matrix differential operator of stresses with elements:

$$T_{nij} = \lambda n_i \frac{\partial}{\partial x_j} + \mu n_j \frac{\partial}{\partial x_i} + \mu \delta_{ij} \frac{\partial}{\partial n},$$
$$\frac{\partial}{\partial n} = \sum_i^2 n_i \frac{\partial}{\partial x_i};$$

n—the direction of the internal normal with directing cosines n_1, and n_2. With $n_1 = 0$, and $n_2 = 1$, we obtain:

$$T_{nij} = \lambda \delta_{2i} \frac{\partial}{\partial x_j} + \mu \left(\delta_{2i} \frac{\partial}{\partial x_i} + \delta_{ij} \frac{\partial}{\partial x_2} \right). \tag{18.101}$$

Furthermore, we assume that as a result of the internal friction, the components of the vector of displacements and stresses tensor become zero at infinity. Let us somehow prolong $\boldsymbol{g}(\boldsymbol{x}, \boldsymbol{a})$ into the extended domain Ω. As the extended domain we take the entire plane. Now let us apply a delta-transformation to the Eq. (18.99) by multiplying on the left side by the two-dimensional delta-function $\boldsymbol{\delta}(\boldsymbol{x}, \boldsymbol{x}^*)$, concentrated at the point $\boldsymbol{x}^* \left(x_1^*, x_2^*\right)$, and integrating over the entire domain Ω. Using the Green's function in the domains of continuity, i.e., in both half-planes $x_2 > 0$ and $x_2 < 0$, and taking into account the properties of a delta-function, we will obtain:

$$\nabla^* \boldsymbol{g}(\boldsymbol{x}^*; \boldsymbol{a}) + \rho\omega^2 \boldsymbol{g}(\boldsymbol{x}^*; \boldsymbol{a}) + \boldsymbol{\delta}(\boldsymbol{x}^*; \boldsymbol{a}) = \int_{-\infty}^{\infty} [\boldsymbol{\delta}(x_0, x^*) \Delta \boldsymbol{T_n g}(x_0, \boldsymbol{a}) - \\ - [\boldsymbol{T_n \delta}(\boldsymbol{x}_0, \boldsymbol{x}^*)]^T \Delta \boldsymbol{g}(x_0, \boldsymbol{a})] dx_1, \tag{18.102}$$

where $\boldsymbol{x}_0 \in \Gamma$—the point on the border of the half-plane; $\boldsymbol{x}^* \in \Omega$; and Δ indicates the difference in values of corresponding vector functions on the inner Γ^+ and outer Γ^- boundary of the half-plane. For solving the Eq. (18.102), we use the known Green's function for the entire plane $\boldsymbol{G}(\boldsymbol{x}^*, \boldsymbol{x})$, which satisfies the Eq. (18.99), with $-\infty < \boldsymbol{x}, \; \boldsymbol{x}^* < \infty$. Then from (18.102) we find:

$$\boldsymbol{g}(\boldsymbol{x}^*, \boldsymbol{a}) = \boldsymbol{G}(\boldsymbol{x}^*, \boldsymbol{a}) - \int_{-\infty}^{\infty} [\boldsymbol{G}(x^*, x_0) \Delta \boldsymbol{T_n g}(x_0, \boldsymbol{a}) - \\ - [\boldsymbol{T_n G}(\boldsymbol{x}_0, \boldsymbol{x}^*)]^T \Delta \boldsymbol{g}(x_0, \boldsymbol{a})] dx_1. \tag{18.103}$$

Thus, we presented the function $\boldsymbol{g}(\boldsymbol{x}, \boldsymbol{a})$ in the entire plane. As it was mentioned above, there are various ways to select jumps of unknown functions on a boundary of a half-plane. Let us pick the following option: $\Delta\boldsymbol{g}(\boldsymbol{x}_0; \boldsymbol{a}) \neq 0$; $\Delta\boldsymbol{T}_n\boldsymbol{g}(\boldsymbol{x}_0, \boldsymbol{a}) \neq 0$. This case corresponds to prolongation of the function $\boldsymbol{g}(\boldsymbol{x}, \boldsymbol{a})$ by zero into the extended domain $\boldsymbol{\Omega}$. Assuming that the outer boundary of the half-plane Γ^- remains undeformed, we obtain the relationships:

$$\boldsymbol{g}(\boldsymbol{x}_0, \boldsymbol{a}) = \boldsymbol{T}_n\boldsymbol{g}(\boldsymbol{x}_0, \boldsymbol{a}) = 0, \quad (\boldsymbol{x}_0 \in \Gamma^-). \tag{18.104}$$

Consequently, $\Delta\boldsymbol{g}(\boldsymbol{x}_0, \boldsymbol{a}) = \boldsymbol{g}(\boldsymbol{x}_0; \boldsymbol{a})$, $\Delta\boldsymbol{T}_n\boldsymbol{g}(\boldsymbol{x}_0, \boldsymbol{a}) = \boldsymbol{T}_n\boldsymbol{g}(\boldsymbol{x}_0, \boldsymbol{a})$, $(\boldsymbol{x}_0 \in \Gamma^+)$.
Then (18.103) will take the form:

$$\begin{aligned} \boldsymbol{g}(\boldsymbol{x}^*, \boldsymbol{a}) = \boldsymbol{G}(\boldsymbol{x}^*, \boldsymbol{a}) - \int_{-\infty}^{\infty} \{\boldsymbol{G}(\boldsymbol{x}^*, \boldsymbol{x}_0)\boldsymbol{T}_n\boldsymbol{g}(\boldsymbol{x}_0, \boldsymbol{a}) - \\ - [\boldsymbol{T}_n\boldsymbol{G}(\boldsymbol{x}_0, \boldsymbol{x}^*)]^T\boldsymbol{g}(\boldsymbol{x}_0, \boldsymbol{a})\}dx_1. \end{aligned} \tag{18.105}$$

Using known properties of the Green's function for a plane:

$$\begin{aligned} \boldsymbol{G}(\boldsymbol{x}, \boldsymbol{x}^*) &= \boldsymbol{G}(\boldsymbol{x}^*, \boldsymbol{x}), \\ \boldsymbol{T}_n\boldsymbol{G}(\boldsymbol{x}, \boldsymbol{x}^*) &= -\boldsymbol{T}_n\boldsymbol{G}(\boldsymbol{x}^*, \boldsymbol{x}), \end{aligned}$$

as well as introducing designations

$$\boldsymbol{G}_q(\boldsymbol{x}, \boldsymbol{x}^*) = -[\boldsymbol{T}_n\boldsymbol{G}(\boldsymbol{x}^*, \boldsymbol{x})]^T \tag{18.106}$$

and using the boundary conditions (18.100), we obtain:

$$\frac{1}{2}\boldsymbol{g}(\boldsymbol{x}_0^*, \boldsymbol{a}) = \boldsymbol{G}(\boldsymbol{x}_0^*, \boldsymbol{a}) - \int_{-\infty}^{\infty} \boldsymbol{G}_q(\boldsymbol{x}_0^*, \boldsymbol{x}_0)\boldsymbol{g}(\boldsymbol{x}_0, \boldsymbol{a})dx_1. \tag{18.107}$$

Here the integral on the right side is understood as having the principal value. For solving the obtained integral in Eq. (18.107) let us apply Fourier transformation to both parts on the coordinate x_1^*

$$\frac{1}{2}\overline{\boldsymbol{g}}(\xi_1, 0; \boldsymbol{a}) = \overline{\boldsymbol{G}}(\xi_1, 0; \boldsymbol{a}) - \iint_{-\infty}^{\infty} \boldsymbol{G}_q(\boldsymbol{x}_0^*, \boldsymbol{x}_0)\boldsymbol{g}(\boldsymbol{x}_0^*, \boldsymbol{a})e^{i\xi_1 x_1^*}dx_1^* dx_1, \tag{18.108}$$

where the feature indicates Fourier transform on the coordinate x_1^*, for example

$$\overline{\boldsymbol{g}}(\xi_1, 0; \boldsymbol{a}) = \int_{-\infty}^{\infty} \boldsymbol{g}(\boldsymbol{x}_0^*, \boldsymbol{a})e^{i\xi_1 x_1^*}dx_1^*.$$

Since $\boldsymbol{G}_q(\boldsymbol{x}_0^*, \boldsymbol{x}_0)$ is the function of the difference of the arguments $\boldsymbol{x}_0^* - \boldsymbol{x}_0$, the dual integral on the right side of (18.108) can be presented in the form of [20].

$$\iint_{-\infty}^{\infty} \boldsymbol{G}_q(\boldsymbol{x}_0^*, \boldsymbol{x}_0)\boldsymbol{g}(\boldsymbol{x}_0^*, \boldsymbol{a})e^{i\xi_1 x_1^*}dx_1^* dx_1 = \widetilde{\boldsymbol{G}}_q(\xi_1)\overline{\boldsymbol{g}}(\xi_1, 0; \boldsymbol{a}), \tag{18.109}$$

where,

$$\widetilde{G}_q(\xi_1) = \int_{-\infty}^{\infty} G_q(R) e^{i\xi_1 R} dR, \qquad (R = x_1^* - x_1). \tag{18.110}$$

Thus, from (18.107) we will obtain:

$$\left[\frac{1}{2}E + \widetilde{G}_q(\xi_1)\right] \overline{g}(\xi_1, 0; a) = \overline{G}(\xi_1, 0; a), \tag{18.111}$$

where E—unit matrix. Designating A as the matrix

$$A = \left[\frac{1}{2}E + \widetilde{G}_q(\xi_1)\right], \tag{18.112}$$

from (18.111) we find

$$\overline{g}(\xi_1, 0; a) = A^{-1} \overline{G}(\xi_1, 0; a). \tag{18.113}$$

Using the Green's function G for a plane and, generated by its tensor of stresses G_q in (18.112) and (18.113), leads to cumbersome calculations. By doing as follows we can bypass these difficulties. We have

$$\begin{aligned} \overline{G}(\xi_1, 0; a) &= \frac{1}{2\pi} \int_{-\infty}^{\infty} \overline{\overline{G}}(\xi_1, \xi_2; a) e^{-i\xi_2 x_2^*} d\xi_2, \\ \widetilde{G}_q(\xi_1, x_2^*) &= \frac{1}{2\pi} \int_{-\infty}^{\infty} \widetilde{\widetilde{G}}_q(\xi_1, \xi_2) e^{-i\xi_2 x_2^*} d\xi_2, \end{aligned} \tag{18.114}$$

where $\overline{\overline{G}}$, $\widetilde{\widetilde{G}}_q$—are dual transforms of Fourier from G and G_q. For determining $\widetilde{\widetilde{G}}_q$ we will use the formula (18.106). Taking into account that $G_q(x^*, x) = -G_q(x, x^*)$, and by applying Fourier transform to its both parts, we obtain

$$\widetilde{\widetilde{G}}_q(\xi_1, \xi_2) = \int_{-\infty}^{\infty}\int [T_n G(x^*, x]^T e^{i(\xi_1 x_1^* + \xi_2 x_2^*)} dx_1^* dx_2^*. \tag{18.115}$$

The elements of the matrix $\widetilde{\widetilde{G}}_q$ will be:

$$\begin{aligned} \left[\widetilde{\widetilde{G}}_q\right]_{jk} = \int_{-\infty}^{\infty}\int &\left[\lambda \delta_{2k} \frac{\partial}{\partial x_m^*} + \mu \left(\delta_{2m} \frac{\partial}{\partial x_k^*} + \delta_{km} \frac{\partial}{\partial x_2^*}\right)\right] \times \\ &\times \, G_{mj}(x^*, x) e^{i(\xi_1 x_1^* + \xi_2 x_2^*)} dx_1^* dx_2^*. \end{aligned} \tag{18.116}$$

Integrating by parts the right side of (18.116), we obtain:

$$\left[\widetilde{\widetilde{G}}_q\right]_{jk} = -i[\lambda \delta_{2k} \xi_m + \mu \xi_k \delta_{2m} + \mu \xi_2 \delta_{km}] \overline{\overline{G}}_{mj}(\xi_1, \xi_2) \tag{18.117}$$

Let us determine $\overline{\overline{G}}_{mj}$ using the dynamic Lame's equations. Let the unit load act in the horizontal (on x_1^*) direction. Lame's equations in the expanded form are taken in the form:

$$\begin{aligned}&(\lambda+\mu)\left(\frac{\partial^2 G_{11}}{\partial x_1^{*2}}+\frac{\partial^2 G_{21}}{\partial x_1^*\partial x_2^*}\right)+\mu\left(\frac{\partial^2 G_{11}}{\partial x_1^{*2}}+\frac{\partial^2 G_{11}}{\partial x_2^{*2}}\right)+\\&+\delta(x_1^*-x_1)\delta(x_2^*-x_2)+\rho\omega^2 G_{11}=0,\\&(\lambda+\mu)\left(\frac{\partial^2 G_{21}}{\partial x_2^{*2}}+\frac{\partial^2 G_{11}}{\partial x_1^*\partial x_2^*}\right)+\mu\left(\frac{\partial^2 G_{21}}{\partial x_1^{*2}}+\frac{\partial^2 G_{21}}{\partial x_2^{*2}}\right)+\rho\omega^2 G_{11}=0.\end{aligned} \tag{18.118}$$

Applying Fourier transform on the coordinate x_1^* to the Eq. (18.118) and having previously designated $x_1^*-x_1=R_1,\quad x_2^*-x_2=R_2$, we will obtain:

$$\begin{aligned}&(\lambda+2\mu)\xi_1^2\tilde{\tilde{G}}_{11}+(\lambda+\mu)\xi_1\xi_2\tilde{\tilde{G}}_{21}+\mu\xi_2^2\tilde{\tilde{G}}_{11}-1-\rho\omega^2\tilde{\tilde{G}}_{11}=0\\&(\lambda+2\mu)\xi_2^2\tilde{\tilde{G}}_{21}+(\lambda+\mu)\xi_1\xi_2\tilde{\tilde{G}}_{11}+\mu\xi_1^2\tilde{\tilde{G}}_{21}-1-\rho\omega^2\tilde{\tilde{G}}_{21}=0\end{aligned} \tag{18.119}$$

where,

$$\tilde{\tilde{G}}_{ij}(\xi_1,\xi_2)=\iint_{-\infty}^{\infty} G_{ij}(R_1,R_2)e^{i(\xi_1 R_1+\xi_2 R_2)}dR_1 dR_2.$$

The multiplier $e^{i(\xi_1 x_1)}e^{i(\xi_2 x_2)}$ in (18.119) was omitted for the simplicity. Using the following designation:

$$\begin{aligned}&k_e^2=\frac{\rho\omega^2}{(\lambda+2\mu)};\qquad k_t^2=\frac{\rho\omega^2}{\mu};\\&\frac{\xi_1^2}{k_e^2}+\frac{\xi_2^2}{k_t^2}-1=a_{11};\\&\left(\frac{1}{k_e^2}-\frac{1}{k_t^2}\right)\xi_1\xi_2=a_{12}=a_{21};\\&\frac{\xi_2^2}{k_e^2}+\frac{\xi_1^2}{k_t^2}-1=a_{22};\end{aligned} \tag{18.120}$$

from (18.119) we will obtain the following system of algebraic equations:

$$\begin{aligned}&a_{11}\tilde{\tilde{G}}_{11}+a_{12}\tilde{\tilde{G}}_{21}=\frac{1}{\rho\omega^2}\\&a_{21}\tilde{\tilde{G}}_{11}+a_{22}\tilde{\tilde{G}}_{21}=0.\end{aligned} \tag{18.121}$$

Hence, we easily find:

$$\tilde{\tilde{G}}_{11}=\frac{1}{\rho\omega^2}\left(\frac{a_{22}}{a_{11}a_{22}-a_{12}a_{21}}\right)=\frac{1}{\rho\omega^2}\left(\frac{\xi_1^2}{\xi_2^2-k_e^2}+\frac{k_t^2-\xi_1^2}{\xi_1^2-k_t^2}\right),$$

$$\tilde{\tilde{G}}_{21}=-\frac{1}{\rho\omega^2}\left(\frac{a_{21}}{a_{11}a_{22}-a_{12}a_{21}}\right)=\frac{1}{\rho\omega^2}\left(\frac{\xi_1\xi_2}{\xi^2-k_e^2}-\frac{\xi_1\xi_2}{\xi^2-k_t^2}\right),$$

where,

$$\xi^2 = \xi_1^2 + \xi_2^2.$$

Under the action of a vertical unit load along the axis x_2^*, analogously by doing likewise, we obtain:

$$\widetilde{\widetilde{G}}_{22} = \frac{1}{\rho\omega^2}\left(\frac{\xi_2^2}{\xi^2 - k_e^2} + \frac{k_t^2 - \xi_2^2}{\xi^2 - k_t^2}\right);$$

$$\widetilde{\widetilde{G}}_{12} = \frac{1}{\rho\omega^2}\left(\frac{\xi_1\xi_2}{\xi^2 - k_e^2} - \frac{\xi_1\xi_2}{\xi^2 - k_t^2}\right).$$

Thus, the formula for the general term, taking into account the omitted multipliers, takes the form

$$\widetilde{\widetilde{G}}_{ij}(\xi_1, \xi_2; x) = \frac{e^{i(\xi_1 x_1 + \xi_2 x_2)}}{\rho\omega^2}\left(\frac{\xi_m\xi_j}{\xi^2 - k_e^2} + \frac{\delta_{mj}k_t^2 - \xi_m\xi_j}{\xi^2 - k_t^2}\right). \tag{18.122}$$

Substituting (18.117) into (18.114) and taking into account (18.122) and the relationships

$$\widetilde{G}_q(\xi_1, x_2^*) = e^{-i\xi_1 x_1}\overline{\overline{G}}_q(\xi_1, x_2^*), \tag{18.123}$$

we obtain:

$$\begin{aligned}2\pi i\rho\omega^2\tilde{G}_{qik}(\xi_1, x_2^*) = \int\limits_{-\infty}^{\infty}&\left(\lambda\delta_{2k}\sum_m^2\frac{\xi_m^2\xi_j}{\xi^2 - k_e^2} + \lambda k_t^2\delta_{2k}\frac{\xi_j}{\xi^2 - k_t^2} -\right.\\ &- \frac{\lambda\delta_{2k}\sum_m^2\xi_m^2\xi_j}{\xi^2 - k_t^2} + \frac{2\mu\xi_2\xi_j\xi_k}{\xi^2 - k_e^2} + \frac{\mu\delta_{2j}k_t^2\xi_j}{\xi^2 - k_t^2} - \frac{2\mu\xi_2\xi_j\xi_k}{\xi^2 - k_e^2} + \\ &\left.+\frac{\mu k_t^2\delta_{kj}\xi_{j2}}{\xi^2 - k_t^2}\right)e^{-i\xi_2 x_2^*}d\xi_2, \qquad (j, k = 1, 2).\end{aligned} \tag{18.124}$$

After calculating integrals [6] let us find:

$$\begin{aligned}2\pi i\rho\omega^2\left[\widetilde{G}_q(\xi_1, x_2^*)\right]_{ik} &= \lambda\xi_1^2\delta_{2k}\left[(\xi_j - \delta_{2j}\xi_2)I_0 + \delta_{2j}I_2\right] + \lambda\delta_{2k}\left[(\xi_j - \delta_{2j}\xi_2)I_3 + \delta_{2j}I_4\right]\\ &+\lambda k_t^2\delta_{2k}\left[(\xi_j - \delta_{2j}\xi_2)I_{0t} + \delta_{2j}I_{1t}\right] + 2\mu\left[(\xi_k - \delta_{2k}\xi_2)(\xi_j - \delta_{2j}\xi_2)I_2\right.\\ &\left.+(\xi_j - \delta_{2j}\xi_2)\delta_{2k}I_\xi + (\xi_k - \delta_{2k}\xi_2)\delta_{2j}I_3 + + \delta_{2k}\delta_{2j}I_4\right]\\ &+\mu k_t^2\left[\delta_{2j}(\xi_k - \delta_{2k}\xi_2)I_{0t} + \delta_{2j}\delta_{2k}I_{1t} + \delta_{kj}I_{1t}\right],\end{aligned} \tag{18.125}$$

where,

$$
\begin{aligned}
&I_{0e} = \frac{\pi}{R_e} e^{-R_e x_2^*}, \qquad I_{0t} = \frac{\pi}{R_t} e^{-R_t x_2^*},\\
&I_0 = I_{0e} - I_{0t},\\
&I_{1e} = -i\pi e^{-R_e x_2^*}, \qquad I_{1t} = -i\pi e^{-R_t x_2^*},\\
&I_2 = I_{1e} - I_{1t} = i\pi\left(e^{-R_t x_2^*} - e^{-R_e x_2^*}\right),\\
&I_{3e} = -\pi R_e e^{-R_e x_2^*}, \qquad I_{3t} = -\pi R_t e^{-R_t x_2^*},\\
&I_3 = I_{3e} - I_{3t} = \pi\left(R_t e^{-R_t x_2^*} - R_e e^{-R_e x_2^*}\right),\\
&I_{4e} = i\pi R_e^2 e^{-R_e x_2^*}, \qquad I_{4t} = i\pi R_t^2 e^{-R_t x_2^*},\\
&I_4 = I_{4e} - I_{4t} = i\pi\left(R_e^2 e^{-R_e x_2^*} - R_t^2 e^{-R_t x_2^*}\right),\\
&R_e = \sqrt{\xi_1^2 - k_e^2} \quad R_t^2 = \sqrt{\xi_1^2 - k_t^2} \qquad x_2^* > 0.
\end{aligned}
\tag{18.126}
$$

Accomplishing a limited transition to $x_2^* \to +0$, we will obtain the boundary value of $\widetilde{G}_q(\xi_1, x_2^*)$

$$
\begin{aligned}
2\pi i\rho\omega^2 \widetilde{G}_{qik}(\xi_1, +0) = &(\lambda + 2\mu)\delta_{2k}\left(\xi_j - \delta_{2j}\xi_2\right)I_3^0 + \lambda\xi_1^2\delta_{2k}\left(\xi_j - \delta_{2j}\xi_2\right)I_{0e}^0 +\\
&+\lambda k_t^2\delta_{2k}\left(\xi_j - \delta_{2j}\xi_2\right)I_{0t}^0 + 2\mu\delta_{2j}(\xi_k - \delta_{2k}\xi_2)I_3^0 +\\
&+\mu k_t^2\left[\delta_{2j}(\xi_k - \delta_{2k}\xi_2)I_{0t}^0 + \delta_{kj}I_{1t}^0\right],
\end{aligned}
\tag{18.127}
$$

where

$$
\begin{aligned}
I_{0e}^0 = \frac{\pi}{R_e}; \quad I_{0t}^0 = \frac{\pi}{R_t};\\
I_{1e}^0 = I_{1t}^0 = -i\pi; \quad I_3^0 = \pi(R_t - R_e).
\end{aligned}
$$

Taking into account the relationship between the boundary and direct singular value of the integral of the stress tensor $\boldsymbol{G}_q$ [23]

$$
\widetilde{G}_q(\xi_1) = \widetilde{G}_q(\xi_1, +0) + \frac{1}{2}\boldsymbol{E}. \tag{18.128}
$$

After simple computations from (18.127) we will find:

$$
\left[\widetilde{G}_q(\xi_1)\right]_{jk} = \frac{1}{2}\left[\delta_{2k}\left(\xi_j - \delta_{2j}\xi_2\right)\alpha(\xi_1) + \delta_{2j}(\xi_k - \delta_{2k}\xi_2)\beta(\xi_1)\right], \tag{18.129}
$$

where,

$$
\begin{aligned}
\alpha(\xi_1) &= i\frac{2(R_t - R_e)R_t + k_t^2}{R_e k_t^2};\\
\beta(\xi_1) &= -i\frac{2(R_t - R_e)R_t + k_t^2}{R_e k_t^2}.
\end{aligned}
\tag{18.130}
$$

Now, according to (18.112) we obtain:

$$
\boldsymbol{A} = \frac{1}{2}\begin{bmatrix} 1 & \xi_1 & \alpha(\xi_1) \\ \xi_1 & \beta(\xi_1) & 1 \end{bmatrix}. \tag{18.131}
$$

The reciprocal (inverse) matrix $[\boldsymbol{A}]^{-1}$ takes the form:

$$[\boldsymbol{A}]^{-1} = \frac{2}{1-\xi_1^2\alpha(\xi_1)\beta(\xi_1)}\begin{bmatrix} 1 & -\xi_1 & \beta(\xi_1) \\ -\xi_1 & \alpha(\xi_1) & 1 \end{bmatrix}^T. \tag{18.132}$$

Now from (18.105) taking into account (18.113) we determine the Green's function for the half-plane.

$$\boldsymbol{g}(\boldsymbol{x}^*,\boldsymbol{a}) = \boldsymbol{G}(\boldsymbol{x}^*,\boldsymbol{a}) + \boldsymbol{B}^*(\boldsymbol{x}^*,\boldsymbol{a}), \tag{18.133}$$

where,

$$\boldsymbol{B}^*(\boldsymbol{x}^*,\boldsymbol{a}) = -\frac{1}{2\pi}\int\limits_{-\infty}^{\infty} \widetilde{G}_q(\xi_1,x_2^*)[\boldsymbol{A}]^{-1}\boldsymbol{G}(\xi_1,0;\boldsymbol{a})e^{-i\xi_1x_1^*}d\xi_1. \tag{18.134}$$

In particular, let the unit load act at the point $\boldsymbol{a} = (0, a_2)$. For this case we write the elements $\overline{\boldsymbol{G}}(\xi_1,0;0,a_2)$

$$\begin{aligned} \overline{\overline{G}}_{11}(\xi_1,0;0,a_2) &= \frac{1}{2\rho\omega^2}\left(e^{-R_ea_2}\frac{\xi_1^2}{R_e} - R_te^{-R_ta_2}\right), \\ \overline{\overline{G}}_{12}(\xi_1,0;0,a_2) &= \overline{\overline{G}}_{21}(\xi_1,0;0,a_2) = \\ &= -\frac{i}{2\rho\omega^2}\left(\xi_1e^{-R_ta_2} - \xi_1e^{-R_ea_2}\right), \\ \overline{\overline{G}}_{22}(\xi_1,0;0,a_2) &= \frac{1}{2\rho\omega^2}\left(\frac{\xi_1^2}{R_t}e^{-R_ta_2} - R_ee^{-R_ea_2}\right). \end{aligned} \tag{18.135}$$

And the elements of $\widetilde{\boldsymbol{G}}_q(\xi_1,x_2^*)$ will be:

$$\begin{aligned} \widetilde{G}_{q11}(\xi_1,x_2^*) &= \frac{\xi_1^2}{k_t^2}\left(e^{-R_tx_2^*} - e^{-R_ex_2^*}\right) - \frac{1}{2}e^{-R_tx_2^*}, \\ \widetilde{G}_{q12}(\xi_1,x_2^*) &= \frac{\xi_1}{i}\left[\frac{R_t}{k_t^2}e^{-R_tx_2^*} - \left(\frac{1}{2R_e} + \frac{R_t^2}{k_t^2R_e}\right)e^{-R_ex_2^*}\right], \\ \widetilde{G}_{q21}(\xi_1,x_2^*) &= \frac{\xi_1}{i}\left[\left(\frac{1}{2R_t} + \frac{R_t}{k_t^2}\right)e^{-R_tx_2^*} - \frac{R_e}{k_t^2}e^{-R_ex_2^*}\right], \\ \widetilde{G}_{q22}(\xi_1,x_2^*) &= \frac{\xi_1^2}{k_t^2}\left(e^{-R_ex_2^*} - e^{-R_tx_2^*}\right) - \frac{1}{2}e^{-R_ex_2^*}. \end{aligned} \tag{18.136}$$

Let us introduce the designations:

$$\widetilde{\boldsymbol{G}}_q(\xi_1,x_2^*)[\boldsymbol{A}]^{-1} = \boldsymbol{C}(\xi_1,x_2^*), \tag{18.137}$$

$$\boldsymbol{B}(\xi_1,x_2^*;0,a_2) = \boldsymbol{C}(\xi_1,x_2^*)\overline{G}(\xi_1,0;0,a_2). \tag{18.138}$$

Thus, we can define:

$$\boldsymbol{g}(x^*,a;0,a_2) = \boldsymbol{G}(x^*;0,a_2) - \frac{1}{2\pi}\int\limits_{-\infty}^{\infty}\boldsymbol{B}(\xi_1,x_2^*;0,a_2)e^{-i\xi_1x_1^*}d\xi_1, \tag{18.139}$$

where the matrix elements $\boldsymbol{B}(\xi_1, x_2^*; 0, a_2)$ have the form:

$$
\begin{aligned}
B_{11} = & \frac{\xi_1^2}{N(\xi_1)\mu k_t^2 R_e} \Big\{ \Big[\big(2\xi_1^2 - k_t^2\big)^2 + 4\xi_1^2 R_e R_t \Big] \Big(\frac{R_t R_e}{2\xi_1^2} e^{-R_t(x_2^* + a_2)} + \\
& + \frac{1}{2} e^{-R_t(x_2^* + a_2)} \Big) - 2\big(2\xi_1^2 - k_t^2\big) R_e R_t \big(e^{-R_t x_2^*} e^{-R_e a_2} + e^{-R_e x_2^*} e^{-R_t a_2} \big) \Big\}, \\
B_{12} = & \frac{i\xi_1}{N(\xi_1)\mu k_t^2} \Big\{ \frac{1}{2} \Big[\big(2\xi_1^2 - k_t^2\big)^2 + 4\xi_1^2 R_e R_t \Big] \Big(e^{-R_t(x_2^* + a_2)} + e^{-R_e(x_2^* + a_2)} \Big) - \\
& - 2\big(2\xi_1^2 - k_t^2\big) R_e R_t e^{-R_t x_2^*} e^{-R_e a_2} - 2\xi_1^2 \big(2\xi_1^2 - k_t^2\big) e^{-R_e x_2^*} e^{-R_t a_2} \Big\}, \\
B_{21} = & \frac{-i\xi_1}{N(\xi_1)\mu k_t^2} \Big\{ \frac{1}{2} \Big[\big(2\xi_1^2 - k_t^2\big)^2 + 4\xi_1^2 R_e R_t \Big] \Big(e^{-R_t(x_2^* + a_2)} + e^{-R_e(x_2^* + a_2)} \Big) - \\
& - 2\big(2\xi_1^2 - k_t^2\big) R_e R_t e^{-R_e x_2^*} e^{-R_t a_2} - 2\xi_1^2 \big(2\xi_1^2 - k_t^2\big) e^{-R_t x_2^*} e^{-R_e a_2} \Big\}, \\
B_{22} = & \frac{\xi_1^2}{N(\xi_1)\mu k_t^2 R_t} \Big\{ \Big[\big(2\xi_1^2 - k_t^2\big)^2 + 4\xi_1^2 R_e R_t \Big] \Big(\frac{R_e R_t}{2\xi_1^2} e^{-R_e(x_2^* + a_2)} + \\
& + \frac{1}{2} e^{-R_t(x_2^* + a_2)} \Big) - 2\big(2\xi_1^2 - k_t^2\big) R_e R_t \big(e^{-R_t x_2^*} e^{-R_e a_2} + e^{-R_e x_2^*} e^{-R_t a_2} \big) \Big\},
\end{aligned}
\tag{18.140}
$$

where

$N(\xi_1) = 2\big(2\xi_1^2 - k_t^2\big)^2 - 4\xi_1^2 R_e R_t$— Rayleigh function.

Substituting $a_2 = 0$ into the obtained solution, we obtain the solution of the Lamb problem of the harmonic force acting on the half-plane and applied to its boundary

$$
g(x_1^*, x_2^*; 0, 0) = G(x_1^*, x_2^*; 0, 0) - \frac{1}{2\pi} \int_{-\infty}^{\infty} B(\xi_1, x_2^*; 0, 0) e^{-i\xi_1 x_1^*} d\xi_1, \tag{18.141}
$$

where

$$
G(x_1^*, x_2^*; 0, 0) = \frac{1}{2\pi} \int_{-\infty}^{\infty} \overline{G}(\xi_1, x_2^*; 0, 0) e^{-i\xi_1 x_1^*} d\xi_1. \tag{18.142}
$$

Substituting (18.142) into (18.141), we will obtain:

$$
g(x_1^*, x_2^*; 0, 0) = \frac{1}{2\pi} \int_{-\infty}^{\infty} \big[\overline{G}(\xi_1, x_2^*; 0, 0) - B(\xi_1, x_2^*; 0, 0)\big] e^{-i\xi_1 x_1^*} d\xi_1. \tag{18.143}
$$

In particular, if a vertical force is acting on the half-plane then

$$
g_i(x_1^*, x_2^*; 0, 0) = \frac{1}{2\pi} \int_{-\infty}^{\infty} \big[G_{i2}(\xi_1, x_2^*; 0, 0) - B_{i2}(\xi_1, x_2^*; 0, 0)\big] e^{-i\xi_1 x_1^*} d\xi_1. \tag{18.144}
$$

Determining the elements of $G_{i2}(\xi_1, x_2^*; 0, 0)$, from (18.143) after simple transformations let us find:

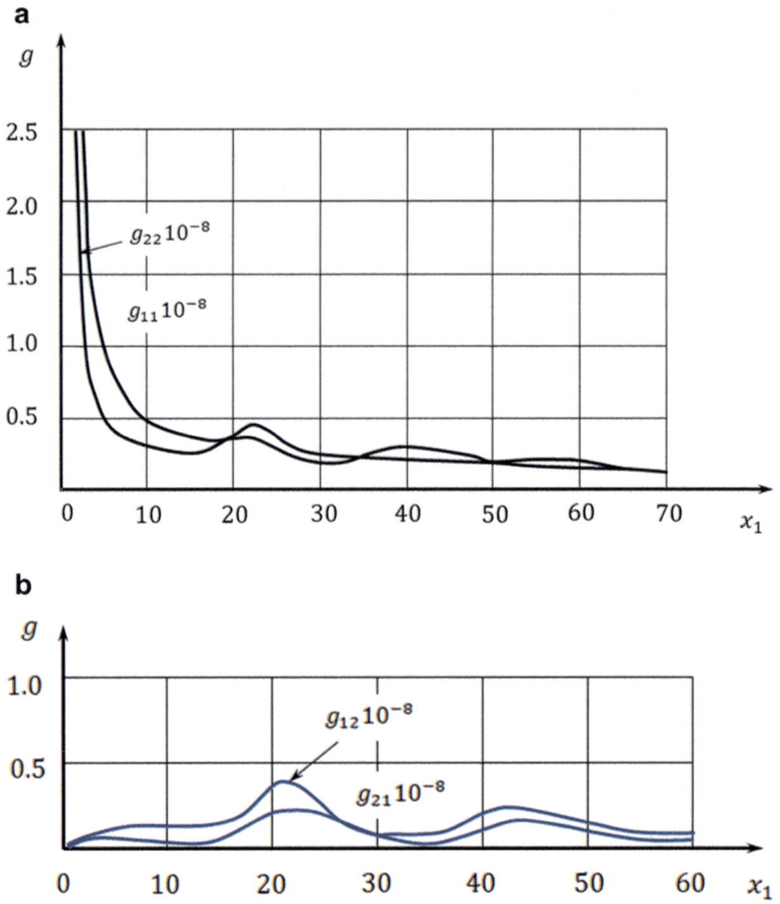

Fig. 18.4 (**a**) Graphs of the amplitudes of the oscillations of the surface of a half-plane under the action of a unit force (**b**) Graphs of the amplitudes of the oscillations of the surface of a half-plane under the action of a unit force

$$g_{12}(x^*;0) = \frac{i}{2\pi\mu}\int_{-\infty}^{\infty} \frac{\xi_1\left[\left(2\xi_1^2 - k_t^2\right)e^{-R_e x_2^*} - 2R_t R_e e^{-R_t x_2^*}\right]e^{-i\xi_1 x_1^*}}{N(\xi_1)} d\xi_1, \tag{18.145}$$

$$g_{22}(x^*;0) = \frac{1}{2\pi\mu}\int_{-\infty}^{\infty} \frac{R_e\left[\left(2\xi_1^2 - k_t^2\right)e^{-R_e x_2^*} - 2\xi_1^2 e^{-R_t x_2^*}\right]e^{-i\xi_1 x_1^*}}{N(\xi_1)} d\xi_1. \tag{18.146}$$

The solutions (18.145) and (18.146) exactly coincide (exactly the same with) with the solution of the Lamb's problem.

Figure 18.4 shows graphs of the amplitudes of oscillations on a surface of a half-plane under the action of a unit force, located at the depth of 10 m (along the vertical coordinate axis) from the surface of the ground and changing in time according to the harmonic low.

The angular (circular) frequency of the force change $\omega = 40$ radian/sec, the modulus of elasticity of the material of half-plane, $E = (1) = 20000 k\Gamma/\text{cm}^2$, the coefficient of loss (losses) $\gamma = 0.2$. The calculations were carried out according to the above given formulas on the mainframe YeS -1080.

18.5 A Frequency-Independent Relative Damping Model

When building linear models of frequency-independent internal frictions, it becomes necessary to observe the fundamental principle of mechanics—the causality principle. From the theory of dynamic systems, it follows that between the real and imaginary parts of the transfer function of a causal system (18.5)

$$\Phi(\omega) = P(\omega) + iQ(\omega) \tag{18.147}$$

must exist the certain dependence (18.6). The transfer function $\Phi(\omega)$, is other than zero only on the positive semi-axis of time. These constrains are associated with the interdependence of real and imaginary parts of the transfer functions. Meanwhile, in the dynamic calculations of structural engineering and mechanical engineering designs, a priori representations of structural behavior with harmonic loads are frequently used, assigning various modes of complex rigidity or pliability.

As indicated above in paragraphs 18.1 and 18.2, the linear model of internal friction must satisfy the principle of causality and the principle of frequency independence of absolute losses. Thus, a linear internal friction model has unrestricted pliability at all frequencies. At the same time, it is possible to create a causal model with frequency-independent relative losses. Such a model can have unlimited pliability (compliance) only at the point $\omega = 0$, that is, under static loading of the structure. The above-mentioned circumstance can be neglected in dynamic calculations. Such a model has been considered in [9, 24]. Below we present the results of Tseitlin and Plotnikov [25].

The frequency independence of the loss factor $\gamma = tg\varphi$ is equivalent to the frequency independence of the phase response of the system.

$$\varphi(\omega) = \varphi_0 = \text{const.} \tag{18.148}$$

As is known, when considering the frequency characteristics on the entire ω- axis, they must be extended to the negative axis in accordance with:

$$A(\omega) = A(-\omega);$$
$$\varphi(\omega) = -\varphi(-\omega)$$

where $A(\omega)$, and $\varphi(\omega)$ are amplitude and phase characteristics, respectively. The amplitude characteristic must be extended in an even way, and the phase characteristic must be extended in an odd way. Similar dependences are also valid for real and imaginary frequency characteristics. Thus, the phase is an odd function of frequency, so it should have the form:

$$\varphi(\omega) = \varphi_0(\omega)\text{sign}\omega = \text{sign}\omega \tan^{-1}\gamma. \tag{18.149}$$

The derivative of the phase characteristic (18.149) has the form:

$$\varphi'(\omega) = 2\varphi_0\delta(\omega). \tag{18.150}$$

In the formula (18.150), the function sign ω has a jump at the point $\omega = 0$ equal to 2. Therefore, a delta function with coefficient 2 appears in (18.150). Let's check whether the causality condition is satisfied. For this, we use the Hilbert transform studied in Sect. 18.1. Dependences similar to (18.6) can also be obtained for the amplitude and phase frequency characteristics. Let's write logarithmic equation for transfer function (18.147). We obtain:

$$\ln\Phi(\omega) = \ln A(\omega) - i\varphi(\omega). \tag{18.151}$$

Since there is a complete analogy between the formula (18.5), and (18.151). In this case, $A(\omega)$ is an even and $\varphi(\omega)$ is an odd function. Then similar calculations can be made for (18.151). As a result of which, we obtain:

$$\ln A(v) = \frac{1}{\pi}\int_{-\infty}^{\infty}\frac{\varphi(\omega)d\omega}{\omega - v} \tag{18.152}$$

$$\varphi(v) = \frac{1}{\pi}\int_{-\infty}^{\infty}\frac{\ln A(\omega)d\omega}{v - \omega}. \tag{18.153}$$

Let's transform the formula (18.152), assuming that $\omega - v = u$, then we obtain:

$$\ln A(v) = \frac{1}{\pi}\int_{-\infty}^{\infty}\frac{\varphi(u + v)du}{u}. \tag{18.154}$$

This integral can be considered as a functional specifying the derivative of the generalized function $\ln|u|$ [12]:

$$\ln A(v) - -\frac{1}{\pi}\int_{-\infty}^{\infty}\ln|u|\varphi'(u + v)du. \tag{18.155}$$

Formulas (18.152) and (18.153) give an unambiguous relationship between $A(\omega)$ and $\varphi(\omega)$ under the condition that $\ln\Phi(\omega)$ has no poles at $Im\omega \leq 0$. [Please see the formulas: (18.36), and (18.37)]. Since the zero of the transfer function is the pole of its logarithm, and then this condition is equivalent to the requirement that $\Phi(\omega)$ does not only have poles, but also zeroes for $Im\omega \leq 0$. Systems are called minimally phased if their transfer functions satisfy these conditions [4]. Such systems, unlike other systems with the same logarithmic amplitude characteristic $\ln A(\omega)$, give a minimum phase shift at any frequency.

Using the transformed formulas (18.152) and (18.153), the following formulas can also be obtained [26]:

$$\begin{aligned}
A(v)\cos\varphi(v) &= -\frac{2}{\pi}\int_0^{\infty}\frac{A(\omega)\sin\varphi(\omega)\omega d\omega}{v^2 - \omega^2};\\
A(v)\sin\varphi(v) &= \frac{2}{\pi}\int_0^{\infty}\frac{A(\omega)\cos\varphi(\omega)d\omega}{v^2 - \omega^2};\\
A^{-1}(v)\cos\varphi(v) &= \frac{2}{\pi}\int_0^{\infty}\frac{A^{-1}(\omega)\sin\varphi(\omega)\omega d\omega}{v^2 - \omega^2};\\
A^{-1}(v)\sin\varphi(v) &= -\frac{2}{\pi}\int_0^{\infty}\frac{A^{-1}(\omega)\sin\varphi(\omega)d\omega}{v^2 - \omega^2};
\end{aligned} \tag{18.156}$$

and

$$\varphi(v)=\frac{1}{\pi}\int_{-\infty}^{\infty}\frac{d\ln A(\omega)}{d\omega}\ \text{ln cot}\frac{\omega}{2}d\omega. \tag{18.157}$$

The formula (18.157) indicates the relationship between ln$A(\omega)$ and $\varphi(\omega)$ [4].

As noted at the beginning of in this chapter, and since the above dependencies connect different characteristics of causal system, these dependences should be taken into account in the a priori construction of linear models of dynamic systems [4].

The derivative of the phase response (18.150) is substituted into (18.155), we obtain:

$$A(\omega)=|\omega|^{\alpha} \tag{18.158}$$

where

$$\alpha=2\varphi_0\pi^{-1}=\frac{2\tan^{-1}\gamma}{\pi}.$$

Since formula (18.154) is valid only for minimum-phase systems. That is, for such systems, the transfer functions do not have poles and zeroes in the lower half-plane, including the real axis. Let's check this condition. The amplitude characteristic has a pole on the real axis. Therefore, the system is not minimum phase. Therefore, the logarithmic amplitude characteristic is determined with the accuracy to the constant summand. Let's assume that in formula (18.157)

$$A(\omega)=c|\omega|^{\alpha}, \tag{18.159}$$

where $c=A(1)=\text{const}$.

Thus, the amplitude characteristic (18.159) and phase characteristic (18.149) define the linear physically realizable element. For this element, we have to find the kernels. To do this, let's move to the basic formulations and the assumptions that characterize linear dynamic systems. We show some basic properties of generalized functions, transfer functions, complex rigidity, and also show the main dependencies between the dynamic characteristics.

As noted at the beginning of this chapter, the displacement of a system at some point at the output can be related to the force acting on it at some other point or at the same point at the input by using a linear operator $y(t)=\boldsymbol{K}P(t)$. Let us assume that, in general, the operator $\boldsymbol{K}$ has a single inverse operator, $\boldsymbol{\Gamma}=\boldsymbol{K}^{-1}$ therefore, $P(t)=\boldsymbol{\Gamma}y(t)$.

The transfer function of a system completely defines its input-output relationship in a steady-state. The impulse transient or impulse response function $y_\delta(t)$ of a causal system, being a Fourier transform of the transfer function $\Phi(\omega)$, is other than zero only on the positive semi-axis of time

$$y_\delta(t)=\frac{1}{2\pi}\int_{-\infty}^{\infty}\Phi(\omega)e^{i\omega t}d\omega=0, \qquad (t<0) \tag{18.160}$$

$y_\delta(t)$ is a function of the movement of the element when the load is applied to it in the form of a delta function. The impulse transfer function is denoted by $y_\delta(t)=K(t)$. Since the considered linear system is also assumed to be stationary, the displacement of the system can be expressed as a convolution of two functions:

$$y(t) = \int_{-\infty}^{\infty} K(t-\tau)P(\tau)d\tau \tag{18.161}$$

where for an elastic inertia-free system $K(t - \tau) = k^{-1}\delta(t - \tau)$, k—stiffness of the element is equal to the force that causes its unit movement.

In the theory of linear dynamical systems, the impulse transfer function is also called the Green's function. Thus, the impulse transfer function of a stable dynamic system must satisfy the following conditions [27]:

$$\int_{0}^{\infty} |K(t)|dt < \infty \tag{18.162}$$

$$K(t) = 0, \quad (t < 0), \tag{18.163}$$

where (18.162) is the stability condition and (18.163) expresses the principle of causality, that is, the system's response cannot precede the impact. Because the cause of movement is action. The integration in (18.161) can be limited to a finite limit, taking into account (18.163):

$$y(t) = \int_{-\infty}^{t} K(t-\tau)P(\tau)d\tau \tag{18.164}$$

The fulfillment of conditions (18.163) for $K(t)$ is equivalent to the fact that the Fourier transform of the function $K(t)$ is an analytic and bounded (limited) function in the lower half-plane.

Usually, loads, including dynamic ones, are represented as a delta function, a Heaviside function. When a harmonic action is suddenly applied, the loads are described by the trigonometric functions of the sine and cosine. At the load (under load, with the load) $P(t) = H(t)$, the movement (displacement) of the element according to (18.5.18) will be:

$$y_H(t) = \int_{-\infty}^{t} K(t-\tau)H(\tau)d\tau = \int_{0}^{t} K(t-\tau)d\tau \tag{18.165}$$

where: $H(t)$ is unit function of Heaviside;

$$H(t) = \begin{cases} 1 & (t \geq 0) \\ 0 & (t < 0) \end{cases},$$

The expression (18.165) is differentiated by t, and we obtain:

$$dy_H(t)/dt = K(t).$$

When the load is applied $P(t) = \cos \omega t$, $P(t) = \sin \omega t$, the displacement of the element (body) changes according to the corresponding law. The shift in the phase of the oscillations should take place in relation to the action, when internal friction is taken into account.

$$\begin{aligned} y_{\cos}(t) &= A(\omega)\cos[\omega t - \varphi(\omega)] \\ y_{\sin}(t) &= A(\omega)\sin[\omega t - \varphi(\omega)]. \end{aligned} \tag{18.166}$$

From the formula (18.166) it follows that at $P(t) = e^{i\omega t}$

$$y_{\exp}(t) = A(\omega)e^{i\,[\omega t - \varphi(\omega)]} = \Phi(\omega)e^{i\omega t}. \tag{18.167}$$

The value of $\Phi(\omega) = A(\omega)e^{-i\varphi(\omega)}$ is called the transfer function and $A(\omega)$, $\varphi(\omega)$ are called amplitude and phase frequency characteristics. The real and imaginary parts of transfer function (18.5) are called real and imaginary frequency characteristics and have the following form:

$$\begin{aligned} P(\omega) &= Re\,\Phi(\omega) = A(\omega)\cos\varphi(\omega) \\ Q(\omega) &= Im\Phi(\omega) = -A(\omega)\sin\varphi(\omega). \end{aligned} \tag{18.168}$$

The real and imaginary parts of the transfer function are related to the amplitude and phase frequency characteristics:

$$\begin{aligned} A(\omega) &= \sqrt{[P(\omega)]^2 + [Q(\omega)]^2}; \\ \varphi(\omega) &= -\tan^{-1}\frac{Q(\omega)}{P(\omega)}. \end{aligned} \tag{18.169}$$

It is known, that when considering the frequency characteristics on the entire frequency axis ω, it is necessary to extend the dependences (18.166) and (18.167) to the negative semi-axis. According to (18.166), herewith $P(t) = e^{-i\omega t}$, $\omega > 0$. From (18.167) when replacing the argument ω with $-\omega$, it follows:

$$y_{exp}(t) = A(-\omega)e^{-i\,[\omega t - \varphi(-\omega)]}.$$

The real and imaginary part of the transfer function through complex rigidity will have the form:

$$\Phi(\omega) = Rek^{-1}(\omega) + Imk^{-1}(\omega);$$

where

$$\Phi^{-1}(\omega) = k(\omega) = A^{-1}(\omega)e^{i\varphi(\omega)}.$$

The relationship between the impulse transient (impulse Response) functions $y_\delta(t)$, $K(t)$, the transfer function $\Phi(\omega)$ [(*Please see the formula* (18.5)], and frequency characteristics $A(\omega)$, $\varphi(\omega)$ can be established using the spectral representation of the delta function [12]. The delta function is defined as a functional acting according to the formula

$$(\delta(x - x_0), \varphi(x)) = \varphi(x_0), \tag{18.170}$$

and at $x_0 = 0$

$$(\delta(x_0), \varphi(x)) = \varphi(0),$$

with the following basic property:

$$\begin{aligned} \delta(x - x_0) &= \begin{cases} \infty & x = x_0 \\ 0 & \text{otherwise} \end{cases}; \\ (\delta(x - x_0),\ \varphi(x)) &= \int_{-\infty}^{\infty} \delta(x - x_0)\varphi(x)dx = \varphi(x_0). \end{aligned} \tag{18.171}$$

The spectral representation of the delta function has the following form:

$$\delta(t) = \frac{1}{2\pi} \int_{-\infty}^{\infty} e^{i\omega t} d\omega; \tag{18.172}$$

$$\delta(t) = \frac{1}{\pi} \int_{0}^{\infty} \cos \omega t d\omega. \tag{18.173}$$

Here we used the formula, known from the theory of the generalized functions:

$$\int_{0}^{\infty} e^{i\omega t} d\omega = it^{-1} + \pi\delta(t), \tag{18.174}$$

where it follows:

$$\int_{0}^{\infty} \sin \omega t d\omega = \frac{1}{t}, \qquad \int_{0}^{\infty} \cos \omega t d\omega = \pi\delta(t). \tag{18.175}$$

(18.147) and (18.160) follow that:

$$y_{\delta}(t) = K(t) = \frac{1}{2\pi} \int_{-\infty}^{\infty} [P(\omega) \cos \omega t - Q(\omega) \sin \omega t\,] d\omega. \tag{18.176}$$

Taking into account the principle of causality (18.163), we find:

$$\frac{1}{2\pi} \int_{-\infty}^{\infty} [P(\omega) \cos \omega t + Q(\omega) \sin \omega t\,] d\omega = 0. \tag{18.177}$$

Further adding and then subtracting (18.176) and (18.177), we obtain for ($t > 0$)

$$K(t) = \frac{2}{\pi} \int_{0}^{\infty} P(\omega) \cos \omega t d\omega = -\frac{2}{\pi} \int_{0}^{\infty} Q(\omega) \sin \omega t d\omega, . \tag{18.178}$$

The transient function is defined as a superposition of the transfer functions that correspond to the decomposition (expansion) of the load $H(t)$ into the integral Fourier:

$$H(t) = \frac{1}{2\pi} \int_{-\infty}^{\infty} F_H(\omega) e^{i\omega t} d\omega, \tag{18.179}$$

here, also taking into account (18.174), we find:

$$F_H(\omega) = \int_{-\infty}^{\infty} H(t) e^{-i\omega t} dt = \int_{0}^{\infty} e^{-i\omega t} d\omega = \pi\delta(\omega) - \frac{i}{\omega}.$$

Since, the load $P(t) = e^{i\omega t}$ corresponds to the displacement $y = \Phi(\omega)e^{i\omega t}$, then the load $P(t) = H(t)$ according to (18.179) will correspond to the displacement:

$$y_H(t) = \frac{1}{2\pi}\int_{-\infty}^{\infty}\left[\pi\delta(\omega) - \frac{i}{\omega}\right]\Phi(\omega)e^{i\omega t}d\omega =$$
$$= \frac{1}{2}P(0) + \frac{1}{\pi}\int_0^{\infty}\frac{P(\omega)\sin\omega t + Q(\omega)\sin\omega t}{\omega}d\omega = \tag{18.180}$$
$$= \frac{1}{2}A(0)\cos\varphi(0) + \frac{1}{\pi}\int_0^{\infty}\frac{A(\omega)\sin[\omega t - \varphi(\omega)]}{\omega}d\omega.$$

In the expression (18.180) it is taken into account that since $Q(\omega)$ is continuous at zero and is odd, the transfer function at zero: $\Phi(0) = P(0)$.

Similar functions can be constructed for the operator $\boldsymbol{\Gamma} = \boldsymbol{K}^{-1}$. Let's represent the dependence $P = \boldsymbol{\Gamma}y$ in the form:

$$P(t) = \int_{-\infty}^{t}\Gamma(t-\tau)y(\tau)d\tau. \tag{18.181}$$

Here $\Gamma(t)$ is the load, that causes a displacement equal to $\delta(t)$. The mechanical interpretation of such a displacement and, accordingly, the operator goes beyond the limits of classical analysis. However, this is not an obstacle to obtaining correct results from a physical and mathematical point of view.

Let's substitute $y(\tau) = \cos\omega\tau$, $y(\tau) = \theta(\tau)$ and also the amplitude and frequency characteristics in (18.181) and transform this expression, we obtain:

$$P_{cos}(t) = A^{-1}(\omega)\cos[\omega t + \varphi(\omega)] \tag{18.182}$$

$$P_{\sin}(t) = A^{-1}(\omega)\sin[\omega t + \varphi(\omega)] \tag{18.183}$$

$$P_{\exp}(t) = A^{-1}(\omega)e^{i[\omega t + \varphi(\omega)]} = [\Phi(\omega)]^{-1}e^{-i\omega t} \tag{18.184}$$

$$P_\delta(t) = \Gamma(t) = \frac{1}{\pi}\int_0^{\infty}A^{-1}(\omega)\cos[\omega t + \varphi(\omega)]d\omega; \tag{18.185}$$

$$P_H(t) = \frac{1}{2}A^{-1}(0)\cos\varphi(0) + \frac{1}{\pi}\int_0^{\infty}\frac{A^{-1}(\omega)\sin[\omega t + \varphi(\omega)]}{\omega}d\omega; \tag{18.186}$$

$$\Gamma(t) = \frac{1}{2\pi}\int_{-\infty}^{\infty}k(\omega)e^{i\omega t}d\omega. \tag{18.187}$$

These dependencies are very general. The characteristics of both ideally elastic and elastic-viscous, for elastically inherited dynamic systems can be described by these dependencies. In general, the functions $K(t) = y_\delta(t)$, and $\Gamma(t) = P_\delta(t)$, along with the regular part, can contain generalized functions in the form of a delta function or its derivatives. The equations of motion of the system will take one form or another depending on this.

According to formulas (18.178) and (18.185), we obtain:

$$K(t) = \begin{cases} \dfrac{ct^{\alpha-1}}{\Gamma(\alpha)} & (t>0) \\ 0 & (t>0) \end{cases}, \tag{18.188}$$

$$\widetilde{\Gamma}(t) = \begin{cases} \dfrac{1}{c\Gamma(-\alpha)t^{\alpha+1}} & (t>0) \\ 0 & (t>0) \end{cases} \tag{18.189}$$

here the values of integrals 3.764 [6] were used. $\Gamma(\alpha)$, $\Gamma(-\alpha)$- gamma function; the kernel $\Gamma(t)$ is denoted by $\widetilde{\Gamma}(t)$. The integral operators with Abelian kernels (18.188), (18.189) act on the positive semi-axis. These kernels are fractional differentiation operators since the transfer function can be represented as a fractional-rational function. In this case, the order of the numerator of transfer function should not exceed the order of the denominator of transfer function. Due to the fact that the transfer function of the dissipative system, in contrast to the same conservative system, is limited at all frequencies. The stability condition of the dynamic system requires that all poles of the transfer function lie in the upper half-plane.

Hereditary models with Abel-type operators are successfully applied in creep theory [18]. The behavior of electrical and mechanical systems, which is described by Abel-type operators, has been studied by many researchers. The dynamic characteristic of a system of the type (18.159) was constructed in the theory of electrical circuits, apparently for the first time [3]. In the theory of creep, the kernel (18.188) was considered by Scott-Blair [28] for the first time. Generalized laws of deformation were introduced by Gerasimov [29]. In this work, operators of fractional differentiation were used.

In works [9, 24], the fractional differentiation operator is used to construct a linear model of frequency-independent internal friction in structural materials. A feature of the proposed model is the operator's description of the fractional differentiation of both components of the resistance forces: elastic and dissipative. In this case, the rigidity (stiffness) of the element, that is, $\cos\varphi(\omega)/A(\omega)$, depends on the frequency of the disturbance. The rigidity (stiffness) of the element at $\omega = 0$ due to unlimited creep is equal to zero. Therefore, very slow processes should be excluded when considering dynamic problems. It is important that a closed expression (18.189) can be obtained for the kernel of the resolvent using the model described by the linear operator with the kernel (18.188). This allows us to analyze equations of motion of systems with frequency-independent internal friction in a simple and comprehensible form. Let us consider the oscillations of a system with single degree of freedom. The system has frequency-independent internal friction. The internal friction is described by the operator obtained above. The equation of motion of such a system can be written in the form:

$$m\ddot{y} + \frac{k_0}{\Gamma(-\alpha)} \int_{-\infty}^{t} \frac{y(\tau)d\tau}{(t-\tau)^{1+\alpha}} = P(t), \tag{18.190}$$

where m is the mass of the system, y is the displacement of the system; P is the load, and $k_0 = A^{-1}$.

If $P(t) \equiv 0$ at $t < 0$, then Eq. (18.190) can be written in differential form

$$mD_t^2y + k_0D_t^\alpha y = P(t), \quad (t > 0).$$

Let us consider harmonic oscillations of the system. Assuming $P = e^{i\omega t}$, we calculate the transfer function. Substituting $y(t) = \Phi(\omega)e^{i\omega t}$ in (18.189) and replacing the integration variable $t - \tau = s$, we obtain:

$$\Phi(\omega) = \left[-m\omega^2 + \frac{k_0}{\Gamma(-\alpha)} \int_0^\infty \frac{\exp(-i\omega s)}{s^{1+\alpha}} ds \right]^{-1} \tag{18.191}$$

The divergent integral (18.191) can be given a meaning if we consider it as the Fourier transform of generalized functions $s_+^{-1-\alpha}$. As is known, the following equality takes place [16]:

$$\int_{-\infty}^{\infty} x_+^\lambda e^{-ix\beta} dx = A(\lambda)\beta_+^{-\lambda-1} + B(\lambda)\beta_-^{-\lambda-1}, \tag{18.192}$$

where

$$A(\lambda) = i\exp\left(i\lambda\frac{\pi}{2}\right)\Gamma(1+\lambda); \quad B(\lambda) = -i\exp\left(-i\lambda\frac{\pi}{2}\right)\Gamma(1+\lambda)$$

$$\beta_+^{-\lambda-1} = \begin{cases} \beta^{-\lambda-1} & (\beta > 0); \\ 0 & (\beta < 0); \end{cases} \quad \beta_-^{-\lambda-1} = \begin{cases} 0 & (\beta > 0); \\ |\beta|^{-\lambda-1} & (\beta < 0). \end{cases}$$

Using the values of the integral (18.192), we find:

$$\int_0^\infty s^{-1-\alpha} e^{-i\omega s} ds = |\omega|^\alpha \,\Gamma(-\alpha) \left(\cos\frac{\pi\alpha}{2} + i\,\mathrm{sign}\omega \cdot \sin\frac{\pi\alpha}{2} \right).$$

Hence, the transfer function is presented in the following form:

$$\Phi(\omega) = \left[-m\omega^2 + k_0|\omega|^\alpha \cdot \left(\cos\frac{\pi\alpha}{2} + i\,\mathrm{sign}\omega \cdot \sin\frac{\pi\alpha}{2} \right) \right]^{-1}. \tag{18.193}$$

Therefore, the amplitude and phase frequency characteristics of the system will be ($\omega \geq 0$):

$$A(\omega) = m^{-1}p^{\alpha-2}\omega^{-\alpha} \left[\left(1 - \frac{\omega^{2-\alpha}}{p^{2-\alpha}} \right) + \gamma^2 \right]^{-1/2}; \tag{18.194}$$

$$\varphi(\omega) = \tan^{-1} \frac{\gamma}{\frac{\omega^{2-\alpha}}{p^{2-\alpha}} - 1}, \tag{18.195}$$

where

$$p = \left(\frac{k_0}{m} \cos\frac{\pi\alpha}{2} \right)^{\frac{1}{2-\alpha}}.$$

The amplitude characteristic (18.194) can also be given the following form:

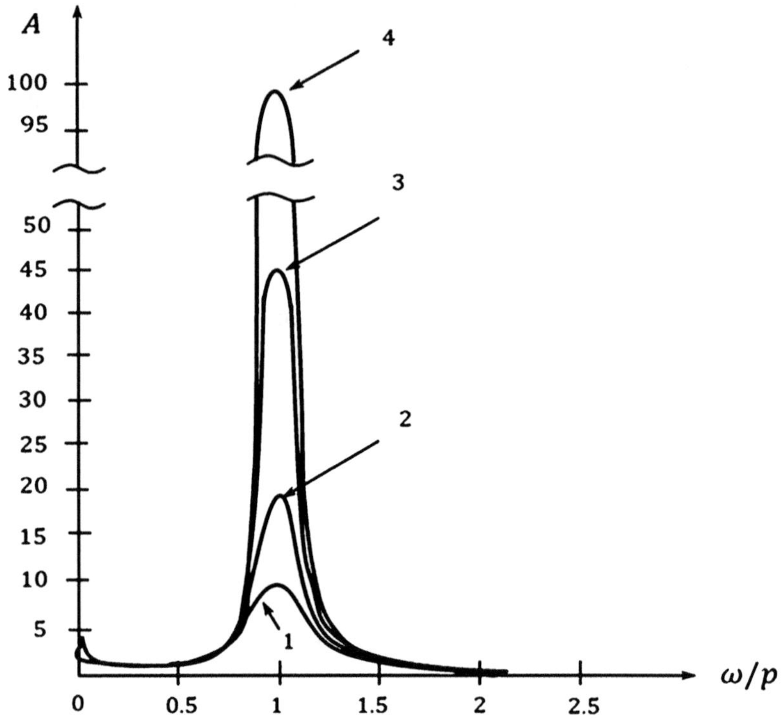

Fig. 18.5 Amplitude-frequency characteristic of an oscillator with frequency-independent relative damping

$$A(\omega) = k^{-1}\left[\left(\frac{\omega^2}{p^2} - \frac{\omega^2}{p^2}\right)^2 + \gamma^2 \frac{\omega^{2\alpha}}{p^{2\alpha}}\right]^{-1/2}, \tag{18.196}$$

where k is the element stiffness at frequency p;

$$k = k_0 p^{\alpha} \cos \frac{\pi\alpha}{2}.$$

For small values of γ, the amplitude characteristic (18.196) does not differ from the same characteristic of a system with constant complex stiffness on the entire axis except for a small neighborhood from the zero-point (Fig. 18.5).

$$1 - \gamma = 0.1; \quad 2 - \gamma = 0.05; \qquad 3 - \gamma = 0.02; \quad 4 - \gamma = 0.01$$

The amplitude characteristic just in this neighborhood turns to infinity. This neighborhood (ω/p) corresponds to very slow processes (close to static) and can be excluded from consideration when solving dynamic problems.

The impulse transient function of the system with frequency-independent internal friction can be obtained by integrating the transfer function (18.193) by the formula:

$$K_c(t) = \frac{1}{2\pi} \int_{-\infty}^{\infty} \Phi(\omega) e^{i\omega t} d\omega$$

As shown in [13], as a result of integration, we come to the following expression:

$$K_c(t) = \begin{cases} \frac{1}{m\lambda} Sc_\alpha(\lambda t), & t \geq 0, \\ 0 & t < 0, \end{cases} \tag{18.197}$$

where

$$\lambda = \left(k_0 m^{-1}\right)^{\frac{1}{2-\alpha}}; \qquad Sc_\alpha(x) = \sum_{n=0}^{\infty} (-1)^n \frac{x^{(2-\alpha)n+1}}{\Gamma((2-\alpha)n+2)}.$$

Impulse transient (impulse response) function of a system with frequency-independent internal friction can also be obtained from the equation:

$$mD_t^2 K_c(t) + k_0 D_t^\alpha K_c(t) = \delta(t); \quad K_c(0) = 0; \quad D_t K_c(0 = 0). \tag{18.198}$$

Using the well-known property of operators of fractional differentiation $D^\alpha D^\beta = D^{\alpha+\beta}$, which allows constructing the algebra of these operators with D^0, we obtain:

$$K_c(t) = m^{-1} D_t^2 \left(D_t^0 + \lambda^{2-\alpha} D_t^{\alpha-2}\right)^{-1} \delta(t).$$

here D^0 plays the role of a unit (here D^0 is assumed to be equal to one). The formal expression in parentheses expanding into a Maclaurin series in powers $\lambda^{2-\alpha} D_t^{\alpha-2}$, we obtain:

$$K_c(t) = \frac{1}{m\lambda} \sum_{n=0}^{\infty} (-1)^n \lambda^{(2-\alpha)n+1} D_t^{(\alpha-2)n-2} \delta(t).$$

Hence, we pass from here to the integral form of the operators of fractional differentiation. Here we take into account the same basic properties of the delta function, as a result we obtain:

$$K_c(t) = \frac{1}{m\lambda} \sum_{n=0}^{\infty} (-1)^n \frac{(\lambda t)^{(2-\alpha)n+1}}{\Gamma[(2-\alpha)n+2]}, \quad (t \geq 0). \tag{18.199}$$

The function $Sc_\alpha(x)$, by analogy with the fractional-exponential functions of Rabotnov [18], can be defined as a fractional sine. Its first derivative

$$Cs_\alpha(x) = \frac{dSc_\alpha(x)}{dx} = \sum_{n=0}^{\infty} (-1)^n \frac{x^{(2-\alpha)n}}{\Gamma[(2-\alpha)n+1]} \tag{18.200}$$

is naturally defined as a fractional cosine. It is obvious that $Sc_0(x) = \sin x$, $Cs_0(x) = \cos x$. The functions $Sc_\alpha(x)$ and $Cs_\alpha(x)$ are closely related to the fractional-exponential Rabotnov's functions:

$$D_x Cs_\alpha(x) = D_x^2 Sc_\alpha(x) = \ni_{1-\alpha}(-1, x),$$

where

$$\ni_\alpha(-\beta, t) = t^\alpha \sum_{n=0}^{\infty} (-1)^n \frac{\beta^n t^{(1+\alpha)n}}{\Gamma[(1+\alpha)(n+1)]}$$

The fractional cosine and fractional sine are expressed through the $\ni_\alpha$ function of the positive parameter for the values of α. The values of α lie in the interval $0 < \alpha < 1$. The function of the negative α parameter $\ni_\alpha$ is widely used in the quasistatic theory of the hereditary theory of elasticity, in contrast to the function of the positive parameter.

Here are some results that relate to the properties of fractional trigonometric functions. These results are presented in [13]. The fractional sine $Sc_\alpha(x)$ can be considered as a special case of the more general Mittag-Leffler function.

$$E_\beta(x;y)=\sum_{n=0}^{\infty}\frac{x^n}{\Gamma(n\beta^{-1}+q)}. \tag{18.201}$$

The properties of this function have been studied in detail. Therefore, the asymptotic representation of the function can be obtained directly by means of the Mittag-Leffler function. In this case, the asymptotic representation of the function $Sc_\alpha(x)$ is valid for large values of the argument.

$$Sc_\alpha(x)\approx-\sum_{n=1}^{\infty}(-1)^n\frac{x^{1-(2-\alpha)n}}{\Gamma[2-(2-\alpha)n]}. \tag{18.202}$$

The representation of the function $Sc_\alpha(x)$ in the form of a series makes it difficult to study the properties of the impulse transient function. Therefore, the exact expression for $K_c(t)$ in the integral form is obtained using the theory of deduction. Let us write the representation for $K_c(t)$ as a Melline integral, assuming $i\omega=s$, and taking into account the features of the integrand:

$$K_c(t)=\frac{1}{2\pi im\lambda}\int_{a-i\infty}^{a+i\infty}\frac{e^{s\lambda t}ds}{s^2+s^\alpha},.\qquad a>0. \tag{18.203}$$

The integrand has a branch point $s=0$, and two poles on the plane $|\arg s|<\pi$. Here the poles are located to the left of the imaginary axis:

$$s_1=\exp\left[i\pi(2-\alpha)^{-1}\right];$$
$$s_2=\exp\left[-i\pi(2-\alpha)^{-1}\right].$$

Let us perform operations similar to those given in Sect. 18.2., i.e., we close the contour of a circle with radius $R\to\infty$. The integration contour is traversed along two sides of the negative semi-axis and a circle of radius $\rho\to0$ around the branch point $s=0$ (see Fig. 18.1). Here, we use the Jordan's lemma and simple estimates for the integral on a small circle, as a result we obtain:

$$K_c(t)=\frac{1}{m\lambda}Sc_\alpha(\lambda t)=\frac{1}{m\lambda}\left[A_0(\lambda t)+Ce^{\eta t}\sin(p_1t-\varphi)\right], \tag{18.204}$$

where

$$\begin{gathered}A_0(x)=\frac{\sin\pi\alpha}{\pi}\int_0^\infty\frac{s^{-\alpha}e^{-sx}}{\left(s^{2-\alpha}+\cos\pi\alpha\right)^2+\sin^2\pi\alpha}ds;\\ C=2(2-\alpha)^{-1};\quad \varphi=\pi\alpha[2(2-\alpha)]^{-1};\quad \eta=\lambda\cos\frac{\pi}{2-\alpha}.\end{gathered} \tag{18.205}$$

This representation allows us to find out the behavior of the fractional cosine over the entire interval of variation of the argument ($0\le x<\infty$). As we can see, the impulse transient (impulse response) function $K_c(t)$ defines a damped harmonic oscillation with a relatively slowly varying (changing) function $A_0(\lambda t)$. Thus, the natural frequency $p_1=\lambda\sin[\pi(2-\alpha)^{-1}]$, and decrement of damped harmonic oscillations $\delta=2\pi|\cot[\pi(2-\alpha)^{-1}]|$, do not depend on the frequency of natural oscillations.

The function $A_0(\lambda t)$ describes the elastic aftereffect. This function is monotonically decreasing and at $t = 0$ takes the following value:

$$A_0(0) = 2(2-\alpha)^{-1} \sin \frac{\pi\alpha}{2(2-\alpha)}.$$

The asymptotic representation of the function $A_0(\lambda t)$ is valid for large t and has the form:

$$A_0(\lambda t) \approx -\sum_{n=1}^{\infty} (-1)^n \frac{\Gamma[(2-\alpha)n-1]}{\Gamma(\alpha n)\Gamma(1-\alpha n)} (\lambda t)^{1-(2-\alpha)n}. \tag{18.206}$$

Small values of α correspond to the usual damping parameters in materials of construction, where we have $A_{0\max} \ll 1$. Therefore, the monotonically decreasing term (addendum) in the expression for the impulse transient function can be neglected. Thus, the oscillatory process decays with a constant decrement in the absence of load. This damping occurs relative to some weakly perturbed position and not relative to the static equilibrium position. The indicated weakly perturbed position of the system asymptotically approaches the statical equilibrium position. As we can see, the proposed model, both for harmonic and free oscillations, provides frequency independence of the parameters of relative damping.

18.6 Oscillations of Beams Lying on an Elastic Base with Frequency- Independent Relative Damping

Let us consider the problems of oscillations/vibrations of beams on an elastic base using the model of frequency-independent relative damping as an example of dynamic calculation. The beam lies on an elastic foundation of the Winkler type with the bed coefficient k . A beam with length l is loaded in section $x = x_1$ with a unit harmonic force. The material of the beam and the base have frequency-independent internal friction. The model of frequency-independent relative damping is used to describe internal friction [9]. The equation of vibrations of the beam is written in the following form:

$$m\ddot{y} + EI\widetilde{\Gamma}_{\alpha_1} y^{IV} + k\widetilde{\Gamma}_{\alpha_2} y = \delta(x-x_1)e^{i\omega t}, \quad (0 \le x, x_1 \le l; \quad -\infty < t < \infty). \tag{18.207}$$

The boundary conditions can be taken arbitrary. Below is the solution of this problem, which belongs to Tseitlin and Kusainov [31].

$$\widetilde{\Gamma}_{\alpha_j} y = \frac{1}{\Gamma(-\alpha_j)} \int_{-\infty}^{t} \frac{y(\tau)d\tau}{(t-\tau)^{1+\alpha_j}}, \qquad (j = 1, 2)$$

Here the standard beam function expansion has been used to solve the inhomogeneous integral-differential Eq. (18.207). The solution of Eq. (18.207) is sought in the following form:

$$y(x,t) = \sum_{r=1}^{\infty} a_r X_r(x) e^{i\omega t} \tag{18.208}$$

where, the normalization beam functions $X_r(x)$ satisfy the given boundary conditions, and are a solution to the equation: $X^{IV}{}_r - \lambda_r^A X_r = 0$.

The right part of (18.207) should also be represented as an expansion in beam functions:

$$\delta(x - x_1) = \sum_{r=1}^{\infty} b_r X_r(x), \tag{18.209}$$

whence

$$b_r = X_r(x_1).$$

Substitute (18.208) and (18.209) in (18.207), then we obtain ($\omega > 0$)

$$\sum_{r=1}^{\infty} a_r \{ [m(\omega^2 - p_r^2 (i\omega)^{\alpha_1}) + k(i\omega)^{\alpha_2}] \} X_r(x) = \sum_{r=1}^{\infty} b_r X_r(x).$$

Now, let's equate the coefficients at X_r. The solution to the problem will take the following form:

$$y(x,t) = \sum_{r=1}^{\infty} A_r(\omega) X_r(x) e^{i[\omega t + \varphi_r(\omega)]}. \tag{18.210}$$

The solution of the problem expressed in terms of the amplitude and phase characteristics of each mode of harmonic oscillations have the form:

$$A_r = \frac{b_r}{m\omega^2 R^{1/2}(\omega)}, \tag{18.211}$$

$$\varphi_r - \tan^{-1}\left(\frac{\gamma_1 \xi + \gamma_2 \zeta}{1 - \xi - \zeta}\right). \tag{18.212}$$

Here it is indicated:

$$\begin{aligned} R(\omega) &= (\xi + \zeta + 1)^2 + (\gamma_1 \xi + \gamma_2 \zeta)^2; \\ \xi &= (p_1/\omega)^{2-\alpha_1}; \quad \zeta = (p_2/\omega)^{2-\alpha_2}; \end{aligned} \tag{18.213}$$

$$p_1 = \left(p_r^2 \cos\frac{\pi\alpha_1}{2}\right)^{\frac{1}{2-\alpha_1}}; \quad p_2 = \left(\frac{k}{m}\cos\frac{\pi\alpha_2}{2}\right)^{\frac{1}{2-\alpha_2}}. \tag{18.214}$$

The extremes of the function (18.211) can be determined from the condition:

$$\frac{d}{d\omega} A_r(\omega) = \frac{b_r}{2m\omega^3 R^{3/2}(\omega)} [4R(\omega) + \omega 4R'(\omega)] = 0.$$

We will exclude singular points $\omega = \infty$ and $\omega = 0$ from the equation and proceed to its solution,

$$4R(\omega) + \omega 4R'(\omega) = 0,$$

which after substitution (18.213) will be presented in the form:

$$\begin{aligned} &\alpha_1(1 + \gamma_1^2)\xi^2 + \alpha_2(1 + \gamma_2^2)\zeta^2 + (\alpha_1 + \alpha_2)(1 + \gamma_1\gamma_2)\xi\zeta - \\ &-(2 + \alpha_1)\xi - (2 + \alpha_2)\zeta + 2 = 0. \end{aligned} \tag{18.215}$$

To evaluate the solutions of this equation, first set $\alpha_1 = \alpha_2 = \alpha$, then from (18.214) we obtain:

$$\zeta = \frac{k}{m}\cos\frac{\pi\alpha_2}{2}\left(\xi / p_r^2 \cos\frac{\pi\alpha_1}{2}\right)^{\frac{2-\alpha_2}{2-\alpha_1}}\Bigg|_{\alpha_1 = \alpha_2} = \frac{k}{m} p_r^{-2} \xi, \tag{18.216}$$

and the Eq. (18.215) is converted to square

$$s_0\xi^2 + s_1\zeta + 2 = 0,$$

where

$$s_0 = \alpha(1+\gamma^2)\left(1+\frac{k}{mp_r^2}\right)^2; \qquad s_1 = -(2+\alpha_1)\left(1+\frac{k}{mp_r^2}\right).$$

Of the two roots of the last equation, we leave one

$$\xi_1 = \frac{(2+\alpha) - \sqrt{(2-\alpha)^2 - 8\alpha\gamma^2}}{2\alpha(1+\gamma^2)(1+k/mp_r^2)}, \tag{18.217}$$

since the second root at $\alpha = 0$ gives a trivial solution $\omega = 0$.

The frequency corresponding to (18.217), is determined by the formula:

$$\omega^* = \left(p_r^2 \cos\frac{\pi\alpha}{2}/\xi_1\right)^{\frac{1}{2-\alpha}}. \tag{18.218}$$

From this expression it can be seen, that at $\alpha_2 > \alpha_1$; $\xi > \xi_1$; $\omega_{\text{rez}} < \omega^*$; and at $\alpha_2 < \alpha_1$; $\xi < \xi_1$; $\omega_{res} > \omega^*$.

The corresponding value is determined from (18.216):

$$\xi_1 = \frac{(2+\alpha) - \sqrt{(2+\alpha)^2 - 8\alpha\gamma^2}}{2\alpha(1+\gamma^2)\left(1+\dfrac{k}{mp_r^2}\right)}.$$

In the absence of damping ($\alpha_j = 0$) we obtain:

$$\xi_1 = \left(1+\frac{k}{mp_r^2}\right)^{-1}, \qquad \omega_{\text{res}} = \left(p_r^2 + \frac{k}{m}\right)^{\frac{1}{2}}.$$

The problem of oscillations of an unlimited beam on an elastic base is further considered [32]. The equation of oscillations of a beam under the action of a concentrated force can be written in a form similar to (18.207). The concentrated force is applied at the beginning of the coordinates and changes according to the harmonic law:

$$m\ddot{y} + EI\widetilde{\Gamma}_{\alpha_1} y^{IV} + k\widetilde{\Gamma}_{\alpha_2} y = \delta(x)e^{i\omega t}, \quad (-\infty \le x, \quad < t \le \infty). \tag{18.219}$$

Equation (18.219) must satisfy the boundary conditions, that is, the disappearance at infinity of the displacement of the beam, and the derivatives of the displacement on coordinate: at $|x| \to \infty$

$$y^{(n)}(x,t) \to 0, \quad n = \overline{0,3}. \tag{18.220}$$

To solve Eq. (18.219), we use the cosine Fourier transform in view of the symmetry of the load. With respect to the cosine transform

$$Y(\xi,t) = \sqrt{2\pi}\int_0^\infty y(x,t),\ \cos(\xi,x)dx$$

we obtain the following equation:

$$m\ddot{Y} + \left(EI\xi^4\Gamma_{\alpha_1} + k\Gamma_{\alpha_2}\right)Y = \sqrt{\frac{1}{2\pi}}e^{i\omega t}. \tag{18.221}$$

The right part of the Eq. (18.221) contains a coefficient ½. This coefficient accounts for the transition to $\delta(x)$ from a symmetrical delta-shaped sequence.

The solution of Eq. (18.221), according to (18.193), can be represented in the form:

$$Y(\xi,t) = \frac{1}{\sqrt{2\pi}}\Big[-m\omega^2 + +EI\xi^4(\omega)^{\alpha_1}\left(\cos\frac{\pi\alpha_1}{2} + i\sin\frac{\pi\alpha_1}{2}\operatorname{sign}\omega\right) + \\ +k(\omega)^{\alpha_2}\left(\cos\frac{\pi\alpha_2}{2} + i\sin\frac{\pi\alpha_2}{2}\operatorname{sign}\omega\right)\Big]^{-1}.$$

The expression for the deflection using the inverse Fourier transform we obtain:
for $\omega > p$

$$y(x,t) = \frac{e^{i\omega t}}{\pi EI\omega^{\alpha_1}\cos\frac{\pi\alpha_1}{2}}\left[\int_0^\infty \frac{\left(\xi^4 - b^4\right)\cos(\xi x)d\xi}{\left(\xi^4 - b^4\right)^2 + \left(\gamma_1\xi^4 + a^4\right)^2} - \right. \\ \left. -i\int_0^\infty \frac{\left(\gamma_1\xi^4 + a^4\right)\cos(\xi x)d\xi}{\left(\xi^4 - b^4\right)^2 + \left(\gamma_1\xi^4 + a^4\right)^2}\right]; \tag{18.222}$$

for $\omega < p$

$$y(x,t) = \frac{e^{i\omega t}}{\pi EI\omega^{\alpha_1}\cos\frac{\pi\alpha_1}{2}}\left[\int_0^\infty \frac{\left(\xi^4 + \tilde{b}^4\right)\cos(\xi x)d\xi}{\left(\xi^4 + \tilde{b}^4\right)^2 + \left(\gamma_1\xi^4 + a^4\right)^2} - \right. \\ \left. -i\int_0^\infty \frac{\left(\gamma_1\xi^4 + a^4\right)\cos(\xi x)d\xi}{\left(\xi^4 + \tilde{b}^4\right)^2 + \left(\gamma_1\xi^4 + a^4\right)^2}\right]. \tag{18.223}$$

The following designations are taken here:

$$p = \left(\frac{k}{m}\cos\frac{\pi\alpha_2}{2}\right)^{\frac{1}{2-\alpha_2}}; \quad a^4 = \frac{k\omega^{\alpha_2}\sin\frac{\pi\alpha_2}{2}}{EI\omega^{\alpha_1}\cos\frac{\pi\alpha_1}{2}}; \\ b^4 = \frac{m\omega^2 - k\omega^{\alpha_2}\cos\frac{\pi\alpha_2}{2}}{EI\omega^{\alpha_1}\cos\frac{\pi\alpha_1}{2}}; \quad \tilde{b}^4 = -b^4. \tag{18.224}$$

The theory of residues used to calculate improper integrals, which are included in the expressions (18.222) and (18.223). Let us first consider the integrals (18.222). The integrand functions have the following poles in Fig. 18.6

$$\xi_1 = Se^{-i\alpha^*}; \qquad \xi_2 = Se^{-i(\alpha^* + \pi)}; \qquad \xi_{3,4} = Se^{-i\left(\pm\alpha^* + \frac{\pi}{2}\right)}, \tag{18.225}$$

where

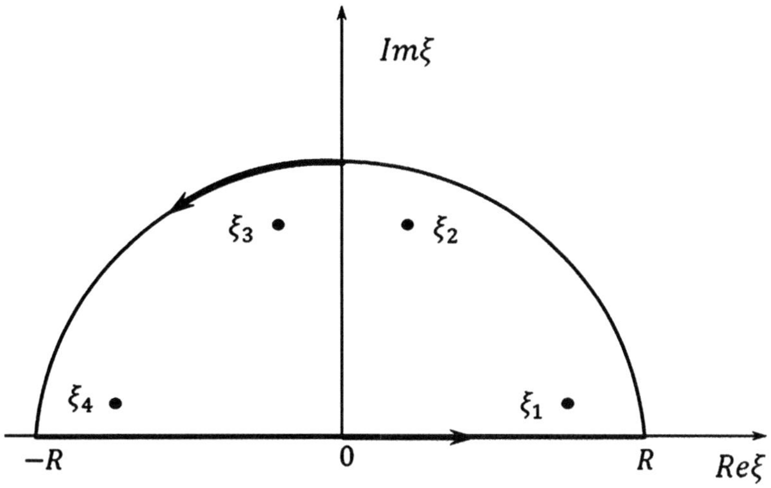

Fig. 18.6 Contour of integration has four poles

$$S=\left(\frac{a^8+b^8}{1+\gamma_1^2}\right)^{1/8};\qquad \gamma_1=\tan\frac{\pi\alpha_1}{2};\qquad \alpha^*=\frac{\pi\alpha_1}{8}+\frac{\theta}{4}.$$

The expression for the deflection will be obtained after simple transformations:

$$y(x,t)=\frac{Se^{-i\omega t}}{TEI\omega^{\alpha}\cos\frac{\pi\alpha_1}{2}}\left[\left(S^4B_1-b^4B_2\right)-i\left(\gamma_1S^4B_1+b^4B_2\right)\right],\tag{18.226}$$

where

$$T=4\left[S^8\left(1-\gamma_1^2\right)^2+2\left(\gamma_1^2a^4-b^4\right)\cos 4\alpha^*S^4\left(1+\gamma_1^2\right)+\left(\gamma_2a^4-b^4\right)^2\right]^{1/2};$$
$$B_1=e^{-\mu^c}\cos(\mu^s-\alpha^*+\nu)+e^{-\mu^s}\sin(\mu^c+\alpha^*-\nu);$$
$$B_2=e^{-\mu^c}\cos(\mu^s+3\alpha^*+\nu)+e^{-\mu^s}\sin(\mu^c-3\alpha^*-\nu);$$
$$\mu^s=xS\sin\alpha^*;\quad \mu^c=xS\cos\alpha^*;$$
$$\nu=\tan^{-1}\left[\frac{S^4\left(1+\gamma_1^2\right)\sin 4\alpha^*}{S^4\left(1+\gamma_1^2\right)\cos 4\alpha^*+\left(\gamma_1a^4-b^4\right)}\right].$$

Thus, when $\omega\geq p$

$$y(x,t)=A(x,\omega)\exp[\omega t-i\varphi(x,\omega)],\tag{18.227}$$

where, the amplitude and phase components of the deflection are determined by the formulas:

$$A(x,\omega)=\left[4S^3EI\omega^{\alpha_1}\right]^{-1}\sqrt{e^{-2\mu^c}+e^{-2\mu^s}+2e^{-2\mu^c-2\mu^s}\sin(\mu^c+\mu^s)};\tag{18.228}$$

$$\varphi(x,\omega)=\tan^{-1}\left\{\frac{\left[\left(\frac{\omega}{p}\right)^{2-\alpha_2}-1\right]B_3+\gamma_2B_1}{\left[\left(\frac{\omega}{p}\right)^{2-\alpha_2}-1\right]B_1+\gamma_2B_3}\right\}.\tag{18.229}$$

Here is the following designation:

$$B_3 = e^{-\mu^c}\sin(\mu^s - \alpha^* + \nu) + e^{-\mu^s}\cos(\mu^c + \alpha^* - \nu).$$

For $\omega \leq p$, the integrands functions have poles:

$$\xi_{1,2} = Se^{-i[\pm\alpha^* + \pi/2]}; \quad \xi_{3,4} = Se^{-i[\pm\alpha^* + 3\pi/4]}.$$

The expressions for the amplitude and phase components of the deflection will take the form:

$$\widetilde{A}(x,\omega) = \left[4S^3EI\omega^{\alpha_1}\right]^{-1}\sqrt{e^{-2\widetilde{\mu}^c} + e^{-2\widetilde{\mu}^s} + 2e^{-2\widetilde{\mu}^c - 2\widetilde{\mu}^s}\sin(\widetilde{\mu}^c + \widetilde{\mu}^s)}; \tag{18.230}$$

$$\varphi(x,\omega) = \tan^{-1}\left\{\frac{\left[1 - \left(\frac{\omega}{p}\right)^{2-\alpha_2}\right]\widetilde{B}_3 + \gamma_2\widetilde{B}_1}{\left[1 - \left(\frac{\omega}{p}\right)^{2-\alpha_2}\right]\widetilde{B}_1 + \gamma_2\widetilde{B}_3}\right\} \tag{18.231}$$

where

$$\begin{aligned}
\widetilde{B}_1 &= e^{-\widetilde{\mu}^c}\cos(\widetilde{\mu}^s - \widetilde{\alpha}^* + \nu) + e^{-\widetilde{\mu}^s}\sin(\widetilde{\mu}^c + \widetilde{\alpha}^* - \nu);\\
\widetilde{B}_3 &= e^{-\widetilde{\mu}^c}\sin(\widetilde{\mu}^s - \widetilde{\alpha}^* + \nu) + e^{-\widetilde{\mu}^s}\cos(\widetilde{\mu}^c + \widetilde{\alpha}^* - \nu)\\
\widetilde{\mu}^c &= xS\cos(\widetilde{\alpha}^* + \pi/4); \qquad \widetilde{\mu}^s = xS\sin(\widetilde{\alpha}^* + \pi/4);\\
\widetilde{\alpha}^* &= \frac{\pi\alpha_1}{8} + \frac{\theta}{4}.
\end{aligned}$$

The amplitude and phase characteristics of the system as a whole are continuous functions of frequency, despite different expressions for $\omega > P$ and $\omega < P$. The continuity of these functions can be easily proved by putting $\omega = p$. in (18.228), (18.229), (18.230) and (18.231). Then $\widetilde{b} = b = 0$; $\theta = \pm\frac{\pi}{2} + \pi k$; $\alpha^* = \frac{\pi\alpha_1}{8} \pm \frac{\pi}{8} + \frac{\pi}{4}k$, $(k = \overline{0,\infty})$ and, therefore, at the point $\omega = p$; $\mu^8 = \widetilde{\mu}^s$; $\mu^c = \widetilde{\mu}^c$, which leads to equivalent amplitudes $A(x,\omega) = \widetilde{A}(x,\omega)$ and phase characteristics:

$$\varphi(x,\omega) = \tan^{-1}\frac{B_1}{B_3} = \tan^{-1}\frac{\widetilde{B}_1}{\widetilde{B}_3} = \widetilde{\varphi}(x,\omega).$$

It should be noted here that the abscissa $\omega = p$ does not correspond to the resonant peak of the amplitude characteristic, although it is close to it. This follows from the fact that $\varphi(x,\omega)|_{\omega = p} \neq \pi/2$, since at this:

$$\widetilde{B}_3 = B_3 = e^{-\widetilde{\mu}^c}\sin(\mu^s - \alpha^* + \nu) + e^{-\mu^s}\cos(\mu^c + \alpha^* - \nu) \neq 0.$$

In the limiting case at $\alpha_1 = \alpha_2 = 0$, we have:
at: $\omega \leq p$

$$\widetilde{A}(x,\omega) = \left(\sqrt{2}/4\widetilde{B}_3EI\right)e^{-\widetilde{b}x/\sqrt{2}}\left(\sin\frac{\widetilde{b}x}{\sqrt{2}} + \cos\frac{\widetilde{b}x}{\sqrt{2}}\right); \tag{18.232}$$

$$\widetilde{\varphi}(x,\omega) = \frac{\pi}{4} + 2\pi k, \left(k = \overline{0,\infty}\right); \tag{18.233}$$

at $\omega \geq p$:

$$A(x,\omega) = \left(4b^3EI\right)^{-1}\left(e^{-2bx} + 2e^{-bx}\sin bx + 1\right)^{1/2}; \tag{18.234}$$

$$\varphi(x,\omega) = \tan^{-1}\left(\frac{\cos bx}{e^{-bx} + \sin bx}\right). \tag{18.235}$$

If we select (separate) the real part of the solution (18.227), then with $\omega \le p$ and $\ge p$, respectively, we obtain:

$$y(x,\omega) = \frac{\cos(\omega t - \pi/4)}{2\sqrt{2}\,EI\tilde{b}^3} e^{-\tilde{b}x/\sqrt{2}}\left(\sin\frac{\tilde{b}x}{\sqrt{2}} + \cos\frac{\tilde{b}x}{\sqrt{2}}\right);$$
$$y(x,t) = -\left(4EIb^3\right)^{-1}\left[e^{-bx}\cos\omega t + \sin(bx - \omega t)\right].$$

The last expressions coincide with those obtained in [1].

Finally, assuming in (18.234) and (18.235) that $k = 0$, we obtain a known solution of the problem of forced harmonic oscillations of a conservative unlimited beam.

Based on the obtained dependencies, the following graphs are plotted: (a) the amplitude-frequency characteristic of an unlimited beam on an elastic foundation (for a central section), taking into account the dissipative properties of only the elastic foundation material ($\gamma_1 = 0$) (Fig. 18.7); (b) the amplitude-coordinate characteristic of an unlimited beam without taking into account the elastic foundation ($\gamma_2 = 0$) at the frequency of forced oscillations $\omega = 10^4$ Fig. 18.8.

It should be noted that the dependence plot (shown with a solid line) is a straight line, not parallel to the abscissa axis. In this case, the damping along the beam length is due to the effect of energy

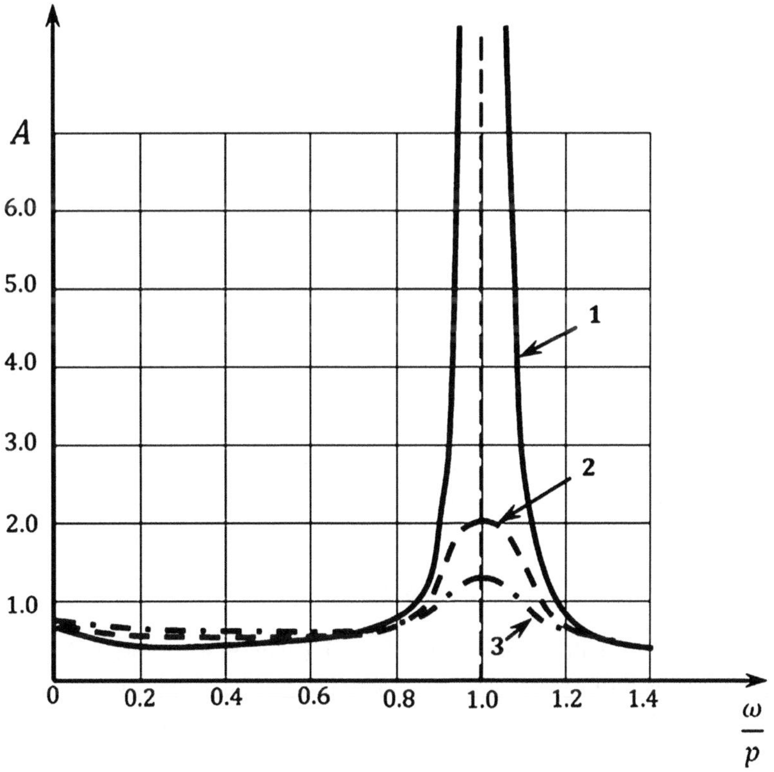

Fig. 18.7 Amplitude and frequency response (characteristic) of an unlimited beam on an elastic base at $\gamma_1 = 0$, ($1 \to \gamma_2 = 0$; $2 \to \gamma_2 = 0.1$; $3 \to \gamma_2 = 0.2$)

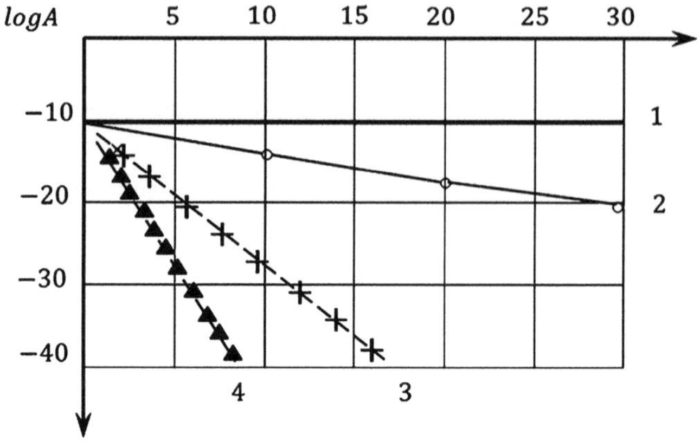

Fig. 18.8 Amplitude-coordinate response (characteristic) of an unlimited beam on an elastic foundation at: $\gamma_2 = 0$, ($1 \to \gamma_1 = 0$; $2 \to \gamma_1 = 0.01$; $3 \to \gamma_1 = 0.05$; $4 \to \gamma_1 = 0.1$)

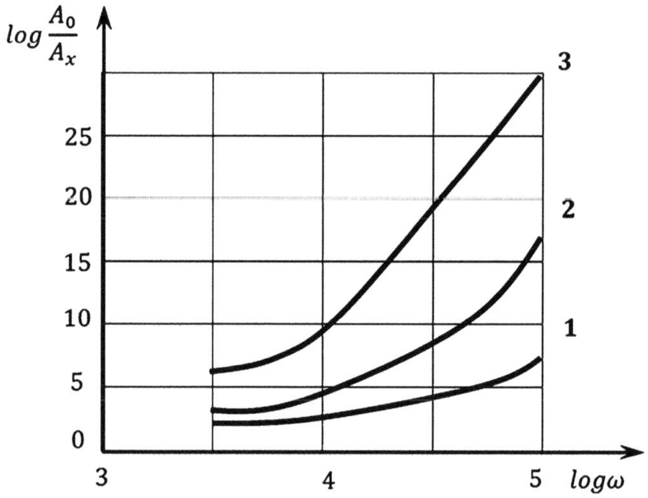

Fig. 18.9 The amplitude-frequency relative characteristic of the section oscillations in logarithmic coordinates ($1 \to x_1 = 2$; $2 \to x = 10$; $3 \to x = 20$)

radiation at infinity. From Fig. 18.8, it is clear that the damping is insignificant compared to the damping due to internal friction.

The decrease in the amplitude of oscillations in the central cross-section of an unlimited beam, which does not lie on the elastic base, is (as a percentage of the amplitude at $\gamma_1 = 0$): 1.5% at $\gamma_1 = 0.01$; 7.7% at $\gamma_1 = 0.05$; 11.8% at $\gamma_1 = 0.1$. This can be explained by putting in (18.211), $k = 0$, $x = 0$, $\omega = 10^4$, so that the solution of the oscillations equation for the central section of the beam without the elastic base will have the following form:

$$A_1(\gamma_1) = \frac{\sqrt{2}}{4S^3 EI \cdot 10^{4\alpha}}; \quad S = \frac{b}{\sqrt{1 + \gamma_1^2}}; \quad \gamma_1 = \tan\frac{\pi \alpha_1}{2}$$

The study of the dependence of spatial damping of harmonic oscillations on the frequency of perturbation is of some practical interest. The graphs of dependence of decimal logarithm of the ratio of oscillations amplitude of the central section A_0 of the beam to the oscillation amplitude of the section with current coordinate A_x on frequency at different values of coordinate x are constructed for this purpose (Fig. 18.9). The analysis of the graphs shows that: (a) the influence of the damping parameters of the elastic foundation on the oscillations of the entire system is greater than the influence of the damping parameters of the beam; (b) the absorption of energy due to dissipation at infinity is less than from the internal friction.

18.7 The Influence (Effect) of Internal Friction On Harmonic Oscillations of the Rigid Die Lying On an Inertial Half-Space

During the oscillation\vibrations of the structures on an elastic foundation, the energy losses are determined mainly by its radiation and internal friction in the base and in the structure itself. Let's try to compare the energy losses on radiation and on micro-plastic deformations of the base, which determine the internal friction in it. To study this issue, consider the harmonic oscillations of a circular rigid die with a plane base.

The rigid die lies on a homogeneous isotropic viscous elastic half-space without coupling (adhesion). Below is the solution of this problem, which belongs to Plotnikov [13]. Let the die be subjected to a vertical load in the type of axial force

$$P(t) = P_{st} + P_0 e^{i\omega t} \tag{18.236}$$

where P_{st} is the static component of the load; P_0 is the amplitude of harmonic load, at the same time: $P_{st} \gg P_0$.

The viscous elastic properties of the base are described by some linear law of hereditary theory. Then, the solution of the problem can be obtained from the solution of a similar problem for an elastic half-space by replacing real constants with complex values [18]. Here the complex values are the Fourier transformants of relaxation kernels.

The problem of oscillations of a circular die on an elastic inertial half-space in a sufficiently precise formulation was first considered in the work of Borodachev [30]. Similar problems for the elastic viscous half-space have been considered by a number of authors [40], however, in these works the influence of internal friction on the general dissipative properties of the base (foundation) was not considered.

The solution of the assigned contact problem for ideally elastic half-space is reduced to the solution of the dual (paired) integral equations [22, 30]:

$$\begin{gathered}
\int_0^\infty G(\xi)\tilde{p}(\xi)J_0(\xi r)d\xi = w_0 \quad (0 \le r \le a), \\
\int_0^\infty \tilde{p}(\xi)J_0(\xi r)d\xi = 0 \quad (r > a)
\end{gathered} \tag{18.237}$$

where

$$G(\xi) = \frac{\rho\omega^2\sqrt{\xi^2 - \rho\omega^2/(\lambda + 2\mu)}}{4\mu^2\left[\left(\xi^2 - \rho\omega^2/2\mu\right)^2 - \xi^2\sqrt{\xi^2 - \rho\omega^2/\mu}\ \sqrt{\xi^2 - \rho\omega^2/((\lambda + 2\mu)}\right]} \tag{18.238}$$

$G(\xi)$ is the transform of the base kernel; $\xi^{-1}\tilde{p}(\xi)$ is the transform of amplitudes of the contact stresses; w_0 is the amplitude of die displacement; a is the radius of die; ρ the density of the base material; λ, μ are Lame's constants; ω the circular frequency of the disturbing load; r is the radial coordinate.

Thus, the solution of the problem of harmonic oscillations of a rigid die on viscous elastic half-space is reduced to the solution of dual (paired) eqs. (18.237) with a kernel transform in the form (18.238). Here, the constant Lamés are replaced by complex parameters:

$$\lambda = \lambda_c + i\lambda_s; \quad \mu = \mu_c + i\mu_s \tag{18.239}$$

were λ_c, μ_c– there are cosine-transforms; λ_s, μ_s– there are sinus-transforms of the corresponding kernels of relaxation $\lambda(t)$ and $\mu(t)$.

The Poisson's coefficient will be considered as non-relaxing value, then $\lambda(t)$ and $\mu(t)$ will be bound by dependence:

$$\lambda(t) = \frac{2\lambda}{1-2\nu}\mu(t).$$

The complex constants (18.239) can be expressed through the coefficient of loss γ of the base material by the following formulas:

$$\lambda = \lambda_0 \exp(i\gamma); \qquad \mu = \mu_0\, exp(i\gamma); \tag{18.240}$$

where

$$\lambda_0 = \sqrt{\lambda_c{}^2 + \lambda_s{}^2}; \qquad \mu_0 = \sqrt{\mu_c{}^2 + \mu_s{}^2}$$
$$\gamma = \tan^{-1}\frac{\mu_s}{\mu_c} = \tan^{-1}\frac{\lambda_s}{\lambda_c}.$$

The behavior of the function $G(\xi)$ then $\xi \to \infty$ is subordinated (subjected) to the ratio:

$$G(\xi) \sim -\left[2\mu\left(1-\varepsilon^2\right)\xi\right]^{-1}[1+0(2)],$$
$$\varepsilon^2 = (1-2\lambda)/2(1-\nu).$$

Therefore

$$\begin{aligned} G(\xi) &= -\left[2\mu\left(1-\varepsilon^2\right)\xi\right]^{-1}\rho^2(\xi); \\ \rho^2(\xi) &= 1 + h(\xi); \\ h(\xi) &= -2\mu\left(1-\varepsilon^2\right)\xi G(\xi) - 1. \end{aligned} \tag{18.241}$$

For solving dual (paired) integral Eq. (18.237), one of the methods usually applied in such cases [33–39] can be used. Here, A.I. Tseitlin's method has been used for solving dual (paired) integral equations, whose accuracy to symbols coincides with the equation obtained by the method of Cook-Lebedev.

The paired integral Eq. (18.237) with the function $G(\xi)$ in the form of (18.241) and the kernel in the form of the Bessel function $J_0(\xi x)$ can be reduced to a Fredholm integral equation of the second kind according to

$$P_1^*(x) + \int_0^1 \Psi(x, x_1) P_1^*(x_1) dx_1 = B^{-2} W_1^*(x) \qquad (0 < x < 1) \tag{18.242}$$

and a symmetric kernel

$$\Psi(x,x_1)=\int_0^\infty h(\xi)J_{\nu-\frac{1}{2}}(\xi x)J_{\nu-\frac{1}{2}}(\xi x_1)\xi d\xi. \tag{18.243}$$

Equation (18.242) with accuracy of symbols coincides with the equation obtained by Cook-Lebedev method.

Let's write down this equation in the form:

$$\varphi(x)+\int_0^1\varphi(y)K(x,y)dy=1, \tag{18.244}$$

$$K(x,y)=\frac{k}{\pi}\left[\widetilde{K}(|x-y|)+\widetilde{K}(|x+y|)\right]. \tag{18.245}$$

Here

$$\widetilde{K}(x)=\int_0^\infty h(\overline{\xi})\cos\overline{\xi}xd\overline{\xi}, \quad k=\omega a\sqrt{\rho/\mu_0}. \tag{18.246}$$

The function $h(\xi)$ is calculated by formula (18.241) taking into account (18.240), and (18.246), then $\overline{\xi}=\xi k/a$

$$h(\overline{\xi})=-1-e^{-i\gamma}\frac{2(1-\varepsilon^2)\overline{\xi}\sqrt{\overline{\xi}^2-\varepsilon^2e^{-i\gamma}}}{\left(2\overline{\xi}^2-e^{-i\gamma}\right)^2-4\overline{\xi}^2\sqrt{\overline{\xi}^2-e^{-i\gamma}}\sqrt{\overline{\xi}^2-\varepsilon e^{-i\gamma}}}. \tag{18.247}$$

The function $h(\overline{\xi})$ has no poles on the real (actual) axis at $\gamma\neq0$, therefore the integral can be calculated on a computer. The amplitudes of contact stresses can be calculated using the formula [30]:

$$\sigma_z(\overline{r},0)=-\frac{4\mu(1-\varepsilon^2)\omega_0}{\pi a}\left[\frac{\varphi(1)}{\sqrt{1-\overline{r}^2}}-\int_{\overline{r}}^1\frac{\varphi'(\eta)d\eta}{\sqrt{\eta^2-\overline{r}^2}}\right] \tag{18.248}$$

where $\overline{r}=r/a$ is the dimensionless radial coordinate.

The amplitude of the total reactive force is determined as a result of integration of the stresses (18.248) over the area occupied by the die:

$$R_0=8\mu(1-\varepsilon^2)w_0a\int_0^1\varphi(\eta)d\eta. \tag{18.249}$$

From the equation of motion of the die

$$m\frac{d^2}{dt^2}\left(w_0e^{i\omega t}\right)+R_0e^{i\omega t}=P_0e^{i\omega t} \tag{18.250}$$

taking into account (18.249) the amplitude of the displacements of the die is determined

$$w_0 = P\mu_0^{-1} a^{-1} \left[8(1-\varepsilon^2) p e^{i\gamma} - \mathfrak{M}^* k^2\right]^{-1} \tag{18.251}$$

where $\mathfrak{M}^* = m/\rho a^3 -$ is dimensionless mass of the die;

$$q = \int_0^1 \varphi(\eta) d\eta. \tag{18.252}$$

Under the static action of force, the settlement of the die will be:

$$w_{st} = P_0 / \left[8\mu_0 a(1-\varepsilon^2)\right]. \tag{18.253}$$

After dividing (division) (18.251) by (into) (18.164) we obtain

$$k_{\text{dyn}} e^{-i\beta} = 8(1-\varepsilon^2) / \left[8(1-\varepsilon^2) p e^{i\gamma} - \mathfrak{M}^* k^2\right]. \tag{18.254}$$

Here k_{dyn} is the dynamic coefficient for displacements; β is the angle of the shift of phases between the displacements of the die and disturbing (excited) force.

The amplitude of displacements of the die (18.251), taking into account (18.164), and (18.254), takes the following form:

$$w_0 = w_{st} k_{\text{dyn}} e^{-i\beta} \tag{18.255}$$

After substitution (18.255) in (18.248), the last expression takes the following form:

$$\sigma_z(\overline{r}, 0) = \sigma_z^{st}(\overline{r}, 0) k_{dyn}(\overline{r}) e^{-i[\theta(\overline{r}) - \beta]} \tag{18.256}$$

where

$$\sigma_z^{st}(\overline{r}, 0) = -P_0 / \left[2\pi a^2 \sqrt{1-\overline{r}^2}\right]$$

is the static contact stresses.

$$k_{\text{dyn}}(\overline{r}) e^{-i\theta(\overline{r})} = k_{\text{dyn}} e^{-i\beta} \left[\varphi(1) - \sqrt{1-\overline{r}^2} \int_{\overline{r}}^1 \frac{\varphi(\eta) d\eta}{\sqrt{\eta^2 - \overline{r}^2}}\right] \tag{18.257}$$

$k_{\text{dyn}}(r)$ is the dynamic coefficient for contact stresses.

As can be seen from the results, the solution of the problem is reduced to the determination of the function $\varphi(x)$ and its derivative $\varphi'(x)$ from the Fredholm integral equation of the second kind (18.244). This equation, after replacement of the integral with a finite sum by one of the quadrature formulas, can be reduced to a system of linear algebraic equations:

$$\boldsymbol{A} \cdot \boldsymbol{\varphi} = \boldsymbol{b}, \tag{18.258}$$

where

$$\begin{aligned} &\boldsymbol{A} = [a_{jm}]; \quad \boldsymbol{\varphi} = [\varphi_j]; \quad \boldsymbol{b} = [b_j]; \\ &b_j = 1; \quad \varphi_j = \varphi\left(\frac{j-1}{n-1}\right); \quad j, m = 1, 2, \ldots n; \\ &a_{jm} = c_m - \frac{k}{\pi}\left[\widetilde{K}\left(\frac{|j-m|}{n-1}\right) + \widetilde{K}\left(\frac{|j+m-2|}{n-1}\right)\right], \end{aligned} \tag{18.259}$$

c_m are coefficients of quadrature formulas, in particular at odd n, and Simpson's formula

$$c_1 = c_n = 1/3(n-1); \quad c_2 = c_4 = \ldots = c_{n-1} = 4/3(n-1); \quad c_3 = c_5 = \ldots = c_{n-2} = 2/3(n-1).$$

After solving the system (18.258), the value q (18.252) is calculated by the following formula

$$q = \sum_{m=1}^{n} c_m \varphi_m. \tag{18.260}$$

For determining the derivatives $\varphi'(x)$, the Eq. (18.244) is differentiated along the x axis and after calculating the integral in parts the equation is reduced to the integral equation of Fredholm of the second kind

$$\varphi'(x) + \int_0^1 \varphi'(y) T(x, y) dy = \varphi(1) T(x, 1) \tag{18.261}$$

with a symmetric kernel

$$T(x, y) = \frac{2h}{\pi} \int_0^\infty h(\xi) \sin(\xi k x) \sin(\xi k y) d\xi. \tag{18.262}$$

The kernel (18.256) can be expressed through the function (18.246)

$$T(x, y) = \frac{k}{\pi}\left[\widetilde{K}(|x-y|) - \widetilde{K}(|x+y|)\right]. \tag{18.263}$$

The Eq. (18.261) is reduced to a system of linear algebraic equations as well as the Eq. (18.244):

$$\boldsymbol{A}_1 \cdot \boldsymbol{\varphi}_1 = \boldsymbol{b}_1, \tag{18.264}$$

where

$$\begin{aligned} &\boldsymbol{A}_1 = \left[a'_{jm}\right]; \quad \boldsymbol{\varphi}_1 = \left[\varphi'_j\right]; \quad \boldsymbol{b} = \left[b'_j\right]; \\ &a'_{jm} = c_{m+1} \frac{k}{\pi}\left[\widetilde{K}\left(\frac{|j-m|}{n-1}\right) - \widetilde{K}\left(\frac{|j+m|}{n-1}\right)\right]; \\ &b'_j = \frac{k}{\pi}\left[\widetilde{K}\left(\frac{|n-j|}{n-1}\right) - \widetilde{K}\left(\frac{|n+j|}{n-1}\right)\right]\varphi(1); \\ &\varphi'_j = \varphi'\left(\frac{j}{n-1}\right); \quad j, m = 1, 2, \ldots n-1. \end{aligned} \tag{18.265}$$

The function $\varphi'(x)$ is approximated by a polynomial

$$\varphi'(x) = \sum_{j=1}^{n-1} \alpha_j x^{2j-1} \tag{18.266}$$

the coefficients of which are determined from equations:

$$\boldsymbol{D} \cdot \boldsymbol{\alpha} = \boldsymbol{\varphi}_1 \tag{18.267}$$

in which

$$\boldsymbol{D} = [d_{jm}]; \quad \boldsymbol{\alpha} = [\alpha_j].$$

$$d_{jm} = \left(\frac{j}{n-1}\right)^{2m-1}; \quad j, m = 1, 2, \ldots n-1.$$

After submission (18.267) into (18.264) we obtain a system of algebraic equations with respect to α_j

$$(\boldsymbol{A}_1 \cdot \boldsymbol{D})\boldsymbol{\alpha} = \boldsymbol{b}_1 \tag{18.268}$$

The integral in (18.257) is easily calculated. For dynamic coefficient the following formula is obtained:

$$k_{\text{dyn}}(\bar{r}) = k_{\text{dyn}} \left| \varphi(1) - \sum_{m=1}^{n-1} \sum_{j=1}^{m} \frac{\alpha_m (m-1)! r^{2(m-j)} (1-r^2)^j}{(j-1)!(m-j)!(2j-1)} \right|. \tag{18.269}$$

The solution of the problem has been obtained on the computer for the following parameters: $\nu = 0;\ 0.25;\ 0.5;\ \gamma = 0.01;\ 0.1;\ 0.2;\ \mathfrak{M}^* = 5;\ 20$. The values of dynamic coefficients depending on the dimensionless frequency k are shown in Figs. 18.10, 18.11, 18.12, 18.13 and 18.14. A comparison of the obtained results at $\gamma = 0.01$, $(n = 5)$ with the results of solving a similar problem for an elastic half-space [5] shows their practical coincidence.

However, at higher values of the coefficient of loss, the effect of internal friction on the maximum values of the dynamic coefficient has a significant influence. It is known that high values of coefficient

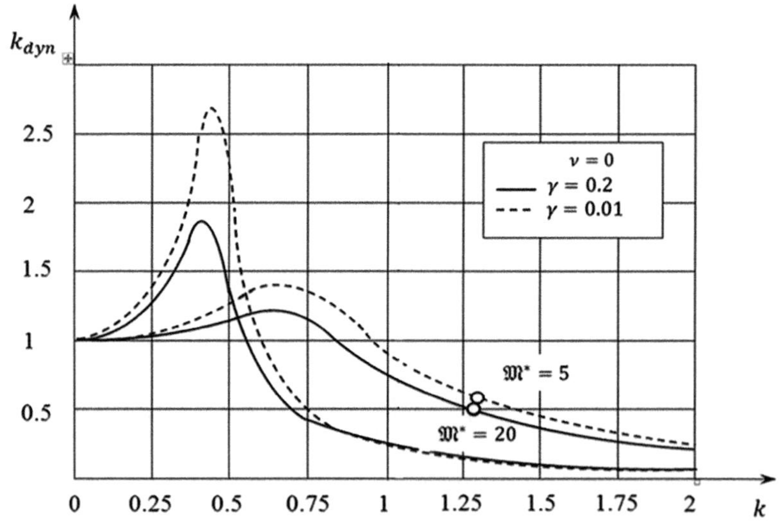

Fig. 18.10 The values of dynamic coefficients depending on the dimensionless frequency k

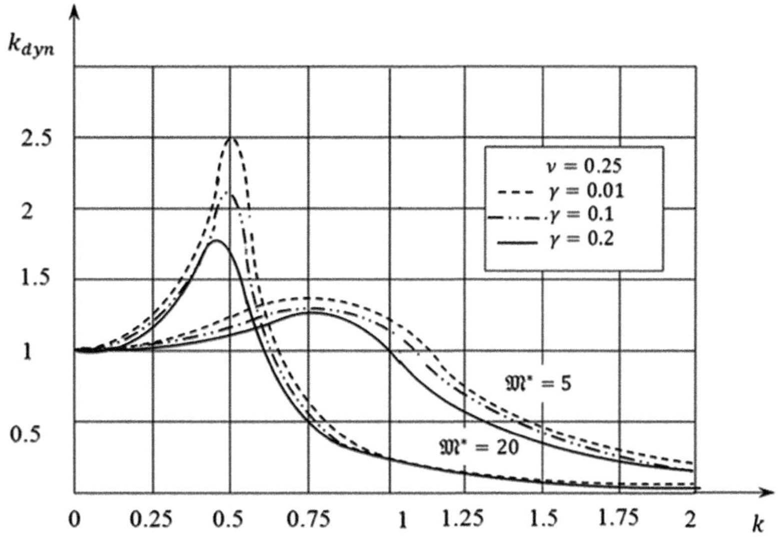

Fig. 18.11 The values of dynamic coefficients depending on the dimensionless frequency k

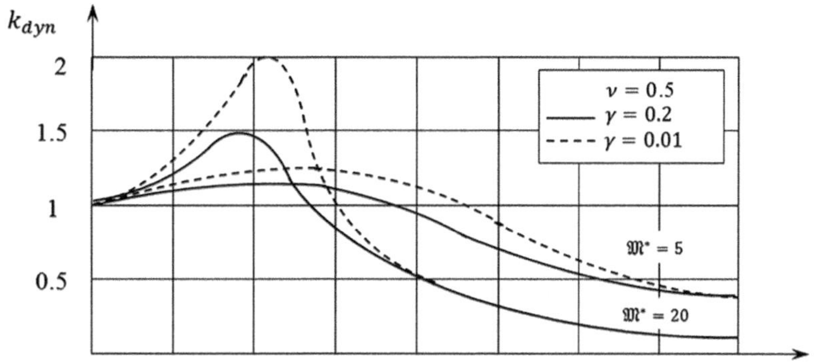

Fig. 18.12 The values of dynamic coefficients depending on the dimensionless frequency k

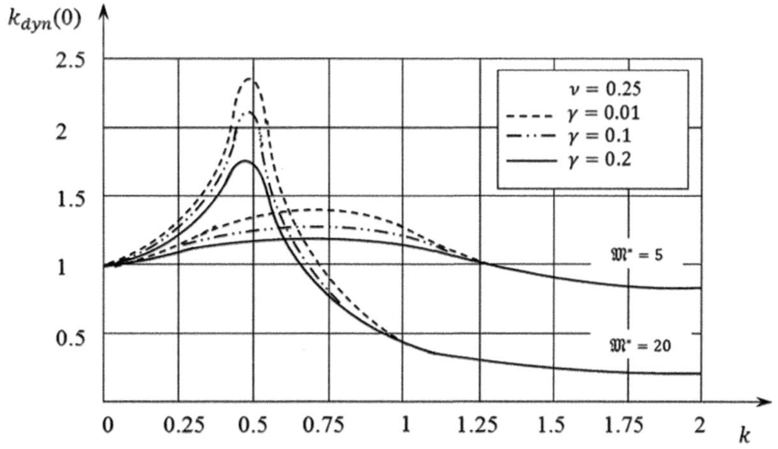

Fig. 18.13 The values of dynamic coefficients depending on the dimensionless frequency k

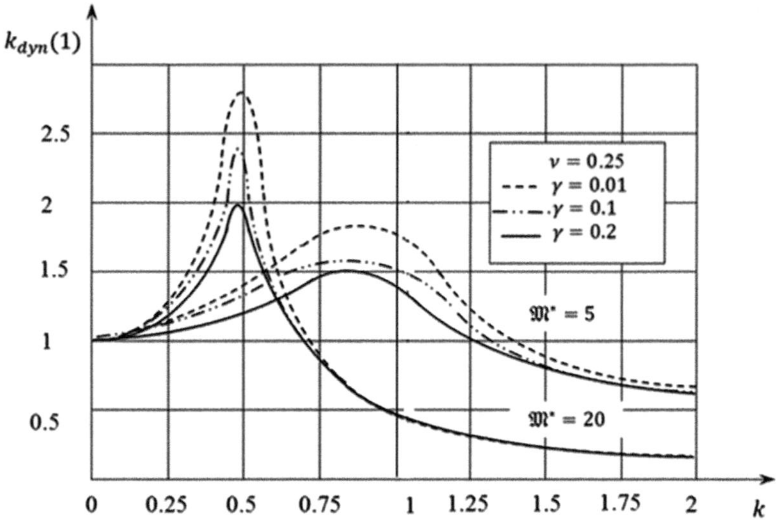

Fig. 18.14 The values of dynamic coefficients depending on the dimensionless frequency k

of loss are typical for soils. In particular, the difference in the maximum values of the dynamic coefficient at $\gamma = 0.01$ and $\gamma = 0.2$ for $\mathfrak{M}^* = 20$ reaches 50%. Thus, taking into account the internal friction in the base material in the dynamic calculation of foundations and structures is necessary.

References

1. Crandall, S. H. (1971). *The role of dumping in the theory of vibration mechanics* (Translation from English) (Vol. 5 (129), pp. 3–22). Mir.
2. Sorokin, E. S. (1960). *To the theory of internal friction with the oscillations of elastic systems* (131pp). Gosstroyizdat.
3. Bode, T. (1948). *Circuit theory and design amplifiers with feedback* (642 p). GITTL.
4. Petrosian, L. G. (1987). About one method of application of Hilbert transformation to the study of dynamic systems. *Izvestia AS Arm SSR, Series: Mechanics, 1*, 17–24.
5. Gelfand, I. M., & Shilov, G. E. (1959). *The generalized functions and operations over them* (470 p). Fizmatgiz.
6. Gradstein, I. S., & Ridjick, I. M. (1991). *Tables of integrals, sums, series, and products* (1100 p). Fizmatgiz.
7. Ditkin, V. A., & Prudnikov, A. P. (1974). *Integral transformations and operational calculus* (542 p). Gostekhizdat.
8. Prudnikov, A. P., Brychkov, Y. A., & Marichev, O. I. (1991). *Integrals and series* (798p). Nauka.
9. Tseitlin, A. I. (1978). About the linear models of the frequency-independent internal friction. *Izvestia AS USSR, MTT, 3*, 18–28.
10. Petrosian, L. G., & Ambartsumian, V. A. (2020). *Static and dynamic analysis of engineering structures: Incorporating the boundary element method* (528p). Wiley.
11. Tseitlin, A. I., Kusainov, A. A., & Petrosian, L. G. (1986). *The free oscillations of dissipative systems* (AS Kazakh SSR, #9) (pp. 50–56). Almaty.
12. Tseitlin, A. I., Petrosian, L. G., Kusainov, A. A., & Mamaeva, G. V. (1988). Design structures on dynamic and seismic impacts, taking into account damping. In *9 WCEE, ninth world conference on earthquake engineering* (Vol. 1, pp. 182–189). Technical Sections.
13. Plotnikov, Yu. G. (1979). *The stationary oscillations of plane and axisymmetric dies on viscose elastic foundations*. The dissertation thesis. Technical sciences, PhD. Moscow, 200p.
14. Zelenev, V. M., Meshkov, S. I., & Rossikhin, Y. A. (1970). *Damped oscillations of elastic—Hereditary system with the weakly singular kernels* (Vol. #2, pp. 14–118). PMTF.
15. Petrosian, L. G. (1986). *The calculation of the internal friction in the one-dimensional dynamic systems* (pp. 40–44). YPTU, Structure Mechanics.
16. Tseitlin, A. I., & Petrosian, L. G. (1987). *The methods of boundary elements in structural mechanics* (200 p). Luis.
17. Christensen, R. (1974). *Introduction to the theory of viscoelasticity* (p. 340p). Mir.
18. Rabotnov Yu, G. *Elements of the hereditary mechanics of solid bodies* (383p). Nauka.

19. Tseitlin, A. I., Atadzhanov, D. R., & Sarkisyan, A. G. (1986). *Grin's function of dynamic problem for the half-plane. TsNIISK named Kucherenko* (9.01.86 # 6446) (14p). DEP VHIIS.
20. Case, K. M., & Hazeltine, R. D. (1970). Elastic radiation in a half space. *Journal of Mathematical Physics, 118*, 2446–2552.
21. Ilichev, V. A., & Shakhter, O. Y. (1976). Determination of dynamic stresses and displacements in the elastic half-plane from the internal source, which imitates the action of the underground metro tunnel of shallow laying. In *Trudi NII Institute of ground and underground structures* (Vol. 67, pp. 42–64).
22. Petrosian, L. G. (1990). *The methods of boundary equations in the problems of calculating the structures on an elastic foundations.* The dissertation on competition of a scientific degree of Doctor of Technical Sciences (Post Doctorate Research). The thesis of Doctor of Technical Sciences. CSRICS, 310p.
23. Kupradze, V. D. (1963). *The methods of potential in the theory of elasticity* (472 p). Fizmatgiz.
24. Caputo, M. (1976). Vibrations of an infinite plate with frequency independent Q. *Journal of the Acoustical Society of America, 60(#3)*, 634–639.
25. Tseitlin, A. I., & Plotnikov, Y. G. (1979). *Free oscillations of a system with frequency-independent internal friction* (pp. 29–35). SM&CC.
26. Tseitlin, A. I., & Guseva, N. I. (1979). *Static methods of calculation of structures for group dynamic impacts.* 176p.
27. Solodovnikov, V. V. (1980). *Statistical dynamics of linear automatic control systems.* 512p.
28. Rogosin, S., & Mainardi, F. (2014). George William Scott Blair—The pioneer of factional calculus in rheology. *Communications in Applied and Industrial Mathematics, 6*(1), 481. https://doi.org/10.1685/journal.caim.481. ISSN 2038-0909.
29. Gerasimov, A. N. (1988). Generalization of liner Laws of deformation and its application to the problems of internal friction. *Applied Mathematics and Mechanics, 12*, 251–260.
30. Borodachev, N. M. (1964). *Some dynamic and temperature contact problems of the theory of elasticity and their applications.* The dissertation on competition of a scientific degree of Doctor of Technical Sciences (Post Doctorate Research). Moscow, 350p.
31. Tseitlin, A. I., & Kusainov, A. A. (1985). *The principle of frequency independence of internal friction in mechanical oscillation systems* (pp. 4–22). Dynamics of Construction.
32. Kusainov, A. A. (1982). *Oscillation of an unlimited beam on an elastic foundation with frequency independent relative damping* (pp. 56–59). SM&CS.
33. Cooke, J. C. (1956). A solution of tranter's dual integral equations problem. *The Quarterly Journal of Mechanics and Applied Mathematics, 9*, 103–110.
34. Lebedev, N. N. (1957). Distribution of electricity on thin parabolic segment. *Dokl AN USSR, 114*, 513–516.
35. Lebedev, N. N. (1963). *Special functions and their applications*, 358p.
36. Aleksandrov, V. M., & Buryak, V. G. (1978). About some dynamic mixed problems of the theory of elasticity. *Public Money & Management, 42*(1), 114–121.
37. Aleksandrov, V. M., & Mkhitarian, C. M. (1983). *Contact problems for bodies with thin coatings and interlayers* (487p). Science.
38. Tseitlin, A. I. (1966). About the method of dual (pair) integral equations and dual (pair) series and its application to problems in mechanics. *Public Money & Management, 30*(32), 259–266.
39. Tseitlin, A. I. (1966). About one class of dual (pair) integral equations. *Differential Equations, 11*(8), 1134–1139.
40. Petrosian, L. G. (1988). The contact problem for a rigid die on the generalized elastic foundation. *Izv. Academy of Science Arm SSR, Applied Mechanics, 3*, 59–64.

[illegible]

19 Nonlinear Vibration of Structural Elements

Abstract

In this chapter presents nonlinear oscillations/vibrations of a system with single-degree-of-freedom (SDOF), multi-degree-of-freedom (MDOF), distributed mass, and oscillations/vibrations of the beam with variable cross section. The values of the periods of free vibration are obtained by various methods, such as direct integration of the equations of motion and the use of asymptotic methods. A study of free vibrations of a beam with an arbitrary law of variation of the cross section is also discussed here.

19.1 The Nonlinear Vibrations of Systems with Single-Degree-of-Freedom

Strictly speaking, the vibrations/oscillations of mechanical systems are always nonlinear because they are described by nonlinear differential equations. In general, linear models are applicable only in a very restrictive domain, that is, when the oscillation amplitude is very small. Thus, to accurately identify and understand the dynamic behavior of a structural system under general loading conditions, it is essential for the nonlinearities present in the system to also be modeled and studied [1]. In structural mechanics, nonlinearities can be broadly classified into the following two general categories: geometric nonlinearity, which exists in systems undergoing large deformations or deflections, and physical nonlinearity, which includes but is not limited to the kinetic energy of the system, which is the source of inertia nonlinearities. For example, damping is essentially a nonlinear phenomenon. Another nonlinear phenomenon is described by material nonlinearity (i.e., stresses and strains are nonlinear, restoring force and displacement are nonlinear). Nonlinearities can also appear in some boundary conditions such as spring supports.

However, in order to avoid difficulties that appear during the solution of nonlinear differential equations, the linearization of these equations is produced, that is, the initial nonlinear differential equation is substituted by a linear equation. Two types of linearization are distinguished; the first is a geometric linearization and it relates to the geometry of the body being deformed. It is assumed that dependences of deformations on the displacements are linear. The second type of linearization is physical linearization, since it concerns the physical properties of the body being deformed. It is assumed that dependences between the strains and stresses are linear.

We will examine vibration/oscillation of "physically nonlinear" mechanical systems, for which the displacements are small, and dependences between the stresses and strains are described by the

L. G. Petrosian, *Dynamics of Structures*, https://doi.org/10.1007/978-3-031-63539-7_19

nonlinear functions. Nonlinear mechanical systems in comparison with the linear systems are characterized by new physical properties. If in the linear system natural frequency is well-defined and independent of the initial conditions value, then in the nonlinear system natural frequency depends on an initial displacement and changes in the dependence on a change of stiffness of the system.

A system is linear if for every element in the system the response is proportional to the excitation. The dynamic properties of each element in the system can be represented by a set of linear differential equations with constant coefficients. For a linear system, using values of natural frequencies and modes of oscillations, we can determine the reactions of the system with the arbitrary forced oscillations. In the nonlinear system that is impossible to do, since the principle of superposition of the separate (individual) modes of oscillations is not applicable. Nevertheless, the determination of frequencies and forms of the oscillations of nonlinear systems represents a significant interest because they characterize systems with single-frequency oscillations, and a similar motion can take place with forced oscillations. On the other hand, the frequencies and modes of free nonlinear oscillations can be used for the "linearization" of nonlinear systems and thus for developing engineering methods of their calculation.

Let us examine the free nonlinear oscillation of a system with single-degree-of -freedom, for which the dependence of restoring force of the displacement takes the form of bilinear diagram. The equation of motion for this system takes the form

$$\begin{gathered} my'' + R(y) = 0, \\ R(y) = \begin{cases} ay & \text{with } y \le y_1 \\ a[y_1 + \gamma(y - y_1)] & \text{with } y \ge y_1, \end{cases} \end{gathered} \tag{19.1}$$

where $m-$ value of the concentrated mass; y, $y''-$ respectively displacement and the acceleration of mass, $R(y)-$ restoring force; $a-$ initial stiffness of system; y_1- given value of displacement at which a change of the rigidity of the system takes place; $\gamma-$ the ratio of the changed (reduced) stiffness (rigidity) of the system to the initial condition. The solution of Eq. (19.1) with the zero initial conditions for both sections will take the following general form:

$$\begin{gathered} y(t) = \frac{c}{\omega^2\gamma} + \left(y(0) - \frac{c}{\omega^2\gamma}\right) \cos\omega\sqrt{\gamma}t + \frac{y'(0)}{\omega\sqrt{\gamma}} \sin\omega\sqrt{\gamma}t, \\ c = \begin{cases} 0, & y \le y_1, \\ -\omega^2(1-\gamma)y_1, & y \ge y_1, \end{cases} \end{gathered} \tag{19.2}$$

where $\omega = \sqrt{\frac{a}{m}}$ is the natural frequency of linear system, $y(0)$, $y'(0)-$ the initial values of displacement and velocity. If we deflect the mass m up to the distance y_2 ($y_2 > y_1$), and then allow it to oscillate without the initial velocity, then for displacing the system from (19.2) we will obtain

$$y(t) = -\frac{(1-\gamma)}{\gamma}y_1 + \left(y_2 + y_1\frac{(1-\gamma)}{\gamma}\right) \cos\omega\sqrt{\gamma}t. \tag{19.3}$$

The time t_{12}, necessary to the return of mass m from $y = y_2$ to $y = y_1$ is obtained from (19.3), if we substitute $y(t)$ with y_1:

$$t_{12} = \frac{1}{\omega\sqrt{\gamma}} \arccos\frac{1}{(\beta-1)\gamma + 1}; \quad \beta = \frac{y_2}{y_1}. \tag{19.4}$$

The time t_{01}, in the course of which the mass m passes the distance from $y = y_1$ to $y = 0$, is obtained from the corresponding expression (19.2) with the substitution of $y(t) = 0,\ y(0) = y_1$, and $y'(0) = y_1{}'$:

$$t_{01} = \frac{1}{\omega}\arctan\frac{\sqrt{\gamma}}{[(\beta-1)\gamma+1]\sin\arccos\frac{1}{(\beta-1)\gamma+1}}. \qquad (19.5)$$

Since the dependence between the force and deflection is symmetrical relative to the center of coordinates, the value of the period of free nonlinear oscillations T_{Nonlin} can be determined according to the formula

$$T_{\text{Nonlinear}} = 4(t_{01} + t_{02}) = \\ = \frac{2\pi}{\omega}\left\{\frac{2}{\pi\sqrt{\gamma}}\arccos\frac{1}{1+(\beta-1)\gamma} + \frac{2}{\pi}\arctan\frac{\sqrt{\gamma}}{[1+(\beta-1)\gamma]\sin\arccos\frac{1}{1+(\beta-1)\gamma}}\right\}. \qquad (19.6)$$

In a similar manner, the period of the free nonlinear oscillation of the system can be determined, for which the loading occurs according to the bilinear law, and the unloading to the linear law (Fig. 19.1). In this case, the formula is obtained:

$$T_{\text{Nonlinear}} = \frac{2\pi}{\omega}\left\{\frac{1}{\pi}\arctan\frac{\sqrt{\gamma}}{[1+(\beta-1)\gamma]\sin\arccos\frac{1}{1+(\beta-1)\gamma}} + \right. \\ \left. + \frac{1}{\pi\sqrt{\gamma}}\arccos\frac{1}{1+(\beta-1)\gamma} + \frac{1}{\pi}\right\}. \qquad (19.7)$$

From the formulas in (19.6) and (19.7) with the substitution of $\beta = 1$, and $\gamma = 1$, the value of the period of linear oscillation is

$$T_{\text{Linear}} = \frac{2\pi}{\omega}.$$

The expressions (19.6) and (19.7) show that the period of nonlinear oscillations is a continuous function of the displacements, and for a nonlinear system, there is no longer a concept of natural frequency since such frequencies have already an infinite set. The values of the relations of the periods of nonlinear and linear oscillations $T_{\text{Nonlinear}}/T_{\text{Linear}}$, depending on the relative displacement of $\beta = y_2/y_1$, are calculated with the aid of these expressions. The calculations were conducted for the values $\gamma = 1.0,\ 0.8,\ 0.6,\ 0.4,\ 0.2,\ 0.1$, and $\beta = 1,\ 2,\ 4,\ 6,\ 8,\ 10$. It follows from the graphs (Fig. 19.1) that an increase in the initial displacement leads to an increase of the value of the periods of nonlinear oscillations. At the same parameters γ and $\beta = \frac{y_2}{y_1}$, this increase is greater with the elastic-nonlinear than with the hysteresis oscillations. The method, with the aid of which the value of the period of oscillations of bilinear system was obtained, is applicable for systems whose stiffness (rigidity) changes according to the piecewise-linear (piece-linear) law.

Let us examine another method of determining the period of nonlinear oscillations with the aid of which precise values for some curvilinear diagrams of deformation [2, 3] were obtained. We will consider the algorithm for solving the problem for an arbitrary dependence $R(y)$ [3]. The equation of motion of the system presets in the form of

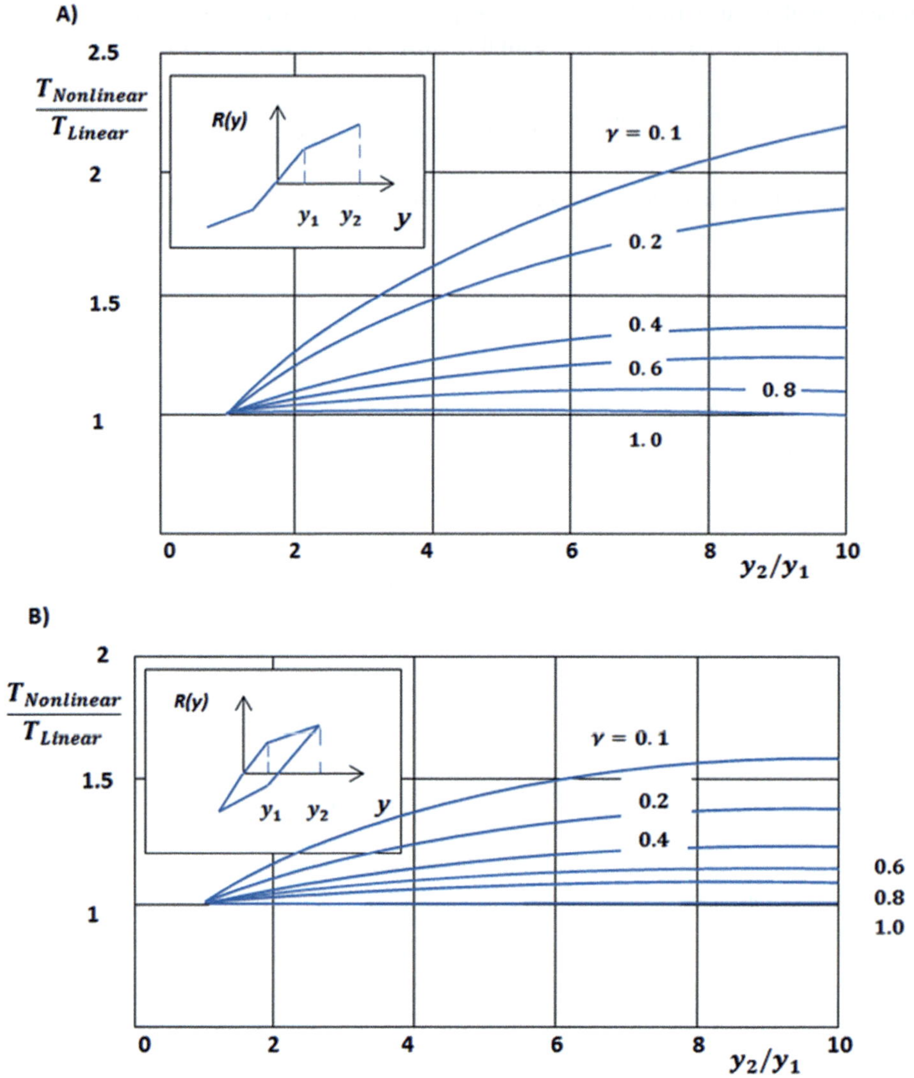

Fig. 19.1 Dependences of the relations of periods of nonlinear and linear oscillation on relative displacements: (**a**) with the bilinear-elastic law of deformation; (**b**) with the bilinear-hysteresis law of the deformation

$$y'' + \omega^2 f(y) = 0, \tag{19.8}$$

$\omega^2 = \frac{a}{m}$ – represents the frequency of free oscillations of linear system that is at $f(y) = y$. The differential Eq. (19.8) can be represented in the form of:

$$\frac{1}{2}\frac{d(y')^2}{dy} + \omega^2 f(y) = 0. \tag{19.9}$$

Integrating the Eq. (19.9) with the initial conditions $y_{(t=0)} = y_0$ and $y'_{(t=0)} = 0,$ we obtain

$$\frac{1}{2}\ y'^2 = \omega^2 \int_y^{y_0} f(y)\,dy = 0. \tag{19.10}$$

Since $y' = \frac{dy}{dt}$ from (19.10) it follows:

$$dt = -\frac{dy}{\sqrt{2\omega^2 \int_y^{y_0} f(y)dy}}. \tag{19.11}$$

By repeated integration of (19.1), we obtain the time t during which the system reaches the position $y = 0$ from the position y_0. The period of the system with the symmetrical diagram of restoring force will be equal to $T = 4t$. For the period $T_{\text{Nonlinear}}$ the following expression is obtained

$$T_{\text{Nonlinear}} = 4 \int_0^{y_0} \frac{dy}{\sqrt{2\omega^2 \int_y^{y_0} f(y)\,dy}} \tag{19.12}$$

With the explicit analytical expression for $T_{\text{Nonlinear}}$ is possible to obtain when $\omega^2 f(y)$ takes the form

$$\omega^2 f(y) = \omega^2 \left(y - \varepsilon y^3\right) \tag{19.13}$$

In this case, the period of oscillations of system $T_{\text{Nonlinear}}$ is expressed through elliptical integral of the first kind, $F\left(\sqrt{\frac{\theta}{2-\theta}}, \frac{\pi}{2}\right)$:

$$T_{\text{Nonlinear}} = \frac{4}{\omega}\sqrt{\frac{2}{2-\theta}} \cdot F\left(\sqrt{\frac{\theta}{2-\theta}}, \frac{\pi}{2}\right), \qquad \theta = \varepsilon y_0^2. \tag{19.14}$$

Using the formula in (19.14) and applying the tables of elliptical integrals we determine the values of the period of nonlinear oscillations $T_{\text{Nonlinear}}$. The ratios $T_{\text{Nonlinear}}/T_{\text{Linear}}$ for values of $\theta = \varepsilon y_0^2$ are given in Table 19.1.

In Fig. 19.2, the dependences of the relation of the periods of nonlinear and linear oscillations $T_{\text{Nonlinear}}/T_{\text{Linear}}$ on the displacement y with different nonlinearity coefficient ε are built. With the determination of the periods of real constructions, we examine the ascent path of function $R(y)$. In Fig. 19.2, this part of the dependence $T_{\text{Nonlinear}}/T_{\text{Linear}} \div y$ is shown by the solid line. For each curve, there is a value y, with which $T_{\text{Nonlinear}}/T_{\text{Linear}} \rightarrow \infty$. This corresponds, as shown in Table 19.1, to the value $\theta = 1$. When considering only the ascending branches of the function, the maximum increase in the period of nonlinear oscillations occurs (takes place at) when $\theta = 1/3$ and comprises 1.15 times. A sinusoidal function is another characteristic of the restoring force, which allows the explicitly precise determination of the oscillatory period [4, 5]. In this case, the function $f(y)$ is taken in the form

$$f(y) = \frac{1}{\varepsilon}\ \sin \varepsilon y \tag{19.15}$$

The equation of motion in this case is the same as in the case of oscillation of the mathematical (simple) pendulum. The difference consists in the fact that in this case, it is possible to use the obtained solution when $-\frac{\pi}{2\varepsilon} \leq y \leq \frac{\pi}{2\varepsilon}$. A similar assumption was used in the previous case. Substituting the

Table 19.1 The ratios $T_{\text{Nonlinear}}/T_{\text{Linear}}$ for values of $\theta = \varepsilon y_0^2$

θ	$T_{\text{Nonlinear}}/T_{\text{Linear}}$	θ	$T_{\text{Nonlinear}}/T_{\text{Linear}}$
0	1	0.5	1.274
0.25	1.115	0.75	1.592
0.33	1.154	1	∞

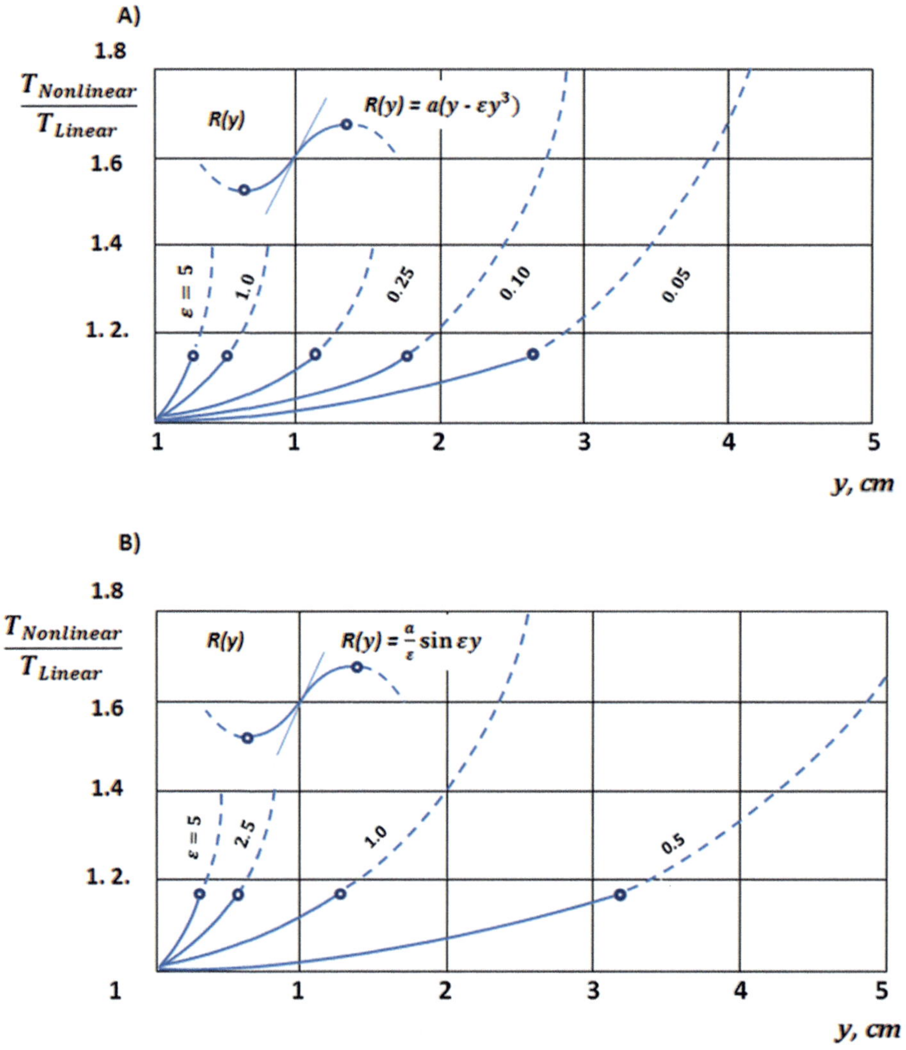

Fig. 19.2 Dependences of relations of periods of nonlinear and linear oscillations on displacements with the laws of the deformation: (**a**) on cubic parabola; and (**b**) on the sinusoid

expression in (19.15) in the formula (19.12), we obtain the value $T_{\text{Nonlinear}}$, expressed through the elliptical integral of the first kind:

$$T_{\text{Nonlinear}} = \frac{4}{\omega} \int_0^{\frac{\pi}{2}} \frac{d\theta}{\sqrt{1 - k^2 \sin^2\theta}}, \tag{19.16}$$

where $k = \sin \frac{\varepsilon y_0}{2}$.

In Fig. 19.2, the dependences for this function are built. If we examine only ascending branches of the function, then the maximum value of the period of nonlinear oscillations will comprise $T_{\text{Nonlinear}}/T_{\text{Linear}} = 1.18$, when $= \pi$, $T_{\text{Nonlinear}}/T_{\text{Linear}} \to \infty$.

The analytic expressions of the restoring force are characterized by special feature; the functions described by them have ascending and descending parts. In the descending part of the diagram, the structural failure occurs, therefore, we introduced limitations at the maximum displacements and maximum values of the coefficient of nonlinearity. But this somewhat limits the framework of the selection of the parameters. Therefore, the application of functions, increasing with the increase in the displacement, represents an interest, as examples of such functions are the following:

$$f(y) = \frac{1}{\varepsilon} \arctan \varepsilon y \tag{19.17}$$

$$f(y) = \frac{1}{\varepsilon} \tan \varepsilon y \tag{19.18}$$

In both cases, ε $(1/cm)$ characterizes the degree of nonlinearity. With $\varepsilon = 0$, $f(y) = y$, that is, a linear dependence occurs. Substituting these values into (19.12) and carrying out integration, we obtain the value of T_{Linear}. For the considered cases it was not possible to obtain the value of the period in an explicit form. The value of $T_{\text{Nonlinear}}$ is obtained in the following form:

For the function (19.17)

$$T_{\text{Nonlinear}} = 4 \int_0^{y_0} \frac{dy}{\sqrt{2\omega^2 \left[y_0 \arctan \varepsilon y_0 - y \arctan \varepsilon y - \frac{1}{2\varepsilon} \ln \frac{1 + \varepsilon^2 y_0^2}{1 + \varepsilon^2 y^2} \right]}}; \tag{19.19}$$

For the function (19.18)

$$T_{\text{Nonlinear}} = 4 \int_0^{y_0} \frac{dy}{\sqrt{2\omega^2 \frac{1}{\varepsilon} \ln \frac{\cosh \varepsilon y_0}{\cosh \varepsilon y}}}. \tag{19.20}$$

The value of $T_{\text{Nonlinear}}$ can be determined by numerical calculation of the integral in each specific case. The application of a characteristic of the restoring force in the form

$$\omega^2 f(y) = \omega_r^2 y^{\frac{1}{r}}, \tag{19.21}$$

where r– positive integer number, is also of interest. With $r = 1$ we have a linear dependence. This function is increasing monotonically. In this case, the initial rigidity of the system is equal to infinity. This dependence does not correspond to the actual properties of deformation of structural materials, but because of the simplicity of expression, it frequently adapts in the engineering calculations, especially for describing the law of the deformation of metals. In the expressions (19.12), (19.15), (19.17), and (19.18), the value ω^2 represents the value of the frequency of free linear oscillations which corresponds to the case $\varepsilon \to 0$. In the expression (19.21), ω_r^2 already does not have a dimensionality of frequency and represents a certain coefficient, dependent on the form of the accepted curve. The number r characterizes the degree of nonlinearity. When r is even, the function $f(y)$ with $y < 0$ should be selected in the form of

$$f(y) = \omega_r^2 |y|^{\frac{1}{r}-1} y$$

We will next assume that r is odd. Substituting the value $f(y)$ into (19.12), we obtain the value of the period of free nonlinear oscillations

$$T_{\text{Nonlinear}} = \frac{4}{\sqrt{2\omega_r^2 \frac{y_0^{\mu-1}}{\mu+1}}} \frac{\sqrt{\pi} \cdot \Gamma\left(\frac{1}{\mu+1}\right)}{\mu+1 \cdot \Gamma\left(\frac{\mu+3}{2\mu+2}\right)}, \tag{19.22}$$

where $\mu = \frac{1}{r}$, $\Gamma(\mu)$– gamma-function. Tables of the values of gamma-functions are seen in [6].

$$\text{With} \quad \mu = 1, \quad T_{\text{Nonlinear}} = 2\pi/\omega_1;$$

with $\mu = 1/3$,

$$T_{\text{Nonlinear}} = \frac{4 \cdot 3.5944}{\omega_3 \sqrt{6}} y_0^{\frac{1}{3}} \tag{19.23}$$

From (19.23) it follows that the period of nonlinear oscillations depends on the amplitude and increases with an increase in the latter. To calculate the values of the period of nonlinear oscillations of a real system it is necessary to know the specific value of the coefficients of nonlinearity ε in the formulas (19.16), (19.19), (19.20) or ω_r and r in the formula (19.22). The values of these coefficients can be obtained with the use of the results of structural testing. However, the dependence of the restoring force on the displacement can also be obtained using the results of material testing under uniaxial linear tension and compression. Let us assume that the dependence between the stresses and strains takes the form of the exponential function:

$$\sigma = B|\varepsilon|^{\mu-1} \cdot \varepsilon, \qquad \mu = \frac{1}{r}, \qquad r = 1, 2, 3 \ \ldots \tag{19.24}$$

where σ– stress, ε– relative deformation, B– the coefficient, characterizing the deformation curve, with $r = 1, B = E$ (E – modulus of elasticity). On the basis of (19.24), let us switch over to obtaining the dependence between the force and displacement. For this, we will consider a one-story frame with the undeformed horizontal element of the frame (beam $EI = \infty$). Using the conventional assumption that the hypothesis of flat sections is also valid for nonlinear deformations, we will obtain a dependency between the displacement and statically loading (acting, impacting) force P:

$$y = \frac{P^r H^{r+2}}{(r+2)2^{r+1}\left(\sum_{i=1}^{s} BI_{ir}\right)^r} \tag{19.25}$$

where H– is the height of the floor, s– the number of floor columns, I_{ir}– generalized moment of inertia of the i column [7, 8]. From (19.25) we have

$$P = a_r y^{\frac{1}{r}} \tag{19.26}$$

where,

$$a_r = \frac{(r+2)^{\frac{1}{r}} \cdot 2^{\frac{r+1}{r}}}{H^{\frac{r+2}{r}}} \sum_{i=1}^{s} BI_{ir}$$

a_r– rigidity of the system in the nonlinear deformation. When (at) $r = 1$

$$a_1 = 12 \sum_{i=1}^{s} \frac{EI_{i1}}{H^3}$$

The value of the generalized moment of inertia for different (various) types of sections is given in [9, 10]. For a rectangular cross section with dimensions $b \times h$, value I_r is determined by the expression

$$I_r = \frac{bh^{2+1/r}}{2^{2+1/r}\left(1 + \frac{1}{2r}\right)},$$

with $r = 1$,

$$I_1 = \frac{bh^3}{12}.$$

If we consider that the examined frame is a system with single-degree-of-freedom and the mass is concentrated at the overlap level, then the equations of motion of this system can be obtained substituting $P = -my''(t)$ into (19.26):

$$my'' + a_r y^{\frac{1}{r}} = 0 \tag{19.27}$$

Above we obtained the value of the period of nonlinear oscillations of a similar system. The period of oscillations of the system, described by Eq. (19.27) can be obtained from (19.22) if instead of ω_r^2 we substitute the value of a_r/m. Thus, knowing the values of B and r (19.24), from (19.26) we will be able to define a_r and from (19.22) define the value of the period with different amplitudes y_0. In order to approximately estimate the values of a_r, we carried out an approximation of the dependency $\sigma - \varepsilon$ for steel of (CT- 3) in the domain $0 \leq \varepsilon \leq 6.10^{-3}$, obtained by uniaxial linear tension [11]. In the first approximation, when $r = 1/3$, we obtained $B = 2.1,\ 10^4$кГс/cm^2.

19.2 The Nonlinear Oscillations of Systems with Multiple-Degrees-of-Freedom

Let us examine free nonlinear oscillations of a system with multiple-degrees-of-freedom. The exact solutions to such problems are very rare in the literature of applied nature. In the works of R.M. Rosenberg [12, 13] the nonlinear oscillations of a multi-mass system with the aid of potential function and use of properties of geodetic lines in space, are investigated. Multi-mass systems, which are design schemas of multistory frame buildings, will be investigated below. The periods and forms of oscillation are defined precisely by the direct integration of the equations of motion. The method used is actually the generalization of Duffing's method, which was adapted for systems with single-degree-of-freedom. Let us examine free oscillations of the multi-story frame, whose mass is concentrated at the overlap levels. The dependence "stress–strain" takes the form (19.24). The equation of motion of this frame takes the form [14].

$$m_i y_i'' + a_{ir}(y_i - y_{i-1})^{\frac{1}{r}} - a_{i+1,r}(y_{i+1} - y_i)^{\frac{1}{r}} = 0 \tag{19.28}$$

where m_i– is the mass, concentrated at the level of *i*- floor, y_i, y_i'' – displacement and acceleration of the *i*-th mass, a_{ir}– generalized rigidity of the *i*-th of floor with the nonlinear deformation with the parameter *r*. The value of a_{ir} is determined by the expression in (19.26). Let us determine the periods and corresponding modes of free oscillations of the nonlinear system. As in linear systems, the modes of oscillations of a nonlinear system are characterized by the following properties: all masses of the system pass through the position of equilibrium simultaneously and oscillate with one and the same period. To solve this problem let us represent (19.28) in the form of

$$\frac{d(y_i')^2}{dy_i} = -2\frac{a_{ir}}{m_i}(y_i - y_{i-1})^{\frac{1}{r}} + 2\frac{a_{i+1,r}}{m_i}(y_{i+1} - y_i)^{\frac{1}{r}}. \tag{19.29}$$

Integrating the Eq. (19.29) with the initial conditions

$$y_i(0) = c_i \quad \text{and} \quad y_i'(0) = 0$$

we will obtain

$$y_i'^2 = -2\frac{a_{ir}}{m_i}\int_{c_i}^{y_i}(y_i - y_{i-1})^{\frac{1}{r}}dy_i + 2\frac{a_{i+1,r}}{m_i}\int_{c_i}^{y_i}(y_{i+1} - y_i)^{\frac{1}{r}}dy_i \tag{19.30}$$

Designating $y_i/y_n = \gamma_i$ and taking into account that with single-frequency oscillations γ_i is taken as constant, we obtain

$$\begin{aligned} y_i'^2 = & -2\frac{a_{ir}}{m_i}\frac{r}{1+r}\left(1 - \frac{\gamma_{i-1}}{\gamma_i}\right)^{\frac{1}{r}}\left(y_i^{\frac{1}{r}+1} - c_i^{\frac{1}{r}+1}\right) + \\ & + 2\frac{a_{i+1,r}}{m_i}\frac{r}{1+r}\left(\frac{\gamma_{i+1}}{\gamma_i} - 1\right)^{\frac{1}{r}}\left(y_i^{\frac{1}{r}+1} - c_i^{\frac{1}{r}+1}\right) + \end{aligned} \tag{19.31}$$

Integrating the expression (19.31) from $t = 0$ to $T/4$, with corresponding displacement from $y_i = c_i$ to $y_i = 0$, we will obtain

$$T'_{j,Nonlinear} = \frac{1}{\sqrt{2\frac{r}{1+r}\left[\frac{a_{ir}}{m_i}\left(1 - \frac{\gamma_{i-1}}{\gamma_i}\right)^{\frac{1}{r}} - \frac{a_{i+1,r}}{m_i}\left(\frac{\gamma_{i+1}}{\gamma_i} - 1\right)^{\frac{1}{r}}\right]c^{\frac{1-r}{2r}}}}\int_0^1 \frac{du}{1 - u^{\frac{1+r}{r}}}$$

After designating

$$\int_0^1 \frac{du}{1 - u^{\frac{1+r}{r}}} = I, \qquad a_{ir} = a_{1r}\alpha_i, \qquad m_i = m_1\mu_i$$

and bearing in mind, that

$$c^{\frac{1-r}{2r}} = \gamma_i^{\frac{1-r}{2r}} c_n^{\frac{1-r}{2r}},$$

we will obtain

$$\frac{T_j\sqrt{2\frac{r}{1+r}}}{4Ic_n^{\frac{r-1}{2r}}}=\frac{1}{\sqrt{\frac{a_{1r}}{m_i}\left\{\frac{\alpha_i}{\mu_i}(\gamma_i-\gamma_{i-1})^{\frac{1}{r}}-\frac{\alpha_{i+1}}{\mu_i}(\gamma_{i+1}-\gamma_i)^{\frac{1}{r}}\right\}\frac{1}{\gamma_i}}} \tag{19.32}$$

Designating

$$\left[\frac{T_j\sqrt{2\frac{r}{1+r}}}{4Ic_n^{\frac{r-1}{2r}}}\right]^r=\frac{1}{\lambda} \tag{19.33}$$

we will obtain

$$\lambda=\left[\frac{\alpha_i}{\mu_i}(\gamma_i-\gamma_{i-1})^{\frac{1}{r}}-\frac{\alpha_{i+1}}{\mu_i}(\gamma_{i+1}-\gamma_i)^{\frac{1}{r}}\right]\frac{1}{\gamma_i} \tag{19.34}$$

We obtain the equation for determining the periods using the following condition:

$$\gamma_0=y_0/y_n=0. \tag{19.35}$$

From (19.34) and with $i = n$, we will obtain

$$\gamma_{n-1}=1-\lambda^r\left(\frac{\mu_n}{\alpha_n}\right)^r,$$

bearing in mind that $\gamma_n = 1$ and $\alpha_{n+1} = 0$. From the system (19.34) and using the recurrent method the following expression is obtained:

$$\gamma_{i-1}=\gamma_i-\left(\frac{\mu_i}{\alpha_i}\right)^r\left[\frac{\alpha_{i+1}}{\mu_i}(\gamma_{i+1}-\gamma_i)^{\frac{1}{r}}+\lambda\gamma_i\right]^r. \tag{19.36}$$

The Eq. (19.35) will take the open form:

$$\gamma_0=\gamma_1-\left(\frac{\mu_1}{\alpha_1}\right)^r\left[\frac{\alpha_2}{\mu_2}(\gamma_2-\gamma_1)^{\frac{1}{r}}+\lambda\gamma_1\right]^r=0 \tag{19.37}$$

where γ_1, γ_2, and γ_i are defined using the expression (19.36). With the aid of the Eq. (19.37), the values of λ_j are determined, through which the value of the period using formula (19.33) is expressed:

$$T_j=4I\sqrt{\frac{1+r}{2r}\frac{m_1}{a_{1r}}\frac{1}{\lambda_j}}c_n^{\frac{r-1}{2r}} \tag{19.38}$$

After the determination of λ_j, the coefficients of form oscillations γ_{ij} are determined by the expression (19.36). The method of determining periods and forms of free oscillations presented is applicable to the linear oscillations, that is, in the case $r = 1$. We determined the values λ_j and γ_{ij}, $j = 1$, 2, 3 for multistory buildings with equal values of concentrated masses and stiffness of floors, that is, with $\alpha_i = 1$, $\mu_i = 1$. The calculations were conducted with the value $r = 3$. The system with n degree of freedom with the linear oscillations has n of values of periods, which are determined using the algebraic equation of n degree (19.37). In the case of linear oscillations, all roots of the Eq. (19.37) are real numbers. With nonlinear oscillations, the degree of an equation for determining the periods is more than n and depends on the parameter value of the nonlinear deformation r. A numerical study of

Table 19.2 The value of roots $\boldsymbol{\lambda}_j$

Number of stories	Roots		
	λ_1	λ_2	λ_3
2	0.50999	1.42000	1
3	0.33280	1.23591	1.47500
4	0.24009	0.90501	1.43870
5	0.18459	0.71180	1.25750
6	0.14809	0.61170	1.17560
7	0.12249	0.53120	0.92461
8	0.10399	0.43590	0.74900
9	0.08911	0.38899	0.64721
10	0.07800	0.31157	0.58998

the Eq. (19.37) showed that there are n valid solutions $\lambda_j,\ j=1,\ 2,\ \ldots n$ and corresponding to them n modes of oscillations. As with linear oscillations, the first form does not have a nodal point, the second mode has one nodal point and so on. The obtained values λ_j, and $\gamma_j,\ j=1,\ 2,\ \ldots n$, for buildings with $n=2-10$ floors are given in Tables 19.2 and 19.3. Let us calculate the values of the periods of nonlinear oscillations by referring to previously used parameters of deformation. For $n=5$, $H/h=10,\ r=3$ are obtained the following values for T_j are obtained

$$T_j = T_{j,\mathrm{Linear}}\left(\frac{c_5}{5H}\right)^{\frac{1}{3}} \cdot k_j \qquad k_1=5.710, \quad k_2=8.461, \quad k_3=9.911 \tag{19.39}$$

As in the case of a system with single-degree-of-freedom the periods of nonlinear oscillations T_j can be greater or smaller than the period of corresponding linear oscillations $T_{j,\ \mathrm{Linear}}$. For example, in this system with $c_5/5H=1/186$, $T_1=T_{1,\ \mathrm{Linear}}$, and with $c_n \lesseqgtr 1/186$, $T_1 \lesseqgtr T_{1,\ \mathrm{Linear}}$. With the second and third modes, the periods of nonlinear oscillations become more than the periods of linear vibrations with $c_5/5H>1/605$ and $c_5/5H>1/973$ respectively.

The modes of the nonlinear vibrations of 10-story buildings are shown in Fig. 19.3. The figure shows that the nature of the deformation of nonlinear and linear systems with single-frequency vibrations of the same tone are identical.

The strain–stress diagrams of the deformation of the materials of structures with dynamic processes are characterized by different laws of loading and unloading. During loading, the restoring force is subordinate to the nonlinear law; while during unloading it is subordinated to the linear. In general, this leads to a significant complication when solving a problem of nonlinear vibrations of structures as compared to nonlinear tasks, when encountered in other areas of technology; for example, in theory of automatic regulation, radio engineering, etc. In the case of different laws of deformation during loading and unloading, the period of nonlinear oscillations can be determined according to the formula:

$$T_{j,\mathrm{Hysteresis}} = \frac{1}{2}\left(T_{j,\mathrm{Linear}} + T_{j,\mathrm{Nonlinear}}\right) \tag{19.40}$$

where $T_{j,\ \mathrm{Hysteresis}}-$ period of $j-$ th modes of the nonlinear hysteresis oscillations of system, $T_{j,\ \mathrm{Linear}}-$ period of $j-$ th modes of the linear oscillations, $T_{j,\ \mathrm{Nonlinear}}-$ period of $j-$ th modes of elastic-nonlinear oscillations.

In the above example, the modes of oscillations of a building during loading and unloading are different. During steady single-frequency hysteresis oscillations, the system will apparently oscillate in

Table 19.3 The value of the coefficient of the modes of oscillations $\gamma_{ij}^{(n)}$

Number of Stories n	i	γ_{i1}	γ_{i2}	γ_{i3}
1	2	3	4	5
	1	0.867	−1.863	
2	2	1	1	
	1	0.684	−0.891	1.946
3	2	0.963	−0.888	2.093
	3	1	1	1
	1	0.553	−1.220	0.822
	2	0.878	−1.220	0.807
4	3	0.986	0.259	1.978
	4	1	1	1
	1	0.460	−1.048	0.869
5	2	0.784	−1.068	0.866
	3	0.944	−0.946	−0.988
	4	0.944	0.639	−0.981
	5	1	1	1
	1	0.393	−0.906	0.903
	2	0.700	−0.976	0.902
	3	0.886	−0.970	−0.649
6	4	0.971	−0.501	−0.711
	5	0.997	0.771	−0.625
	6	1	1	1
	1	0.344	−0.999	1.017
	2	0.628	−1.000	0.171
	3	0.824	−0.957	−1.050
7	4	0.935	−0.532	−1.081
	5	0.983	0.397	−0.980
	6	0.998	0.926	0.627
	7	1	1	1
	1	0.317	−0.806	1.063
	2	0.576	−0.997	1.074
	3	0.769	−1.000	0.875
8	4	0.894	−0.975	−1.027
	5	0.961	−0.606	−1.130
	6	0.990	0.337	−1.077
	7	0.999	0.918	0.580
	8	1	1	1
	1	0.284	−1.000	0.999
	2	0.524	−0.999	0.833
	3	0.714	−0.987	−0.293
	4	0.848	−0.848	−0.947
9	5	0.931	−0.402	−0.977
	6	0.975	0.279	−0.958
	7	0.994	0.790	−0.387
	8	0.999	0.976	0.796
	9	1	1	1
	1	0.257	−0.956	0.892
	2	0.480	−0.999	0.951
	3	0.664	−1.000	0.940
	4	0.802	−0.988	0.394

Number of Stories n	i	γ_{i1}	γ_{i2}	γ_{i3}
1	2	3	4	5
10	5	0.896	−0.852	−0.825
	6	0.953	−0.410	−0.985
	7	0.983	0.273	−0.624
	8	0.996	0.788	0.742
	9	0.999	0.976	0.742
	10	1	1	1

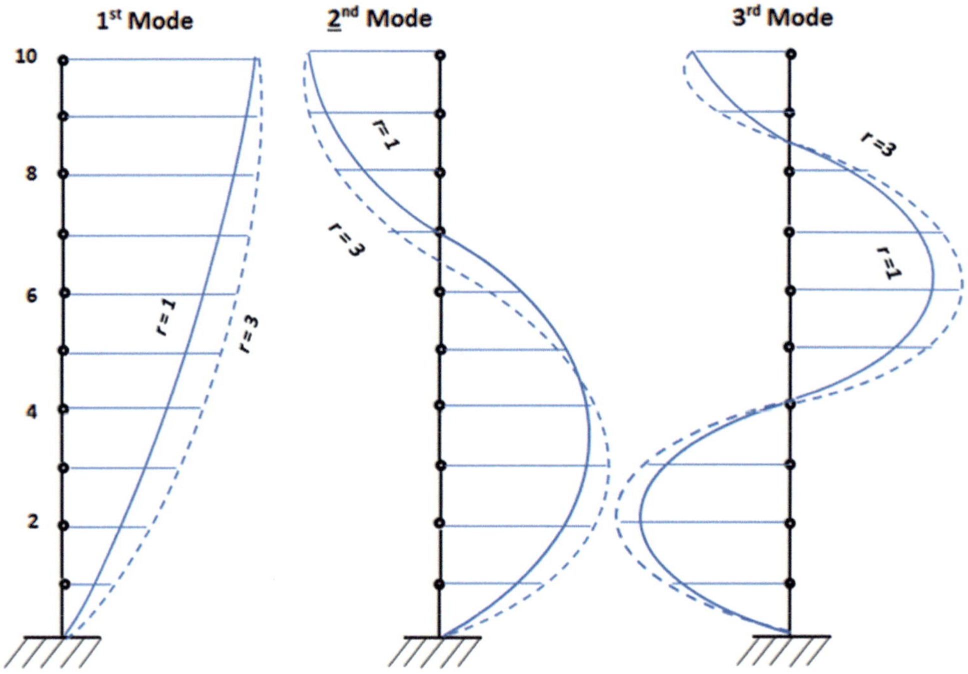

Fig. 19.3 Modes of the free linear and nonlinear vibrations of the 10-story building

a mode occupying the mid-position with respect to the modes of linear and elastic-nonlinear modes of oscillations.

In the case of the power law of deformation, we succeeded in obtaining the precise value of periods and modes of nonlinear oscillations for a multiple-degree-of-freedom system. In the case of other dependences, obtaining the exact solution of the problems proved to be impossible. The effective methods of study of nonlinear systems with the "small nonlinearity" are the methods of approximations; the asymptotic method of Krylov-Bogoliubov [15] in particular. Let us consider free nonlinear oscillations of multi-story frames with the undeformed horizontal elements of frame. The equation of motion of this frame with the power law of deformation takes the form of (19.28). When $r = 1$ the linear oscillations take place. The equations of the linear oscillations for a frame are written in the form presented by E. E. Khachian, particularly for analyzing an elasto-plastic system [16].

$$\sum_{k=1}^{s} m_k y_k'' + a_s(y_s - y_{s+1}) = 0, \qquad s = 1, 2, .. \ n \tag{19.41}$$

Here, the concentrated masses are numbered on top, for example, m_1- is concentrated mass of the uppermost floor. The presence in each equation of one elastic reaction, corresponding to a given floor, is the characteristic property of the Eq. (19.41). Let us examine a case when dependency between the restoring force and relative floor deformation takes the form of a cubic parabola. In this case, the equations of elastic-nonlinear oscillations take the form [17]

$$\sum_{k=1}^{s} m_k y_k'' + a_s\left[(y_s - y_{s+1}) - \varepsilon_s(y_s - y_{s+1})^3\right] = 0, \qquad s = 1, 2, .. \ n \tag{19.42}$$

where ε_s- is a small positive parameter, characterizing the nonlinearity of deformation $(n - s + 1)$ th floor. In a multistory frame, the lower and upper levels are compressed to a different degree. This means that in the Eq. (19.42), the value of normal forces for the ground floors must be more as compared to the upper. However, we further neglect the influence of the normal forces and accept the following:

$$\varepsilon_1 = \varepsilon_2 = \cdots = \varepsilon_s = \varepsilon$$

On the basis of the physical representations, the Eq. (19.42) can be used for such values $(y_s - y_{s+1})$, with which the restoring force grows in its absolute value. From the conditions

$$R_s'(y_s - y_{s+1}) = a_s\left[1 - 3\varepsilon(y_s - y_{s+1})^2\right] \geq 0,$$

we see that

$$\varepsilon \leq \frac{1}{3(y_s - y_{s+1})^2_{\max}}. \tag{19.43}$$

For finding the frequency of the period of a nonlinear system, described by the Eq. (19.42), we will use the asymptotic method of Krylov-Bogoliubov [15]. We search for the general solution of Eq. (19.42), in the form of decomposition

$$\begin{aligned} y_s = {} & C_s^{(1)} Y \cos\left(\omega_l^{(1)} t + \theta\right) + \varepsilon U_s^{(1)}\left(Y, \omega_\wedge^{(1)} t + \theta\right) + \\ & + \varepsilon^2 U_s^{(2)}\left(Y, \ \omega_l^{(1)} t + \theta\right) + \cdots, \end{aligned} \tag{19.44}$$

where $Y-$ maximum displacement of the free end of the frame, $C_s^{(1)}$ – the particular solution of the linear system of the Eq. (19.41); $\omega_l^{(1)}$ – the angular frequency of the main form of linear oscillations; $U_s^{(1)}, \ U_s^{(2)}$ – the periodical functions of angle $\psi = \omega_l^{(1)} t + \theta$ with period 2π. The values of Y and ψ, as the functions of time, are determined by differential equations:

$$\begin{aligned} \frac{dY}{dt} &= \varepsilon A_1(Y) + \varepsilon^2 A_2(Y) + \varepsilon^3 \ \cdots, \\ \frac{d\psi}{dt} &= \omega_l^{(1)} + \varepsilon B_1(Y) + \varepsilon^2 B_2(Y) + \varepsilon^3 \ \cdots \end{aligned}$$

As the first approximation, we assume:

$$y_s = C_s^{(1)} Y \cos\psi, \quad s = 1, 2 \cdots n, \tag{19.45}$$

where

$$\begin{aligned} \frac{dY}{dt} &= \varepsilon A_1(Y) \\ \frac{d\psi}{dt} &= \omega(Y) = \omega_l^{(1)} + \varepsilon B_1(Y). \end{aligned} \tag{19.46}$$

Here $\omega(Y)$– the desired frequency of nonlinear oscillations. To find $B_1(Y)$, we will use one of the equations of harmonic balance [17]:

$$\int_0^{2\pi} \sum_{s=1}^{n} C_s^{(1)} \left\{ \sum_{k=1}^{s} m_k y_k'' + a_s(y_s - y_{s+1}) - a_s\varepsilon(y_s - y_{s+1})^3 \right\} \cos\psi d\psi = 0 \tag{19.47}$$

where: $s = 1,\ 2,\ \cdots n$

Differentiating (19.45) and accounting for (19.46), with the accuracy to the values of the first-order of smallness, we find

$$\begin{aligned} y_s{}^{(1)} = C_s^{(1)} \Big[&- 2\varepsilon\omega_l^{(1)} \sin\psi A_1(Y) - Y\omega_l^{(1)^2} \cos\psi - \\ &- 2\varepsilon Y\omega_l^{(1)} \cos\psi B_1(Y) \Big]. \end{aligned} \tag{19.48}$$

After substituting (19.45) and (19.48) into the Eq. (19.47), we will obtain:

$$\begin{aligned} &\int_0^{2\pi} \sum_{r=1}^{n} C_r^{(1)} \left\{ \sum_{k=1}^{s} - m_k C_s{}^{(1)} \Big[2\varepsilon\omega_l^{(1)} \sin\psi A_1(Y)\omega_l^{(1)^2} + Y\cos\psi + \right. \\ &+ 2\varepsilon Y\omega_l^{(1)} \cos\psi B_1(Y) \Big] + a_s Y \cos\ \psi \left(C_s{}^{(1)} - C_{s+1}{}^{(1)} \right) - \\ &\left. - \varepsilon a_s Y^3 \cos^3\psi \left(C_s{}^{(1)} - C_{s+1}{}^{(1)} \right)^3 \right\} \cos\ \psi d\psi = 0 \end{aligned} \tag{19.49}$$

After integration and appropriate conversion, we will obtain.

$$\begin{aligned} &- 2Y\omega_l^{(1)} \varepsilon B_1(Y)\pi \sum_{r,k=1}^{n,\ s} m_k C_r{}^{(1)} C_s{}^{(1)} + Y\pi \sum_{r,k=1}^{n,s} C_r{}^{(1)} \Big[- m_k \omega_l^{(1)^2} C_s{}^{(1)} + \\ &+ a_s \left(C_s{}^{(1)} - C_{s+1}{}^{(1)} \right) \Big] = \frac{3}{4} \pi\varepsilon Y^3 \sum_{r,s=1}^{n} a_s C_r^{(1)} \left(C_s^{(1)} - C_{s+1}^{(1)} \right)^3. \end{aligned} \tag{19.50}$$

Assuming that all concentrated masses are equal and the rigidity (stiffness) of all floors is identical:

$$a_1 = a_2 = \cdots = a, \qquad m_1 = m_2 = \cdots = m.$$

Keeping in mind that

$$\sum_{r,k=1}^{n,s} \left\{ - m\omega_l^{(1)^2} C_s{}^{(1)} + a \left(C_s^{(1)} - C_{s+1}^{(1)} \right) \right\} = 0, \qquad s = 1,\ 2,\ \ldots\ n,$$

and designating

$$\sum_{r,k=1}^{n,\ s} C_r^{(1)} C_s^{(1)} = M_1^{(n)}$$

and

$$\sum_{r,s=1}^{n} C_r^{(1)} \left(C_s^{(1)} - C_{s+1}^{(1)}\right)^3 = \Delta_1^{(n)}$$

we will obtain

$$\varepsilon B_1 = -\frac{3}{8} \frac{\varepsilon Y^2}{\omega_l^{(1)}} \frac{a}{m} \frac{\Delta_1^{(n)}}{M_1^{(n)}}. \tag{19.51}$$

The values of $C_s^{(1)}$ and $\lambda_1 = \omega_l^{(1)^2} \frac{m}{a}$ for $n = 2 - 20$ are presented in [18]. Using these data and calculating $\Delta_1^{(n)}$ and $M_1^{(n)}$, we will obtain εB_1. Further substituting εB_1 into the second equation of the system (19.46), we obtain the value of the frequency of nonlinear oscillations $\omega(Y)$ in the first approximation. Calculations were conducted for $n = 2,\ 10,\ 15$, and 20 stories. The obtained values $M_1^{(n)}$, $\Delta_1^{(n)}$, and $\omega(Y)$ are given in Table 19.4. The solution of the Eq. (19.42) in the second approximation has the form

$$y_s = C_s^{(1)} Y \cos\ \psi + \varepsilon U_s^{(1)}, \qquad s = 1, 2\ \cdots\ n \tag{19.52}$$

where Y and ψ are determined from the equations of the second approximation:

$$\frac{dY}{dt} = \varepsilon A_1(Y) + \varepsilon^2 A_2(Y), \qquad \frac{d\psi}{dt} = \omega_l^{(1)} + \varepsilon B_1(Y) + \varepsilon^2 B_2(Y). \tag{19.53}$$

The value $U_s^{(1)}$ in the Eq. (19.52) is obtained as the amplitude of the forced oscillations, excited in the system (19.42) by the forces $a\varepsilon(y_s - y_{s+1})^3$, in which y_s and y_{s+1} are sinusoidal:

Table 19.4 Values of $M_1^{(n)}$, $\Delta_1^{(n)}$, and relations on nonlinear and linear frequencies for frame buildings up to 20 floors (stories)

n	$M_1^{(n)}$	$\Delta_1^{(n)}$	$\omega(Y)/\omega_l^{(1)}$
1	1	1	1–0.375
2	4.235	0.869	1–0.109
3	11.328	0.316	1–0.0528
4	24.089	0.247	1–0.0317
5	43.542	0.197	1–0.0209
6	71.373	0.164	1–0.01479
7	109.336	0.141	1–0.01107
8	158.886	0.123	1–0.00912
9	221.977	0.111	1–0.00687
10	299.713	0.100	1–0.00562
15	961.594	0.068	1–0.00256
20	2224.004	0.051	1–0.00147

$$U_s^{(1)} = \sum_{j=1}^{n} C_s^{(j)} \frac{\sum_{p=1}^{n} f_p^{(0)} C_p^{(j)}}{mM_j\omega_j^2} + \sum_{j=2}^{n} C_s^{(j)} \frac{\sum_{p=1}^{n}\left(f_p^{(1)}\cos\psi + g_p^{(1)}\sin\psi\right)C_p^{(j)}}{mM_j\left(\omega_j^2 - \omega_l^{(1)^2}\right)} + \\ + \sum_{i=2}^{\infty}\sum_{j=1}^{n} C_s^{(j)} \frac{\sum_{p=1}^{n}\left(f_p^{(i)}\cos\ i\psi + g_p^{(i)}\sin i\psi\right)C_p^{(j)}}{mM_j\left(\omega_j^2 - i^2\omega_l^{(1)^2}\right)} \tag{19.54}$$

where

$$\sum_{r,k=1}^{n,\ s} C_r^{(j)} C_s^{(j)} = M_j \ (s,j=1,\ 2\cdots n)$$

$f_p^{(i)},\ g_p^{(i)}$ – the coefficients of the decomposition function $a(y_s - y_{s+1})^3$ in the Fourier series. When calculating $U_s^{(1)}$ in the sum of the third addend, three members were taken into account. Differentiating (19.52) and taking into account (19.53) with the accuracy to the values of the second-order of smallness inclusively, and substituting y_s and y_s'' in (19.47), we find $B_2(Y)$. After substituting the latter into the second equation of the system (19.53), we find value $\omega(Y)$ by conducting two approximations.

The values $B_2(Y)$ are calculated for $n = 1,\ 2,\ 3$ floors (stories). The values $\omega(Y)$ taking into account two approximations take the following form:

when $n = 1$ $\omega(Y) = \omega_l^{(1)}\left(1 - 0.375\varepsilon Y^2 - 0.058\varepsilon^2 Y^4\right)$,
when $n = 2$ $\omega(Y) = \omega_l^{(1)}\left(1 - 0.109\varepsilon Y^2 - 0.0039\varepsilon^2 Y^4\right)$,
when $n = 3$ $\omega(Y) = \omega_l^{(1)}\left(1 - 0.058\varepsilon Y^2 - 0.0008\varepsilon^2 Y^4\right)$.

For systems with degrees of freedom higher than three, the second approximation is not produced because of the bulkiness, of calculations. Furthermore, as can be seen from Table 19.5, in the given interval of variation of coefficient of nonlinearity ε, the first approximation gives results with the accuracy sufficient for practical purposes. Quantitative change of the frequency of nonlinear

Table 19.5 Values of relations of nonlinear and linear frequencies depending on coefficient of nonlinearity and numbers of stories

	$\omega(Y)/\omega_l^{(1)}$ First approximation		$\omega(Y)/\omega_l^{(1)}$ Second approximation	
n	$\varepsilon = 0.16$	$\varepsilon = 0.33$	$\varepsilon = 0.16$	$\varepsilon = 0.33$
1	0.93	0.8763	0.9384	0.8700
2	0.9302	0.8562	0.9286	0.8501
3	0.9239	0.8432	0.9222	0.8362
4	0.9188	0.8327	–	–
5	0.9164	0.8276	–	–
6	0.9147	0.8242	–	–
7	0.9137	0.8222	–	–
8	0.9124	0.8195	–	–
9	0.9109	0.8164	–	–
10	0.9100	0.8446	–	–
15	0.9078	0.8100	–	–
20	0.9060	0.8060	–	–

oscillations $\omega(Y)$, depending on the number of floors and the coefficient of nonlinearity, is illustrated by the data of Table 19.5, where $(y_s - y_{s+1})_{\max} = 1$ is set for the sake of simplicity.

From the conditions of (19.43), it turns out that $\varepsilon \leq 0.33$. To compare the influence of height of the building on the value of $\omega(Y)/\omega_l^{(1)}$ in the first approximation, we take that $Y = n(y_s - y_{s+1})_{\max}$. As can be seen in Table 19.5, with increasing height of the building, the ratio of $\omega(Y)/\omega_l^{(1)}$ decreases, and the difference between the frequencies of nonlinear and linear oscillations is more essential at the large values of ε. Data analysis of Table 19.5 showed that the values of $\omega(Y)/\omega_l^{(1)}$ were well approximated by the function

$$\omega(Y)/\omega_l^{(1)} = 1 - 0.375n^{-1.84}\varepsilon Y^2,$$

and for relation of the periods of oscillations $T(Y)/T_l^{(1)}$ with the accuracy of ε, we will have

$$\frac{T(Y)}{T_l^{(1)}} = 1 + 0.375n^{-1.84}\varepsilon Y^2. \tag{19.55}$$

After substituting $T_l^{(1)}$ in (19.35) with its value as given in [16], we will obtain a similar formula for determining the period of single-frequency elastic-nonlinear oscillations:

$$T(Y) = 2\pi(0.367 + 0.633n)\sqrt{\frac{m}{a}}\left(1 + 0.375n^{-1.84}\varepsilon Y^2\right).$$

If $\varepsilon = 0$, then we obtain the value of the period of the main (fundamental) tone of free linear oscillations.

Now let us examine the case when restoring force under load is subordinate to the nonlinear law:

$$R_s = a\left[\left(y_s - y_{s+1}\right) - \varepsilon\left(y_s - y_{s+1}\right)^3\right],$$

and unloading occurs linearly. The angle of the slope of the unloading line with the axis $y_s - y_{s+1}$ is equal $\alpha = \arctan a$. Using (19.40), we obtain the formula for determining the period of hysteresis oscillations:

$$T_{\text{Hysteresis}} = T_l^{(1)}\left[1 + \frac{1}{2}0.375n^{-1.84}\varepsilon Y^2\right]. \tag{19.56}$$

When determining the period of hysteresis oscillations, it was accepted that the rigidity of the system during the unloading was equal to the greatest initial rigidity of the system when loading. If we continuously reduce the rigidity of the system during the unloading, then it is possible to ascertain that the period of hysteresis oscillations will approach the period of linear oscillations.

In calculating the periods of nonlinear oscillations of reinforced concrete structures according to the formulas (19.55) and (19.56), it is possible to use the values of coefficient of nonlinearity ε, given in [19]. This work is dedicated to the experimental studies of the forced oscillations of the reinforced concrete bent elements. The patterns were tested with different values of perturbing force, until destruction. As the design scheme of the tested elements with adequate accuracy shows, it is possible to accept a system with single-degree-of-freedom. According to the data of these experiments, the coefficient of nonlinearity ε for the frame columns with the initial period $T = 0.15. - 0.40\ s.$ was equal $0.0255 - 0.066\ cm^{-2}$, and for the simple beam $T \leq 0.1s. - 0.10^{-2}$. The results of vibrational tests of two reinforced concrete monolithic and buildup frame buildings given in [20] make it possible to calculate coefficients of nonlinearity in this case for a building as a whole. The values of these coefficients were equal to $\varepsilon = 1$ and $\varepsilon = 1.8$ respectively. Comparatively large values of the coefficient of nonlinearity can be explained not only by nonlinear deformation of reinforced concrete but also by the pliability of joints and the failure of the load-bearing elements of the framework filling.

19.3 The Nonlinear Oscillations of Systems with Distributed Mass

Let us examine free nonlinear oscillations of systems with distributed mass. As an example of such systems, we will examine the oscillations of bent beams with different supporting fastenings and shifted cantilever beam. To solve these problems, we will use a method used in the work of G. Kauder [21]. In a rectangular coordinate system, a prismatic bent bar with a symmetrical relative to y and z cross section is examined. It is well established that the dependency between the normal stress σ_x and relative strain θ_x takes the form

$$\sigma_x = E\theta_x\left(1 - \varepsilon E^2 {\theta_x}^2\right) \tag{19.57}$$

where $E-$ modulus of elasticity of material, $\varepsilon-$ coefficient of nonlinearity of deformation.

To this oscillatory system, we will apply Hamilton's principle, according to which in an interval of real motion of a system in the time interval $t_1 \div t_2$, the integral $\int_{t_1}^{t_2}(P-K)\,dt$ takes the extreme value

$$\delta\int_{t_1}^{t_2}(P-K)\,dt = 0 \tag{19.58}$$

where $P-$ potential energy and $K-$ kinetic energy of the oscillatory system.

The potential energy of the bar is equal to

$$P = \frac{EI_0}{2}\int_0^l\left[1 - \frac{1}{2}\varepsilon E^2\frac{I_2}{I_0}\left(\frac{\partial^2\eta}{\partial x^2}\right)^2\right]\left(\frac{\partial^2\eta}{\partial x^2}\right)^2 dx. \tag{19.59}$$

and the kinetic energy is equal to

$$K = \frac{m}{2}\int_0^l\left(\frac{\partial\eta}{\partial t}\right)^2 dx. \tag{19.60}$$

In (19.59) and (19.60), the following designations are accepted: $m-$ the mass of the unit length of the bar (rod); $I_0 = \iint_F y^2 dydz$, $I_2 = \iint_F y^4 dydz$, $-$ the area of cross section; $\eta = \eta(x, t)-$ deflection.

We will compose Hamilton's integral and move to dimensionless parameters according to the formulas

$$\xi = \pi\frac{x}{l}; \quad \tau = \omega t,$$

where $\omega-$ is the desired frequency of nonlinear oscillations. Hamilton's integral takes the form

$$F = \frac{mF_1}{2\pi\omega}\int_0^{2\pi}\int_0^{\pi}\left\{a^2\left[1 - \frac{\lambda}{6}\left(\frac{\partial^2\eta}{\partial x^2}\right)^2\right]\left(\frac{\partial^2\eta}{\partial x^2}\right)^2 - \frac{\omega^2\partial^2\eta}{\partial t^2}\right\} d\xi d\tau,$$

where the following designations are accepted

$$a^2 = \frac{\pi^4 EI}{ml^4}; \qquad \lambda = \frac{3\pi^4 \varepsilon E^2}{l^4} \frac{I_2}{I_0}.$$

It is assumed that the desired function $\eta(\xi, \tau)$ can be found in the following way

$$\eta(\xi, \tau) = X(\xi) \cdot Y(\tau)$$

Let us introduce the designations:

$$b_0 = \int_0^\pi X^2 d\xi; \qquad \nu^2 = \frac{1}{b_0} \int_0^\pi X''^2 d\xi; \qquad b_2 = \frac{1}{\nu^2 b_0} \int_0^\pi X''^4 d\xi. \tag{19.61}$$

Integral F takes the form

$$F = \frac{mF_1}{2\pi\omega} b_0 \int_0^{2\pi} \left[\nu^2 a^2 Y^2 \left(1 - \frac{\lambda}{6} b_2 Y^2\right) - \omega^2 \left(\frac{\partial Y}{\partial \tau}\right)^2 \right] d\tau. \tag{19.62}$$

Function $Y(\tau)$ must satisfy the Euler's equation, composed for the functional in (19.62), and having the form:

$$\frac{\partial^2 Y}{\partial \tau^2} + \frac{\nu^2 a^2}{\omega^2} \left(1 - \frac{1}{3} \lambda b_2 Y^2\right) Y = 0. \tag{19.63}$$

The solution of the Eq. (19.63) is given in [21]. Using it, we find the value of the period of nonlinear oscillations T

$$T = \frac{1}{\nu a} \left[1 + \frac{3}{8} \frac{\lambda b_2}{3} Y_i^2 + \frac{57}{265} \left(\frac{\lambda b_2}{3}\right)^2 Y_i^4 + \frac{315}{2048} \left(\frac{\lambda b_2}{3}\right)^3 Y_i^6 + \cdots \right], \tag{19.64}$$

where Y_i– the fixed value of initial maximum displacement.

From the foregoing, it is clear that in order to solve the problem it is necessary to set the function of $X(\xi)$. The best approximation for the "weakly nonlinear" systems can be obtained if we proceed from the corresponding fundamental functions, obtained at (with) $\varepsilon = 0$. Let us examine several types of beams with different supporting fastenings, that is, with different boundary conditions.

Simply Supported Beams The function which determines the main form (first mode) of free linear oscillations has the form

$$X(\xi) = \sin \frac{k\xi}{\pi}, \qquad \text{where} \quad k = \pi$$

Constants (19.61) will take the following values:

$$b_0 = \frac{\pi}{2}; \qquad \nu^2 = 1; \quad b_2 = 0.75$$

The period of nonlinear oscillations according to (19.64) will be:

$$T = T_l \left[1 + 0.09375\lambda Y_i^2 + 0.013916\lambda^2 Y_i^4 + 0.002403\lambda^3 Y_i^6 + \cdots\right], \tag{19.65}$$

where $T_l = \frac{2\pi}{a}$ – period of free nonlinear oscillations. This result is obtained by G. Kauder [21].

Beams with Built-In Ends As $X(\xi)$, we assign a corresponding fundamental function, determining the main mode of free linear oscillations [10], with the value of maximum amplitude equal to one:

$$X(\xi) = \frac{1}{1.61643}\left[\sin\frac{k\xi}{\pi} - 1.0178\cos\frac{k\xi}{\pi} - \sinh\frac{k\xi}{\pi} + 1.0178\cosh\frac{k\xi}{\pi}\right];$$
$$k = 4.73$$

After calculating integrals (3.30), we will obtain the following values:

$$b_0 = 1.24542;\quad \nu^2 = 5.1384;\quad b_2 = 3.77213.$$

From (19.64), we will have

$$T = T_l\left[1 + 0.47152\lambda Y_i^2 + 0.35203\lambda^2 Y_i^4 + 0.30577\lambda^3 Y_i^6 + \cdots\right],$$
$$T_l = \frac{2\pi}{2.267a}. \tag{19.66}$$

Beams with One End Hinged Support and Another End Built-In Support For function $X(\xi)$, we have

$$X(\xi) = \frac{1}{1.06676}\left[\sin\frac{k\xi}{\pi} + 0.02785\sinh\frac{k\xi}{\pi}\right],\qquad k = 3.927.$$

The maximum value of the function $X(\xi)$ is equal to one, and takes place at the distance $0.42l$ from the supported end.

$$b_0 = 1.37911;\quad \nu^2 = 2.4423;\quad b_2 = 1.8116$$

$$T = T_l\left[1 + 0.22645\lambda Y_i^2 + 0.081193\lambda^2 Y_i^4 + 0.03387\lambda^3 Y_i^6 + \cdots\right],$$
$$T_l = \frac{2\pi}{1.563a} \tag{19.67}$$

The Cantilever Beam

$$X(\xi) = \frac{1}{2.7242}\left[\sin\frac{k\xi}{\pi} - 1.3622\cos\frac{k\xi}{\pi} - \sinh\frac{k\xi}{\pi} + 1.3622\cosh\frac{k\xi}{\pi}\right]$$
$$k = 1.875;\quad b_0 = 0.78536;\quad \nu^2 = 0.1269;\quad b_2 = 0.074514$$
$$T = T_l\left[1 + 0.009314\lambda Y_i^2 + 0.000137\lambda^2 Y_i^4 + 0.0000023\lambda^3 Y_i^6 + \cdots\right],$$
$$T_l = \frac{2\pi}{0.362a}. \tag{19.68}$$

As an illustration, we conduct calculations of the periods of nonlinear oscillations of the reinforced concrete bent beams. Let us determine the periods of the nonlinear oscillations of beams, with the following geometric dimensions: $l = 300\ cm;\ b = h = 40\ cm$. On the basis of physical considerations, it is possible to use the dependency (19.57) with such values of θ, for which $\sigma'(\theta) > 0$, that is,

$$\sigma'(\theta) = E\left(1 - 3\varepsilon^2\theta^2\right) > 0.$$

Hence, we have:

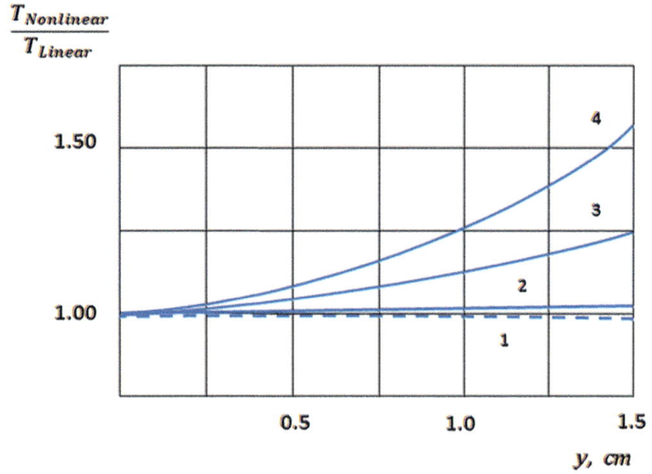

Fig. 19.4 Dependencies of periods of nonlinear and linear oscillations of bent beams and displacement: 1–cantilever beam; 2–simply hinge-supported beam; 3–beam with one end hinged support and another with end built-in support; and 4–beam with built-in ends

$$\varepsilon_{\max} \leq \frac{1}{3E^2\theta^2{}_{\max}}$$

Assuming that the maximum relative deformation in the reinforced concrete bent beams equals to $\theta = 0.003$, we obtain the maximum value of coefficient λ:

$$\lambda_{\max} \leq \frac{\pi^4}{l^4\theta^2{}_{\max}} \frac{12}{80} h = 0.32$$

Based on this data, Fig. 19.4 shows the dependencies of periods T/T_l and the initial displacement Y_i maximum value of which is equal to $Y_{i,\max} = \frac{l}{200} = 1.5$ cm. As can be seen in Fig. 19.4, the same initial displacement the nonlinearity of deformation has the greatest effect on the period of oscillations of the beam with the built-in ends.

If in the simple beam with $Y_i = 1.5$ cm. , the period of nonlinear oscillations increases by 1.08 times, then in the beam with the built-in ends, it increases by 1.6 times. Similarly, a dependency "period–displacement" can be obtained for a shifted cantilever beam with distributed mass m. The relationship of sheering stress τ and angular deflection γ is written in the form of:

$$\tau = \gamma G\left(1 - \varepsilon\gamma^2 G^2\right)$$

where $G-$ shear modulus, $\varepsilon-$ coefficient of nonlinearity of deformation.

We compose Hamilton's integral

$$F = \frac{1}{2}\int_0^{\frac{2\pi}{\omega}}\int_0^l \left\{kF_1 G\left[1 - \frac{\varepsilon G^2}{2}\left(\frac{\partial\eta}{\partial x}\right)^2\right]\left(\frac{\partial\eta}{\partial x}\right)^2 - \left(\frac{\partial\eta}{\partial t}\right)^2\right\} dxdt, \qquad (19.69)$$

where $k-$ coefficient for non-uniform distribution of shearing stresses on the height of the cross section.

We look for the unknown displacement in the form of

$$\eta(x,\ t) = X(x) \cdot T(t) \qquad (19.70)$$

For the cantilever beam, the function $X(x)$ is set as one corresponding to the $i-$ th mode of the free linear oscillations, that is,

$$X(x) = \sin\frac{2i-1}{2}\pi\frac{x}{l}; \quad i = 1,\ 2\ \ldots$$

Let us introduce the following designations:

$$a_0 = \int_0^l X^2 dx; \qquad a_1 = \int_0^l X'^2 dx; \qquad a_2 = \int_0^l X'^4 dx; \tag{19.71}$$

The calculations of coefficients for the cantilever beam gives:

$$a_0 = \frac{l}{2}; \quad a_1 = \frac{1}{2}\left(\frac{2i-1}{2}\pi\right)^2; \quad a_2 = \frac{3}{4l^3}\left(\frac{2i-1}{2}\pi\right)^4. \tag{19.72}$$

Substituting (19.70) in (19.40), then conducting integration for X, taking into account (19.72), and composing the Euler's equation for this functional, we obtain the following:

$$Y'' + G\frac{kF_1}{m}\left(\frac{2i-1}{2}\frac{\pi}{l}\right)^2 Y\left[1 - \frac{3}{4}\varepsilon G^2\left(\frac{2i-1}{2}\frac{\pi}{l}\right)^2 Y^2\right] = 0. \tag{19.73}$$

Let us designate:

$$\chi_i^2 = \frac{kGF_1}{m}\left(\frac{2i-1}{2}\frac{\pi}{l}\right)^2; \qquad \mu_i = \varepsilon G^2\left(\frac{2i-1}{2}\frac{\pi}{l}\right)^2\frac{kGF_1}{m}. \tag{19.74}$$

The approximate solution of the nonlinear differential equation given above is known [21]. With the aid of that solution, the following expression is obtained for the period of $i-$ th mode

$$T_i = \frac{2\pi}{\chi_i}\left[1 + 0.28125\mu_i Y_i^2 + 0.125244\mu_i Y_i^4 + 0.064888\mu_i^3 Y_i^6 + \right]. \tag{19.75}$$

As in [21], here, too, it is possible to show that the relation of the periods of nonlinear and linear oscillations with the identically stressed states does not depend on the number of the form of oscillations. If, in the first approximation, we use linear theory, then the maximum shearing stress τ_{max} can be determined according to the formula:

$$\tau_{\max} \approx kGX'(x)_{\max} = \frac{(2i-1)}{2}\frac{\pi}{2}xY_i. \tag{19.76}$$

Hence, we find

$$Y_i = \frac{\tau_{\max}}{kG}\frac{2l}{(2i-1)\pi}.$$

For the second term of the series in (19.75), we will obtain the expression:

$$0.28125\mu Y_i^2 = \varepsilon\frac{GF_1}{km}\tau_{\max}. \tag{19.77}$$

The expression (19.77) shows that with the identical stressed state, an increase in the period of nonlinear oscillations, as compared to the period of linear oscillations of the corresponding form, occurs to the identical degree.

If, in the examined systems with the distributed parameters hysteresis oscillations occur, then their periods are determined from the formula (19.40).

19.4 The Oscillations of the Beam of the Variable Cross Sections

The solution to the problem of free oscillation of the wedge-shaped beam in the Bessel functions was obtained by Kirchhoff. Subsequently, the exact solutions for beams of variable sections, expressed through the special functions, were obtained in works [22–24]. However, the set of problems that have exact solutions is very small. Therefore, when studying free oscillations of a bar (rod) of variable cross sections, the methods of approximations are more frequently applied in particular the Ritz method and Galerkin method, as well as other methods of approximations. It is practically impossible to provide a review of all the research in which these methods are used. We will note only some basic analysis which led to the mathematical foundation of these methods for more complete bibliography, see [3, 7, 23]. The enumerated methods, being approximate, under specific conditions allow for obtaining solutions with predetermined accuracy. So, for example, with the aid of Ritz's method it is possible to obtain the exact solution of a problem if as the basic functions we select a system of linear independent functions, which satisfy boundary conditions and condition of completeness of energy [25]. In [26] the solution of the problem of free vibrations of hinge-supported beam with linearly changing height of the cross section is obtained by this method. Below, with the aid of Ritz's method, we examine the vibrations of the hinge-supported beam with the arbitrary changing height of the cross section [5] (Fig. 19.5).

The equation of the free oscillations of the beam with variable cross section has the form

$$\frac{\partial^2}{\partial x^2}\left[EI(x)\frac{\partial^2 y}{\partial x^2}\right]+\rho F(x)\frac{\partial^2 y}{\partial t^2}=0. \tag{19.78}$$

Assuming harmonic oscillations with a frequency ω,

$$Y(x,\ t)=Y(x)e^{-i\omega t},$$

and replacing the variable in the formula $\xi=\frac{x}{l}$, we will obtain

$$\frac{d^2}{d\xi^2}\left[I(\xi)\frac{d^2 Y}{d\xi^2}\right]-\lambda F(\xi)Y(\xi)=0,\quad \lambda=\frac{\rho\omega^2}{E}l^4. \tag{19.79}$$

The boundary conditions for the hinge-supported beam are the following:

$$Y(0)=Y(l)=Y''(0)=Y''(l)=0. \tag{19.80}$$

The functional for which the Euler equation is the equation of motion (19.79) has the form

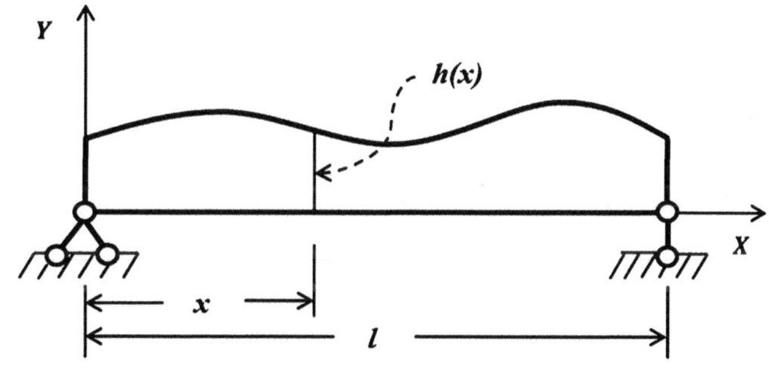

Fig. 19.5 Scheme of the beam

$$S = \frac{1}{2}\int_0^l \left\{ I(\xi)[Y''(\xi)]^2 - \lambda F(\xi)[Y(\xi)]^2 \right\} d\xi. \tag{19.81}$$

The problem is solved for the case of the arbitrary law of variation in the cross section. It is accepted that cross sectional area $F(\xi)$ and the moment of inertia of the cross section $I(\xi)$ are given in the form of Fourier series:

$$\begin{aligned} F(\xi) &= \frac{a_0}{2} + \sum_{j=1}^{\infty} a_j \cos j\pi\xi + b_j \sin j\pi\xi \\ I(\xi) &= \frac{c_0}{2} + \sum_{j=1}^{\infty} c_j \cos j\pi\xi + d_j \sin j\pi\xi. \end{aligned} \tag{19.82}$$

The solution of the Eq. (19.79) is sought in the form of the Fourier series with members, satisfying the boundary conditions of the problem:

$$Y(\xi) = \sum_{k=1}^{\infty} A_k \sin k\pi\xi \tag{19.83}$$

Substituting the expression for $Y(\xi)$ and $Y''(\xi)$, and also $F(\xi)$, and $I(\xi)$ in (19.81), we will obtain

$$\begin{aligned} 2S = \int_0^l \Bigg\{ & \sum_{i=1}^{\infty}\sum_{k=1}^{\infty} \left(\frac{c_0}{2} i^2k^2\pi^4 - \lambda\frac{a_0}{2}\right) A_iA_k \sin\ k\pi\xi\ \sin\ i\pi\xi + \\ & + \sum_{j=1}^{\infty}\sum_{i=1}^{\infty}\sum_{k=1}^{\infty} \left(c_j i^2k^2\pi^4 - \lambda a_j\right) A_iA_k \sin\ k\pi\xi\ \sin\ i\pi\xi\ \cos\ j\pi\xi + \\ & + \sum_{j=1}^{\infty}\sum_{i=1}^{\infty}\sum_{k=1}^{\infty} \left(d_j i^2k^2\pi^4 - \lambda b_j\right) A_iA_k \sin\ k\pi\xi\ \sin\ i\pi\xi\ \sin\ j\pi\xi \Bigg\}\ d\xi. \end{aligned} \tag{19.84}$$

Let us calculate the integrals, entering in (19.84)

$$I_0(k,\ i) = \int_0^l \sin\ k\pi\xi\ \sin\ i\pi\xi d\xi = \begin{cases} 0, & i \neq k \\ 1/2, & i = k \end{cases} \tag{19.85}$$

$$I_1(k,i,j) = \int_0^l \sin\ k\pi\xi\ \sin\ i\pi\xi\ \cos\ j\pi\xi\ d\xi = \begin{cases} 0, & j \neq \pm(k \mp i) \\ 1/4, & j = \pm(k - i) \\ -1/4, & j = \pm(k + i) \end{cases}$$

$$I_2(k,i,j) = \int_0^l \sin\ k\pi\xi\ \sin\ i\pi\xi\ \sin\ j\pi\xi\ d\xi = \begin{cases} \dfrac{j}{2\pi}\left[\dfrac{1-(-1)^{k-i+j}}{j^2-(k-i)^2} - \dfrac{1-(-1)^{k+i+j}}{j^2-(k+i)^2}\right] \\ k \mp i \mp j \neq 0 \\ 0, \quad j = \pm(k \mp i) \end{cases}$$

Taking into account (19.85), the functional $2S$ is represented in the form of

$$2S=\frac{1}{4}\sum_{i=1}^{\infty}\left(c_0 i^4\pi^4-\lambda a_0\right)A_i^2+\sum_{j=1}^{\infty}\sum_{i=1}^{\infty}\sum_{k=1}^{\infty}I_1(k,i,j)\left(c_j i^2k^2\pi^4-\lambda a_j\right)A_iA_k+ \\ +\sum_{j=1}^{\infty}\sum_{i=1}^{\infty}\sum_{k=1}^{\infty}I_2(k,i,j)\left(d_j i^2k^2\pi^4-\lambda b_j\right)A_iA_k. \tag{19.86}$$

According to the Ritz method, the conditions of the minimum (minimal conditions) of functional (19.86) will be

$$\frac{\partial S}{\partial A_i}=0 \qquad i=1,\ 2,\ 3\ \cdots \tag{19.87}$$

The Eq. (19.87) in the expanded form will be

$$\frac{1}{2}\left(c_0 i^4\pi^4-\lambda a_0\right)A_i+\frac{\partial}{\partial A_i}\sum_{i=1}^{\infty}f_{ji}{}^{(1)}A_i+\frac{\partial}{\partial A_i}\sum_{i=1}^{\infty}f_{ji}{}^{(2)}A_i=0,$$

where

$$f_{ji}{}^{(1)}=\sum_{j=1}^{\infty}\sum_{k=1}^{\infty}I_1(j,i,k)\left(c_j i^2k^2\pi^4-\lambda a_j\right)A_k, \\ f_{ji}{}^{(2)}=\sum_{j=1}^{\infty}\sum_{k=1}^{\infty}I_2(j,i,k)\left(d_j i^2k^2\pi^4-\lambda b_j\right)A_k, \tag{19.88}$$

Bearing in mind that

$$\frac{\partial}{\partial A_i}\sum_{i=1}^{\infty}f_{ji}{}^{(1,2)}A_i=f_{ji}{}^{(1,2)}+\sum_{k=1}^{\infty}A_k\frac{\partial f_{ji}{}^{(1,2)}}{\partial A_i}, \qquad I_{1,2}(j,i,k)=I_{1,2}(j,k,i),$$

we will obtain an infinite system of homogeneous equations, which has the form:

$$\sum_{k=1}^{\infty}A_k\mu_{ik}=0, \quad i=1,\ 2,\ 3\cdots \\ \mu_{ik}=2\sum_{j=1}^{\infty}I_1(j,i,k)\left[c_j i^2k^2\pi^4-\lambda a_j\right]+I_2(j,i,k)\left[d_j i^2k^2\pi^4-\lambda b_j\right], \\ \mu_{ii}=2\sum_{j=1}^{\infty}\left\{I_1(j,i,i)\left[c_j i^4\pi^4-\lambda a_j\right]+I_2(j,i,i)\left[d_j i^4\pi^4-\lambda b_j\right]\right\}+ \\ +\frac{1}{2}\left(c_0 i^4\pi^4-\lambda a_0\right). \tag{19.89}$$

The natural frequencies are determined by the solution of the infinite determinant of the form

$$|\mu_{ik}|=0, \quad i,k=1,\ 2.\ 3\ \cdots \tag{19.90}$$

Since in solving the problem, the basis was taken as complete (total, full) energy functions, then according to [25], at i, $k\to\infty$, the dimensionless frequencies λ have to approach their exact values.

After determining eigen frequencies, the forms of oscillations can also be determined. If dimensionless frequencies λ are determined from the truncated determinant of order P, the coefficients A_i can be determined from the system of equations of $P-1$ order, in which one of the coefficients is taken as the given one (for example $A_1 = 1$). The system of such equations will take the form:

$$\sum_{k=2}^{P} A_k \mu_{ik} = -\mu_{i1}, \qquad i = 2, 3, \cdots P. \tag{19.91}$$

As an example, let us examine the free oscillations of a beam of a constant width b and variable height $h(\xi)$:

$$h(\xi) = h_0 + h_1 \sin \pi\xi \tag{19.92}$$

where h_0, h_1– heights of the sections at $\xi = 0,\ 1/2$.

In this case, $F(\xi)$, and $I(\xi)$– are represented in the form of:

$$\begin{aligned} &F(\xi) = F_0(1 + \gamma \sin \pi\xi) \\ &I(\xi) = I_0\{1 + 1.5\gamma^2 - 1.5\gamma^2 \cos\ 2\pi\xi + (3\gamma + 0.75\gamma^3)\ \sin\ \pi\xi - 025\gamma^3 \sin 3\pi\xi\} \end{aligned} \tag{19.93}$$

where

$$F_0 = bh_0, \qquad I_0 = \frac{bh_0{}^3}{12}, \qquad \gamma = \frac{h_1}{h_0}.$$

Comparing (19.93) and (19.82), we find that the coefficients of decomposition of the functions $F(\xi)$ and $I(\xi)$ in the Fourier series take the form

$$\begin{aligned} &\frac{a_0}{2} = F_0; \quad a_1 = a_2 = \cdots = 0; \quad b_1 = F_0\gamma; \quad b_2 = b_3 = \cdots = 0; \\ &\frac{c_0}{2} = I_0(1 + 1.5\gamma^2); \quad c_1 = 0; \quad c_2 = -1.5\gamma^2 I_0; \quad c_3 = c_4 = \cdots = 0; \\ &d_1 = I_0(3\gamma + 0.75\gamma^3); \quad d_2 = 0; \quad d_3 = -0.25\gamma^3 I_0; \quad d_4 = d_5 = \cdots = 0. \end{aligned}$$

The Eq. (19.90) was solved at $i, k = 1,\ 3,\ 5$. The speed of the convergence of the solution can be judged by the data in Table 19.6, where the values of dimensionless frequency $\sqrt{\lambda_i}$ were given $\left(\sqrt{\lambda_i} = \omega_i \frac{l^2}{\pi^2}\sqrt{\frac{m}{EI_0}}\right)$.

As one would expect, the convergence of the process is correct, especially when determining the fundamental frequency and for small values of γ. The accuracy of the obtained results can be judged by the proximity of the results for the same frequencies, obtained during the solution of the determinants of gradually increasing order.

Figure 19.6 shows the dependencies of the first three natural frequencies of the beams, cross sections of which change according to the law (19.92), depending on the parameter γ.

As evident in Fig. 19.6, the values of natural frequencies increase with an increase in γ.

In the examined interval of the variation γ the obtained dependences are very close to the linear dependency.

Table 19.6 The values of $\sqrt{\lambda_i}$

γ	$i, k = 1$	$i, k = 3$			$i, k = 5$				
	$\sqrt{\lambda_1}$	$\sqrt{\lambda_1}$	$\sqrt{\lambda_2}$	$\sqrt{\lambda_3}$	$\sqrt{\lambda_1}$	$\sqrt{\lambda_2}$	$\sqrt{\lambda_3}$	$\sqrt{\lambda_4}$	$\sqrt{\lambda_5}$
0	1	1	4	9	1	4	9	16	25
0.1	1.085	1.084	4.274	9.599	1.084	4.268	9.585	17.053	26.638
0.25	1.215	1.209	4.691	10.528	1.209	4.660	10.453	18.693	29.192
0.75	1.666	1.623	6.122	13.798	1.618	5.910	13.301	24.494	38.251

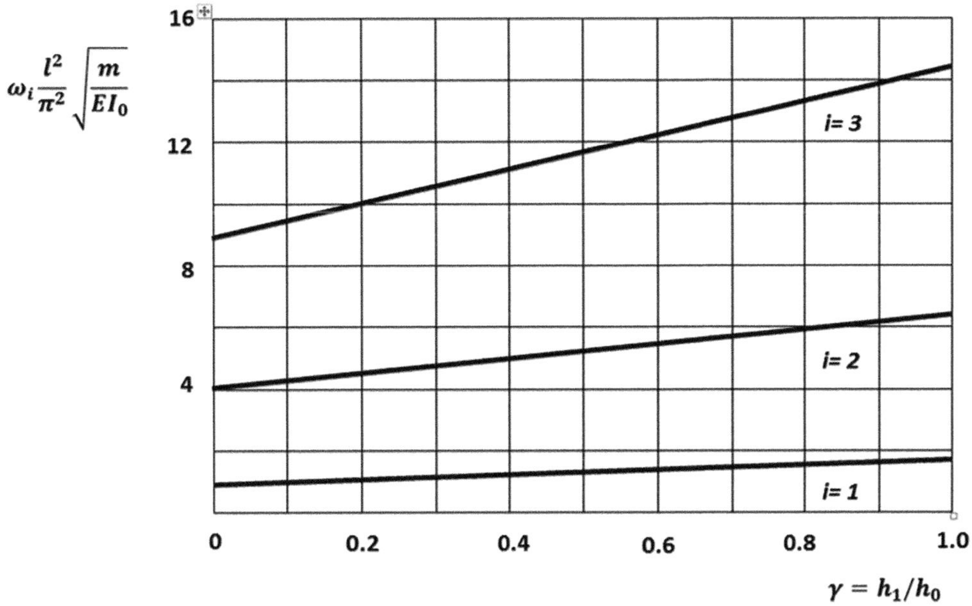

Fig. 19.6 Dependencies of frequencies of free oscillation of a beam from a parameter characterizing changes of cross sections

References

1. Clough, R. W., & Penzien, G. (1975). *Dynamics of structures* (345p).
2. Timoshenko, S. (1956). *Strength of materials, part II, advanced theory and problems* (3rd ed.). Van Nostrand.
3. Timoshenko, S. P. (1967). *Oscillations in engineering* (444p). Nauka.
4. Ambartsumian, V. A. (1972). *To the question of calculation of bar Bent Systems to the seismic impacts in non-linear elasticity law* (p. 19). Soobshenya ArmNIISA vip.
5. Petrosian, L. G., & Ambartsumian, V. A. (2020). *Static and dynamic analysis of engineering structures: Incorporating the boundary element method* (528p). Wiley.
6. Yanke, E., Edme, P., & Lesh, P. (1968). *Special functions. Formulas, tables and graphs, 344.* Nauka.
7. Smirnov, A. P. (1974). *Construction stability and oscillations* (571p). Transzheldorizdat.
8. Smirnov, V. A. (1975). *Suspension bridges of large spans* (369p). Vishaya Shkola.
9. Strength, Stability, and Oscillations. (1968). *Handbook of mechanical engineering* (Vol. I, 831p). Mashinostroenie.
10. Strength, Stability, and Oscillations. (1968). *Handbook of mechanical engineering* (Vol. II, 567p). Mashinostroenie.
11. Ponomarev, S. D., & Biderman, V. L. (1952). *Strength calculations in mechanical engineering* (947c).
12. Rosenberg, R. M. (1961). About the modes of oscillations of the normal type of nonlinear system with two degrees of freedom. *Mechanics, 5*, 3–15.
13. Rosenberg, R. M. (1962). The normal mode shapes of oscillation of nonlinear systems with *n* degree of freedom. *Transactions of the American Society of Mechanical Engineers. Series E, Journal of Applied Mechanics, 5*, 78–84.

14. Ambartsumian, V. A. (1971). Periods of the free non-linear oscillations of frame buildings. *Izv Academy of Sciences of the Armenian SSR, Series of Technical Sciences, XXIV*(1), 16–22.
15. Bogolyubov, N. N., & Mitropolsky, Y. A. (1958). *Asymptotic methods in the theory of nonlinear oscillations* (472p). Fizmatgiz.
16. Khachiyan, E. E. (1973). *Seismic influence on tower buildings and structures* (328p). Hayastan.
17. Khachiyan, E. E., & Ambartsumyan, V. A. (1981). *Dynamic models of construction in the theory of seismic stability* (204p). Hauka.
18. Khachiyan, E. E. (1985). *Recommendations—Guidelines for determining dynamic characteristics and seismic loads for buildings and structures using Accelerograms of earthquakes* (107p). ArmNIISA.
19. Murusidze, R. X. (1965). Experimental study of the nonlinearity of deformation with the oscillations of reinforced concrete elements. In *Seismic resistance of structures* (pp. 129–139). Tbilisi.
20. Zakaryan, V. A. (1971). Experimental study of the dynamic characteristics of high-rise buildings. *Izv AH Arm SSR, Series of Technical of Nauk, XXIV*(3), 33–44.
21. Kauder, G. (1961). *Nonlinear mechanics* (776p). Inostranaya Literatura.
22. Dinnik, A. N. (1952). *The application of the Bessel functions to the problems of theory of elasticity*. Izdatelstvo AH USSR.
23. Korenev, B. G. (1960). *Some tasks of the theory of elasticity and thermal conductivity solving in Bessel functions* (458p). Fizmatgiz.
24. Lardner, T. J. (1968). Vibration of beams with exponentially varying properties. *Acta Mechanica, 6*(2–3), 197–202.
25. Mikhlin, S. G. (1957). *Variational methods in mathematical physics* (476p). Gostexizdat.
26. Krynitsky, E., & Mazurkevich, Z. (1962). The natural oscillations of a hinge-supported beam with the linearly varying height of the cross section. *Transactions of the American Society of Mechanical Engineers. Series E, Applied Mechanics, 28*(3), 314–321.

20 Vibration Damage and Vibration Control Measures

Abstract

This chapter deals with vibrational damage, criteria and limits, short-term and long-term vibrations, vibration control measures, and reduction of dynamic impact on a structure. Ways to reduce vibrations and resonant phenomena, are discussed.

20.1 Vibrational Damage

Vibration damage to structures has been a known phenomenon for a long time and is largely a consequence of the transition to the era of industrialization, the development of modern technologies, and the emergence of new sources of vibration.

Usually, vibrations arise from wind or seismic forces, influence of blasts, exposures, or man-made sources such as construction activities, movement of large masses of people, traffic operations (vehicles, railways, streetcar, subways, etc.), heavy machine operations, etc. Vibrations are oscillatory motions and can be experienced in buildings, usually through floors. Vibrations can affect the strength of structures or even lead to a complete failure or collapse of the structure itself.

Numerous cases of destruction of structures caused by the action of wind at very moderate wind speeds are known. A similar problem arose in the aircraft industry at the beginning of the nineteenth century, when the destruction of the wings of an aircraft occurred from unexpected vibrations at certain flight speeds. This problem <<flatter>> has been successfully solved with advancement of both aircraft engineering and the theory of mechanical vibrations.

Today, vibrations from an external source such as subway or streetcar operation can transfer through soil and propagate through foundation, walls, floors, etc. of buildings and can cause cracks in concrete, bricks, or other masonry structures.

Vibration damage to building structures is associated, as a rule, with the strength of disorders (violations). This occurs due to the brittle fracture of structural elements, the development of plastic deformations, the accumulation of fatigue damage, the appearance of cracks, settlement (sedimentation) of structure supports, distortions, and other damages. The abovementioned damages create the risk of destruction of the structure and disruption of its normal operation.

Resonant vibrations of machine elements, mechanisms, and structures, often lead to their damage, as well as to very expensive shutdowns of the production process. Therefore, when designing vibration-active equipment and building structures, the most important principle is the prevention of

L. G. Petrosian, *Dynamics of Structures*, https://doi.org/10.1007/978-3-031-63539-7_20

resonance. Resonance prevention is achieved by calculating and by comparing the frequencies of natural and forced oscillations. The improvement of computer technology has ensured the creation of modern high-speed computers through which it is possible to carry out adequate calculations not only of individual elements but also of very complex structures and machines as a whole and to give a reasonable forecast of vibration. At the same time, the development of vibration protection means: vibration isolators, dampers, vibration dampers, and so on made it possible to successfully combat vibration, including in resonant modes. The problems of vibration damage began to shift from traditional machines and mechanisms (hammers, presses, saws, compressors, turbine units, etc.) towards more complex equipment, including high tech. At present, the protection of high-precision technologies (microelectronics, radio engineering, optics, instrumentation engineering) from vibration damage is a very urgent problem of the dynamics of structures.

Problems that are associated with vibration damage can be divided into the following main groups:

- Industrial interference in technological processes
- Damage to machinery and equipment
- Adverse (harmful) effect on the human body
- Damage to buildings and structures, foundations, communications, utility lines, etc.
- Vibration interference outside the industrial area and transport routes
- Structural noises that are caused by the propagation of vibrations, infrasound

It should be noted that vibration interference in technological processes can affect the deterioration of the quality of products and reduce their production volume. Vibration interference outside the industrial area and transport routes can lead to long disputes and complaints from the residents of nearby houses and representatives of health care. In general, humans are able to feel vibration at much less lower intensity than that which might result in property damage. Some levels of vibrations can have detrimental effects on human bodies and human minds. Usually vibration is accompanied by noise of varying intensity. The influence of noise can also have a negative impact on people on their physiological and mental state. As an example, some noise control data are presented below. According to the District of Columbia (Washington DC) Noise Control Act, the maximum sound pressure level for a commercial zone in a city environment should not exceed 65dBA (decibels) during daytime (7:00AM – 9:00PM), and 60dBA in the nighttime; for an industrial zone accordingly 70dBA and 65dBA. (Note: dBA takes into account human perception. dBA is an expression of the relative loudness of sounds as perceived by the human ear. The sound pressure level in decibels (dB) is the absolute amount of pressure change exerted in the air by a sound and is characterized by the energy of acoustic wave). The maximum sound level for a residential zone should not exceed 60dBA during daytime, and 55dBA in the nighttime. Please consider any noise exceeding 70bBA as disturbing.

Construction activities in a city environment and other urban areas frequently require excavations or use of heavy vibrating tools and equipment close to existing buildings and structures. These types of work require careful planning, organization, and adherence to engineering and work zone standards, Occupational Safety and Health Administration (OSHA) standards, and safety regulations in order to prevent damage to nearby buildings.

As noted in Chap. 1, vibrations can be divided into two large groups, depending on how the oscillatory process is described:

- Deterministic vibration
- Random vibration

Deterministic vibration can be determined and predicted at any time. Deterministic vibration can be described by any function of time and spatial coordinates.

Random vibration cannot be strictly defined at any given time. Random vibration can only be described by statistical methods using some averaged motion characteristics. An example of random vibrations can be attributed to non-deterministic and stochastic: seismic and wind loads. Vibrations caused by such loads must be considered as random since the loads cannot be determined in advance. The motion parameters of the system caused by such loads will also be random.

Deterministic processes that can be divided into periodic and non-periodic are described in detail in the first paragraphs of this book. For this reason, here we will proceed to the consideration of random vibration. Many dynamic impacts and, respectively, the responses of structures to these actions can be described using the theory of random processes. Here, let us limit ourselves to the basic characteristics of these processes. The most important characteristics of oscillation in the interval of time $t_2 - t_1$ are the peak values $A_{\max} = |\max y(t)|$, $(t_1 \leq t \leq t_2)$, as well as the range of oscillations in the interval $t_2 - t_1$ equal to: $A_{\max} + A_{\min}$.

To describe random processes, some of their averaged characteristics are used, in particular, the average (mean) value

$$y_{\text{mean}} = \frac{1}{t_2 - t_1} \int_{t_1}^{t_2} y(t)dt, \tag{20.1}$$

and the root-mean-square value

$$y_{\text{mean sq.}} = \sqrt{\frac{1}{t_2 - t_1} \int_{t_1}^{t_2} y^2(t)dt}. \tag{20.2}$$

The root-mean-square value is an energy measure of the mechanical oscillatory process.

To describe almost periodic and nonperiodic stationary processes

$$y_{\text{mean sq.}} = \sqrt{\frac{1}{2}\sum_n A_n}, \quad c = \frac{1}{2}\sum_n A_n, \tag{20.3}$$

it is assumed here that $t_1 = 0$, $t_2 = \infty$. In particular, the amplitude values and the root-mean-square value for harmonic oscillations will take the following form:

$$A_{\max} = A_{\min} = A, \quad y_{\text{mean sq.}} = \frac{A}{\sqrt{2}} \tag{20.4}$$

Peak and root-mean-square (RMS) values of oscillatory processes are used to assess the level of vibrations. Peak and RMS values of oscillatory processes affect people and vibration-sensitive technological processes. Since the frequency composition of vibrations significantly influences a person's response to his physiological perception and also affects precision instruments, high-precision equipment, and high-tech processes, the level of vibrations is assessed in certain frequency ranges (octave and one-third octave frequency bands). In an octave band, the top boundary frequency is twice the bottom boundary frequency, in the one-third octave band it is 1.26 times greater.

The geometric mean frequencies of the bands are determined by the formula:

$$\omega_{\text{mean sq.}} = \sqrt{\omega_{\text{bot.}} \cdot \omega_{\text{top}}} \tag{20.5}$$

where $\omega_{\text{bot.}}$, ω_{top} are the lower and upper boundary frequencies of the band.

Modern regulatory documents for assessing the level in octave or third-octave frequency bands use the RMS values of vibration parameters of buildings and structures (displacements, velocities, or accelerations). At the same time, taking into account significant changes in vibration levels at the frequency range of interest to us (1–100 Hz.), it is convenient to use a logarithmic scale. Logarithmic level of vibration displacements, vibration velocities, and vibration accelerations is measured in bels or decibels (tenths of a bel). Increasing the level by one bel means increasing the initial parameter ten times and increasing by one decibel $10^{0.1} = 1.57$ times. Logarithmic level is defined by the formula:

$$L = 10 \log \frac{y}{y_0}, \tag{20.6}$$

where y is the measured value of the parameter and y_0 is some threshold value. If the determined value depends on the square of some oscillation parameter a, as is the case when determining the oscillation/vibration energy, then the logarithmic level will be:

$$L = 20 \log \frac{a}{a_0}. \tag{20.7}$$

The following values are accepted as thresholds in sanitary norms:

- For vibration speeds $v_0 = 5 \cdot 10^{-8}$m/c.
- For vibration accelerations $v_0' = 3 \cdot 10^{-4}\text{m/c}^{2}$.

20.2 Vibration Measurement of Buildings and Structures

Vibration normalization (vibration standards) of buildings and structures is based on vibration measurements using seismic scales. Most of these scales were developed to assess the seismic resistance of buildings during earthquakes. One of the most common scales is the MCS scale. This scale was developed by Mercalli, Cancani, and Sieberg. The other scales were developed later. To assess the risk of damage to buildings and structures from explosions, the scales developed by Langefors and Kihlstrom are the most used.

Permissible vibration levels indicated in the scales are empirical. They are based on the results of measurements of vibrations caused by explosions. It should be noted that measurements of vibrations caused by explosions were carried out in many countries of the world. The best known are vibration measurements carried out in Europe and in the United States. In these studies, the vibration velocity is a criterion for damage to buildings and structures. In the 1970s, many national standards and recommendations appeared. Among the main and proven in practice, we should point to (Deutsches Institute for Normung) *DIN 4150* [1]. This standard specifies methods of determining and evaluating the effects of vibration on structures designed for predominantly static loading. It covers structures that do not need to be designed to specific standards or codes of practice on dynamic loading.

In the 1980s, research work on the international standard (International Organization for Standardization) *ISO 4866* [2] measuring and evaluating the impact of vibrations on buildings was carried out. In the *ISO 4866* standard, as well as in DIN, the criterion for damage to buildings is the vibration velocity.

ISO is a worldwide federation of national standards bodies (ISO member bodies). The work of preparing International Standards is normally carried out through ISO technical committees. Each member body interested in a subject for which a technical committee has been established has the right to be represented on that committee. International organizations, governmental and non-governmental, in liaison with ISO, also take part in the work. ISO collaborates closely with the International Electrotechnical Commission (IEC) on all matters of electrotechnical standardization.

The reason for this practice is that the stresses in individual structural elements are mainly directly proportional to the vibration velocity. To accurately determine the effect of vibration, it is necessary to use direct methods of measuring stresses or deformations or to measure vibration with appropriate dynamic calculations.

Below is the concise information about standard specified methods of determining and evaluating the effects of vibration on structures *DIN 4150* [1], *ISO 4866* [2]

20.2.1 Definition

Building vibration is defined by the Deutsches Institute for Normung (*DIN 4150-3*) standard as a mechanical oscillation of solid bodies that have the potential to cause discomfort to persons or damage to structures. Following the DIN standard, the damage is a permanent effect of vibration that reduces the serviceability of a structure or one of its components.

20.2.2 Building and Ground Vibrations Effects

- Damage to buildings structure, including fatigue damage such as cracking and breakage
- Damage to buildings' contents, including sensitive electronic equipment and work of art
- Disturbance to human comfort
- Audible effects such as structure-borne noise and secondary rattling of building elements or contents

In each case, the response to vibration should be evaluated using appropriate equipment and compared to required standards [3].

20.2.3 Vibration Sources

The International Organization for Standardization (*ISO—4866*), [2] standard defines a vibration source as a simple or multiple solid, liquid, or gaseous body causing vibration in its environment. Common vibration sources are:

- *Construction sites* and excavation equipment
- Rail and road traffic (most common source)
- Industrial machinery-induced vibrations
- Mechanical equipment
- Low-frequency, airborne pressure waves emitted by heavy vehicles, aircraft, mechanical equipment, or blasting

How to Measure the Vibration of a Building?

Vibration occurring in a structure can be detected directly using an accelerometer, a displacement transducer, or a velocity transducer. The measure is a Peak Particle Velocity, which is the maximum absolute value of the unweight signal (particle velocity signal) throughout the measurement.

Following *DIN 4150-3* [1] measurement of the structure as a whole shall be based on the topmost floor vibration velocity, measured in two horizontal directions. As an alternative to evaluating the topmost floor horizontal vibrations, measurements can also be taken at the foundation.

How to Measure Building Content Vibration?

Vibration influencing sensitive objects in buildings is assessed using an accelerometer or a velocity transducer measuring walls or floor vibrations. Following the Institute of Environmental Science and Technology (*IEST*) standard, the RMS of vibration velocity in 1/3 octave bands is used to measure the floor vibration effects on sensitive equipment in buildings.

How Can the Human Perception of Vibrations in Buildings Be Measured?

The human exposure to floor vibration is assessed using an accelerometer or a velocity transducer. Following ISO standards, common measures of human response to vibrating floors are the *PEAK* and *RMS* of vibration acceleration or velocity or Vibration Dose Value (*VDV*) of vibration acceleration [4]

20.2.4 Criteria and Limits

The standard DIN 4150-3 [1] defines the values that must be met to avoid damage and impact on structures. In general, do different Peak Particle Velocity (PPV) limits apply depending on the building types in the frequency range from 1 Hz to 100 Hz:

- From 20 mm/s up to 50 mm/s—for buildings used for commercial and industrial buildings.
- From 4 mm/s up to 20 mm/s for residential buildings.
- From 3 mm/s up to 10 mm/s for building sensitive to vibration that has great intrinsic value.

Vibration Criterion (VC) Curves expressed as 1/3 octave bands of RMS vibration velocity, are vibration limits used by the American standard IEST for the measurement of vibration-sensitive equipment.

To meet American standards Institute of Environmental Sciences and Technology (IEST) that use 1/3 octave bands *Vibration Criterion VC* curves expressed in RMS vibration velocity, the 1/3 octave band method compares RMS results in each band to the limit curve [4]. So, in fact, the whole spectrum is used for comparison, unlike in the fast Fourier Transform (FFT) method where the point PPV, dominant frequency (DF) is positioned versus the curve. Vibration occurs in the time domain, meaning it oscillates over time. After collecting vibration data, it is often advantageous to transform it into a different domain for analysis. Patterns not apparent in the time domain are visible in others, giving engineers a clearer understanding of their product's vibrational behavior.

The fast Fourier transform (FFT) is an efficient algorithm used to compute a discrete Fourier transform (DFT). This Fourier transform outputs vibration amplitude as a function of frequency so that the analyzer can understand what is causing the vibration.

The frequency resolution in an FFT is directly proportional to the signal length and sample rate. To improve the resolution, the time of the recording must be extended, but be careful of a changing vibration environment. Engineers often analyze vibration as a function of frequency. The fast Fourier transform (FFT) transforms time-domain data into the frequency domain by deconstructing the signal into sine and cosine waves. It is a computational algorithm that efficiently implements a mathematical operation called the discrete-time Fourier transform (DTFT).

The FFT helps engineers determine the excitation frequencies in a complex signal and their amplitude. It also highlights changes in frequency and amplitude and harmonic excitation in the selected frequency range.

Short-Term and Long-Term Vibrations

The standard *DIN 4150-3* [1] divides vibrations into:

- Short-term vibrations, which do not occur frequently enough to cause fatigue and do not cause resonance in the assessed structures.
- Long-term vibrations, on the other hand, are all vibration types that cannot be classified as short-term vibrations.

British Standard *BS 7385-2*

BS 7385-2 [3, 8] guides the assessment of the possibility of vibration-induced damage in buildings due to a variety of common sources. Vibration from common sources that are considered include blasting, demolition, piling, ground treatments (e.g., compaction), construction equipment, tunneling, road and rail traffic, and industrial machinery-induced vibrations. It gives guidance on the levels of vibration above which building structures could be damaged. Only the direct effect of vibration on buildings is considered. The indirect effects on the building structure due to ground movement, the movement of loose objects within buildings, the possibility of damage to sensitive equipment, and the effect of vibration on people are outside the scope of this standard. The lowest frequency originating from man-made common sources is 1 Hz and the highest frequency expected from either machinery or close-in construction blasting on hard ground is 1000 Hz, however, to select guide values more limited range of 4 Hz to 250 Hz is taken into consideration. BS standards indicate two types of buildings:

- Reinforced or framed structures Industrial and heavy commercial buildings
- Unreinforced or light framed structures Residential or light commercial type buildings

What Are VC Curves for Vibration?

VC Vibration Criterion Curves expressed as 1/3 octave bands of RMS vibration velocity, are vibration limits used by the American standard IEST for the measurement of vibration-sensitive equipment.

To meet American standards (IEST) that use 1/3 octave bands *VC Vibration Criterion* curves expressed in RMS vibration velocity, the 1/3 octave band method compares RMS results in each band to the limit curve. So in fact the whole spectrum is used for comparison, unlike in the FFT method, where the point (PPV, DF) is positioned versus the curve.

What Are Vibration Modes?

A vibration mode is a method of vibrating, or a pattern of vibration when applied to a system or structure with many points and varying amplitudes of movement. The two elements of a vibration mode are a time variation of the vibration and, second, a spatial variation of the amplitude of the motion across the structure [3].

The normalized vibration levels of structures for the frequency range from 0 to 50 Hz are shown in Fig. 20.1 [5]. Curve 1 indicates the maximum permissible level of vibration in structures. Exceeding the vibration level can lead to accidents and even destruction of structures. The area below curve 2 indicates the vibration of structures in the safe zone. The area of the graph between curve 1 and curve 2 corresponds to the vibration level of the threatening structures. According to the recommendations of some studies, the safe vibration zone for structures is below curve 3 and curve 4.

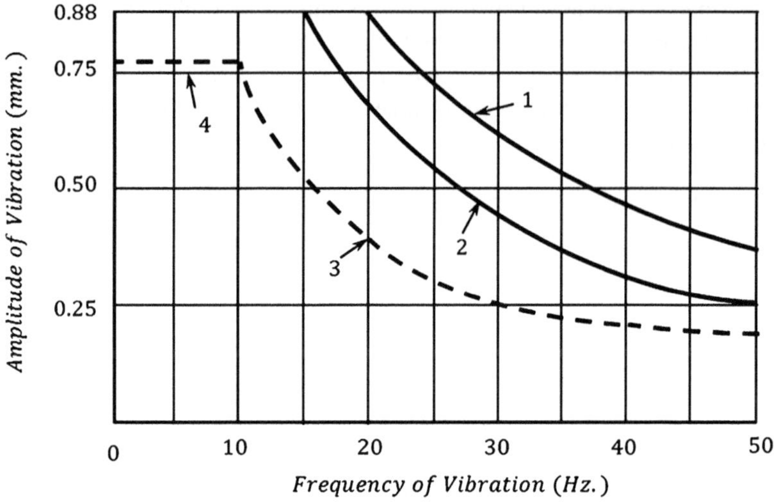

Fig. 20.1 Normalized vibration levels

According to the recommendations of some studies, the safe vibration zone for structures is below curve 3 and curve 4.

20.3 Ways to Reduce Vibrations and Resonant Phenomena

In cases where, as a result of the dynamic design of the structure, the requirements of strength and stiffness are not satisfied, or when the vibration characteristics of the impact on people and equipment are not satisfied, vibration control measures should be taken. These measures are quite diverse and effective, provided that they are chosen correctly in each particular case. When choosing vibration protection measures, the following factors should be taken into account: cost-effectiveness, efficiency, ease of implementation, and reliability of vibration protection; maximum improvement of working conditions, and increase of production productivity.

Vibration protection measures can be divided into the following main groups:

- Reducing the dynamic impact on structures.
- Changing the location of the sources of dynamic impact in the structure.
- Changing the elastic, inertial, and dissipative characteristics of the part of structure or whole building structure.
- Protection of people and workplaces. (Controlling vibration at work).
- Changing the conditions of the interaction of structures with the base (foundation).

An analysis of the causes and nature of vibrations, as well as an assessment of the parameters of a dynamic system, should always be carried out before choosing vibration protection measures. At the same time, the evaluation of the parameters of the dynamic system should take into account the source of vibration, vibration characteristics, as well as the main dynamic characteristics of the vibrations of the building or structure itself. The main purpose of the analysis is to identify the source of vibration and establish the relationship between the vibrations of the system and the dynamic forces of excitation. The experience and ability of an engineer to choose a relatively simple and at the same time reliable calculation model of the system is of great importance when analyzing a dynamic system.

A thorough (comprehensive) analysis of the dynamic system could give an adequate picture of the excitation and propagation of vibrations in the structure and in the environment. Once the vibration source and the design of the part of structure or whole building structure have been identified, the choice of vibration protection measures can be initiated. Vibration protection measures can be implemented in different ways, depending on the type of vibration source, foundation, as well as the properties of the structure and the base. In particular, the force of inertia can be changed by increasing or decreasing the mass of the structure and foundation. The force of excitation can be changed by balancing the rotating parts of machines and mechanisms, by balancing the forces of inertia of machine mechanisms, and by changing the rotational frequency of individual parts of machines. The elastic forces can be changed by changing the geometry of the structure, by changing the area of the foundation, by changing the rigidity of the structure, and so on. Here, let's recall that the damping force is important mainly only in the resonant region, while the elastic force is decisive in the vibrations of the structures and foundations outside the resonant (resonance) region. Thus, the above scheme or method of vibration analysis can be attributed mainly to the foundations of machines as systems with single-degree-of-freedom. This scheme can be considered as the simplest for analyzing the vibration of foundations for machines.

20.4 Reduction of Dynamic Impact on the Structure

Expedient arrangement of machines by this or that rearrangement of the machines that are the source of vibrations, it is possible to reduce or even avoid resonant phenomena. So, for example, by moving the source of vibrations closer to the support of a building (column) or a load-bearing beam, it is possible to increase the frequency of natural vibrations of the beam. Machines, mechanisms, that create horizontal forces of inertia should be positioned so that these forces act in the direction for which the stiffness (rigidity) of the building is greater, or the frequency of the natural vibrations of the building is significantly different from the frequency of rotation of the machines, or mechanisms.

Changing the rigidity of the structure Increasing the rigidity of the structure by increasing the dimensions of its cross sections by reducing the spans leads to an increase in the frequencies of natural vibrations and can be used to remove the lowest of these vibration frequencies up from the frequency of exciting loads. Reducing the rigidity of the structure may be acceptable only in rare cases, since it leads to a decrease in strength and creates the possibility of resonance with higher tones of vibrations of the altered structure.

Changing the mode of operation of machines Lowering or increasing the speed of the vibration source can lead to the same effect as increasing or, respectively, lowering the rigidity of the structure, that is, to a decrease in the dynamic coefficient. Vibrations generated by machines with rotating masses with reciprocating rotating masses can be significantly reduced by balancing the inertial forces. The balancing of the machine that is the source of vibration can be achieved by balancing its moving masses. The most common way to reduce dynamic loads from machines is the power (force) vibration isolation of the vibration source. The principle of operation of power vibration isolation is based on the use of the inertial forces of the machine to counteract the external load. For this purpose, the vibration source, that is, the machine together with the substation or support frame is turned into a vibrating system by supporting it on elastic elements or by using an elastic suspension. Springs, rubber and rubber-metal gaskets, rubber insulators, etc. are used as elastic elements.

The phase of movement of a system with a single-degree-of-freedom in the off-resonance region is opposite to the phase of the disturbing force. Therefore, vibration isolation is carried out in such a way that the vibration-isolated system oscillates in the off-resonance zone, that is, that the frequency of natural oscillations of the system is less than the frequency of the external force.

It follows from formula (6.74) that the amplitude of the force that transmits to the base $R = kA$, with the amplitude of the external force P will be:

$$R = \frac{P}{\sqrt{\left(1 - \xi_k^2\right)^2 + \gamma^2 \xi_k^2}} \tag{20.8}$$

The ratio of this force to the amplitude of the external load $R/P = \mu$ is called the transmission coefficient. The inverse of the transmission coefficient $k_{\text{eff.}} = 1/\mu = P/R$ is called the vibration isolation efficiency coefficient. Thus, the greater the ratio of the frequency of the disturbing force to the natural frequency of the installation $\xi_k = \omega/\omega_0$, the higher the efficiency of vibration isolation. It is usually recommended to take $\xi_k \geq 3$ for vertical vibrations and $\xi_k \geq 2.5$ for horizontal and rotational vibrations.

In practical implementation of vibration isolation, the frequency of natural oscillations of the installation must be calculated or experimentally determined. Moreover, in order to optimally select the characteristics of vibration isolators and avoid possible resonances for high-vibration modes, it is also necessary to determine the natural vibration frequencies for these modes.

The use of vibration dampers A dynamic vibration damper is an additional system that is installed on a structure that performs forced vibrations. The purpose of using a dynamic vibration damper is to neutralize vibrations from a perturbing load by the reactions of the vibration damper itself. For this, the frequency of natural vibration of the damper must be equal to the frequency of the exciting load.

The most common types of dampers that are used in structural engineering and in construction industry are dynamic dampers: spring dynamic dampers, rubber dynamic dampers; rubber-spring combined dampers; multi-layers of steel and rubber antivibration mounts; shock absorbers; shock/impact dynamic dampers; dynamic dampers with rotating pendulums; dry friction dampers; dampers with viscous friction, vibration isolation pads, etc.

Vibration isolation Vibration isolation is a very effective way of vibration control. Correct calculation, design, as well as strict implementation of design requirements are necessary conditions that ensure the effectiveness of vibration isolation. Usually, a vibration-isolated object is considered as an absolutely solid body, as a system with a single-degree-of-freedom under harmonic vibrations of the object itself or its base. Vibration isolation is one of the simplest ways to reduce the harmful effects of vibrations. The effectiveness of vibration isolation can be evaluated by the transmission coefficient. There are two types of vibration isolation: active vibration isolation and passive vibration isolation. The transmission coefficient of active vibration isolation is equal to the ratio of the amplitude of the force that transmits the displacement through pliable springs (vibration isolators, vibration dampers) on the base, to the amplitude of force acting on the vibration-isolated object. The effectiveness of passive vibration isolation is determined by the ratio of the amplitude of movement of the vibration-isolated object to the amplitude of the displacement of the base. In most cases, the calculation of passive vibration is more complex than the calculation of active vibration isolation. While single-frequency disturbances are typical in active vibration isolation, multi-frequency vibrations of the base (foundation) are typical in passive vibration isolation.

The purpose of active vibration isolation is to reduce the dynamic forces acting on the structure. Vibration is detected by the sensor mounted above the instrument, triggering the actuator to output a force in the opposite direction to cancel the vibration. This makes it possible to attenuate vibration amplification due to resonance, which cannot be avoided with a passive isolator. Usually, active vibration isolation (AVI) system is applied for high-performance tools, which require accurate posture control and higher throughput.

The purpose of passive vibration isolation is to isolate certain places of work of personnel as well as the locations of devices from vibrations of load-bearing or enclosing structures. Passive vibration isolation investigated countermeasures base isolation of the office building, where the superstructure is dynamically isolated from the foundation.

In those cases when vibration does not threaten the strength of the structure and the normal use of technological equipment but only has a harmful or adverse effect on people, local vibration protection of people and workplaces can be carried out. For vertical vibrations, when the human body weight and movement coincide, the human body is sensitive to vibrations between 4 and 8 Hz. For horizontal vibrations, the human body is sensitive to vibrations between 1 and 2 Hz, whereas the hand/arm system is most affected by vibration frequencies between 4 and 16 Hz [6].

Spring steel vibration dampers allow a large static settlement and make it possible to reduce the frequency of natural vibrations of the isolated object. Spring steel vibration isolators are effective vibration dampers in the frequency range from 5 to 25 Hz.

Practice shows that high-quality installation of vibration isolation equipment and regular inspection during the operation is an essential condition for ensuring the effectiveness of vibration isolation. Skewing/misalignment of support frames, tightening of bolts that limit the pliability of vibration dampers, loss of spring stability and resistance, contamination, aging of the rubber elements of isolators, and dampers can all dramatically reduce the effectiveness of vibration protection and should be immediately eliminated.

According to the International Organization for Standardization (*ISO 2631-1*) [7]. People are more sensitive in perceiving vibration than structures are to damage from vibration. Humans begin to perceive vibration at about 0.005in/s peak particle velocity (PPV), while one of the lowest proposed allowable ground vibration velocities for structures is at 0.05 in/s.

A vibration with a velocity level greater than 1.0in/s should be avoided. An acceptable vibration level would be less than 0.16in/s (Peak) or 2.8mm/s (RMS). The vibration perception threshold (VPT) is the lowest vibrational intensity that can be tolerated. Humans are generally sensitive to vibrations with VPT between 0.14mm/s and 0.3mm/s (0.055in/s and 0.118in/s). The perceptible vibration levels for humans, the threshold perception is 0.015 m/s^2 (peak acceleration as an acceptable vibration level for building occupants). If the floor vibrations are only slightly above this level of vibrations, then building occupants are likely to complain.

The following categories can be attributed to personal local means of vibration protection: vibration protection stands, pad layers, mats; vibration protection chairs and seats; as well as anti-vibration shoes and anti-vibration gloves.

This local protective equipment against general vibrations that affect a person cannot provide effective protection against low-frequency vibrations. Since this is due to the relatively small mass of a person, the use of personal protective equipment is limited.

Let us consider the idea of active vibration isolation using the example of vibrations of an isolated foundation on a pliable base as a system with a single degree of freedom. In this case, the mass of the foundation is under the action of a vibration harmonic force $P = P_0 \sin \omega t$, that arises from the operation of an electric motor. From this load, we assume that the mass of the foundation makes/performs only vertical vibrations with the following frequency ω. (Fig. 20.2).

One of the effective ways to reduce vibration is the use of vibration dampers. To reduce vibration, we install a dynamic vibration spring damper with mass m_1 and spring stiffness (Fig. 20.3)

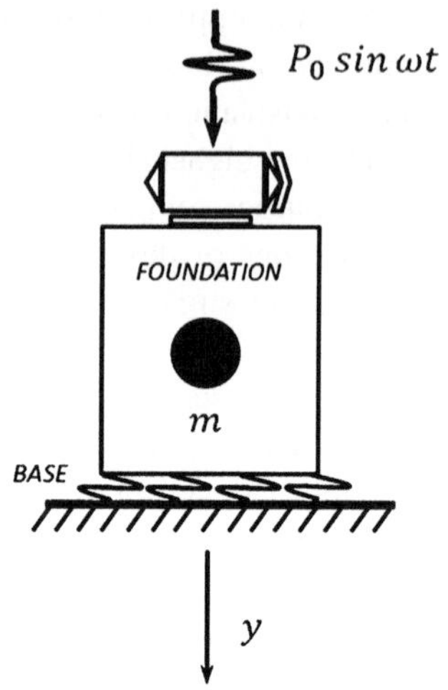

Fig. 20.2 Isolated foundation base

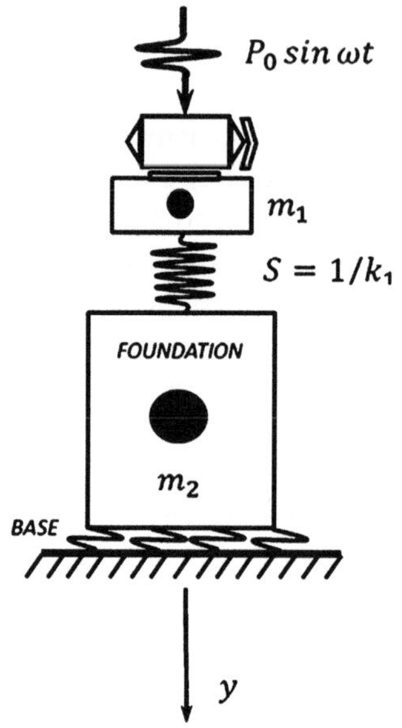

Fig. 20.3 Vibration spring damper

Masses m_1 and m_2 will perform vertical oscillations with frequencies ω_1 and ω_2 respectively. The challenge is to select such a stiffness for the spring that the amplitudes of masses m_1 and m_2 would be less than before the spring is introduced to the system.

The stiffness of spring S connecting the vibration damper to the foundation mass is called the value $\frac{1}{k_1}$.

The increment of the spring length during the vibrations is represented as:

$$\Delta y_1 = k_1 R_1(t) \tag{20.9}$$

where

Δy_1 is the increment of the spring length during the vibration; $R_1(t)$ is the resistance force of the spring when it is stretched or compressed by Δy_1; k_1 is the coefficient included in the expression of the increment of the spring length during the vibration (20.9). Based on the same reasoning, let's write down the expression of the dependence of vertical displacement on the resistance force for the base:

$$y_2(t) = kR(t) \tag{20.10}$$

Let us find the relationship between the amplitudes of the oscillations/vibrations of masses m_1 and m_2 and the stiffness of the spring. For this purpose, we will compose equations of motion of masses m_1 and m_2 .

Differential equations of oscillations/vibrations with a dynamic damper are as follows:

$$m_1 \frac{d^2 y_1}{dt^2} + R_1(t) = P_0 \sin \omega t \tag{20.11}$$

$$m_2 \frac{d^2 y_2}{dt^2} + R(t) - R_1(t) = 0$$

Based on (20.9) and (20.10)

$$R_1(t) = \Delta y_1 / k_1 = (y_1 - y_2)/k_1 \tag{20.12}$$

$$R(t) = y_2/k \tag{20.13}$$

The differential equations of motion of the two parts of the foundation are obtained by substituting expressions (20.12) and (20.13) into (20.11):

$$\begin{aligned} m_1 \frac{d^2 y_1}{dt^2} + \frac{1}{k_1}(y_1 - y_2) = P_0 \sin \omega t \\ m_2 \frac{d^2 y_2}{dt^2} + \frac{y_2}{k} - \frac{1}{k_1}(y_1 - y_2) = 0 \end{aligned} \tag{20.14}$$

The solution of this system of differential equations corresponding to the steady-state oscillations/vibrations is determined by the expressions:

$$y_1 = A_1 \sin \omega t; \quad y_2 = A_2 \sin \omega t \tag{20.15}$$

The following equations for amplitudes A_1, and A_2 are obtained by substituting (20.15) into (20.14):

$$-m_1 \omega^2 A_1 + \frac{1}{k_1}(A_1 - A_2) = P_0;$$

$$-m_2 \omega^2 A_2 + \frac{1}{k} A_2 - \frac{1}{k_1}(A_1 - A_2) = 0,$$

or

$$\begin{aligned} A_1(-m_1\omega^2 + s_1) - s_1 A_2 = P_0 \\ A_1 s_1 + A_2(-m_2\omega^2 + s + s_1) = 0 \end{aligned} \tag{20.16}$$

where

$$\frac{1}{k_1} = s_1; \quad \frac{1}{k} = s$$

Let us find the amplitude values from the system (20.16):

$$\begin{aligned} A_1 &= P_0 \frac{s_1 + s - m_2\omega^2}{(s_1 - m_1\omega^2)(s + s_1 - m_1\omega^2) - s_1^2}; \\ A_2 &= P_0 \frac{s_1}{(s_1 - m_1\omega^2)(s + s_1 - m_1\omega^2) - s_1^2}. \end{aligned} \quad (20.17)$$

Thus, the amplitude A_1 will turn to zero, if the stiffness of the spring (vibration damper) s_1 is equal to $m_2\omega^2 - s$, and then the spring will completely dampen the vibrations of mass m_1.

Mass m_2 will not vibrate/oscillate if the circular natural frequency of the attached damper is equal to the circular frequency of the change in the disturbing force. The use of a damper is especially effective if the system is in a state of resonance or in a near-resonance domain.

References

1. DIN 4150-3, 2016 Edition, Vibrations in buildings – Part 3: Effects on structures, December 2016, 23p.
2. International Standard ISO 4866. ISO (4866:2010(E). Mechanical vibration and shock—Vibration of fixed structures — Guidelines for the measurement of vibrations and evaluation of their effects on structures. 33p.
3. Svantek. Building vibration. Vibration damage to buildings. https://svantek.com/applications/building-vibrations/
4. Svantek. Building vibration, measurement. Measure. https://svantek.com/applications/building-vibrations/
5. Vainer, D., & Tseitlin, A. I. (1994). *Vibration Damage in industrial construction*. Swedish Council for Research in Construction Buildings Protection Center. 336p.
6. Malerbi, B. (1989). Vibration. In *Occupational health practice (Third Edition)*. Butterworth – Heinemann. Evaluation of vibration. Whole-body vibration (WBV).
7. ISO 2631-1:1997. Mechanical vibration and shock. Evaluation of human exposure to whole-body vibration. Part 1: General requirements. 6p.
8. BS 7385-2 Evaluation and measurement for vibration in buildings Guide to damage levels from ground-borne vibration. An item dealing with the vibratory loading of structures, ISO/FDIS 10317, is being processed within ISO Technical Committee ISO/TC 98/SC/2, Safety of structures. 16p.

Appendix A

Certificate of Essential Building Data

Proposed Format for a "Certificate of Essential Building Data"

Levon G. Petrosian
Visiting Scholar, Building and Fire Research Laboratory,
National Institute of Standards and Technology, Gaithersburg, Maryland
Formerly, Director, Armenian Earthquake Engineering Research Institute
Andrew W. Taylor
Research Engineer, Building and Fire Research Laboratory
National Institute of Standards and Technology, Gaithersburg, Maryland
Diana Todd
Research Engineer, Building and Fire Research Laboratory
National Institute of Standards and Technology, Gaithersburg, Maryland

In the aftermath of a natural or man-made disaster, rescue workers, engineers, and public officials need quick access to accurate information about the essential structural features of damaged buildings. A minimal set of essential data is required to make judgments about the suitability of a structure for continued occupancy, to estimate the cost of damage and the feasibility of repairs, and to assess the vulnerability of the structure to further damage (e.g., from aftershocks). Experiences in past earthquakes, particularly the recent Northridge and Kobe events, have confirmed that even basic information about most damaged buildings is not readily available. Fundamental data such as the date of construction, the version of the building code under which the structure was designed, the type of structural framing system, and a history of significant modifications and repairs to the structure are almost never available on short notice. Such information, if recorded at all, usually resides in the files of government planning offices or consulting engineering firms. Therefore, it is proposed that a systematic method of recording, in a brief format, the fundamental information about a building be developed. This "Certificate of Essential Building Data" (CEBD) would be kept on file at the building

L. G. Petrosian, *Dynamics of Structures*, https://doi.org/10.1007/978-3-031-63539-7

site (much like an elevator certificate) and at a central municipal repository. Possible features of such a certificate and its proposed users are described below.

The CEBD would be an abstract of data essential to making informed decisions about the condition and use of structures following a natural disaster, such as an earthquake. The CEBD would be about two pages long and would contain information in the following five categories:

1. Building Identification (name, address, owner, etc.)
2. General Description (occupancy class, number of stories, sketch of plan, and elevation)
3. Structural System (structural frame type, dynamic and static structural characteristics)
4. Foundation System (foundation type, site soil conditions)
5. Building History (dates of design and constructions, record of significant modifications and repairs)

The information would be arranged in a format that corresponds to post-disaster reconnaissance survey forms, such as the one currently being developed by the Applied Technology Council in project ATC-38, and to other building databases, such as the one currently being developed for Federal buildings by the Interagency Committee on Seismic Safety in Construction (ICSSC 1995).

Perhaps the main questions related to the development of a CEBD are not the content of the CEBD form itself, but rather how the CEBD would be implemented and the information used. Some of these issues are outlined below.

- Who would use the CEBD?

Although the main purpose of the CEBD is to assist rescue workers, engineers, and government officials in post-disaster recovery efforts, there are many other uses for data contained in the CEBD. First, the information would be invaluable to researchers in earthquake engineering. If CEBDs were available for a broad range of buildings in an area struck by an earthquake, much could be learned about the vulnerability of various structure types. Furthermore, if CEBDs are updated over the life of a building, the effects of previous weak, moderate, and strong earthquakes and structural modifications or repairs, could be monitored. Second, the CEBD could be used by disaster relief planners to estimate the extent and cost of damage resulting from hypothetical disaster scenarios. Third, the information could be used by owners, buyers, and insurers of buildings as a capsule summary of the essential structural features of the building. The CEBD could not take the place of a detailed structural evaluation, but it would provide an overview of the essential characteristics of a structure.

- Which buildings would have a CEBD on file?

In a perfect world, nearly every public building would have a CEBD on file. However, practically speaking, this would be impossible. The most likely candidates for CEBDs are new structures, existing structures that undergo significant modifications, and government buildings. The most qualified person to complete the CEBD is the structural engineer responsible for the design of a new building or the construction permit process. This would ensure that major modifications to the building, which require a building permit, are recorded on the CEBD. Regarding government buildings, local jurisdictions that undertake a program of evaluating the vulnerability of their building stock to natural hazards could include completion of CEBDs as part of the evaluation process.

- Where would the CEBD be stored?

One copy of the CEBD would be kept on the building site with a building manager. Another copy would be kept at a central repository, such as a city building department or disaster response agency.

In short, the CEBD would fulfill two purposes. First, and foremost, the CEBD would be a readily available record of essential structural information which could be used in post-disaster relief and recovery efforts. Second, over time, as the CEBD database grew, the CEBD would become an invaluable tool for studying the seismic vulnerability of various classes of structures and for developing improved retrofit and repair measures for buildings.

NIST

Proposed Format for

Certificate of Essential Building Data

Levon G. Petrosian Visiting Scholar, Ph. D., Professor, Building and Fire Research Laboratory,
National Institute of Standards and Technology
Andrew W. Taylor Research Structural Engineer, Ph.D., Building and Fire Research Laboratory,
National Institute of Standards and Technology
Diana R. Todd Research Structural Engineer, Ph.D., Building and Fire Research Laboratory,

What is it?

The Certificate of Essential Building Data (CEBD) is a systematized, concise record of the most important structural features of a building, including basic geometry, structural system type, foundation type, etc.

Who would use it?

- **Emergency workers (police, fire, rescue, building inspectors) immediately following a natural or man-made disaster, to make safety evaluations of damaged buildings.**
- **Disaster Planners, for community risk analysis and emergency response planning.**
- **Insurers, to estimate disaster risk exposure in a given area.**
- **Researchers, to learn about the actual performance of various types of buildings in natural disasters.**

How would it be implemented?

- **For existing Federal buildings, the CEBD could be completed as part of the implementation of Executive Order 12941 (seismic safety of existing federal buildings)**
- **For new construction, the CEBD could be completed as part of the building permit process, since most of the information required for the CEBD can be supplied by the designer.**
- **For significant renovations, the CEBD could also be completed as part of the building permit process, since significant renovations require the designer to compile most of the information on the CEBD.**
- **For private buildings, the CEBD could be completed as part of local, state, or federal programs to evaluate significant buildings in a community.**

Where would the information be stored?

- **At the building site**
- **In a central repository (e.g. city building department)**
- **On a publicly accessible electronic data base**

What information is in the CEBD?

Examples of key data listed in the proposed CEBD:

- **Identification of the structure**
  - **Name of structure**
  - **Address**
  - **Use and Occupancy**
  - **Map location**
- **Structure configuration**
  - **Overall plan and elevation dimensions**
  - **Story heights and weights**
  - **Locations of seismic separation joints**
- **Structural System**
  - **Gravity load system description**
  - **Lateral load system description**
  - **Special structural conditions**
  - **Foundation system description**
  - **Experimentally obtained dynamic properties**
  - **Dates of design and construction**
  - **Dates of remodeling or reinforcement**
  - **Seismic design coefficients**
  - **Codes used in design**
- **Estimated or insured value of structure**

NATIONAL INSTITUTE OF STANDARDS AND TECHNOLOGY
BUILDING AND FIRE RESEARCH LABORATORY

(Proposed Format for)

CERTIFICATE OF ESSENTIAL BUILDING DATA

1. IDENTIFICATION OF STRUCTURE

1.1 Name of Structure (if any) ______________________________

____

____

1.2 Address

______________________________City______________________________

State ____________________ Zip Code ________ Building Name ______________

____

1.3 County ______________________________

____

1.4 Name of Building development, of which the structure is part (if any) ______________

____

____

1.5 Building Owner / Manager Contact ______________________________

____

______________________________ Phone #______________________________

1.6 Jurisdiction Governing the Structure (name of city, county, state, or federal agency)

____

____

1.7 Structure Use and Occupancy:

1.7.1 Function of the Structure (e.g. residential, commercial office, light manufacturing, heavy manufacturing, health care, auditorium, etc.)

____

____

1.7.2 Maximum Occupancy (number of persons)

____

____

____

1.7.3 If hazardous materials are stored in the structure, type and location

____

____

____

1.8 Population setting (urban, suburban, or rural) ______________________

____

1.9 Photograph (general view of structure)

1.10 Map and plans:

1.10.1 State Map Number, and coordinates in state plane coordinates system

____

____

1.10.2 County or city map number, and coordinates

____

____

1.10.3 Project – Specific map number, and structure identification number

____

____

1.10.4 Structure plan number ______________________

____

____

1.10.5 Latitude and longitude coordinates ______________________

____

2. STRUCTURE CONFIGURATION

2.1 Number of Stories (identify below-grade and above –grade stories, including penthouse levels), and the height of each story ______________________

____

____

____

2.2. Basic plan shape of the structure (sketch with dimensions sufficient to describe the overall plan shape of the structure) ______________________

____

2.3 Basic elevations of the structure (sketches with dimensions sufficient to describe the overall elevation shape of the structures on all sides). Include the number of stories (below-grade and above-grade stories, including penthouse levels), and the height of each story ____

____

____

____

2.4 Seismic separation joints (sketches showing the locations of joints) ______________

____

____

____

____

3. STRUCTURAL SYSTEM

3.1 Structure system identification

3.1.1 Gravity Load Structural System (using Applied Technology Council ATC -33 standard structural system types)

____

____

____

____

____

____

3.1.2 Lateral load structural system, if distinct from the gravity load structural system (using either Applied Technology Council ATC -33 or NEHRP Recommended Provisions standard structural system types) ____________________________

____

____

____

____

____

3.1.2.1 Lateral load structural system along the longitudinal building axis

____

____

____

____

____

3.1.2.2 Lateral load structural system along the transverse building axis

____

____

____

____

3.1.3 Approximate weight of each floor, excluding contents of the structure

____
____
____
____

3.1.4 Secondary elements ______________________________
____
____
____
____

3.1.4.1 Structural raming system of floors ________________
____
____
____
____

3.1.4.2 Structural framing system of roof ________________
____
____
____
____

3.1.4.3 Exterior cladding system ______________________
____
____
____
____

3.1.4.4 Description of interior, non-loadbearing partitions __________
____
____
____
____

3.1.5 Special structural conditions (e.g. mixed framing types significant modifications, unusual geometry, unusual mass distribution, unusual materials, etc.)

____
____
____
____
____
____
____

3.1.6 Is the structure essentially a standardized design? ______________________

3.1.7 Are there special seismic resistant elements in the structure (e.g. base isolators, supplemental passive dampers, semi-active, active dampers)? ______________

3.1.8 Basic material grades ______________________

3.1.8.1 Predominant superstructure steel grade ______________________

3.1.8.1.1 Columns ______________________

3.1.8.1.2 Girders and floors ______________________

3.1.8.2 Predominant superstructure concrete strength ______________________

3.1.8.2.1 Columns

3.1.8.2.2 Girders and floors

3.1.8.2.3 Walls

3.1.8.3 Predominant superstructure concrete reinforcement grade

3.1.8.4 Other important superstructure materials and characteristics

3.2 Foundation system

3.2.1 Foundation types employed (spread footing, pile and pile caps, mat, etc.)

3.2.2 Are there changes of surface elevation on the site which significantly affect the foundation design?

3.2.3 Soil conditions

____
____
____

3.2.3.1 Soil category (NEHRP Recommended Provisions categories)

____
____
____

3.2.3.2 Is a geotechnical engineering report available? ______________

____
____
____

3.2.3.3 Are special ground hazards present? (e. g. high potential for liquefaction, landslide) ______________________________

____
____
____

3.2.4 Have dynamic properties of the soil been obtained through field measurements? If so, what values were obtained?

____
____
____
____
____

3.3 Design

3.3.1 Dates of design (beginning to end) ______________________________

____
____
____

3.3.2 Name of design organization (commercial firm, government agency)

____
____
____

3.3.3 Engineer of record______________________________________

____
____
____

3.3.4 Architect of record ______________________________________

____

____

____

3.3.5 Building codes under which the structure was designed, and code version

____

____

3.3.6 Design seismic acceleration ____________________________

____

____

3.3.7 Have dynamic properties of the structure been obtained by field experiment? If so, what values were obtained? ____________________________

____

____

3.4 Construction ____________________________________

3.4.1 Dates of construction (beginning to end) ____________________

____

____

3.4.2 Name of general contractor ____________________________

____

3.5 Date when the structure was first approved for occupancy ________________

____

____

4. STRUCTURE VALUE

4.1 Estimated or insured value of the structure, excluding contents. Include method of valuation ____________________________________

____

4.2 Estimated or insured value of structure contents. Include method of valuation

____

____

5. ASSESSMENT OF THE STRUCTURE

5.1 Assessment of the disaster resistance of the structure. What criteria were used?

____

___

Provide possible rating:

5.1.1 Does not satisfy code requirements in existence at the time of design

___

___

5.1.2. Does not satisfy the current code requirements, which are different from the code requirements in effect at the time of design? ___

___

___

5.1.3 Does not satisfy the current code requirements because of previous damage?

___

___

5.2 Assessment of fitness for occupancy. What criteria were used? ___

___

___

Provide possible rating: ___

___

___

5.2.1. Fully fit for occupancy ___

___

___

5.2.2. Fit for occupancy only after retrofit or repair ___

___

___

5.2.3 Not fit for occupancy, and recommended for demolition ___

___

___

5.3 Overall assessment of the structure ___

___

___

Notes:

___

___

___

____
____
____

Inspector (s):

Name ____
____

Title ____
____

Address ____
____
____

Signatures

Date: ______________________ Time: ______________________

In addition the assessment of the structure of building must be provided. Here we are using the same format:

5. Assessment of the structure

5.1 Assessment of the disaster resistance of the structure. What criteria were used?
(Possible rating:
A. Does not satisfy code requirements in existence at the time of design.
B. Does not satisfy the current code requirements, which are different from the code requirements in effect at the time of design.
C. Does not satisfy the current code requirements because of previous damage.

5.2 Assessment of fitness for occupancy. What criteria were used?
(Possible rating:
A. Fully fit for occupancy.
B. Fit for occupancy only after retrofit or repair.
C. Not fit for occupancy, and recommended for Demolition

5.3 Overall assessment of the structure

Appendix B

Amplitudes of Dynamic Reactions

Values of the Functions Those Are Included in the Formulas of the Displacement Method

Correction Functions and Their Numerical Values (Table B.1)

Table B.1 Amplitudes of dynamic reactions

##	Design scheme and diagram of bending moments	Calculation formulas for reaction amplitudes
1	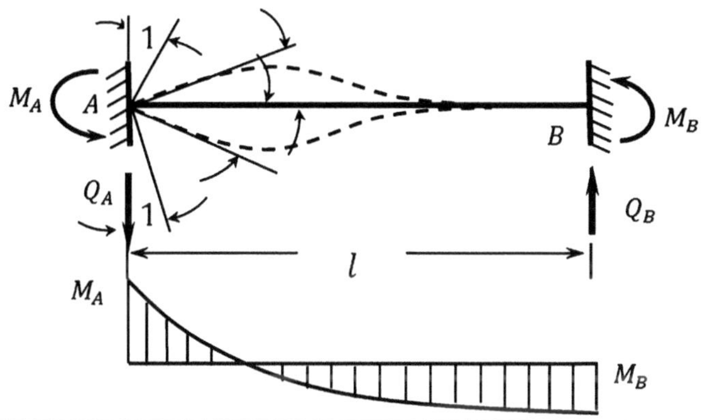	$M_A = 4i\psi_2(u)$ $M_B = 2i\psi_3(u)$ $Q_A = \frac{6i}{l}\psi_5(u)$ $Q_B = \frac{6i}{l}\psi_6(u)$
2	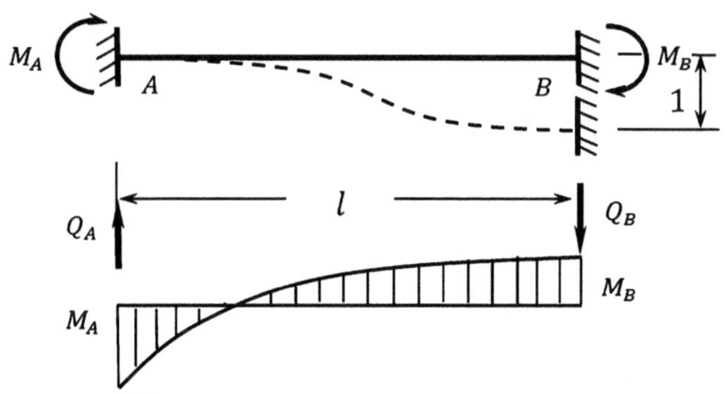	$M_A = \frac{6i}{l}\psi_5(u)$ $M_B = \frac{6i}{l}\psi_6(u)$ $Q_A = \frac{12i}{l^2}\psi_{10}(u)$ $Q_B = \frac{12i}{l^2}\psi_{11}(u)$

L. G. Petrosian, *Dynamics of Structures*, https://doi.org/10.1007/978-3-031-63539-7

##	Design scheme and diagram of bending moments	Calculation formulas for reaction amplitudes
3	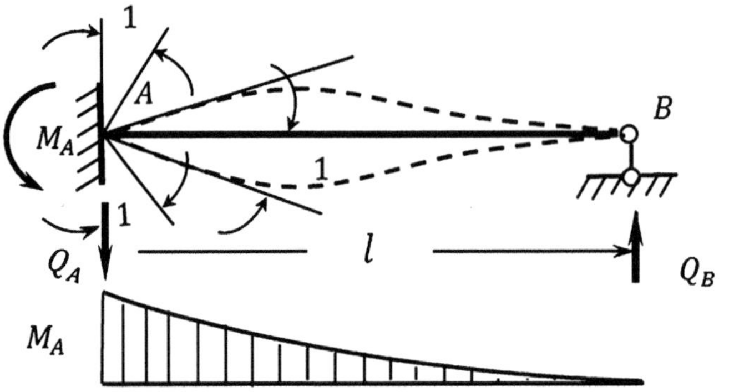	$M_A = 3i\psi_1(u)$ $Q_A = \frac{3i}{l}\psi_4(u)$ $Q_B = \frac{3i}{l}\psi_7(u)$
4	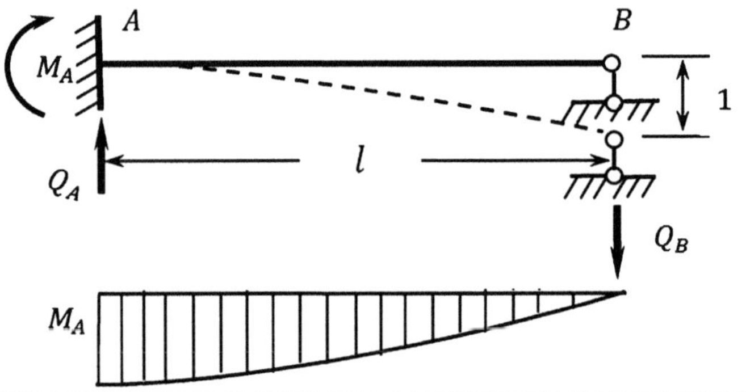	$M_A = \frac{3i}{l}\psi_7(u)$ $Q_A = \frac{3i}{l^2}\psi_9(u)$ $Q_B = \frac{3i}{l^2}\psi_{12}(u)$
5	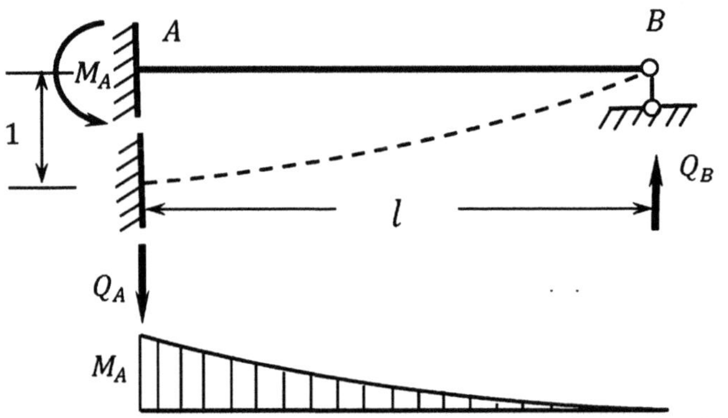	$M_A = \frac{3i}{l}\psi_4(u)$ $Q_A = \frac{3i}{l^2}\psi_8(u)$ $Q_B = \frac{3i}{l^2}\psi_9(u)$
6	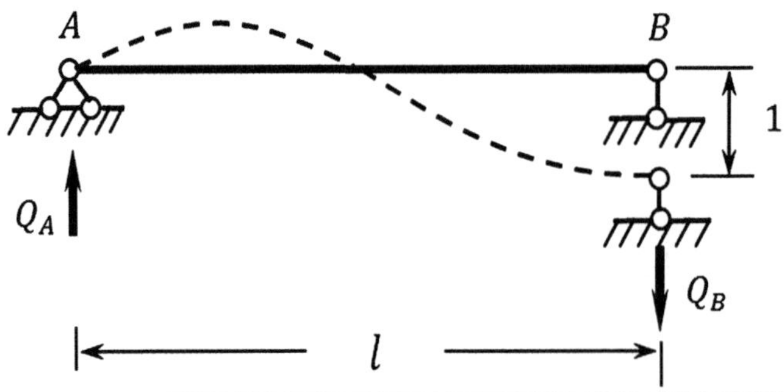	$Q_A = \frac{i}{l^2}\psi_{13}(u)$ $Q_B = \frac{i}{l^2}\psi_{14}(u)$

##	Design scheme and diagram of bending moments	Calculation formulas for reaction amplitudes
7	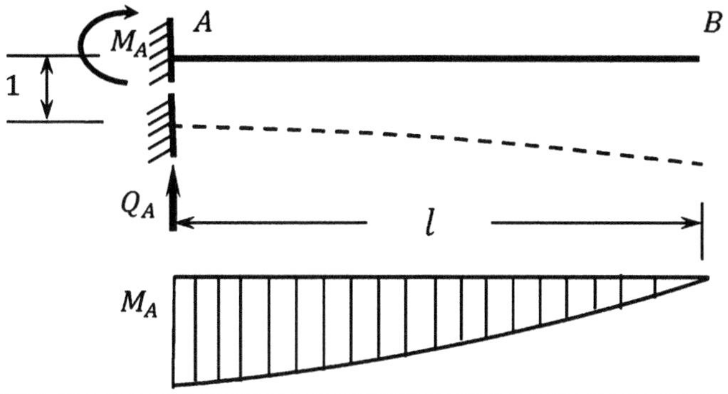	$M_A = \frac{i}{l}\psi_{15}(u)$ $Q_A = \frac{i}{l^2}\psi_{15}(u)$
8	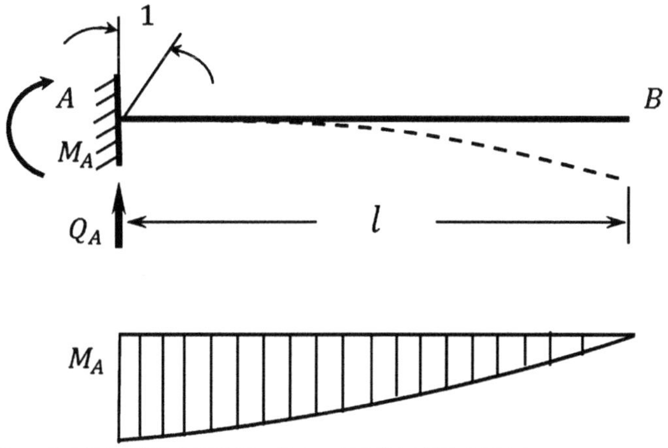	$M_A = i\psi_{17}(u)$ $Q_A = \frac{i}{l}\psi_{15}(u)$
9	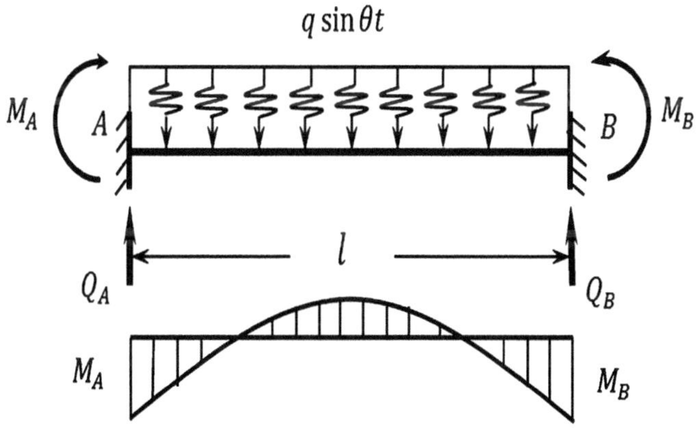	$M_A = M_B = \frac{ql^2}{12}\psi_{1q}(u)$ $Q_A = Q_B = \frac{ql}{2}\psi_{2q}(u)$

##	Design scheme and diagram of bending moments	Calculation formulas for reaction amplitudes
10	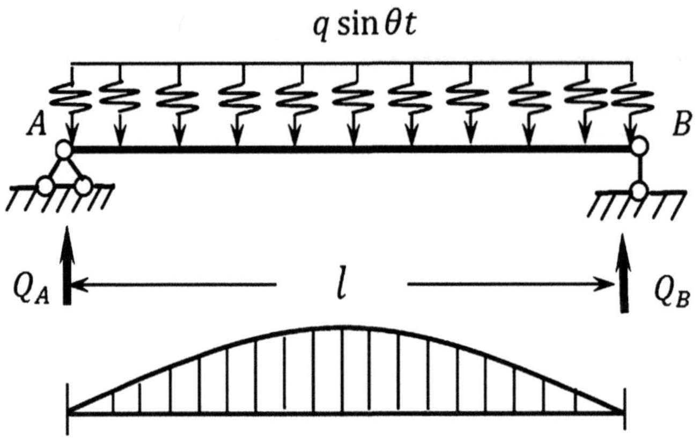	$Q_A = Q_B = \frac{ql}{2}\psi_{3q}(u)$
11	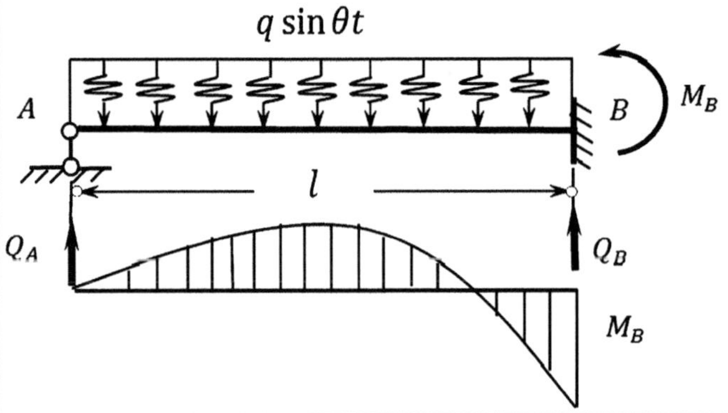	$M_B = \frac{ql^2}{8}\psi_{4q}(u)$ $Q_A = \frac{3ql}{8}\psi_{5q}(u)$ $Q_B = \frac{5ql}{8}\psi_{6q}(u)$
12	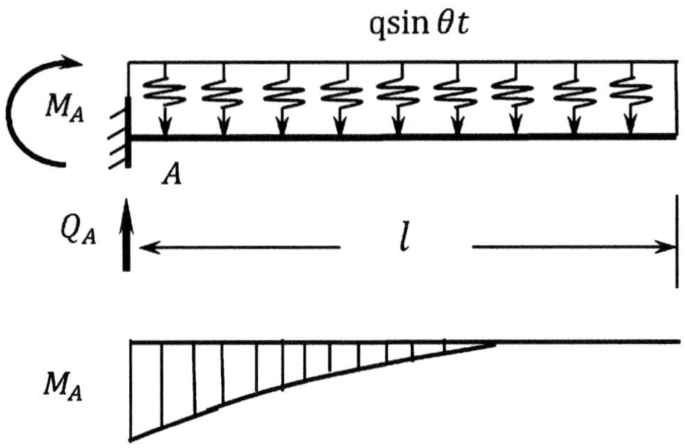	$M_A = \frac{ql^2}{2}\psi_{7q}(u)$ $Q_A = ql\psi_{8q}(u)$

##	Design scheme and diagram of bending moments	Calculation formulas for reaction amplitudes
13	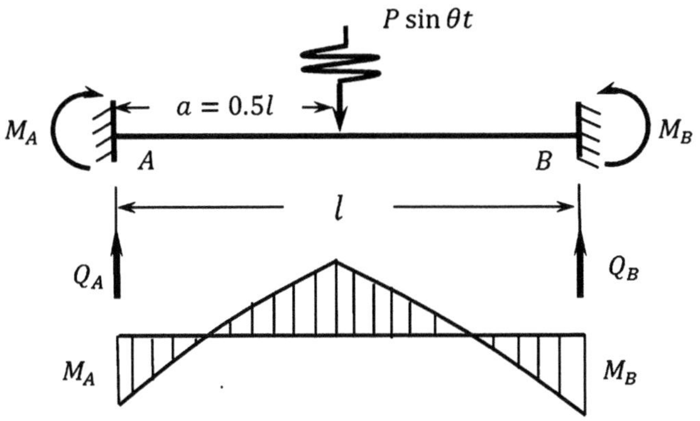	$M_A = M_B = \frac{Pl}{8}\psi_{1p}(u)$ $Q_A = Q_B = \frac{P}{2}\psi_{2p}(u)$
14	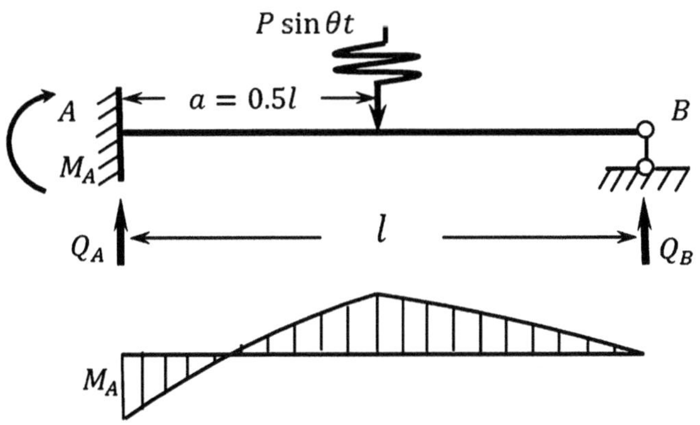	$M_A = \frac{3}{16}Pl\psi_{3p}(u)$ $Q_A = \frac{11}{16}P\psi_{4p}(u)$; $Q_B = \frac{5}{16}P\psi_{5p}(u)$
15	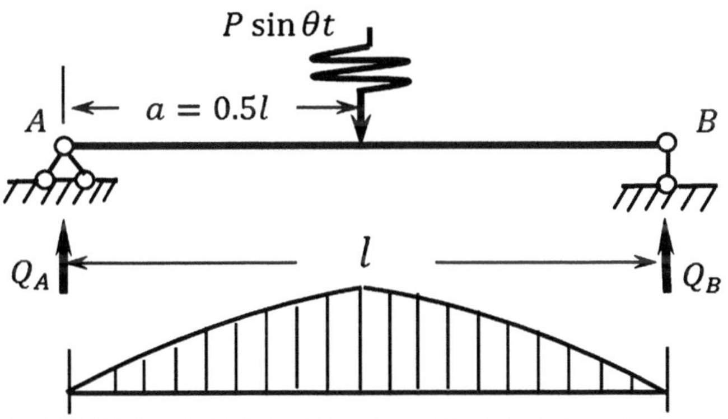	$Q_A = Q_B = \frac{P}{2}\psi_{6p}(u)$

##	Design scheme and diagram of bending moments	Calculation formulas for reaction amplitudes
16	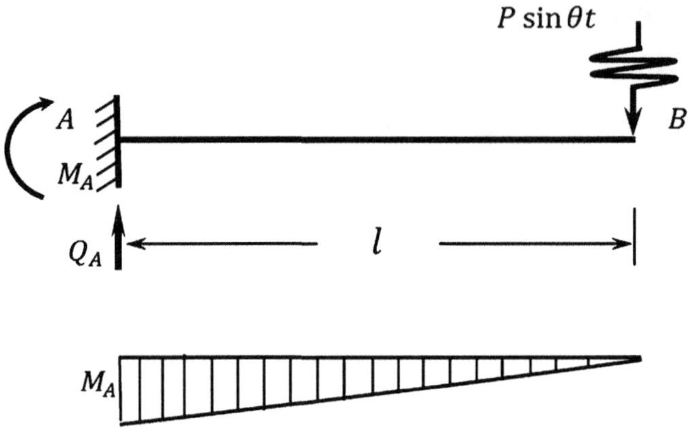	$M_A = Pl\psi_{3p}(u)$ $Q_A = P\psi_{8p}(u)$
17	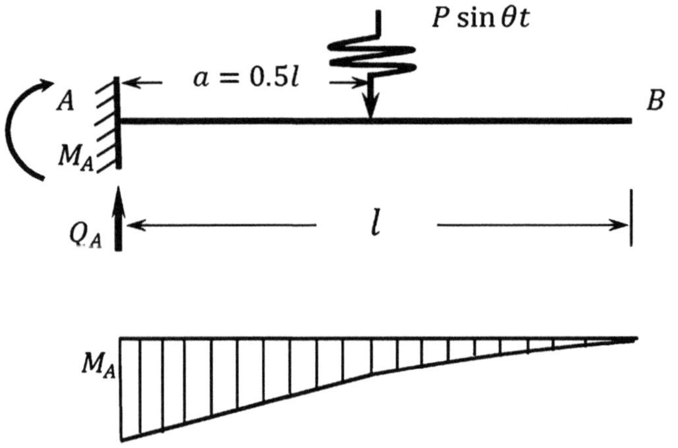	$M_A = \frac{l}{2}Pl\psi_{9p}(u)$ $Q_A = P\psi_{10p}(u)$

The Values of the Functions That Are Included in the Formulas of the Displacement Method

$$\psi_1(u) = \frac{u}{3} \cdot \frac{2\sinh u \sin u}{\cosh u \sin u - \sinh u \cos u};$$

$$\psi_2(u) = \frac{u}{4} \cdot \frac{\cosh u \sin u - \sinh u \cos u}{1 - \cosh u \cos u};$$

$$\psi_3(u) = \frac{u}{2} \cdot \frac{\sinh u - \sin u}{1 - \cosh u \cos u};$$

$$\psi_4(u) = \frac{u^2}{3} \cdot \frac{\cosh u \sin u + \sinh u \cos u}{\cosh u \sin u - \sinh u \cos u};$$

$$\psi_5(u) = \frac{u^2}{6} \cdot \frac{\sinh u \ \sin u}{1 - \cosh u \cos u};$$

$$\psi_6(u) = \frac{u^2}{6} \cdot \frac{\cosh u - \cos u}{1 - \cosh u \cos u};$$

$$\psi_7(u) = \frac{u^2}{3} \cdot \frac{\sinh u + \sin u}{\cosh u \sin u - \sinh u \cos u};$$

$$\psi_8(u) = \frac{u^3}{3} \cdot \frac{2 \cosh u \cos u}{\cosh u \sin u - \sinh u \cos u};$$

$$\psi_9(u) = \frac{u^3}{3} \cdot \frac{\cosh u + \cos u}{\cosh u \sin u - \sinh u \cos u};$$

$$\psi_{10}(u) = \frac{u^3}{12} \cdot \frac{\cosh u \sin u + \sinh u \cos u}{1 - \cosh u \cos u};$$

$$\psi_{11}(u) = \frac{u^3}{12} \cdot \frac{\sinh u + \sin u}{1 - \cosh u \cos u};$$

$$\psi_{12}(u) = \frac{u^3}{3} \cdot \frac{1 + \cosh u \cos u}{\cosh u \sin u - \sinh u \cos u}.$$

$$\psi_{13}(u) = \frac{u^3}{3} \cdot \frac{\sinh u - \sin u}{\Delta_3};$$

$$\psi_{14}(u) = \frac{u^3}{2} \cdot \frac{\cosh u \sin u - \sinh u \cos u}{\Delta_3};$$

$$\psi_{15}(u) = u^2 \cdot \frac{\sinh u \cdot \sin u}{\Delta_4};$$

$$\psi_{16}(u) = u^3 \cdot \frac{\cosh u \sin u - \sinh u \cos u}{\Delta_4};$$

$$\psi_{17}(u) = u \cdot \frac{\cosh u \sin u - \sinh u \cos u}{\Delta_{4.}}$$

$$\psi_{1p}(u) = \frac{2}{\left(\frac{u}{2}\right)} \cdot \frac{\cosh \frac{u}{2} + \sin \frac{u}{2}}{\cosh \frac{u}{2} \sin \frac{u}{2} + \sinh \frac{u}{2} \cos \frac{u}{2}};$$

$$\psi_{2p}(u) = \frac{\sinh \frac{u}{2} - \cos \frac{u}{2}}{\cosh \frac{u}{2} \sin \frac{u}{2} + \sinh \frac{u}{2} \cos \frac{u}{2}};$$

$$\psi_{3p}(u) = \frac{3}{\left(\frac{u}{2}\right)^2} \cdot \frac{\cosh \frac{u}{2} \sin \frac{u}{2} - \sinh \frac{u}{2} \cos \frac{u}{2}}{\cosh \frac{u}{2} \sin \frac{u}{2} + \sinh \frac{u}{2} \cos \frac{u}{2}};$$

$$\psi_{4p}(u) = \frac{2}{u} \cdot \frac{-2 \sinh \frac{u}{2} \sin \frac{u}{2}}{\cosh \frac{u}{2} \sin \frac{u}{2} + \sinh \frac{u}{2} \cos \frac{u}{2}}.$$

$$\psi_{5p}(u) = \frac{16}{10} \cdot \left[\frac{(\sinh u + \sin u)(\cosh 0.5u - \cos 0.5u) -}{\Delta_1}\right.$$

$$\left.\frac{(\cosh u + \cos u)(\sinh 0.5u - \sin 0.5u)}{1}\right];$$

$$\psi_{6p}(u) = \frac{\sinh u \cdot \sin 0.5u + \sinh 0.5u \sin u}{\Delta_3};$$

$$\psi_{7p}(u) = \frac{1}{u} \cdot \frac{\sinh u + \sin u}{\Delta_4}; \quad \psi_{8p}(u) = \frac{\cosh u + \cos u}{\Delta_4};$$

$$\psi_{9p}(u) = \frac{1}{u} \cdot \left[\frac{(\cosh 0.5u + \cos 0.5u)(\sinh u + \sin u) -}{\Delta_4} \right.$$

$$\left. \frac{(\sinh 0.5u + \sin 0.5u)(\cosh u + \cos u)}{1} \right];$$

$$\psi_{10p}(u) = \left[\frac{(\cosh 0.5u + \cos 0.5u)(\cosh u + \cos u) -}{2\Delta_4} \right.$$

$$\left. \frac{(\sinh 0.5u + \sin 0.5u)(\sinh u - \sin u)}{1} \right];$$

where:

$$\Delta_1 = \cosh u \cdot \sin u - \sinh u \cdot \cos u; \quad \Delta_2 = 1 - \cosh u \cdot \cos u;$$

$$\Delta_3 = \sinh u \cdot \sin u; \quad \Delta_4 = 1 + \cosh u \cdot \cos u;$$

$$\Delta_5 = \cosh 0.5u \cdot \sin 0.5u + \sinh 0.5u \cdot \cos 0.5u;$$

$$\psi_{1q}(u) = \frac{12}{u^2} \cdot \frac{\cosh u - \sinh u \cdot \sin u - \cos u}{\Delta_2};$$

$$\psi_{2q}(u) = \frac{2}{u} \cdot \frac{\sinh u + \sin u - \cosh u \cdot \sin u - \sinh u \cdot \cos u}{\Delta_2};$$

$$\psi_{3q}(u) = \frac{1}{u} \cdot \frac{\cosh u \cdot \sin u - \sinh u \cdot \cos u + \sinh u \cdot \sin u}{\Delta_3};$$

$$\psi_{4q}(u) = \frac{8}{u^2} \cdot \frac{\sinh u + \sin u - \cosh u \cdot \sin u - \sinh u \cdot \cos u}{\Delta_1};$$

$$\psi_{5q}(u) = \frac{8}{3u} \cdot \frac{\cosh u + \cos u - \cosh u \cdot \cos u - 1}{\Delta_1};$$

$$\psi_{6q}(u) = \frac{8}{5u} \cdot \frac{\cosh u + \cos u - 2\cosh u \cdot \cos u}{\Delta_1};$$

$$\psi_{7q}(u) = \frac{2}{u^2} \cdot \frac{\sinh u \cdot \sin u}{\Delta_4};$$

$$\psi_{8q}(u) = \frac{1}{u} \cdot \frac{\cosh u \cdot \sin u + \sinh u \cdot \cos u}{\Delta_4}.$$

The correction functions and their numerical values are presented in the following table (Table B.2)

Table B.2 Correction functions and their numerical values

u	$\psi_1(u)$	$\psi_2(u)$	$\psi_3(u)$	$\psi_4(u)$	$\psi_5(u)$	$\psi_6(u)$
0	1.0000	1.0000	1.0000	1.0000	1.0000	1.0000
0.1	1.0000	1.0000	1.0000	0.9999	0.9999	1.0000
0.2	0.9999	1.0000	1.0000	0.9999	0.9999	1.0000
0.3	0.9999	0.9999	1.0001	0.9998	0.9998	1.0001
0.4	0.9998	0.9999	1.0001	0.9993	0.9997	1.0002
0.5	0.9996	0.9999	1.0002	0.9982	0.9995	1.0003
0.6	0.9992	0.9997	1.0005	0.9964	0.9989	1.0007
0.7	0.9985	0.9994	1.0009	0.9931	0.9979	1.0012
0.8	0.9974	0.9990	1.0015	0.9883	0.9964	1.0021
0.9	0.9958	0.9984	1.0023	0.9812	0.9943	1.0034
1.0	0.9936	0.9976	1.0036	0.9313	0.9913	1.0052
1.1	0.9906	0.9965	1.0052	0.9580	0.9872	1.0076
1.2	0.9867	0.9950	1.0074	0.9403	0.9818	1.0108
1.3	0.9817	0.9932	1.0103	0.9176	0.9750	1.0148
1.4	0.9752	0.9908	1.0138	0.8888	0.9663	1.0200
1.5	0.9672	0.9878	1.0182	0.8529	0.9555	1.0265
1.6	0.9573	0.9842	1.0237	0.8086	0.9422	1.0343
1.7	0.9453	0.9798	1.0304	0.7546	0.9262	1.0439
1.8	0.9306	0.9745	1.0384	0.6892	0.9069	1.0555
1.9	0.9130	0.9682	1.0479	0.6107	0.8840	1.0693
2.0	0.8919	0.9608	1.0592	0.5170	0.8569	1.0857
2.1	0.8667	0.9521	1.0725	0.4055	0.8252	1.1051
2.2	0.8368	0.9419	1.0885	0.2733	0.7881	1.1278
2.3	0.8012	0.9300	1.1064	0.1168	0.7451	1.1544
2.4	0.7589	0.9162	1.1277	−0.0684	0.6953	1.1853
2.5	0.7085	0.9002	1.1525	−0.2879	0.6379	1.2215
2.6	0.6484	0.8819	1.1813	−0.5488	0.5718	1.2635
2.7	0.5761	0.8606	1.2146	−0.8604	0.4958	1.3122
2.8	0.4886	0.8362	1.2534	−1.2350	0.4086	1.3691
2.9	0.3817	0.8080	1.2984	−1.6895	0.3084	1.4352
3.0	0.2494	0.7754	1.3509	−2.2482	0.1934	1.5124
3.1	0.0826	0.7377	1.4121	−2.9464	0.0609	1.6028
3.2	−0.1325	0.6940	1.4840	−3.8388	0.0920	1.7091
3.3	−0.4185	0.6430	1.5687	−5.0147	−0.2691	1.8349
3.4	−0.8150	0.5832	1.6693	−6.6306	−0.4753	1.9844
3.5	−1.3990	0.5126	1.7896	−8.9890	−0.7172	2.1640
3.6	−2.3415	0.4284	1.9349	−12.762	−1.0032	2.3816

3.7	−4.1148	0.3269	2.1127	−19.808	−1.3453	2.6487
3.8	−8.6838	0.2027	2.3335	−37.845	−1.7603	2.9817
3.9	−47.556	0.0478	2.6131	−190.69	−2.2730	3.4048
4.0	19.468	−0.1501	2.9758	72.589	−2.9217	3.9557
4.1	9.1701	−0.4110	3.4615	32.016	−3.7687	4.6961
u	$\psi_1(u)$	$\psi_2(u)$	$\psi_3(u)$	$\psi_4(u)$	$\psi_5(u)$	$\psi_6(u)$
4.2	6.3934	−0.7700	4.1402	20.984	−49232	5.7342
4.3	5.0927	−1.2950	5.1472	15.743	−6.5952	7.2795
4.4	4.3307	−2.1357	6.7817	12.607	−9.2489	9.7956
4.5	3.8236	−3.7021	9.8635	10.460	−14.155	14.552
4.6	3.4560	−7.6655	17.734	8.8476	−26.492	26.725
4.7	3.1721	−37.947	78.238	7.5480	−120.37	120.43
4.8	2.9412	18.304	−34.332	6.4399	53.839	−53.975
4.9	2.7452	8.3439	−14.483	5.4496	22.905	−23.252
5.0	2.5722	5.7486	−9.3710	4.5288	14.786	−15.362
5.1	2.4142	4.5445	−7.0495	3.6424	10.971	−11.797
5.2	2.2652	3.8417	−5.7383	27666	8.7024	−9.8001
5.3	2.1207	3.3749	−4.9080	1.8767	7.1570	−8.5522
5.4	1.9765	3.0368	−4.3454	0.9537	6.0024	−7.7233
5.4	1.9765	3.0368	−4.3454	0.9537	6.0024	−7.7233
5.5	1.8292	2.7759	−3.9483	−0.0221	5.0778	−7.1556
5.6	1.6752	2.5639	−3.6619	−1.0721	4.2951	−6.7650
5.7	1.5105	2.3842	−3.4546	−2.2202	3.6012	−6.5032
5.8	1.3306	2.2260	−3.3067	−3.4958	2.9618	−6.3409
5.9	1.1300	2.0819	−3.2061	−4.9360	2.3526	−6.2605
6.0	0.9016	1.9465	−3.1450	−6.5901	1.7551	−62514
6.1	0.6356	1.815	−3.1186	−8.5259	1.1542	−6.3082
6.2	0.3118	1.6861	−3.1245	−10.841	0.5364	−6.4291
6.3	−0.7184	1.5542	−3.1618	−13.683	−0.1116	−6.6159
6.4	−0.5673	1.4170	−3.2315	−17.284	−0.8038	−6.8738
6.5	−1.2242	1.2710	−3.3360	−22.041	−1.5559	−7.2115
6.6	−2.1462	1.1122	−3.4798	−28.685	−2.3871	−7.6424
6.7	−3.5490	0.9359	−3.6699	−38.742	−3.3215	−8.1858
6.8	−5.9688	0.7357	−3.9165	−56.001	−4.3912	−8.8699
6.9	11.214	0.5030	−42349	−93.244	−5.6407	−9.7356
7.0	−31.636	0.2255	−4.6482	−237.78	−7.1341	−10.844
7.1	77.672	−0.1155	−5.1922	53467	−8.9685	−12.289
7.2	20.557	−0.5497	−5.9252	130.73	−11.300	−14.224
7.3	12.760	−1.1281	−6.9480	75.383	−14.395	−16.914
7.4	9.6349	−1.9466	−8.4512	53.045	−18.755	−20.858
7.5	7.9308	−3.2090	−10842	40.731	−25.450	−27.122
7.6	6.8430	−5.4410	−15.170	32.754	−37.233	−38.457
7.7	6.0765	−10.540	−25.229	27.025	−64.044	−64.802
7.8	5.4970	−34.665	−73.337	22.597	−190.55	−190.82
7.9	5.0348	44.151	84.441	15.972	222.29	222.53
8.0	4.6496	15.528	27.347	15.863	72.197	72.981
u	$\psi_7(u)$	$\psi_8(u)$	$\psi_9(u)$	$\psi_{10}(u)$	$\psi_{11}(u)$	$\psi_{12}(u)$
0	1.0000	1.0000	1.0000	1.0000	1.0000	1.0000
0.1	1.0000	0.9999	1.0001	0.9999	1.0000	0.9999
0.2	1.0001	0.9997	1.0002	0.9999	1.0000	0.9999
0.3	1.0002	0.9987	1.0007	0.9997	1.0001	0.9994

0.4	1.0004	0.9958	1.0014	0.9992	1.0004	0.9980
0.5	1.0008	0.9899	1.0029	0.9981	1.0007	0.9951
0.6	1.0017	0.9790	1.0060	0.9960	1.0014	0.9898
0.7	1.0031	0.9611	1.0111	0.9926	1.0026	0.9811
0.8	1.0054	0.9336	1.0190	0.9873	1.0044	0.9678
0.9	1.0086	0.8936	1.0306	0.9797	1.0070	0.9484
1.0	1.0132	0.8377	1.0467	0.96910	1.0107	0.9213
1.1	1.0193	0.7621	1.0685	0.9546	1.0157	0.8846
1.2	1.0274	0.6626	1.0973	0.9357	1.0223	0.8363
1.3	1.0379	0.5345	1.1346	0.9113	1.0308	0.7741
1.4	1.0512	0.3724	1.1820	0.8806	1.0416	0.6955
1.5	1.0679	0.1705	1.2414	0.8425	1.0549	0.5976
1.6	1.0886	−0.0777	1.3150	0.7958	1.0714	0.4772
1.7	1.1139	−03794	1.4054	0.7393	1.0914	0.3309
1.8	1.1447	−0.7430	1.5155	0.6716	1.1155	0.1547
1.9	1.1819	−0.1775	1.6489	0.5913	1.1444	−0.0559
2.0	1.2267	−1.6936	1.8098	0.4967	1.1787	−0.3059
2.1	1.2806	−2.3035	2.0034	0.3861	1.2192	−0.6012
2.2	1.3450	−3.0212	2.2362	0.2574	1.2668	−0.9487
2.3	1.4222	−3.8668	2.5160	0.1087	1.3227	−1.3563
2.4	1.5148	−4.8513	2.5830	−0.0626	1.3879	−1.8337
2.5	1.6263	−6.0086	−3.2601	−0.2592	1.4641	−2.3928
2.6	1.7610	−7.3665	3.7543	−0.4840	1.5530	−3.0482
2.7	1.9248	−8.9647	4.3582	−0.7405	1.6565	−3.8189
2.8	2.1257	−10.855	5.1028	−1.0327	1.7775	−4.7296
2.9	2.3747	−13.108	6.0312	−1.3651	1.9187	−5.8136
3.0	2.6880	−15.823	7.2056	−1.7432	2.0842	−7.1176
3.1	3.0891	−19.142	8.7185	−2.1736	2.2789	−8.7095
3.2	3.6150	−23.284	10.714	−2.6641	2.5087	−10.693
3.3	4.3262	−28.605	13.430	−3.2245	2.7817	−13.236
3.4	5.3294	−35.725	17.285	−3.8671	3.1083	−16.631
3.5	6.8316	−45.833	23.091	−4.6079	3.5020	−22.442
3.6	9.2938	−61.587	32.6567	−5.4678	3.9819	−28.918
3.7	13.991	−90.280	50.994	−6.4756	4.5742	−42.510
3.8	26.227	−162.26	98.945	−76716	5.3165	−76.545
3.9	131.08	−764.08	510.82	−9.1145	6.2652	−360.80
4.0	−50.020	269.20	−201.00	−10.894	7.5072	127.06
4.1	−22.351	108.35	−92.452	−13.157	9.1857	51.053
4.2	−15.002	63.467	−63.776	−16.158	11.552	29.793
u	$\psi_7(u)$	$\psi_8(u)$	$\psi_9(u)$	$\psi_{10}(u)$	$\psi_{11}(u)$	$\psi_{12}(u)$
4.3	−11.654	41.210	−50.990	−20.388	15.092	19.210
4.4	−9.7781	27.086	−43.734	−26.925	20.883	12.461
4.5	−8.6096	16.701	−39.428	−38.725	31.873	7.4703
4.6	−7.8395	8.2554	−36.721	−67.821	60.093	3.3879
4.7	−7.3194	−0.8683	−35.036	−286.43	277.75	0.2033
4.8	−6.9700	−5.9531	−34.067	117.88	−127.58	−3.5364
4.9	−6.7457	−12.451	−33.642	45.470	−56.284	−6.7568
5.0	−6.6193	−19.024	−33.661	26.034	−38.052	−9.9639
5.1	−6.5744	−25.638	−34.071	16.557	−29.877	−13.233
5.2	−6.6016	−32.485	−34.847	10.628	−25.361	−16.625
5.3	−6.6962	−39.681	−35.977	6.3337	−22.599	−20.198

5.4	−6.8577	−47.339	−37.507	2.8963	−20.826	−24.007
5.5	−7.0888	−55.582	−39443	−0.06147	−19.678	−27.686
5.6	−7.3956	−64.544	−41.850	−2.7487	−18.962	−32.580
5.7	−7.7884	−74.388	−44.808	−5.2934	−18.569	−37.492
5.8	−8.2820	−85.315	−48.430	−7.7815	−18.435	−42.949
5.9	−8.8980	−97.584	−52.874	−10.276	−18.524	−49.080
6.0	−9.6672	−111.54	−58.361	−12.828	−18.818	−56.059
6.1	−10.634	−127.67	−65.207	−15.4813	−19.309	−64.126
6.2	−11.864	−146.66	−73.881	−18.279	−20.004	−73.628
6.3	−13.459	−169.55	−85.098	−21.266	−20.918	−85.087
6.4	−15.5807	−197.99	−100.01	−24.491	−22.078	−99.331
6.5	−18.507	−234.81	−120.57	−28.013	−23.522	−117.77
6.6	−22.753	−285.16	−150.43	−31.904	−25.307	−142.99
6.7	−29.396	−359.82	−197.20	−36.258	−27.512	−180.40
6.8	−41.116	−485.62	−279.83	−41.200	−30.250	−243.43
6.9	−66.960	−752.89	−462.24	−46.904	−33.682	−377.38
7.0	−168.74	−1778.8	−1181.4	−53.623	−38.051	−891.56
7.1	378.71	3676.85	2688.6	−61.736	−43.728	1842.86
7.2	93.356	816.85	671.98	−71.859	−51.315	409.43
7.3	54.827	420.63	400.06	−85.042	−61.852	210.96
7.4	39.711	257.46	283.70	−103.26	−77301	129.09
7.5	31.739	164.86	237.89	−130.71	−101.85	82.692
7.6	26.891	102.60	204.23	−178.21	−146.32	51.506
7.7	23.963	55.918	182.32	−284.84	−249.75	28.124
7.8	21.487	18.071	167.47	−783.31	−744.85	9.1728
7.9	19.923	−14.470	157.27	837.61	879.63	−7.1185
8.0	18.811	−43.762	150.37	246.32	292.09	−21.780

u	$\psi_{13}(u)$	$\psi_{14}(u)$	$\psi_{15}(u)$	$\psi_{16}(u)$	$\psi_{17}(u)$
0	0.0000	0.0000	0.0000	0.0000	0.0000
0.1	0.0000	0.0000	0.0001	0.0001	0.0000
0.2	0.0003	0.0005	0.0008	0.0016	0.0005
0.3	0.0014	0.0027	0.0041	0.0081	0.0027
0.4	0.0043	0.0085	0.0128	0.0256	0.0086
0.5	0.0104	0.0208	0.0314	0.0627	0.0209
0.6	0.0216	0.0432	0.0654	0.1304	0.0436
0.7	0.0401	0.0801	0.1222	0.2430	0.0816
0.8	0.0686	0.1369	0.2111	0.4183	0.1411
0.9	0.1102	0.2196	0.3445	0.6788	0.2306
1.0	0.1687	0.3355	0.5393	1.0544	0.3618
1.1	0.2485	0.4926	0.8198	1.5856	0.5517
1.2	0.3546	0.7005	1.2233	2.3316	0.8265
1.3	0.4933	0.9698	1.8109	3.3855	1.2298
1.4	0.6718	1.3130	2.6934	4.9095	1.8412
1.5	0.8992	1.7447	4.0971	7.2255	2.8239
1.6	1.1868	2.2819	6.5736	11.104	4.5777
1.7	1.5486	2.9453	11.929	19.045	8.4136
1.8	2.0032	3.7601	31.579	46.774	22.622
1.9	2.5747	4.7581	−106.41	−142.36	−77.704
2.0	3.2961	5.9799	−23.322	−27.038	−17.433
2.1	4.2129	7.4796	−14.017	−13.117	−10.782
2.2	5.3906	9.3317	−10.331	−6.7495	−82309

(continued)

Table B.2 (continued)

2.3	6.9258	11.643	−8.2656	−2.4110	−6.8777
2.4	8.9685	14.572	−6.8656	1.2373	−6.0311
2.5	11.763	18.377	−5.7836	4.7004	−5.4417
2.6	15.735	23.493	−4.8601	8.2280	−4.9972
2.7	21.699	30.749	−4.0085	11.973	−4.6386
2.8	31.425	41.930	−3.1751	16.050	−4.3320
2.9	49.624	61.758	−2.3222	20.555	−4.0553
3.0	94.316	108.27	−1.4189	25.585	−3.7934
3.1	356.89	372.88	−0.4377	31.242	−3.5345
3.2	−282.01	−263.76	0.6498	37.644	−3.2688
3.3	−115.24	−94.465	1.8748	44.932	−2.9867
3.4	−78.217	−54.654	3.2745	53.280	−2.6785
3.5	−62.409	−35.753	4.8958	62.911	−2.3329
3.6	−53.992	23.911	6.7998	74.123	−1.9360
3.7	−49.054	−15.182	9.0707	87.324	−1.4696
3.8	−46.069	−80039	11.828	103.09	−0.9080
3.9	−44.326	1.6216	15.249	122.29	−0.2138
4.0	−43.456	4.3833	19.611	146.25	0.6716
4.1	−43.256	10.272	25.379	177.21	1.8451
4.2	−43.614	16.223	33.388	219.17	3.4815
4.3	−44.470	22.377	45.318	280.19	5.9323
u	$\psi_{13}(u)$	$\psi_{14}(u)$	$\psi_{15}(u)$	$\psi_{16}(u)$	$\psi_{17}(u)$
4.4	−45.804	28.849	65.130	379.21	10.0261
4.5	−47.622	35.749	104.94	574.19	18.2974
4.6	−49.955	43.185	228.38	1169.3	44.0537
4.7	−52.860	51.277	−3807.5	−18119.7	−800.19
4.8	−56.419	60.160	−220.75	966.67	−50.035
4.9	−60.751	69.998	−117.11	−464.95	−28.439
5.0	−66.019	80.994	−80.673	−284.08	−20.909
5.1	−72.449	93.409	61.713	186.26	−17.042
5.2	−80.354	107.59	49.812	121.67	−14.660
5.3	−90.184	124.03	41.423	73.315	−13.022
5.4	−102.59	143.40	35.004	33.780	−11.806
5.5	−118.59	166.75	29.772	−0.72089	−10.850
5.6	−139.75	195.69	25.283	−32.361	−10.062
5.7	−168.77	232.95	21.264	−62.510	−9.3851
5.8	−210.57	283.50	17.530	−92.110	−8.7829
5.9	−275.22	357.43	13.950	−121.86	−8.2296
6.0	−387.06	479.13	10.422	−152.35	−7.7062
6.1	−623.53	726.09	6.8622	−184.09	−7.1972
6.2	−1434.7	1548.4	3.1920	−217.57	−66896
6.3	−7435.1	−7309.5	−0.66501	−253.32	−6.1713
6.4	1124.27	−985.87	−4.7906	−291.93	−5.6301
6.5	637.90	−486.05	−9.2781	−334.09	−5.0526
6.6	481.02	−294.70	−14.241	−380.66	−4.4234
6.7	371.08	−189.27	−19.822	−432.76	−3.7235
6.8	317.83	−119.41	−26.213	−491.88	−2.9278
6.9	283.63	−67.379	−33.678	−560.07	−2.0022
7.0	260.73	−25.298	−42.598	−640.36	−0.89767
7.1	245.20	10.906	−53.552	−737.27	0.45965

(continued)

Table B.2 (continued)

7.2	234.86	43.576	−67.466	−858.09	2.1879
7.3	228.45	74.186	−85.926	−1015.3	4.4894
7.4	225.20	103.74	−111.91	−1232.2	7.7431
7.5	224.65	132.99	−151.73	−1558.5	12.754
7.6	226.54	162.51	−221.62	−2121.6	21.591
7.7	230.79	192.84	−379.76	−3378.0	41.664
7.8	237.43	224.46	−1109.1	−9118.6	134.51
7.9	246.60	257.87	1377.4	10380.5	−182.39
8.0	258.58	293.65	437.20	2983.2	−62.686

u	$\psi_{1q}(u)$	$\psi_{2q}(u)$	$\psi_{3q}(u)$	$\psi_{4q}(u)$	$\psi_{5q}(u)$	$\psi_{6q}(u)$
0	1.0000	1.0000	1.0000	1.0000	1.0000	1.0000
0.1	1.0729	1.0006	1.0000	1.0025	0.9993	1.0001
0.2	1.0059	1.0000	1.0000	1.0001	1.0000	1.0000
0.3	0.9993	1.0000	1.0001	1.0000	1.0000	1.0000
0.4	0.9999	1.0000	1.0002	1.0001	1.0001	1.0001
0.5	0.9997	1.0001	1.0005	1.0003	1.0002	1.0002
0.6	1.0002	1.0002	1.0011	1.0005	1.0004	1.0004
0.7	1.0004	1.0003	1.0020	1.0009	1.0008	1.0007
0.8	1.0007	1.0006	1.0034	1.0015	1.0014	1.0012
0.9	1.0012	1.0009	1.0055	1.0025	1.0022	1.0020
1.0	1.0018	1.0014	1.0084	1.0038	1.0034	1.0030
1.1	1.0026	1.0020	1.0124	1.0056	1.0050	1.0044
1.2	1.0037	1.0029	1.0177	1.0079	1.0071	1.0062
1.3	1.0051	1.0040	1.0245	1.0109	1.0098	1.0086
1.4	1.0069	1.0054	1.0333	1.0147	1.0132	1.0116
1.5	1.0091	1.0071	1.0445	1.0195	1.0174	1.0154
1.6	1.0119	1.0092	1.0585	1.0254	1.0227	1.0201
1.7	1.0152	1.0118	1.0761	1.0326	1.0292	1.0258
1.8	1.0191	1.0149	1.0980	1.0414	1.0370	1.0327
1.9	1.0239	1.0186	1.1254	1.0520	1.0465	1.0410
2.0	1.0295	1.0230	1.1595	1.0647	1.0579	1.0510
2.1	1.0361	1.0281	1.2024	1.0798	1.0715	1.0630
2.2	1.0439	1.0341	1.2569	1.0979	1.0877	1.0773
2.3	1.0529	1.0412	1.3271	1.1195	1.1070	1.0943
2.4	1.0634	1.0493	1.4191	1.1453	1.1301	1.1146
2.5	1.0756	1.0588	1.5431	1.1761	1.1577	1.1389
2.6	1.0898	1.0698	1.7169	1.2131	1.1909	1.1681
2.7	1.1061	1.0825	1.9738	1.2578	1.2309	1.2035
2.8	1.1251	1.0972	2.3869	1.3122	1.2797	1.2462
2.9	1.1470	1.1143	3.1496	1.3791	1.3398	1.2989
3.0	1.1725	11341	5.0022	1.4626	1.4146	1.3646
3.1	1.2021	1.1571	15.804	1.5685	1.5097	1.4480
3.2	2.2367	1.1840	−10.410	1.7060	1.6332	1.5563
3.3	1.2773	1.2155	−3.5365	1.8904	1.7988	1.7014
3.4	1.3253	1.2527	−1.9886	2.1480	2.0304	1.9041
3.5	1.3824	1.2970	−1.3083	2.5302	2.3741	2.2047
3.6	1.4510	1.3503	−0.9276	3.1515	2.9328	2.6933
3.7	1.5344	1.4150	−0.6855	4.3280	3.9913	3.6182
3.8	1.6374	1.4949	−0.5186	7.3747	6.7330	6.0130
3.9	1.7671	1.5955	−0.3972	33.380	30.139	26.452

(continued)

Table B.2 (continued)

4.0	1.9343	1.7252	−0.3053	−11.495	−10.252	−8.8163
4.1	2.1568	1.8977	−0.2335	−4.6174	−4.0626	−3.4108
4.2	2.4660	2.1373	−0.1760	−2.7756	−2.4056	−1.9628
4.3	2.9220	2.4907	−0.1292	−19233	−1.6394	−1.2925
u	$\psi_{1q}(u)$	$\psi_{2q}(u)$	$\psi_{3q}(u)$	$\psi_{4q}(u)$	$\psi_{5q}(u)$	$\psi_{6q}(u)$
4.4	3.6584	3.0613	−0.0905	−1.4334	−1.1994	−0.9070
4.5	5.0405	4.1320	−0.0579	−1.1161	−0.9149	−0.6570
4.6	8.5578	6.8565	−0.0302	−0.8945	−0.7166	−0.4822
4.7	35.531	27.749	−0.0065	0.7312	−0.5711	−0.3532
4.8	−14.623	−11.098	0.0141	−0.6063	−0.4601	−0.2542
4.9	−5.7649	−4.2363	0.0321	−0.5077	−0.3731	−0.1759
5.0	−3.4732	−2.4609	0.0479	−0.4281	−0.3033	−0.1124
5.1	−2.4231	—1.6473	0.0620	−0.3625	−0.2464	−0.0598
5.2	−1.8220	−1.1813	0.0745	−0.3075	−0.1994	−0.0155
5.3	−1.4335	−0.88003	0.0858	−0.2608	−0.1601	0.0225
5.4	−1.1622	−0.66965	0.0960	−0.2205	−0.1270	0.0555
5.5	0.96256	−0.51449	0.1053	−0.1853	−0.0991	0.0847
5.6	−0.80973	−0.395666	0.1138	−0.1543	−0.0754	0.1108
5.7	−0.30183	−0.30183	0.1216	−0.1266	−0.0554	0.1345
5.8	−0.59188	−0.22595	0.1289	−0.1015	−0.0387	0.1564
5.9	−0.51178	0.16337	0.1357	−0.0785	−0.0250	0.1771
6.0	−0.44481	−0.11092	0.1421	−0.0570	−0.0142	0.1970
6.1	−0.38805	−0.06635	0.1481	−0.0365	−0.00625	0.2165
6.2	−0.33940	−0.02802	0.1539	−0.0166	−0.00137	0.2364
6.3	−0.29728	0.00529	0.1595	0.00341	−0.00006	0.2573
6.4	−0.26050	0.03452	0.1649	0.02436	−0.00323	0.2804
6.5	−0.22812	0.06038	0.1701	0.04751	`-0.01258	0.3072
6.6	−0.19941	0.0834	0.1753	0.07503	−0.03140	0.3408
6.7	−0.17380	0.10417	0.1804	0.11131	−0.06671	0.3874
6.8	−0.15082	0.12292	0.1856	0.16708	−0.13617	0.4620
6.9	−0.13007	0.13999	0.1908	0.27830	−0.29953	0.6155
7.0	0.11125	0.15564	0.1961	0.69018	−0.96561	1.1944
7.1	−0.09408	0.17008	0.2016	−1.4729	2.6626	−1.8666
7.2	−0.07836	0.18347	0.2072	−0.33379	0.78157	−0.2587
7.3	−0.06388	0.19599	0.2131	−0.17373	0.53301	−0.0348
7.4	−0.05049	0.20775	0.2194	−0.10672	0.43914	0.03800
7.5	−0.03804	0.21887	0.2261	−0.06820	0.39240	0.1108
7.6	−0.02642	0.22946	0.2332	−0.04217	0.36622	0.1462
7.7	−0.01552	0.23960	0.2410	−0.02273	0.35091	0.1726
7.8	−0.00524	0.24939	0.2496	−0.00719	0.34213	0.1937
7.9	0.00450	0.25889	0.2690	0.00586	0.33764	0.2116
8.0	0.01378	0.26819	0.2696	0.01727	0.33623	0.2275
u	$\psi_{7q}(u)$	$\psi_{8q}(u)$	$\psi_{1p}(u)$	$\psi_{2p}(u)$	$\psi_{3p}(u)$	$\psi_{4p}(u)$
0	1.0000	1.0000	1.0000	1.0000	1.0000	1.0000
0.1	1.0000	1.0000	1.0000	1.0000	1.0000	1.0000
0.2	1.0001	1.0001	1.0000	1.0000	1.0000	1.0000
0.3	1.0006	1.0004	1.0000	1.0000	1.0000	1.0000
0.4	1.00019	1.0013	1.0001	1.0001	1.0001	1.0001
0.5	1.00045	1.0031	1.0001	1.0002	1.0003	1.0003
0.6	1.00095	1.0065	1.0003	1.0003	1.0005	1.0006

(continued)

Table B.2 (continued)

0.7	1.0177	1.0122	1.0005	1.0006	1.0010	1.0011
0.8	1.0306	1.0212	1.0009	1.0011	1.0017	1.0019
0.9	1.0500	1.0346	1.0015	1.0017	1.0028	1.0030
1.0	1.0786	1.0544	1.0023	1.0026	1.0043	1.0046
1.1	1.1199	1.0830	1.0033	1.0038	1.0062	1.0067
1.2	1.1799	1.1244	1.0047	1.0054	1.0089	1.0095
1.3	1.2681	1.1854	1.0065	1.0075	1.-123	1.0132
1.4	1.4023	1.2780	1.0087	1.0101	1.0166	1.0178
1.5	1.6186	1.4273	1.0115	1.0133	1.0219	1.0178
1.6	2.0061	1.6944	1.0150	1.0173	1.0286	1.0308
1.7	2.8566	2.2803	1.0192	1.0221	1.0367	1.0395
1.8	6.0163	4.4557	1.0242	1.0279	1.0466	1.0501
1.9	−16.331	−10.924	1.0302	1.0349	1.0585	1.0629
2.0	−2.9153	−1.6899	1.0373	1.0431	1.0727	1.0783
2.1	−1.4415	−0.6745	1.0457	1.0527	1.0898	1.0966
2.2	−08820	−0.21881	1.0555	1.0640	1.1101	1.1185
2.3	−0.5907	−0.0862	1.0669	1.0772	1.1344	1.1446
2.4	−0.4139	0.0373	1.0802	1.0926	1.1634	1.1758
2.5	−0.2961	0.1203	1.0957	1.1104	1.1980	1.2130
2.6	−0.2127	0.1801	1.1136	1.1311	1.2396	1.2577
2.7	−0.1509	0.2253	1.1343	1.1550	1.2898	1.3117
2.8	−0.1033	0.2611	1.1582	1.1827	1.3509	1.3774
2.9	−0.0657	0.2906	1.1860	1.2149	1.4260	1.4581
3.0	−0.0350	0.3159	1.2183	1.2521	1.5197	1.5587
3.1	−0.0095	0.3383	1.2558	1.2956	1.6385	1.6864
3.2	0.0124	0.3590	1.2997	1.3463	1.7929	1.8521
3.3	0.0316	0.3789	1.3512	1.4059	1.9996	2.0740
3.4	0.0490	0.3987	1.4120	1.4763	2.2885	2.3839
3.5	0.0652	0.4192	1.4844	1.5601	2.7171	2.8436
3.6	0.0810	0.4413	1.5715	1.6609	3.4134	3.5903
3.7	0.0968	0.4659	1.6775	1.7837	4.7315	5.0036
3.8	0.1134	0.4944	1.8083	1.9353	8.1445	8.6619
3.9	0.1318	0.5286	1.9731	2.1264	37.272	39.879
u	$\psi_{7q}(u)$	$\psi_{8q}(u)$	$\psi_{1p}(u)$	$\psi_{2p}(u)$	$\psi_{3p}(u)$	$\psi_{4p}(u)$
4.0	0.1532	0.5713	2.1857	2.3729	−12.989	−13.986
4.1	0.1796	0.6271	2.4688	2.7013	−5.2853	−5.7289
4.2	0.2146	0.7043	2.8622	3.1573	−3.2215	−3.5162
4.3	0.2651	0.8195	3.4427	3.8317	−2.2660	−2.4913
4.4	0.3475	1.0117	4.3805	4.9205	1.7163	−1.9019
4.5	0.5118	1.4003	6.1410	6.9653	−1.3598	−1.5176
4.6	1.0201	2.6116	10.622	12.171	−1.1102	−1.2488
4.7	−15.605	−37.133	44.993	52.103	−0.9260	−1.0497
4.8	−0.8317	−1.8210	−18.919	−22.152	−0.7844	−0.8961
4.9	−04063	−0.8065	−7.6316	−9.0392	−0.6721	−0.7737
5.0	−0.2582	−0.4545	—47123	−5.6485	−0/5808	−0.6735
5.1	−0.18244	−0.2753	−3.3753	−4.0961	−0.5049	−0.5895
5.2	−0.13625	−0.1664	−2.6104	−3.2085	−0.4406	−0.5174
5.3	−0.10499	−0.0929	−21164	−2.6358	−0.3850	−0.4543
5.4	−0.08233	−0.0397	−1.7721	−2.2370	−0.3362	−0.3978
5.5	−0.06507	−0.0008	−1.5190	−1.9444	−0.2925	−0.3461

(continued)

Table B.2 (continued)

5.6	−0.05142	0.0329	−1.3257	−1.7212	−0.2527	−0.2978
5.7	−0.04029	0.0592	−1.1737	−1.5461	−0.2156	−0.2513
5.8	−0.03098	0.0814	−1.0513	−1.4055	−0.1803	−0.2054
5.9	−0.02302	0.1006	−0.9510	−1.2907	−0.1458	−0.1587
6.0	−0.1608	0.1176	−0.8676	−1.1956	−0.1111	−0.1096
6.1	−0.00991	0.1330	−0.7973	−1.1159	−0.0751	−0.0562
6.2	−0.00432	0.1472	−0.7374	−1.0484	−0.0361	−0.0043
6.3	0.00084	0.1608	0.6860	−0.9910	−0.0079	0.0757
6.4	0.00571	0.1740	−0.6416	−0.9417	0.0600	0.1639
6.5	0.01040	0.1872	−0.6029	−0.8993	0.1256	0.2789
6.6	0.01501	0.2006	−0.5690	−0.8627	0.2141	0.4390
6.7	0.01967	0.2148	−0.5393	−0.8311	0.3454	0.6821
6.8	0.02452	0.2300	−0.5132	−0.8038	0.5685	1.1028
6.9	0.02971	0.2471	−0.4901	−0.7802	1.0485	2.0193
7.0	0.03548	0.2667	0.4696	−0.7600	2.9136	5.6062
7.1	0.04215	0.2901	−0.4515	−0.7282	−7.0692	−13.647
7.2	0.05021	0.3193	−0.4356	−0.7283	−1.8554	−3.6037
7.3	0.06052	0.3575	−0.4214	−0.7162	−1.1462	−2.2442
7.4	0.07464	0.4109	−0.4090	−0.7064	−0.8646	−1.7090
7.5	0.09591	0.4926	−0.3981	−0.6987	−0.7138	−1.4259
7.6	0.13286	0.6359	−0.3886	−0.6930	−0.6203	−1.2532
7.7	0.21606	0.9609	−0.3805	−0.6893	−0.5572	−1.1393
7.8	0.59928	2.4635	−0.3736	−0.6874	−0.5125	−1.0697
7.9	−0.70727	−2.6651	−0.3678	−0.6873	−0.4797	−1.0054
8.0	−0.21347	−0.7283	−0.3631	−0.6891	−0.4552	−0.9666
u	$\psi_{5p}(u)$	$\psi_{6p}(u)$	$\psi_{7p}(u)$	$\psi_{8p}(u)$	$\psi_{9p}(u)$	$\psi_{10p}(u)$
0	1.000	1.000	1.000	1.000	1.000	1.000
0.1	1.000	1.000	1.000	1.000	1.000	1.000
0.2	1.000	1.000	1.0001	1.0002	1.0001	1.0001
0.3	1.0001	1.0001	1.0007	1.0010	1.0005	1.0004
0.4	1.0002	1.0003	1.0024	1.0032	1.0016	1.0011
0.5	1.0004	1.0008	1.0058	1.0079	1.0040	1.0028
0.6	1.0009	1.0017	1.0120	1.0164	1.0083	1.0058
0.7	1.0016	1.0031	1.0224	1.0306	1.0154	1.0108
0.8	1.0027	1.0054	1.0388	1.0530	1.0267	1.0187
0.9	1.0044	1.0086	1.0635	1.0867	1.0436	1.0306
1.0	1.0067	1.0132	1.0998	1.1361	1.0685	1.0481
1.1	1.0098	1.0194	1.1523	1.2079	1.1046	1.0733
1.2	1.0139	1.0276	1.2285	1.3121	1.1568	1.1098
1.3	1.0191	1.0383	1.3408	1.4657	1.2336	1.1634
1.4	1.0259	1.0521	1.5115	1.6995	1.3503	1.2446
1.5	1.0343	1.0695	1.7871	2.07775	1.5385	1.3752
1.6	1.0447	1.0915	2.2811	2.7557	1.8751	1.6083
1.7	1.0574	1.1190	3.3663	4.2472	2.6134	2.1181
1.8	1.0728	1.1533	7.4005	9.7976	5.3546	4.0067
1.9	1.0914	1.1960	−21.142	−29.494	−14.025	−9.3290
2.0	1.1138	1.2494	−4.0099	−5.9157	−2.3891	−1.3177
2.1	1.1406	1.3166	−2.1298	−3.3321	−1.1096	−0.4337
2.2	1.1725	1.4020	−1.4177	−2.3572	−0.6227	−0.0947
2.3	1.2107	1.5118	−1.0486	−1.8551	−0.3681	0.0851

(continued)

Table B.2 (continued)

2.4	1.2563	1.6560	−0.8261	−1.5559	−0.2124	0.1975
2.5	1.3109	1.8504	−0.6797	−1.3625	−0.1076	0.2758
2.6	1.3766	2.1229	−0.5777	−1.2316	−0.0322	0.3349
2.7	1.4560	2.5259	−0.5040	−1.1412	0.0250	0.3826
2.8	1.5528	3.1742	−0.4494	−1.0789	0.0704	0.4234
2.9	1.6721	4.3716	−0.4085	−1.0374	0.1079	0.4604
3.0	1.8211	7.2810	−0.3776	−1.0124	0.1400	0.4955
3.1	2.0107	24.247	−0.3547	−1.0010	0.1687	0.5304
3.2	2.2574	−16.930	−0.3381	−1.0020	0.1954	0.5664
3.3	2.5886	−6.1342	−0.3269	−1.0147	0.2210	0.6050
3.4	3.0525	−3.7039	−0.3205	−1.0393	0.2468	0.6475
3.5	3.7422	−2.6364	−0.3186	−1.0769	0.2737	0.6956
3.6	4.8653	−2.0398	−0.3214	−1.1293	0.3028	0.7514
3.7	6.9953	−1.6608	−0.3291	−1.1996	0.3354	0.8176
3.8	12.519	−1.4003	−0.3426	−1.2926	0.3733	0.8982
3.9	59.702	−1.2112	−0.3634	−1.4160	0.4192	0.9989
4.0	−21.734	−1.0686	−0.3937	−1.5819	0.4767	1.1287
u	$\psi_{5p}(u)$	$\psi_{6p}(u)$	$\psi_{7p}(u)$	$\psi_{8p}(u)$	$\psi_{9p}(u)$	$\psi_{10p}(u)$
4.1	−9.2615	−0.9578	−0.4378	−1.8109	0.5523	1.3026
4.2	−5.9270	−0.8698	−0.5035	−2.1407	0.6572	1.5475
4.3	−4.3890	−0.7986	−0.6067	−2.6471	0.8144	1.9179
4.4	−3.5094	−0.7402	−0.7847	−3.5096	1.0775	2.5420
4.5	−2.9439	−0.6917	−1.1525	−5.2779	1.6106	3.8121
4.6	−2.5529	−0.6512	−2.3140	−10.839	3.2758	7.7888
4.7	−2.2692	−0.6170	36.008	172.36	−51.416	−122.95
4.8	−2.0562	−0.5881	1.9709	9.6332	−2.8270	−6.8060
4.9	−1.8926	−0.5635	0.9984	4.9789	−1.4329	−3.4766
5.0	−1.7649	−0.5426	0.6643	3.3783	−0.9506	−2.3263
5.1	−1.6646	−0.5248	0.4968	2.5748	−0.7062	−1.7446
5.2	−1.5846	−0.5248	0.4968	2.5748	−0.7062	−1.3944
5.3	−1.5244	−0.4969	0.3315	1.7817	−0.4602	−1.1611
5.4	−1.4777	−0.4861	0.2857	1.5624	−0.3899	−0.9949
5.5	−1.4437	−0.4773	0.2522	1.4031	−0.3372	−0.8710
5.6	−1.4214	−0.4701	0.2270	1.2845	−0.2964	−0.7753
5.7	−1.4100	−0.46444	0.2077	1.1951	−0.2639	−0.6994
5.8	−1.4095	−0.4601	0.1928	1.1276	−0.2375	−0.6379
5.9	−1.4204	−0.4571	0.1813	1.0773	−0.2156	−0.5873
6.0	−1.4436	−0.4554	0.1724	1.0411	−0.1973	−0.5450
6.1	−1.4813	−0.4549	0.1658	1.0169	−0.1817	−0.5093
6.2	−1.5367	−0.4555	0.1611	1.0034	−0.1683	−0.4789
6.3	−1.6151	−0.4572	0.1582	1.0001	−0.1567	−0.4527
6.4	−1.7250	−0.4602	0.1569	1.0068	−0.1466	−0.4301
6.5	−1.8817	−0.4642	0.1572	1.0238	−0.1377	−0.4103
6.6	−2.1131	−0.4695	0.1591	1.0521	−0.1299	−0.3929
6.7	−2.4782	−0.4760	0.1630	1.0932	−0.1229	−0.3776
6.8	−3.1238	−0.4838	0.1689	1.1495	−0.1166	−0.3640
6.9	−4.5455	−0.4930	0.1774	1.2249	−0.1109	−0.3517
7.0	−10.130	−0.5038	0.1893	1.3251	−0.1057	−0.3404
7.1	19.855	−0.5161	0.2055	1.4590	−0.1008	−0.3299
7.2	4.2072	−0.5303	0.2280	1.6413	−0.0962	−0.3197

(continued)

Table B.2 (continued)

7.3	2.0807	−0.5464	0.2600	1.8973	−0.0917	−0.3092
7.4	1.2344	−0.5648	0.3076	2.2751	−0.0869	−0.2973
7.5	0.7767	−0.5858	0.3838	2.8768	−0.0814	−0.2821
7.6	0.4870	−0.6098	0.5221	3.9652	−0.0737	−0.2582
7.7	0.2844	−0.6371	0.8425	6.4827	−0.0592	−0.2070
7.8	0.1323	−0.6685	2.3425	18.258	−0.0015	0.0254
7.9	0.0113	−0.7047	−2.7988	−22.093	−0.1996	−0.7692
8.0	−0.0896	−0.7466	−0.8657	−6.9040	−0.1223	−0.4712

Index

L. G. Petrosian, *Dynamics of Structures*, https://doi.org/10.1007/978-3-031-63539-7

GPSR Compliance

The European Union's (EU) General Product Safety Regulation (GPSR) is a set of rules that requires consumer products to be safe and our obligations to ensure this.

If you have any concerns about our products, you can contact us on ProductSafety@springernature.com

In case Publisher is established outside the EU, the EU authorized representative is:

Springer Nature Customer Service Center GmbH
Europaplatz 3
69115 Heidelberg, Germany

Batch number: 10732805

Printed by Printforce, the Netherlands